D0076064

# Integrated Electronics: Operational Amplifiers and Linear ICs with Applications

# Integrated Electronics: Operational Amplifiers and Linear ICs with Applications

Joseph J. Carr
*MSEE*
*Annandale, Virginia*

**HARCOURT BRACE JOVANOVICH, PUBLISHERS**
**Technology Publications**

San Diego New York Chicago Austin Washington, D.C.
London Sydney Tokyo Toronto

ISBN: 0-15-541360-0
Library of Congress Number: 89-85309
Printed in the United States of America

# Contents

## Chapter 15
## IC DATA CONVERTER CIRCUITS AND THEIR APPLICATION 395



## Chapter 16
## AUDIO APPLICATIONS OF LINEAR IC DEVICES 420



## Chapter 17
## COMMUNICATIONS APPLICATIONS OF LINEAR IC DEVICES 451

## Chapter 18
## ANALOG MULTIPLIERS AND DIVIDERS 489



## Chapter 19
## ACTIVE FILTER CIRCUITS 513

## Chapter 20
## DESIGN AND CONSTRUCTION OF DC POWER SUPPLIES 552



## Chapter 21
## TROUBLESHOOTING DISCRETE AND IC SOLID-STATE CIRCUITS 583

# Preface

Electronics technology has always been exciting, and people outside the industry sometimes impute a degree of wizardry to those on the inside. In the early to mid-1960s that impression gained a tremendous foothold on the public imagination with the invention of the *integrated circuit* (IC). Although many ICs today are digital devices, linear ICs are very common, growing in both range and scope of applications, and form one of bases of knowledge required of anyone engaged in electronics. Linear devices were also among the very first IC devices on the market: uA-709 was an early operational amplifier (see Chapter 2) and uA-703 was an RF/IF chip.

The linear IC market is indisputably large and growing. Such growth will continue to require well-trained technicians who understand how these devices work and how to apply them. This textbook provides in-depth coverage of the devices and their operation, but not at the expense of practical applications in which linear IC devices figure prominently.

Chapter 1 offers a general introduction that will provide students with the foundations of linear IC technology. Starting in Chapter 2, students will find thorough coverage of the operational amplifier—perhaps the most common of all linear IC devices. The book continues to develop the theme of op-amps over several chapters and then switches to non-op-amp forms. Finally, because microwave linear IC devices (*MMIC* chips) are becoming increasingly important, a chapter is devoted to high-frequency (VHF and up) devices.

No textbook on linear ICs is complete without an in-depth look at applications circuits. This textbook covers timers, waveform generators, signal processing circuits (such as logarithmic amplifiers, integrators, and differentiators), audio circuits, and physical instrumentation circuits.

Each chapter opens with a set of objectives that will alert both the student and instructor to those important topics to be mastered in that chapter. An initial, brief "Pre-Quiz" allows the student to test his or her knowledge of the material before reading the chapter. This serves two important functions: correct answers will increase confidence and incorrect answers will point out specific areas that require attention. At the conclusion of the chapter a point-by-point summary of significant topics is provided. Because the points are given individually, the summary can also serve as a helpful resource guide. Finally, because the terms, equations, and technology presented must be reinforced, the book presents a large number of problems and questions at the end of each chapter.

The author wishes to express his gratitude to the reviewers who provided insightful comments: Weston Beale, Camden County College; Glenn Blackwell, Purdue University; Russell Puckett, Texas A & M University; and Robert Reaves, Durham Technical College.

*Joseph J. Carr, MSEE*
*Annandale, Virginia*

# CHAPTER 1

# Introduction to Linear Integrated Circuit Devices

## OBJECTIVES

1. Learn the advantages of IC devices compared with discrete circuitry.
2. Be able to identify IC package types.
3. Learn the differences between linear and digital devices.
4. Understand the power requirements and methods of construction for linear IC devices.

## 1-1 PREQUIZ

These questions test your prior knowledge of the material in this chapter. Try answering them before you read the chapter. Look for the answers (especially those you answered incorrectly) as you read the text. After you have finished studying the chapter, try answering these questions again and those at the end of the chapter (see Section 1-12).

1. List four common types of IC packages.
2. A(n) ____________ signal is continuous in both range and domain, and has a transfer function of the form $V = F(t)$.
3. ____________ amplifiers normally operate with bipolar DC power supplies even though no ground terminal exists on the device.
4. The transfer function of an operational transconductance amplifier is of the form ____________ .

## 1-2 HISTORY OF THE INTEGRATED CIRCUIT DEVICE

The modern era of "solid-state" electronics began in the late 1940s with the invention of the bipolar transistor. Only a decade later engineers were working to build a device containing multiple transistors and resistors formed onto a single semiconductor substrate. By the early 1960s the "integrated circuit" (IC) was created and the modern chip revolution began. Now literally hundreds of millions of IC devices are sold annually and the IC semiconductor industry is a major contributor to the U.S. economy.

Two of the earliest commercially successful IC devices were the uA-703 RF/IF gain block and the uA-709 operational amplifier. The uA-703 device was a simple high frequency gain block and was frequently used in the mid-1960s. These devices were often used as the FM intermediate frequency (IF) amplifier in broadcast and communications receivers. Typically, three to five uA-703 devices were used to provide the 60- to 80-dB gain typically required of an IF amplifier in consumer FM broadcast receivers.

The uA-709 was an operational amplifier, or "op-amp." Invented in 1948 (in vacuum tube form) by George Philbrick, the operational amplifier gets its name from the fact that it was originally intended to perform mathematical operations in analog computers. The unique property of the op-amp that makes it so useful and so versatile is its circuit transfer function ($V_o/V_{in}$) is entirely controlled by the feedback network between output and input. This makes the op-amp one of the most powerful IC devices on the market.

There is a well-known production "learning curve" in the semiconductor industry under which costs drop dramatically after an initial high-priced period. In 1964, for example, electronic distributor catalogs listed the uA-709 operational amplifier for $110; two years later it was $79; today the uA-709 costs less than $1 if it is available at all. Devices with superior performance compared to the uA-709 now cost only a few dollars and are easily available. Production yields and device availability have improved since the early years. Where designers in 1963 had to wait for rationed parts, operational amplifiers are available today from distributor stock inventories in almost any quantity. Even electronic hobbyists can buy common operational amplifiers at low prices in retail stores.

Integrated circuit electronics became successful so quickly because it allows immense circuit density, permitting many circuits to be implemented with fewer overall discrete components. A large number of devices can be put together in a small package. Figure 1-1A shows a photomicrograph of a typical integrated circuit, while Fig. 1-1B shows the circuitry contained in a common IC operational amplifier.

Performance can be dramatically improved in an IC device compared to a discrete component version of the same circuit. Consider thermal drift of DC amplifiers, for example. All DC amplifiers suffer from an undesired change in output DC offset voltage which is a function of temperature ($\Delta V_{off} = F(T)$). Even low-cost modern IC operational amplifiers are several orders of magnitude better with respect to drift than traditional vacuum tube or discrete transistor models because all of the semiconductors and internal resistors (the main sources of drift) share the same thermal environment on the common substrate in IC devices. In discrete circuits, on the other hand, these components are spread out over several square inches of circuit board and thus do not share the same thermal environment so drift will be more pronounced.

Also consider issues such as stray capacitance, inductance and inadvertent series resistances due to conductor placement and length. These stray parameters are less of a factor in IC devices because of the small sizes and short distances involved in IC design.

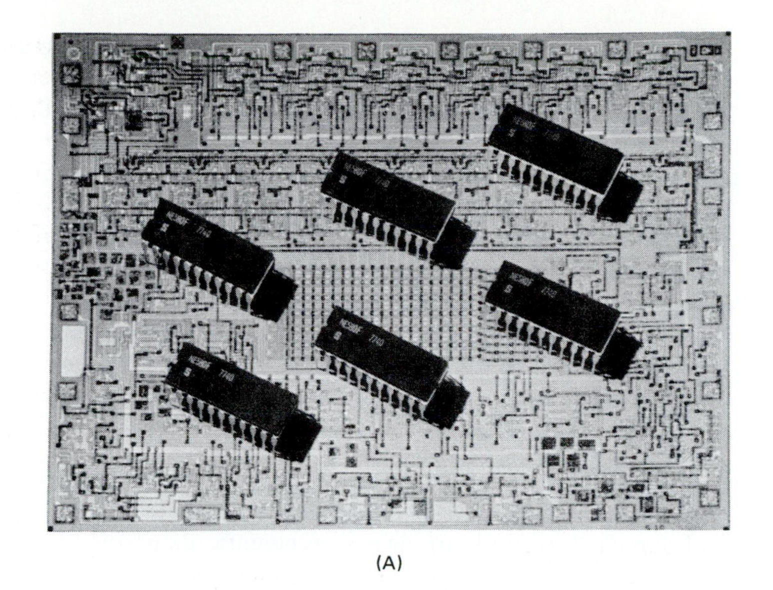

(A)

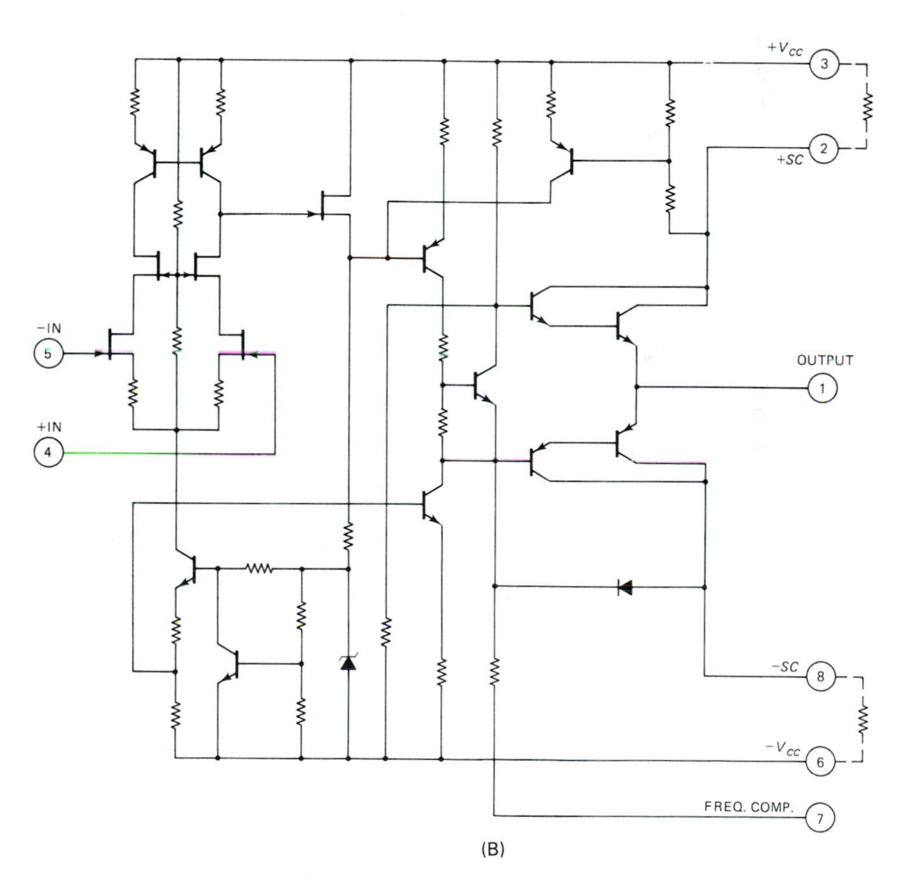


(B)

FIGURE 1-1

Cost is another great advantage of the integrated circuit. Early transistor and vacuum tube operational amplifiers were not only larger and ran hotter than their modern IC counterparts, but they were more costly as well. In addition, those early amplifiers poorly approximated the performance of the ideal textbook version of the amplifier (see Chapter 3). Modern IC operational amplifiers come very close to the ideal, especially in input impedance and open-loop gain.

Most of the spectacular benefits of modern electronics derive from integrated circuit electronics technology. From consumer devices to commercial instruments to military equipment, the IC either improves the product or simply makes it possible. Many of the marvels of modern electronics could not exist without the integrated circuit. This book is designed to give you an overview of the field, as well as detailed technical information on the workings and applications of the most commonly available devices.

## 1-3 INTEGRATED CIRCUIT SYMBOLS

Manufacturers and users of modern integrated electronics devices have adopted several standard symbols for use in schematic diagrams. Figure 1-2 shows the most common versions of these symbols as they apply to linear IC devices.

The standard single-ended amplifier symbol is shown in Fig. 1-2A. In this type of amplifier the input signal is applied between the single input terminal and common terminal (or "ground" in less rigorous terminology). The symbol shown in Fig. 1-2A has a $V+$ DC power supply terminal and a ground or common terminal. In some actual schematic circuit diagrams these terminals are not shown for sake of simplicity. Do not assume, however, that they are not used in such cases.

A *differential amplifier* symbol is shown in Fig. 1-2B. This type of amplifier produces an output signal that is proportional to the difference between the two input signals. If $A_v$ is the voltage gain of the amplifier, and $V1$ and $V2$ are the two input signals, then the output voltage is defined by $A_v(V2\text{-}V1)$.

There are always two inputs on a differential amplifier. The *inverting input* (−IN) produces an output signal that is 180° out of phase with the input signal. The *noninverting input* (+IN) produces an output that is in phase with the input signal.

Figures 1-2C and 1-2D show two alternate symbols used for the operational amplifier. The supposed standard version is shown in Fig. 1-2C. But in common industry practice the linear amplifier symbol of Fig. 1-2D is the *de facto* standard.

The standard operational amplifier symbol in Fig. 1-2D has several features. First, there are two inputs (inverting and noninverting) because most op-amps are differential amplifier devices. In fact, only the differential amplifier gives true, full-range, operational amplifier performance. Second, there are two DC power supply terminals. The "$V+$" terminal requires a potential that is positive with respect to common, while the "$V-$" terminal requires a potential that is negative with respect to common. Third, the output is single-ended, so the output signal is taken between the output terminal and common.

Notice what is missing on this symbol? There is no "ground" or "common" terminal on the standard operational amplifier. The common connection is established by connecting together the "cold" ends of the $V-$ and $V+$ DC power supplies (see Section 1-8).

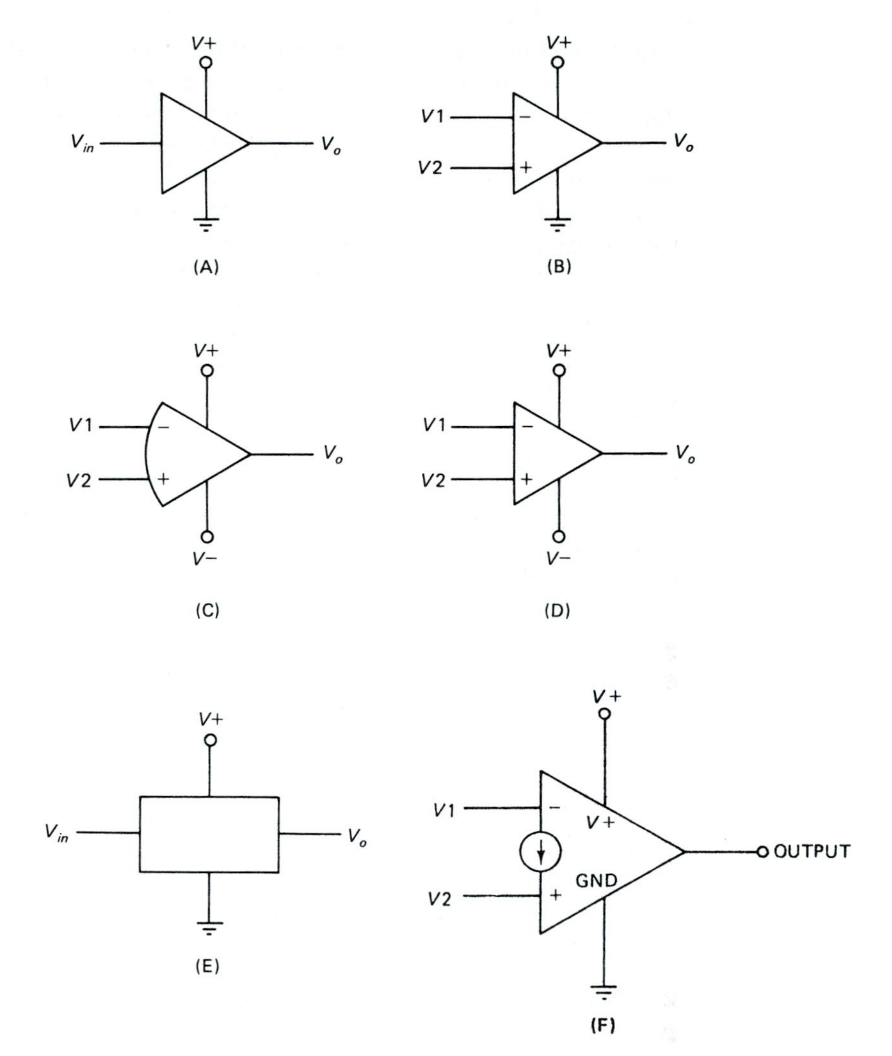


FIGURE 1-2

Some manufacturers use the universal or generic IC symbol shown in Fig. 1-2E for their products. Although these devices tend to be special-function products, or limited-use amplifiers, the same generic symbol is sometimes also used for other purposes.

Figure 1-2F is the symbol used for a special type of amplifier called a *current difference amplifier* (CDA), or *Norton amplifier*. The CDA produces an output voltage that is proportional to the difference between two input currents. The operation of the CDA is not exactly analogous to the op-amp (with the input voltages replaced by input currents), but it is similar. The symbol for the CDA differs from the normal op-amp symbol in order to distinguish its unique operation. The CDA symbol shown in Fig. 1-2F is the regular differential amplifier symbol with a current source symbol added along one edge to let the reader know that current mode operation is intended.

Another form of linear IC amplifier, different from either op-amp or CDA, is the *operational transconductance amplifier*, or OTA. This type of amplifier has a transfer function that relates output current to input voltage ($\Delta I_o/\Delta V_{in}$). Because the transfer function expression has the units amperes/volts (or subunits thereof), the transfer function gain can be expressed in the units of conductance ("mhos" or the subunits millimhos or micromhos; Siemens is the modern unit). Since these are units of conductance we call the amplifier a "transconductance amplifier." The name "operational" conveys the idea that some of the functions are similar to those of the operational amplifier. The OTA symbol is the same as the op-amp shown in Fig. 1-2D, sometimes with the letters "OTA" superimposed inside the triangle.

Although there is some variation in schematic symbols used by manufacturers, the versions shown in Fig. 1-2 represent the majority of modern IC device applications.

## 1-4 LINEAR VS. DIGITAL IC DEVICES

The integrated circuit revolution brought us both linear and digital devices. It is therefore appropriate to consider the differences between these devices, especially because there are digital applications of linear devices and linear applications of digital devices.

### 1-4.1 Analog vs. Digital Signals

It is instructive to examine the differences between analog and digital signals. Although the formal mathematical difference is more rigorously defined, the diagram of Fig. 1-3 shows the concept intuitively.

*An analog signal (Fig. 1-3A) is continuous in both range and domain.*

The range of an analog signal will be either a current ($I$) or voltage ($V$), while the domain is typically time ($t$). Thus, an analog signal is a continuous function of the form $V = F(t)$. The actual waveshape of the analog signal is not significant for this definition, but it is often critical for applications purposes.

*A digital signal is discrete (noncontinuous) in range and possibly also domain.*

In a digital signal the range is a voltage or current that can take on only specified values (called VLOW and VHIGH in Fig. 1-3B). Although a digital signal may be continuous in domain ($t$), many digital signals are discrete in domain as well as range. The key identifying feature of the digital signal is the fact that the amplitude (range) can only take on specific discrete values. The overwhelming majority of digital signals are *binary* signals in which there are only two permissible states (VLOW and VHIGH in Fig. 1-3B). Modern digital computers use this type of signal. In Chapter 10, however, there is a discussion of a circuit with tristate outputs.

The actual values of VLOW and VHIGH depend on the particular family of devices selected. In *transistor transistor logic* (TTL) devices VLOW nominally ranges from zero to 0.80 volts. VHIGH is nominally +5 volts with a tolerance range of +2.4 to +5.2 volts. In *complementary symmetry metal oxide semiconductor* (CMOS) devices VLOW can be any voltage from zero to −15 volts, while VHIGH can be any voltage from zero to +15 volts. A constraint in the CMOS system is that the condition VLOW = VHIGH = zero is not permitted. Furthermore, the absolute values of VLOW and VHIGH need not be equal. In other words, while VLOW = −10 volts and VHIGH

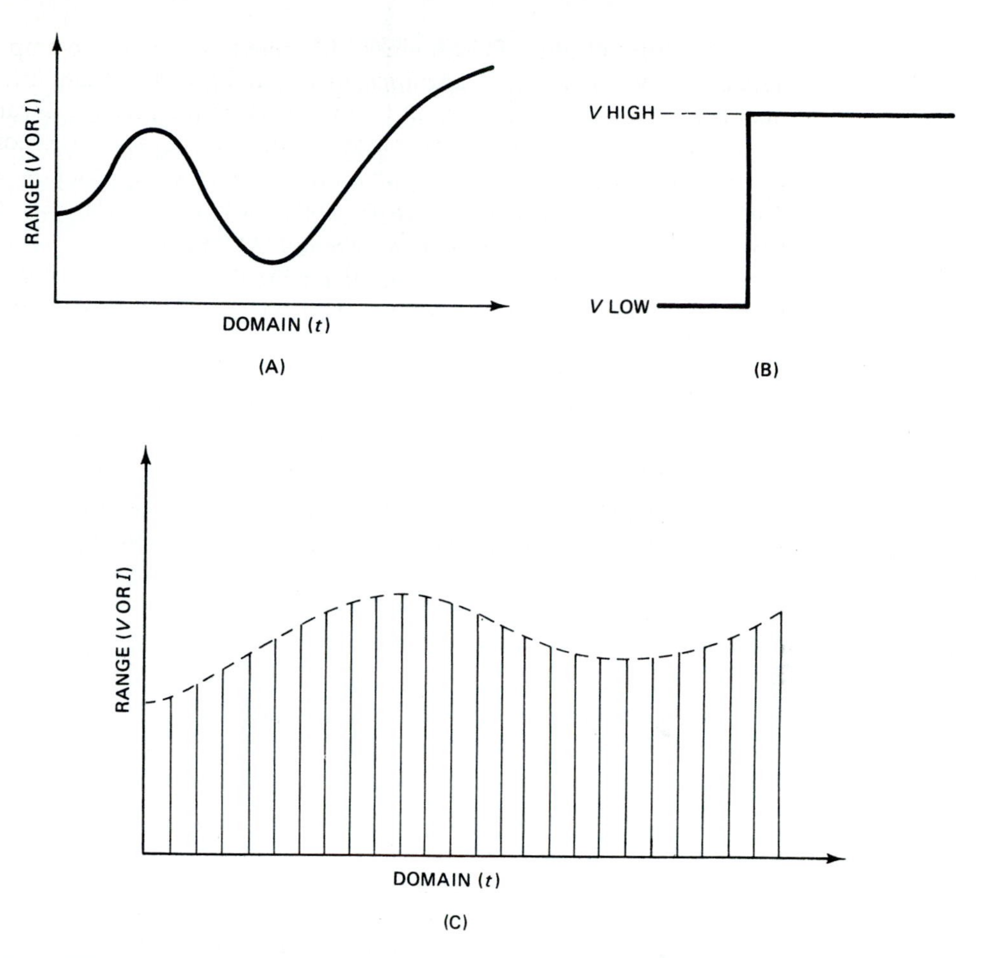


FIGURE 1-3

= +10 volts is a permissible condition, so is VLOW = −5 volts and VHIGH = +10 volts. The transition point from VLOW to VHIGH or from VHIGH to VLOW, is one-half the difference between the two voltage levels.

A third category of signal is the *sampled analog signal* shown in Fig. 1-3C. This type of signal is *continuous in range but discrete in domain.* The sampled signal is found extensively in instrumentation applications, especially in computerized (microprocessor-based) analog instruments.

*A "linear circuit" is one that shows a simple scalar relationship between output and input signals.* For example, the gain of ten analog amplifier is "linear" because the output has the same waveshape as the input but is ten times larger. In contrast, both input and output signals of a digital circuit can take on only one of two values, VLOW or VHIGH. These are "nonlinear" circuits. For purposes of our discussion, a linear IC device is one that processes analog signals. However some linear circuits may also process sampled signals as well as analog signals. Also included in this admittedly loose definition are certain nonlinear circuits discussed in Chapters 10, 11, and 12.

### 1-4.2 Applications of Linear vs. Digital Devices

The inherent differences between linear and digital IC devices make most applications unambiguous. There are areas, however, where the situation is not easily determined. For example, some CMOS digital gates can be biased in a way that makes them operate as linear amplifiers from DC up to AC frequencies exceeding 5 MHz. Similarly, operational amplifiers can work as clock oscillators in digital circuits, comparators, and coincidence detectors (for instance, AND gates) in a manner that suggests a binary condition normally found in digital circuits. While it is true the linear device output voltage must match the system voltage levels (VLOW and VHIGH) of the particular digital logic family it works with, it is usually not difficult to interface such devices.

We will also see cases where an inherently linear IC device, such as the operational amplifier, is used to produce some decidedly nonlinear output signals because the output waveshape does not resemble the input waveshape. Examples of this type of circuit are integrators, differentiators, and logarithmic amplifiers. These are, nonetheless, applications of linear IC devices in modern terminology.

## 1-5 COMMON PACKAGE TYPES

The integrated circuit is formed on a tiny chip of silicon material by a photolithographic process. Typical chip die sizes range around 100 mils (0.100 inch). The die is mounted inside a package and connected to the package pins by fine wires.

Figure 1-4A shows a die with wire attached, while Fig. 1-4B shows a packaged die with a see-through window for illustration purposes. The connecting wires between

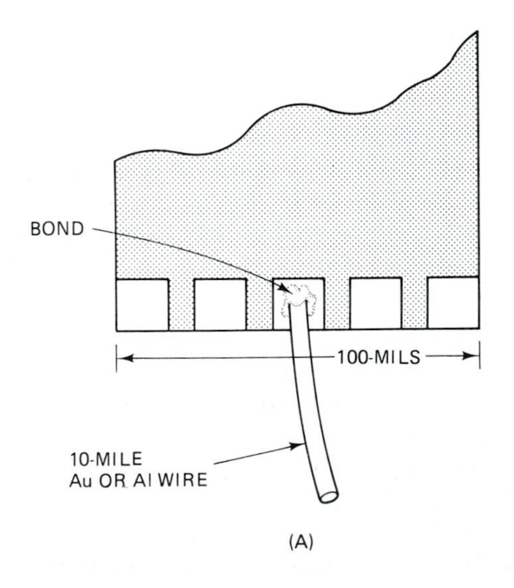


(A)

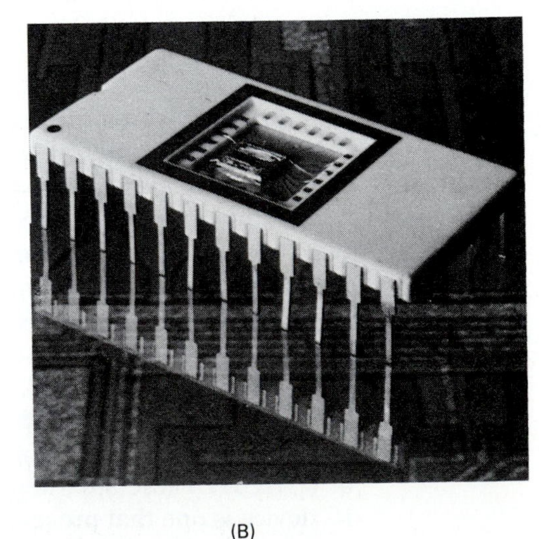

(B)

FIGURE 1-4

package pins and solder pads on the die are around 10 mils (0.010 inch) diameter, and are usually made of either gold or aluminum. Either an electric current or a thermosonic process is used to melt the end of the wire onto the die's connecting pad.

The particular package style selected for any given IC device depends in part on the intended application and the number of pins required. For many IC devices several different packages are available. The earliest IC packages were the 6-, 8-, 10-, and 12-lead metal can devices (Fig. 1-5A). These packages were redesigns of (and similar to) the TO-5 metal transistor package.

When viewed from the bottom of the package, the keyway marks the highest number lead or pin (pin 8 in Fig. 1-5A). Pin 1 is the next pin clockwise from the keyway. One must be careful when looking at IC base diagrams to know whether a top or bottom view is depicted.

Perhaps the largest number of IC devices on the market today are sold in *dual in-line packages* (DIP), examples of which are shown in Fig. 1-5B. DIP packs are available in a wide variety of sizes from four to more than 48 pins. Although many devices are available in other size DIP packs, most linear devices are found in 8-, 14-, or 16-pin DIP packs.

The DIP pack is symmetrical with respect to pin count regardless of the package size, so some other means is needed to designate pin 1; Fig. 1-5B shows several common methods. In all cases the IC DIP pack is viewed from the top. In some devices a paint dot, called a "pimple" or "dimple," will mark pin 1. In other cases, a semicircular or square notch marks where pin 1 is located. When viewed from the top, with the notch pointed away from you, pin 1 is to the left of the notch while the highest number pin is to the right of the notch.

Both plastic and ceramic materials are used in DIP pack construction. In general, plastic packages are used in consumer and noncritical commercial or industrial equipment. Ceramic packs are used in military and critical commercial equipment. The principal difference between the plastic and ceramic packages is the intended temperature range. Although exceptions exist, typical temperature range specifications are 0° to +70°C for commercial plastic devices, and −10° to +80°C for ceramic. Military and some critical commercial ceramic devices are rated from −55° to +125°C. The principal difference between military chips and the highest grade commercial or industrial chips is the amount of testing, burn-in, and documentation that accompanies each device.

An example of an IC flatpack is shown in Fig. 1-5C. This type of package is typically used where very high component density is required. Most flatpack devices are digital ICs, although a few linear devices are also available. DIP package IC devices are mounted either in sockets or by insertion of the pins through holes drilled in a printed circuit board (PCB). Flatpacks, on the other hand, are mounted on the surface by direct soldering to the conductive track of the PCB.

A relatively new style of package is the *surface-mounted device* (SMD), an example of which is shown in Fig. 1-5D. The SMD technology represents a significant improvement in packaging density. SMD components can be mounted closer together and are amenable to automatic PCB production methods. The SMD will probably replace other types of packages, especially in VLSI applications.

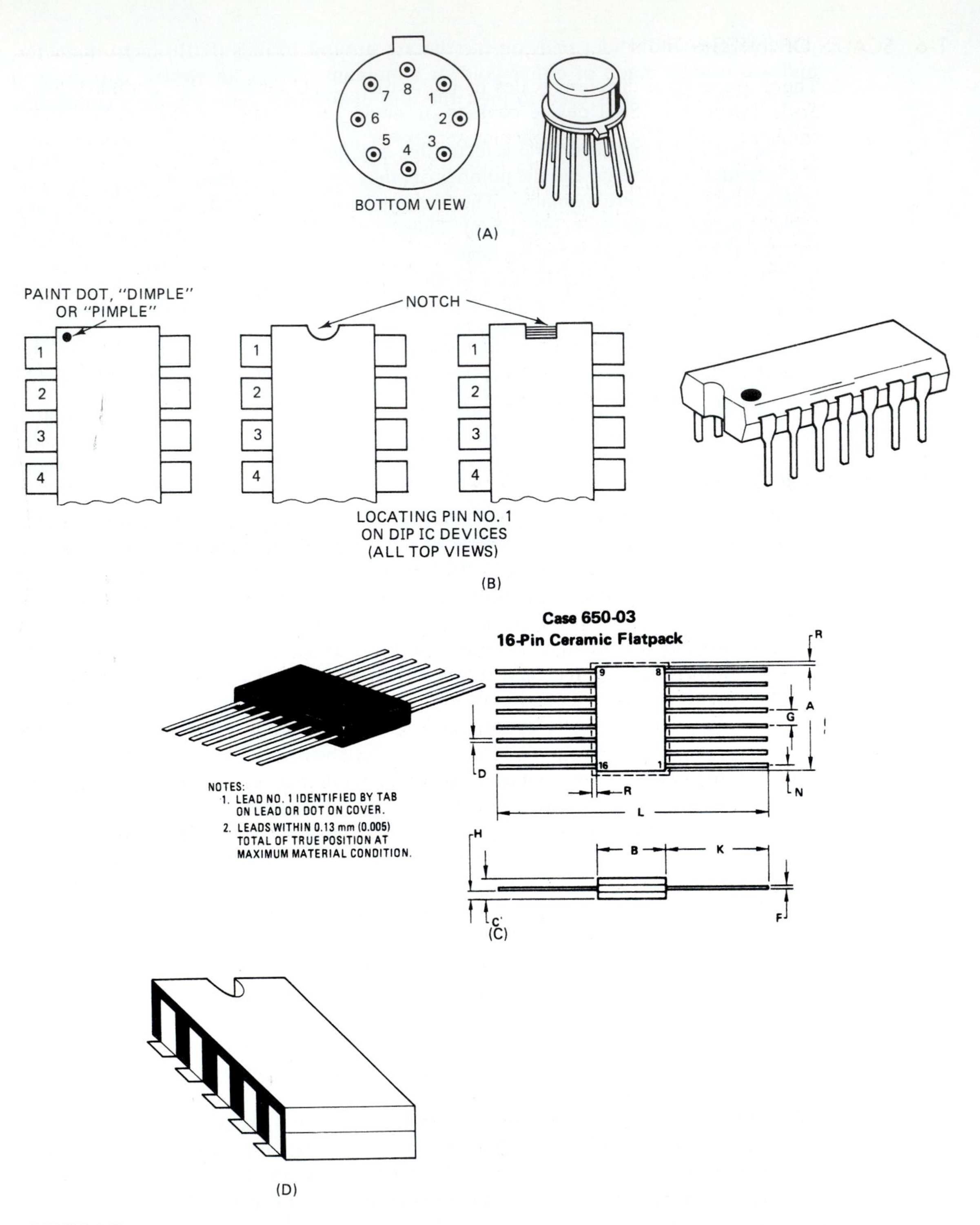

8
7
1
6
2
5
4
3
BOTTOM VIEW
(A)
PAINT DOT, "DIMPLE"
OR "PIMPLE"
NOTCH
1
2
3
4
LOCATING PIN NO. 1
ON DIP IC DEVICES
(ALL TOP VIEWS)
(B)
Case 650-03
16-Pin Ceramic Flatpack
NOTES:
1. LEAD NO. 1 IDENTIFIED BY TAB ON LEAD OR DOT ON COVER.
2. LEADS WITHIN 0.13 mm (0.005) TOTAL OF TRUE POSITION AT MAXIMUM MATERIAL CONDITION.
9
8
16
1
R
A
G
D
N
L
H
B
K
C
F
(C)
(D)


FIGURE 1-5

## 1-6 SCALES OF INTEGRATION

There are several different scales of integration in IC devices. The ordinary *Small Scale Integration* (SSI) device consists of single gates, small amplifiers, and other smaller circuits. The number of components on each chip is usually 20 or less. *Medium Scale Integration* (MSI) devices have a higher degree of complexity and may have about 100 or so components on the chip. Devices such as operational amplifiers, shift registers, and counters are usually classed as MSI devices. *Large Scale Integration* (LSI) devices are mostly digital ICs and include functions such as calculators and microprocessors. Typical LSI devices may contain 100 to 1000 components. Some newer devices are called *Very Large Scale Integration* (VLSI) and include some of the latest computer chips.

The numbers and descriptions listed for SSI, MSI, and LSI devices are only approximate and are meant to serve as guidelines. Most linear IC devices are either SSI or MSI, with the latter predominating.

## 1-7 DISCRETE, MONOLITHIC IC, AND HYBRID CIRCUITS

The generic term *microelectronics* describes two distinctly different categories of solid-state devices: *monolithic integrated circuits*, which includes most devices discussed in this text, and *hybrid circuits*. We will see how monolithic and hybrid circuits differ from each other, and how they differ from regular discrete solid-state circuits.

The monolithic IC device is made using photolithographic and other processes on a single crystal of semiconductor material. It is called monolithic because only a single piece of semiconductor material is used in the fabrication of the device (see Fig. 1-6A). A hybrid circuit, on the other hand, is made using a cross between monolithic and printed circuit methods. Although larger than most monolithic devices, tight packaging makes the hybrid very different from the printed circuit board, which it superficially resembles. The macro-view of the hybrid makes it usable as an integrated circuit, even though the internal structure is quite different from IC devices.

The substrate diode shown in Fig. 1-6A is formed by an interface between the P-type substrate and the regions of semiconductor crystal used for other components. This diode is a natural (if undesired) feature of IC devices. The substrate diode is normally reverse-biased. If circuit conditions force the diode to be forward-biased, the device may be damaged. Although some designers have cleverly used the substrate diode in certain applications, it is normally ignored.

The typical hybrid consists of a multi-layer (usually ceramic) substrate on which conductive tracks are printed by several deposition methods. The integrated circuits, transistors, and other semiconductor devices used in the hybrid circuit are either cemented or otherwise bonded to the ceramic substrate (Fig. 1-6B). A critical difference between the hybrid and other printed circuit devices is that unpackaged semiconductors are used in the hybrid. The chips diodes and transistors are used in die form. In other words, unpackaged dies are installed directly onto the ceramic substrate of the hybrid. Tiny (about 10 mil) gold or aluminum wires connect the die electrical contacts either to the conductors of the hybrid substrate or to the hybrid package pinouts.

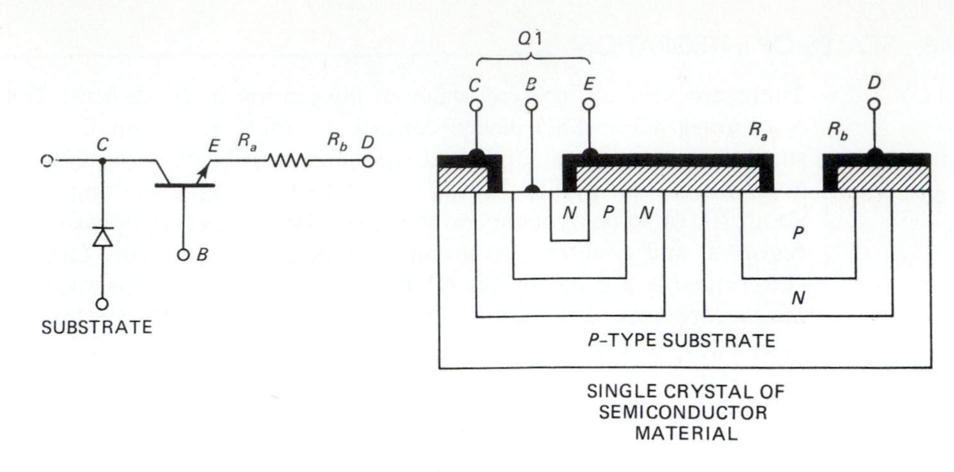


(A)

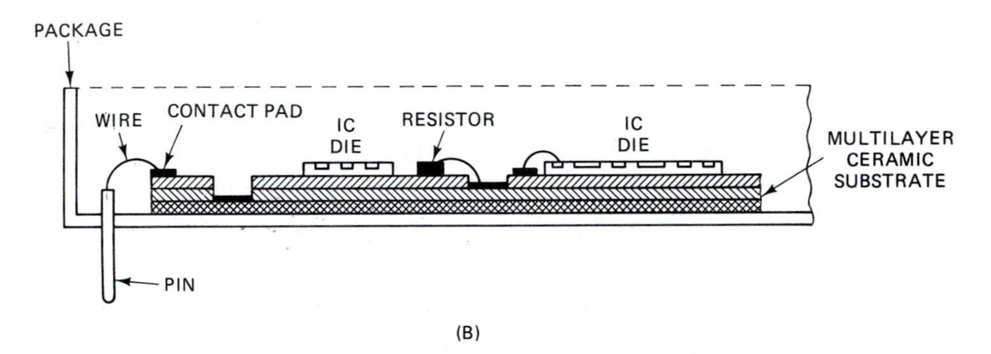


(B)

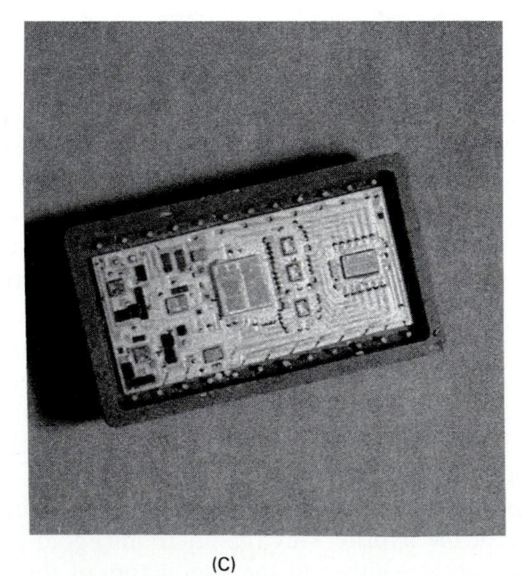

(C)

FIGURE 1-6A-C

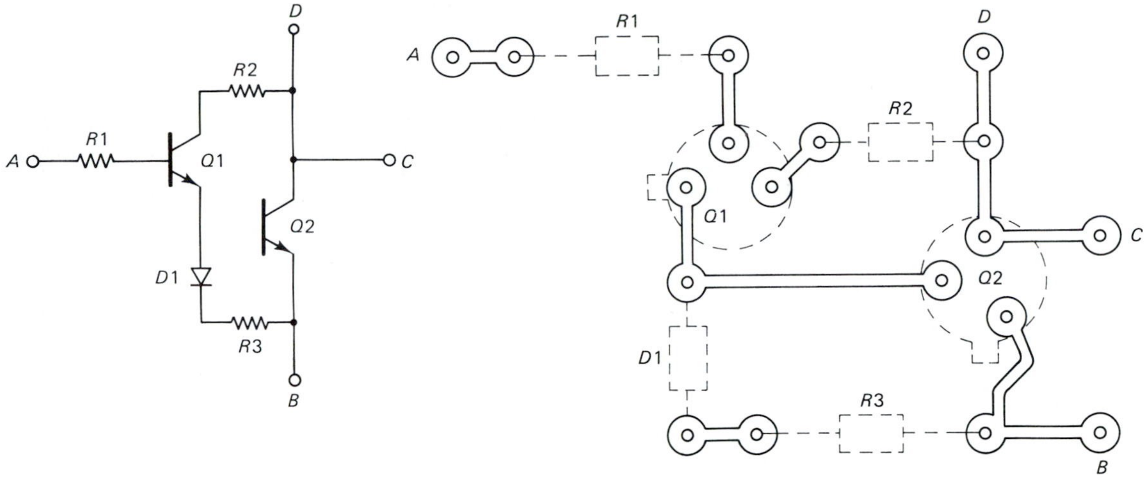

(D)

FIGURE 1-6D

The hybrid circuit may contain a variety of components which are either difficult, expensive, or impossible to implement in monolithic IC form. Once the components are installed and the circuit is tested, the package lid is attached and the device is sealed, evacuated, and prepared for final testing. Once completed, the hybrid can be treated like a large integrated circuit. A photo of a "de-lidded" hybrid is shown in Fig. 1-6C.

A principal advantage of the hybrid is its IC-like packaging densities, while being relatively easy to design and manufacture without a semiconductor foundry. The hybrid is used when either complexities or small quantities make monolithic IC implementation difficult or uneconomical.

Discrete electronic circuits are formed on printed circuit boards (PCB) of individually packaged active solid-state and passive components interconnected by photo-etched copper tracks. The simple discrete circuit shown in Fig. 1-6D would easily fit onto a 100-mil (or less) IC die, or a 250-mil hybrid substrate. At normal discrete component packaging densities this same circuit would require a 1-inch (1000-mil) square area on the PCB. By comparing the ratios of $(1000\text{-mil})^2$ to $(250\text{-mil})^2$ and $(100\text{-mil})^2$ we can see how IC and hybrid construction offer greatly reduced size.

## 1-8 LINEAR IC POWER SUPPLY REQUIREMENTS

Operational amplifiers and other linear IC devices normally operate from bipolar DC power supplies (Fig. 1-7). The pin numbers in this figure are for the so-called "industry standard" type 741 operational amplifier. This circuit shows that the two DC power supplies are completely independent of each other. The $V+$ power supply is positive with respect to common and the $V-$ supply is negative with respect to the common. The operational amplifier manufacturer will specify minimum and maximum values for $V-$ and $V+$. Typically, the maximum voltages will be on the order of $\pm 18$ volts with some able to handle up to $\pm 22$ volts and in at least one case $\pm 40$ volts.

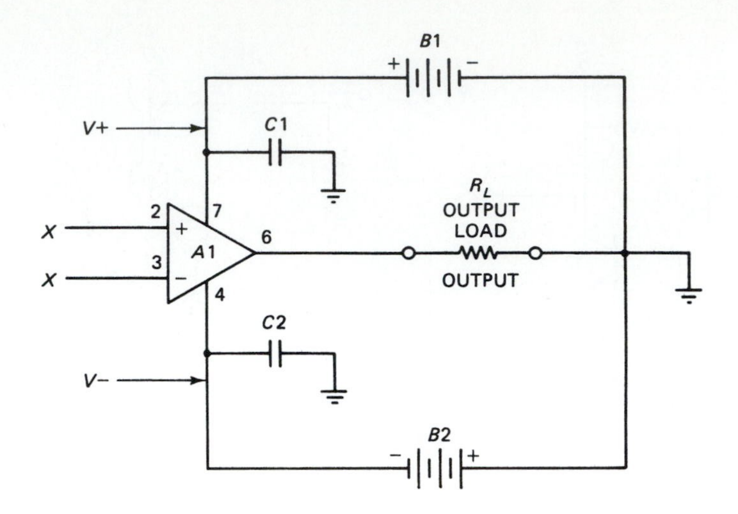


FIGURE 1-7

There may be certain specified limitations on the maximum permissible supply voltages for any given device. For example, one such limitation occasionally seen in data sheets gives $(V-)$ and $(V+)$ potentials. The manufacturer will specify the quantity $[(V+) - (V-)]$ cannot exceed a certain critical voltage. On some older 741 devices, for example, the maximum $V-$ and $V+$ ratings are both specified at 18 volts. But that does not mean the permissible algebraic sum of the two is 36 volts! The pin-to-pin supply voltage was specified "not to exceed 30 volts." Since $[(+18\ Vdc) - (-18\ Vdc)] = 36$ volts, operating both supply terminals at their maximum voltage is not permitted. Let's look at a practical example.

## EXAMPLE 1-1

An operational amplifier must operate with $V+$ at 18 volts. What is the maximum value permitted for $V-$ that will not exceed a 30-volt pin-to-pin limit?

$$[(V+) - (V-)] = 30$$
$$[(+18\ Vdc) - (V-)] = 30$$
$$-(V-) = 30 - 18$$
$$-(V-) = 12$$
$$(V-)_{max} = -12 \text{ volts}$$

■

Fortunately, most modern IC devices do not suffer such limitations. One should however, check manufacturer data sheets before assuming otherwise.

Because most practical circuits operate with equal bipolar DC power supplies, operational amplifiers with the 30-volt pin-to-pin limit also limit the $V-$ and $V+$ DC power supplies to not more than 15 volts each.

**Single DC Supply Operation.** The operational amplifier and many other linear IC devices are designed for operation from "split," dual, or bipolar power supplies (all three terms describe the same situation). There are, however, many applications where

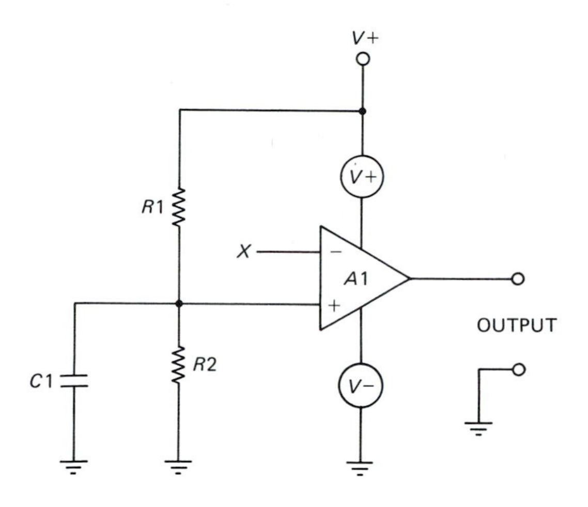


FIGURE 1-8

only a single polarity DC power supply is available. In order to operate the op-amp in these cases we must either supply the missing potential (with an on-board DC-to-DC converter), or eliminate the need for the missing potential.

It is easy to supply a missing potential. All that is needed is a DC-to-DC converter circuit that provides the needed voltage from the existing DC supply voltage. There are several devices on the market that will produce either −15 volts from +15 volts (or +12 volts, as the case may be), or will produce isolated ±15 volt potentials from an existing nonisolated +12 to +15 volt potential.

Another method for using a single DC power supply is shown in Fig. 1-8. A resistor voltage divider, $R1/R2$, is used to bias the noninverting input of the operational amplifier to some potential between ground and $V+$. The $V-$ terminal of the device is grounded. The bias voltage on the noninverting input also appears on the output terminal as a DC offset potential. Unless the circuit following the amplifier is not affected by this offset potential, the output terminal must be capacitor-coupled. The value of the bias voltage is found from

$$V1 = \frac{(V+)\,R2}{R1 + R2} \tag{1-1}$$

The capacitor shown in Fig. 1-8 is used to place the noninverting input at or near ground potential for AC signals, while retaining the DC level produced by the resistor voltage divider. This capacitor sometimes causes noisy operation of the device, especially when significant ground loop or ground plane noise is present, so it is often omitted in practical circuits. The value of the capacitor is such that it has a capacitive reactance of less than $R2/10$ at the lowest frequency of operation. For example, if the amplifier is designed to operate down to a frequency of 10 Hz and the value of $R2$ is 2200 ohms (a typical value in real circuits), then $C1$ must have a reactance of 220 ohms or less at 10 Hz. This requirement evaluates to

$$C1 = 1{,}000{,}000/(2\ \pi\ F\ X)$$

$$C1 = 1{,}000{,}000/((2)(3.14)(10\text{ Hz})(220\text{ ohms}))$$

$$C1 = 1{,}000{,}000/13{,}816 = 73\text{ uF}$$

Because 100 uF is the next higher standard value capacitor, most designers will select 100 uF for $C1$ instead of 73 uF.

### 1-8.1 Protecting Linear IC Amplifiers

Operational and other linear IC amplifiers are sensitive to problems on the DC power supply lines. For example, the amplifier may oscillate if the DC lines are not properly decoupled. Furthermore, voltage variations, noise, and transients coupled to the DC power supply lines in one stage can affect the other stages in the same circuit, especially if power supply rejection is poor.

Reversed polarity in DC power supplies can also be a problem, especially when breadboarding, troubleshooting, or using portable (battery-operated) equipment. The results can be catastrophic if the DC power supply potentials are reversed. An operational amplifier with reversed DC power supplies will probably be destroyed instantly.

There are certain remedies available for each of the defects discussed above.

The problems of noise and oscillation due to cross-coupling between stages can be cured by using decoupling capacitors on the amplifier power supply terminals. Capacitors $C1$ and $C2$ (each 0.1 uF) in Fig. 1-9A are used to decouple high frequencies, while the low frequency decoupling is provided by $C3$ and $C4$ (each typically 1 uF or higher).

Why are two forms of capacitors needed at each op-amp power supply terminal? The higher value capacitors ($C3$ and $C4$) are typically aluminum or tantalum electrolytics. The performance of these capacitors drops drastically as frequency increases and may be very poor at higher frequencies which are nonetheless within the device's range. At those high frequencies the typical electrolytic capacitor is ineffective. For this reason we sometimes use a smaller value capacitor, but one that will work at higher frequencies (mylar, mica, or ceramic). This situation is changing, however, because certain new forms of capacitors are available which offer high frequency operation and high capacitance.

Figures 1-9B and 1-9C show the output waveform of a high frequency operational amplifier in both nondecoupled (Fig. 1-9B) and decoupled (Fig. 1-9C) configurations. In Fig. 1-9B we see a 400-Hz sinusoidal waveform that has a high frequency oscillation superimposed on it. In Fig. 1-9C, however, we see the same waveform after the decoupling capacitors were added to the $V-$ and $V+$ pins of the operational amplifier used in this experiment. Note that the oscillation is removed. In Fig. 1-9D we see the actual oscillation (with 400-Hz sinewave turned off) in an expanded oscilliscope timebase range.

One practical rule of thumb ensures the success of the circuit in regard to noise and oscillation: *place those capacitors as close as physically possible to the body of the amplifier*. The 0.1 uF capacitors ($C1$ and $C2$) are more important than the higher value capacitors. So they should be closest to the IC amplifier body.

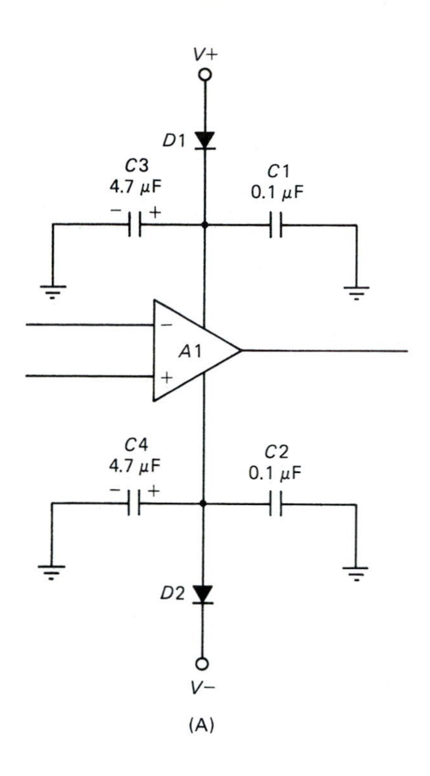

(A)

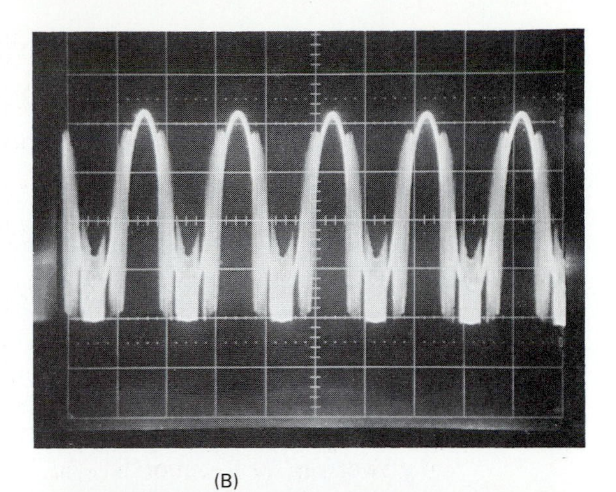
(B)

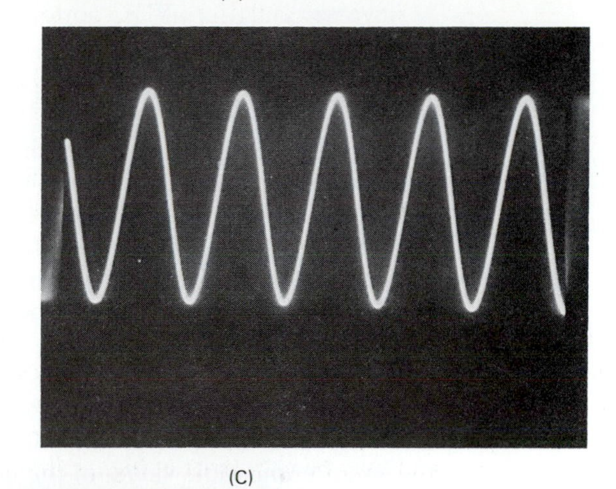
(C)

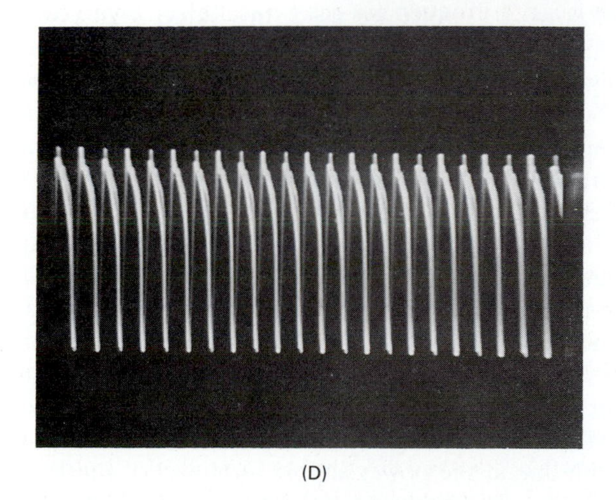
(D)

FIGURE 1-9

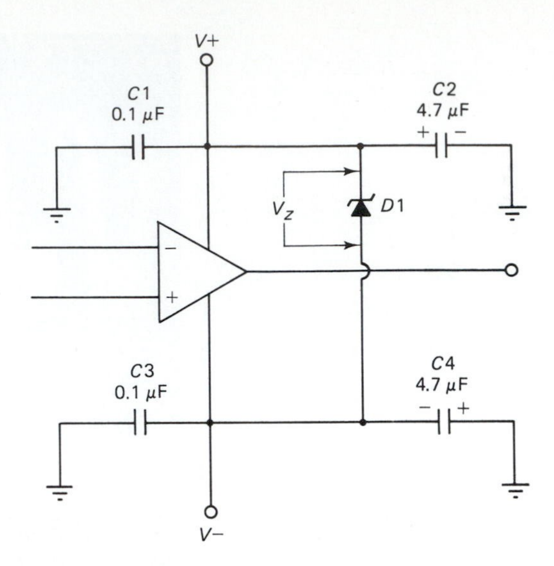


FIGURE 1-10

Protection against reverse polarity conditions is shown in Fig. 1-9 also. Diodes $D1$ and $D2$ are placed in series with each DC power supply line. Under normal operation these diodes are forward-biased, so they will conduct current to the amplifier. If someone accidentally connects one or both DC power supplies backwards, these diodes are reverse-biased and will not conduct current. Thus, the series diodes protect the amplifier IC from incorrect power supply connection. Typical diodes for this application are any of the 1N400x series (1N4001 through 1N4007).

Another method for protecting the device is shown in Fig. 1-10. In this case, a zener diode is placed across the two DC power supply terminals of the amplifier. The zener potential must be greater than the maximum actual value, but less than the maximum permissible value, of the quantity $[(V+) - (V-)]$. In the case where DC power supplies of $\pm 12$ volts are used, this value is 24 volts. A 28-volt zener diode would be adequate if the maximum permissible value is 30 volts (provided the power supply voltages are reasonably stable). Under these conditions, with $V_z$ greater than the voltage between the terminals, zener diode $D1$ is reverse-biased at a voltage lower than the zener potential. Thus, it is not conducting in normal operation. In reverse polarity operation, diode $D1$ becomes forward-biased in the normal nonzener mode. It will harmlessly pass current around the amplifier. If excessive $(V-)$-to-$(V+)$ voltage is applied, the zener diode will conduct and clamp the voltage to $V_z$.

The protection of the multiple stage amplifier is shown in Fig. 1-11. There are two alternatives shown in this figure. In one case, 1N400*x*-series diodes in reverse-bias state are placed across the DC power supply lines and current-limiting resistors in series to prevent them from burning up. The diodes are normally reverse-biased, but when one or both DC power supplies are reversed, these diodes become forward-biased and short the line to ground. The second alternative is to place the diodes in series with the line at the power supply terminals of the IC device (shown in dotted lines in Fig. 1-11). This method is analogous to the method shown in Fig. 1-9, except it serves more than one amplifier.

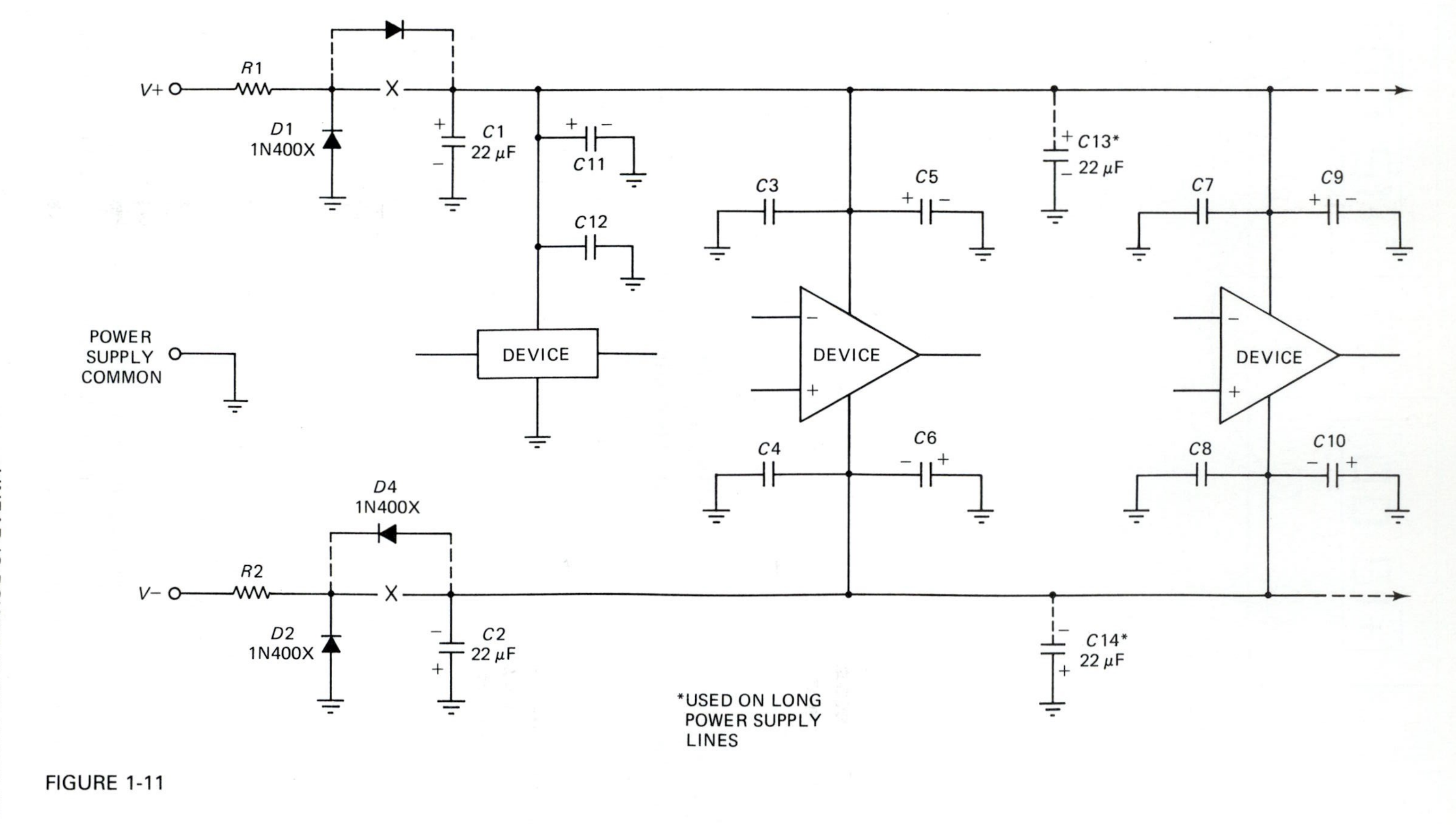
V+
R1
D1
1N400X
C1
22 μF
C11
C12
DEVICE
POWER
SUPPLY
COMMON
C3
C5
C4
C6
DEVICE
C13*
22 μF
C7
C9
C8
C10
DEVICE
D4
1N400X
V−
R2
D2
1N400X
C2
22 μF
C14*
22 μF
*USED ON LONG
POWER SUPPLY
LINES

FIGURE 1-11

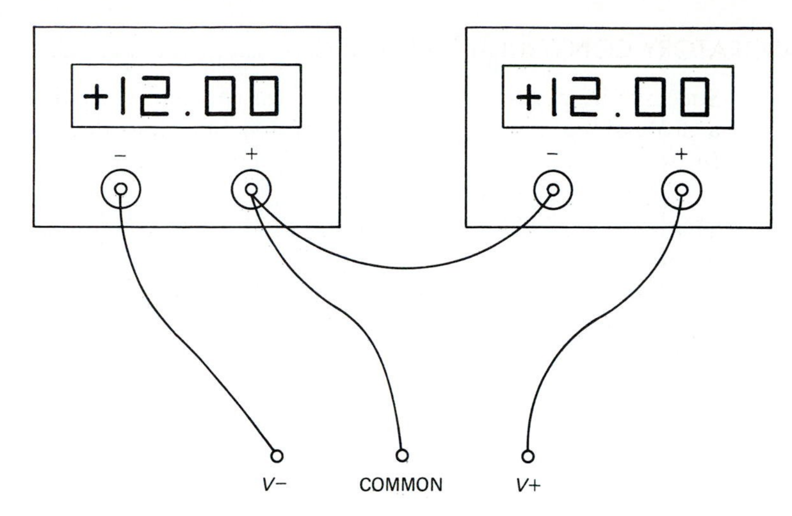


FIGURE 1-12

### 1-8.2 Power Supplies for Laboratory Experiments

A power supply must be selected (or built) to perform laboratory experiments in this book. Unless otherwise specified, the experiments are designed to use ±12 volt DC-regulated DC power supplies. The power supply should offer either a single bipolar power supply or two independent 12-volt DC supplies that are not ground referenced. The nongrounded feature will allow you to create a bipolar supply by connecting the positive output terminal of one supply to the negative output terminal of the other (see Fig. 1-12). Students who plan to build their own power supply are encouraged to see Appendix A for pertinent details.

Desirable features to have on a bench power supply include the following:

1. Output voltages fixed at ±12 volts DC (or ±15 volts), or
2. Output voltages adjustable from zero to greater than 12 Vdc
3. Current available from each polarity not less than 100 milliamperes
4. Metered outputs (I and V)
5. Voltage regulation
6. Current limiting for output short circuit protection and
7. Overvoltage protection

Although selections will not have all these features, those that do are clearly superior for most laboratory applications. All of the experiments in this book can be run on a power supply that meets these requirements.

Successfully performing student experiments and designing or debugging circuits on a professional level requires certain skills and knowledge of construction practices. In this section we will review some of these practical matters.

There are four major categories of preproduction construction practice: (1) partial breadboard, (2) complete breadboard, (3) brassboard, and (4) full-scale development model.

The two breadboard categories involve using special construction methods to check circuit validity and make certain preliminary measurements. The breadboard is in no way usable in actual equipment. It is a bench or laboratory device.

The difference between partial and complete breadboards is merely one of scale. The partial breadboard is used to check out a partial circuit or a small group of circuits. The complete breadboard will be the full circuit and is much more extensive than the partial breadboard. There are fundamental differences in the type of breadboarding hardware used for both types, especially in large circuits.

The two commonly used types of hardware are shown in Figs. 1-13 and 1-14. The device shown in Fig. 1-13 is an IC pin socket breadboard. This model is the Heath/Zenith Educational Systems ET-3300. It, like others of its type, offers three built-in DC power supplies: +12 volts at 100 mA, −12 volts at 100 mA and +5 volts at 1.5 amperes (for TTL digital logic IC devices). Located between the large multipin IC sockets are bus sockets in which all pins in a given row are strapped together. These sockets are generally used for power distribution and ground busses. Interconnections between sockets, power sources, and components are made using #22 insulated hookup wire. The bared ends of the wires are pressed into the socket holes.

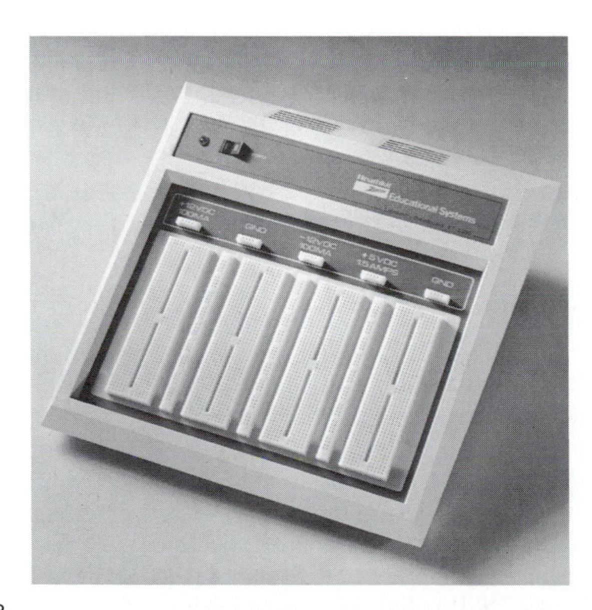


FIGURE 1-13

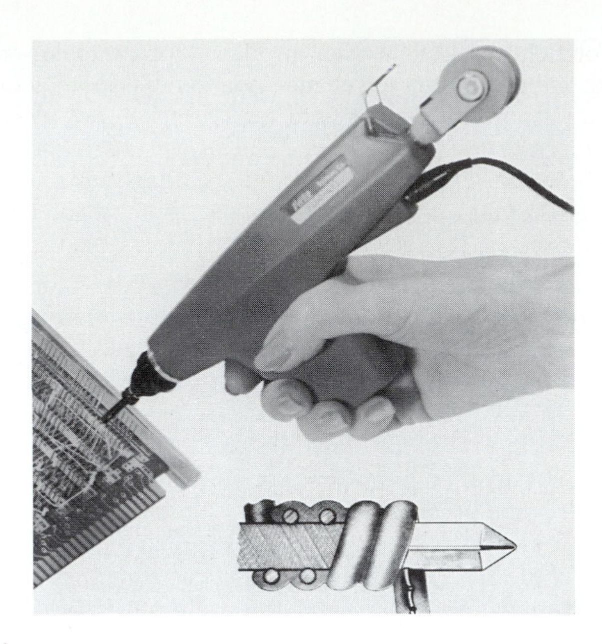

FIGURE 1-14

The other method of breadboarding is wirewrapping (Fig. 1-14). Sockets with special square or rectangular wirewrap pins are mounted on a piece of universal printed circuit board. These boards are perforated to accept wirewrap IC sockets, and usually have printed power distribution and ground busses. Using a special tool, the wire interconnections are strapped from point-to-point as needed by the circuit. The wire is wrapped around each contact tightly enough to break through the insulation to make the electrical connection.

In general, socketed breadboards are typically used for student applications, for partial breadboarding in professional laboratories, and for small design projects that do not justify the cost of wirewrap board. Wirewrapping is used for larger, more extensive, projects. Although the distinctions between types of breadboard and the respective supporting hardware are quite flexible, they hold true for a wide range of situations.

*Brassboards* are full-up model shop versions of the circuit which will plug into the actual equipment cabinet where the final circuit will reside. Some brassboards are wirewrapped, but most are printed circuits. It is important to have the brassboard in its nearly final configuration so it can be tested in the actual equipment. In contrast, the breadboard is purely a laboratory model.

The brassboard is usually made using model shop handwork methods and not routine automated production methods. The brassboard differs from the final version in that it might contain dead bug and kluge card modifications in which components are informally mounted on the board for purposes of testing circuit changes.

The full-scale development (FSD) model is the highest preproduction model. It might look like the first article production model that is eventually made and is intended for field tests of the final product. For example, an FSD model of a two-way landmobile radio transceiver may be mounted in an actual vehicle and used for its intended purpose in tests or field trials. For aircraft electronics, the FSD model must be built and tested according to flight-worthiness criteria. The FSD model is built as near to regular production methods as possible.

There are several principles to remember when breadboarding electronic circuits. While it is possible to get away with ignoring these rules, they constitute good practice and ignoring them is risky. The rules are:

1. Insert and remove components only with the power turned off.
2. Wiring and changes to wiring are done with the power turned off.
3. Check all wiring prior to applying power for the first time.
4. Power distribution and ground wiring is always done first, and then checked before any other wiring is done. (In many schematics the $V-$ and $V+$ wiring is omitted, but that doesn't mean these connections are not to be made.)
5. Use single point (or star) grounding wherever possible. A ground plane or ground bus should not be used unless signal frequencies are low, signal voltage levels are relatively large, and current drain from the DC power supply is low. Otherwise, ground noise and ground loop voltage drop problems will exist.
6. Applying a signal to IC input pins when there is not DC applied can cause the substrate diode to be forward-biased, with potential for damage to the device being very high. Therefore, do not apply signal until the DC power is turned on, turn off the signal source prior to turning off the DC power supply, and never apply a signal with a positive peak that exceeds $V+$ or a negative peak that exceeds $V-$.
7. All measurements are to be made with respect to common or ground, unless special test equipment is provided. For example, in Fig. 1-15 differential voltage $V_d$ is composed of two ground referenced voltages, $V1$ and $V2$. Measure these voltages separately and then take the difference between them ($V2$-$V1$).

In this chapter we introduced the integrated circuit, discussed the power supply for the IC, and demonstrated some construction practices and techniques. In the next chapter we will introduce the ideal and practical operational amplifier, and examine the applications of an op-amp with no feedback network.

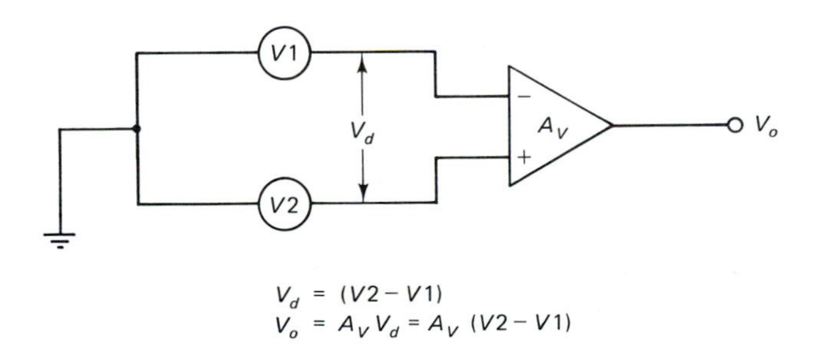


FIGURE 1-15

## 1-10 SUMMARY

1. Universal IC amplifiers include devices such as the operational amplifier, current difference or Norton amplifier, and the operational transconductance amplifier.
2. Analog signals are continuous in both range and domain; digital signals are discrete in range and possibly also domain; sampled signals are continuous in range but discrete in domain.
3. Common forms of integrated circuit packages include the metal can, dual inline package (DIP), flatpack, and surface-mounted device.
4. There are several scales of integration available: Small Scale Integration (less than 20 internal devices), Medium Scale Integration (approximately 100 internal devices), Large Scale Integration (100 to 1000 internal devices), and Very Large Scale Integration (more than 1000 internal devices).
5. Hybrid devices are a cross between integrated and discrete technology. Treated as if they are an IC device, these hybrids are actually packaged miniature printed circuits.
6. Most linear IC devices operate from bipolar power supplies (a supply in which $V-$ is negative to common while $V+$ is positive to common). By special biasing methods, the linear device can operate from single polarity supplies.
7. Both socketed breadboards and wirewrapping techniques are used to make prototype circuits.

## 1-11 RECAPITULATION

Now return to the objectives and Prequiz questions at the beginning of the chapter and see how well you can answer them. If you cannot answer certain questions, mark them and review the appropriate parts of the text. Next, try to answer the questions and work the problems below, using the same procedure.

## 1-12 QUESTIONS AND PROBLEMS

1. In an ____________ amplifier device the voltage amplifier transfer function is controlled entirely by the feedback loop characteristics.
2. List two principal advantages of integrated vs. discrete electronics components.
3. A ____________ amplifier has two inputs: inverting and noninverting. This type of amplifier forms the basis for the popular operational amplifier.
4. A Norton amplifier is also called a ____________ ____________ amplifier.
5. The ____________ ____________ amplifier has a transfer function of the basic form $\Delta I_o/\Delta V_{in}$.
6. A signal is found to be continuous in both range and domain. This is a(n) ____________ signal.
7. A signal is found to be discrete in both range and domain. This is a(n) ____________ signal.
8. A signal is found to be continuous in range, but discrete in domain. This is a(n) ______ ____________ signal.
9. List three types of package used for IC linear devices.

10. True or false: All circuits involving linear IC devices have an output signal which is related to the input signal by a linear scaler factor.
11. A device contains 12 internal components. This device is __________ scale integration.
12. A device contains 100 internal components. This device is __________ scale integration.
13. A device contains 684 internal components. This device is __________ scale integration.
14. The substrate diode must be kept normally __________ biased.
15. An operational amplifier has specified voltages as follows: maximum $V- = -18$ volts, maximum $V+ = +18$ volts, and maximum pin-to-pin [$(V-)$ to $(V+)$] voltage of 28 Vdc. If $V+ = 15$ volts, what is the maximum permissible value of $V-$?
16. What is the maximum permissible input voltage peak value when the maximum applied $V+ = 18$ volts?
17. What is the value of voltage $V1$ in Fig. 1-16?

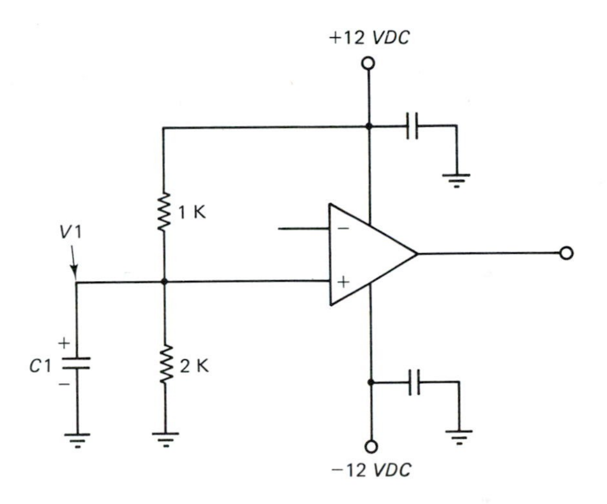


FIGURE 1-16

18. Using the standard engineering practice "rule of thumb," what is the minimum allowable value of $C1$ in Fig. 1-16 if the amplifier must operate down to a frequency of 30 Hz?
19. The military device temperature range is __________ °C to __________ °C.
20. Two general forms of microelectronic linear device are __________ and __________. Which of these are commonly called "integrated circuits?"
21. A __________ is a microelectronic device that is analogous to a miniature printed circuit, but has IC devices in "die" form, chip resistors and other generally unpackaged components mounted to a ceramic substrate.
22. Draw the circuit for an operational amplifier operating from a single positive $(V+)$ DC power supply. Select bias resistors to set the noninverting input to one-third of the $V+$ supply.
23. An operational amplifier circuit similar to Fig. 1-8 uses resistor values as follows: $R1 =$ 2.7 kohms, $R2 = 6.8$ kohms. Calculate the bias applied to the noninverting input if $V+$ is $+15$ volts.

24. Draw the circuit for a protection scheme that will protect an operational amplifier from reversed DC power supplies.
25. A cascade chain of three operational amplifiers connected as inverting followers operates from a single DC power supply. Draw the circuit of the power distribution network for this amplifier in which provision is made for preventing oscillation due to cross-coupling between stages.
26. List desirable features in a DC power supply for use on the bench where operational amplifier and other linear IC devices are used.
27. Define the differences between a *breadboard* and *brassboard*.

CHAPTER 2

# The IC Operational Amplifier

## OBJECTIVES

1. Learn the basic pin functions of the standard operational amplifier.
2. Learn the most common package styles used for IC operational amplifiers.
3. Describe the electrical parameters associated with each basic pin on an operational amplifier.
4. Learn the properties of the ideal operational amplifier.

## 2-1 PREQUIZ

These questions test your prior knowledge of the material in this chapter. Try answering them before you read the chapter. Look for the answers (especially those you answered incorrectly) as you read the text. After you have finished studying the chapter, try answering these questions again and those at the end of the chapter (see Section 2-11).

1. List the seven properties of the ideal operational amplifier.
2. List the five basic pins found on most operational amplifiers.
3. Describe the difference between inverting and noninverting inputs.
4. Write the transfer function for a differential amplifier in which $A_v$ is the gain, and $V1$ and $V2$ are the input signals applied to the inverting and noninverting inputs, respectively.

## 2-2 INTRODUCTION

The original operational amplifiers, first built in 1948, were designed to perform mathematical operations in analog computers. Their use in other applications followed because the op-amp is a very good DC differential amplifier with an extremely high gain. The immense benefits and extreme flexibility of the device derive from this one simple fact. Applications for the operational amplifier are found in instrumentation, process monitoring and control, servo control systems, signals processing, communications, measuring and testing circuits, alarm systems, medicine, science, and even in some digital computers. The operational amplifier might be called a universal linear amplifier.

## 2-3 OPERATIONAL AMPLIFIERS

Operational amplifiers are probably the most widely used linear integrated circuits on the market today. They are the most flexible linear IC devices available because we can manipulate the overall forward transfer function by controlling the feedback network properties. Other universal amplifiers (CDA and OTA) have failed to overcome the op-amp in either sales or usefulness.

Figure 2-1 shows a simplified schematic diagram of the internal circuitry of the popular and inexpensive 741 operational amplifier. There are three main sections to this circuit: *input amplifier*, *gain stages*, and *output amplifier.*

The output stage is a complementary symmetry push-pull DC power amplifier. Typical designs produce from 50 to 500 milliwatts of output power. The output stage operates as a push-pull amplifier because transistors $Q9$ and $Q10$ have opposite polarities: $Q9$ is NPN and $Q10$ is PNP. An output signal is taken from the junction of the $Q9/Q10$ emitters. If the two transistors conduct an equal amount, the net output voltage is zero.

The intermediate gain stages are shown here in simplified block diagram form. These stages provide the high gain required, some level translation, and (in the case of the 741) some internal frequency compensation.

The input stage is a DC differential amplifier made from bipolar transistors. Although there are a few single-input devices on the market sold as operational amplifiers, they are in reality high gain DC amplifiers, not true op-amps. The reason is that a true *operational* amplifier must be able to perform a wide range of mathematical operations and that ability requires both inverting and noninverting input functions. This also explains why bipolar DC power supplies must be used in op-amp circuits because output voltages may be either zero, positive, or negative, and the device must be able to accommodate all three possibilities.

### 2-3.1 DC Differential Amplifiers

Figure 2-2A shows a simplified DC differential amplifier circuit. Two transistors ($Q1$ and $Q2$) are connected at their emitters to a single *constant current source* (CCS), $I3$. Because current $I3$ cannot vary, changes in either $I1$ or $I2$ will also affect the other current. For example, an increase in current $I1$ means a necessary decrease in current $I2$ in order to satisfy Eq. 2-1

$$I3 = I1 + I2 \tag{2-1}$$

where

$I3$ is the constant current
$I1$ is the C-E current of transistor $Q1$
$I2$ is the C-E current of transistor $Q2$

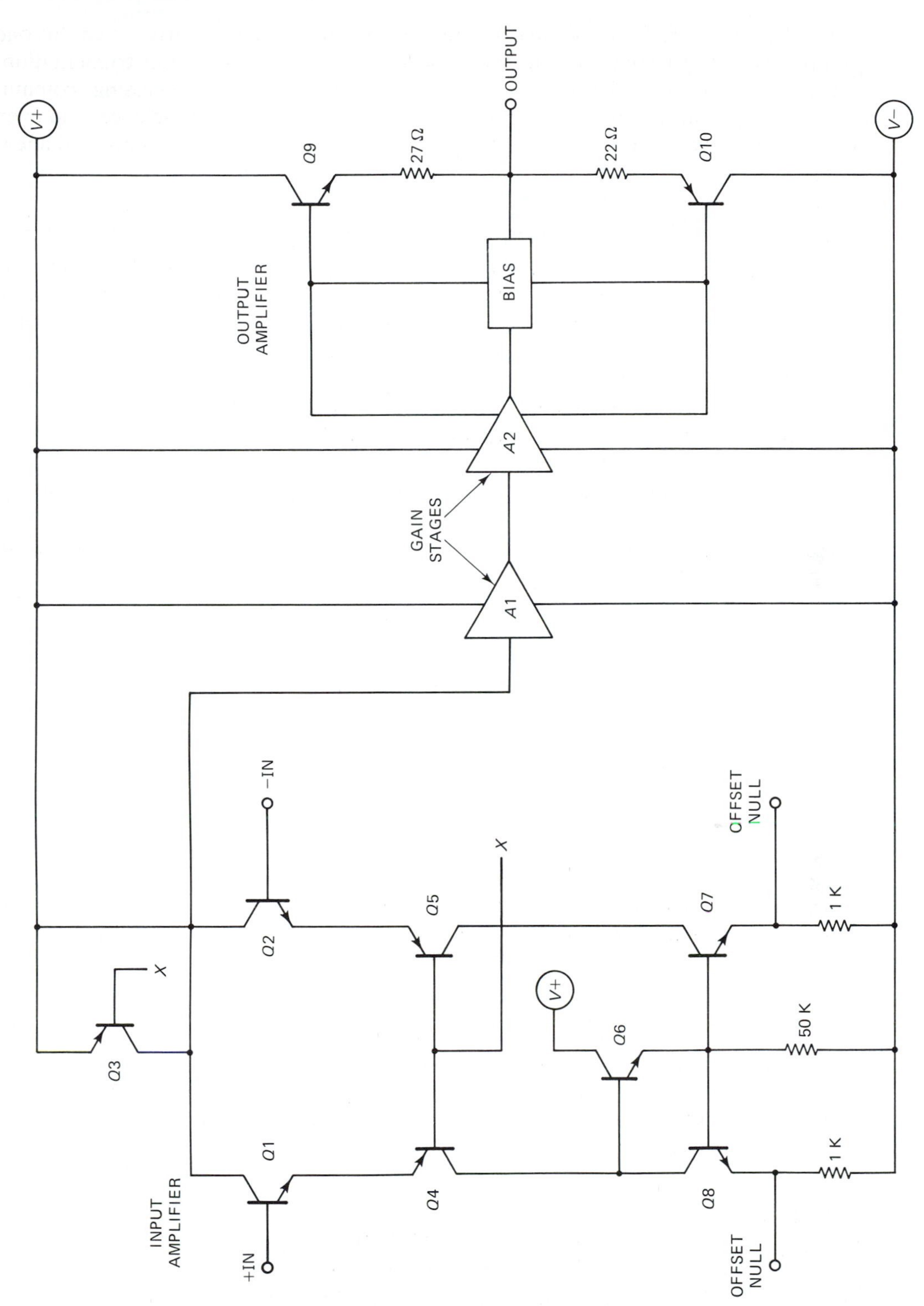
V+
OUTPUT
V−
Q9
27 Ω
22 Ω
Q10
OUTPUT
AMPLIFIER
BIAS
A2
GAIN
STAGES
A1
−IN
OFFSET
NULL
X
Q5
Q7
1 K
Q2
Q3
V+
Q6
50 K
INPUT
AMPLIFIER
Q1
1 K
+IN
Q4
Q8
OFFSET
NULL

FIGURE 2-1

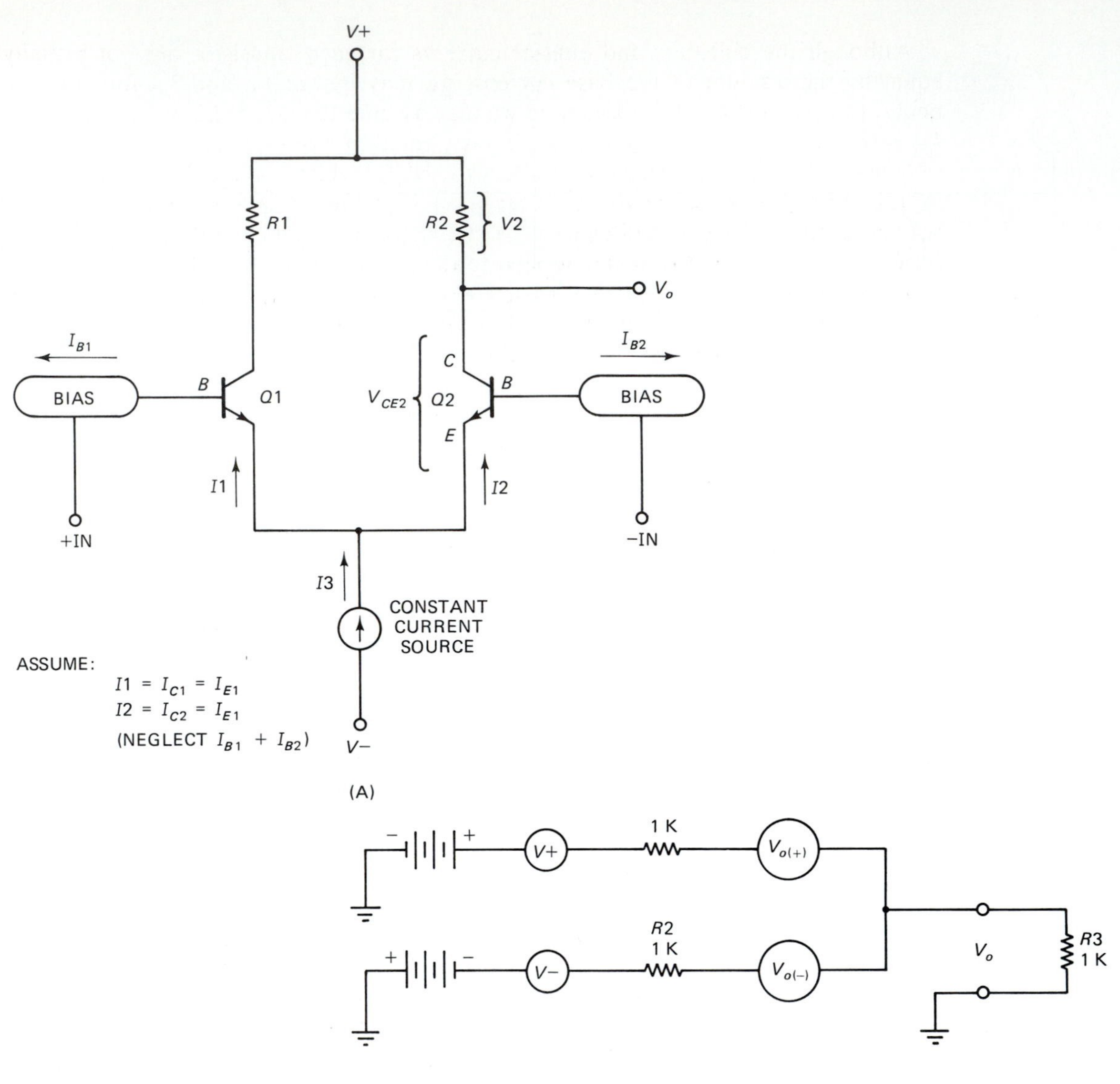


ASSUME:

$$V+ = +10\ VDC$$
$$V- = -10\ VDC$$
$$V_O = V_{O(+)} + V_{O(-)}$$
$$R1 = R2 = R3 = 1\ \text{K}\Omega$$

A) $$V_{O(-)} = \frac{(V-)(R3)}{(R2 + R3)} = \frac{(-10\ Vdc)(1\ \text{K})}{(1\ \text{K} + 1\ \text{K})} = \frac{(-10\ Vdc)(1)}{2} = -5\ Vdc$$

B) $$V_{O(+)} = \frac{(V+)(R3)}{(R1 + R3)} = \frac{(+10\ Vdc)(1\ \text{K})}{(1\ \text{K} + 1\ \text{K})} = \frac{(+10\ Vdc)(1)}{2} = +5\ Vdc$$

C) $$V_O = V_{O(-)} + V_{O(+)} = (-5\ Vdc) + (+5\ Vdc) = 0\ Vdc$$

(B)

FIGURE 2-2

Although the collector and emitter currents for each transistor are not actually equal by the amount of the base current, we may ignore $I_{B1}$ and $I_{B2}$ for the time being. For purposes of this discussion we may assume that $I1 = I_{C1} = I_{E1}$, and $I2 = I_{C2} = I_{E2}$, even though they are not, in fact, equal in real circuits.

Consider two voltage drops created by current $I2$ (which is the C-E current flowing in transistor $Q2$). $V2$ is the voltage drop $I2 \times R2$, while $V_{CE2}$ is the collector-emitter voltage drop across $Q2$. Voltage $V_{CE2}$ depends on the conduction of $Q2$, which in turn is determined by the signal applied to the noninverting input (+IN).

If the voltage applied to both −IN and +IN inputs are equal, $I1$ and $I2$ are equal. In this case, the quiescent values of $V2$ and $V_{CE2}$ are approximately equal (as determined by internal bias networks) so the relative contributions of $V-$ and $V+$ to output potential $V_o$ are equal.

A model circuit is shown in Fig. 2-2B to demonstrate how $V_o$ is formed. The output voltage $V_o$ is the sum of two contributors, $V_{o(-)}$ is the contribution from $V-$ and $V_{o(+)}$ is the contribution from $V+$. These voltages are derived from the voltage drops across $R1$ and $R2$ respectively, and in our model represents the voltage drops $V2$ and $V_{CE2}$ above. As long as $R1$ and $R2$ are balanced, the sum $V_o = V_{o(-)} + V_{o(+)}$ is equal to zero. But if either $R1$ or $R2$ change, then $V_o$ will not be zero. If you wish to prove this fact, repeat the arithmetic shown in Fig. 2-2B using a different value than 1 kohms for either $R1$ or $R2$. For example, when $R1$ is changed to 2 kohms, then the output voltage $V_o$ is −1.667 Vdc instead of zero.

Now let's consider the operation of the inverting input (−IN), which is the base terminal of transistor $Q2$. Recall that an inverting input produces an output signal that is 180° out of phase with the input signal. In other words, as $V_{-IN}$ goes positive, $V_o$ goes negative; as $V_{-IN}$ goes negative, $V_o$ goes positive.

If a positive signal voltage is applied to the −IN input and +IN = 0, then NPN transistor $Q2$ is turned on harder. This increases current $I2$ because the collector-emitter resistance of $Q2$ drops. We now have an inequality between $V2$ and $V_{CE2}$, voltage $V2$ increases, while $V_{CE2}$ decreases. The result is the contribution of $V-$ to $V_o$ is greater, so $V_o$ goes negative. The base of $Q2$ is therefore the inverting input because *a positive input voltage produced a negative output voltage.*

If the signal voltage applied to the −IN input is negative instead of positive, the situation changes. Then $Q2$ starts to turn off, so $I2$ drops. Voltage $V_{CE2}$ therefore increases and $V2$ decreases. The relative contribution of $V+$ to $V_o$ is now greater than the contribution of $V-$, so $V_o$ goes positive. Again we see inverting behavior because *a negative input voltage produced a positive output voltage.*

Now consider the noninverting input, which is the base of transistor $Q1$. Recall that a noninverting input produces an output signal that is in phase with the input signal. A positive going input signal produces a positive going output signal, and a negative going input signal produces a negative going output.

Suppose −IN = 0 and a positive signal voltage is applied to +IN. In this case $I1$ increases. Because Eq. 2-1 must be satisfied, an increase in $I1$ results in a decrease of $I2$ in order to keep $I3 = I1 + I2$ constant. Reducing $I2$ reduces $V2$ (which is $I2 \times R2$), so the contribution of $V+$ to $V_o$ goes up, or $V_o$ *goes positive in response to a positive input voltage.* This is noninverting input.

Now suppose that a negative signal voltage is applied to +IN. Now there is a decrease in the conduction of $Q1$, so $I1$ drops. Again, to satisfy Eq. 2-1 current $I2$ increases. This condition increases $V2$, reducing the contribution of $V+$ to $V_o$, forcing $V_o$ negative. Because *a negative input voltage produced a negative output voltage*, we may again say +IN is a noninverting input.

### 2-3.2 Categories of Operational Amplifiers

Now we shall widen our discussion beyond the basic operational amplifier to a larger selection of devices. Table 2-1 shows a hierarchy of commonly available devices. Some of these devices have been on the market for a long time, while others are relatively new. This list is intended to be representative, rather than exhaustive. Let's take a brief look at each category.

**General Purpose.** These are "garden variety" operational amplifiers that are neither special purpose nor premium devices. Most of these devices are frequency compensated, so designers trade off bandwidth for inherent stability. As such, the general purpose devices can be used in a wide range of applications with few external components. Devices are usually selected from this category unless a property of another class brings a unique advantage to some particular application.

**Voltage Comparators.** These devices are not true operational amplifiers, but they are based on op-amp circuitry. While all op-amps can be used as voltage comparators, the reverse is not true. IC comparators (like the LM-311) usually cannot be used as op-amps (Section 2-7).

**Low Input Current.** Although ideal op-amps (Section 2-4) have zero input bias current, real devices have a small current due to input transistor biasing or leakage. This class of devices typically uses MOSFET, JFET, or superbeta (Darlington) transistors for the input stage instead of NPN/PNP bipolar devices. The manufacturer may choose to use a nulling or selection technique in process that will reduce input bias current. Low input current devices typically have picoampere level currents, rather than microampere or milliampere input currents found in other devices.

**Low Noise.** These devices are optimized to reduce internally generated noise.

**Low Power.** This category of op-amp optimizes internal circuitry to reduce power consumption. Many of these devices also operate at very low DC power supply potentials ($\pm 1.5$ Vdc).

**Low Drift.** All DC amplifier circuits experience an erroneous change of output voltage as a function of temperature. Devices in this category are internally compensated to minimize temperature drift. These devices are typically used in instrumentation circuits where drift is an important concern, especially when handling low level input signals.

**Wide Bandwidth.** Also called video op-amps in some literature, these devices have a very high gain-bandwidth level. One device, for example, has a G-B product of 100 MHz, compared with 0.300 to 1.2 MHz for various 741 devices.

**Single DC Supply.** These devices provide op-amp-like behavior from a monopolar (typically $V+$) DC power supply. However, not all op-amp performance will be

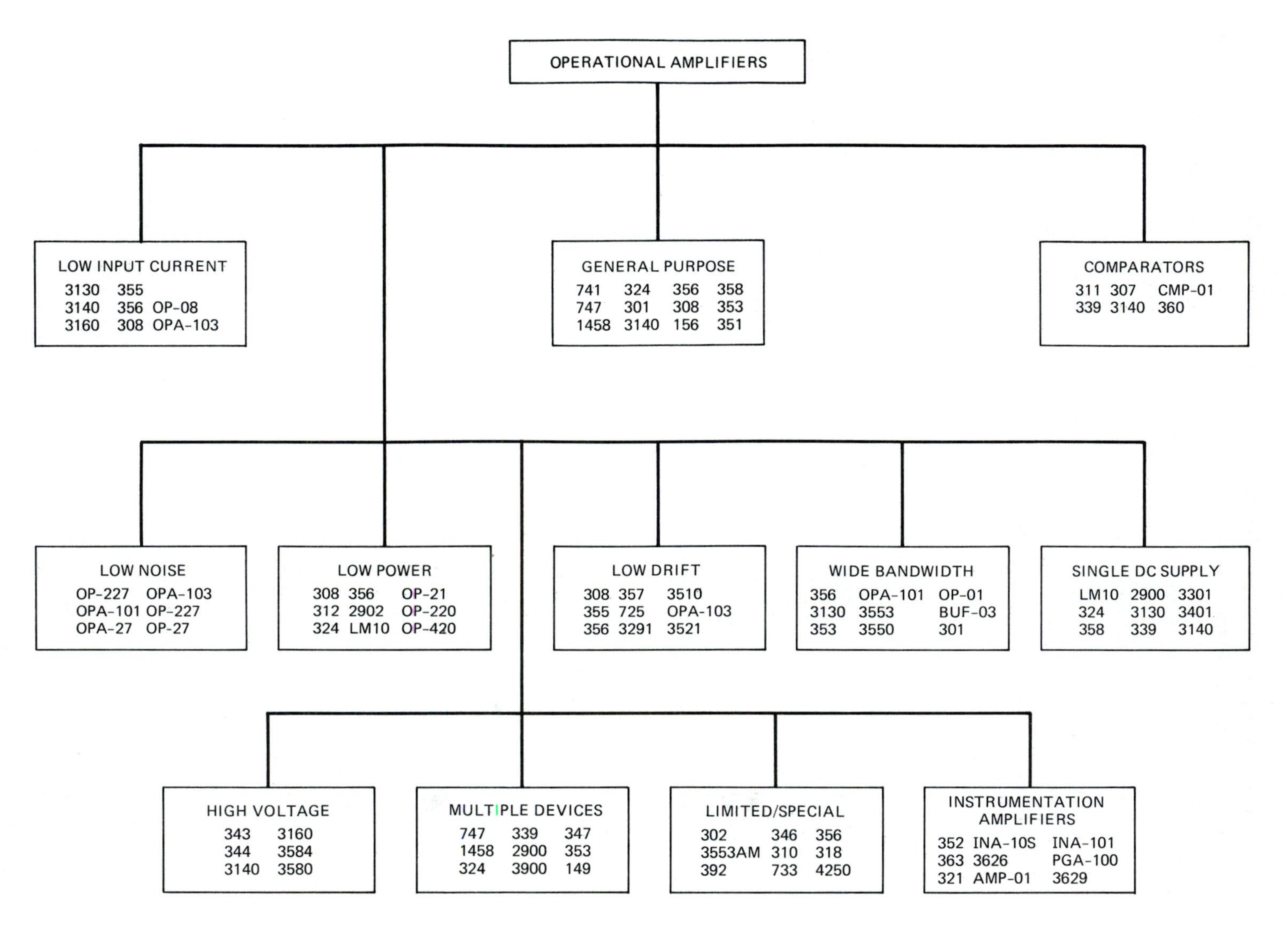
OPERATIONAL AMPLIFIERS
LOW INPUT CURRENT
3130 355
3140 356 OP-08
3160 308 OPA-103
GENERAL PURPOSE
741 324 356 358
747 301 308 353
1458 3140 156 351
COMPARATORS
311 307 CMP-01
339 3140 360
LOW NOISE
OP-227 OPA-103
OPA-101 OP-227
OPA-27 OP-27
LOW POWER
308 356 OP-21
312 2902 OP-220
324 LM10 OP-420
LOW DRIFT
308 357 3510
355 725 OPA-103
356 3291 3521
WIDE BANDWIDTH
356 OPA-101 OP-01
3130 3553 BUF-03
353 3550 301
SINGLE DC SUPPLY
LM10 2900 3301
324 3130 3401
358 339 3140
HIGH VOLTAGE
343 3160
344 3584
3140 3580
MULTIPLE DEVICES
747 339 347
1458 2900 353
324 3900 149
LIMITED/SPECIAL
302 346 356
3553AM 310 318
392 733 4250
INSTRUMENTATION AMPLIFIERS
352 INA-10S INA-101
363 3626 PGA-100
321 AMP-01 3629

Table 2.1

available from some of these devices because the output voltage may not be able to assume negative values.

**High Voltage.** Most op-amps operate from ±6 to ±22 Vdc power supplies. A few devices in the high voltage category operate from ±44 Vdc supplies, and at least one proprietary hybrid model (not listed here) operates from ±100 Vdc power supplies.

**Multiple Devices.** This category of devices has two, three, or four operational amplifiers in a single package. The 1458 device, for example, contains two 741 devices in either 8-pin metal can or miniDIP packages.

**Limited/Special Purpose.** The devices in this category are designed for either a limited range of uses, or highly specialized uses. The type 302 buffer, for example, is connected internally into the noninverting unity gain follower configuration (see Chapter 3). Some consumer audio devices also fit into this category.

**Instrumentation Amplifiers.** Although the instrumentation amplifier is arguably a special purpose device, it is sufficiently universal to warrant a class of its own. The IA is a DC differential amplifier made of either two or three internal op-amps. Voltage gain can be set by either one or two external resistors. Chapter 6 deals with these amplifiers in detail.

## 2-4 THE "IDEAL" OPERATIONAL AMPLIFIER

When you study any type of electron device, be it vacuum tube, transistor, or integrated circuit, it is wise to start with an ideal representation of that device and then proceed to practical devices. In some cases, the practical and ideal devices are so far apart you might wonder at the wisdom of this approach. In engineering and technical schools the students often do laboratory experiments to put into practice what was learned in the theory classes. When students study their lab workbook sheets they may wonder why the experiments fail to come close to predicted theoretical behavior. But IC operational amplifiers, even low-cost commercial products, so nearly approximate the ideal op-amp of textbooks that the lab experiments actually work. The ideal model analysis method thus becomes extremely useful for understanding the technology, learning to design new circuits, or figuring out how someone else's circuit works.

In Chapter 3 we will discuss the inverting and noninverting amplifier configurations of the op-amp. There we will derive design equations that describe the operation of real circuits from both the ideal model and a feedback amplifier model. The usefulness of this simplified approach stems from the correspondence of the ideal and practical operational amplifier IC devices.

### 2-4.1 Properties of the Ideal Operational Amplifier

The ideal op-amp is characterized by seven properties. From this short list of properties we can deduce circuit operation and design equations. Also, the list gives us a basis for examining nonideal operational amplifiers and their defects (plus solutions to the problems caused by those defects). The basic properties of the op-amp are:

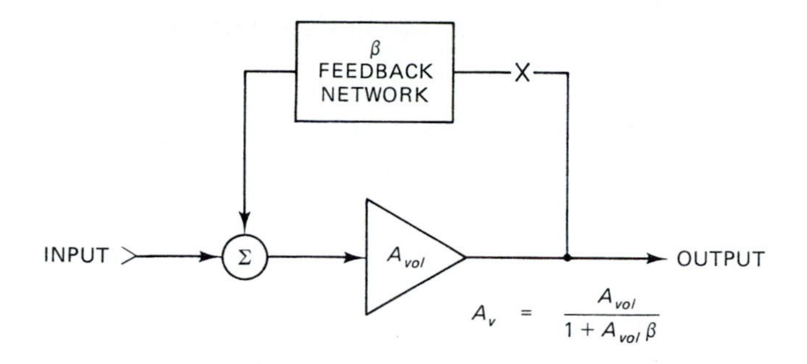


FIGURE 2-3

1. Infinite open-loop voltage gain
2. Infinite input impedance
3. Zero output impedance
4. Zero noise contribution
5. Zero DC output offset
6. Infinite bandwidth
7. Both differential inputs stick together

Let's look at these properties to determine what they mean in practical terms. You will find that some real op-amps only approximate these ideals, but for others the approximation is extremely good.

**Property No. 1—Infinite Open-Loop Gain.** The open-loop gain of any amplifier is its DC gain without negative or positive feedback. By definition, negative feedback is a signal fed back to the input 180° out of phase. In operational amplifier terms this means feedback between the output and the inverting input.

Negative feedback has the effect of reducing the open loop gain ($A_{vol}$) by a factor (called $B$) which depends on the transfer function and properties of the feedback network. Figure 2-3 shows the basic configuration for any negative feedback amplifier. The transfer equation for any circuit is the output function divided by the input function. The transfer function of a voltage amplifier is, therefore, $A_{vol} = V_o/V_{in}$. In Fig. 2-3 the term $A_{vol}$ represents the gain of the amplifier element only, that is the gain with the feedback network disconnected. The term B represents the transfer function of the feedback network without the amplifier. The overall transfer function of this circuit, with both amplifier element and feedback resistor in the loop, is defined as

$$A_v = \frac{A_{vol}}{1 + A_{vol}\text{B}} \tag{2-2}$$

where

$A_v$ is the closed-loop gain

$A_{vol}$ is the open-loop gain

B is the transfer equation for the feedback network

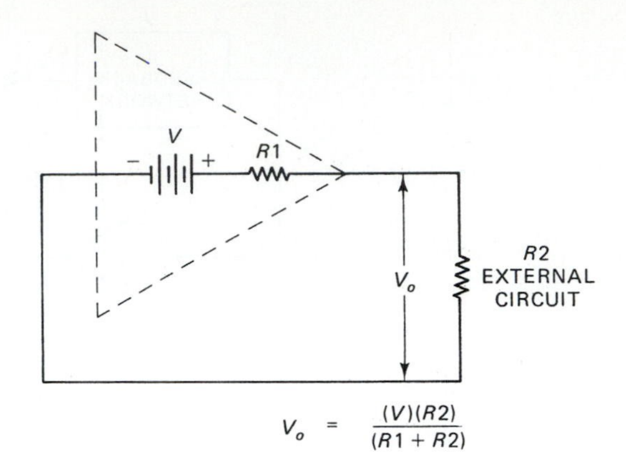


FIGURE 2-4

In the ideal op-amp $A_{vol}$ is infinite, so the voltage gain is a function only of the feedback network. In real op-amps, the value of the open-loop gain is not infinite, but it is quite high. Typical values range from 20,000 in low-grade consumer audio models to more than 2,000,000 in premium units (typically 200,000 to 300,000).

**Property No. 2—Infinite Input Impedance.** This property implies the op-amp input will not load the signal source. The input impedance of any amplifier is definable as the ratio of the input voltage to the input current, $Z_{in} = V_{in}/I_{in}$. When the input impedance is infinite, we must assume the input current is zero. Thus, an important implication of this property is the operational amplifier inputs neither sink nor source current. In other words, it will neither supply current to an external circuit, nor accept current from an external circuit. We will use an implication of this property ($I_{in} = 0$) to perform the circuit analysis in Chapter 3.

Real operational amplifiers have some finite input current other than zero. In low-grade devices this current can be substantial (for example, 1 milliampere) and will cause a large output offset voltage error in high gain circuits. The primary source of this current is the base bias currents from the NPN and PNP bipolar transistors used in the input circuits. Certain premium grade op-amps which feature bipolar inputs reduce this current to nanoamperes or picoamperes. However, in op-amps that use field effect transistors (FET) in the input circuits, the input impedance is quite high due to the very low leakage currents normally found in FET devices. The JFET input devices are typically called BiFET op-amps, while the MOSFET input models are called BiMOS devices. The CA-3140 device is a BiMOS op-amp in which the input impedance approaches 1.5 terraohms (i.e., $1.5 \times 10^{12}$ ohms)—which is near enough to infinite to make the inputs of those devices approach the ideal.

**Property No. 3—Zero Output Impedance.** A voltage amplifier, of which class the op-amp is a member, ideally has a zero output impedance. All real voltage amplifiers, however, have a nonzero impedance. Figure 2-4 represents any voltage source, including amplifier outputs, and its load or external circuit. Potential $V$ is a perfect internal voltage source with no internal resistance. Resistor $R1$ represents the internal resistance of the source and $R2$ is the load. Because the internal resistance, which in amplifiers is usually called output resistance, is in series with the load resistance, the

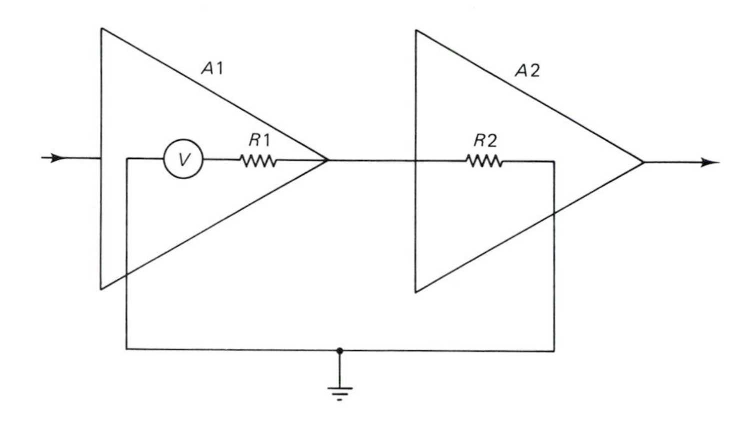


FIGURE 2-5

output voltage $V_o$ available to the load is reduced by the voltage drop across $R1$. Thus, the output voltage is given by

$$V_o = \frac{VR2}{R1 + R2} \tag{2-3}$$

It is clear from the above equation that the output voltage will equal the internal source voltage only when the output resistance of the amplifier ($R1$) is zero. In that case, $V_o = V \times (R2/R2) = V$. Thus, in the ideal voltage source we get the maximum output voltage and the least error because no voltage is dropped across the internal resistance of the amplifier.

Real operational amplifiers do not have a zero output impedance. The actual value is typically less than 100 ohms, with many being in the neighborhood of 30 to 50 ohms. So for typical devices the operational amplifier output can be treated as if it were ideal.

A rule of thumb used by designers is to set the input resistance of any circuit which is driven by a nonideal voltage source output to be at least ten times the previous stage output impedance. We see this situation in Fig. 2-5. Amplifier $A1$ is a voltage source driving the input of amplifier $A2$. Resistor $R1$ represents the output resistance of $A1$ and $R2$ represents the input resistance of amplifier $A2$. In practical terms, we find that the circuit where $R2 > 10R1$ will yield results acceptably close to ideal for many purposes. In some cases, however, the rule $R2 > 100R1$ must be followed where greater precision is required.

**Property No. 4—Zero Noise Contribution.** All electronic circuits, even simple resistor networks, produce noise signals. A resistor creates noise due to the movement of electrons in its internal resistance element material. For all resistive circuits, the noise contribution is:

$$V_{\text{noise}} = \sqrt{4\ K\ T\ B\ R} \tag{2-4A}$$

$$P_{\text{noise}} = 4\ K\ T\ B\ R \tag{2-4B}$$

where

$V_{noise}$ = the noise voltage

$P_{noise}$ = the noise power

$K$ = Boltzman's constant ($1.38 \times 10^{-23}$ J/°K)

$T$ = the temperature in degrees Kelvin

$B$ = the bandwidth in hertz

$R$ = the resistance in ohms

As you see from Eqs. 2-4 A and B, the noise voltage produced by a resistor varies with the square root of its resistance, while noise power varies directly with resistance.

In the ideal operational amplifier, zero noise voltage is produced internally. Thus, any noise in the output signal must have been present in the input signal as well. Except for amplification, the output noise voltage will be exactly the same as the input noise voltage. In other words, the op-amp contributed nothing to the output noise. This is one area where practical devices depart from the ideal. Practical op-amps do not approximate the ideal, except for certain high cost premium low-noise models.

Amplifiers use semiconductor devices which create not only the resistive noise described above, but they also create a special noise of their own. There are a number of internal noise sources in semiconductor devices. Any good text on transistor theory will give you more information on them. For present purposes, however, assume the noise contribution of the op-amp to be considerable in low signal level situations. Premium op-amps are available in which the noise contribution is very low. These devices are usually advertised as premium low-noise types. Others, such as the RCA/GE metal can CA-3140 device, offer relatively low-noise performance when the DC supply voltages are limited to ±5 Vdc and the metal package of the op-amp is fitted with a flexible "TO-5" style heatsink.

**Property No. 5—Zero Output Offset.** The output offset voltage of any amplifier is the output voltage that exists when it should be zero. The voltage amplifier sees a zero input voltage when the inputs are both grounded. This should produce a zero output voltage. If the output voltage is nonzero, there is an output offset voltage present. In the ideal op-amp, this offset voltage is zero volts. But real op-amps exhibit at least some amount of output offset voltage. In the real IC operational amplifier the output offset voltage is nonzero, although in some cases it can be quite low. Several methods for lessening the output offset voltage are discussed in Chapter 5.

**Property No. 6—Infinite Bandwidth.** The ideal op-amp will amplify all signals from DC to the highest AC frequencies. In real op-amps, however, the bandwidth is sharply limited. The "gain-bandwidth (G-B) product," which is symbolized by $F_t$, is the frequency at which the voltage gain drops to unity (1). The maximum available gain at any frequency is found by dividing the maximum required frequency into the gain-bandwidth product. If the value of $F_t$ is not sufficiently high, the circuit will not behave in classical op-amp fashion at some frequencies.

Some op-amps have G-B products in the 10 to 20 MHz range. Others are more limited. The 741 devices are very limited. They will perform as an op-amp only to frequencies of a few kilohertz. Above that range, the gain drops off considerably. But in return for this apparent limitation, unconditional stability is obtained. Such op-amps are said to be "frequency compensated." The frequency compensation of these devices reduces the G-B product and provides the inherent stability. Noncompensated op-amps yield wider frequency response, but only at the expense of a tendency to spontaneously oscillate.

**Property No. 7—Differential Inputs Stick Together.** Most operational amplifiers have two inputs: one inverting (−IN) input and one noninverting (+IN) input. "Sticking together" means that a voltage applied to one of these inputs also appears at the other input. This voltage is real; it is not merely a theoretical device used to evaluate circuits. If you apply a voltage to the inverting input, for example, and then connect a voltmeter between the noninverting input and the power supply common, the voltmeter will read the same potential on the noninverting as it did on the inverting input. The implication of this property is that we must treat both inputs the same mathematically. This fact will be important when we discuss the concept of virtual grounds as opposed to actual grounds and again when we examine the noninverting follower circuit.

## 2-5 STANDARD OPERATIONAL AMPLIFIER PARAMETERS

Understanding operational amplifier circuits requires knowledge of the parameters given in specification sheets. The list below represents the most commonly needed parameters. Methods of measuring some of these parameters are discussed in Chapter 5.

**Open-Loop Voltage Gain ($A_{vol}$).** Voltage gain is defined as the ratio of output voltage to input signal voltages ($V_o/V_{in}$), which is a dimensionless quantity. The open-loop voltage gain is the gain of a circuit without feedback (with the feedback loop open). In an ideal operational amplifier $A_{vol}$ is infinite, but in practical devices it will range from about 20,000 for low-cost devices to over 1,000,000 in premium devices.

**Large Signal Voltage Gain.** This is the ratio of the maximum allowable output voltage swing (usually one to several volts less than $V-$ and $V+$) to the input signal required to produce a swing of ±10 volts (or some other standard).

**Slew Rate.** This specifies the ability of the amplifier to change from one output voltage extreme to another while delivering full rated output current to the external load. The slew rate is measured in terms of voltage change per unit of time. The 741 operational amplifier, for example, is rated for a slew rate of 0.5 volts per microsecond (0.5 V/μS). Slew rate is usually measured in the unity gain noninverting follower configuration.

**Common Mode Rejection Ratio (CMRR).** A common mode voltage is one that is presented simultaneously to both inverting and noninverting inputs. In an ideal operational amplifier, the output signal resulting from the common mode voltage is zero, but in real devices it is nonzero. The common mode rejection ratio (CMRR) is the measure of the device's ability to reject common mode signals, and is expressed as the ratio of the differential gain to the common mode gain. The CMRR is usually expressed in decibels, with common devices having ratings between 60 dB and 120 dB. The higher the number, the better the device.

**Power Supply Rejection Ratio (PSRR).** Also called "power supply sensitivity," the PSRR is a measure of the operational amplifier's insensitivity to changes in the power supply potentials. The PSRR is the change of the input offset voltage (see below) for a 1-volt change in one power supply potential while the other is held constant. Typical values are in microvolts or millivolts per volt of power supply potential change.

**Input Offset Voltage.** The voltage required at the input to force the output voltage to zero when the input signal voltage is zero. The output voltage of an ideal operational amplifier is zero when $V_{in}$ is zero.

**Input Bias Current.** The current flowing into or out of the operational amplifier inputs. In some sources, this is defined as the average difference between currents flowing in the inverting and noninverting inputs.

**Input Offset (Bias) Current.** The difference between inverting and noninverting input bias current when the output voltage is held at zero.

**Input Signal Voltage Range.** The range of permissible input voltages measured in the common mode configuration.

**Input Impedance.** The resistance between the inverting and noninverting inputs. This value is typically very high: 1 megohm in low-cost bipolar operational amplifiers and over $10^{12}$ ohms in premium BiMOS devices.

**Output Impedance.** This parameter refers to the "resistance looking back" into the amplifier's output terminal. It is usually modeled as a resistance between output signal source and output terminal. Typically the output impedance is less than 100 ohms.

**Output Short-Circuit Current.** The current that flows in the output terminal when the output load resistance external to the amplifier is zero ohms (a short to common).

**Channel Separation.** This parameter is used on multiple operational amplifier integrated circuits (devices in which two or more operational amplifiers sharing the same package with common power supply terminals). The separation specification describes part of the isolation between the op-amps inside the same package. It is measured in decibels (dB). The 747 dual operational amplifier, for example, offers 120 dB of channel separation. From this specification we may imply that a 1-microvolt change will occur in the output of one of the amplifiers when the other amplifier output changes by 1 volt (1 V/1μV = 120 dB).

### 2-5.1 Minimum and Maximum Parameter Ratings

Operational amplifiers, like all electronic components, are subject to maximum ratings. If these ratings are exceeded, the user can expect either premature—often immediate—failure or unpredictable operation. The ratings described below are the most commonly used.

**Maximum Supply Voltage.** This is the maximum voltage that can be applied to the operational amplifier without damaging the device. The operational amplifier uses $V+$ and $V-$ DC power supplies which are typically $\pm 18$ Vdc, although some exist with much higher maximum potentials. The maximum rating for either $V-$ or $V+$ may depend on the value of the other (see below).

**Maximum Differential Supply Voltage.** This is the algebraic sum of $V-$ and $V+$, namely $[(V+) - (V-)]$. Quite often this rating is not the same as the summation of the maximum supply voltage ratings. For example, one 741 operational amplifier specification sheet lists $V-$ and $V+$ at 18 volts each, but the maximum differential supply voltage is only 30 volts. Thus, when both $V-$ and $V+$ are at maximum (18 volts DC each), the actual differential supply voltage is $[(+18\text{ V}) - (-18\text{ V})] = 36$. This is 6 volts over the maximum rating. Therefore when either $V-$ or $V+$ is at maximum value, the other must be proportionally lower. For example, when $V+$ is $+18$ volts, the maximum allowable value of $V-$ is $[30\text{ V} - 18\text{ V}] = 12$ Vdc.

**Power Dissipation, $P_d$.** This rating is the maximum power dissipation of the operational amplifier in the normal ambient temperature range (80° Celsius in commercial devices and 125°C in military devices). A typical rating is 500 milliwatts (0.5 watts).

**Maximum Power Consumption.** The maximum power dissipation, usually under output short circuit conditions, that the device will survive. This rating includes both internal power dissipation and device output power requirements.

**Maximum Input Voltage.** This is the maximum that can be simultaneously applied to both inputs. Thus, it is also the maximum common-mode voltage. In most bipolar operational amplifiers the maximum input voltage is nearly equal to the power supply voltage. There is also a maximum input voltage that can be applied to either input when the other input is grounded.

**Differential Input Voltage.** This is the maximum differential-mode voltage that can be applied across the inverting (−IN) and noninverting (+IN) inputs.

**Maximum Operating Temperature.** The maximum temperature is the highest ambient temperature at which the device will operate according to specifications with a specified level of reliability. The usual rating for commercial devices is 70 or 80°C, while military components must operate to 125°C.

**Minimum Operating Temperature.** The lowest temperature at which the device operates within specifications. Commercial devices operate down to either 0 or −10°C, while military components operate down to −55°C.

**Output Short-Circuit Duration.** This is the length of time the operational amplifier will safely sustain a short circuit of the output terminal. Many modern operational amplifiers are rated for indefinite output short-circuit duration.

**Maximum Output Voltage.** The maximum output potential of the op-amp is related to the DC power supply voltages. Operational amplifiers have one or more bipolar PN junctions between the output terminal and either $V-$ or $V+$ terminals. The voltage drop across these junctions reduces the maximum achievable output voltage. For example, if there are three PN junctions between the output and power supply terminals, the maximum output voltage is $[(V+) - (3 \times 0.7)]$, or $[(V+) - 2.1]$ volts. If the maximum $V+$ voltage permitted is 15 volts, then the maximum allowable output voltage is [(15 V) − (2.1 V)], or 12.9 volts. It is not always true, especially in older devices, that the maximum negative output voltage is equal to the maximum positive output voltage.

**Maximum Output Voltage Swing.** The absolute value of the voltage swing from maximum negative to maximum positive.

## 2-6 PRACTICAL OPERATIONAL AMPLIFIERS

Now let's turn our attention to practical devices. Because of its popularity and low cost, we will concentrate on the 741 device. The 741 family also includes the 747 and 1458 dual 741 devices. Although there are better operational amplifiers on the market, the 741 class of amplifiers is considered the industry standard.

Figure 2-6 shows the two most popular packages used for the 741. Figure 2-6A is the 8-pin miniDIP package and Fig. 2-6B is the 8-pin metal can package. The 741 is also available in flatpacks and 14-pin DIP packages, although these are becoming rare. The miniDIP pin-outs for a 1458 dual op-amp are shown in Fig. 2-6C.

The 741 has the following pins:

**−IN Inverting Input (Pin No. 2).** The output signals produced from this input are 180° out of phase with the input signal applied to −IN.

**+IN Noninverting Input (Pin No. 3).** Output signals are in phase with signals applied to the +IN input terminal.

**Output (Pin No. 6).** On most op-amps, the741 included, the output is single-ended. This means the output signals are taken between this terminal and the power supply common (see Fig. 2-7). The output of the 741 is said to be short-circuit proof because it can be shorted to common indefinitely without damage to the IC.

**V+ Power Supply (Pin No. 7).** The positive DC power supply terminal.

**V− Power Supply (Pin No. 4).** The negative DC power supply terminal.

**Offset Null (Pins 1 & 5).** These are used to accommodate external circuitry which compensates for offset (error) voltages (see Chapter 4).

The pin-out scheme shown in Fig. 2-6 is the industry standard for generic single operational amplifiers. There are numerous examples of amplifiers using different pin-outs, but a large percentage of the available devices use this scheme.

### 2-6.1 Standard Circuit Configuration

The standard circuit configuration for 741 family of operational amplifiers is shown in Fig. 2-7. The pin-outs are industry standard. The output signal voltage is impressed across load resistor $R_L$ connected between the output terminal (pin 6) and the power supply common. Most manufacturers recommend a 2 kohm minimum value for $R_L$. Also, note that some operational amplifier parameters shown in Table 2-2 are based on a 10 kohm load resistance. Because it is referenced to common, the output is single-ended.

The ground symbol shown in Fig. 2-7 indicates it is optional. The point of reference for all measurements is the common connection between the two DC power supplies ($V-$ and $V+$). Whether or not this point is physically connected to a ground, the

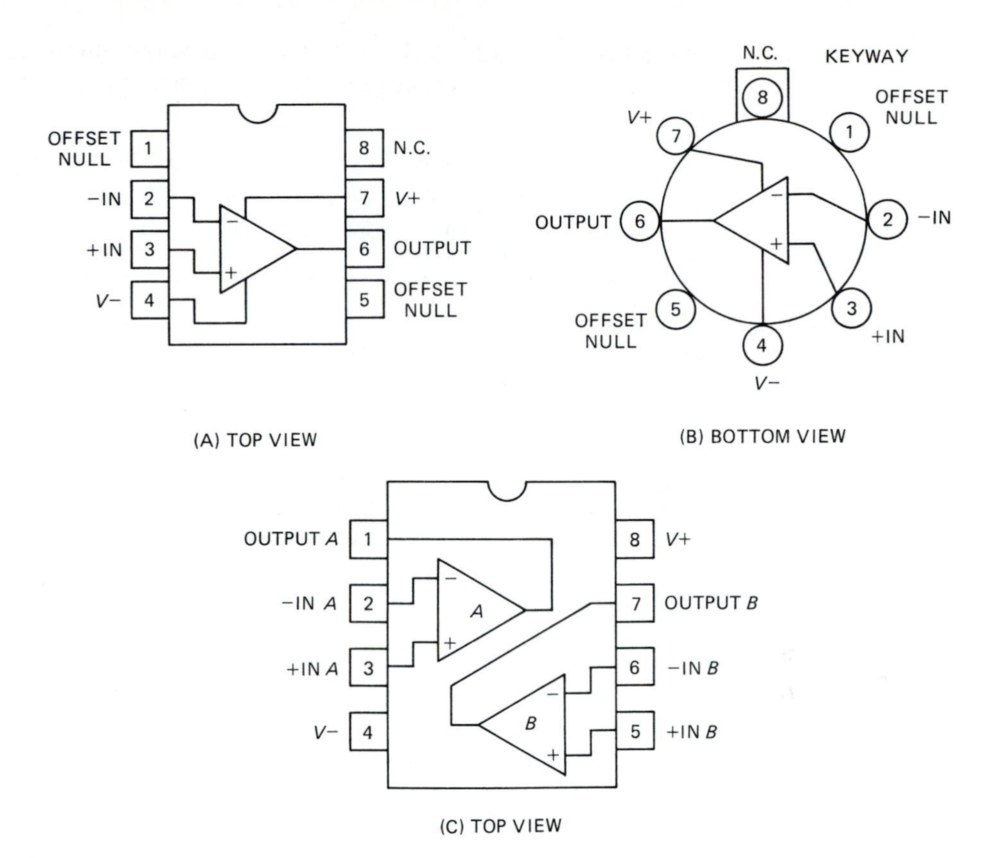


FIGURE 2-6

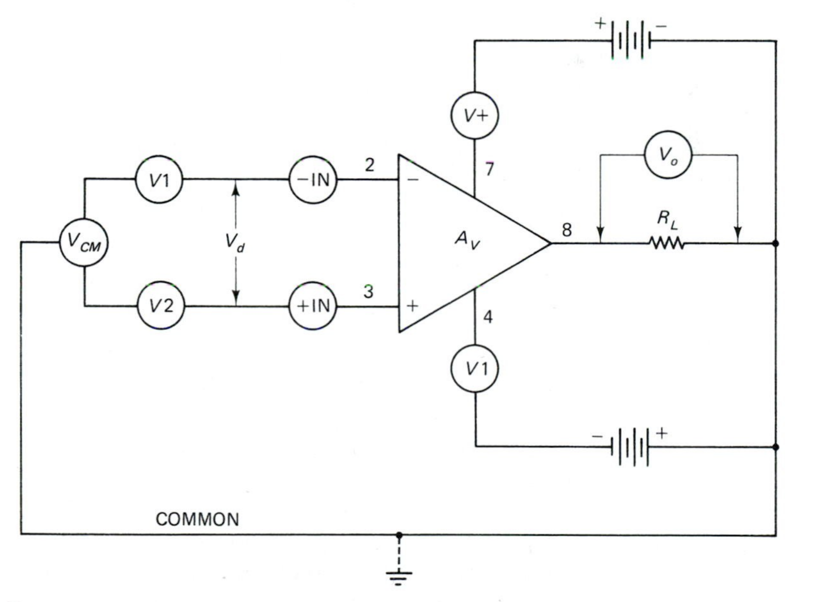


FIGURE 2-7

**TABLE 2-2**

| *Parameter* | *Min.* | *Typical* | *Max.* | *Units* |
|---|---|---|---|---|
| Input offset voltage $V_{io}$ | — | 1 | 5 | mV |
| Input bias current | — | 80 | 500 | nA |
| Input resistance | 0.3 | 2.0 | — | MΩ |
| Input voltage range | ±12 | ±13 | — | Volts |
| Large signal voltage gain | 50 | 200 | — | V/mV[1] |
| Output voltage swing | | | | |
| $R_L$ = 10 K | ±12 | ±14 | — | Volts |
| $R_L$ = 2 K | ±10 | ±13 | — | Volts |
| Output short circuit | — | 25 | — | MA |
| CMRR | 80 | 90 | — | dB |
| PSRR | 77 | 96 | — | dB[2] |
| Rise time ($A_v$ = 1) | — | 0.3 | — | uS |
| BW | — | 1.16 | — | MHz |
| Power consumption | — | 50 | 85 | mW |
| Power supply | — | ±22 | | Vdc |
| Power dissipation | — | 500 | — | mW |
| Diff. input voltage | — | ±30 | — | Volts |
| Input voltage | — | ±15 | — | Volts |

[1] $V_s = \pm 15$ Vdc, $V_o = \pm 10$ V
[2] $R_L \leq 10$ KΩ

equipment chassis or a dedicated system ground bus is optional in some cases and required in others. Whether it is required or not, however, the determination is made on the basis of circuit factors other than the basic nature of the op-amp.

The $V-$ and $V+$ DC power supplies are independent of each other. Do not make the mistake of assuming that these terminals are merely different ends of the same DC power supply. In fact, $V-$ is negative with respect to common while $V+$ is positive with respect to common.

The two input signals in Fig. 2-7 are labelled $V1$ and $V2$. Signal voltage $V1$ is the single-ended potential between common and the inverting input (−IN). $V2$ is the single-ended potential between common and the noninverting input (+IN). The 741 operational amplifier is *differential* because both −IN and +IN are present. Any differential amplifier produces an output proportional to the difference between the two input potentials. In Fig. 2-7 the *differential input potential* ($V_d$) is the difference between $V1$ and $V2$

$$V_d = V2 - V1 \tag{2-5}$$

### EXAMPLE 2-1

The table shown below shows six situations for input voltages to an operational amplifier. Calculate the differential input voltage, $V_d$, for each case.

| | $V1$ | $V2$ | $V_d$ |
|---|---|---|---|
| a. | +2 V | +1.5 V | ? |
| b. | −2 V | +1.5 V | ? |
| c. | +3 V | +5 V | ? |
| d. | −3 V | +5 V | ? |
| e. | +2 V | +2 V | ? |
| f. | +1.6 V | +1.75 V | ? |

**Solutions**

(a) $V_d = V2\text{-}V1 = (+2\text{ V}) - (+1.5\text{ V}) = \mathbf{+0.50\ V}$
(b) $V_d = V2\text{-}V1 = (-2\text{ V}) - (+1.5\text{ V}) = \mathbf{-0.50\ V}$
(c) $V_d = V2\text{-}V1 = (+3\text{ V}) - (+5\text{ V}) = \mathbf{-2\ V}$
(d) $V_d = V2\text{-}V1 = (-3\text{ V}) - (+5\text{ V}) = \mathbf{-8\ V}$
(e) $V_d = V2\text{-}V1 = (+2\text{ V}) - (+2\text{ V}) = \mathbf{0\ V}$
(f) $V_d = V2\text{-}V1 = (+1.6\text{ V}) - (+1.75\text{ V}) = \mathbf{-0.15\ V}$ ■

Signal voltage $V_{cm}$ in Fig. 2-7 is the *common mode* signal, that is, a potential common to both −IN and +IN inputs. This potential is equivalent to $V1 = V2$. In an ideal operational amplifier there will be no output response to a common mode signal. In real devices, however, there is some small response to $V_{cm}$. The freedom from such responses is called the *common mode rejection ratio* (CMRR).

### 2-6.2 Open-Loop Gain ($A_{vol}$)

The flexibility of the operational amplifier is due in large part to its extremely high open-loop DC voltage gain. By definition, the open-loop voltage gain ($A_{vol}$) is the gain of the amplifier without feedback. If the feedback network in Fig. 2-3 had been interrupted at point X, the gain of the circuit becomes $A_{vol}$. Negative feedback reduces overall circuit gain to something less than $A_{vol}$.

The open-loop voltage gain of operational amplifiers is always very high. Some audio amplifier devices intended for consumer electronics applications offer $A_{vol}$ of 20,000. Some premium operational amplifiers offer gains to 1,000,000 and more. Depending upon the specific device, the 741 op-amp will typically exhibit $A_{vol}$ values in the 200,000 to 300,000 range.

A consequence of such high values of $A_{vol}$ is that very small differential input signal voltages will cause the output to saturate. On the 741 device the value of the maximum permissible output voltage, $\pm V_{sat}$, is typically about 1 volt (or a little less) below the power supply potential of the same polarity. Certain BiMOS devices, such as the CA-3130 or CA-3140, operate to within a few tenths of a volt of the supply rail. For ±15 Vdc supplies, the maximum 741 output potential is ±14 Vdc.

Let's consider the maximum input potential that *will not* cause saturation of the op-amp at four popular values of power supply potential, assuming the "1 volt less" rule. The calculation is

$$V_{in(max)} = \frac{\pm V_{sat}}{A_{vol}} \tag{2-6}$$

$A_{vol} = 300{,}000$

| *Power Supply* | $\pm V_{sat}$ | $\pm V_{in(max)}$ |
|---|---|---|
| ± 6 Vdc | ± 5 Vdc | 17 uV |
| ± 10 Vdc | ± 9 Vdc | 30 uV |
| ± 12 Vdc | ±11 Vdc | 37 uV |
| ± 15 Vdc | ±14 Vdc | 47 uV |

One of the consequences of high $A_{vol}$ is that op-amps usually saturate at either the $V-$ or $V+$ power supply rails when the input lines are either shorted to common or floating open. This is due to tiny imbalances in the input bias conditions internal to the device, often said to be random in nature. Accordingly, one might expect half of a group of op-amps to saturate at $+V_{sat}$, and half to saturate at $-V_{sat}$. This situation is probably true for a large number of devices procured from various lots and various manufacturers. However, a collection of 100 devices purchased at the same time from the same manufacturer will show a marked tendency toward either $-V_{sat}$ or $+V_{sat}$, not the expected random distribution. The reason is the input bias current imbalances tend to be design- and process-related. So they are generally uniform from device to device within a given lot. For example, in a lot of 20 741 devices of the same brand tested, 19 flipped to $+V_{sat}$ and only 1 flipped to $-V_{sat}$ at turn-on. Some sources might say we should have expected ten to fall into each group, but that does not usually happen in real situations.

## 2-7 VOLTAGE COMPARATORS

A voltage comparator is an operational amplifier with no negative feedback network (Fig. 2-8A). With the open-loop gain on the order of 200,000 to 300,000 for many common devices and with no negative feedback, the operational amplifier functions as a high gain DC amplifier with an output which saturates at a small input potential (see Example 2-1).

So what use is an amplifier which saturates with only a few microvolts of input signal voltage? Such an amplifier can be used as a voltage comparator. The voltage comparator compares two input voltages and issues an output signal that indicates their relationship ($V1 = V2$, $V1 > V2$, or $V1 < V2$). In Fig. 2-8A potential $V1$ is applied to the inverting input and $V2$ is applied to the noninverting input. If $V1 = V2$, then $V_o = 0$. Otherwise, the output voltage adheres to the relationships shown in Fig. 2-8B, which is the transfer function of the comparator.

According to the normal rules for operational amplifiers, making $V1$ larger than $V2$ (see Fig. 2-8A) looks like a positive input to the inverting input, so the output potential is saturated just below $V-$. Alternatively, when $V1$ is smaller than $V2$ it looks like a negative input potential, so the output is saturated just below $V+$.

In Fig. 2-8B there is a small hysteresis band around zero where no output changes occur. This is an unfortunate defect in practical operational amplifiers. It is possible to measure the hysteresis of operational amplifiers and IC comparators in a laboratory experiment (see Section 2-10). In one such experiment involving operational amplifier and IC comparator (LM-311) devices, hysteresis bands of 1 to 25 mV were found. Not surprisingly, the low-cost 741 family of operational amplifiers had high hysteresis levels (on the order of 25 mV). The LM-311 devices had 8–10 mV hysteresis. Certain other devices had 10–20 mV of hysteresis. The overall best device in the experiment was the RCA/GE CA-3140, a BiMOS operational amplifier. The CA-3140 device uses the industry standard 741 pinouts, which are shown in Fig. 2-8A.

The LM-311 device (Fig. 2-9A) is a low-cost voltage comparator in IC form. Although based internally on op-amp circuitry, this device is specifically designed as a voltage comparator. Contrary to op-amp practice, it has a ground terminal (pin 1)

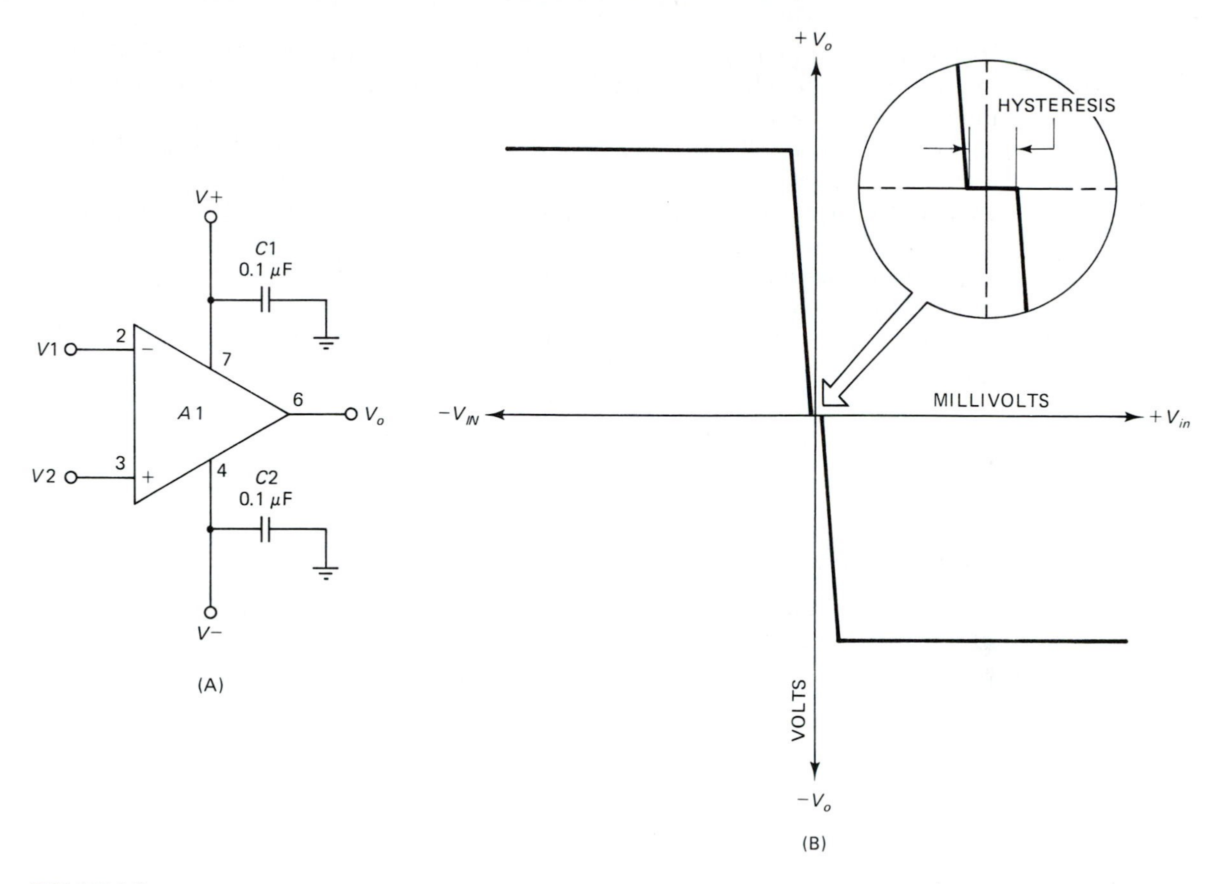


FIGURE 2-8

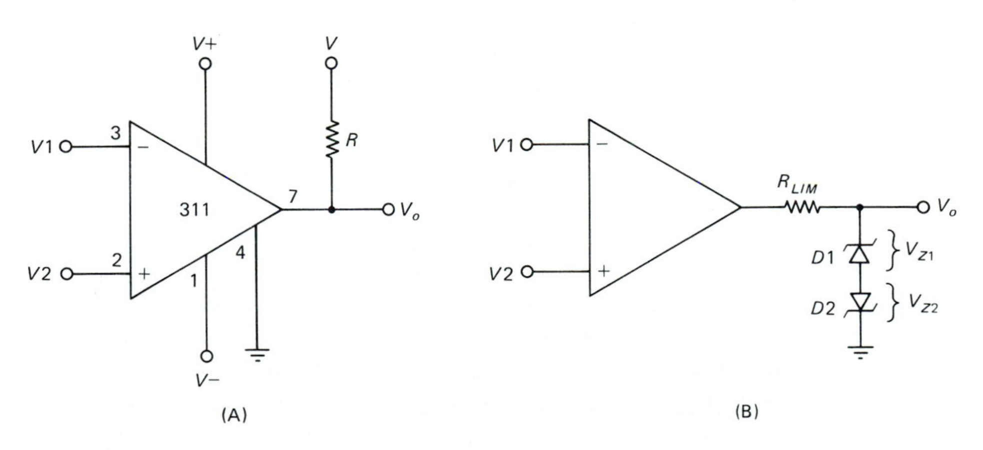


FIGURE 2-9

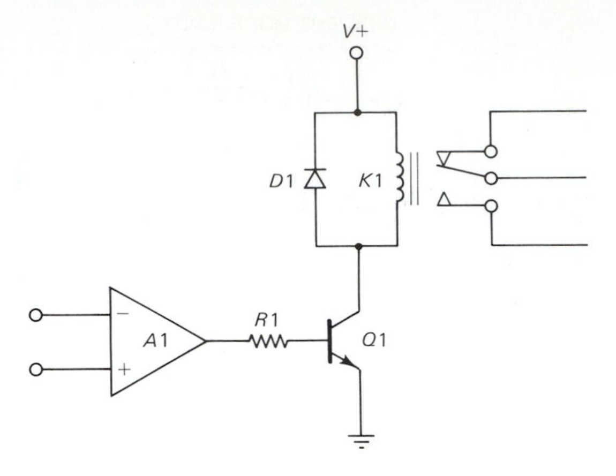


FIGURE 2-10

and requires an output pull-up resistor ($R$) to a positive power supply voltage. The output terminal can drive loads like relay coils, lamps and LEDs operated at potentials up to 40–50 volts (depending on the category of device) and current loads of 50 milliamperes. If the LM-311 is operated to be compatible with TTL digital logic, the pullup resistor is terminated in a +5 Vdc potential and usually has a value of 1.5 to 3.3 kohms.

A means for limiting the output level, improving the sharpness of the transfer function corners (see Fig. 2-8B), and improving speed by reducing latchup problems, is shown in Fig. 2-9B. In this circuit a pair of back-to-back zener diodes is connected across the output line. When the output voltage is HIGH it is limited to $V_{Z1} + 0.7$ volts; and when LOW it is limited to $V_{Z2} + 0.7$ volts. These potentials represent the reverse-bias zener voltages of $D1$ and $D2$, plus the normal forward-bias voltage drop of the other diode (which is forward-biased).

### 2-7.1 High Drive Capacity Comparators

Figure 2-10 shows a way to increase the drive capacity of the comparator. In this circuit a bipolar transistor (2N3704, 2N2222, etc.) is used to control a larger load than the device could normally handle such as the relay coil shown here. The output voltage ($V_o$) of the comparator is used to set up the DC bias for the NPN transistor. When the comparator output is HIGH, the transistor is biased hard-on and the load is grounded through the transistor's collector-emitter path. Alternatively, when the comparator output is LOW, the transistor is reverse biased and the load remains ungrounded.

The diode across the relay coil is essential for any inductive load. When the magnetic field surrounding an inductor like a relay coil collapses, the counterelectromotive force (CEMF) generates a high-voltage spike which is capable of damaging components or interrupting circuit operation (especially digital circuits). The diode is normally reverse-biased, but for the CEMF spike it is forward-biased. The diode therefore clamps the high voltage spike to about 0.7 volts.

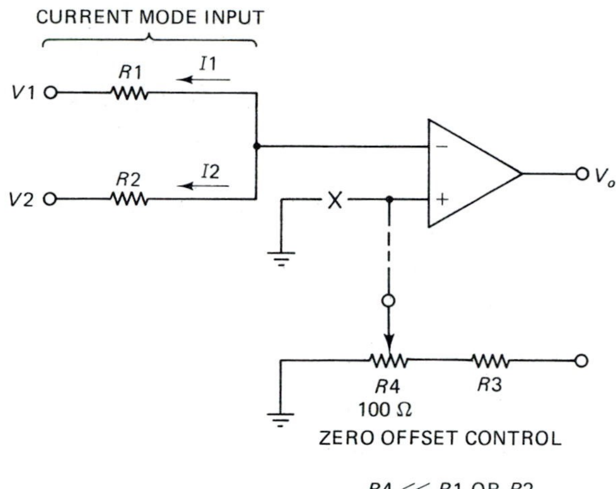


FIGURE 2-11

## 2-7.2 Current Mode Comparator

Figure 2-11 shows two comparator circuit techniques applied to the same circuit. One technique is a *zero offset control* used to reduce the effects of the hysteresis band. The other technique is the *current mode* configuration. The offset control ($R4$) biases one input to a small but nonzero level so it is ready to trip when the other input is also nonzero. In this case, the noninverting input is grounded ($V2 = 0$), but could easily be connected to a nonzero voltage.

Current mode operation is usually faster and less prone to latchup than voltage mode operation. For this reason, current mode comparators are sometimes used in high speed analog to digital converters (A/D).

Assume that the noninverting input is grounded. In this case, the output potential $V_o$ will reflect the relationship of the two currents. If $I1 = I2$, then $V_o = 0$. To the outside observer, this circuit is a voltage comparator because $I1 = V1/R1$ and $I2 = V2/R2$. The circuit is also useful for accommodating current output devices such as the LM-334 temperature monitor IC as well as voltages as shown.

## 2-7.3 Zero-Crossing Detectors

Figure 2-12A shows a *zero-crossing detector* circuit. In this case a comparator is connected with its noninverting input grounded. When $V_{in}$ is nonzero, the output will also be nonzero. But when the input voltage crosses zero, the output briefly goes to zero, producing the differentiated output pulse shown. These relationships are shown in Fig. 2-12B.

## 2-7.4 Window Comparators

A *window comparator* is shown in Fig. 2-13. This circuit consists of two voltage comparators connected so one or the other input is activated when the input voltage ($V_{in}$) exceeds either positive or negative limits. The limits are set by setting $V1$ or $V2$

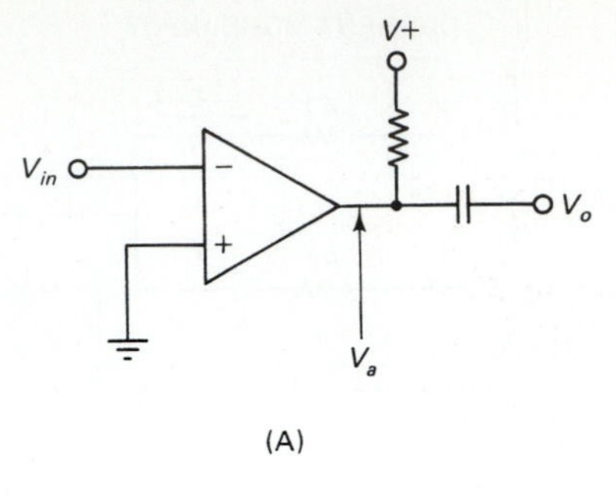


(A)

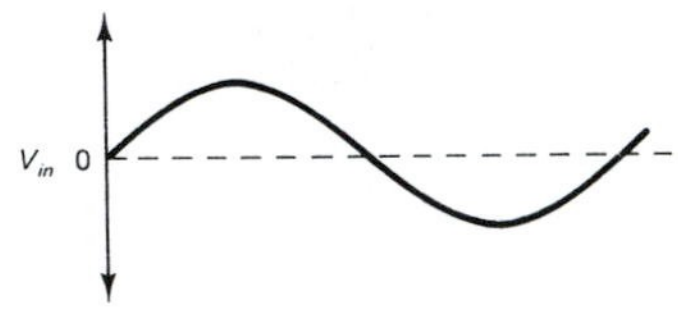


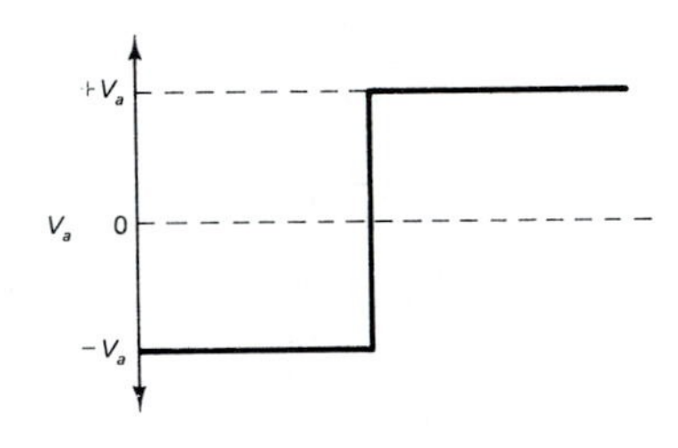


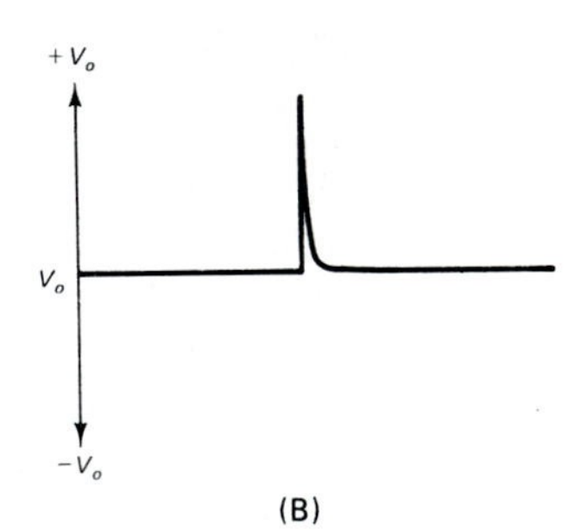


FIGURE 2-12

(B)

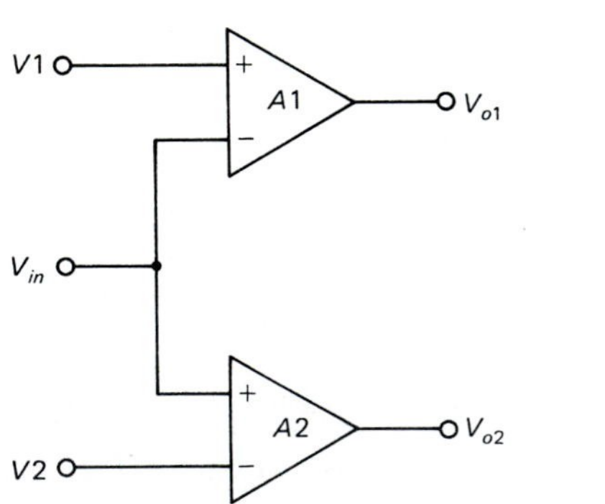


FIGURE 2-13

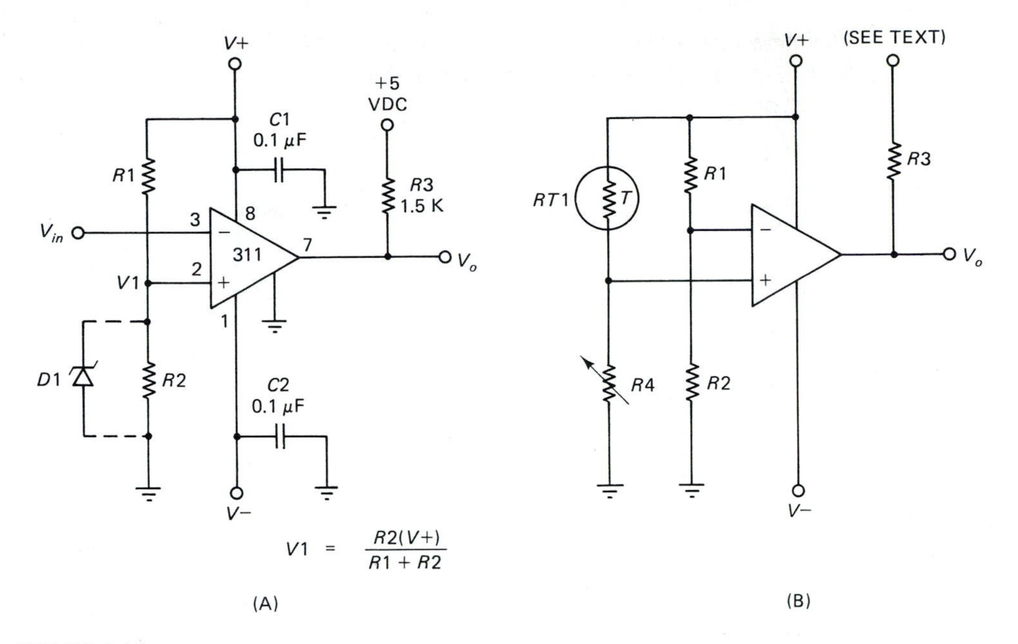


FIGURE 2-14

reference voltages. A possible application for this circuit is in alarm systems (for example, over- and under-temperature alarms) and other applications where a range of permissible values exists between two separate regions.

### 2-7.5 Pre-Biased Comparator (Voltage Level Detector)

Figure 2-14A shows a method for biasing either comparator input to a specific reference voltage. This circuit is called a *voltage level detector.* Although in this case the noninverting input is biased and the inverting input is active, the roles can easily be reversed. Two methods of biasing are used, resistor voltage divider and zener diode. If $R2$ is replaced with a zener diode, the reference potential is the zener potential. In this case, $R1$ is the normal current limiting resistor needed to protect the zener from self-destruction. In the case where a resistor voltage divider is used, the bias voltage $V1$ is set by the voltage divider equation

$$V1 = \frac{R2\ (V+)}{R1 + R2} \tag{2-7}$$

For example, suppose $R1 = R2 = 10$ kohms, and $V+ = 12$ Vdc

$$V1 = \frac{R2\ (V+)}{R1 + R2}$$

$$V1 = \frac{(10 \text{ kohms})\ (+12 \text{ Vdc})}{(10 \text{ kohms} + 10 \text{ kohms})}$$

$$V1 = \frac{120 \text{ volts}}{20 \text{ kohms}} = 6 \text{ volts}$$

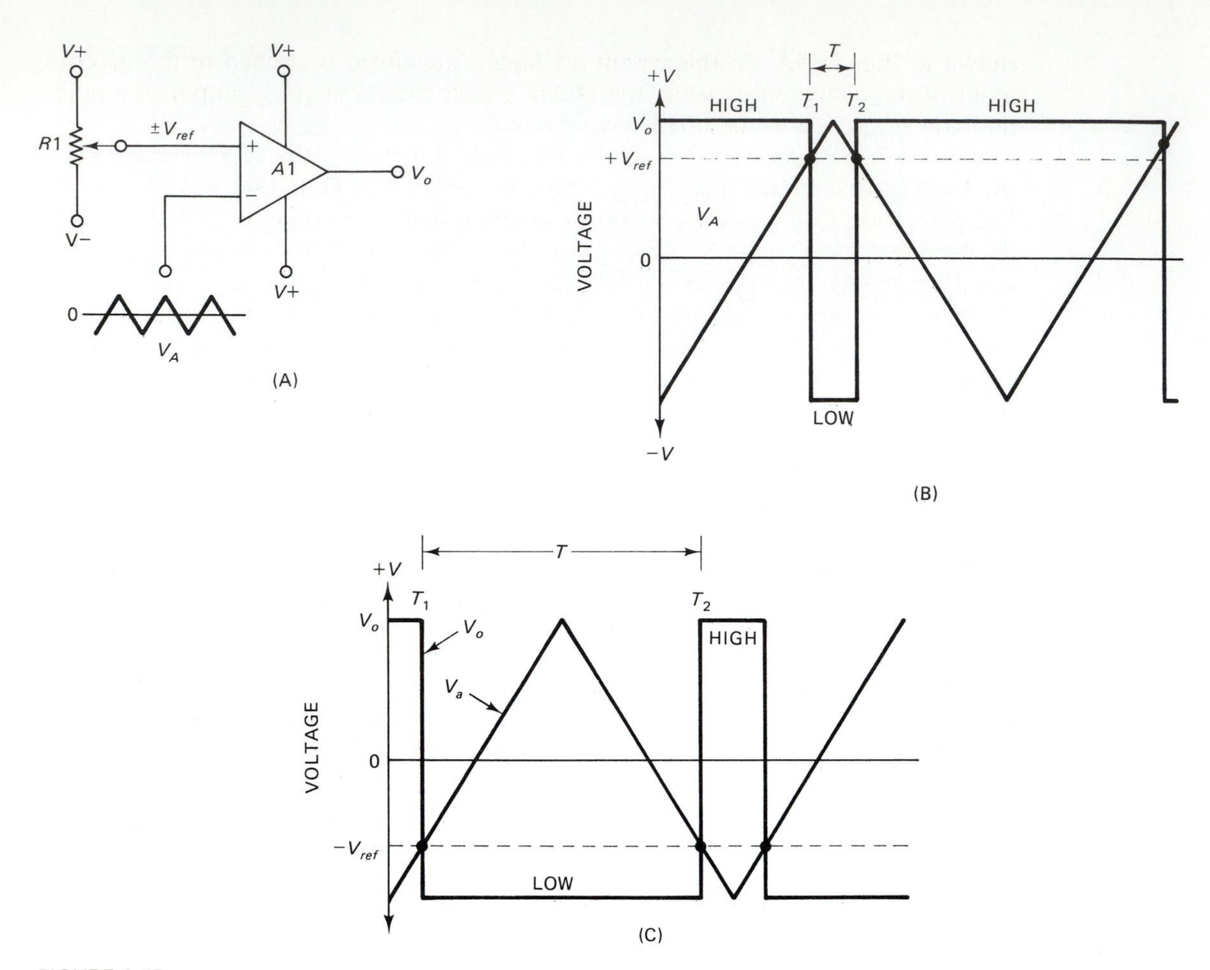


FIGURE 2-15

## 2-7.6 Temperature Alarm

Figure 2-14B shows an over-temperature circuit based on Fig. 2-14A. In this circuit the inverting input is biased by voltage divider *R*1/*R*2. The noninverting input is set by another voltage divider, *R*4/*RT*1. Resistance *RT*1 is a thermistor, which has a resistance proportional (or inversely proportional in some types) to the temperature. Potentiometer *R*4 is used to set the trip-point temperature. The values of the resistors depend upon the trip-point desired and the resistance of the thermistor over the range of temperatures being monitored.

## 2-7.7 Pulse Width Controller

Pulse width modulation is used in many communications systems, motor and load controllers, switching DC power supplies, and other applications. A pulse width modulator will vary the width of an output pulse proportionally to an applied input voltage. We can use a voltage comparator to form a basic pulse width modulator, as shown in Fig. 2-15A. In this circuit a triangle waveform is applied to the inverting input of the comparator, while the DC reference potential ($V_{ref}$) output of a potentiometer is applied to the noninverting input.

Figures 2-15B and 2-15C show the relationship between the output terminal and the two input signals, $V_a$ and $V_{ref}$. Examine Fig. 2-15B. Note that $V_{ref}$ is a positive DC potential. As long as the applied triangle waveform is more or less than $+V_{ref}$, the output of the comparator ($V_o$) is high. But at the time $T1$ the two voltages become equal, so the output flips as the triangle waveform increases in value above $+V_{ref}$. During the period $T2-T1$, signal $V_a$ is greater than $+V_{ref}$, so $V_o$ is low. At time $T2$, however, the situation again reverts back to where $V_a$ is less than $+V_{ref}$ so $V_o$ drops to low.

Now consider a slightly different situation in Fig. 2-15C. In this case the value of $V_{ref}$ is readjusted to a negative value, $-V_{ref}$. As in the previous case, the output $V_o$ is low during the period $T2-T1$, but note that this segment is much longer than it was in Fig. 2-15B. This difference is caused by the relationship between $V_{ref}$ and $V_a$ in the two different situations.

### 2-7.8 Analog-to-Digital Converter

An analog-to-digital converter (A/D or ADC) converts an analog voltage into a binary number which is proportional to that voltage. For example, in an eight-bit unipolar ADC we might find that binary number 00000000 represents zero volts, while 11111111 represents the full-scale potential (2.56 volts, 5 volts, 10 volts). A digital-to-analog converter (DAC) is exactly the opposite. It converts a binary number into an analog voltage. In both cases, the minimum step of output change is called the "one least significant bit" (1-LSB) value, or the amount of change caused by a change in the least significant bit of the binary number (xxxxxxx0 to xxxxxxx1, or vice versa).

Figure 2-16 shows one type of ADC circuit that can be implemented with a voltage comparator and a DAC circuit. This type of ADC circuit is known both as the *binary ramp ADC* and the *servo ADC*.

The comparator noninverting input in the circuit of Fig. 2-16A is connected to the unknown voltage, $V_x$, while the inverting input is connected to the output of the DAC, $V_a$. The output state of the comparator indicates the relationship of $V_a$ and $V_x$, that is, $V_a < V_x$ or $V_a = V_x$. These relationships are graphed in Fig. 2-16B. The DAC output voltage is a stepped potential which is raised by the 1-LSB amount every time the binary counter is clocked. During this time the comparator output ($V_b$) is HIGH. At the point where the DAC output voltage equals (or exceeds) the unknown input voltage, the comparator output switches to the LOW condition, indicating the end of conversion. The binary number appearing on the output of the binary counter represents the unknown voltage.

A variant of this circuit is often used in microprocessor-based instrumentation. The binary counter is replaced with an eight-bit output port on the computer, while the comparator output is read by one bit of an input port. Software replaces the control logic. The program will output a binary number starting with 00000000, test the comparator output for HIGH or LOW, and then branch according to the results of the test. For example, if the output is HIGH, it will increment the binary number and output this new number to the DAC. This process continues in an iterative fashion

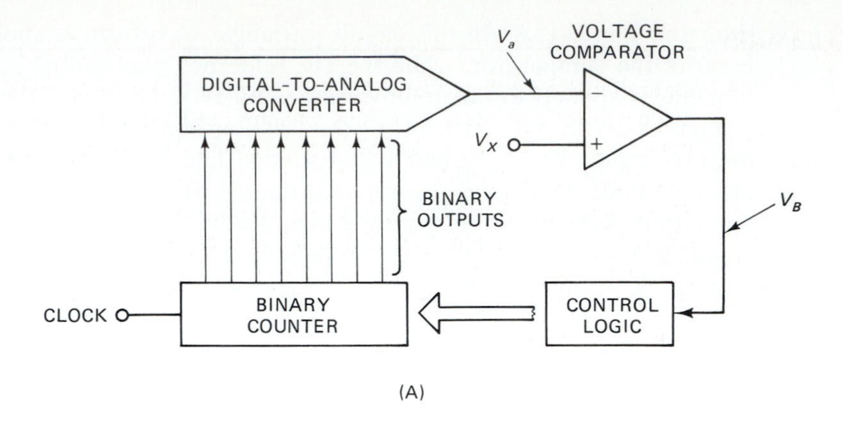


(A)

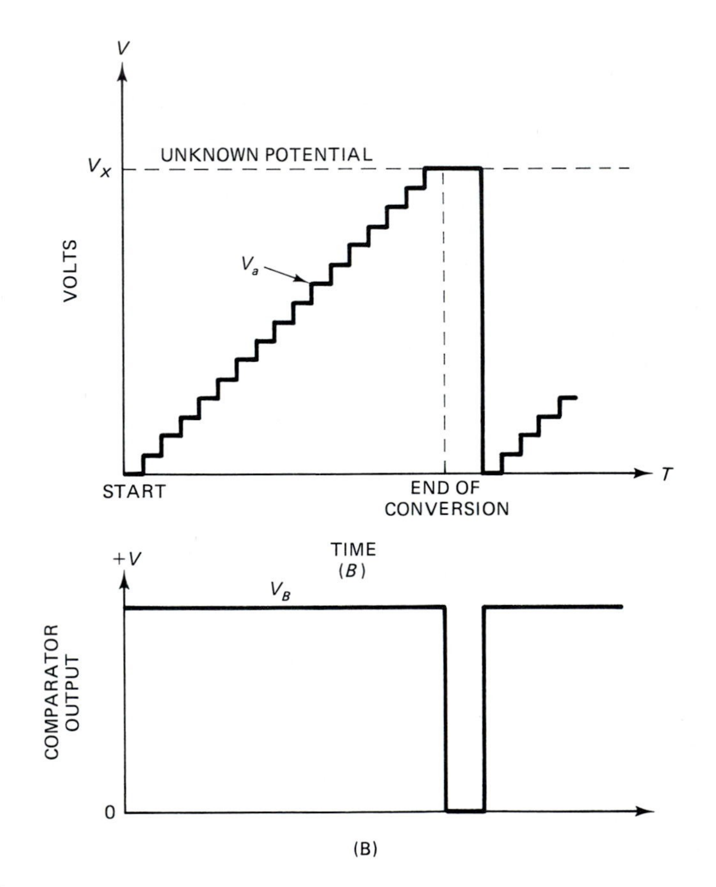


(B)

FIGURE 2-16

until the comparator output drops LOW indicating the condition $V_a = V_x$. At this point the computer program branches to store the binary number in a memory location or use it in a calculation.

## 2-8 SUMMARY

1. Operational amplifiers were initially designed to perform mathematical operations in analog computers. Because of this op-amps use bipolar power supplies to accommodate both polarity outputs, and need both inverting and noninverting inputs. Most op-amps are differential amplifiers.
2. There are a number of categories of operational amplifiers: general purpose, voltage comparators, low input current, low noise, low power, low drift, wide bandwidth, single DC supply, high voltage, multiple devices, limited/special purpose, and instrumentation amplifiers.
3. The ideal operational amplifier has seven properties (1) infinite open-loop voltage gain, (2) infinite input impedance, (3) zero output impedance, (4) zero noise contribution, (5) zero DC output offset, (6) infinite bandwidth, (7) both differential inputs stick together.
4. There are two configurations for the operational amplifier, inverting follower and noninverting follower.
5. The differential input voltage is the difference between single-ended voltages applied to the −IN ($V1$) and +IN ($V2$) inputs [$V_d$ = (V2-V1)].
6. A common-mode voltage ($V_{cm}$) is one that applies equally to both −IN and +IN inputs. It is equivalent to the situation where $V1 = V2$. The common mode rejection ratio (CMRR) is a measure of the ability of the differential amplifier to reject common-mode voltages.
7. Open-loop voltage gains range from 20,000 to more than 1,000,000, with 200,000 to 300,000 being common in 741 devices.
8. A voltage comparator is an operational amplifier with no feedback. It compares the two input voltages, and issues an output signal that indicates whether $V1 = V2$, $V1 > V2$, or $V1 < V2$.

## 2-9 RECAPITULATION

Now return to the objectives and Prequiz questions at the beginning of the chapter and see how well you can answer them. If you cannot answer certain questions, mark them and review the appropriate parts of the text. Next, try to answer the questions and work the problems below, using the same procedure.

## 2-10 LABORATORY EXERCISES

1. Connect the circuit of Fig. 2-E1, using a 741 operational amplifier. The output indicator is a digital voltmeter. (A) ground both −IN and +IN inputs (pins 2 and 3). Note what happens to the output voltage when $V-$ and $V+$ are applied simultaneously. (B) repeat the experience using a large variety of 741 devices from different manufacturers. (C) repeat the experiment with −IN and +IN floating (note: break the circuit at points X).
2. Connect the circuit of Fig. 2-E2, initially with the inverting input connected to the wiper terminal of potentiometer $R1$. Connect the noninverting input to a positive voltage between +1 Vdc and +10 Vdc (the +5 Vdc shown in Fig. 2-E2 was used because it is commonly available on laboratory breadboard power supplies). (A) Starting with point "B" at zero volts, slowly increase the potential by adjusting $R1$ until a transition of output potential "C" occurs. Measure and record voltages present at A and B. (B) repeat the experiment starting

with the voltage at point A at +12 Vdc. (C) disconnect the inverting input from $R1$ and reconnect to the output of a function generator. Monitor point C with an oscilloscope and note the relationship between input and output signals.

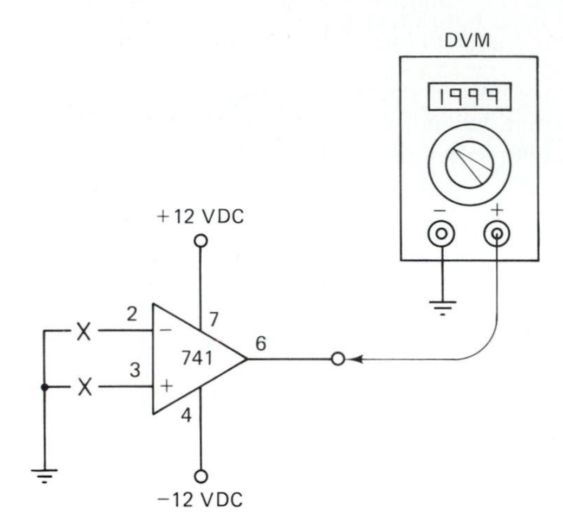


FIGURE 2-E1

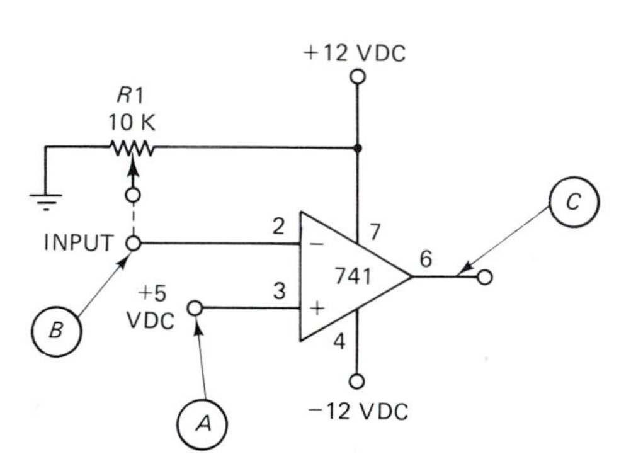


FIGURE 2-E2

## 2-11 QUESTIONS AND PROBLEMS

1. List several areas where operational amplifiers are typically found.
2. List three major sections of the internal circuitry of an operational amplifier.
3. A DC differential amplifier has two inputs: ____________ , which produces out of phase output signals, and ____________ which produces in phase output signals.
4. List five different categories of operational amplifiers.
5. List the seven properties of the ideal operational amplifier.
6. The open-loop gain of an operational amplifier is 300,000. Calculate the closed-loop gain if the feedback factor ($B$) is 0.01.
7. A potential of +5 Vdc is applied to the noninverting input of an operational amplifier. A voltmeter connected between the inverting input and common will read: __________ Vdc.

8. The ____________ ____________ rejection ratio relates output change to a change in either $V-$ or $V+$, while the other is held constant.

9. A dual operational amplifier has a channel separation of 100 dB. Calculate the output voltage change at amplifier B if the output of amplifier A changes 2.5 volts.

10. The industry standard operational amplifier pin-outs are the same pin-outs as found on the ____________ operational amplifier.

11. There is no ____________ terminal on the operational amplifier. Output signal is taken between the output terminal and the power supply ____________ line.

12. Calculate the differential input voltage ($V_d$) if an operational amplifier has 1.83 Vdc applied to the $-$IN input, and 4.5 Vdc applied to the $+$IN input.

13. Calculate the differential input voltage ($V_d$) if an operational amplifier has 1.05 Vdc applied to the $-$IN input and $-2.5$ Vdc applied to the $+$IN input.

14. Calculate the input voltage that will cause saturation at the positive output limit if $+V_{sat} = 13.4$ Vdc and $A_{vol} = 250{,}000$.

15. Calculate the input voltage that will cause saturation at the negative output limit if $-V_{sat} = -14$ Vdc and $A_{vol} = 300{,}000$.

16. A constant current source used in a transistor DC differential amplifier provides 12 mA. If $I1 = 4.2$ mA, then $I2 =$ ____________ .

17. A Darlington input amplifier is characteristic of a ____________ operational amplifier.

18. ____________ ____________ operational amplifiers are internally compensated to minimize thermal variations of the output voltage.

19. A DC voltmeter connected to the noninverting input of an operational amplifier measures $+4.5$ Vdc. What is the DC voltage at the inverting input if $V_{in} = 0$?

20. An operational amplifier has an open-loop voltage gain of 500,000. Calculate the closed-loop voltage gain of a simple resistor voltage divider consisting of a 120-kohm resistor from the output to the inverting input, and a 27-kohm resistor from the inverting input to ground.

21. Calculate the noise power dissipated in a 2.2-megohm resistor that is operating at a temperature of 45°C. Assume a bandwidth of 100 KHz.

22. Calculate the noise voltage across the resistor in question 21 above.

23. Calculate the noise voltage present across a 1.5-megohm resistor operated at room temperature (25°C) when a 100-pF capacitor is shunted across the resistor.

24. Define *virtual ground.*

25. Define *slew rate* and give the units normally used to measure slew rate.

26. Define *CMRR.*

27. Define *PSRR.*

28. Draw the circuit diagram for an operational amplifier operated as a voltage comparator.

29. Draw the circuit diagram for a current-mode comparator based on an operational amplifier.

30. Draw the circuit of a voltage comparator used as a voltage crossing detector.

31. Draw the circuit for a window voltage comparator.

32. Draw the circuit for a prebiased comparator used as a voltage level detector.

CHAPTER 3

# Inverting and Noninverting Operational Amplifier Configurations

## OBJECTIVES

1. Learn to recognize the three principal operational amplifier configurations.
2. Be able to set the gain of either inverting or noninverting amplifiers and use them in practical applications.
3. Learn the advantages and disadvantages of the principal configurations.
4. Be able to describe the response of linear amplifiers to common DC, AC, and composite AC/DC signals.

## 3-1 PREQUIZ

These questions test your prior knowledge of the material in this chapter. Try answering them before you read the chapter. Look for the answers (especially those you answered incorrectly) as you read the text. After you have finished studying the chapter, try answering these questions again and those at the end of the chapter (see Section 3-10).

1. An inverting follower must have a voltage gain of $-10$ and exhibit an input impedance of not less than 22 kohms. Find the value of feedback resistor $R_f$ that will allow the required gain with the specified minimum value of input impedance.
2. An inverting amplifier has an input resistor of 10 kohms and a feedback resistor of 1 megohms. Find the output voltage if an input signal of $+100$ mVdc is applied to the input.
3. Calculate the value of a capacitor needed in the amplifier described in question 2 to shunt $R_f$ and thus limit the upper AC frequency response to 100 Hz.
4. A noninverting follower uses a 1 kohm input resistor and a 100 kohm feedback resistor. What is the voltage gain of this circuit?

## 3-2 INVERTING FOLLOWER CIRCUITS

The inverting follower is an operational amplifier circuit in which the output signal is 180° out of phase with the input signal. Figure 3-1A is a cathode ray oscilloscope (CRO) rendering which shows the relationship between input and output signals for an inverting follower with a gain of $-2$. Note the phase reversal present in the output signal with respect to the input signal. In order to achieve this inversion the inverting input ($-$IN) of the operational amplifier is active and the noninverting input ($+$IN) is grounded.

Figure 3-1B shows the basic configuration for the inverting follower (also called inverting amplifier) circuits. The noninverting input is not used so it is set to ground potential. There are two resistors in this circuit. Resistor $R_f$ is the negative feedback path from the output to the inverting input while $R_{in}$ is the input resistor. We will examine the $R_f/R_{in}$ relationship to determine how gain is fixed in this type of circuit. But first, let's take a look at the implications of grounding the noninverting input in this circuit.

### 3-2.1 What's a Virtual Ground?

A "virtual ground" is a ground that only acts like a ground. While this definition sounds strange at first, it's not an unreasonable description. But the terminology is confusing and leads to the erroneous conclusion the virtual ground does not really function as a ground. So let's examine the concept in detail.

In Chapter 2 we discussed the properties of the ideal operational amplifier. Property 7 was differential inputs stick together. Put another way, this means a voltage applied to one input appears on the other input also. Therefore in the arithmetic of op-amps, we must treat both inputs as if they are at the same potential. This fact is not merely a theoretical device. If you actually apply a potential, 1 Vdc for example, to the noninverting input the same 1 Vdc potential can also be measured at the inverting input.

In Fig. 3-1A the noninverting input is grounded, so it is at 0 Vdc potential. Considering the properties of the ideal op-amp, this means the inverting input of the op-amp is also at the same 0 Vdc ground potential. Since the inverting input is at ground potential, but has no physical ground connection, it is said to be at virtual (as opposed to physical) ground. *A virtual ground is a point which is fixed at ground potential (0 Vdc), even though it is not physically connected to the actual ground or common of the circuit.*

### 3-2.2 Developing the Transfer Equation for the Inverting Amplifier

The transfer equation of any circuit is the output function divided by the input function. For an operational amplifier used as a voltage amplifier, the transfer function describes the voltage gain

$$A_v = \frac{V_o}{V_{in}} \tag{3-1}$$

where

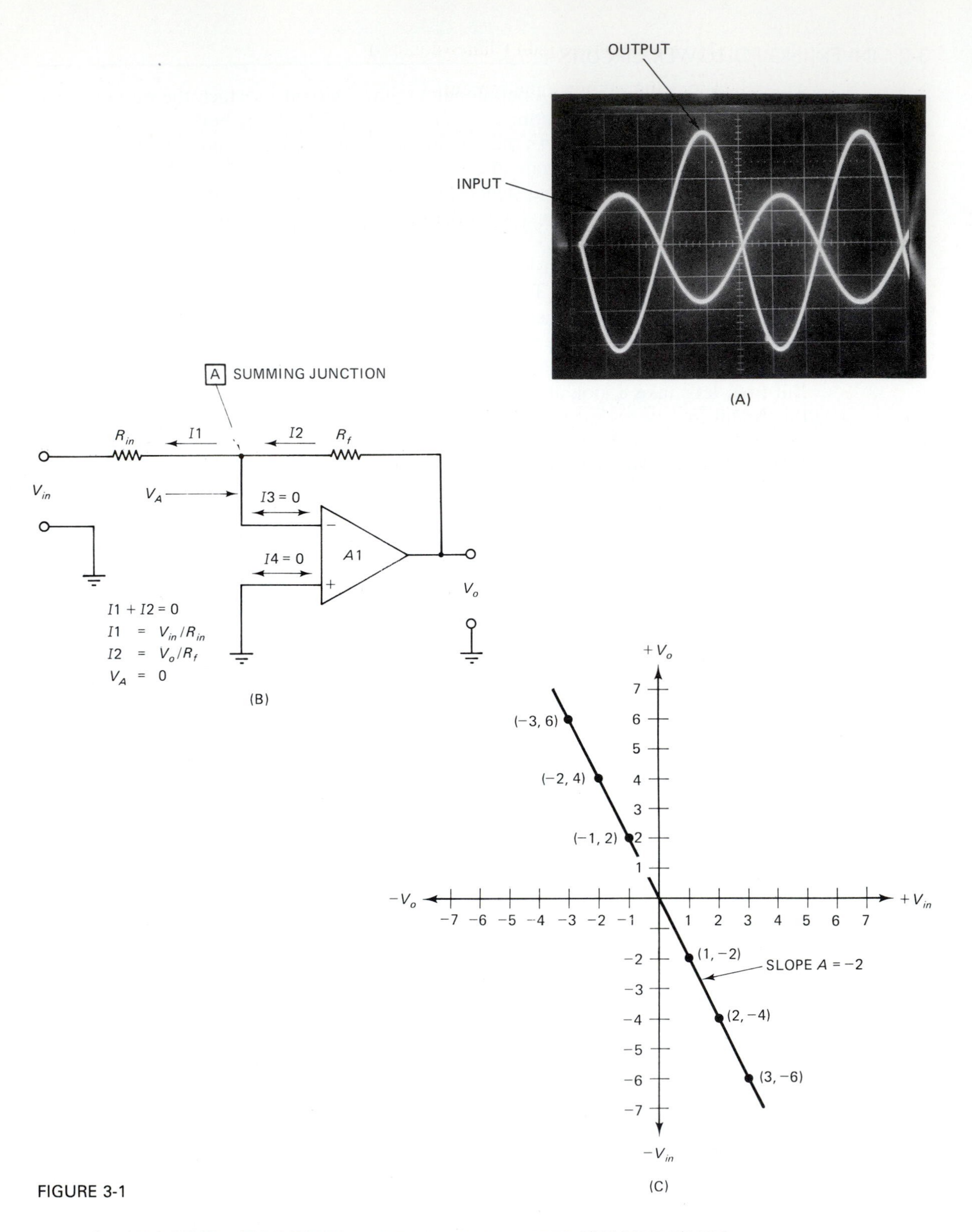
OUTPUT
INPUT
(A)
A SUMMING JUNCTION
$R_{in}$
$I1$
$I2$
$R_f$
$V_{in}$
$V_A$
$I3 = 0$
$I4 = 0$
A1
$V_o$
$I1 + I2 = 0$
$I1 = V_{in}/R_{in}$
$I2 = V_o/R_f$
$V_A = 0$
(B)
$+V_o$
$-V_o$
$+V_{in}$
$-V_{in}$
(−3, 6)
(−2, 4)
(−1, 2)
(1, −2)
(2, −4)
(3, −6)
SLOPE $A = -2$
(C)

FIGURE 3-1

$A_v$ = the voltage gain (dimensionless)

$V_o$ = the output signal potential

$V_{in}$ = the input signal potential ($V_o$ and $V_{in}$ are in the same units)

### EXAMPLE 3-1

What is the voltage gain of an amplifier if +100 mVdc (0.100 Vdc) at the input produces a −10 Vdc output potential?

$$A = V_o/V_{in}$$

$$A = -10/0.100$$

$$A = -100$$ ■

In the inverting follower circuit (Fig. 3-1A) the gain is set by the ratio of two resistors, $R_f$ and $R_{in}$. Let's make a step-by-step analysis to see if we can find this relationship. Consider the currents flowing in Fig. 3-1B. The input bias currents, $I3$ and $I4$, are assumed to be zero for purposes of analysis. This is a reasonable assumption because our model is an *ideal* operational amplifier. In a real op-amp these currents are nonzero and have to be accounted for, but in the analysis case we use the ideal model. Thus, in the analysis below we can ignore bias currents (assume that $I3 = I4 = 0$).

Remember that the *summing junction* (point A) is a virtual ground and is at ground potential because the noninverting input is grounded. Current $I1$ is a function of the applied input voltage, $V_{in}$, and the input resistance $R_{in}$. By Ohm's law then, the value of $I1$ is

$$I1 = \frac{V_{in}}{R_{in}} \tag{3-2}$$

Further, we know that current $I2$ is also related by Ohm's law to the output voltage, $V_o$, and the feedback resistor $R_f$ (again, because the summing junction is at 0 Vdc)

$$I2 = \frac{V_o}{R_f} \tag{3-3}$$

How are $I1$ and $I2$ related? These two currents are the only currents entering or leaving the summing junction (recall that $I3 = 0$), so by Kirchoff's Current Law (KCL) we know that

$$I1 + I2 = 0 \tag{3-4a}$$

so

$$I2 = -I1 \tag{3-4b}$$

We can arrive at the transfer function by substituting Eq. (3-2) and Eq. (3-3) into Eq. (3-4b)

$$I2 = -I1 \tag{3-4b}$$

$$\frac{V_o}{R_f} = \frac{-V_{in}}{R_{in}} \tag{3-5}$$

Algebraically rearranging Eq. (3-5) yields the transfer equation in standard format

$$\frac{V_o}{V_{in}} = \frac{-R_f}{R_{in}} \tag{3-6}$$

According to Eq. (3-1), the gain ($A_v$) of the circuit is $V_o/V_{in}$, so we may also write Eq. (3-6) in the form

$$A_v = \frac{-R_f}{R_{in}} \tag{3-7}$$

**EXAMPLE 3-2**

What is the gain of an op-amp inverting follower if the input resistor ($R_{in}$) is 10 kohms and the feedback ($R_f$) resistor is 1 megohms?

$$A = -R_f/R_{in}$$

$$A = -(1{,}000{,}000)/10{,}000$$

$$A = -100$$

■

We have shown above that the voltage gain of an op-amp inverting follower is merely the ratio of the feedback resistance to the input resistance ($-R_f/R_{in}$). The minus sign indicates that a 180° phase reversal takes place. Thus, a negative input voltage produces a positive output voltage and vice versa.

We often see the transfer equation [Eq. (3-6)] written to express output voltage in terms of gain and input signal voltage. The two expressions are

$$V_o = -A_v\, V_{in} \tag{3-8}$$

and

$$V_o = -V_{in}\,\frac{R_f}{R1} \tag{3-9}$$

**EXAMPLE 3-3**

What is the output voltage from an inverting follower in which $V_{in}$ = 100 mV (0.100 V), $R_f$ = 100k and $R_{in}$ = 2k?

**Solution**

$$V_o = -V_{in}(R_f/R_{in})$$

$$V_o = -(0.100\text{ V})(100\text{k}/2\text{k})$$

$$V_o = -(0.100)(50)$$

$$V_o = -5\text{ volts}$$

■

In the example above the voltage gain ($A_v$) is $-R_f/R_{in} = -100\text{k}/2\text{k} = -50$.

The transfer function ($A_v = V_o/V_{in}$) can be plotted on graph paper in terms of input and output voltage. Figure 3-1C shows the plot $V_o$ against $V_{in}$ for an inverting amplifier with a gain of $-2$. Straight lines plotted as in Fig. 3-1C have the mathematical form $Y = aX + b$, where $a$ is the slope of the line and $b$ is the $Y$-intercept. In the case of a perfect amplifier the $Y$-intercept is 0 volts. Given the nature of Fig. 3-1C the basic form becomes for our purposes $V_o = A_v V_{in} \pm V_{offset}$.

### 3-2.3 Inverting Amplifier Transfer Equation by Feedback Analysis

In Section 3-2.2 we developed the inverting amplifier transfer equation from the ideal model of the operational amplifier. Now let's consider the same matter from the point of view of the generic feedback amplifier to see if Eq. (3-7) is valid. When used in a closed-loop circuit the operational amplifier is a feedback amplifier, so feedback analysis will result in the same transfer equation as the ideal model analysis.

Figure 3-2 shows an operational amplifier with its feedback network. The overall gain of this type of amplifier is defined by

$$A_v = \frac{A_{vol}\,C}{1 + A_{vol}B} \tag{3-10}$$

where

$A_v$ = the closed-loop voltage gain

$A_{vol}$ = the open-loop voltage gain

$C$ = the transfer equation of the input network

$B$ = the transfer equation of the feedback network

Two networks must be considered in this analysis, the input network ($C$) and the feedback network ($B$). Both networks are resistor voltage divider attenuators so we

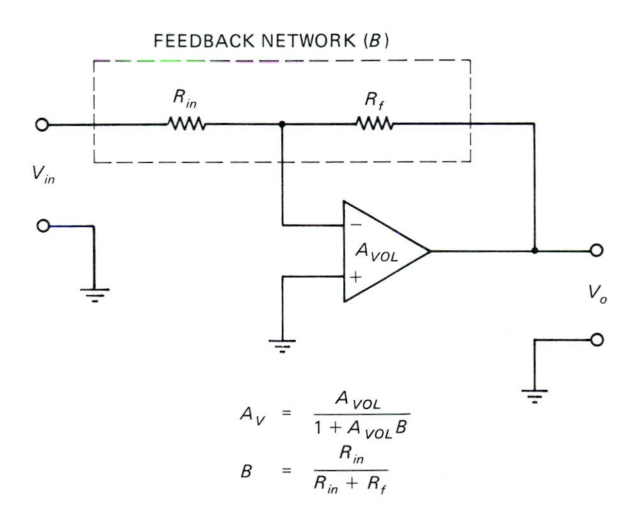


FIGURE 3-2

can expect $B$ and $C$ to be fractions. The expression for the input network in Fig. 3-2 is

$$C = \frac{R_f}{R_f + R_{in}} \tag{3-11}$$

The C term is needed because the input signal is attenuated by the $R_{in}/R_f$ voltage divider network. If the signal is applied directly to the inverting input, as it might be in other feedback amplifiers, this input attenuation term is unity, so it disappears from Eq. (3-10).

The feedback transfer equation is defined by the feedback voltage divider, $R_f/R_{in}$

$$B = \frac{R_{in}}{R_f + R_{in}} \tag{3-12}$$

We can now substitute the expressions for $B$ [Eq. (3-12)] and $C$ [Eq. (3.11)] into the equation for the standard feedback amplifier

$$A_v = \frac{A_{vol}\,C}{1 + A_{vol}B}$$

so

$$A_v = \frac{A_{vol}\left[\dfrac{R_f}{(R_f + R_{in})}\right]}{1 + A_{vol}\left[\dfrac{R_{in}}{(R_f + R_{in})}\right]} \tag{3-13}$$

We may legally divide both numerator and denominator by the open-loop gain which yields

$$A_v = \frac{\left[\dfrac{R_f}{(R_f + R_{in})}\right]}{\left(\dfrac{1}{A_{vol}}\right) + \left[\dfrac{R_{in}}{(R_f + R_{in})}\right]} \tag{3-14}$$

Because $A_{vol}$ is infinite in ideal devices and very high in practical devices, the term $1/A_{vol} \rightarrow 0$, so we may write Eq. (3.14) in the form

$$A_v = \frac{\left[\dfrac{R_f}{(R_f + R_{in})}\right]}{\left[\dfrac{R_{in}}{(R_f + R_{in})}\right]} \tag{3-15}$$

Earlier we discovered that $A_v = R_f/R_{in}$. If the feedback analysis is correct, then Eq. (3-15) will be equal to $R_f/R_{in}$. Solving this relationship we invert and multiply:

$$\left[\frac{R_f + R_{in}}{R_{in}}\right] \times \left[\frac{R_f}{R_f + R_{in}}\right] = \frac{R_f}{R_{in}} \tag{3-16}$$

$$\frac{R_f}{R_{in}} = \frac{R_f}{R_{in}} \tag{3-17}$$

Equation (3-17) demonstrates the equality of the two methods, proving that the transfer equation [Eq. (3-7)] derived earlier is valid.

The following equations apply to inverting followers

$$A_o = -R_f/R_{in} \tag{3-18}$$

$$V_o = -A\ V_{in} \tag{3-19}$$

$$V_o = -V_{in}\left(\frac{R_f}{R_{in}}\right) \tag{3-20}$$

### 3-2.4 Multiple Input Inverting Followers

We can accommodate multiple signal inputs on an inverting follower by using a circuit such as Fig. 3-3. There are a number of applications of these circuits such as summers, audio mixers, and instrumentation. The multiple input inverter of Fig. 3-3 can be

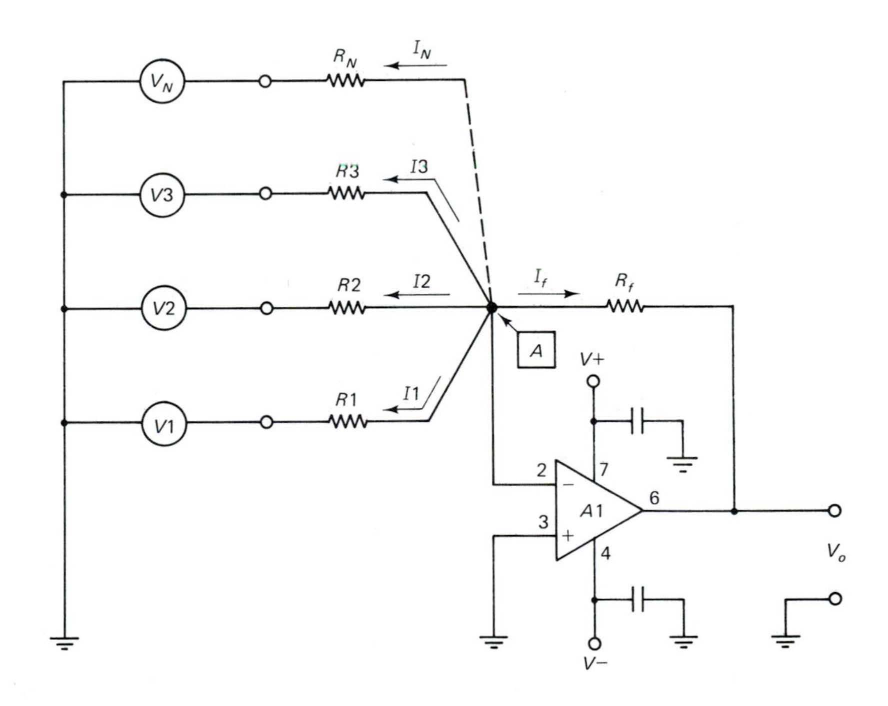


FIGURE 3-3

evaluated exactly like Fig. 3-1B except we have to account for more than one input. Again appealing to KCL, we know that

$$I1 + I2 + I3 + \ldots + I_n = I_f \tag{3-21}$$

Also by Ohm's law, considering that summing junction A is virtually grounded, we know that

$$I1 = \frac{V1}{R1} \tag{3-22}$$

$$I2 = \frac{V2}{R2} \tag{3-23}$$

$$I3 = \frac{V3}{R3} \tag{3-24}$$

$$I_n = \frac{V_n}{R_n} \tag{3-25}$$

$$I_f = \frac{V_o}{R_f} \tag{3-26}$$

Substituting Eq. (3-22) through Eq. (3-26) into Eq. (3-21)

$$\frac{V1}{R1} + \frac{V2}{R2} + \frac{V3}{R3} + \ldots + \frac{V_n}{R_n} = \frac{V_o}{R_f} \tag{3-27}$$

or, algebraically rearranging Eq. (3-27) to solve for $V_o$

$$V_o = R_f \left[ \frac{V1}{R1} + \frac{V2}{R2} + \frac{V3}{R3} + \ldots + \frac{V_n}{R_n} \right] \tag{3-28}$$

Equation (3-28) is the transfer equation for the multiple input inverting follower. The terms $V_n$ and $R_n$ refer to the *nth* voltage and *nth* resistance, respectively. Let's consider a three-input example ($n = 3$).

### EXAMPLE 3-4

A circuit such as Fig. 3-3 has a 100k feedback resistor ($R_f$), and the following input resistors: $R1 = 10\text{k}$, $R2 = 50\text{k}$, and $R3 = 100\text{k}$. Find the output voltage when the following input voltages are present: $V1 = 0.100$ volts, $V2 = 0.200$ volts and $V3 = 1$ volt.

$$V_o = R_f \left[ \frac{V1}{R1} + \frac{V2}{R2} + \frac{V3}{R3} \right]$$

$$V_o = (100\text{k}) \left[ \frac{(0.100)}{10\text{k}} + \frac{(0.200)}{50\text{k}} + \frac{(1.00)}{100\text{k}} \right]$$

$$V_o = [(10)(0.100)] + [(2)(0.200)] + [(1)(1.00)]$$

$$V_o = (1 \text{ volt}) + (0.4 \text{ volt}) + (1 \text{ volt})$$

$$V_o = 2.4 \text{ volts}$$

■

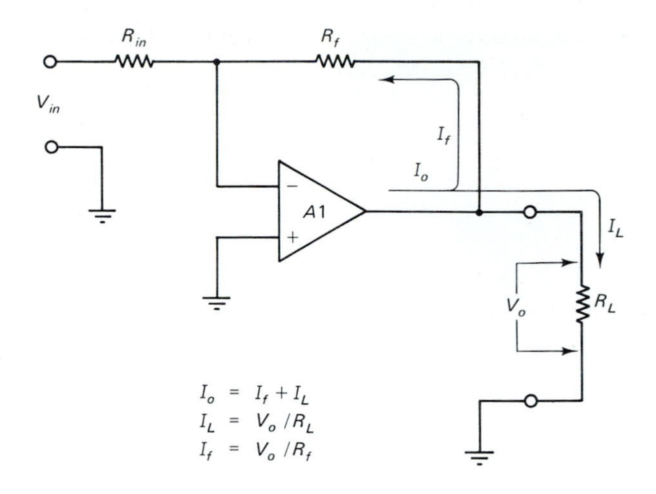


FIGURE 3-4

### 3-2.5 Output Current

The output current $I_o$ must be supplied by the output terminal of the op-amp. Typically small-signal op-amps supply 5 to 25 mA of current depending on the device, while power op-amps such as the Burr-Brown OPA-511 supply up to 5 amperes at potentials of $\pm 30$ volts. The output current ($I_o$) splits into two paths (Fig. 3-4); a portion of the output current flows into the feedback path ($I_f$) and a portion flows into the load ($I_L$). The total current is

$$I_o = I_f + I_L \tag{3-29}$$

where

$I_o$ = the output current

$I_f$ = the feedback current ($V_o/R_f$)

$I_L$ = the load current ($V_o/R_L$)

In normal voltage amplifier service both $I_f$ and $I_L$ tend to be very small compared to the available output current. But in applications where load and feedback resistances are low, the output currents may approach the maximum specified value. To determine whether this limit is exceeded, divide output potential $V_o$ by the parallel combination of $R_f$ and $R_L$

$$\frac{(V_{o(max)})(R_f + R_L)}{R_f R_L} < I_{o(max)} \tag{3-30}$$

where

$V_{o(max)}$ = the maximum expected output voltage

$I_{o(max)}$ = the maximum allowable output current

$R_f$ = the feedback resistance in ohms

$R_L$ = the load resistance in ohms

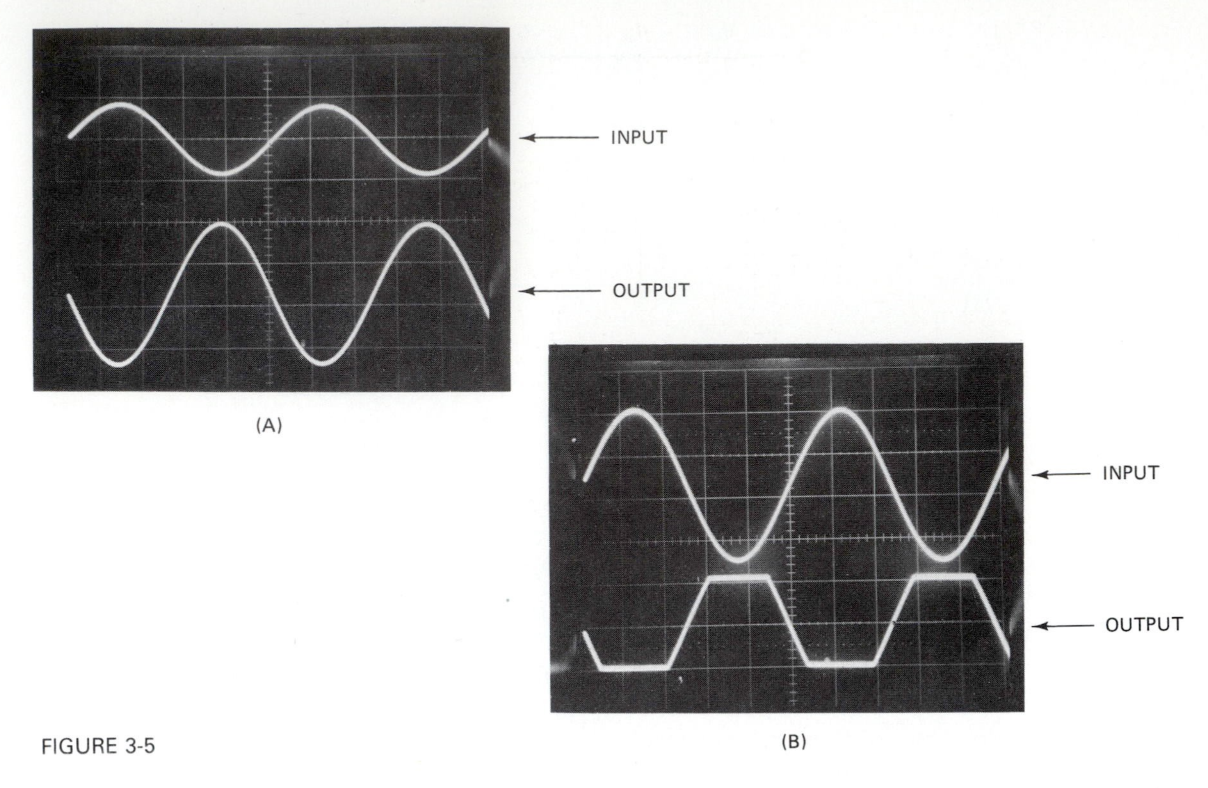


FIGURE 3-5

In general, the output current limit is not approached in ordinary devices unless load and/or feedback resistances are less than 1000 ohms. But power devices can drive a lower load and/or feedback resistance combination.

### 3-2.6 Response to AC Signals

Thus far in this discussion of inverting amplifiers we have assumed a DC input signal voltage. The behavior of the circuit in response to AC signals (sinewaves, squarewaves, and triangle waves) is similar. Recall the rules for the inverter: positive input signals produce negative output signals, and negative input signals produce positive output signals. These relationships mean that a 180° phase shift occurs between input and output. The relationship is shown in Fig. 3-5A.

Although the DC-coupled op-amp will respond to AC signals, there is a limit that must be recognized. If the peak value of the input signal is too great, output clipping (Fig. 3-5B) will occur. The peak output voltage will be

$$V_{o(peak)} = A_v V_{in(peak)} \tag{3-31}$$

where

$V_{o(peak)}$ = the peak output voltage

$V_{in(peak)}$ = the peak input voltage

$A_v$ = the voltage gain

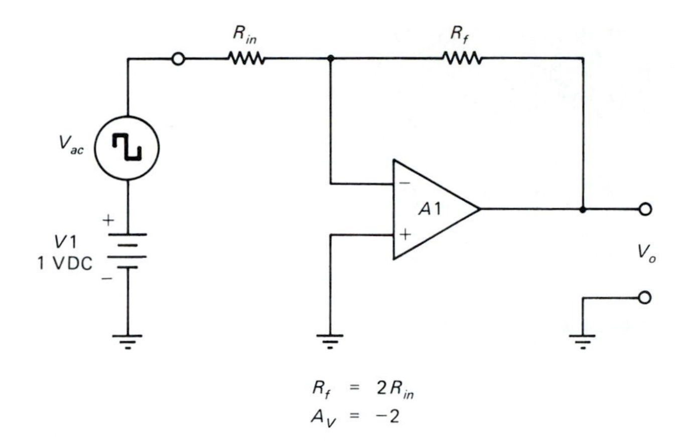


FIGURE 3-6

For every value of $V-$ and $V+$ power supply potentials there is a maximum attainable output voltage, $V_{o(max)}$. As long as the peak voltage is less than the maximum allowable output potential, the input waveform will be reproduced in the output, except amplified and inverted. But if the value of $V_{o(peak)}$ determined by Eq. (3-31) is greater than $V_{o(max)}$, clipping will occur.

In a linear voltage amplifier clipping is generally undesirable. The maximum output voltage can be used to calculate the maximum input signal voltage

$$V_{in(max)} = \frac{V_{o(max)}}{A_v} \tag{3-32}$$

There are occasions when clipping is desired. For example, in radio transmitter circuits called modulation limiters are often simple clippers followed by an audio low-pass filter which removes the harmonic distortion created by clipping. Clipping is also desired in generating squarewaves from sinewaves. The goal here is to drive the input so hard that sharp clipping occurs. Although there are better methods to achieve this the "overdriven clipper" squarewave generator does work.

### 3-2.7 Response to AC Input Signals with DC Offset

The case we considered in the previous section had an assumed waveform which is symmetrical close to the zero volts baseline. In this section we will examine a case where an AC waveform is superimposed on a DC voltage. Figure 3-6 shows an inverting amplifier circuit with an AC signal source in series with a DC source. In Fig. 3-7A there is a 4 volt peak-to-peak squarewave superimposed on a 1 Vdc fixed potential. Thus, the nonsymmetrical signal will swing between +3 volts and −1 volt.

The output waveform is shown in Fig. 3-7B. With the 180° phase inversion and the gain of −2 depicted in Fig. 3-6, the waveform will be a nonsymmetrical oscillation between −6 volts and +2 volts. Because of this gain ($A_v = -2$), the degree of asymmetry has also doubled to 2 Vdc.

Dealing with AC signals that have a DC component can lead to problems at high gain and high input signal levels. The output may saturate at either $V-$ or $V+$ power supply rails as in the case of the high amplitude symmetrical signal. If this limit is

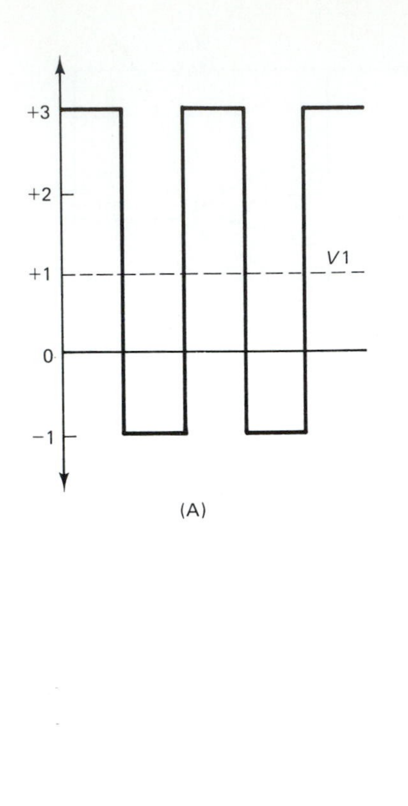


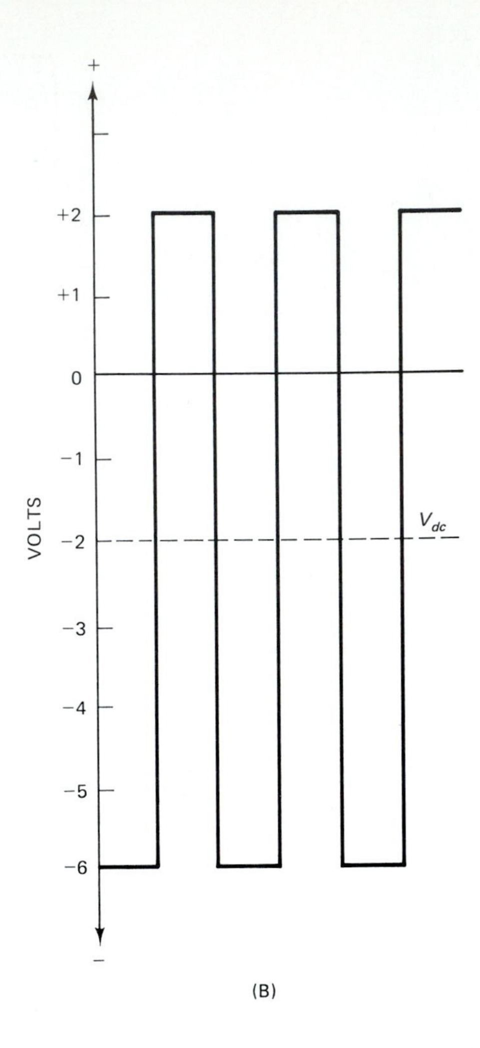


FIGURE 3-7

reached, clipping will result. The DC component is seen as a valid input signal so it will drive the output to one power supply limit or the other. For example, if the op-amp in Fig. 3-6 has a $V_{o(max)}$ value of $\pm 10$ volts, a $+4$ volt positive input signal will saturate the output (with $A_v = -2$), while the negative excursion can reach $-7$ volts before causing output saturation.

### 3-2.8 Response to Squarewaves and Pulses

Most amplifiers respond efficiently to sinusoidal and triangle waveforms. Some amplifiers, however, will exhibit problems with fast risetime waveforms such as squarewaves and pulses. The source of these problems is the high frequency content of the waveforms.

All continuous mathematical functions (including electronic waveforms) are made of a series of harmonically related sine and cosine constituent waves (with a possible DC component). The sine wave consists of a single frequency, or fundamental sin-

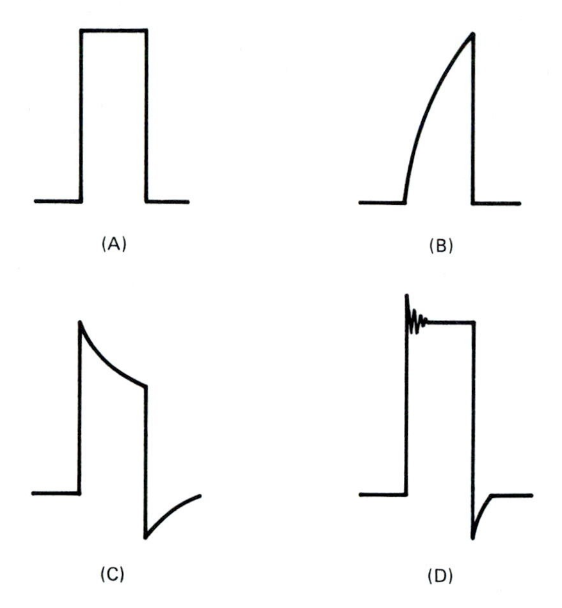


FIGURE 3-8

usoidal wave. All nonsinusoidal waveforms, however, are made up of a fundamental sinewave plus its harmonics. The actual waveshape is determined by the number of harmonics present, which particular harmonics are present (odd or even), the relative amplitudes of those harmonics, and their phase relationship with respect to the fundamental. These factors can be deduced from the quarter wave or halfwave symmetry of the wave.

The listing of the constituent frequencies forms a *Fourier series* and determines the bandwidth of the system required to process the signal. For example, the symmetrical squarewave is made up of a fundamental frequency sinewave (F), plus *odd* harmonics (3F, 5F, 7F . . .), theoretically up to infinity. As a practical matter, most squarewaves are "square" if the first 100 harmonics are present. Furthermore, if the squarewave is truly symmetrical, all of the harmonics are in phase with the fundamental. Other waveshapes have different "Fourier spectrums."

In general, the risetime of a pulse is related to the highest significant frequency in the Fourier spectrum by the rule-of-thumb approximation

$$F = \frac{0.35}{T_r} \tag{3-33}$$

where

$F$ = the highest Fourier frequency in hertz

$T_r$ = the pulse risetime in seconds

Because pulse shape is a function of the Fourier spectrum for that wave, the frequency response characteristic of the amplifier has an effect on the waveshape of the reproduced signal. Figure 3-8 shows an input pulse signal (Fig. 3-8A) and two possible responses.

The response shown in Fig. 3-8B results from attenuation of the high frequencies. The rounding shown will be either moderate or severe depending on the −3 dB bandwidth of the amplifier. Or in other words, by how many harmonics are attenuated by the amplifier frequency response characteristic, and to what degree. This problem becomes especially severe when the fundamental frequency (or pulse repetition rate) is high, the risetime is very fast, and the amplifier bandwidth is low.

Frequency compensated operational amplifiers achieve their claimed unconditional stability by rolling off the high frequency response drastically above a few kilohertz. A type 741AE is a frequency compensated op-amp with a gain-bandwidth product of 1,250,000 Hz and an open-loop gain of 250,000. Frequency response at maximum gain is 1,250,000 Hz/250,000, or 5000 Hz. Thus, we can expect good squarewave response only at relatively low frequencies. A rule of thumb for squarewaves is to make the amplifier bandwidth at least 100 times the fundamental frequency. As with all such rules, however, this one should be applied with caution even though it will be part of your "standard engineering wisdom."

The other class of problems is shown in Figs. 3-8C and 3-8D. In this case we see *peaking* and *ringing* of the pulse. There are three principal causes of these phenomena. First, a skewed bandpass characteristic in which either the low frequencies are attenuated or the high frequencies are amplified. Second, there are LC resonances in the circuit that give rise to ringing. Although not generally a problem at low frequencies, video operational amplifiers may have this problem. Third, there are both significant harmonics present at frequencies where circuit phase shifts add up to 180° and the loop gain is unity or greater. When combined with the 180° phase shift inherent in inverting followers, we have Barkhausen's criteria for oscillation.

Those criteria are

1. Total phase shift of 360° at the frequency of oscillation;
2. Output-to-input coupling (may be accidental); and
3. Loop gain of unity or greater.

Under some conditions the device will break into sustained oscillation. In other cases, however, oscillation will occur only on fast risetime signal peaks as shown in Fig. 3-8D.

### 3-2.9 Some Basic Rules

We must consider two important factors when designing inverting follower amplifiers. First, we must determine the voltage gain required by the application. Second, we must consider the input impedance of the circuit. That specification is needed in order to prevent the amplifier input from loading down the driving circuit. In the case of the inverting follower, the input impedance is the value of the input resistor ($R_{in}$), and a simple design rule is in effect: *The input resistor (hence the input impedance) should be equal or greater than ten times the source resistance of the previous circuit.*

The implication of this rule is that we must determine the source resistance of the driving circuit and make the input impedance of the operational amplifier inverting follower at least ten times larger. When the driving source is another operational amplifier we can assume the source impedance (the output impedance of the driving op-amp) is 100 ohms or less. For these cases, make the value of $R_{in}$ at least 1000 ohms (10 × 100 ohms = 1000 ohms). This value is based on the available output load current.

In other cases, however, we have a slightly different problem. Some transducers, for example a thermistor for measuring temperature, have a much higher source resistance. One *thermistor* has an advertised resistance which varies 10k to 100k over the temperature range of interest, so a minimum input impedance of 1 megohm (10 × 100 kohm) is required. When the input impedance gets this high the designer might want to consider the noninverting follower rather than the inverting follower.

In the inverting follower circuit the choice of input impedance drives the design. As part of the design procedure

1. Determine the minimum allowable input resistance (ten or more times the source impedance).
2. If the source resistance is 1000 ohms or less, try 10 kohms as an initial trial input resistance ($R_{in}$). This value might be lowered if the feedback resistor ($R_f$) becomes too high for the required gain. The value of $R_{in}$ is the input resistance or 10k whichever is higher.
3. Determine the amount of gain required. In general, the closed-loop gain of a single inverting follower should be less than 500. For higher gains use a multiple op-amp cascade circuit. Some low-cost op-amps should not be operated at closed-loop gains greater than 200, because of problems found in real versus ideal devices (see Chapter 4). In these cases the distributed gain of a cascade amplifier may prove easier to tame in practical situations.
4. Determine the frequency response (the frequency at which the gain drops to unity). From steps 3 and 4 we can calculate the minimum gain bandwidth product of the op-amp required (G-B = GAIN × FREQUENCY).
5. Select the operational amplifier. If the gain is high, e.g. over 100, you might want to select a BiMOS or BiFET operational amplifier in order to limit the output offset voltage caused by the input bias currents. Select a 741-family device if (a) you don't need more than a few kilohertz frequency response and (b) the unconditionally stable characteristics of the 741 are valuable for the application. Also look at the package style. For most applications the 8-pin miniDIP package is probably the easiest to handle. The 8-pin metal can is also useful, and it can be made to fit 8-pin miniDIP positions by correctly bending the leads.
6. Select the value of the feedback resistor

$$R_f = ABS(A_v) \times R_{in} \tag{3-34}$$

   $ABS(A_v)$ means the absolute value of voltage gain.

7. If the value of the feedback resistor is higher than the range of standard values (about 20 megohms or so), or too high for the input bias currents, try a lower input resistance.

### 3-2.10 Altering AC Frequency Response

The natural bandwidth of an amplifier is sometimes too great for certain applications. Noise power, for example, is a function of bandwidth as indicated by the expression $P_n$ = KTBR. Thus, it is possible the signal-to-noise ratio will suffer in some applications if bandwidth is not limited to just what is needed to process the expected

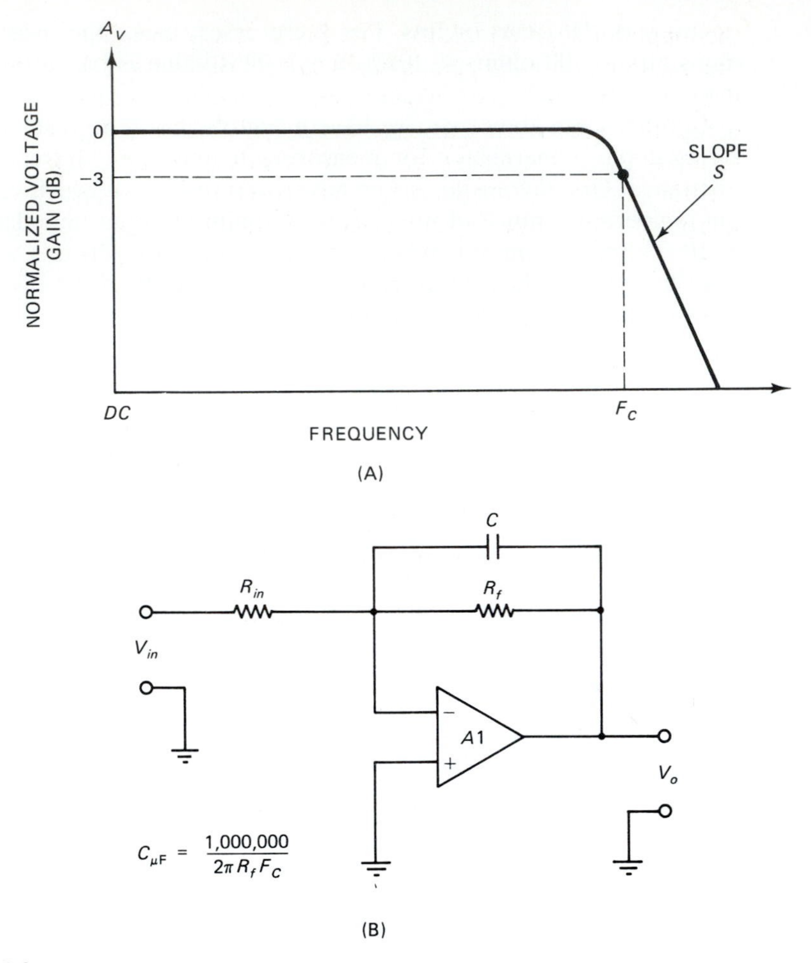


FIGURE 3-9

waveform. In other cases we find the rejection of spurious signals suffers if the bandwidth of an amplifier circuit is not tailored to the required bandwidth of the applied input signal. Amplifier stability is improved if the loop gain of the circuit is reduced to less than 1 at the frequency the circuit phase shifts (including internal amplifier phase shift) reach 180°. When the distributed phase shift is added to the 180° phase shift seen normally on inverting amplifiers, Barkhausen's criteria for oscillation are satisfied and the amplifier will oscillate.

If these criteria are satisfied at any frequency, the operational amplifier will oscillate at that frequency. We will discuss this topic further in Section 3-4, in Chapter 4 when we discuss operational amplifier problems and again in Chapter 15 when we discuss audio and telecommunications uses for operational amplifiers.

The design goal in tailoring the AC frequency response is to roll off the voltage gain at the frequencies above a certain critical frequency, $F_c$. This frequency is determined by evaluating the application, and is defined as the frequency at which the gain of the circuit drops off $-3$ dB from its in-band voltage gain. The response of the amplifier should look like Fig. 3-9A. It is shown here in normalized form where the maximum in-band gain is taken to be 0 dB. Above the critical frequency the gain drops off −6 dB/octave (an octave is a 2:1 change in frequency) by shunting a capacitor across the feedback resistor, as shown in Fig. 3-9B. The reactance of the capacitor is shunted across the resistance of $R_f$, so reduces the gain. The low-pass filter characteristic is achieved because the capacitive reactance becomes lower as frequency increases. The value of the capacitor is found from

$$C = \frac{1}{2 \pi R_f F_c} \tag{3-35}$$

where

$C$ = the capacitance in farads

$R_f$ = the feedback resistance in ohms

$F_c$ = the −3 dB frequency in hertz

Alternatively, to calculate the capacitance of $C$ in microfarads (μF) we use

$$C\mu F = \frac{1{,}000{,}000}{2 \pi R_f F_c} \tag{3-36}$$

**EXAMPLE 3-5**

An inverting follower with a gain of −100 has a 470 kohm feedback resistor. Calculate the capacitance (in μF) if the −3 dB point in the response curve is 200 Hz.

**Solution**

$$C2 = \frac{1{,}000{,}000}{2 \pi R_f F_c}$$

$$C2 = \frac{1{,}000{,}000}{(2)(3.14)(470{,}000 \text{ ohms})(200 \text{ Hz})}$$

$$C2 = \frac{1{,}000{,}000}{5.9 \times 10^8}$$

$$C2 = 1.69 \times 10^{-3}\ \mu\text{F} = 0.00169\ \mu\text{F}$$

■

In the example above, a standard value of either 0.0015 μF or 0.002 μF would probably be used as a practical matter.

## 3-3 NONINVERTING FOLLOWERS

The next standard op-amp circuit is the *noninverting follower.* This type of amplifier uses the noninverting input of the operational amplifier to apply signal and the output signal is in phase with the input signal (Fig. 3-10). There are two basic noninverting configurations, *unity gain* and *greater-than-unity gain.*

Figure 3-11 shows the circuit for a unity gain noninverting follower. The output terminal is connected directly to the inverting input, resulting in 100% negative feed-

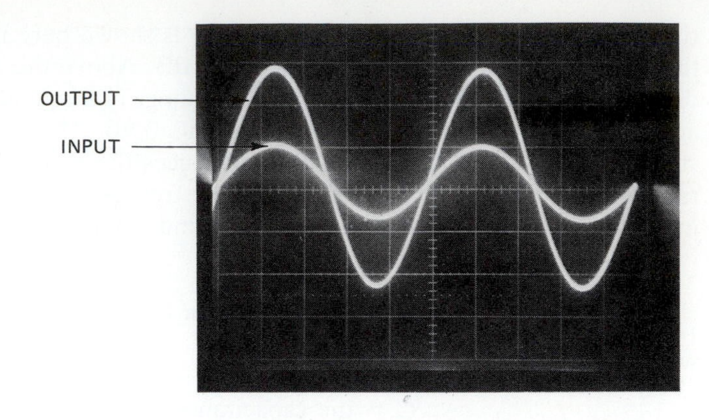


FIGURE 3-10

back. Recall the voltage gain expression for all feedback amplifiers

$$A_v = \frac{A_{vol}\, C}{1 + A_{vol}B} \tag{3-37}$$

where

$A_v$ = the closed-loop voltage gain (gain with feedback)

$A_{vol}$ = the open-loop voltage gain (gain without feedback)

$B$ = the feedback factor

$C$ = the input attenuation factor

In this circuit the input signal is applied directly to +IN, so $C = 1$, and can therefore be ignored. The feedback factor $B$ represents the transfer function of the feedback network. When that network is a resistor voltage divider network, the value of $B$ is a decimal fraction representing the attenuation of the op-amp output voltage before it is applied to the op-amp inverting input. In the unity gain follower circuit the value of $B$ is also 1 so it too is ignored. Eq. (3-24) therefore reduces to

$$A_v = \frac{A_{vol}}{1 + A_{vol}} \tag{3-38}$$

## EXAMPLE 3-6

A 741 device has an open-loop gain of 300,000. For this device, the voltage gain of a circuit such as Fig. 3-11 is

$$A_v = \frac{A_{vol}}{1 + A_{vol}}$$

$$A_v = \frac{300{,}000}{1 + 300{,}000} \tag{3-37}$$

$$A_v = 300{,}000/300{,}001 = 0.9999967$$

■

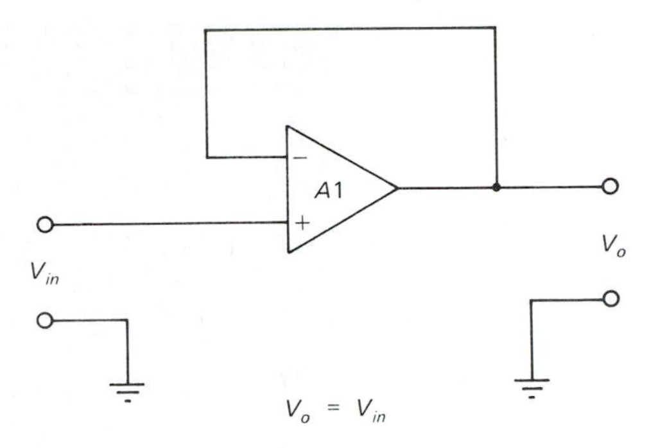


FIGURE 3-11

A gain of 0.9999967 is close enough to 1.0 to justify calling the circuit of Fig. 3-11 a unity gain follower.

### 3-3.1 Applications of Unity Gain Followers

What use is an amplifier that does not amplify? There are three principal uses of the unity gain noninverting follower, *buffering*, *power amplification*, and *impedance transformation*.

A buffer amplifier is placed between a circuit and its load to improve the isolation between the two. An example is a buffer amplifier used between an oscillator or waveform generator and its load. The buffer is especially useful where the load exhibits a varying impedance which could result in "pulling" of the oscillator frequency. Such unintentional frequency modulation of the oscillator is annoying because it causes some oscillator circuits to malfunction.

Another common use for buffer amplifiers is isolation of an output connection from the main circuitry of an instrument. An example might be an instrumentation circuit which uses multiple outputs, perhaps one to a digital computer A/D converter input and another to an analog oscilloscope or strip chart recorder. By buffering the analog output to the oscilloscope, we prevent short circuits in the display wiring from affecting the signal to the computer and vice versa.

A special case of buffering is represented by using the unity gain follower as a power driver. A long cable run may attenuate low-power signals. To overcome this problem we could use a low impedance power source to drive a long cable. This application shows a unity gain follower does have power gain ("unity gain" refers only to the voltage gain). If the input impedance is typically much higher than the output impedance, but $V_o = V_{in}$, then by $V^2/R$ it stands to reason the delivered power output is much greater than the input power. Thus, the circuit of Fig. 3-11 is unity gain for voltage signals and greater than unity gain for power. It is therefore a power amplifier.

The impedance transformation capability is obtained from the op-amp's high input impedance and low output impedance. Let's illustrate this application by a practical example. Figure 3-12A is a generic equivalent of a voltage source driving a load ($R2$).

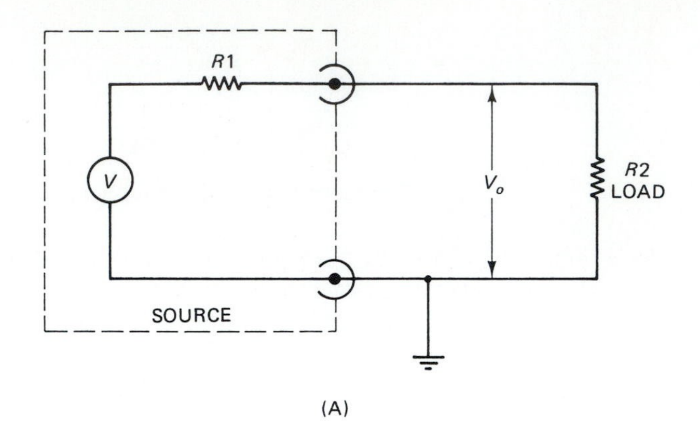


(A)

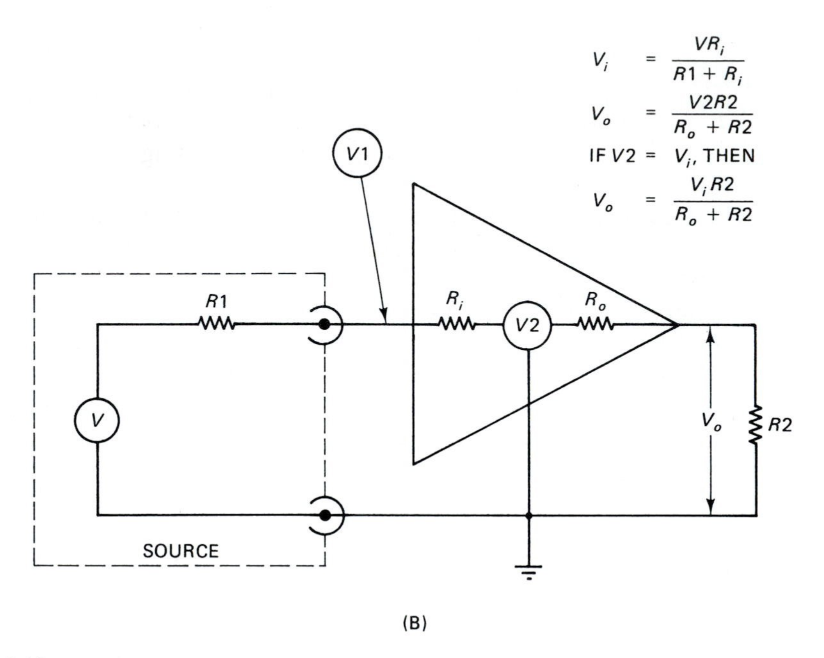


(B)

FIGURE 3-12

The resistance $R1$ represents the internal impedance of the signal source (usually called source impedance). The signal voltage, $V$, is reduced at the output ($V_o$) by whatever voltage is dropped across source resistance $R1$. The output voltage is found from

$$V_o = \frac{VR2}{R1 + R2} \tag{3-39}$$

## EXAMPLE 3-7

Assume that a high impedance source is driving a low impedance load similar to Fig. 3-12A. Calculate the loss in a case where $R1 = 10$ kohms, $R2 = 1$ kohm, and $V = 1$ volt.

$$V_o = \frac{V\,R2}{R1 + R2}$$

$$V_o = \frac{(1)(1\text{k})}{(10\text{k} + 1\text{k})}$$

$$V_o = 1/11 = 0.091 \text{ volt}$$ ■

In Example 3-7, 90% of the signal voltage was lost by the voltage divider action. But if we transform the circuit impedance with a unity gain noninverting amplifier, as in Fig. 3-12B, we change the situation entirely. If the amplifier input impedance is much larger than the source resistance, and the amplifier output impedance is much lower than the load impedance, there is very little loss and $V$ will closely approximate $V_o$.

### 3-3.2 Noninverting Followers with Gain

Figure 3-13A shows the circuit for a noninverting follower with gain. In this circuit, the signal ($V_{in}$) is applied to the noninverting input, while the feedback network ($R_f/R_{in}$) is almost the same as it was in the inverting follower circuit. The difference is one end of $R_{in}$ is grounded.

We can evaluate this circuit using the same general method used with the inverting follower. We know from Kirchoff's Current, and the fact Law that the op-amp inputs neither sink nor source current, that $I1$ and $I2$ are equal. Thus, the Kirchoff expression for these currents at the summing junction (point A) can be written as

$$I1 = I2 \tag{3-40}$$

We know from the properties of the ideal op-amp that any voltage applied to the noninverting input ($V_{in}$) also appears at the inverting input. Therefore

$$V1 = V_{in} \tag{3-41}$$

From Ohm's law we know that the value of current $I1$ is

$$I1 = V1/R_f \tag{3-42}$$

or, because $V1 = V_{in}$

$$I1 = V_{in}/R_f \tag{3-43}$$

Similarly, current $I2$ is equal to the voltage drop across resistor $R_f$ divided by the resistance of $R_f$. The voltage drop across resistor $R_f$ is the difference between output voltage $V_o$ and the voltage found at the inverting input, $V1$. By Ideal Property No. 7 $V1 = V_{in}$. Therefore

$$I2 = \frac{(V_o - V_{in})}{R_f} \tag{3-44}$$

We can derive the transfer equation of the noninverting follower by substituting Eqs. (3-43) and Eq. (3-44) into Eq. (3-40)

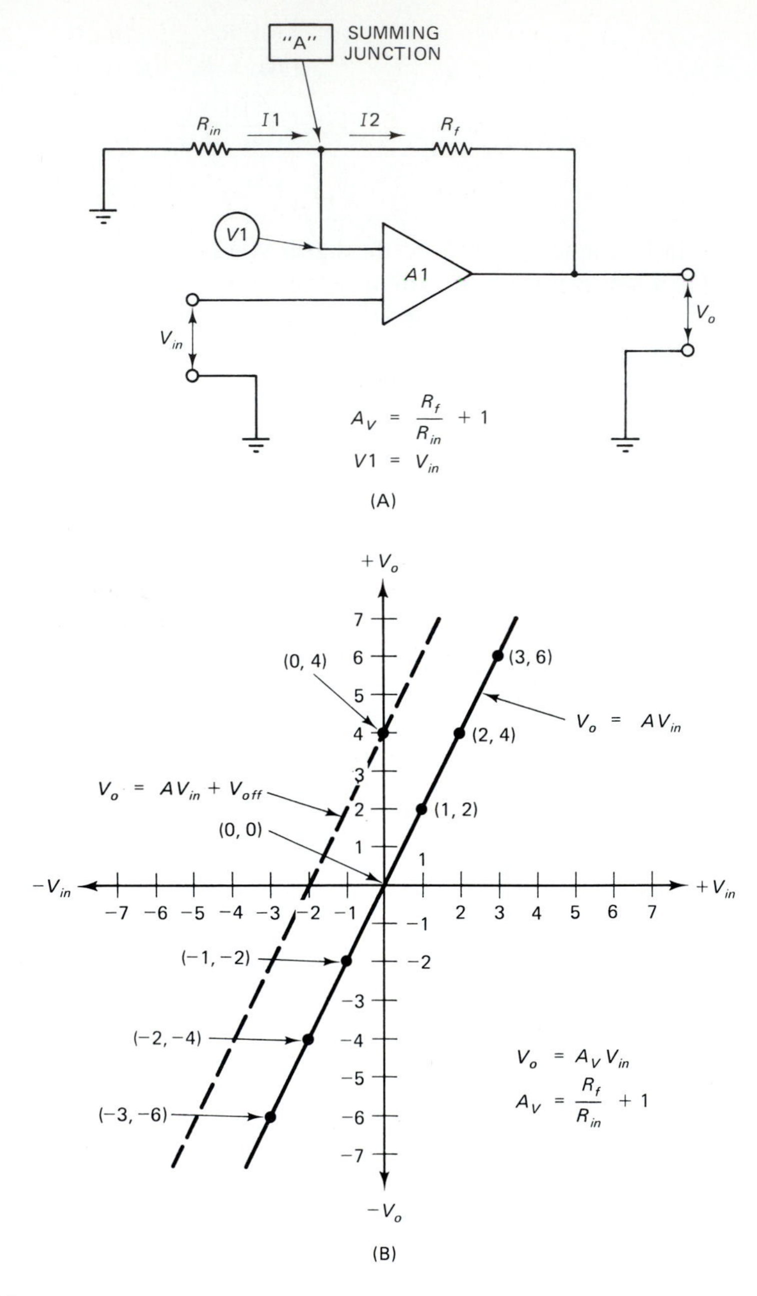


FIGURE 3-13

$$\frac{V_{in}}{R_{in}} = \frac{(V_o - V_{in})}{R_f} \tag{3-45}$$

We must now solve Eq. (3-45) for output voltage $V_o$

$$\frac{V_{in}}{R_{in}} = \frac{(V_o - V_{in})}{R_f} \tag{3-46}$$

$$\frac{R_f V_{in}}{R_{in}} = V_o - V_{in} \tag{3-47}$$

$$\frac{R_f V_{in}}{R_{in}} + V_{in} = V_o \tag{3-48}$$

Factoring out $V_{in}$

$$V_{in} \times \left(\frac{R_f}{R_{in}} + 1\right) = V_o \tag{3-49}$$

or, reversing the order to the conventional style

$$V_o = V_{in} \times \left(\frac{R_f}{R_{in}}\right) + 1 \tag{3-50}$$

Equation (3-50) is the transfer equation for the noninverting follower.

The transfer function $V_o/V_{in}$ for a gain of $-2$ noninverting amplifier is shown in Fig. 3-13B. The solid line assumes no output offset voltage is present ($V_o = A_v V_{in} + 0$), while the dotted line represents a case where the offset voltage is nonzero.

**EXAMPLE 3-8**

A noninverting amplifier has a feedback resistor of 100 kohms, and an input resistor of 20 kohms. What is the voltage gain?

**Solution**

$$A_v = \frac{R_f}{R_{in}} + 1$$

$$A_v = \frac{100 \text{ kohms}}{20 \text{ kohms}} + 1$$

$$A_v = 5 + 1 = 6$$

■

### 3-3.3 Advantages of Noninverting Followers

The noninverting follower offers several advantages. In our discussion of the unity gain configuration we mentioned buffering, power amplification, and impedance transformation were advantages. Also, the gain noninverting amplifier can provide voltage gain with no phase reversal.

The input impedance of the noninverting followers shown thus far is very high, being essentially the input impedance of the op-amp itself. In the ideal device, this impedance is infinite, while in practical devices it may range from 500,000 to more than $10^{12}$ ohms. Thus, the noninverting follower is useful for amplifying signals from any high impedance source, regardless of whether or not impedance transformation is a circuit requirement.

When the gain required is known (as it usually is in practical situations) we select a trial value for $R_{in}$, and then solve Eq. (3-50) to find $R_f$. This new version of the equation is:

$$R_f = R_{in}(A_v - 1) \tag{3-51}$$

We can solve many design problems using Eq. (3-51).

**EXAMPLE 3-9**

Design a noninverting amplifier with a gain of 100.

1. Select a trial value of resistance for $R_{in}$ (typically 100 to 5000 ohms): $R_{in}$ = 1200 ohms.
2. Use Eq. (3-51) to determine the value of $R_f$:

$$R_f = R_{in}(A_v - 1)$$

$$R_f = (1200)(100 - 1)$$

$$R_f = (1200)(99)$$

$$R_f = 118{,}800$$

■

Determine if the trial result obtained from this operation is acceptable by evaluating the application. If the result is not acceptable, work the problem again using a new trial value.

What does acceptable mean? If the value of $R_f$ is exactly equal to a standard resistor value, then all is well. But, as in the case above, the value 118,800 ohms is not a standard value. What we have to determine, therefore, is whether the nearest standard values result in an acceptable gain error. This is determined from the application. Both 118 kohm and 120 kohm are standard values, with 120 kohms being somewhat easier to obtain from distributor stock inventories. Both of these standard values are less than 1% from the calculated value, so this result is acceptable if a 1% gain error is within reasonable tolerance limits for the application.

### 3-3.4 The AC Response of Noninverting Amplifiers

The noninverting amplifier circuits discussed in the preceding sections are DC amplifiers. Nonetheless, as with the inverting amplifiers considered earlier, the noninverting amplifier will also respond to AC signals up to the upper frequency response limit of the circuit. (You may wish to review Sections 3-2.6 and 3-2.7.)

Figure 3-14 shows the input signal circuit for a noninverting follower. In this case there is an AC signal source in series with a DC potential ($V1$). These are applied to the noninverting input of the operational amplifier. A squarewave input signal (Fig. 3-15A) is applied to the input, but it is offset by a DC component (Fig. 3-15B). If the amplifier has a gain of +2, the output signal will be as shown in Fig. 3-15C. This signal swings from +1 volt to +5 volts. The offset of 1.5 Vdc is amplified by two, and becomes a 3 Vdc offset, with the AC signal swinging about this level.

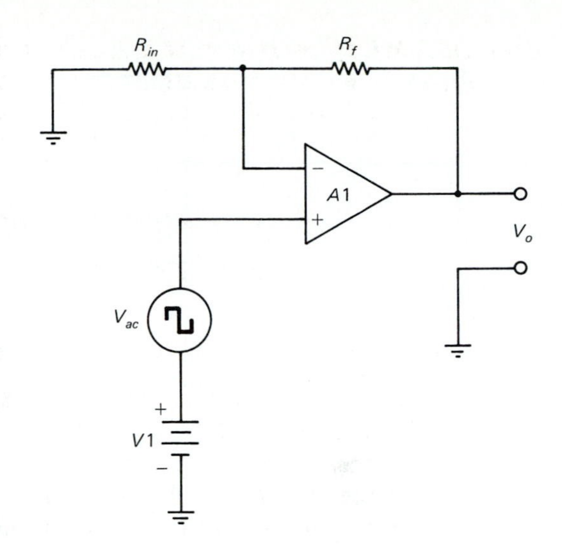


FIGURE 3-14

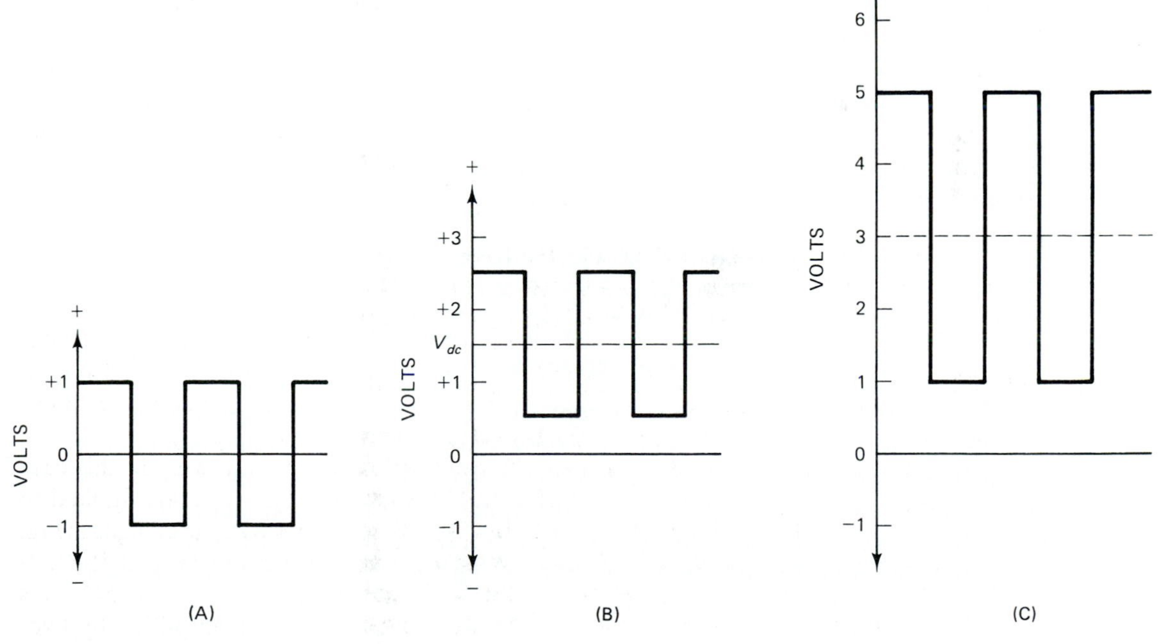


FIGURE 3-15

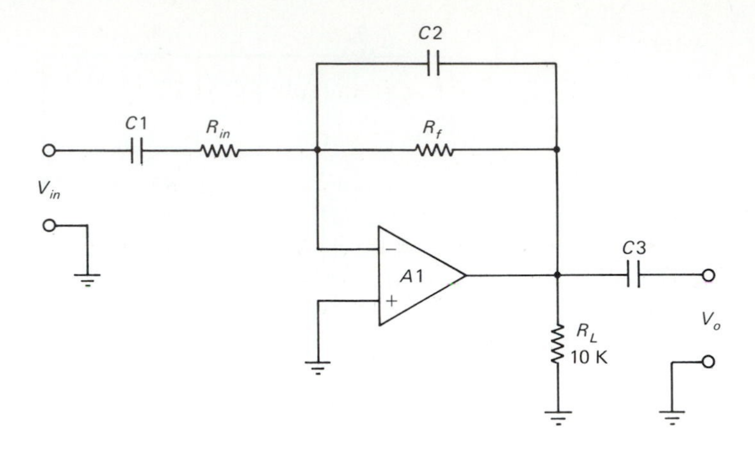


FIGURE 3-16

## 3-4 FREQUENCY RESPONSE TAILORING

In Section 3-2.10 we learned one can tailor the upper-end frequency response of the inverting follower operational amplifier with a capacitor shunting the feedback resistor. The same method also works for the noninverting follower. In this section we will expand the subject and discuss not only tailoring of the upper −3 dB frequency response, but also a lower −3 dB limit as well.

The capacitor across the feedback resistor (Fig. 3-16) sets the frequency at which the upper end frequency response falls off −3 dB below the low end in-band gain. The gain at frequencies higher than this −3 dB frequency falls off at a rate of −6 dB/octave, or −20 dB/decade (a decade is a 10:1 frequency ratio).

The value of capacitor $C2$ is found by

$$C2_{\mu F} = \frac{1,000,000}{2\ \pi\ R_f\ F_c} \tag{3-52}$$

where

$C2_{\mu F}$ = the capacitance in microfarads (μF)

$R_f$ = in ohms

$F_c$ = the upper −3 dB frequency in hertz (Hz)

The low frequency response is controlled by placing a capacitor in series with the input resistor, which makes the inverting follower an AC-coupled amplifier. Figure 3-16 is the circuit for an inverting follower that uses AC coupling at both input and output circuits. Capacitor $C2$ limits the upper −3 dB frequency response point. Its value is set by the method discussed above. The lower −3 dB point is set by the combination of $R_{in}$ and input capacitor $C1$. This frequency is set by the equation

$$C1 = \frac{1,000,000}{2\ \pi\ R_{in}\ F} \tag{3-53}$$

where

$$C1 = \text{in microfarads } (\mu F)$$

$$R_{in} = \text{in ohms}$$

$$F = \text{the lower } -3 \text{ dB point in hertz (Hz)}$$

In some cases we will want to AC-couple the output circuit. Capacitor $C3$ is used to AC-couple the output, thus preventing any DC component present on the op-amp output from affecting the following stages. Resistor $R_L$ is used to keep capacitor $C3$ from being charged by the offset voltage from op-amp $A1$. The value of capacitor $C3$ is set to retain the lower $-3$ dB point, using the resistance of the stage following as the $R$ in the equations above.

## 3-5 AC-COUPLED NONINVERTING AMPLIFIERS

The noninverting amplifiers discussed thus far have all been DC-coupled. They will respond to signals from either DC or near-DC up to the frequency limit of the amplifier selected. Sometimes, however, we do not want the amplifier to respond to DC or slowly varying near-DC signals. For these applications we can select an AC-coupled noninverting follower circuit. In this section we will examine several AC-coupled noninverting amplifiers.

Figure 3-17A shows a capacitor input AC-coupled amplifier circuit. It is essentially the same as the previous circuits, except for the input coupling network, $C1/R3$. The capacitor in Fig. 3-17A serves to block DC and very low frequency AC signals. If the op-amp has zero or very low input bias currents, then we can safely delete resistor $R3$. For all but a few commercially available devices, however, resistor $R3$ is required if closed-loop gain is high. Input bias currents will charge capacitor $C1$, creating a voltage offset which is recognized by the op-amp as a valid DC signal and amplified to form an output offset voltage. In some devices the output saturates from the $C1$ charge shortly after turn-on and resistor $R3$ keeps $C1$ discharged.

Resistor $R3$ also sets the input impedance of the amplifier. Previous circuits had a very high input impedance determined by the high op-amp input impedance. In Fig. 3-17A, however, the input impedance seen by the source is equal to $R3$.

Resistor $R3$ and capacitor $C1$ also limit the low frequency response of the circuit. Filtering occurs because $R3C1$ forms a high-pass filter (see Fig. 3-17B). The $-3$ dB frequency is found from:

$$F = \frac{1,000,000}{2\,\pi\,R1\,C1} \tag{3-54}$$

where

$$F = \text{the } -3 \text{ dB frequency in hertz (Hz)}$$

$$R1 = \text{the resistance in ohms}$$

$$C1 = \text{the capacitance in microfarads}$$

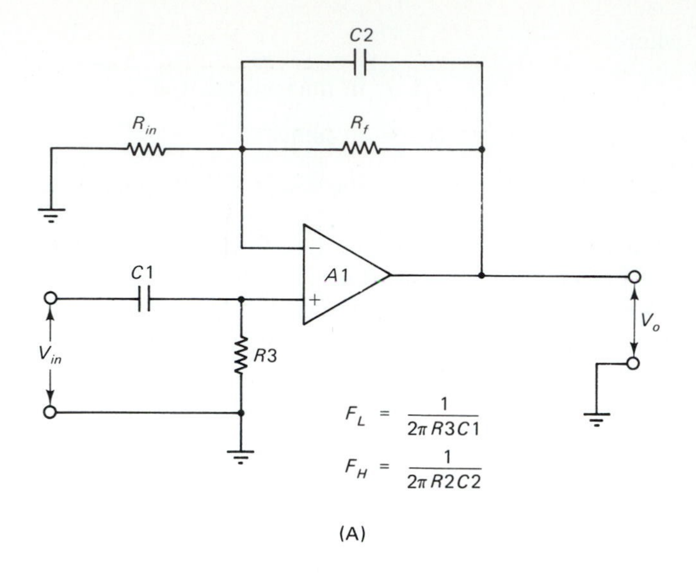


(A)

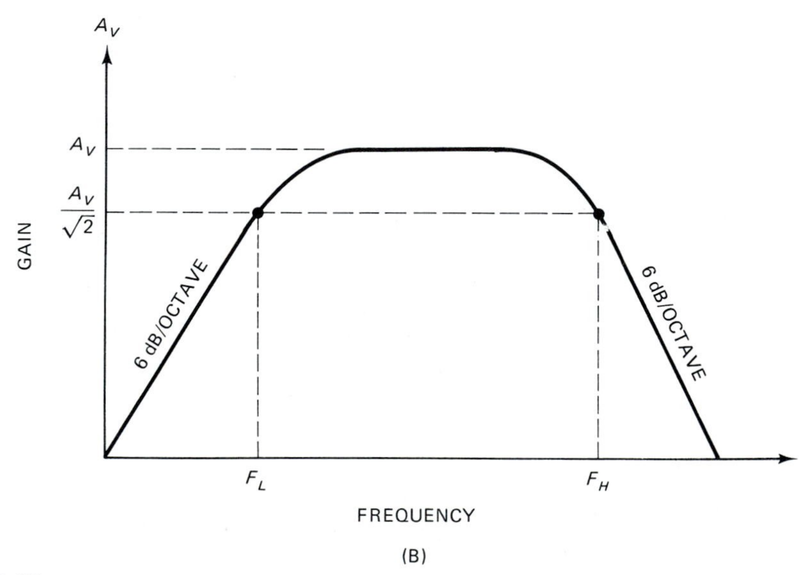


(B)

FIGURE 3-17

The form of Eq. (3-54) is reversed from the standpoint of practical circuit design problems. In most cases, we know the required frequency response limit from the application. We also know from the application what the minimum value of $R3$ should be (an implication from source impedance), and often set it as high as possible as a practical matter (10 megohms). Thus, we want to solve the equation for $C$, as shown below

$$C = \frac{1{,}000{,}000}{2\ \pi\ R1\ F} \tag{3-55}$$

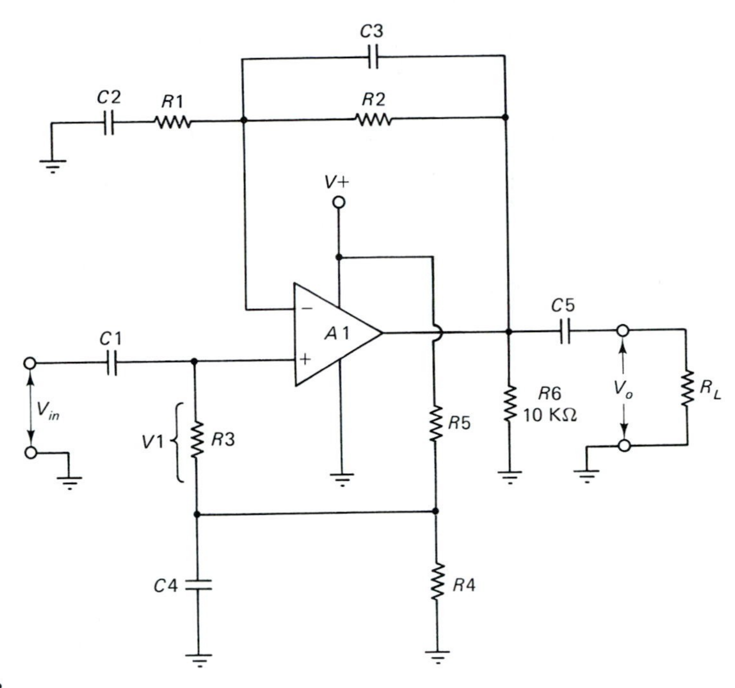


FIGURE 3-18

The technique of Fig. 3-17A works well for dual-polarity DC power supply circuits. In single-polarity DC power supply circuits, however, the method fails because of the large DC offset voltage present on the output. For these applications we use a circuit like the one shown in Fig. 3-18.

The circuit in Fig. 3-18 is operated from a single $V+$ DC power supply (the $V-$ terminal of the op-amp is grounded). In order to compensate for the $V-$ supply being grounded, the noninverting input is biased to a potential of

$$V1 = (V+)\,\frac{R4}{R4 + R5} \tag{3-56}$$

If $R4 = R5$, then $V1$ will be $(V+)/2$. Because the noninverting input typically sinks very little current, the voltage at both ends of $R3$ is the same ($V1$).

The circuit of Fig. 3-18 does not pass DC and some low AC frequencies because of the capacitor coupling. Also, because capacitor $C3$ shunts feedback resistor $R2$, there is also a rolloff of the higher frequencies. The high frequency rolloff $-3$ dB point is found from

$$F = \frac{1{,}000{,}000}{2\,\pi\,R2\,C3} \tag{3-57}$$

where

$F$ = the $-3$ dB frequency in hertz (Hz)

$R2$ = in ohms

$C3$ = in microfarads

We can restate Eq. (3-57) in a form which takes into account the fact we usually know the value of $R2$ (from setting the gain) and the nature of the application sets the minimum value of frequency $F$. We can rewrite Eq. (3-57) in a form which yields the value of $C3$ from these data:

$$C3 = \frac{1{,}000{,}000}{2\ \pi\ R3\ F} \tag{3-58}$$

The lower $-3$ dB frequency is set by any or all of several $RC$ combinations within the circuit

1. $R1/C2$
2. $R3/C1$
3. $R3/C4$
4. $R_L/C5$

Resistor $R1$ is part of the gain-setting feedback network. Capacitor $C2$ is used to keep the cold end of $R1$ above ground at DC, while keeping it grounded for AC signals.

Resistor $R3$ is the input resistor and serves the same purpose as the similar resistor in the previous circuit. At midband the input impedance is set by resistor $R3$, although at the extreme low end of the frequency range the reactance of $C4$ becomes a significant feature. In general, $X_{C4}$ should be less than or equal to $R3/10$ at the lowest frequency of operation.

Capacitor $C1$ is in series with the input signal path and serves to block DC and certain very low frequency signals. The value of $C1$ should be

$$C1 = \frac{1{,}000{,}000}{2\ \pi\ F\ R3} \tag{3-59}$$

where

$$C1 = \text{in microfarads}$$
$$F = \text{in hertz (Hz)}$$
$$R3 = \text{in ohms}$$

Capacitor $C5$ is used to keep the DC output offset from affecting succeeding stages. The 10 kohm output load resistor ($R6$) keeps $C5$ from being charged by the DC offset voltage. The value of $C5$ should be greater than

$$C5 > \frac{1{,}000{,}000}{2\ \pi\ F\ R_L} \tag{3-60}$$

where

$$C5 = \text{in microfarads}$$
$$F = \text{the low end } -3 \text{ dB frequency in hertz (Hz)}$$
$$R_L = \text{the load resistance in ohms}$$

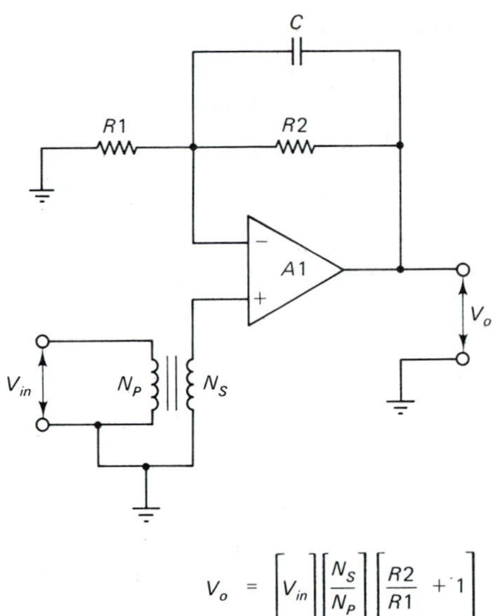


FIGURE 3-19

## 3-6 TRANSFORMER COUPLED NONINVERTING AMPLIFIERS

Figure 3-19 shows the circuit for a transformer coupled noninverting follower. This type of circuit is often used in audio and broadcasting applications where audio signals might pass over a 600 ohm balanced line. The important point here is this circuit is an AC-only amplifier, with upper and lower −3 dB points determined by the frequency response of the transformer ($T1$), the limitations of the operational amplifier, and any capacitance shunting feedback resistor $R2$.

The gain of the amplifier in Fig. 3-19 is given in Eq. (3-61) below:

$$A_v = (V_{in}) \left[\frac{N_s}{N_p}\right] \left[\frac{R2}{R1} + 1\right] \tag{3-61}$$

where

$A_v$ = the voltage gain

$N_s$ = the number of turns in the secondary of $T1$

$N_p$ = the number of turns in the primary of $T1$

$R2$ = the op-amp feedback resistor

$R1$ = the op-amp input resistor

**EXAMPLE 3-10**

Find the gain of an amplifier such as Fig. 3-19 if transformer $T1$ has a turns ratio ($N_p/N_s$) of 1:4.5 ($N_s/N_p$ = 4.5:1), $R1$ = 1000 ohms, and $R2$ = 3900 ohms.

$$A_v = \left[\frac{N_s}{N_p}\right]\left[\frac{R2}{R1} + 1\right]$$

$$A_v = \left[\frac{4.5}{1}\right]\left[\frac{3900}{1000} + 1\right]$$

$$A_v = (4.5)(3.9 + 1)$$

$$A_v = (4.5)(4.9)$$

$$A_v = 22.1$$

■

## 3-7 SUMMARY

1. There are three principal configurations for the operational amplifier: inverting follower, unity gain noninverting follower, and noninverting follower with gain.
2. The inverting amplifier produces an output signal which is 180° out of phase with the input signal.
3. A noninverting amplifier produces a signal that is in phase with the input signal.
4. A virtual ground is any point where the potential is zero volts, even though the point is not physically connected to ground or the power supply common.
5. The gain of an operational amplifier circuit is set by the external feedback resistor network according to the equations given in Section 3-9.

## 3-8 RECAPITULATION

Now return to the objectives and Prequiz questions at the beginning of the chapter and see how well you can answer them. If you cannot answer certain questions, mark them and review the appropriate parts of the text. Next, try to answer the questions and work the problems below, using the same procedure.

## 3-9 LABORATORY EXERCISES

1. Connect the circuit of Fig. 3-20 using values of $R_f$ selected from the range 22 kohms to 56 kohms (try several). With each value of feedback resistor make a graph of $V_o$ against $V_{in}$ for five values of $+V_{in}$ and five values of $-V_{in}$.
2. Using the circuit of Fig. 3-20, and a value of $R_f$ of 150 kohms, make a plot of $V_o$ against $V_{in}$ similar to the previous exercise. Explain the difference between the two plots.
3. Replace the DC voltage source in Fig. 3-20 with a 100 Hz sinewave source. Observe the input and output waveforms on a cathode ray oscilloscope (CRO).
4. Design a noninverting amplifier with a gain of (A) 2 (B) 101.

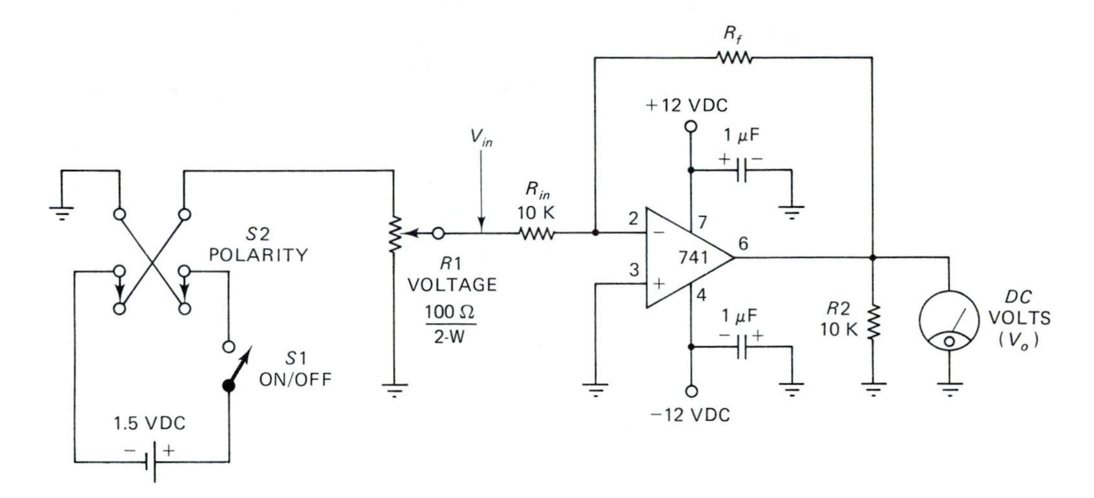


FIGURE 3-20

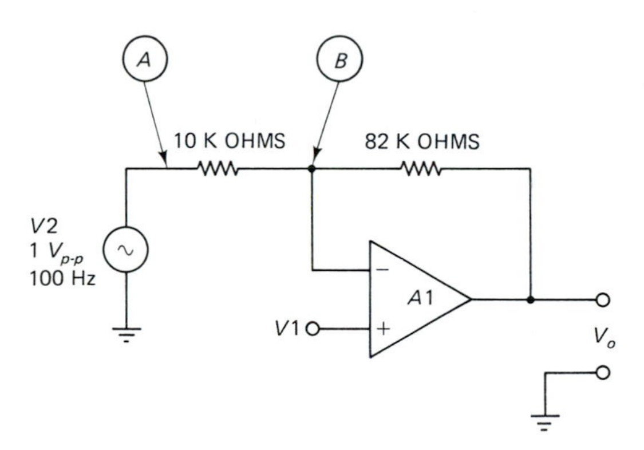


FIGURE 3-21

## 3-10 QUESTIONS AND PROBLEMS

1. The potential measured at a properly functioning virtual ground is: ______________ −Vdc.
2. An operational amplifier is connected as an inverting follower. The feedback resistor is 100 kohms, and the input resistor is 15 kohms. (A) What is the gain of this circuit? (B) What is the output voltage if a 0.800 Vdc input signal voltage is applied?
3. Consider the circuit of Fig. 3-21. Sketch one cycle of the output voltage, $V_o$, if the input signal is 1 volt (peak-to-peak) at 100 Hz, and $V1$ is 1.5 Vdc. (Hint: $V1$ and $V2$ use different operational amplifier configurations.)
4. In the circuit of Fig. 3-21 point A is grounded and a +2.5 Vdc potential is applied to the noninverting input ($V1$). (A) which amplifier configuration is this new circuit? (B) what is the output voltage? (C) what voltage is measured at point B?
5. Consider Fig. 3-22. Calculate $V_o$ for the following conditions: $V1$ = 1 Vdc, $V2$ = 0.4 Vdc, $V3$ = 2 Vdc, and $V4$ = −1.5 Vdc.

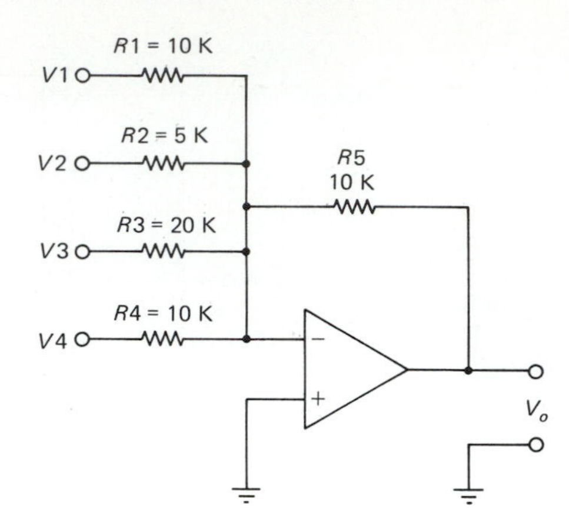


FIGURE 3-22

6. Calculate $V_o$ in Fig. 3-22 if $V1 = 2$ Vdc, $V2 = 1$ Vdc, $V3 = 560$ mV, and $V4 = 2.75$ Vdc.
7. A 741 operational amplifier is connected to $\mp$ 12 Vdc power supply. In this amplifier the maximum value of $V_o$ is 1 volt less than the DC power supply potential of the same polarity. If $R_f = 100$ kohms, and $R_{in} = 10$ kohms, will this amplifier saturate with input signals of (a) 100 mV, (b) 1 Vdc, (c) 1 volt p-p 100-Hz AC, (d) 5 Vdc, (e) $-5$ Vdc?
8. What is the maximum allowable input signal voltage if the closed-loop gain of an inverting follower is 150 and the maximum allowable output voltage is 14 volts?
9. Sketch the output waveform from a gain of $+2$ noninverting follower if the input waveform is a 2 volt peak-to-peak squarewave with a 1 Vdc offset.
10. Calculate the approximate highest Fourier frequency in a squarewave with a 1.6 μsec risetime.
11. State Barkhausen's criteria for oscillation, and explain how they apply to self-oscillation of an operational amplifier.
12. A signal source has an internal impedance of 15 kohms. What is the minimum rule-of-thumb input impedance of an inverting follower that is used to amplify the output signal of this signal source?
13. A noninverting follower has a 470 kohm feedback resistor. Calculate the value of shunt capacitor needed to limit the $-3$ dB frequency response to 100 Hz.
14. List three uses for a unity gain noninverting follower.
15. Consider Fig. 3-23. (A) calculate $V1$, (B) calculate $V_o$ when $V_{in}$ is 0-Vdc, (C) sketch the output waveform when $V_{in}$ is a 100-Hz, 5-volt peak-to-peak sinewave.
16. Consider $V_o$ compared to $V_{in}$ characteristic plotted in Fig. 3-24. (A) What is the gain of the amplifier depicted as curve A? (B) What is the gain of the amplifier depicted as curve B? (C) Design amplifiers that will exhibit these characteristics.
17. Using Fig. 3-24 as a guide draw characteristics for the following gains: (A) $+5$, (B) $-5$, (C) $+10$, (D) $-1$, and (E) $-6$.
18. Draw the circuit of an inverting follower in which a method is used to prevent self-oscillation when an uncompensated operational amplifier is used.
19. Derive the transfer equation for an inverting follower using Eq. (3-37).

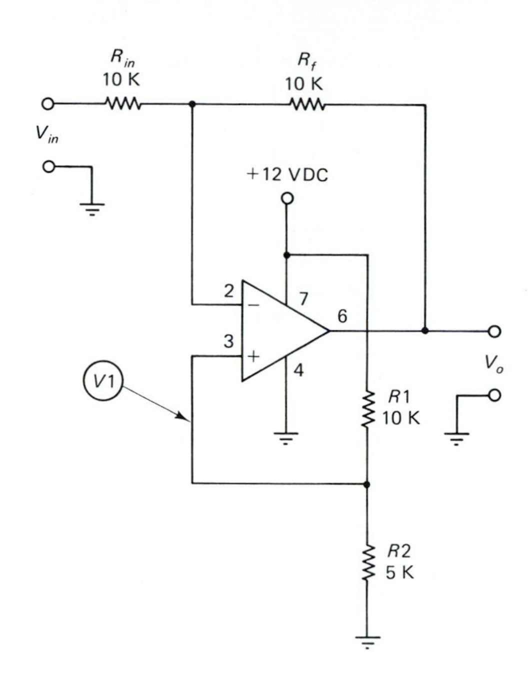


FIGURE 3-23

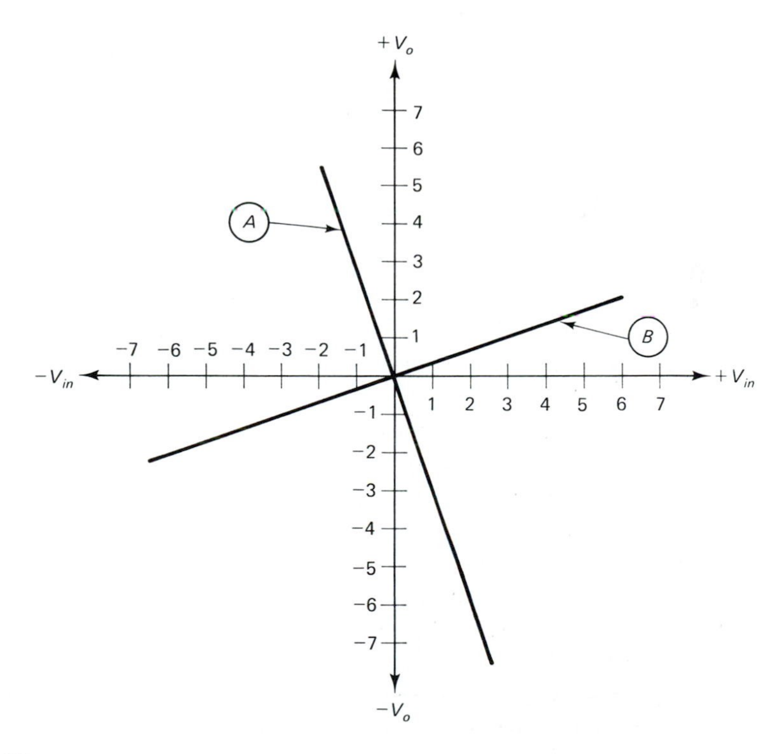


FIGURE 3-24

20. Derive the transfer function for a noninverting follower using Eq. (3-37). What is the value of the term $C$ in Eq. (3-37).
21. A unity gain noninverting follower is based on an operational amplifier that has an open-loop gain of 20,000. Calculate the exact gain of this circuit.
22. In question 21, the operational amplifier is replaced with a newer model that has an open-loop gain of 250,000. Calculate the exact gain of this new amplifier.
23. A noninverting amplifier has a gain of 114, and an input resistance of 8.2-kohms. Calculate the capacitance that must be shunted across the feedback resistance in order to limit the frequency response to 2200 Hz.
24. Find the voltage gain of a transformer coupled AC amplifier (Fig. 3-19) in which the turns ratio of the transformer is 1:6.5, and the resistance ratio of the feedback network is 3:1.
25. Find the voltage gain of the circuit in question 24 if the transformer is reversed so that the turns ratio is 6.5:1.

## CHAPTER 4

# Dealing with Practical Operational Amplifiers

## OBJECTIVES

1. Understand problems existing in real versus ideal operational amplifiers.
2. Understand how to deal with offset voltages in real operational amplifiers.
3. Learn how to compensate the frequency in operational amplifiers.
4. Learn how to measure the parameters of operational amplifiers.

## 4-1 PREQUIZ

These questions test your prior knowledge of the material in this chapter. Try answering them before you read the chapter. Look for the answers (especially those you answered incorrectly) as you read the text. After you have finished studying the chapter, try answering these questions again and those at the end of the chapter (see Section 4-9).

1. An input bias offset current of 3.7 mA flows in the inputs of a 741 operational amplifier. Calculate the input offset voltage on the inverting input if $R_f$ if 100 kohms and $R_{in}$ is 10 kohms.
2. What are the implications of not decoupling the $V-$ and $V+$ power supply lines of an uncompensated IC linear amplifier?
3. Why are two different types of capacitors sometimes used in power supply decoupling of IC linear amplifiers?
4. The unity gain capacitance for an operational amplifier is 68 pF. Find the value of a compensation capacitor if $R_f$ is 100 kohms, and $R_{in}$ is 2.7 kohms.

## 4-2 OPERATIONAL AMPLIFIER PROBLEMS AND THEIR SOLUTIONS

In Chapter 2 we discussed the ideal operational amplifier. Such a hypothetical device is a good learning tool, but it doesn't really exist. It makes our analysis easier, but it cannot actually be purchased and used in practical circuits. All real operational amplifiers depart from the ideal.

We find, for example, that open-loop gain is not really infinite, but rather high values in the range from about 20,000 to over 1,000,000 are found. Similarly, real operational amplifiers do not have infinite bandwidth, and in fact most are intentionally made with severe bandwidth limitations. Such amplifiers are said to be unconditionally stable or frequency compensated devices (an example is the 741 device). While stability is a highly desirable feature for many applications, it is obtained at the expense of frequency response. In this chapter we will discuss some of the common problems in real devices, and provide solutions. We will also examine how various amplifier parameters are measured.

## 4-3 MEASUREMENT OF OPERATIONAL AMPLIFIER PARAMETERS

In this section you will learn how to make your own measurements on real devices in the laboratory. Second, and perhaps most important, you will understand specification parameters and how they are derived. Many op-amp parameters can only be understood in the context of the measurement situation. Comparison of devices becomes impossible if different test methods are used for each device. For example, thermal drift can be compared in two devices only if both are measured at the same $V-$ and $V+$ power supply potentials.

### 4-3.1 Voltage Gain

Voltage gain is the ratio of signal output voltage ($V_o$) to signal input voltage ($V_{in}$), or

$$A_v = \frac{V_o}{V_{in}} \tag{4-1}$$

Measuring the closed-loop voltage gain is a simple matter of applying an input signal of known amplitude and measuring the output level generated. The voltage gain is calculated from Eq. (4-1). The process is simple, but watch for saturation of the output signal. An AC input signal such as a sinewave or triangle wave is preferred for this measurement because flat-topping (the indicator of saturation) is easier to spot.

The input signal must have an amplitude that is high enough to produce meaningful output level, one which is easy to measure accurately but low enough to prevent saturating the amplifier. When the input signal must be low, as might occur when the overall closed-loop gain is high, a means must be provided to ensure that any DC input offset does not affect the measurement. In those cases it is necessary to null the input offset using one of the methods discussed later in this chapter (Section 4-4.1). In some cases these offsets can approach or even exceed the amplitude of the signal voltage.

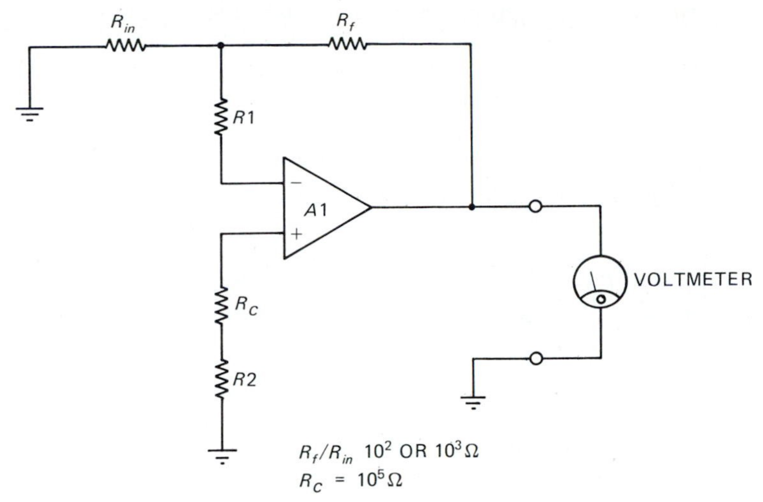


FIGURE 4-1

### 4-3.2 Input Offset Current

Input offset current is measured in a test circuit like the one shown in Fig. 4-1. Input offset current can be specified by the relationship of two different output voltages taken under different input conditions

$$I_{io} = \frac{R_{in}(V_{o1} - V_{o2})}{R2(R_{in} + R_f)} \tag{4-2}$$

The first output voltage, $V_{o1}$, is measured with $R1$ and $R2$ connected in the circuit. Voltage $V_{o2}$ is then measured with resistors short-circuited and all other conditions the same. With $V_{o1}$, the resulting output voltage can be used in Eq. (4-2) to determine input offset current.

### 4-3.3 Input Offset Voltage

The input offset voltage is the voltage required to force the output voltage ($V_o$) to zero when the input voltage is also zero. The operational amplifier is connected in an inverting amplifier configuration such as Fig. 4-2. To make the measurement, the

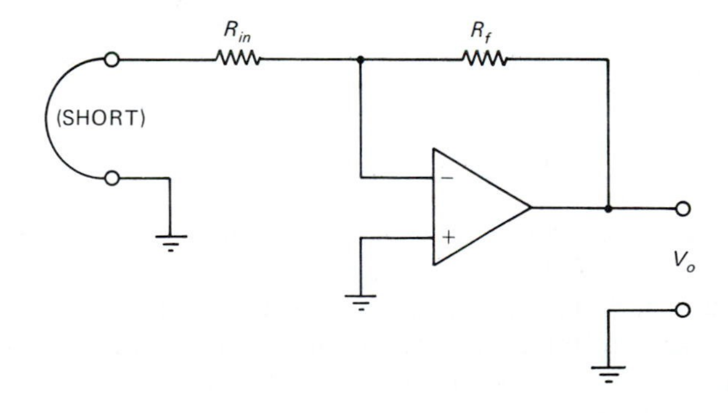


FIGURE 4-2

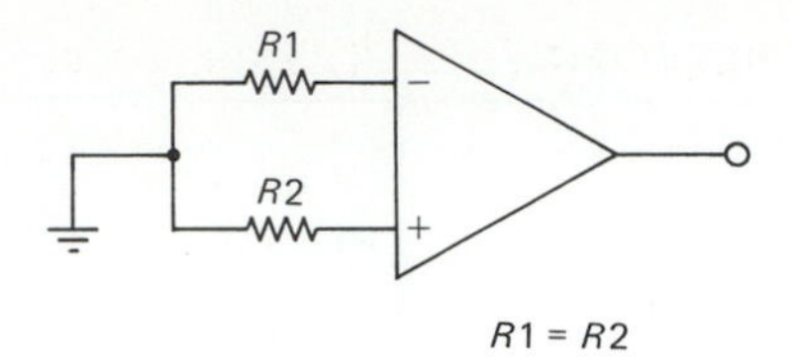


FIGURE 4-3

input terminal is connected to ground. The input offset voltage is found by measuring the output voltage (when $V_{in} = 0$) and then using the standard voltage divider equation

$$V_{io} = \frac{V_o R_{in}}{R_f + R_{in}} \tag{4-3}$$

Greater accuracy is achieved if the gain of the amplifier is 100, 1000, or even higher, provided such gains can be accommodated without saturating the amplifier.

### 4-3.4 Input Bias Current

This test requires a pair of closely matched resistors connected between the op-amp inputs (−IN and +IN) and ground (Fig. 4-3). Power is applied to the operational amplifier and the voltage is measured at each input. The value of resistors $R1$ and $R2$ must be high enough to create a measurable voltage drop at the level of current anticipated. Although the actual value of these resistors is not too critical, the match between them ($R1 = R2$) is critical to the success of the measurement. The definition of readable voltage drop depends on the available instrumentation. For example, assume an actual input bias current of 5 microamperes (μA) is flowing and the resistors are each 10 kohms. In this case the measured voltage will be (by Ohm's law)

$$V = IR$$

$$V = (5 \times 10^{-6} \text{ amperes}) \times (10{,}000 \text{ ohms})$$

$$V = 5 \times 10^{-2} \text{ volts} = 50 \text{ mV}$$

If your voltage-measuring equipment is not capable of measuring these levels, higher resistor values will be required. If the two inputs had ideally equal input bias currents, only one measurement would be needed. Since real devices usually have unequal bias currents, however, it is sometimes necessary to measure both and use the higher input bias current. Alternatively, the root sum squares (RSS) value can be used.

### 4-3.5 Power Supply Sensitivity (PSS)

Power supply sensitivity is the worst-case change of input offset voltage for a 1.0 Vdc change of *one* DC power supply voltage (either $V-$ or $V+$), with the other supply potential being held constant. The test to measure input offset voltage (Fig. 4-2) is also used to measure this parameter. First, the two power supply voltages are set to equal levels and the input offset voltage is measured. One of the power supply voltages is then changed by precisely 1.00 Vdc and the input offset voltage is again measured.

The power supply sensitivity (PSS) is given by

$$\psi = \frac{\Delta V_{io}}{\Delta V_o} \tag{4-4}$$

**EXAMPLE 4-1**

An operational amplifier has an initial offset voltage of 5.0 mV. A value of 6.2 mV is found to exist after the $V+$ power supply voltage is changed from +12 Vdc to +13 Vdc. Calculate the power supply sensitivity for a +1 Vdc change in $V+$.

**Solution**

$$\psi = \frac{\Delta V_{io}}{\Delta V_o}$$

$$\psi = \frac{(6.2 - 5.0)\ \text{mV}}{1.0\ \text{V}}$$

$$\psi = 1.2\ \text{mV}/1.0\ \text{V} = 1.2\ \text{mV/V}$$ ■

The actual power supply sensitivity is the worst case when this measurement is made under four conditions: (1) $V+$ *increased 1 Vdc*, (2) $V+$ *decreased 1 Vdc*, (3) $V-$ *increased 1 Vdc*, and (4) $V-$ *decreased 1 Vdc*. The worst case of these four measurements is taken to be the true power supply sensitivity.

### 4-3.6 Slew Rate

Slew rate measures the operational amplifier's ability to shift between the two possible opposite output voltage extremes while supplying full output power to the load. This parameter is usually specified in terms of volts per unit of time (V/μS).

A saturating squarewave is usually used to measure the slew rate of an operational amplifier. The squarewave must have a rise time which substantially exceeds the expected slew rate of the operational amplifier. The value of rise time is found from examination of the leading edge of the output waveform on an oscilloscope while the input is over-driven by the squarewave. The time measured is that which is required for the output to slew from 10% of the final value to 90% of the final value. Slew rate can be affected by gain, so the value at unity gain will not match either the slew rate under open-loop or high gain closed-loop conditions. Once the switching time is known, the slew rate ($S_r$) is closely approximated by:

$$S_r = \frac{(V+) + ABS(V-)}{T_s} \tag{4-5}$$

where

$S_r$ = the slew rate in volts per microsecond (V/μS)

$V+$ = the positive supply voltage

$ABS(V-)$ = the absolute value of the negative supply voltage

$T_s$ = the switching time

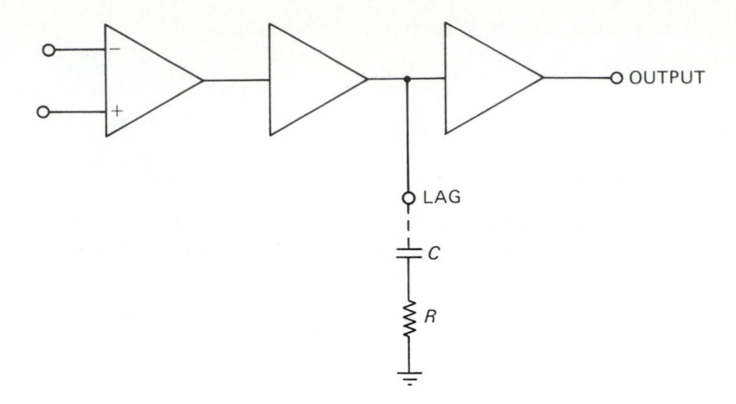


FIGURE 4-4

Because most manufacturers specify slew rate for the open-loop configuration in their data sheets, we can use this relationship to approximate the switching times of specific operational amplifier "digital" circuits.

It is possible to improve the closed-loop slew rate at any given gain figure through the use of appropriate lag compensation techniques (Fig. 4-4). Keep the values of $R_f$ and $R_{in}$ low when trying to improve slew rates; values under 10 kohms will be best. The compensation capacitor will have a value of:

$$C = \left[\frac{R_f + R_{in}}{4\ \pi\ R_f R_{in}}\right] \left[\frac{F_{oi}}{10^m}\right] \tag{4-6}$$

where

$F_{oi}$ = the −3 dB half-power point

$R_f$ = the feedback resistance

$R_{in}$ = the input resistance

$m$ = the quantity $[A_{vol(dB)} - A_{v(dB)}]/20$

The resistor value is found from

$$R = \frac{1}{2\ \pi\ F_{oi}\ C} \tag{4-7}$$

It is the usual practice to measure slew rate in the noninverting unity gain voltage follower configuration because generally it has the poorest slew rate in most op-amp devices. As in the previous test, the worst case figure is used as a matter of standard practice.

The unity gain follower is driven by a squarewave of sufficient amplitude to drive the device well beyond the full saturation point. This is necessary to eliminate the rounded curves which will exist at points just below full saturation. The output waveform can then be examined with a wideband oscilloscope which has a time base fast

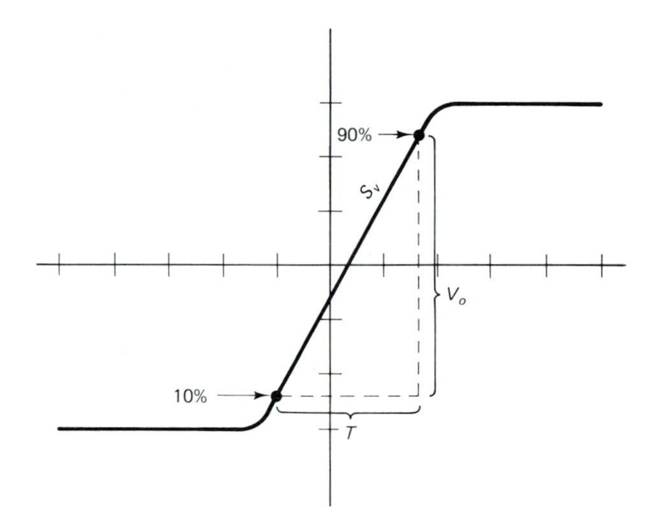


FIGURE 4-5

enough to allow for a meaningful examination. The trace (Fig. 4-5) will be a straight line with a certain slope. It is standard practice to measure rise time as the time of transition from 10% of full amplitude to 90% of full amplitude. Adjust the time base triggering of the oscilloscope so the slope covers several horizontal scale divisions. The slew rate ($S_r$) is then found from the slope of the trace on the oscilloscope

$$S_r = \frac{V_o}{t} \tag{4-8}$$

### 4-3.7 Phase Shift

The phase shift ($\phi$) of an operational amplifier can be measured using a sinewave and an oscilloscope. In one version of this test, an X-Y oscilloscope (or a dual-channel model with an X-Y capability) is used. The input signal is applied to the vertical channel of the CRO and the operational amplifier output is applied to the horizontal channel. The gains of the two channels are set to produce equal beam deflections. The points marked $Y1$ through $Y4$ (Fig. 4-6A) are measured and the phase shift is calculated from

$$\phi = \text{SIN}^{-1}\left[\frac{Y1 - Y2}{Y4 - Y3}\right] \tag{4-9}$$

An alternative approach uses a dual-trace CRO in which the input signal is applied to one channel and the output signal is applied to the other channel. The noninverting unity gain operational amplifier configuration is used. The CRO channel gains are adjusted to be identical, and the traces are superimposed (Fig. 4-6B). The phase shift is found from

$$\phi = \frac{360\ B}{A} \tag{4-10}$$

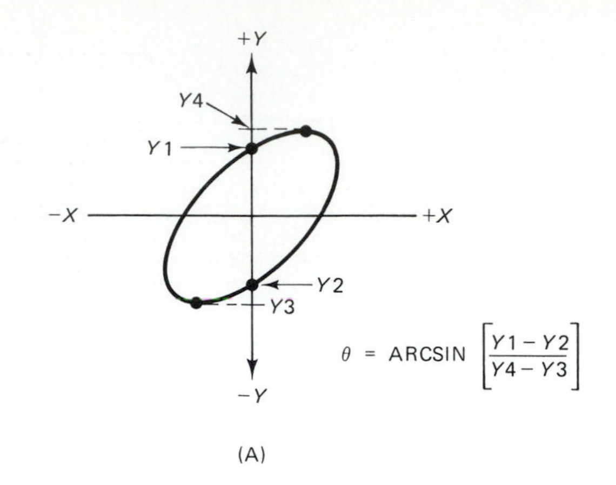


(A)

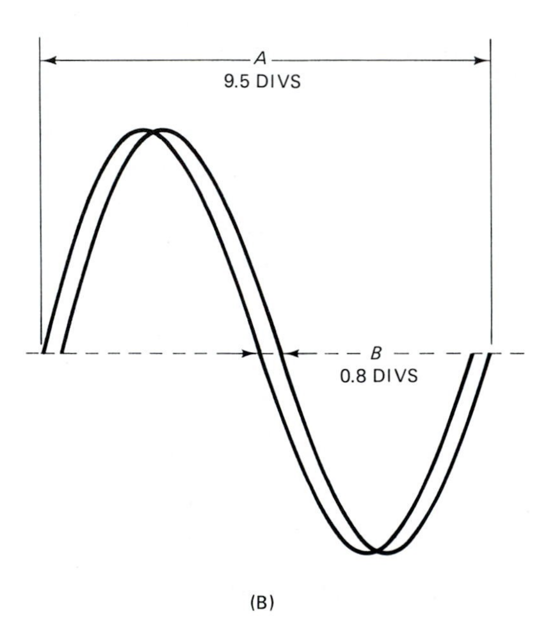


(B)

FIGURE 4-6

## EXAMPLE 4-2

In Fig. 4-6B the overall trace is adjusted to produce a horizontal deflection of 9.5 divisions on the CRO screen. When the traces are overlapped, dimension B, the time difference between zero crossings, is found to be 0.8 divisions. Calculate the phase shift.

### Solution

$$\phi = 360B/A$$

$$\phi = [(360)(0.8)]/(9.5)$$

$$\phi = 288/9.5 = 30.32°$$

■

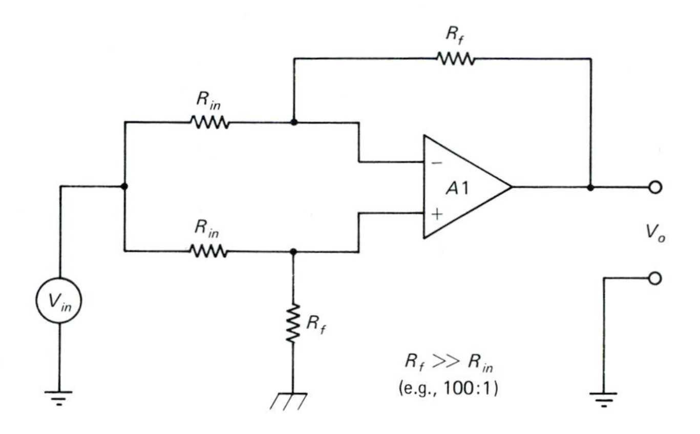


FIGURE 4-7

This test only works if both channels of the CRO are identical to each other, and thus have the same internal phase shift. In addition, the CRO must have a high internal chopping frequency in the dual beam mode. It must be recognized that instrument limitations limit the usefulness of this test.

### 4-3.8 Common Mode Rejection Ratio (CMRR)

The *common mode rejection ratio* is defined as the ratio of the differential gain to the common mode gain

$$CMRR = \frac{A_{vd}}{A_{vcm}} \tag{4-11}$$

where

$CMRR$ = the common mode rejection ratio

$A_{vd}$ = the differential voltage gain

$A_{vcm}$ = the common mode voltage gain

The common mode voltage gain is not always a linear function of common mode input voltage ($V_{cm}$), so it is usually specified at the maximum allowable value for $V_{cm}$. The amplifier is connected into a circuit such as Fig. 4-7 and CMRR determined from

$$CMRR = \frac{(R_f + R_{in})\, V_{in}}{R_{in} V_o} \tag{4-12}$$

Provided that $R_f >> R_{in}$ (100:1).

## 4-4 DC ERRORS AND THEIR SOLUTIONS

DC errors in operational amplifiers are usually in the form of current and voltage levels that combine and force the output voltage to differ from the theoretical value. In many cases the error will be in terms of an actual output voltage which is nonzero at a time when it should be zero (when $V_{in} = 0$). We will also propose some circuit tactics to minimize or eliminate certain errors.

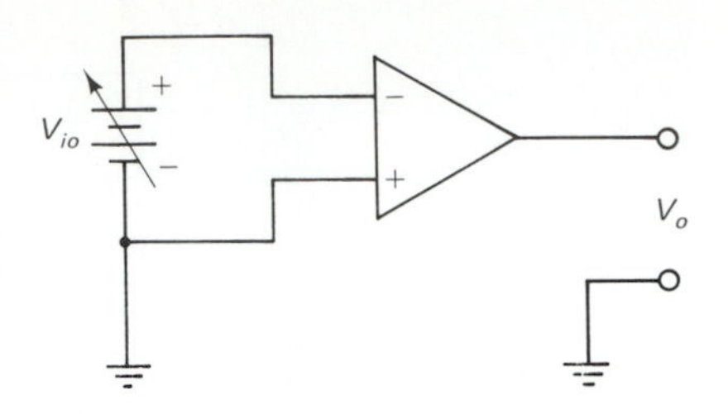


(A)

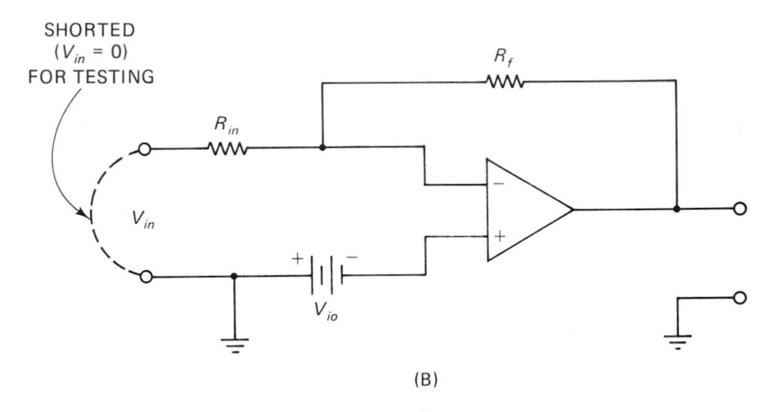


(B)

FIGURE 4-8

### 4-4.1 Output Offset Compensation

The DC error factors result in an output offset voltage $V_{oo}$ which exists between the output terminal and ground at a time when $V_o$ should be zero. This phenomenon helps explain discrepancies between voltages which actually exist and those which equations say should exist. One method for classifying output offset voltages is by causes, *input offset voltage* and *input bias current.*

Input offset voltage $V_{io}$ (Fig. 4-8A) is defined as the differential input voltage required to force the output to zero when no other signal is present ($V_{in} = 0$). A reasonably good model for the input offset voltage phenomenon (Fig. 4-8B) is a voltage source with one end connected to ground and the other end connected to the noninverting input. Although voltage source polarity is shown here, the actual polarity found in any given situation may be either positive or negative to ground, depending on the device being tested. Values for input offset are typically from 1 to several millivolts. The popular type 741 operational amplifier is specified to have a 1 to 5 mV input offset voltage, with 2 mV being listed as typical.

The value of the output offset voltage ($V_{oo}$) caused by an input offset voltage is given by

$$V_{oo} = R_f V_{io} \tag{4-13}$$

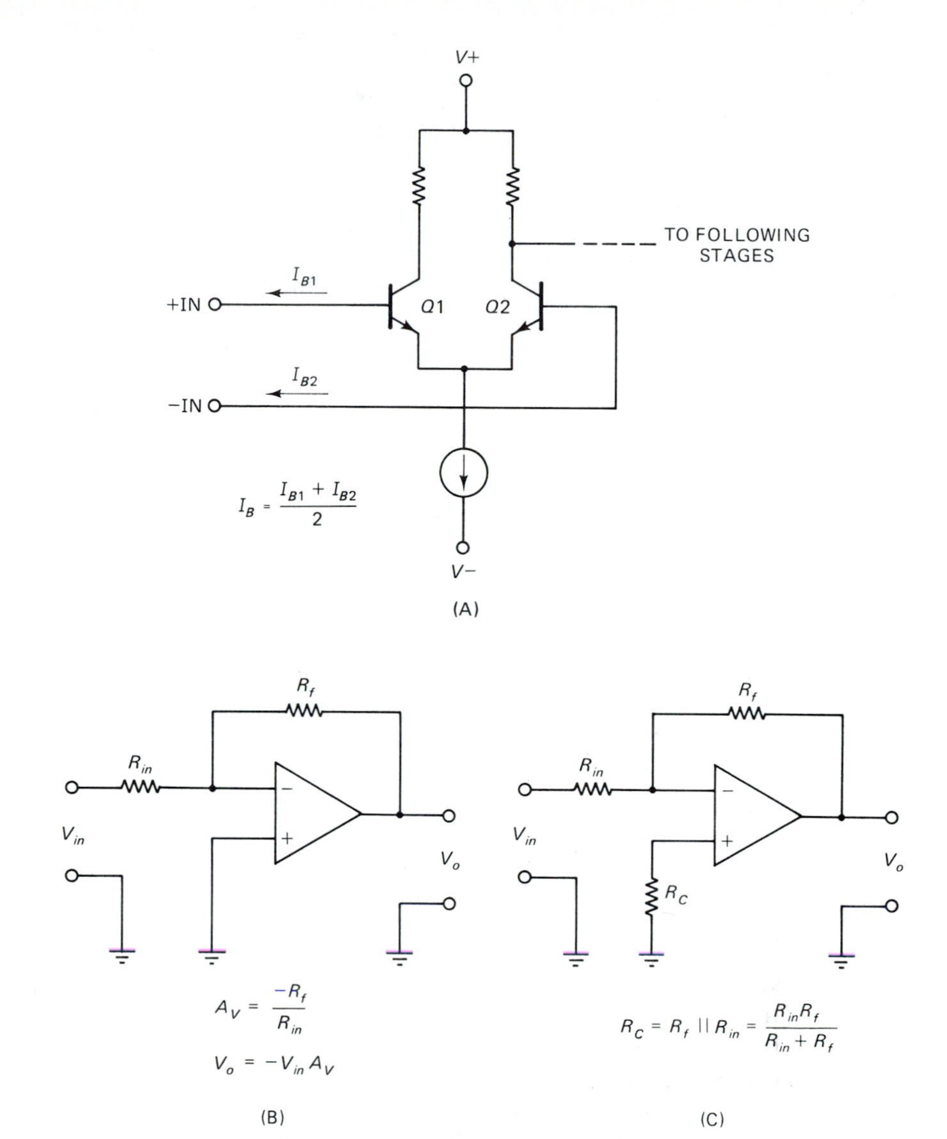


FIGURE 4-9

If the circuit gain is low and $V_{in}$ remains at relatively high values, the input offset voltage may be of little practical consequence. But where either high values of $A_v$ or low values of $V_{in}$ are encountered, the input offset voltage becomes a problem. These conditions are often encountered together.

The second major cause of output offset voltage is spurious input current. This can be further subdivided into two more classes, *normal input bias* and *input offset current.*

Figure 4-9A shows a typical operational amplifier differential input stage. Whenever bipolar (NPN or PNP) transistors are used in this stage, some small input bias current

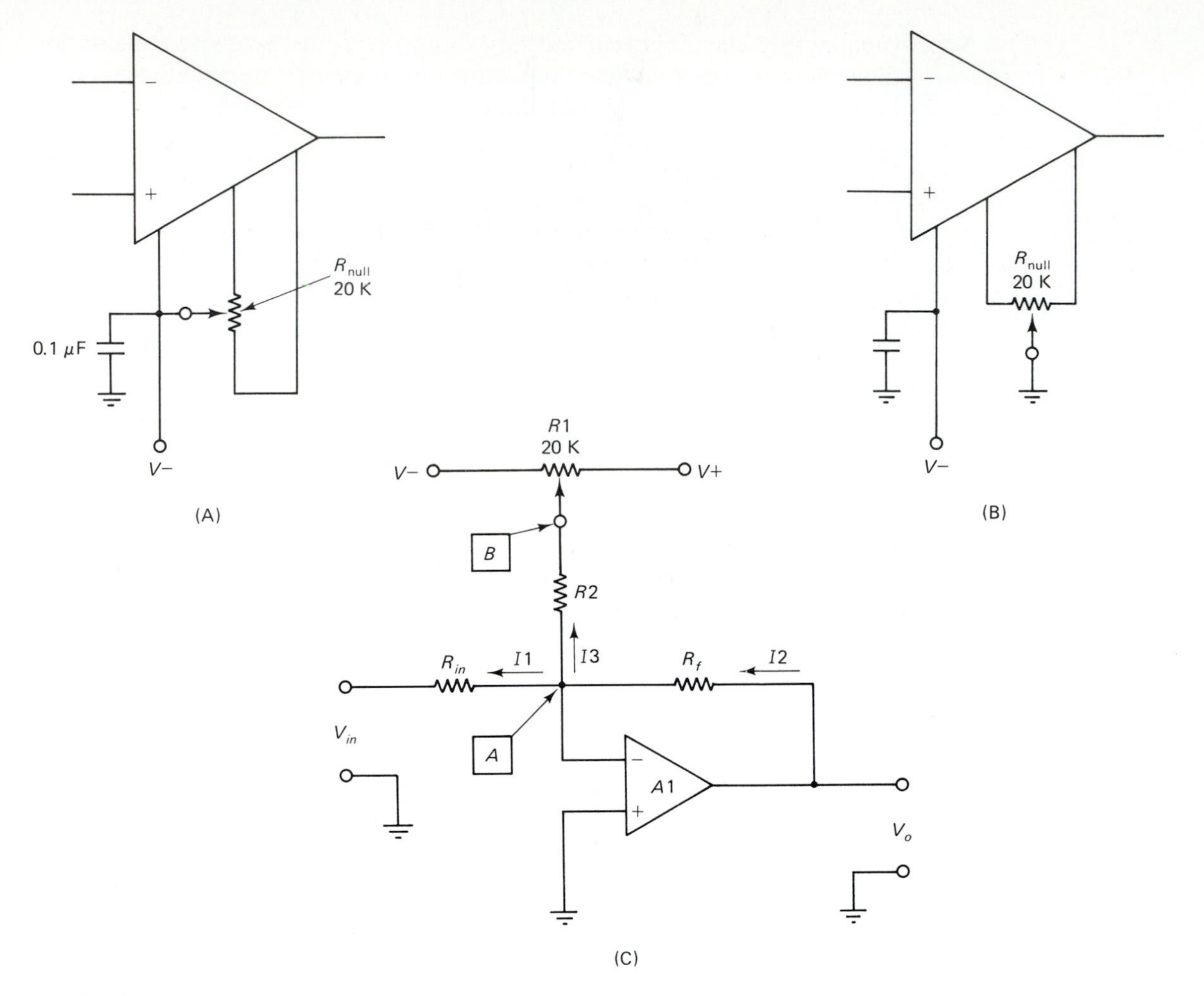


FIGURE 4-10

will be required for normal operation. This is one of those unavoidable conditions inherent in the transistor rather than any deficiency in the internal op-amp circuitry. The problem input bias currents cause becomes acute when high values of input and feedback resistances are used (Fig. 4-9B). When these resistors are in a circuit, the bias current causes a voltage drop across the resistances even when $V_{in} = 0$, causing an output voltage equal to

$$V_{oo} = I_b \times R_f \tag{4-14}$$

Figure 4-9C shows the use of a compensation resistor, $R_c$, to reduce the offset potential due to input bias currents. This same resistor also improves thermal drift. This resistor has a value equal to the parallel combination of the other two resistors:

$$R_c = \frac{R_f R_{in}}{R_f + R_{in}} \tag{4-15}$$

Approximately the same bias current flows in both inputs, so the compensation resistor will produce the same voltage drop at the noninverting input and the inverting input. Because the operational amplifier inputs are differential, the net output voltage is zero.

The method shown in Fig. 4-9C is used when the source of the DC offset potential in the output signal is due to input bias currents.

If the input signal contains an undesired DC component, the DC component will also create an amplifier output error. Depending on the situation, we may need a greater range of control than that offered by Fig. 4-9C. For this, we turn to one of several external offset null circuit techniques.

Figure 4-10 shows two methods for nulling output offsets whether the source is bias currents, other op-amp defects, or an input signal DC component. Figure 4-10A shows the use of offset null terminals found on some operational amplifiers. A potentiometer is placed between the terminals, while the wiper is connected to the $V-$ power supply. The potentiometer is adjusted to produce the null required, the input terminals are shorted together at ground potential, and the potentiometer is adjusted to produce 0 Vdc output. In the LM-101/201/301 family of devices the wiper terminal of the potentiometer is connected to common rather than $V-$ (see Fig. 4-10B).

Figure 4-10C shows a nulling circuit which can be used on any operational amplifier, inverting or noninverting, except the unity gain noninverting follower. A counter current ($I3$) is injected into the summing junction (point A) of a magnitude and polarity enough to cancel the output offset voltage. The voltage at point B is set to produce a null offset at the output. The output voltage component due to this voltage is

$$V'_o = -V_B \times (R_f/R2) \tag{4-16}$$

If finer control over the output offset is required, the potentiometer can be replaced with another resistor voltage divider consisting of a low-value potentiometer (1 kohm) in series with two fixed resistors, one connected to each end of the potentiometer. Thus, the potentiometer resistance is only a fraction of the total voltage divider resistance. In some cases, narrower limits are set by using zener diodes or three-terminal IC voltage regulators at each end of the potentiometer to reduce the range of $V-$ and $V+$.

### 4-4.2 Thermal Drift

Another category of DC error is *thermal drift.* This parameter is usually given in the data sheets relative to input conditions. It is usually related to *the change of input offset voltage per degree Celsius of temperature change.* Typical figures for common operational amplifiers are in the range 1 to 5 microvolts per degree Celsius (μV/°C), with typical values around 3 μV/°C. Keep in mind, however, this specification is usually an expression of drift in steady state circuits and may not accurately represent drift under dynamic situations. These conditions may be considerably higher than the drift given in the specification sheet.

A potential source of increased thermal drift is in the use of resistor networks to null offset voltages (Section 4-4.1 and Fig. 4-10C). The current passing through the network adds to the drift because, as the resistors change value in response to temperature changes, the op-amp interprets varying current as a valid input signal. Thus the overall drift performance increases. Use of low temperature coefficient resistors helps this situation. In some cases, an op-amp manufacturer will recommend an offset null circuit which is different from those described in the previous section. Such circuits are preferred because they are generally designed to compensate the temperature for a specific device. Although the null methods shown previously will work well and are reasonable in stable temperature or noncritical situations, they may fail where temperature varies because they unbalance input offset currents.

The input offset current of an op-amp can result in an offset voltage (Section 4-3). A solution to the problem is the use of a compensation resistor between the noninverting input and ground. That resistor has a value equal to the parallel combination of feedback and input resistances. Although couched in terms of the inverting follower circuit in the previous discussion, the actual goal is to ensure that both inputs (+IN and −IN) look out to the same resistance to ground. Thus, the differential nature of the op-amp inputs tends to cancel out the effects of the voltage drop created by bias currents flowing in external resistances.

The same method also works to improve thermal drift. Bias current is a function of temperature, so thermal drift is due to bias current changes as a function of temperature change. These effects are smoothed out when both inputs have the same resistance. It is sometimes important to use identical, low temperature coefficient resistors in the op-amp input and feedback circuits. In other cases, thermistors or PN junction devices are used to compensate the temperature of circuits.

Another means for reducing overall thermal drift is to reduce thermoelectric sources in the circuit. A phenomenon called the *Seebeck effect* exists when dissimilar metals are joined together. If two metals with different quantum work functions are joined together in a junction, a voltage is generated between them which is proportional to the junction temperature. The Seebeck effect is used to form temperature sensors called *thermocouples.* Ordinary copper wire or printed wiring board tracks produce thermoelectric voltages of 40 to 60 μV/°C when brought into contact with the *Kovar*® leads used on integrated circuits. Lead/tin solder reduces the effect to 1 to 5 μV/°C, and cadmium-based solders to less than 0.5 μV/°C.

For low level signals, especially where high gain is also present, several steps can be taken to reduce thermal drift. First, operate the device at the lowest possible internal junction temperature. Two methods are useful in this respect. One is to use a heatsink (where possible) on the IC device. The other is to operate the temperature sensitive stage at the lowest practical DC power supply voltages; for example, at ∓6 Vdc instead of ∓15 Vdc. Second ensure both amplifier inputs have the same resistance to ground (use a compensation resistor). Third, make all resistors in the circuit low temperature coefficient types. Fourth, eliminate thermoelectric sources. Fifth, stabilize the ambient temperature. Finally, if necessary use thermistor or PN junction devices to compensate the temperature of the circuit.

Although drift may not be important in some circuits, it can be critical where low-level signals are being processed. If an amplifier has a drift of 10 μV/°C and the circuit is expected to maintain its performance over a large temperature range, a significant drift component may exist when low input signals are processed. For example, a temperature change of 20°C will result in a 200 $\mu$V error, or 20% of a 1 mV input signal.

## 4-5 NOISE, SIGNALS, AND AMPLIFIERS

Although gain, bandwidth, and the shape of the passband are important amplifier characteristics, we must also be concerned about circuit *noise.*

At any temperature above Absolute Zero (0°K or about −273°C) electrons in any material are in constant random motion. Because of the inherent randomness of that motion, there is no detectable current in any one direction. In other words, electron drift in any single direction is cancelled over short times by equal drift in the opposite direction. There is, however, a continuous series of random current pulses generated in the material, and those pulses are seen by the outside world as a noise signal. This signal is known as *thermal agitation noise*, *thermal noise*, or *Johnson noise.*

Johnson noise is a so-called white noise because it has a broadband gaussian spectral density. The term white noise is a metaphor developed from white light, which is composed of all visible color frequencies. The expression for Johnson noise is

$$(V_n)^2 = 4KTRB \text{ Volts}^2/\text{Hz} \qquad (4\text{-}17)$$

where

$V_n$ = the noise voltage

$K$ = Boltzmann's constant ($1.38 \times 10^{-23}$ J/°K)

$T$ = the temperature in degrees Kelvin

$R$ = the resistance in ohms

$B$ = the bandwidth in hertz

With the constants collected, and the expression normalized to 1 kohm, Eq. [4-17] reduces to:

$$V_n = 4\sqrt{\frac{R}{1 \text{ kohms}}}\,\frac{nV}{\sqrt{Hz}} \qquad (4\text{-}18)$$

The evaluated solution of Eq. (4-18) is normally read *nanovolts of AC noise per square root hertz.*

### EXAMPLE 4-3

Calculate the spectral noise for a 1 megohm resistor.

### Solution

$$V_n = 4\left[\frac{R}{1 \text{ kohm}}\right]^{1/2} \frac{nV}{[\text{Hz}]^{1/2}}$$

$$V_n = 4\,[1000 \text{ kohms}/1 \text{ kohm}]^{1/2}\ \text{nV}/[\text{Hz}]^{1/2}$$

$$V_n = (4)\,(31.6) = 126.4 \text{ nV}/[\text{Hz}]^{1/2}$$

■

Several other forms of noise are present in linear ICs. For example, because current flow at the quantum level is not smooth and predictable, an intermittent burst phenomenon is sometimes seen. This noise is called popcorn noise. It consists of pulses of many milliseconds duration. Another form of noise is shot noise (also called Schottky noise). The name "shot" is derived from the fact that the noise sounds like a handful of BB shot thrown against a metal surface. Shot noise is a consequence of DC current flowing in any conductor. It is found from

$$(I_n)^2 = 2qIB \text{ amps}^2/\text{Hz} \tag{4-19}$$

where

$I_n$ = the noise current

$q$ = the elementary electric charge ($1.6 \times 10^{-19}$ coulombs)

$I$ = the current in amperes

$B$ = the bandwidth in hertz

Finally, we see flicker noise, also called pink noise or 1/F noise. The latter name applies because flicker noise is predominantly a low-frequency (<100 Hz) phenomenon. This type of noise is found in all conductors, and becomes important in IC devices because of manufacturing defects.

Amplifiers are evaluated on the basis of *signal-to-noise ratio* (S/N or SNR). The goal of a designer is to enhance the SNR as much as possible. Ultimately, the minimum signal detectable at the output of an amplifier is that which appears above the noise level. Therefore, the lower the system noise, the smaller the *minimum allowable signal.* Although usually thought of as a radio receiver parameter, SNR is applicable in other amplifiers where signal levels are low and gains are high. This situation occurs in scientific, medical, and engineering instrumentation, as well as other applications.

Noise resulting from thermal agitation of electrons is measured in terms of *noise power* ($P_n$), and carries the units of power (watts or its sub-units). Noise power is found from

$$P_n = KTB \tag{4-20}$$

where

$P_n$ = the noise power in watts

$K$ = Boltzmann's constant ($1.38 \times 10^{-23}$ J/°K)

$B$ = the bandwidth in hertz

Notice there is no center frequency term in Eq. (4-20), only a bandwidth. True thermal noise is *gaussian* (or near-gaussian) so frequency content, phase, and amplitudes are equally distributed across the entire spectrum. Thus, in bandwidth limited systems, such as a practical amplifier or network, the total noise power is related only to temperature and bandwidth. We can conclude that a 200-Hz bandwidth centered on 1 KHz produces the same thermal noise level as a 200-Hz bandwidth centered on 600 Hz or some other frequency.

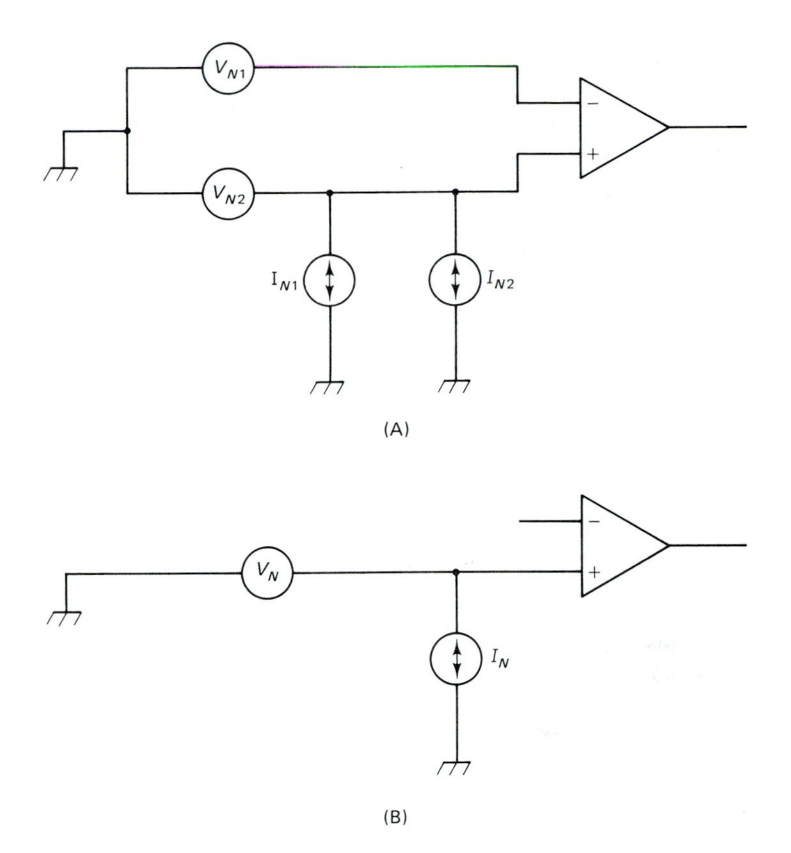


FIGURE 4-11

Noise sources can be categorized as either *internal* or *external.* The internal noise sources are due to thermal currents in the semiconductor material resistances. It is the noise component contributed by the amplifier. If noise, or S/N ratio, is measured at both input and output of an amplifier, the output noise is greater. The internal noise of the device is the difference between output noise level and input noise level.

External noise is the noise produced by the signal source and is often called "source noise." This noise signal is due to thermal agitation currents in the signal source, and even a simple zero-signal input termination resistance has some amount of thermal agitation noise.

Figure 4-11A shows a circuit model showing that several voltage and current noise sources exist in an op-amp. The relative strengths of these noise sources vary with the type of op-amp. In an FET-input op-amp, for example, the current noise sources are small, but voltage noise sources are large. The opposite is true in bipolar op-amps.

All of the noise sources in Fig. 4-11A are unrelated so one cannot simply add noise voltages. Only noise *power* can be added. To characterize noise voltages and currents they must be added in the root sum squares (RSS) manner.

Models such as Fig. 4-11A are more complex than most situations, so it is standard practice to lump all of the voltage noise sources into one source and all of the current noise sources into another. The composite sources have a value equal to the RSS voltage (or current) of the individual sources. Figure 4-11B is such a model where only a single current source and a single voltage source are used. The *equivalent AC noise* in Fig. 4-11B is the overall noise which is given a specified value of source resistance ($R_s$). It is found from the RSS value of $V_n$ and $I_n$

$$V_{ne} = \sqrt{(V_n)^2 + (I_n R_s)^2} \tag{4-21}$$

### 4-5.1 Noise Factor, Noise Figure, and Noise Temperature

The noise of a system or network can be defined in three different but related ways, *noise factor* ($F_n$), *noise figure* ($NF$), and *equivalent noise temperature* ($T_e$). These properties are definable as a ratio, decibel, or temperature, respectively.

**Noise Factor ($F_n$).** The noise factor is the ratio of output noise power ($P_{no}$) to input noise power ($P_{ni}$)

$$F_n = \left. \frac{P_{no}}{P_{ni}} \right|_{T = 290°K} \tag{4-22}$$

In order to make comparisons easier the noise factor is always measured at the standard temperature ($T_o$) 290°K (approximately room temperature).

The input noise power $P_{ni}$ can be defined as the product of the source noise at standard temperature ($T_o$) and the amplifier gain

$$P_{ni} = GKBT_o \tag{4-23}$$

It is also possible to define noise factor $F_n$ in terms of output and input S/N ratio

$$F_n = \frac{SNR_{in}}{SNR_{out}} \tag{4-24}$$

which is also

$$F_n = \frac{P_{no}}{KT_oBG} \tag{4-25}$$

where

$SNR_{in}$ = the input signal to noise ratio

$SNR_{out}$ = the output signal to noise ratio

$P_{no}$ = the output noise power

$K$ = Boltzmann's constant ($1.38 \times 10^{-23}$ J/°K)

$T_o$ = 290°K

$B$ = the network bandwidth in hertz

$G$ = the amplifier gain

The noise factor can be evaluated in a model which takes into account the amplifier

ideal and therefore only amplifies the noise produced by the input noise source through gain $G$

$$F_n = \frac{KT_oBG + \Delta N}{KT_oBG} \tag{4-26}$$

or

$$F_n = \frac{\Delta N}{KT_oBG} \tag{4-27}$$

where

$\Delta N$ = the noise added by the network or amplifier

All other terms are as defined above

**Noise Figure ($NF$).** The noise figure is frequently used to express an amplifier's "goodness" or its departure from "idealness." Thus, it is a figure of merit. The noise figure is the noise factor converted to decibel notation

$$NF = 10 \text{ LOG } F_n \tag{4-28}$$

where

$NF$ = the noise figure in decibels (dB)

$F_n$ = the noise factor

LOG refers to the system of base-10 logarithms

**Noise Temperature ($T_e$).** The noise temperature is a means for specifying noise in terms of an equivalent temperature. Evaluating Eq. (4-23) shows noise power is directly proportional to temperature in degrees Kelvin, and also that noise power collapses to zero at the temperature of Absolute Zero (0°K).

Note the equivalent noise temperature $T_e$ is *not* the physical temperature of the amplifier, but rather a theoretical construct *equivalent* temperature which produces that amount of noise power. The noise temperature is related to the noise factor by

$$T_e = (F_n - 1)\ T_o \tag{4-29}$$

and to noise figure by

$$T_e = \left[\text{ANTILOG}\left(\frac{NF}{10}\right) - 1\right] K\ T_o \tag{4-30}$$

Now that we have noise temperature $T_e$, we can also define noise factor and noise figure in terms of noise temperature

$$F_n = \frac{T_e}{T_o} + 1 \tag{4-31}$$

and

$$NF = 10 \text{ LOG} \left[\frac{T_e}{T_o} + 1\right] \tag{4-32}$$

The total noise in any amplifier or network is the sum of internally generated and externally generated noise. In terms of noise temperature

$$P_{n(\text{total})} = GKB(T_o + T_e) \tag{4-33}$$

where

$P_{n(\text{total})}$ = the total noise power

All other terms are as previously defined

### 4-5.2 Noise in Cascade Amplifiers

A following amplifier interprets a noise signal as a valid input signal. Thus, in a cascade amplifier the final stage acts on an input signal which consists of the original signal and the noise amplified by each successive stage. Each stage in the cascade chain both amplifies signals and noise from previous stages. Each also contributes some noise of its own. The overall noise factor for a cascade amplifier can be calculated from *Friis' noise equation*

$$F_n = F_1 + \frac{F_2-1}{G1} + \frac{F_3-1}{G1G2} + \ldots + \frac{F_{n-1}}{G1G2\ldots G_{n-1}} \tag{4-34}$$

where

$F_n$ = the overall noise factor of $N$ stages in cascade

$F_1$ = the noise factor of stage-1

$F_2$ = the noise factor of stage-2

$F_n$ = the noise factor of the *nth* stage

$G1$ = the gain of stage-1

$G2$ = the gain of stage-2

$G_{n-1}$ = the gain of stage (n−1).

From Eq. (4-34), we see the noise factor of the entire cascade chain is dominated by the noise contribution of the first stage or two. Typically, high gain amplifiers use a low noise device for only the first stage or two in the cascade chain.

## 4-6 AC FREQUENCY STABILITY

Operational amplifiers are subject to spurious oscillations, especially those which are not internally frequency-compensated. Figure 4-12 shows the plot of the open-loop phase shift against the frequency for a typical operational amplifier. From DC to a certain frequency there is essentially zero phase-shift error, but above that breakpoint the phase error increases rapidly. This change is due both to the internal resistances and capacitances of the amplifier acting as an RC phase-shift network, and the phase shift of the feedback network. This shift is called the "propagation phase shift."

At some frequency ($F$), the propagation phase-shift error reaches 180°, which when

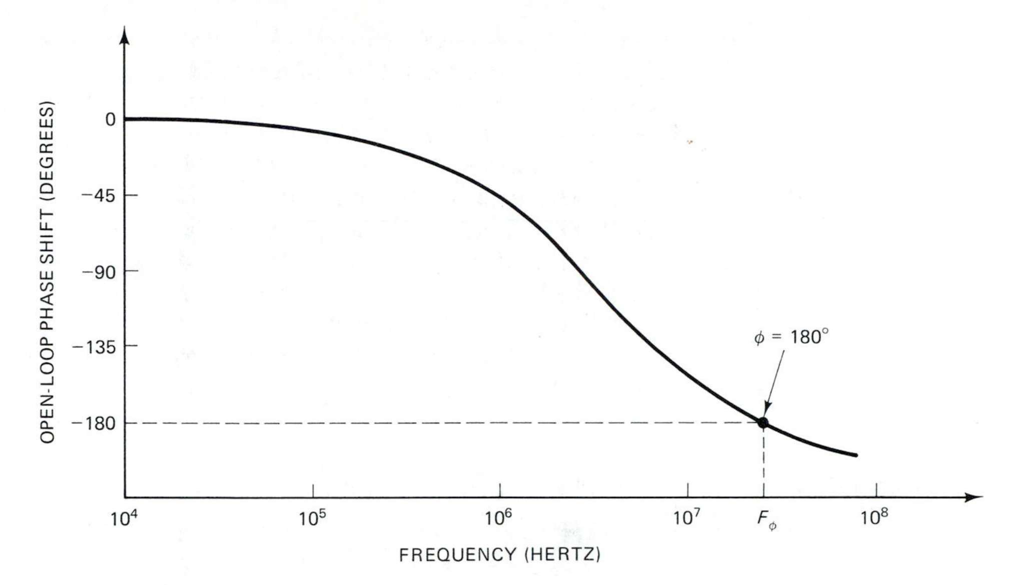


FIGURE 4-12

added to the 180° inversion normal in an inverting follower amplifier, adds up to the 360° phase shift. This satisfies Barkhausen's criteria for oscillation. At this frequency the amplifier will become an oscillator.

In some cases, we may need to use a variant of the methods shown in Fig. 4-13. In Fig. 4-13A we see lead compensation. If the operational amplifier is equipped with compensation terminals (usually either pins 1 and 8, or 1 and 5 on standard packages), connect a small value capacitor (20 to 100 pF) as shown. An alternate scheme is to connect the capacitor from a compensation terminal to the output terminal.

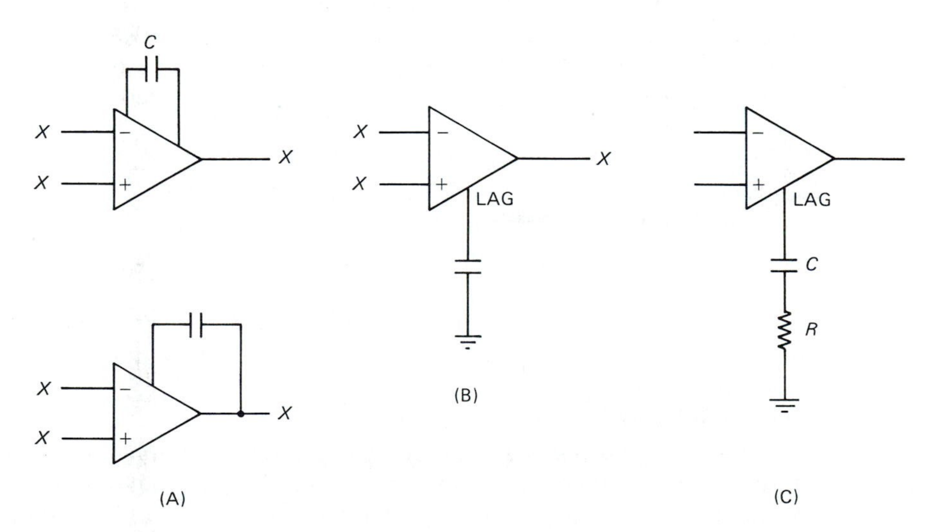


FIGURE 4-13

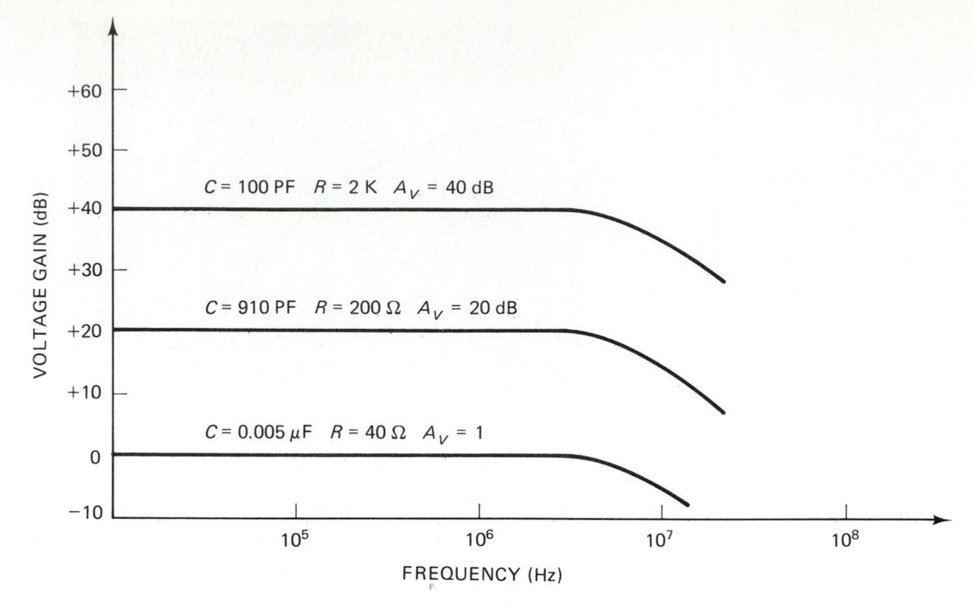


FIGURE 4-14

The recommended capacitance in manufacturers' specification sheets is for the unity gain noninverting follower configuration. For a gain follower the capacitance is reduced by the feedback factor, B

$$C = C_m B \tag{4-35}$$

where

$C$ = the required capacitance

$C_m$ = the recommended unity gain capacitance

$B$ = the feedback factor $R_{in}/(R_{in} + R_f)$

Lag compensation is shown in Fig. 4-13B and Fig. 4-13C. Here, either a single capacitor (Fig. 4-13B) or a resistor-capacitor network (Fig. 4-13C) is connected from the compensation terminal to ground. A related method places the resistor-capacitor series network between the inverting and noninverting input terminals.

The object of these methods (see Fig. 4-14) is to reduce the high frequency loop gain of the circuit to a point where the total loop gain is less than unity at the frequency where the 180° phase shift occurs. The amount of compensation required to accomplish this determines the maximum amount of feedback which can be used without violating the stability requirement.

Several factors cause an operational amplifier to oscillate, and this is highly undesirable. Often these oscillations occur at frequencies far in excess of the passband of the associated circuit. Two of these factors, both of which can be overcome, are positive feedback via the DC power supply and spurious internal phase shift.

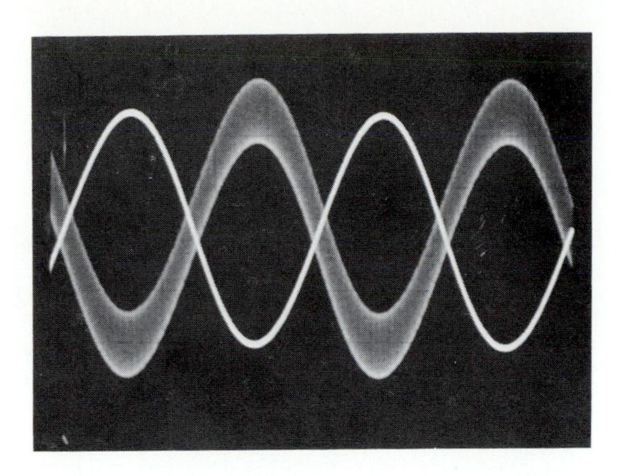

FIGURE 4-15

Figure 4-15 shows the input and output waveforms from a unity gain inverting follower amplifier ($R_{in} = R_f$) which used no decoupling of $V-$ and $V+$ terminals. The output waveform shows a high frequency oscillation superimposed on the output sinewave. The cause of this problem is a high AC impedance to ground through the DC power supply terminals. The use of power supply decoupling helps to solve this problem. It is considered poor engineering practice to use an uncompensated operational amplifier without these decoupling capacitors.

### 4-6.1 Diagnosing and Fixing Instability

Finding and fixing instability sources in linear IC amplifier circuits is sometimes erroneously believed to rely on a little magic. But in reality the process is relatively straightforward. However, there may be a certain amount of cut-and-try involved.

But first be sure the problem you are trying to solve is indeed instability in the circuit and not an external problem. Three kinds of external problems can mimic op-amp instability. One is a defective source signal. For example, a low-frequency (subhertzian) "motorboating" oscillation may be due to a high AC impedance to ground in the amplifier DC power supply, but it may also be due to an oscillation in the signal source. Another problem of this sort is 60-Hz AC hum caused by a broken shield or ground wire on the input line. In fact, don't look for any other cause if the oscillation is exactly 60 Hz (the AC power line frequency) until this possibility is checked.

The second external problem is noisy or defective power supply lines. Examine the $V-$ and $V+$ lines on an oscilloscope to be sure that the power is clean. A 120 Hz oscillation could indicate an excessive ripple condition in a DC power supply. Additional ripple filtering or a voltage regulator in each DC line ($V-$ and $V+$) usually are successful solutions.

Finally, we sometimes see electromagnetic interference in IC amplifiers. Strong local fields, such as RF interference from a local broadcasting station, may get into the circuit and appear to be high frequency oscillations. Such interference may be direct radiation or may be carried into the circuit on an input line, output line, or a DC power supply line. Figure 4-16 shows an op-amp circuit containing several anti-

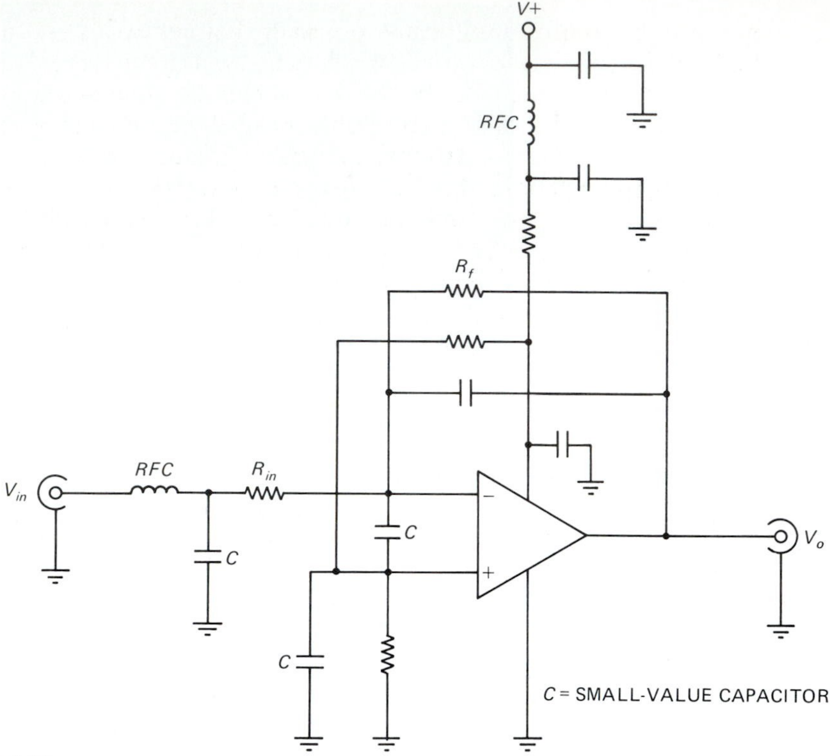


FIGURE 4-16

interference techniques. Small value capacitors should be used with caution because they may attenuate signals at higher frequencies. Also, they may add to the phase shift of signals in the feedback network and cause an oscillation even while eliminating the interference.

Once the obvious noninstabilities have been eliminated, it is necessary to concentrate on the causes of actual instabilities. First, examine the circuit for any *unintended common impedances* between input and output circuits, or between stages in a cascade chain of amplifiers. Two areas of concern are often found, the DC power supply lines and ground. The power supply circuits are usually handled by bypass capacitors.

Common impedances in the ground system, usually called ground loops, are suppressed by using large conductive ground planes (normally on the component side of the printed circuit board) rather than thin point-to-point conductors. Diagnosis and resolution of ground loops is subject to the particular design and a few rules.

1. Use large ground plane surfaces where possible.
2. Use single-point "star" grounding rather than either random grounding or daisy-chain grounding.
3. Keep the DC power supply and AC signal lines separate except at the single "star" point.

Beyond this a cut-and-try approach is advised. The cut-and-try method involves moving ground connections and actual components around the board to find "cold" spots on the ground plane.

Another source of difficulty, especially but not exclusively in high gain circuits, is erroneous circuit layout. Keep input and output circuits separated as much as possible. A source of unintended feedback is sometimes found when several stages are in cascade. To overcome the constraints of small printed wiring boards, some designers double a circuit back on itself on the board. This arrangement places the output circuit components next to either the input circuit components or the intermediate stages. Such a layout induces radiation feedback. This is especially likely if the total propagation feedback is 360°, as it is in noninverting amplifiers or circuits with an even number of inverting amplifiers.

Once layout problems, common impedances, and other causes are ruled out, it is time to evaluate the linear IC amplifier circuit itself for instability problems. Two areas must be investigated, feedback network and the rest of the circuit. Although not an absolute indicator, the frequency of oscillation often tells you where a problem is located. Measure the oscillation frequency ($F_o$) and compare it to the frequency at which the amplifier gain drops to unity ($F_t$, which is found in the specifications sheet or data book). If $F_o$ is close to $F_t$, it is probable the problem is in the feedback network. One diagnostic aid is temporarily increasing the amplifier closed-loop gain by a factor of X2 to X10 and observing the effect on stability. If the oscillation ceases or $F_o$ drops appreciably, the problem is probably in the feedback network. If neither of these events occurs, the problem is most likely in another part of the circuit.

Only a few linear IC devices (and practically no op-amps) operate into the VHF region. Yet op-amps and other low frequency devices sometimes oscillate in the 50- to 200-MHz region, way beyond the bandwidth of the device. The output power amplifier in the IC device is the cause of these VHF parasitic oscillations. This is especially likely when resonances are present.

The wrong capacitor type on the DC power supply terminals is one source of stray resonance which leads to parasitics. In Fig. 4-17A, for example, we see disk ceramic capacitors used for $V-$ and $V+$ bypassing. These have significant stray capacitance and inductance (see inset) that tend to resonate in the VHF region. Fixes for test problems are shown in Fig. 4-17B. The $V+$ lead bypass capacitor uses a 2- to 12-ohm snubber resistor in series to lower the $Q$ of the stray resonant LC circuit elements. If the $Q$ is lowered sufficiently, the LC circuit elements are unable to cause an oscillation. In the $V-$ lead of the same op-amp we see an alternate fix. A ferrite bead is slipped over the lead to the bypass capacitor. These beads act like RF chokes at VHF frequencies, but are practically transparent to low frequencies.

A hidden source of feedback problems can be capacitance in the load of the amplifier. Such capacitance adds to the propagation phase shift of the feedback network, possibly causing oscillation. If a load is known to be capacitive, identification of the problem is easy. But often other sources of capacitance are found in a circuit. For example, shielded or coaxial cables have a high value of capacitance per unit of length. Similarly, some chassis or in-line connectors offer significant capacitances. Circuit stray capacitance may also be significant.

A circuit fix which isolates a capacitance load from an IC amplifier output is shown in Fig. 4-18. A small feedback capacitor reduces the closed-loop gain at frequencies where oscillation is likely to occur. Isolation is obtained by the series snubber resistor, $R_{sn}$.

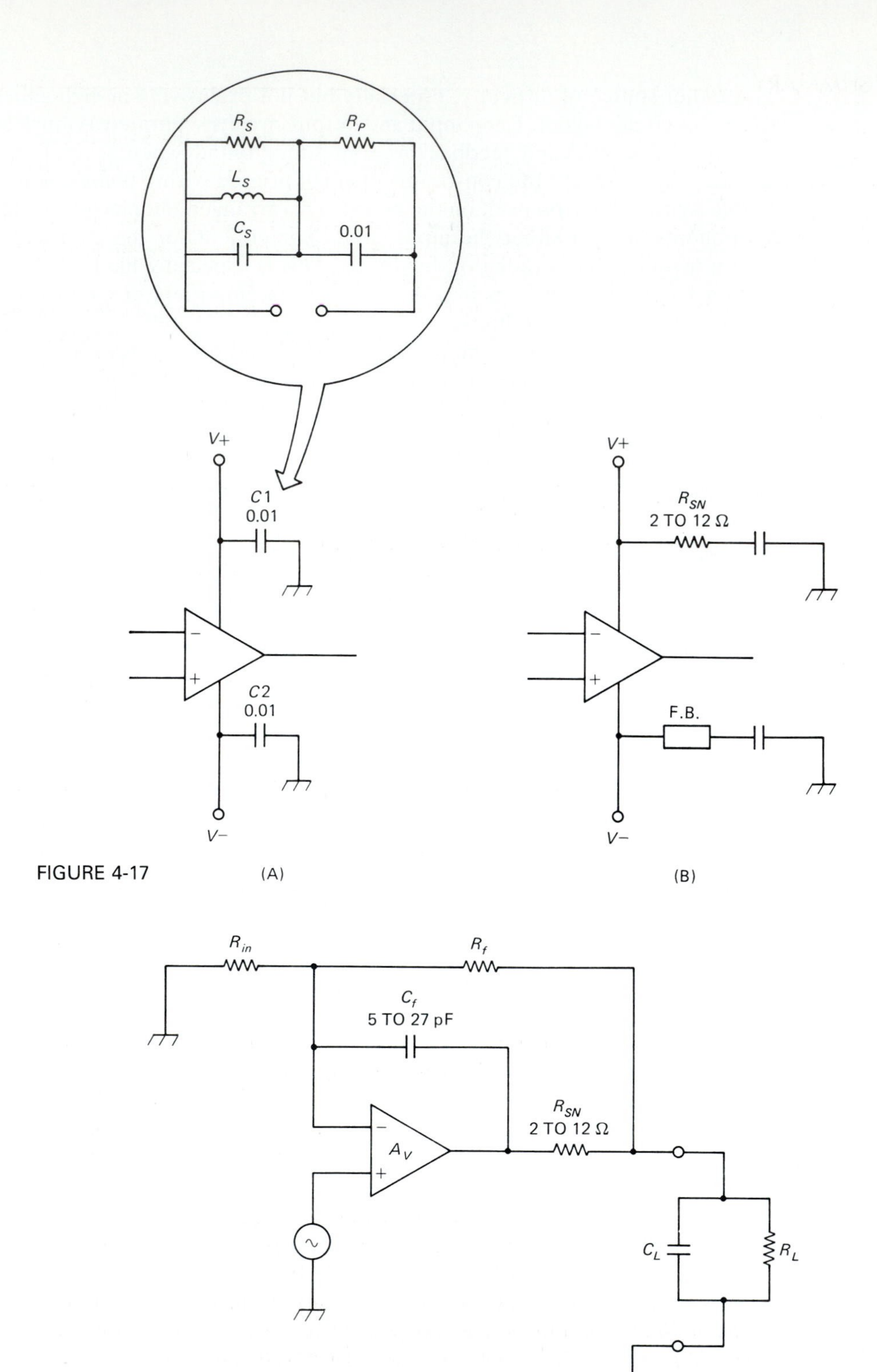


FIGURE 4-17

FIGURE 4-18

## 4-7 SUMMARY

1. Real operational amplifiers diverge considerably from the ideal devices discussed in textbooks. In many cases, the specifications seem nearly ideal, but actually differ considerably from the ideal under test conditions.
2. Slew rate is a measure of an op-amp's ability to shift between two possible opposite output voltage extremes while supplying full output power to the load. This parameter is usually expressed in volts per microsecond.
3. DC errors include input offset voltages, input offset currents and output offset voltages. These errors can be corrected by using a compensation resistor, a potentiometer attached to the offset null terminals of the device, or an external universal null circuit.
4. Thermal drift is a change in the input offset voltage per degree Celsius. Use of a compensation resistor reduces thermal drift.
5. Noise generated inside the op-amp and in its external circuitry obscures weak signals. Low-noise operation of a cascade chain is possible if the input stage is of low-noise design (see Friis' equation).
6. Oscillation in an operational amplifier occurs when the propagation phase shift and any inherent inversion phase shift add up to 360° at any frequency where loop gain is unity or more. Oscillation can be overcome by compensating the circuit with a capacitor or RC network which reduces the gain at higher frequencies.
7. Oscillation can often be diagnosed by examining the frequency. If the frequency of oscillation is close to the natural gain-bandwidth product of the amplifier, the problem is most likely in the feedback loop. Otherwise, look elsewhere for the problem.

## 4-8 RECAPITULATION

Now return to the objectives and Prequiz questions at the beginning of the chapter and see how well you can answer them. If you cannot answer certain questions, mark them and review the appropriate parts of the text. Next, try to answer the questions and work the problems below, using the same procedure.

## 4-9 QUESTIONS AND PROBLEMS

1. If the operational amplifier has an initial offset voltage of 4.0 mV and a value of 5.6 mV is found to exist after the $V+$ power supply voltage is changed from +12 Vdc to +13 Vdc, calculate the power supply sensitivity for a +1 Vdc change in $V+$.
2. If the operational amplifier has an initial offset voltage of 5.0 mV and a value of 12 mV is found to exist after the $V+$ power supply voltage is changed from +14 Vdc to +15 Vdc, calculate the power supply sensitivity for a +1 Vdc change in $V+$.
3. The trace on an oscilloscope is adjusted to produce a horizontal deflection of 8.5 divisions on the CRO screen. When the traces are overlapped dimension B, the time difference between zero-crossings, is found to be 0.96 divisions. Calculate the phase shift.
4. An oscilloscope is adjusted to produce a horizontal deflection of 8.8 divisions on the CRO screen. When the traces are overlapped dimension B, the time difference between zero-crossings, is found to be 0.86 divisions. Calculate the phase shift.
5. Calculate the spectral noise for a 150-kohm resistor.

6. Calculate the spectral noise for a 10-megohm resistor.
7. State Friis' equation for noise in a cascade amplifier and describe how it applies to creating a low-noise amplifier system.
8. An operational amplifier specification sheet lists a compensation capacitor of 47 pF in unity gain circuits. Calculate the capacitance required in a noninverting follower with a gain of 11.
9. An operational amplifier specification sheet lists a compensation capacitor of 82 pF in unity gain circuits. Calculate the capacitance required in a noninverting follower with a gain of 15.
10. An inverting follower amplifier uses a feedback resistor of 1.5 megohms and an input resistor of 100 kohms. Calculate the value of a compensation resistor which will cancel input bias currents.
11. An inverting follower amplifier uses a feedback resistor of 150 kohms and an input resistor of 100 kohms. Calculate the value of a compensation resistor which will cancel input bias currents.
12. Sketch the output waveform of an amplifier that is in slight saturation so that it shows a small amount of clipping. Assume a sinewave input.
13. A 10-volt p-p, 100-Hz sinewave is applied to the input of an inverting follower that has a gain of 1000. Sketch the output waveform if the maximum allowable output voltages are $\pm$ 10 volts.
14. Sketch the output of the circuit in question 13 if the input voltage is reduced to 10-mV p-p.
15. An operational amplifier in a comparator configuration switches from $-V_{max}$ to $+V_{max}$ in 135 nanoseconds. Assume DC power supply voltages of $\pm$ 12 Vdc.
16. A DC differential amplifier has a voltage gain of 500, and a measured CMRR of 90 dB. Calculate the common-mode voltage gain of this amplifier.
17. A 50-$\mu$A input offset current flows in an op-amp inverting input circuit in which the feedback resistor is 150 kohms. Calculate the output offset voltage due to this current.
18. Sketch the circuit for a universal output offset voltage null control. If the feedback resistor is 100 kohms, and the input resistor is 10 kohms, select values for the nulling circuit that will null the output voltage over a range of $\pm$ 10-volts.
19. An amplifier with a gain of 500 and a bandwidth of 1000 Hz is operated at room temperature. Calculate the output noise power of this circuit if the noise factor is 20.
20. Calculate the noise factor of an amplifier if the input S/N ratio is 4.5 dB and the output S/N ratio is 8.9 dB.
21. Calculate the noise added to the signal by an amplifier that has a noise factor of 10 when the bandwidth of 10 KHz and a gain of 200.
22. Calculate the equivalent noise temperature of an amplifier that has a noise factor of 4.5.
23. Calculate the noise figure of an amplifier that has an equivalent noise temperature of 270°K.
24. A cascade chain of three amplifiers (A1, A2, and A3) has the following stage gains: $G1 = 40$, $G2 = 5$, and $G3 = 2$. Calculate the total noise factor if $NF1 = 3$, $NF2 = 7$, and $NF3 = 10$.
25. Discuss the most likely cause of oscillation in an operational amplifier circuit if the frequency of oscillation is near the gain-bandwidth product of the device.

CHAPTER 5

# DC Differential Operational Amplifier Circuits

## OBJECTIVES

1. Learn the properties of the DC differential amplifier.
2. Be able to design DC differential amplifier circuits.
3. Understand the range of applications for the DC differential amplifier.
4. Understand the limitations of the DC differential amplifier.

## 5-1 PREQUIZ

These questions test your prior knowledge of the material in this chapter. Try answering them before you read the chapter. Look for the answers (especially those you answered incorrectly) as you read the text. After you have finished studying the chapter, try answering these questions again and those at the end of the chapter (see Section 5-11).

1. Describe the difference between common-mode gain and differential gain.
2. Solve the voltage gain transfer equation for a DC differential amplifier in which both input resistors are 10 kohms and both feedback resistors are 200 kohms.
3. Define *Common Mode Rejection Ratio* (CMRR).
4. Draw the circuit for a single op-amp DC differential amplifier which incorporates a CMRR ADJUST control.

Most operational amplifiers have differential inputs; that is, there are two inputs (−IN and +IN). Each provides the same amount of gain but has opposite polarities. The inverting input (−IN) of the operational amplifier provides an output 180° out of phase with the input signal. In other words, a positive-going input signal will produce a negative-going output signal, and vice versa. The noninverting input (+IN) produces an output signal that is in phase with the input signal. For this type of input, a positive-going input signal will produce a positive-going output signal. We will use these properties to understand a class of amplifiers in which both inputs are used. But before going further with our discussion of differential amplifiers, let's reexamine the basic differential input stage of an operational amplifier.

## 5-2 DIFFERENTIAL INPUT STAGE

Figure 5-1 shows a hypothetical input stage for a differential input operational amplifier. Transistors $Q1$ and $Q2$ are a matched pair and share common $V-$ and $V+$ DC power supplies through separate collector load resistors. The collector of transistor $Q2$ serves as the output terminal for the differential amplifier. Thus, collector voltage $V_o$ is the output voltage of the stage. The exact value of the output voltage is the difference between supply voltage ($V+$) and the voltage drop across resistor $R2$ ($V_{R2}$).

The emitter terminals of the two transistors are connected and fed from a *constant current source* (CCS). For purposes of analysis, we can assume the collector and emitter currents ($I1$ and $I2$) are equal. (Even though these currents are not actually equal, they are very close to each other which makes this convention useful for our present purpose.) Because of Kirchoff's Current Law (KCL), we know that

$$I3 = I1 + I2 \tag{5-1}$$

Because current $I3$ is a constant current, a change in either $I1$ or $I2$ will change the other. For example, an increase in current $I1$ must force a decrease in current $I2$ in order to satisfy Eq. (5-1).

Now let's consider how the stage works. First, the case where $V1 = V2$. In this case, both $Q1$ and $Q2$ are equally biased, so the collector currents ($I1$ and $I2$) are equal. In a normal situation, this will result in $V_{R2}$ being equal to $V_o$ and also equal to $(V+)/2$.

Next, let's consider the case where $V1$ is greater than $V2$. Here transistor $Q1$ is biased on, harder than $Q2$, so $I1$ will increase. Because $I1 + I2$ is always equal to $I3$, an increase in $I1$ will force a decrease in $I2$. A decrease of $I2$ will also reduce $V_{R2}$ and increase $V_o$. Thus, an increase in $V1$ increases $V_o$, so the base of transistor $Q1$ is the noninverting input of the differential amplifier.

Now let's consider the case where $V2$ is greater than $V1$. Transistor $Q2$ is biased on, harder than $Q1$, so $I2$ will increase and $I1$ will decrease. An increase in $I2$ forces a larger voltage drop ($V_{R2}$) across resistor $R2$, so $V_o$ will go down. Thus, an increase in $V2$ forces a decrease in $V_o$, so the base of transistor $Q2$ is the inverting input of the differential amplifier.

Voltage $V3$ is a common mode signal, so it will affect transistors $Q1$ and $Q2$ equally. For this kind of signal the output voltage ($V_o$) will not change.

The base bias currents required to keep transistors $Q1$ and $Q2$ operating become the input bias currents which make the practical op-amp nonideal. In order to make the input impedance high we need to make these currents very low. Some manufacturers offer operational amplifiers with MOSFET transistors (BiMOS op-amps) or JFET transistors (BiFET op-amps). These transistors inherently have lower input bias currents than bipolar NPN or PNP transistors. In fact, the bias currents approach mere leakage currents.

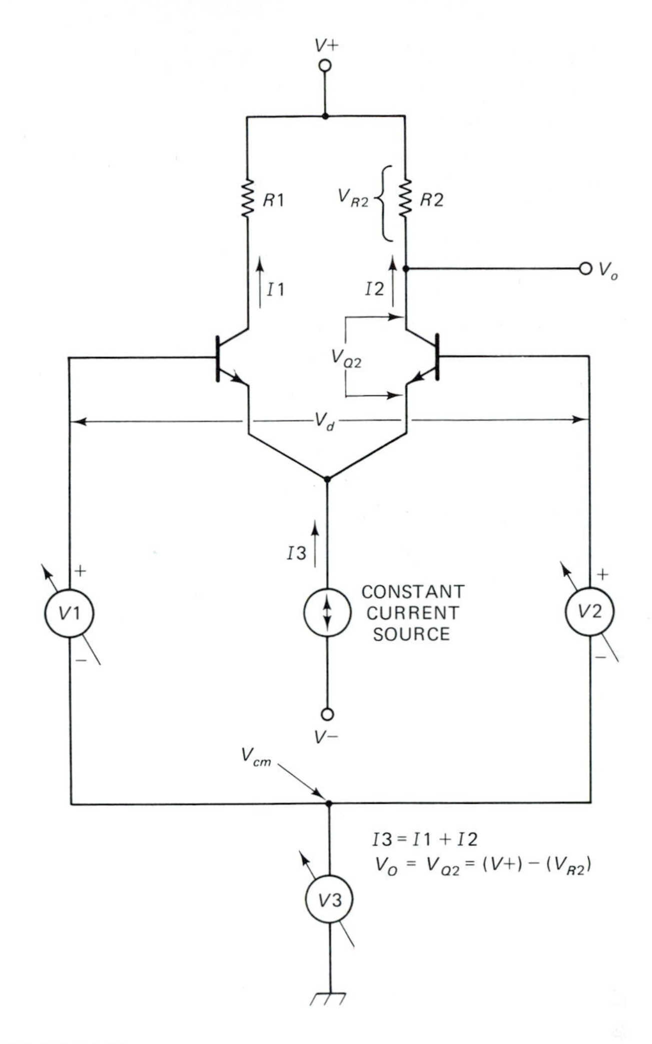


FIGURE 5-1

## 5-3 INPUT SIGNAL CONVENTIONS

Figure 5-2 shows a generic differential amplifier with the standard signals applied. Signals $V1$ and $V2$ are single-ended potentials applied to the −IN and +IN inputs, respectively. The *differential input signal* ($V_d$) is the difference between the two single-ended signals, $V_d = (V2 - V1)$. Signal $V_{cm}$ is a *common-mode signal.* That is, it is applied equally to both −IN and +IN inputs. These signals are described in detail below.

**Common Mode Signals.** A common-mode signal ($V_{cm}$) is one which is simultaneously applied to both inputs. Such a signal might be either voltage such as $V_{cm}$, or where voltages $V1$ and $V2$ are equal and of the same polarity ($V1 = V2$). When the common-mode signal is applied equally to inverting and noninverting inputs, the output voltage is zero. Because the two inputs have equal but opposite polarity gains for common-mode signals, the net output signal in response to a common-mode signal is zero.

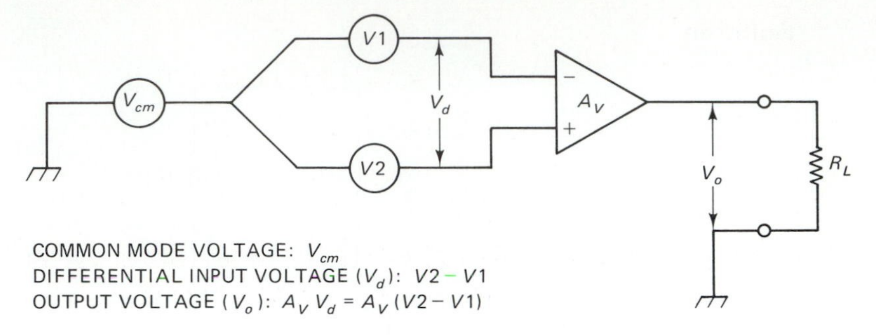


FIGURE 5-2

The operational amplifier with differential inputs cancels common-mode signals. An example of the usefulness of this property is in the performance of the differential amplifier with respect to 60-Hz hum pickup from local AC power lines. Almost all input signal cables for practical amplifiers will pick up 60-Hz radiated energy and convert it to a voltage which is interpreted by the amplifier as a genuine input signal. In a differential amplifier, however, the 60-Hz field will affect both inverting and noninverting input lines equally, so the 60-Hz artifact signal will disappear in the output. This feature of common-mode signals is used to good advantage in Chapter 13 when we discuss physical measurement and instrumentation circuits.

The practical operational amplifier will not exhibit perfect rejection of common mode signals. A specification called the Common Mode Rejection Ratio (CMRR) expresses the op-amp's ability to reject such signals. The CMRR is usually specified in decibels (dB) and is defined as

$$CMRR = \frac{A_{vd}}{A_{cm}} \tag{5-2}$$

or, in decibel form

$$CMRR_{dB} = 20 \text{ LOG} \left[\frac{A_{vd}}{A_{cm}}\right] \tag{5-3}$$

where

$CMRR$ = the common mode rejection ratio

$A_{vd}$ = the voltage gain to differential signals

$A_{cm}$ = the voltage gain to common mode signals

In general, the higher the CMRR the better the operational amplifier. Typical low-cost devices have CMRR ratings of 60 dB or more.

## EXAMPLE 5-1

An operational amplifier exhibits an open-loop differential gain of 100, and a common-mode gain of 0.001. Calculate the CMRR in decibels.

**Solution**

$$CMRR_{dB} = 20 \text{ LOG } (A_{vd}/V_{cm})$$

$$CMRR_{dB} = 20 \text{ LOG } (100/0.001)$$

$$CMRR_{dB} = (20)(5) = 100 \text{ dB}$$

■

**Differential Signals.** Signals $V1$ and $V2$ in Fig. 5-2 are single-ended signals. The total differential signal seen by the operational amplifier is the difference between the single-ended signals

$$V_d = V2 - V1 \tag{5-4}$$

The output signal from the differential operational amplifier is the product of the differential voltage gain and the difference between the two input signals (hence the term "differential" amplifier). Thus, the transfer equation for the operational amplifier is

$$V_o = A_v (V2 - V1) \tag{5-5}$$

### EXAMPLE 5-2

A DC differential amplifier has a gain of 200. Calculate the output voltage if $V1 = 2.06$ and $V2 = 2.03$.

**Solution**

$$V_o = A_v(V2 - V1)$$

$$V_o = (200)(2.03 \text{ Vdc} - 2.06 \text{ Vdc})$$

$$V_o = (200)(-0.03 \text{ Vdc}) = -6 \text{ Vdc}$$

■

## 5-4 DIFFERENTIAL AMPLIFIER TRANSFER EQUATION

The basic circuit for the DC differential amplifier is shown in Fig. 5-3A. This circuit uses only one operational amplifier, so it is the simplest possible configuration. In Chapter 6 you will see additional circuits based on two or three operational amplifiers. In its most common form the circuit of Fig. 5-3A is balanced so $R1 = R2$ and $R3 = R4$.

Consider the redrawn differential amplifier circuit shown in Fig. 5-3B. Assume source resistances $R_{S1}$ and $R_{S2}$ are zero. Further assume that $R1 = R2 = R$ and that $R3 = R4 = kR$, where $k$ is a multiplier of $R$.

1. Set $V2 = 0$. In this case $V_o$ is found from

$$V_{O1} = \frac{-kR}{R} V1 \tag{5-6}$$

$$V_{O1} = -k\, V1 \tag{5-7}$$

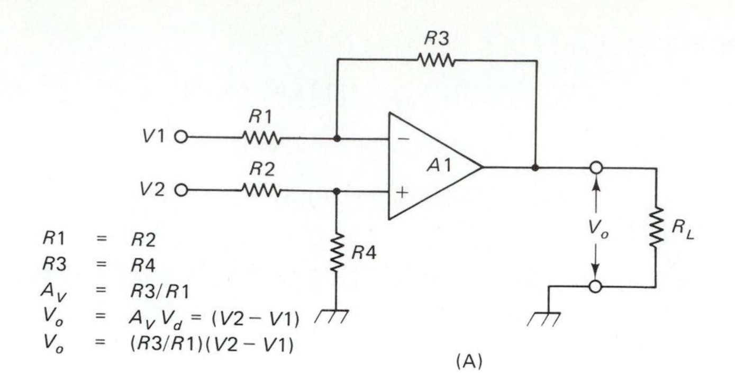


(A)

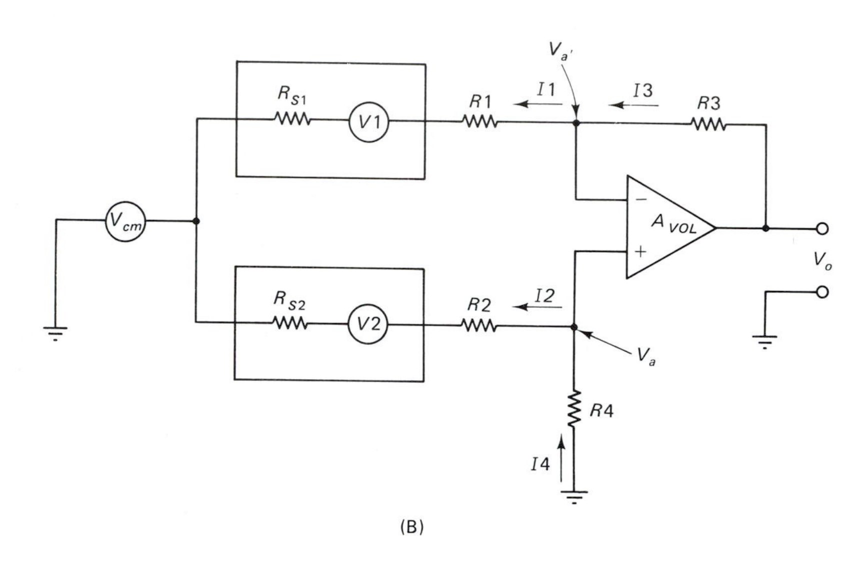


(B)

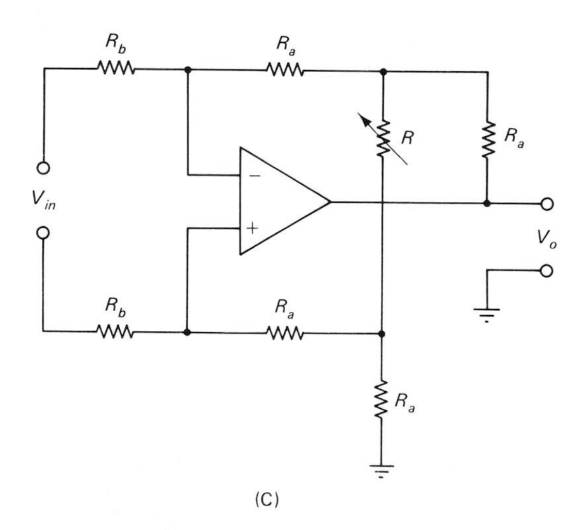


(C)

FIGURE 5-3

2. Now assume that $V1 = 0$ instead

$$V_a = \frac{V2\,k\,R}{kR + R} \tag{5-8}$$

$$V_a = \frac{V2\,k}{k + 1} \tag{5-9}$$

$$V_{O2} = \left[\frac{kR}{R} + 1\right] V_a \tag{5-10}$$

$$V_{O2} = \left[\frac{kR}{R} + 1\right]\left[\frac{V2\,k}{k + 1}\right] \tag{5-11}$$

$$V_{O2} = (k + 1)\left[\frac{V2\,k}{k + 1}\right] \tag{5-12}$$

$$V_{O2} = V2\,k \tag{5-13}$$

3. Now, by superimposing the two expressions for $V_o$

$$V_o = V_{O2} + V_{O1} \tag{5-14}$$

$$V_o = (V2\,k) - (V1\,k) \tag{5-15}$$

$$V_o = k\,(V2 - V1) \tag{5-16}$$

According to Eq. (5-16), the output voltage is the product of the difference between single-ended input potentials $V1$ and $V2$, and a factor $k$. The differential input voltage ($V_d$) is

$$V_d = V2 - V1 \tag{5-17}$$

While factor $k$ is the differential voltage gain, $A_{vd}$. Thus, the output voltage is

$$V_o = V_d\,A_{vd} \tag{5-18}$$

We may also use a less parametric analysis by appealing to Fig. 5-3B. We know the following relationships are true (assuming $R_{S1} = R_{S2} = 0$)

$$V_a = \frac{(V_{cm} + V2)\,R4}{R2 + R4} \tag{5-19}$$

$$I1 = -I3 \tag{5-20}$$

$$I1 = \frac{V_{cm} + V1 - V_{a'}}{R1} \tag{5-21}$$

$$I3 = \frac{V_o - V_{a'}}{R3} \tag{5-22}$$

From the properties of the ideal op-amp we know that voltage $V_{a'} = V_a$, so

$$I1 = \frac{V_{cm} + V1 - V_a}{R1} \tag{5-23}$$

$$I3 = \frac{V_o - V_a}{R3} \tag{5-24}$$

Combining equations and solving for $V_o$

$$V_o = V_{cm}\left[\frac{R3R4 + R1R4 - R2R3 - R3R4}{R1(R2 + R4)}\right] - [R3\ V1/R1] + \left[\left(\frac{R4}{R2}\right)\left(\frac{1 + R3/R1}{1 + R4/R2}\right)\right] V2 \tag{5-25}$$

Assuming $R1 = R2$ and $R3 = R4$, Eq. (5-25) resolves to

$$V_o = \frac{R3}{R1}(V2 - V1) \tag{5-26}$$

Equation (5-26) is similar to Eqs. (5-5) and (5-16). In this case, $A_{vd}$ is $R3/R1$ and $V_d$ is $(V2 - V1)$. The standard transfer equation for the single op-amp DC differential amplifier is:

$$V_o = V_d\, A_{vd} \tag{5-27}$$

$$V_o = V_d \frac{R3}{R1} \tag{5-28}$$

It is difficult to build a DC differential amplifier with variable gain control. It is, for example, very difficult to get two ganged potentiometers (used to replace $R3$ and $R4$ in Fig. 5-3A) to track well enough to vary gain while maintaining required balance. Figure 5-3C shows one attempt to solve the problem. In this case, a potentiometer is connected between the midpoint of the two feedback resistances. This circuit works, but the gain control is a nonlinear function of the potentiometer setting. It is generally assumed to be better practice to use a post-amplifier stage following the differential amplifier and perform gain control there. Another alternative is to use one of the differential amplifier circuits shown in Chapter 6 which have more than one operational amplifier.

## 5-5 COMMON-MODE REJECTION

Figure 5-4 shows two situations. Figure 5-4A is a single-ended amplifier, while Fig. 5-4B is a differential amplifier in a similar situation. In these circuits a noise signal, $V_n$, is placed between the input ground and the output ground. This noise signal might be either AC or DC noise.

In Fig. 5-4A we see a case where the noise signal is applied to a single-ended input amplifier. The input signal interpreted by the amplifier is the algebraic sum of the two independent signals: $V_{in} + V_n$. Because of this, the amplifier output signal will see a noise artifact equal to the product of the noise signal amplitude and the amplifier gain, $-A_v V_n$.

Now consider the situation of a differential amplifier depicted in Fig. 5-4B. The noise signal in this case is common-mode, so is essentially cancelled by the common mode rejection ratio. Of course, in nonideal amplifiers the input signal ($V_{in}$) is subject

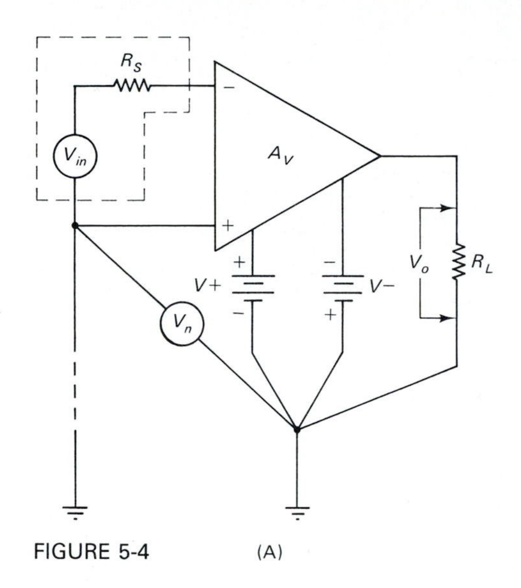


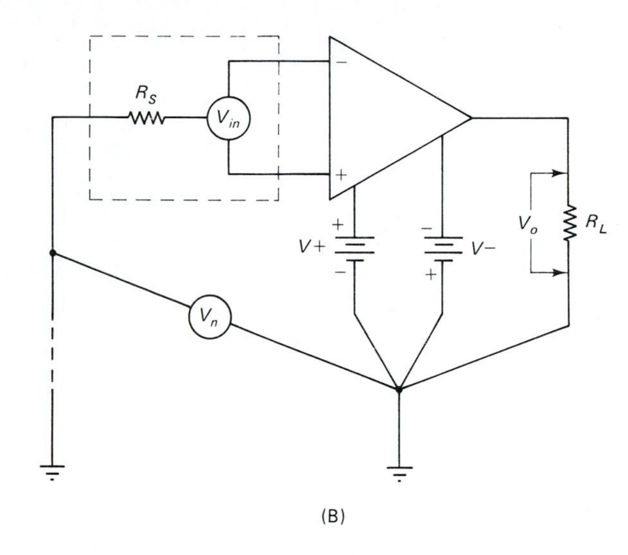


FIGURE 5-4 (A) (B)

to the differential gain, while the noise signal ($V_n$) is subject to the common-mode gain. If an amplifier has a CMRR of 90 dB, for example, the gain of the noise signal will be 90 dB down from the differential gain.

## EXAMPLE 5-3

A DC differential amplifier has a common-mode rejection ratio of 100 dB. Calculate the output potential caused by a common-mode signal of 1 Vdc.

### Solution

$$100 \text{ dB} = 20 \text{ LOG} \left[\frac{V_{cm}}{V_o}\right]$$

$$100 \text{ dB} = 20 \text{ LOG} \left[\frac{1 \text{ Vdc}}{V_o}\right]$$

$$\frac{100 \text{ dB}}{20} = \text{LOG} \left[\frac{1 \text{ Vdc}}{V_o}\right]$$

$$5 \text{ dB} = \text{LOG} \left[\frac{1 \text{ Vdc}}{V_o}\right]$$

$$10^5 \text{ dB} = \frac{1 \text{ Vdc}}{V_o}$$

$$100{,}000 = \frac{1 \text{ Vdc}}{V_o}$$

Solving for $V_o$

$$V_o = \frac{1 \text{ Vdc}}{100{,}000} = 0.00001 \text{ Vdc}$$

■

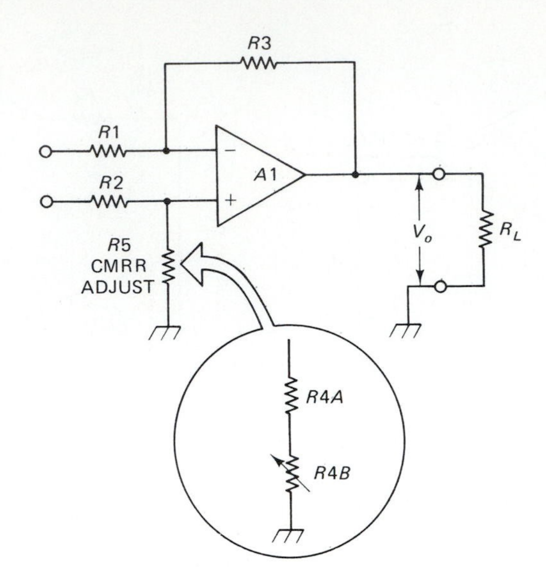


FIGURE 5-5

The common-mode rejection ratio (CMRR) of the operational amplifier DC differential circuit is principally dependent on two factors. First, the natural CMRR of the operational amplifier used as the active device. Second, the balance of the resistors $R1 = R2$ and $R3 = R4$. Unfortunately, the balance is often difficult to obtain with fixed resistors. We can use a circuit such as Fig. 5-5. In this circuit, $R1$ through $R3$ are the same as in previous circuits. The fourth resistor, however, is a potentiometer (see $R4$). The potentiometer will adjust out the CMRR errors caused by resistor and related mismatches.

A version of the circuit with greater resolution is shown in the inset to Fig. 5-5. In this version the single potentiometer is replaced by a fixed resistor and a potentiometer in series, the sum resistance ($R4A + R4B$) is equal to approximately 20% more than the normal value of $R3$. Ordinarily, the maximum value of the potentiometer is 10 to 20% of the overall resistance.

The adjustment procedure for either version of Fig. 5-5 is the same (see Fig. 5-6):

1. Connect a zero-center DC voltmeter to the output terminal (M1 in Fig. 5-6).
2. Short together inputs $A$ and $B$, and connect them to either a signal voltage source, or ground.
3. Adjust potentiometer $R4$ (CMRR ADJ) for zero volts output.
4. If the output indicator (meter $M1$) has several ranges, switch to a lower range and repeat stages 1–3 until no further improvement is possible.

Alternatively, connect the output to an audio voltmeter or oscilloscope, and connect the input to a 1 volt to 5 volt peak-to-peak AC signal which is within the frequency range of the particular amplifier. For audio amplifiers, a 400 to 1000 Hz 1-volt signal is typically used.

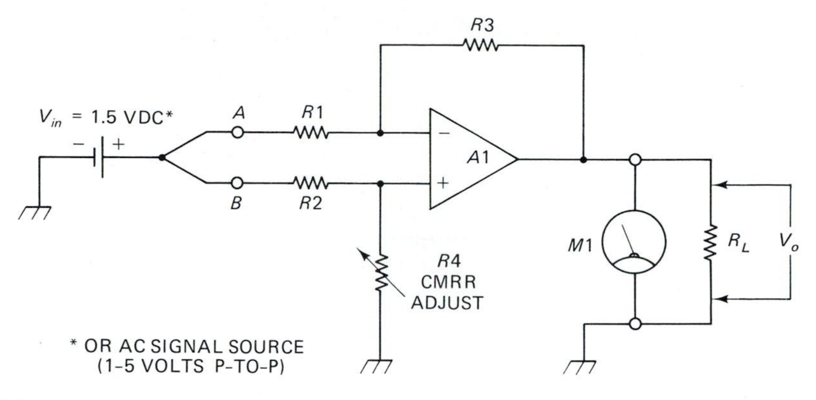


FIGURE 5-6

## 5-6 PRACTICAL DIFFERENTIAL AMPLIFIER CIRCUITS

Figure 5-7 shows the circuit of the simple DC differential amplifier based on the operational amplifier. The gain of the circuit is set by the ratio of two resistors

$$A_v = R3/R1 \tag{5-29}$$

or

$$A_v = R4/R2 \tag{5-30}$$

Provided that

$$R1 = R2$$

$$R3 = R4$$

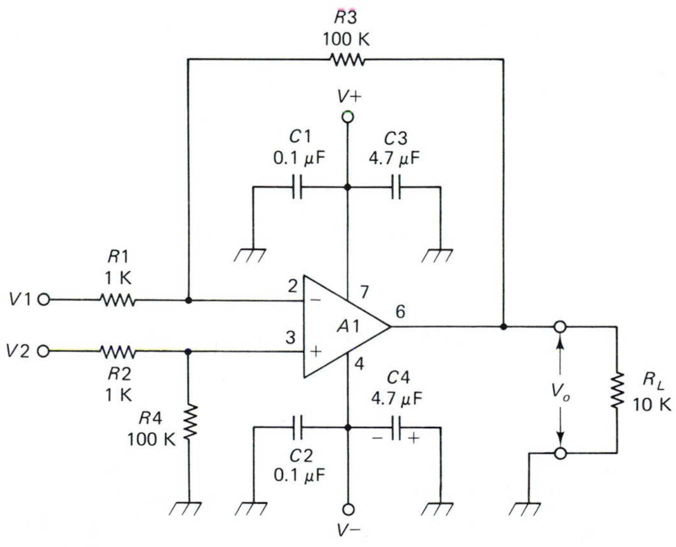


$A_V = \frac{R3}{R1} = \frac{100\text{ K}}{1\text{ K}} = 100$

FIGURE 5-7

The output voltage of the DC differential amplifier is given by

$$V_o = (R3/R1)(V2 - V1) \tag{5-31}$$

or

$$V_o = A_v(V2 - V1) \tag{5-32}$$

where

$V_o$ = the output voltage

$R3$ = the feedback resistor

$R1$ = the input resistor

$A_v$ = the differential gain

$V1$ = the signal voltage applied to the inverting input

$V2$ = the signal voltage applied to the noninverting input

Again, the constraint is $R1 = R2$, and $R3 = R4$. These two balances must be maintained or the common-mode rejection ratio (CMRR) will deteriorate rapidly. In many common applications, the CMRR can be maintained within reason by specifying 5% tolerance resistors for $R1$ through $R4$. But where superior CMRR is required, especially where the differential voltage gain is high, closer tolerance resistors (1% or better) are required.

### DESIGN EXAMPLE 5-1

Design a DC differential amplifier with a gain of 100. Assume the source impedance of the preceding stage is about 100 ohms (refer to Fig. 5-7).

1. Because the source impedance is 100 ohms, we need to make the input resistors of the DC differential amplifier ten times larger (or more). Thus, the input resistors ($R1$ and $R2$) must be at least 100 ohms, $R1 = R2 = 1000$ ohms.
2. Set the value of the two feedback resistors (keep in mind $R3 = R4$). The value of these resistors is found from rewriting Eq. (5-29)

$$R3 = A_v\, R1 \tag{5-33}$$

$$R3 = (100)(1000) \tag{5-34}$$

$$R3 = 100{,}000 \text{ ohms} \tag{5-35}$$

■

Figure 5-7 shows the finished circuit of this amplifier. The values of the resistors are

$R1$ = 1 kohm

$R2$ = 1 kohm

$R3$ = 100 kohms

$R4$ = 100 kohms

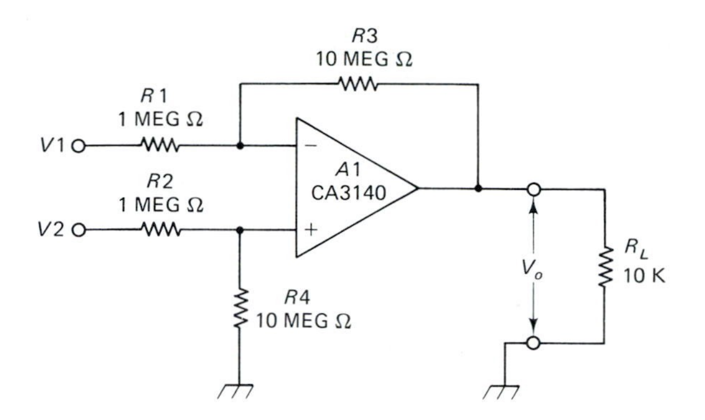


FIGURE 5-8

For best results, either make $R1$, $R2$, $R3$, or $R4$ 1% precision resistors.

The pinouts shown in Fig. 5-7 are for the industry standard 741 family of operational amplifiers. These same pinouts are found on many other op-amp products as well.

The DC power supply voltages are usually either ∓ 12 volts DC, or ∓ 15 volts DC. Lower potentials can be accommodated, however, if a corresponding reduction in the output voltage swing is tolerable. A typical lower grade op-amp will produce a maximum output voltage approximately 3 volts lower than the supply voltage. For example, when $V+$ is 12 volts DC the maximum positive output voltage permitted is $(12 - 3)$ or 9 volts.

The decoupling bypass capacitors shown in Fig. 5-7 are used to keep the circuit stable, especially in those cases where the same DC power supplies are used for several stages. The low value 0.1 μF capacitors ($C1$ and $C2$) are used to decouple high frequency signals. These capacitors should be physically placed as close as possible to the body of the operational amplifier. The high value capacitors ($C3$ and $C4$) are needed to decouple low frequency signals.

The reason we need two values of capacitors is the high value capacitors needed for low frequency decoupling are typically electrolytics (tantalum or aluminum), some of which are ineffective at high frequencies. Thus, we must provide smaller value capacitors which have a low enough capacitive reactance to do the job and are effective at high frequencies.

A different situation is shown in Fig. 5-8. Most differential amplifiers have relatively low input impedances. The amplifier in Fig. 5-8 uses a high input impedance by virtue of the high values of resistors $R1$ and $R2$. In order to attain this input impedance, however, we need to specify an operational amplifier for $A1$ that has a very low input bias current (a very high natural input impedance). The GE/RCA BiMOS devices which use MOSFET input transistors, and the BiFET which uses JFET input transistors, are good selections for DC differential amplifier circuits when high input impedances are needed.

## 5-7 ADDITIONAL DIFFERENTIAL AMPLIFIERS

The simple DC differential amplifier circuits shown in this chapter are useful for low gain applications and for applications where a low to moderate input impedance is permissible (300 to 200,000 ohms). Where a higher gain is required, a more complex circuit called the operational amplifier instrumentation amplifier (IA) must be used.

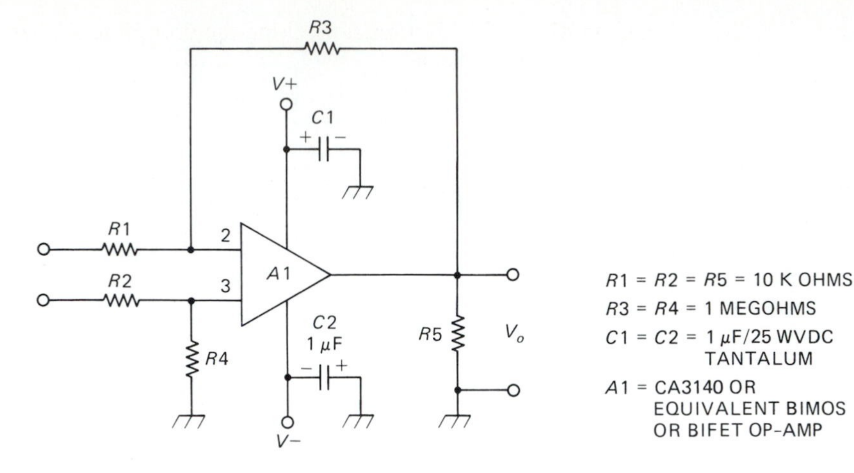


FIGURE 5-9

## 5-8 SUMMARY

1. A differential voltage amplifier produces an output voltage signal equal to the product of the differential voltage gain and the difference between the potentials applied to the −IN and +IN inputs.
2. A single op-amp differential amplifier such as Fig. 5-9 has a gain equal to $R3/R1$ or $R4/R2$, assuming that $R1 = R2$ and $R3 = R4$.
3. The *common-mode gain* of a differential amplifier is the signal gain simultaneously applied to both −IN and +IN inputs. The common-mode rejection ratio (CMRR) is the ratio of differential gain to common mode gain.
4. The input impedance of the simple DC differential amplifier (such as Fig. 5-9) is limited to the sum of the two input resistors.

## 5-9 RECAPITULATION

Now return to the objectives and Prequiz questions at the beginning of the chapter and see how well you can answer them. If you cannot answer certain questions, mark them and review the appropriate parts of the text. Next, try to answer the questions and work the problems below, using the same procedure.

## 5-10 STUDENT EXERCISES

1. Construct a DC differential amplifier using the circuit of Fig. 5-8. The gain is 100. Short both inputs together and connect this junction to the output of an audio signal generator. Apply a 3 volt peak-to-peak sinewave signal (frequency between 100 and 1000 Hz). Measure the output signal amplitude using an oscilloscope.
2. In the exercise above, change resistor $R4$ to 100 kohms and repeat the experiment.
3. In the first exercise above, replace $R4$ with a series combination of a 1-megohm potentiometer

and an 820-kohm fixed resistor. Using the procedure similar to that in the text, adjust for minimum output signal.

4. In all three cases, calculate the common mode rejection ratio, and compare them with each other.

## 5-11 QUESTIONS AND PROBLEMS

1. Define differential voltage gain.
2. Define common-mode voltage gain.
3. What is the gain of a DC differential amplifier (Fig. 5-9) in which $R1 = R2 = 100$ kohms and $R3 = R4 = 1$ megohms?
4. What is the gain of a DC differential amplifier (Fig. 5-9) in which $R1 = R2 = 100$ kohms and $R3 = R4 = 10$ kohms?
5. What is the gain of a DC differential amplifier (Fig. 5-9) in which $R1 = R2 = 500$ ohms and $R3 = R4 = 1$ kohms?
6. Draw a circuit for a DC differential amplifier in which a fine CMRR ADJUST control covers the range of approximately 80 to 120% of the value of the feedback resistor. Assume a gain of 100 and an input resistance of 1000 ohms.
7. A DC differential amplifier has a differential voltage gain of 120 and a common mode gain of 0.001. Calculate the CMRR in decibels.
8. In the amplifier in the previous question, assume that $V1 = 3.04$ and $V2 = 3.06$. Calculate the output voltage.
9. A 3-Vdc level is applied to a DC differential amplifier as a common mode signal. If the differential gain of the circuit is 200 and the CMRR is 90 dB, what is the output voltage? Assume that no differential signal is present.
10. A DC differential amplifier must be designed with an input impedance of 1 megohm and a gain of 10. Using only one operational amplifier, design a suitable circuit.
11. A DC differential amplifier must be designed with an input impedance of 1 kohm and a gain of 1000. Using only one operational amplifier, design a suitable circuit.
12. A DC differential amplifier must be designed with an input impedance of 100 kohms and a gain of 50. Using only one operational amplifier, design a suitable circuit.
13. A DC differential amplifier must be designed with an input impedance of 10 kohms and a gain of 200. Using only one operational amplifier, design a suitable circuit.
14. What is the gain of a DC differential amplifier (Fig. 5-9) in which $R1 = R2 = 10$ kohms and $R3 = R4 = 10$ kohms?
15. What is the gain of a DC differential amplifier (Fig. 5-9) in which $R1 = R2 = 100$ kohms and $R3 = R4 = 150$ kohms?
16. What is the gain of a DC differential amplifier (Fig. 5-9) in which $R1 = R2 = 470$ kohms and $R3 = R4 = 1$ megohm.
17. An operational amplifier has a differential voltage gain of 300,000 in the open-loop configuration. If the CMRR is rated at 120 dB, find the common mode voltage gain.
18. Calculate the CMRR in decibels if the common-mode voltage gain is 0.5 and the differential gain is 220,000.
19. Calculate the CMRR in decibels if the common mode voltage gain is 0.02 and the differential gain is 150,000.

20. A differential input operational amplifier has a measured CMRR of 80 dB, and a differential voltage gain of 100 dB. Calculate the output voltage that would occur if a common-mode voltage of 1 volt is applied to both inputs simultaneously.
21. A common-mode voltage of 10 volts results in an output voltage of 100 mV, while a 10 mV differential signal produces a 2.5 volt output. Calculate the common mode rejection ratio (CMRR) in decibels.
22. Sketch the circuit for a single op-amp differential amplifier that has a means for adjusting the CMRR. Set the differential voltage gain at 100 and the input impedance to at least 10 kohms. Select values for the resistors used in this circuit.
23. Draw the circuit for a differential DC amplifier with a gain of 1000 and a minimum input impedance of 2.2 kohms.
24. A DC differential amplifier based on an op-amp has a common mode rejection ratio of 120 dB. Calculate the differential voltage gain if the common mode gain is 0.001.
25. Draw the circuit for a DC differential amplifier that has a gain of 500 and an input impedance of more than 50 megohms. More than one active device will be required in order to avoid using resistors of excessive value.

CHAPTER 6

# Instrumentation Amplifiers

## OBJECTIVES

1. Know the basic properties of the instrumentation amplifier.
2. Know how to design instrumentation amplifier circuits.
3. Understand the range of applications for, and uses of, instrumentation amplifiers.
4. Become familiar with various forms of IC instrumentation amplifiers (ICIA).

## 6-1 PREQUIZ

These questions test your prior knowledge of the material in this chapter. Try answering them before you read the chapter. Look for the answers (especially those you answered incorrectly) as you read the text. After you have finished studying the chapter, try answering these questions again and those at the end of the chapter (see Section 6-13).

1. List three advantages of the IC instrumentation amplifier.
2. Write the transfer equation for the classical three-device instrumentation amplifier.
3. Why are guard shield connections sometimes used in instrumentation circuits?
4. Calculate the gain of an INA-101 ICIA circuit in which the gain setting resistor ($R_g$) is 500 ohms.

## 6-2 INTRODUCTION

The simple DC differential amplifier discussed in Chapter 5 suffers from several drawbacks. First, the input impedance ($Z_{in}$) is limited, approximately equal to the sum of the two input resistors. Second, there is a practical limitation on the gain of the simple single-device DC differential amplifier. If we attempt to obtain high gain, we find either the input bias current tends to cause large output offset voltages, or the input impedance becomes too low.

In this chapter we will demonstrate a solution to these problems in the form of the instrumentation amplifier (IA). All of these amplifiers are differential amplifiers, but offer superior performance over the simple DC differential amplifiers of the last chapter. The instrumentation amplifier can offer higher input impedance, higher gain, and better common-mode rejection than the single-device DC differential amplifier.

## 6-3 SIMPLE IA CIRCUIT

The simplest form of instrumentation amplifier circuit is shown in Fig. 6-1. In this circuit the input impedance is improved by connecting inputs of a simple DC differential amplifier ($A3$) to two input amplifiers ($A1$ and $A2$) which are unity gain, noninverting followers. (Their use here is as buffer amplifiers.) The input amplifiers offer a large input impedance (a result of the noninverting configuration) while driving the input resistors of the actual amplifying stage ($A3$). The overall gain of this circuit is the same as that for any simple DC differential amplifier

$$A_v = R3/R1 \tag{6-1}$$

where

$$A_v = \text{the voltage gain}$$

assuming

$$R1 = R2, \text{ and } R3 = R4.$$

It is considered best practice if $A1$ and $A2$ are identical operational amplifiers. In fact, it is advisable to use a dual operational amplifier for both $A1$ and $A2$. The common thermal environment of the dual amplifier will reduce thermal drift problems. The very high input impedance of superbeta (Darlington), BiMOS, and BiFET operational amplifiers makes them ideal for use as the input amplifiers in this type of circuit.

One of the biggest problems with the circuit of Fig. 6-1 is it wastes two good operational amplifiers. The most common instrumentation amplifier circuit uses the input amplifiers to provide voltage gain in addition to a higher input impedance. Such amplifier circuits are discussed in the next section.

## 6-4 STANDARD INSTRUMENTATION AMPLIFIERS

The standard instrumentation amplifier (IA) is shown in Fig. 6-2A. Like the simple circuit discussed above, this circuit uses three operational amplifiers. The biggest difference is the input amplifiers ($A1$ and $A2$) are used in the noninverting follower with gain configuration. Like the circuit of Fig. 6-1, the input amplifiers are ideally BiMOS, BiFET, or superbeta input types for maximum input impedance. Again, for best thermal performance use a dual, triple, or quad operational amplifier for this application. The signal voltages shown in Fig. 6-2A follow the standard pattern, voltages $V1$ and $V2$ form the differential input signal ($V2$-$V1$) and voltage $V_{cm}$ represents the common-mode signal because it affects both inputs equally.

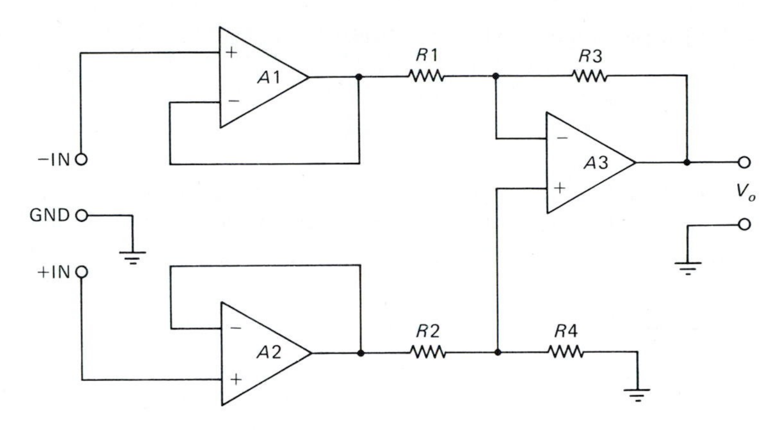


FIGURE 6-1

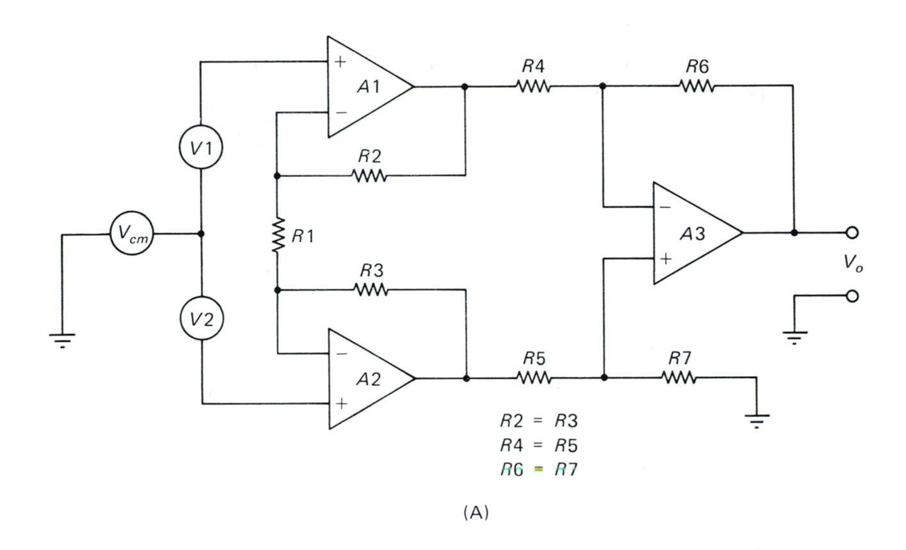


(A)

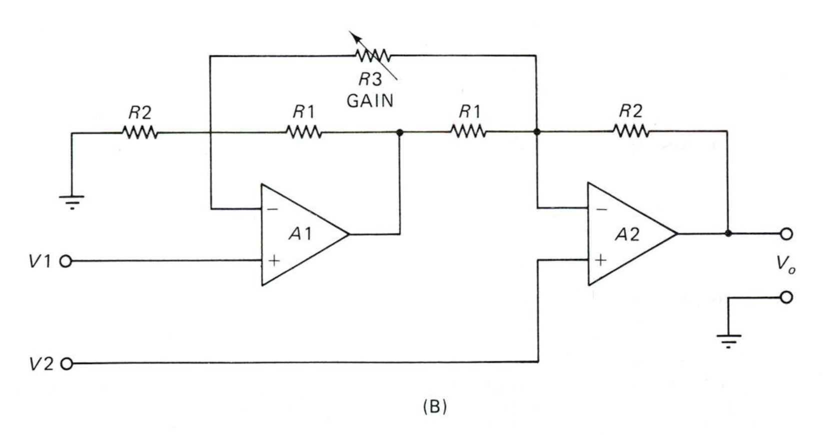


(B)

FIGURE 6-2

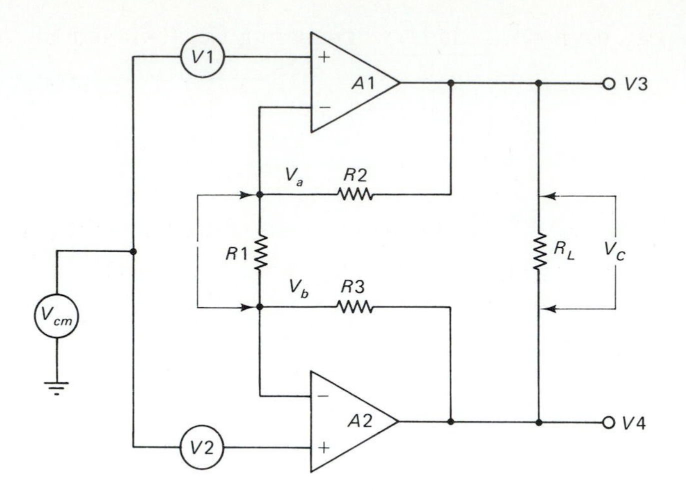


FIGURE 6-3

Let's evaluate Fig. 6-2A by first examining the behavior of the input stages, $A1$ and $A2$. In the partial circuit of Fig. 6-3, the output voltage $V_c$ is the difference between the output potential of $A1$ ($V3$) and the output potential of $A2$ ($V4$). Because resistor $R1$ is shared by both $A1$ and $A2$, we count its value as $R1/2$ for each calculation. Our method is to calculate $V3$ when $V2 = 0$ and $V4$ when $V1 = 0$. Then superimpose the result to find $V_c = V4 - V3$.

1.
$$V4 = V2\left[\frac{R3}{R1/2} + 1\right] \tag{6-2}$$

2.
$$V3 = V1\left[\frac{R2}{R1/2} + 1\right] \tag{6-3}$$

3. Therefore
$$V4 - V3 = V2\left[\frac{R3}{R1/2} + 1\right] - V1\left[\frac{R2}{R1/2} + 1\right] \tag{6-4}$$

4. Because $V_c = V4 - V3$, and $R2 = R3 = R$, we may rewrite Eq. (6-4) in the form
$$V_c = V2\left[\frac{R}{R1/2} + 1\right] - V1\left[\frac{R}{R1/2} + 1\right] \tag{6-5}$$

5.
$$V_c = (V2 - V1)\left[\frac{R}{R1/2} + 1\right] \tag{6-6}$$

6.
$$V_c = (V2 - V1)\left[\frac{R2}{R1} + 1\right] \tag{6-7}$$

7. Or, in the form that identifies specific circuit components
$$V_c = (V2 - V1)\left[\frac{2\,R2}{R1} + 1\right] \tag{6-8}$$

In the form of a standard differential amplifier
$$V_c = V_d\,A_{V12} \tag{6-9}$$

So, we may conclude by comparing Eq. (6-8) and Eq. (6-9)

$$V_d = V2 - V1 \tag{6-10}$$

$$A_{V12} = \frac{2\,R2}{R1} + 1 \tag{6-11}$$

From our discussions in Chapter 5 we know the gain of $A3$ in Fig. 6-2A is

$$A_{V3} = \frac{R6}{R4} \tag{6-12}$$

The overall gain of Fig. 6-2A is

$$A_{V13} = A_{V12} \times A_{V3} \tag{6-13}$$

By substituting Eq. (6-12) into Eq. (6-13), we arrive at the transfer equation for the instrumentation amplifier

$$A_v = \frac{2\,R2}{R1} + 1\,\frac{R6}{R4} \tag{6-14}$$

Provided, $R2 = R3$, $R4 = R5$, $R6 = R7$.

### EXAMPLE 6-1

Find the gain of the instrumentation amplifier of Fig. 6-2A if the following values of resistors are used, $R1 = 220$ ohms, $R2 = R3 = 2200$ ohms, $R4 = R5 = 10$ kohms, and $R6 = R7 =$ 82 kohms.

#### Solution

$$A_v = \frac{2\,R2}{R1} + 1\,\frac{R6}{R4}$$

$$A_v = \frac{(2)\,(2200)}{(220)} + 1\,\frac{(82{,}000)}{(10{,}000)}$$

$$A_v = \frac{(4400)}{(220)} + 1$$

$$A_v = (20 + 1)\,(8.2)$$

$$A_v = (21)\,(8.2)$$

$$A_v = 172$$

■

An alternate instrumentation amplifier circuit is shown in Fig. 6-2B. This circuit also offers the advantage of high input impedance, but it uses only two operational amplifier devices rather than three. The gain of this circuit is given by

$$A_{vd} = \frac{R2\,(2R1 + R3)}{R1R3} + 1 \tag{6-15}$$

In most practical situations the problem will be to select a value for the gain-ranging resistor which is consistent with the required differential voltage gain and the values of the remaining resistors ($R1$ and $R2$). For this application we use the following equation

$$R3 = \frac{2\,R2}{A_v - 1 - (R2/R1)} \tag{6-16}$$

Gain control is a severe problem in the simple DC differential amplifier. That problem is easily solved in instrumentation amplifier circuits as you will see in Section 6-5.

## 6-5 GAIN CONTROL FOR THE IA

It is difficult to provide a gain control for a simple DC differential amplifier without adding an extra amplifier stage (for example, an inverting follower with a gain of 0 to −1). For the instrumentation amplifier, however, resistor $R1$ can be used as a gain control provided the resistance does not go to a value near zero ohms. Figure 6-4 shows a revised circuit with resistor $R1$ replaced by a series circuit consisting of fixed resistor $R1A$ and potentiometer $R1B$. This circuit prevents the gain from rising above the level set by $R1A$. Do not use a potentiometer alone in this circuit because it can have disastrous effect on the gain. Note in Eq. (6-14) the term $R1$ appears in the denominator. If the value of $R1$ gets close to zero, the gain goes very high (supposedly to infinity if $R1 = 0$). The maximum gain of the circuit is limited by using the fixed resistor in series with the potentiometer. The gain of the circuit in Fig. 6-4 varies from a minimum of 167 (when $R1B$ is set to 2000 ohms) to a maximum of 1025 (when $R1B$ is zero). The gain expression for Fig. 6-4 is

$$A_v = \left[\frac{2\,R2}{R1A + R1B} + 1\right]\left[\frac{R6}{R4}\right] \tag{6-17}$$

or, rewriting Eq. (6-17) to take into account that $R1A$ is fixed,

$$A_v = \left[\frac{2\,R2}{390 + R1B} + 1\right]\left[\frac{R6}{R4}\right] \tag{6-18}$$

where

$R1B$ varies from 0 to 2000 ohms

## 6-6 COMMON-MODE REJECTION RATIO ADJUSTMENT

The instrumentation amplifier is no different from any other practical DC differential amplifier in that there will be imperfect balance for common-mode signals. The operational amplifiers are not ideally matched, so there will be a gain imbalance. This gain imbalance is enhanced by a mismatch of the resistors. The result is the instrumentation amplifier will respond to some extent to common-mode signals. As in the simple DC differential amplifier, we can provide a common-mode rejection ratio adjustment by making resistor $R7$ variable (see Fig. 6-5).

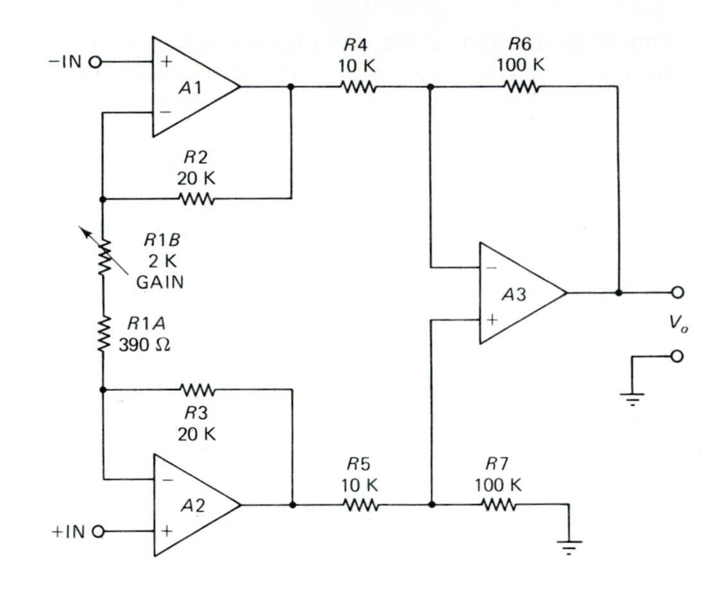


FIGURE 6-4

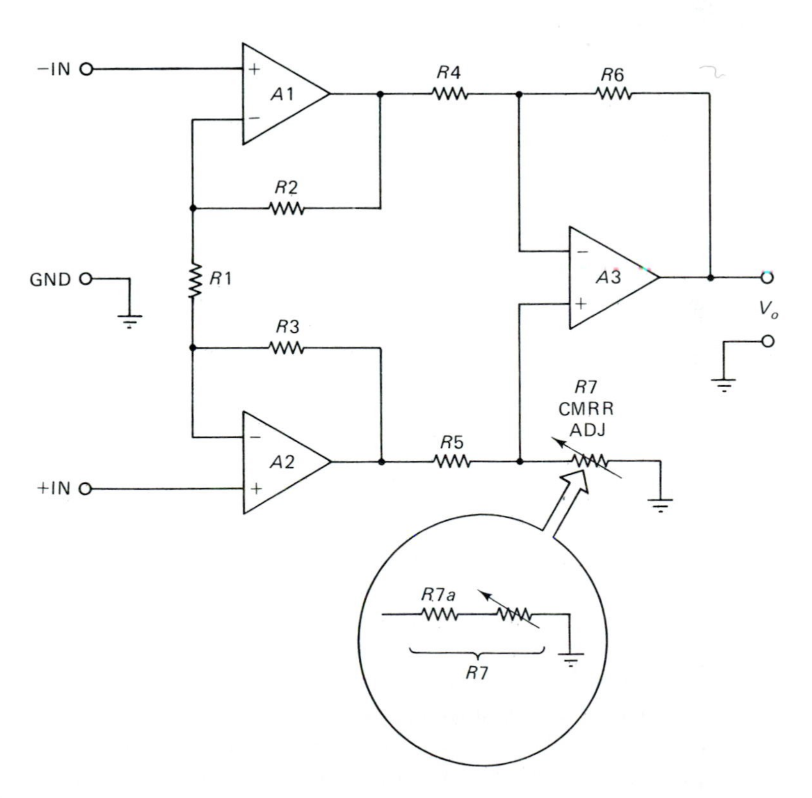


FIGURE 6-5

One configuration of Fig. 6-5 uses a single potentiometer ($R7$) which has a value 10 to 20% larger than the required resistance of $R6$. For example, if $R6$ is 100 kohms, $R7$ should be 110 to 120 kohms. Unfortunately, these values are somewhat difficult to obtain, so we pick a standard value for $R7$ (100 kohms) and then select a value for $R6$ which is somewhat lower (82 kohms or 91 kohms).

The second configuration of Fig. 6-5 uses a fixed resistor in series with a potentiometer. The general rule is to make $R7A$ approximately 80% of the total required value and $R7B$ 40% of the required value. As was true in the other configuration, the sum of $R7A$ and $R7B$ is approximately 110 to 120% of the value of resistor $R6$. Resistance values of this type permit the total resistance to vary from less than to greater than the nominally required value. The adjustment of the CMRR ADJUST control follows the same procedure as is given in Chapter 5 for all differential amplifiers.

## 6-7 AC INSTRUMENTATION AMPLIFIERS

What is the principal difference between DC amplifiers and AC amplifiers? A DC amplifier will amplify both AC and DC signals up to the frequency limit of the particular circuit being used. The AC amplifier, on the other hand, will not pass or amplify DC signals. In fact, AC amplifiers will not pass AC signals of frequencies from close to DC to as low as a $-3$ dB bandpass limit. The gain in the region between near-DC and the full-gain frequencies within the passband rises at a rate determined by the design, usually $+6$ dB/octave. The standard low-end point in the frequency response curve is defined as the frequency at which the gain drops off $-3$ dB from the full gain.

Figure 6-6A shows a modified version of the instrumentation amplifier designed as an AC amplifier. The input circuitry of $A1$ and $A2$ is modified by placing a capacitor in series with each op-amp's noninverting input. Resistors $R8$ and $R9$ are used to keep the input bias currents of $A1$ and $A2$ from charging capacitors $C1$ and $C2$. In some modern low input current operational amplifiers these resistors are optional because of the extremely low levels of bias current which are normally present.

The $-3$ dB frequency of the amplifier in Fig. 6-6A is a function of the input capacitors and resistors (assuming that $R8 = R9 = R$ and $C1 = C2 = C$):

$$F = \frac{1{,}000{,}000}{2\,\pi\,R\,C_{\mu F}} \tag{6-19}$$

where

$$F = \text{the } -3 \text{ dB frequency in Hz}$$

$$R = \text{in ohms}$$

$$C_{\mu F} = \text{in microfarads}$$

### EXAMPLE 6-2

Find the lower $-3$ dB breakpoint frequency if $R9 = R10 = 10$ megohms and $C1 = C2 = 0.1\ \mu F$.

$$A_v = \frac{1{,}000{,}000}{2\,\pi\,R\,C_{\mu F}}$$

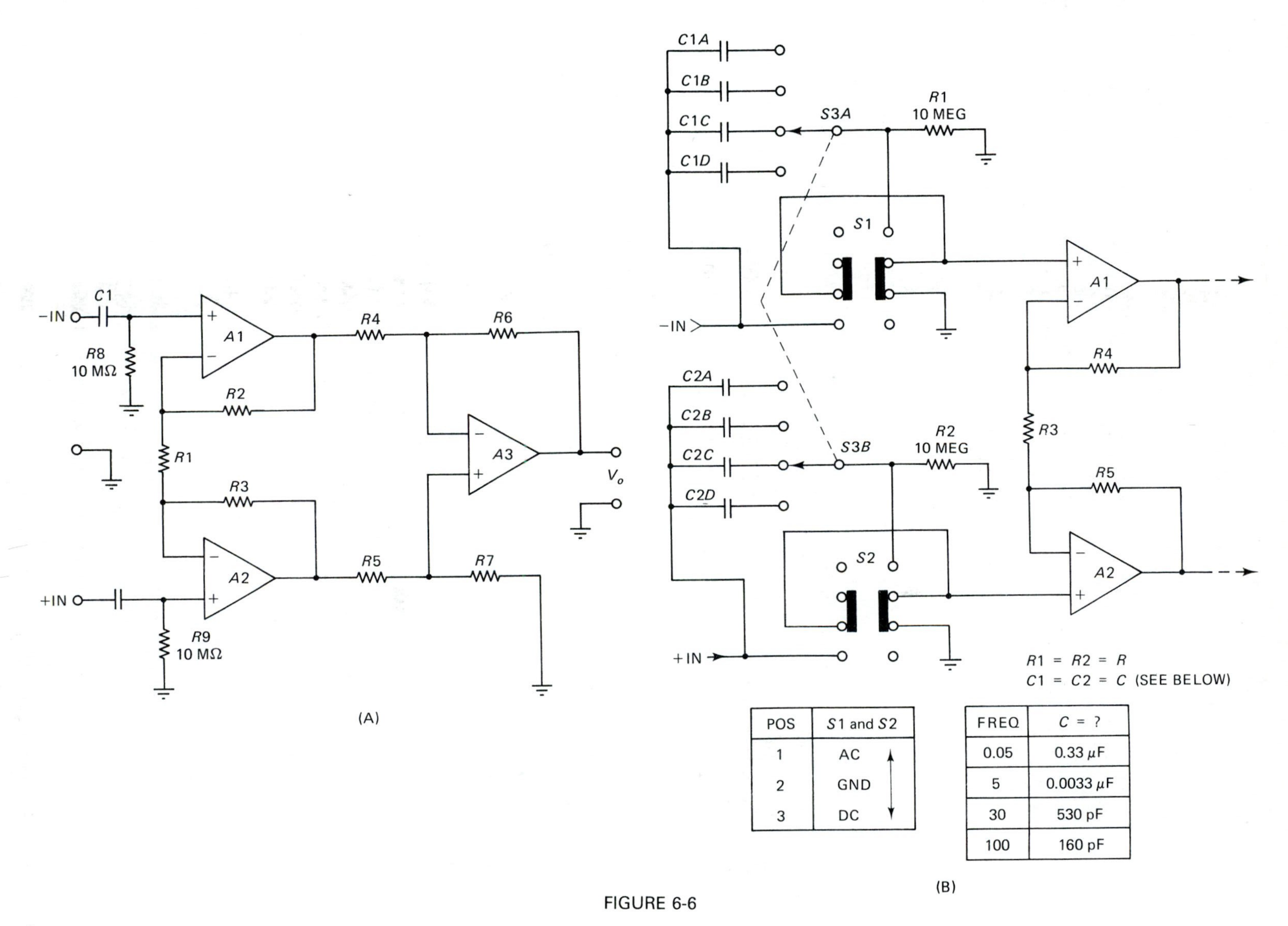


| POS | S1 and S2 |
|---|---|
| 1 | AC |
| 2 | GND |
| 3 | DC |

| FREQ | C = ? |
|---|---|
| 0.05 | 0.33 μF |
| 5 | 0.0033 μF |
| 30 | 530 pF |
| 100 | 160 pF |

FIGURE 6-6

$$A_v = \frac{1{,}000{,}000}{(2)(3.14)(10{,}000{,}000)(0.1)}$$

$$A_v = \frac{1{,}000{,}000}{6{,}280{,}000} = 0.16 \text{ Hz}$$

■

The equation given above for frequency response is not the most useful. In most practical cases you will know the required frequency response from evaluation of the application. Furthermore, you will know the value of the input resistors ($R9$ and $R10$) because they are selected for high input impedance and are by convention either $>10\times$ or $>100\times$ the source impedance depending on the application. Typically, these resistors are selected to be 10 megohms. You will therefore need to select the capacitor values from Eq. (6-20)

$$C_{\mu f} = \frac{1{,}000{,}000}{2\,\pi\,R\,F} \tag{6-20}$$

where

$C_{\mu f}$ = the capacitance of $C1$ and $C2$, in microfarads

$R$ = the resistance of $R9$ and $R10$ in ohms

$F$ = the $-3$ dB frequency in hertz (Hz)

The AC instrumentation amplifier can be adapted to all the other modifications of the basic circuit discussed earlier in this chapter. We may, for example, use a gain control (replace $R1$ with a fixed resistor and a potentiometer), or add a CMRR ADJ control. In fact, these adaptations are probably necessary in most practical AC IA circuits.

In many instrumentation amplifier applications it is desirable to provide selectable AC or DC coupling and the ability to ground the input of the amplifiers. The latter is especially desirable in circuits where an oscilloscope, strip-chart paper recorder, digital data logger or computer is used to receive the data. By grounding the input of the amplifier (without also grounding the source, which could be dangerous), it is possible to set (or at least determine) the $V_d = 0$ baseline. Figure 6-6B shows a modified input circuit which uses a switch to select AC-GND-DC coupling. In addition, a second switch sets the low-end $-3$ dB frequency response point according to Eq. (6-20). A table of popular frequency response limits is shown inset to Fig. 6-6B.

### 6-7.1 Designing an ECG Amplifier

The heart in humans and animals produces a small electrical signal that can be recorded through skin surface electrodes and displayed on an oscilloscope or paper strip-chart recorder. This signal is called an electrocardiograph or ECG signal. The peak values of the ECG signal are on the order of one millivolt (1 mV). In order to produce a 1-volt signal which will display on a recorder or oscilloscope, we need a gain of 1000 mV/1 mV, or 1000. Therefore, our ECG amplifier must provide a gain of 1000 or more. Furthermore, because skin has a relatively high electrical resistance (1 to 20 kohms), the ECG amplifier must have a very high input impedance.

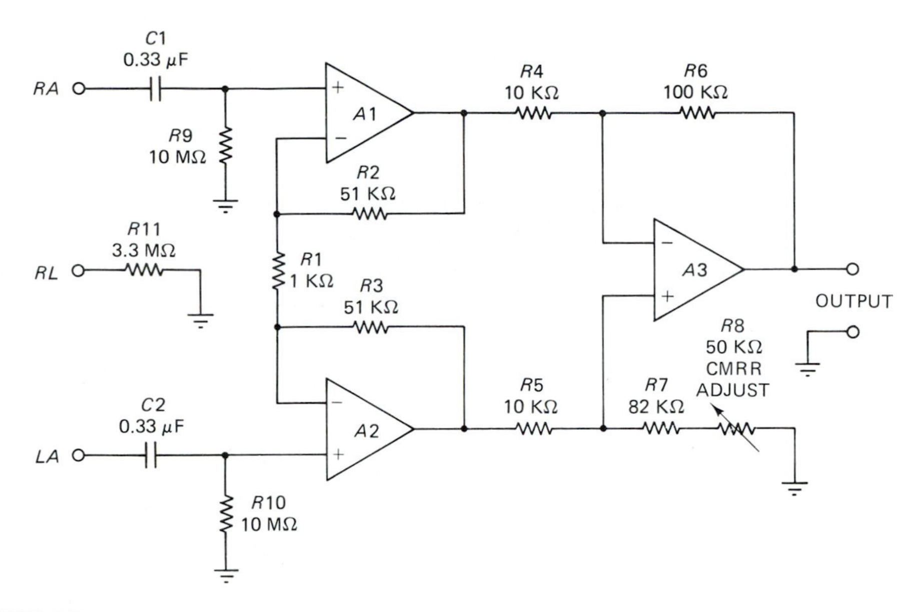


FIGURE 6-7

The ECG amplifier must also be an AC amplifier. The reason is metallic electrodes applied to the electrolytic skin produce a halfcell potential. This potential tends to be on the order of 1 to 2 volts, so it is more than 1000 times higher than the signal voltage. By making the amplifier respond only to AC, we eliminate the DC halfcell potential.

The frequency selected for the −3 dB point of the ECG amplifier must be very low, close to DC, because the standard ECG waveform contains very low frequency components. The typical ECG signal has significant frequency components in the range 0.05 to 100 Hz, which is the industry standard frequency response for diagnostic ECG amplifier. (Some clinical monitoring ECG amplifiers use 0.05 Hz to 40 Hz to eliminate muscle artifact due to patient movements.)

The typical ECG amplifier has differential inputs because the most useful ECG signals are differential in nature. They also suppress 60-Hz hum picked up on the leads and the patient's body. In the most simple case, the right arm (RA) and left arm (LA) electrodes form the inputs to the amplifier, with the right leg (RL) defined as the common (see Fig. 6-7). The basic configuration of the amplifier in Fig. 6-7 is the AC-coupled instrumentation amplifier discussed earlier. The gain for this amplifier is set to slightly more than ×1000, so a 1 Mv ECG peak signal will produce a 1-volt output from this amplifier. Because of the high gain it is essential the amplifier be well balanced. This requirement suggests the use of a dual amplifier for $A1$ and $A2$. An example might be the RCA CA-3240 which is a dual BiMOS device made up of two CA-3140's in a single 8-pin miniDIP package. Also in the interest of balance, 1% or less tolerance resistors should be used for the equal pairs.

The lower end −3 dB frequency response point is set by the input resistors and capacitors. In this case, the combination forms a response of

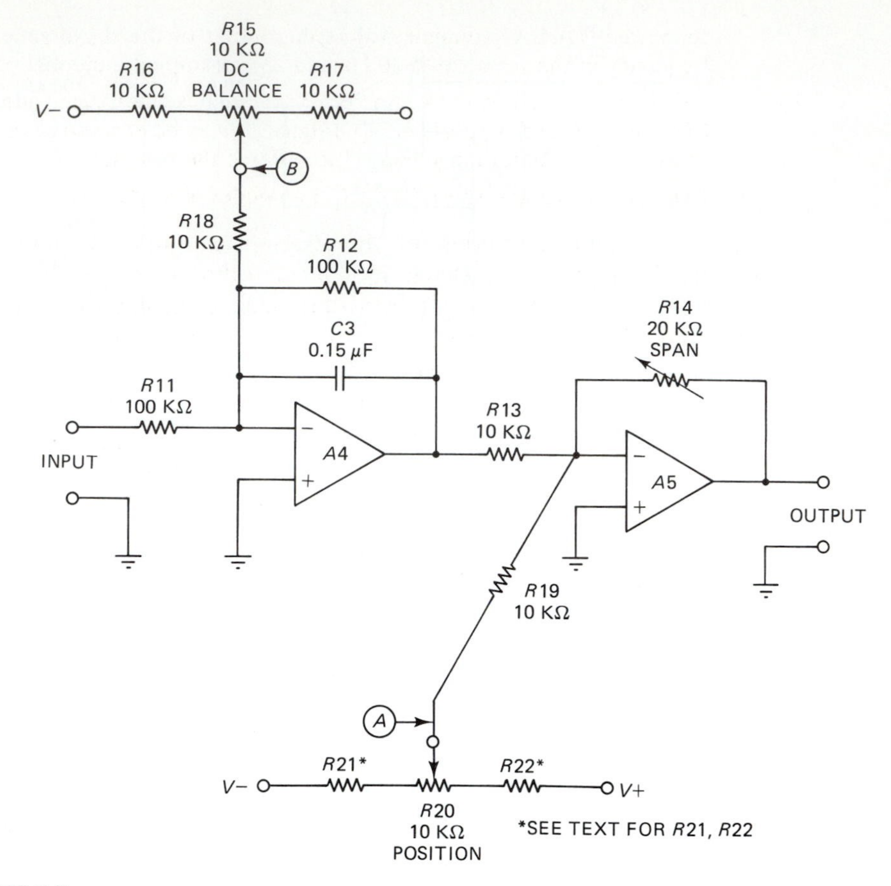


FIGURE 6-8

$$F_{\text{Hz}} = \frac{1{,}000{,}000}{2\ \pi\ R\ C}$$

$$F_{\text{Hz}} = \frac{1{,}000{,}000}{(2)(3.14)(10{,}000{,}000\text{ ohms})(0.33\ \mu\text{F})}$$

$$F_{\text{Hz}} = \frac{1{,}000{,}000}{20{,}724{,}000} = 0.048\text{ Hz}$$

The CMRR ADJUST control in Fig. 6-7 is usually a 10-to-20-turn trimmer potentiometer. It is adjusted in the following manner.

1. Short together the RA, LA, and RL inputs.
2. Connect a DC voltmeter to the output. It should be either a digital voltmeter or an analog meter with a 1.5 volt DC scale, or a DC-coupled oscilloscope (but be sure to identify the zero baseline).
3. Adjust CMRR ADJUST control (*R*7) for zero volts output.
4. Disconnect the RL input and connect a signal generator between RL and the still-

connected RA/LA terminal. Adjust the output of the signal generator for a sinewave frequency in the range 10 to 40 Hz and a peak-to-peak potential of 1 volt to 3 volts.

5. Using an AC scale on the voltmeter (or an oscilloscope), again adjust CMRR ADJUST (*R*7) for the smallest possible output signal. It may be necessary to readjust the voltmeter or oscilloscope input range control to observe the best null.
6. Remove the RA/LA short. The ECG amplifier is ready to use.

A suitable post-amplifier for the ECG preamplifier is shown in Fig. 6-8. This amplifier is placed in the signal line between the output of Fig. 6-7 and the input of the oscilloscope or paper-chart recorder used to display the waveform. The gain of the postamplifier is variable from 0 to +2. It will produce a 2-volt maximum output when a 1-millivolt ECG signal provides a 1-volt output from the preamplifier. Because of the high level signals used, this amplifier can use ordinary 741 operational amplifiers.

The frequency response of the amplifier is set to an upper −3 dB point of 100 Hz, with the response dropping off at a −6 dB/octave rate above that frequency. This frequency response point is determined by capacitor *C*3 operating with resistor *R*12

$$F_{Hz} = \frac{1{,}000{,}000}{2\ \pi\ R\ C}$$

$$F_{Hz} = \frac{1{,}000{,}000}{(2)(3.14)(100{,}000)(0.015\ \mu F)}$$

$$F_{Hz} = \frac{1{,}000{,}000}{9420} = 106\ Hz$$

There are three controls in the post-amplifier circuit, *span*, *position*, and *DC balance*. The span control is the 0 to 2 gain control. The label "span" is instrumentation users' language, rather than electronics language. The position control sets the position of the output waveform on the display device. Resistors *R*21 and *R*22 are selected to limit the travel of the beam or pen to full scale. Set these resistors so the maximum potential at the end terminals of *R*20 corresponds to full-scale deflection of the display device.

The DC BALANCE control is used to cancel the collective effects of offset potentials created by the various stages of amplification. This control is adjusted as follows

1. Follow the CMRR ADJUST procedure and then reconnect the short-circuit at RA, LA, and RL. The voltmeter is moved to the output of Fig. 6-8.
2. Adjust the position control for zero volts at point A.
3. Adjust the DC BALANCE control for zero volts at point B.
4. Adjust the SPAN control (*R*14) through its entire range from zero to maximum while monitoring the output voltage. If the output voltage does not shift, no further adjustment is needed.
5. If the output voltage in the previous step varied as the span control varied, adjust DC balance until varying the span control over its full range does not produce an output voltage shift. Repeat this step several times until no further improvement is possible.
6. Remove the RA, LA, RL short and the amplifier is ready for use.

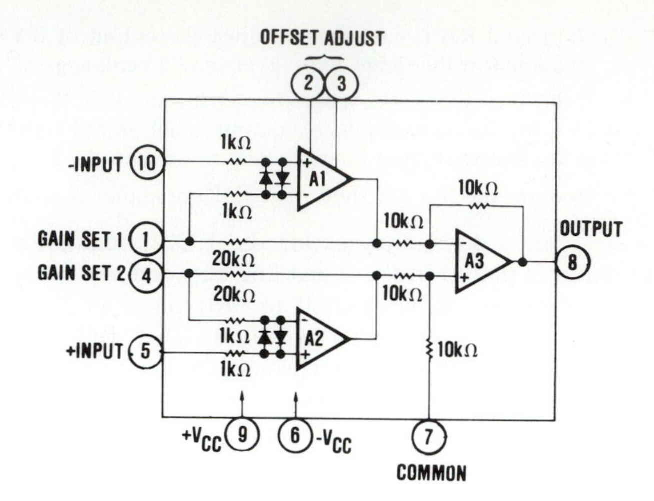


FIGURE 6-9

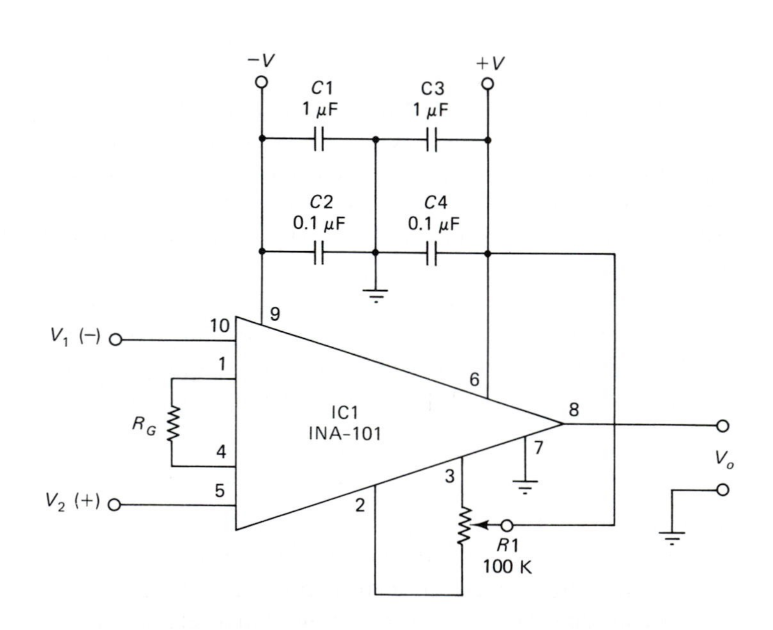


$$A_{VD} = \frac{40\text{ K}}{R_G} + 1$$

$R_G$ IS IN KOHMS

FIGURE 6-10

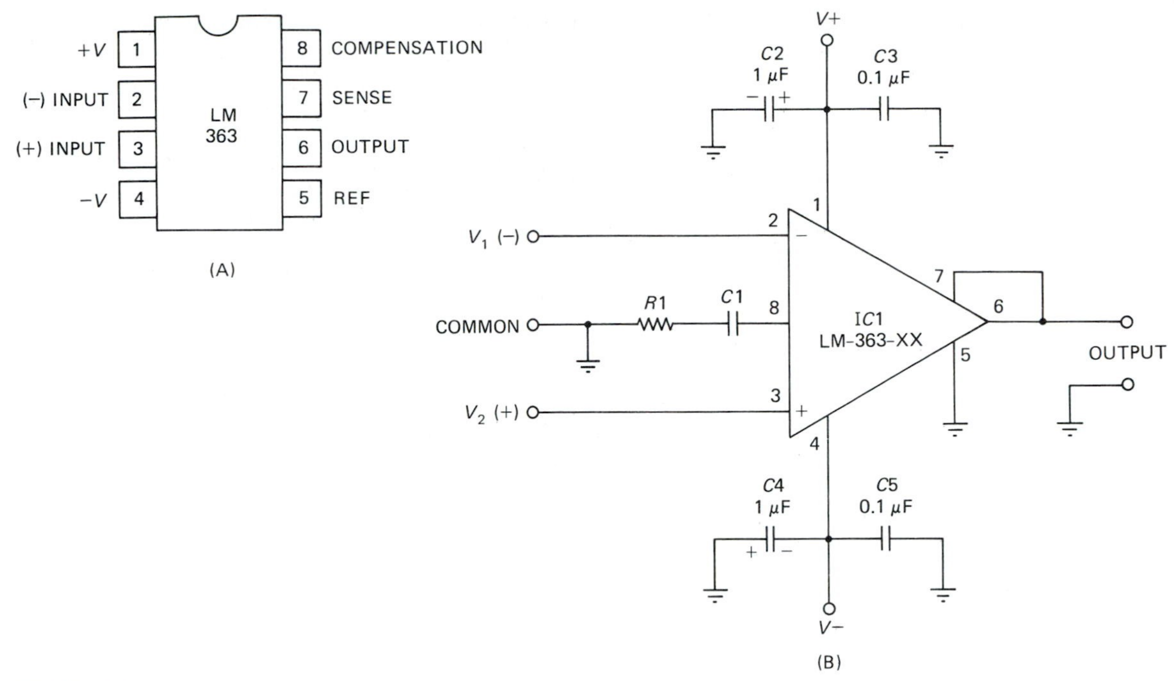


FIGURE 6-11

## 6-8 IC INSTRUMENTATION AMPLIFIERS: SOME COMMERCIAL EXAMPLES

The operational amplifier revolutionized analog circuit design. For a long time, the only additional advances were their gradual improvement (they approached ideal op-amp of textbooks). The next big breakthrough came when the analog device designers made an IC version of the instrumentation amplifier (Fig. 6-2A), the integrated circuit instrumentation amplifier (ICIA). Today, several manufacturers offer substantially improved ICIA devices.

The Burr-Brown INA-101 (Fig. 6-9) is a popular ICIA device. A sample INA-101 circuit is shown in Fig. 6-10. This ICIA amplifier is simple to connect and use. There are only DC power connections, differential input connections, offset adjust connections, ground, and an output. The gain of the circuit is set by

$$A_{vd} = (40\ k/R_g) + 1 \tag{6-21}$$

The INA-101 is basically a low-noise, low-input bias current integrated circuit version of the IA of Fig. 6-2A. The resistors $R2$ and $R3$ in Fig. 6-2A are internal to the INA-101 and are 20 kohms each, hence the "40 k" term in Eq. (6-21).

Potentiometer $R1$ in Fig. 6-10 is used to null the offset voltages appearing at the output. An offset voltage is a voltage which exists on the output at a time when it should be zero (when $V1 = V2$, so V1-V2 = 0). The offset voltage might be internal to the amplifier, or it might be a component of the input signal. DC offsets in signals are common, especially in biopotentials amplifiers like ECG and EEG, and chemical transducers such as pH, pO2, and pCO2.

Another ICIA is the LM-363-*xx* device shown in Fig. 6-11. The miniDIP version is shown in Fig. 6-11A (an 8-pin metal can is also available). A typical circuit is shown in Fig. 6-11B. The LM-363-*xx* device is a fixed-gain ICIA. There are three versions of the LM-363-*xx* listed according to gain

| *Designation* | *Gain ($A_v$)* |
|---|---|
| LM-363-10 | × 10 |
| LM-363-100 | × 100 |
| LM-363-500 | × 500 |

The LM-363-*xx* is useful where a standard gain is required and there is minimum space. For example, the LM-363-*xx* can be used as a transducer preamplifier, especially in noisy signal areas. The LM-363-*xx* can be built onto (or into) the transducer to build up its signal before sending it to the main instrument or signal acquisition computer. (A design example of this form of amplifier is given in Chapter 13.) Another possible use is in biopotentials amplifiers. Biopotentials are typically very small, especially in lab animals. The LM-363-*xx* can be mounted on the subject and a higher level signal sent to a main instrument.

A selectable gain version of the LM-363 device is shown in Fig. 6-12A. The 16-pin DIP package is shown in Figure 6-12A and a typical circuit is shown in Fig. 6-12B. The type number of this device is LM-363-AD, which distinguishes it from the LM-363-*xx* devices. The gain can be ×10, ×100, or ×1000 depending on the programming of the gain setting pins (2, 3, and 4). The programming protocol is as follows:

| *Gain Desired* | *Jumper Pins* |
|---|---|
| × 10 | (All Open) |
| × 100 | 3 & 4 |
| × 1000 | 2 & 3 |

Switch *S*1 in Fig. 6-12B is the GAIN SELECT switch. This switch should be mounted close to the IC device, but it is quite flexible in mechanical form. The switch could also be made from a combination of CMOS electronic switches (for example, 4066).

The DC power supply terminals are treated in a manner similar to the other amplifiers. Again, the 0.1-μF capacitors need to be mounted as close as possible to the body of the LM-363-AD.

Pins 8 and 9 are guard shield outputs. These pins are a feature that makes the LM-363-AD very useful for many instrumentation problems. By outputting a signal sample back to the shield of the input lines, we can increase the common-mode rejection ratio. This feature is frequently used in biopotentials amplifiers and in other applications where a low-level signal must pass through a strong interference (high noise) environment. Guard shield theory will be discussed in Section 6-9.

The LM-363 devices will operate with DC supply voltages of ± 5 volts to ± 18 volts DC, with a common-mode rejection ratio (CMRR) of 130 dB. The 7 nV/(SQR(Hz)) noise figure makes the device useful for low noise applications (a 0.5 nV model is also available).

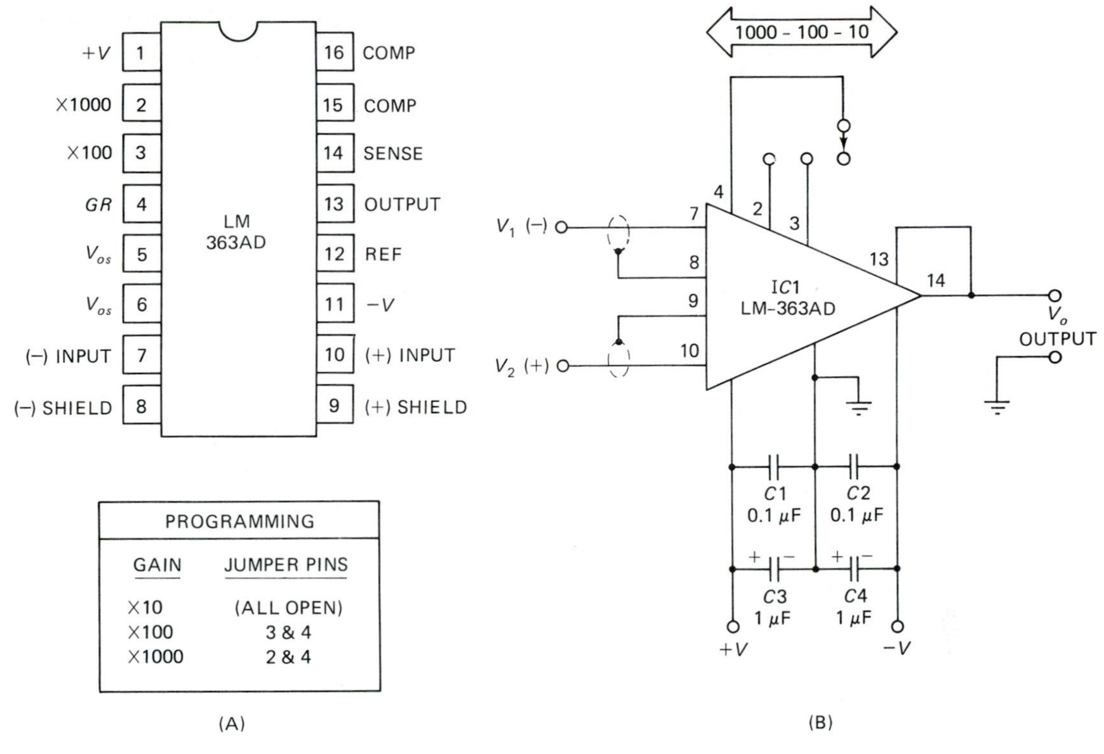


FIGURE 6-12

## 6-9 GUARD SHIELDING

Differential amplifiers, including instrumentation amplifiers, tend to suppress interfering signals from the environment. The common-mode rejection process is the basis of this capability. When an amplifier is connected to an external signal source through wires, those wires are subjected to strong local 60 Hz AC fields from nearby power line wiring. Fortunately, in the case of the differential amplifier the field affects both lines equally, so the induced interfering signal is cancelled out by the common-mode rejection property of the amplifier.

However, the cancellation of interfering signals is not total. There may be, for example, imbalances in the circuit that tend to deteriorate the CMRR of the amplifier. These imbalances may be either internal or external to the amplifier circuit. Figure 6-13 shows a common scenario. In this figure we see the differential amplifier connected to shielded leads from the signal source, $V_{in}$. Shielded lead wires offer some protection from local fields, but there is a flaw in the standard wisdom regarding shields. This is that it is possible for shielded cables to create a valid differential signal voltage from a common mode signal!

Figure 6-13B shows an equivalent circuit that demonstrates how a shielded cable pair can create a differential signal from a common mode signal. The cable has capacitance between the center conductor and the shield conductor surrounding it.

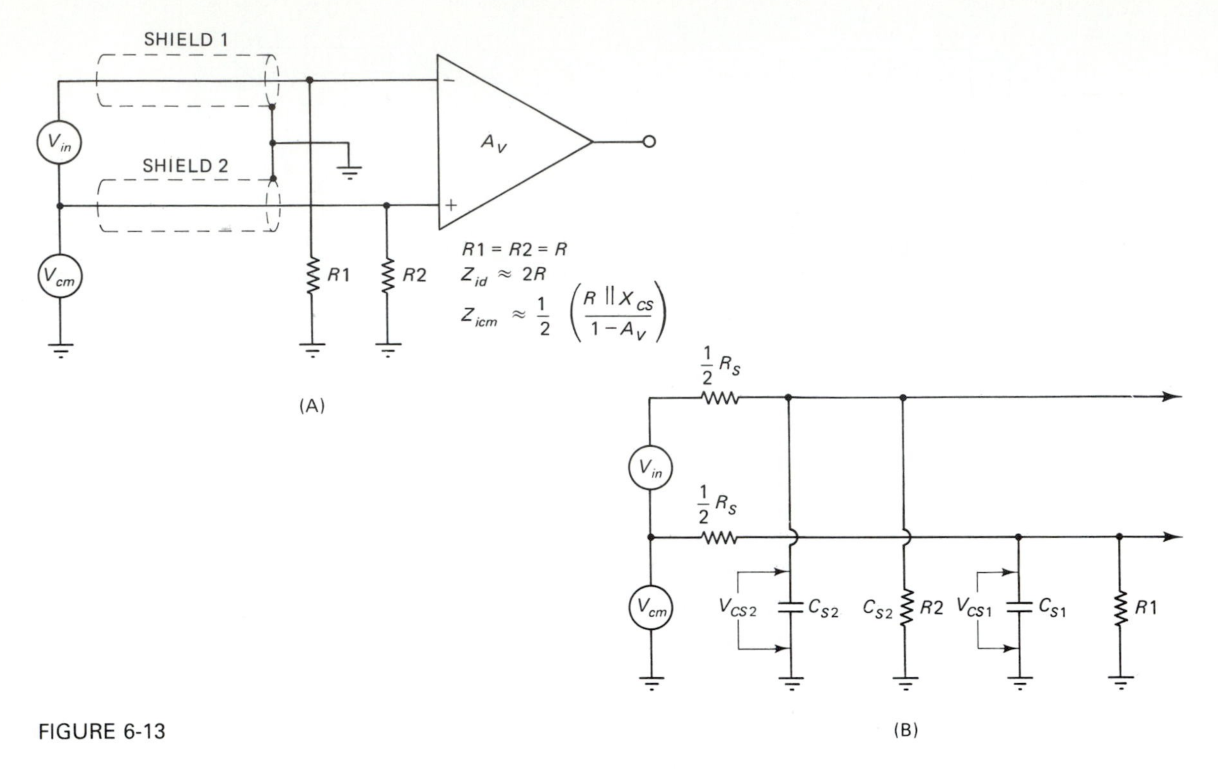


FIGURE 6-13

In addition, input connectors and the amplifier's internal wiring also exhibit capacitance. These capacitances are lumped together in the model of Fig. 6-13B as $C_{S1}$ and $C_{S2}$. As long as the source resistances and shunt resistances are equal, and the two capacitances are equal, there is no problem with circuit balance. But inequalities in any of these factors (which are commonplace) create an unbalanced circuit in which common mode signal $V_{cm}$ can charge one capacitance more than the other. As a result, the difference between the capacitance voltages ($V_{CS1}$ and $V_{CS2}$) is seen as a valid differential signal.

A low-cost solution to the problem of shield-induced artifact signals is shown in Fig. 6-14A. In this circuit a sample of the two input signals is fed back to the shield, which in this situation is not grounded. Alternatively, the amplifier output signal is used to drive the shield. This type of shield is called a *guard shield.* Either double shields (one on each input line) or a common shield for the two inputs can be used.

An improved guard shield example for the instrumentation amplifier is shown in Fig. 6-14B. In this case a single shield covers both input lines, but it is possible to use separate shields. In this circuit a sample of the two input signals is taken from the junction of resistors *R*8 and *R*9, and fed to the input of a unity gain buffer/driver guard amplifier (*A*4). The output of *A*4 is used to drive the guard shield.

Perhaps the most common approach to guard shielding is the arrangement shown in Fig. 6-14C. Here we see two shields used. The input cabling is double-shielded insulated wire. The guard amplifier drives the inner shield, which serves as the guard shield for the system. The outer shield is grounded at the input end in the normal manner and serves as an electromagnetic interference suppression shield.

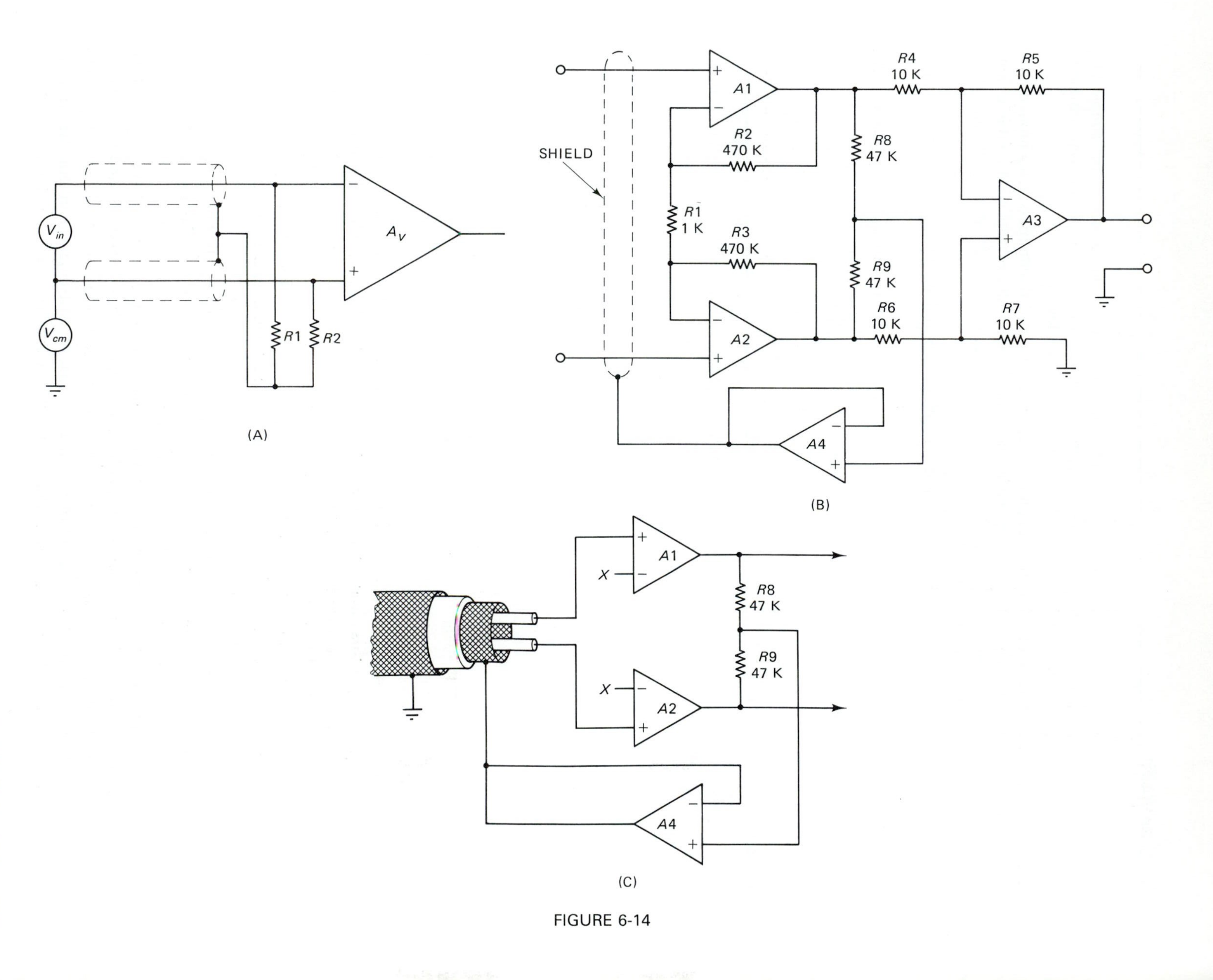

$V_{in}$
$V_{cm}$
$A_V$
R1
R2
(A)
SHIELD
A1
A2
A3
A4
R1 1 K
R2 470 K
R3 470 K
R4 10 K
R5 10 K
R6 10 K
R7 10 K
R8 47 K
R9 47 K
(B)
X
(C)


FIGURE 6-14

## 6-10 SUMMARY

1. The instrumentation amplifier circuit overcomes several difficulties of the simple DC differential amplifier. It provides higher input impedance levels, improved common-mode rejection ratio (CMRR), and higher achievable gain levels.
2. The basic instrumentation amplifier uses three operational amplifiers. Two input amplifiers (*A*1 and *A*2) are connected in the noninverting follower with gain configuration, while the third (*A*3) is an output amplifier connected as a DC differential amplifier. Amplifiers *A*1/*A*2 drive *A*3
3. Gain control on the instrumentation amplifier is controllable by adjusting a single input resistor which is common to both input amplifiers.
4. AC-coupled instrumentation amplifiers are formed by using capacitors in series with the inputs. In most circuits each input will require a very high value resistor to common in order to prevent the input capacitors from being charged by the op-amp input bias currents.

## 6-11 RECAPITULATION

Now return to the objectives and Prequiz questions at the beginning of the chapter and see how well you can answer them. If you cannot answer certain questions, mark them and review the appropriate parts of the text. Next, try to answer the questions and work the problems below, using the same procedure.

## 6-12 STUDENT EXERCISES

1. Using a circuit like Fig. 6-2A, design and build an instrumentation amplifier with a gain of (a) 100, (b) 500, (c) 1000. Measure the gain and CMRR in each case.
2. Using a circuit like Fig. 6-6A, design an AC-coupled instrumentation amplifier which has a lower −3 dB point of 10 Hz in its frequency response characteristic. Measure the frequency response from near DC to 1000 Hz, using at least 10 data points between 0.1 Hz and 100 Hz.
3. In the exercise above change the input resistance values and/or the input capacitor values to form frequency responses of 0.1 Hz, 1 Hz, or 100 Hz. Measure the frequency response from near DC to 1000 Hz.

## 6-13 QUESTIONS AND PROBLEMS

1. Find the gain of the instrumentation amplifier of Fig. 6-2A if the following values of resistors are used: $R1 = 330$ ohms, $R2 = R3 = 20{,}000$ ohms, $R4 = R5 = 10$ kohms, and $R6 = R7 = 56$ kohms.
2. Find the gain of the instrumentation amplifier of Fig. 6-2A if the following values of resistors are used: $R1 = 1$ kohm, $R2 = R3 = 10$ kohms, $R4 = R5 = 10$ kohms, and $R6 = R7 = 100$ kohms.
3. Find the lower −3 dB breakpoint frequency of an AC instrumentation amplifier such as Fig. 6-6A if $R8 = R9 = 10$ megohms, and $C1 = C2 = 0.33$ μF.

4. Find the lower $-3$ dB breakpoint frequency in an AC instrumentation amplifier such as Fig. 6-6A if $R8 = R9 = 10$ megohms, and $C1 = C2 = 0.01$ μF.
5. Design an instrumentation amplifier for a Wheatstone bridge transducer that will drive an oscilloscope trace full scale (8 cm) when the input selector is set to 1 Vdc/cm. Assume that the full-scale output of the transducer is 10 mV.
6. Design an AC instrumentation amplifier that has a gain of 2000, an input impedance of 10 megohms or more, and a lower end AC frequency response of 0.5 Hz.
7. Design an AC instrumentation amplifier that has a gain of 1 to 1000, an input impedance of 10 megohms or more, and a lower end AC frequency response of 10 Hz.
8. Sketch the circuit for a standard instrumentation amplifier circuit based on three operational amplifiers.
9. List at least three advantages of the instrumentation amplifier over the standard DC differential op-amp circuit.
10. What advantage is gained by making at least the two input op-amps part of the same IC package?
11. Derive the transfer function for the standard three op-amp instrumentation amplifier using the Kirchoff's law method used in this book. Show your work.
12. Sketch the circuit for an instrumentation amplifier that uses only two op-amp devices.
13. Calculate the differential gain of a standard three op-amp instrumentation amplifier if all resistors are 10 kohms except the resistor connected between the inverting inputs of the two input amplifiers ($R1$), which is 100 ohms.
14. Calculate the differential voltage gain of a two op-amp instrumentation amplifier if $R1 =$ 10 kohms, $R2 = 150$ kohms and $R3 = 10$ kohms.
15. Select a value for $R3$ in the two op-amp instrumentation amplifier if the other resistors are all 10 kohms and the required voltage gain is 100.
16. Sketch the circuit for a three op-amp instrumentation amplifier that has a fine resolution CMRR ADJ control. Select appropriate resistor values for a fixed voltage gain of 500. Show your calculations to justify the selections.
17. Write out a brief procedure for adjusting the CMRR in the amplifier above (a) when the input signal is 1 Vdc, and (b) when the input signal is a 10 volt peak-to-peak 100-Hz sinewave.
18. Draw the circuit for a bioelectric instrumentation amplifier that has a lower $-3$ dB frequency response point of 0.05 Hz or less (but not DC), an upper $-3$ dB response of 100 Hz, an input impedance of at least 10 megohms, and a voltage gain that is variable from about 200 to more than 2,000.
19. Draw the circuit for an ECG preamplifier that has a gain of 1,000 and provides some protection against electrical shock in the common line for the patient.
20. Why must ECG and other bioelectric preamplifiers use AC-coupled input circuitry?
21. Draw the circuit for an instrumentation amplifier input circuit that offers a 10-megohm input impedance (or more), with the following options: DC coupling, AC coupling and input grounded (but not the input voltage!).
22. Draw the circuit for a postamplifier suitable for use with an ECG preamplifier. Label component values.
23. A Burr-Brown INA-101 is used in a circuit that must offer a gain of 100, and input impedance

of 10 megohms or more, and an AC-coupled lower $-3$ dB point of 0.1 Hz or less. Calculate the values of (a) the gain setting resistor ($R_g$) and (b) the coupling capacitors.

24. What is the differential voltage gain of an LM-363-100?
25. Why are guard shield circuits sometimes used in instrumentation amplifier circuits?
26. Draw the circuit for a universal guard shield driver that will work on a three op-amp instrumentation amplifier.
27. Draw the use of guard shield connections in the LM-363-AD ICIA device.

CHAPTER 7

# Isolation Amplifiers

## OBJECTIVES

1. Understand the applications where isolation amplifiers are required.
2. Learn the different approaches to isolation amplifier design.
3. Examine typical applications for isolation amplifiers.
4. Learn to design simple instrumentation circuits based on isolation amplifiers.

## 7-1 PREQUIZ

These questions test your prior knowledge of the material in this chapter. Try answering them before you read the chapter. Look for the answers (especially those you answered incorrectly) as you read the text. After you have finished studying the chapter, try answering these questions again and those at the end of the chapter (see Section 7-9).

1. List two potential applications for isolation amplifiers and why.
2. List three approaches to isolation amplifier design.
3. Under what circumstances is a battery powered amplifier considered "isolated?"
4. From what are the inputs of an isolation amplifier isolated?

## 7-2 INTRODUCTION

There are a number of applications in which ordinary solid-state amplifiers are either in danger themselves or present a danger to users. An example of the former is an amplifier in a high voltage experiment such as a biochemist's electrophoresis system (Sec. 7-4.3). The latter might be cardiac monitors and other devices used in clinical

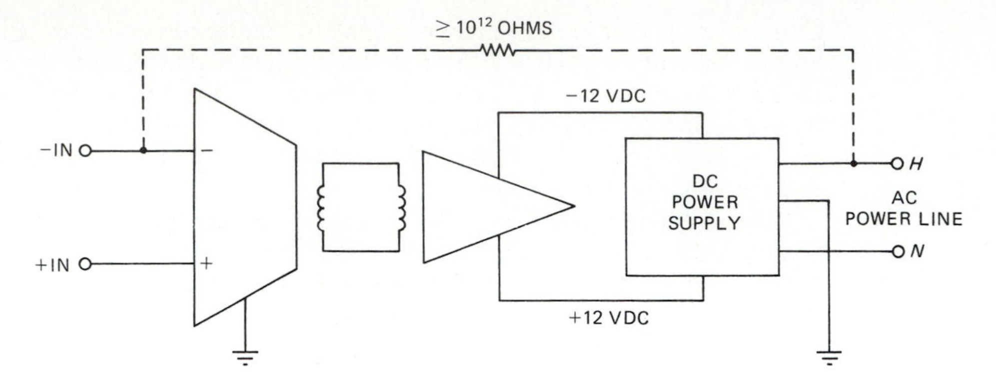


FIGURE 7-1

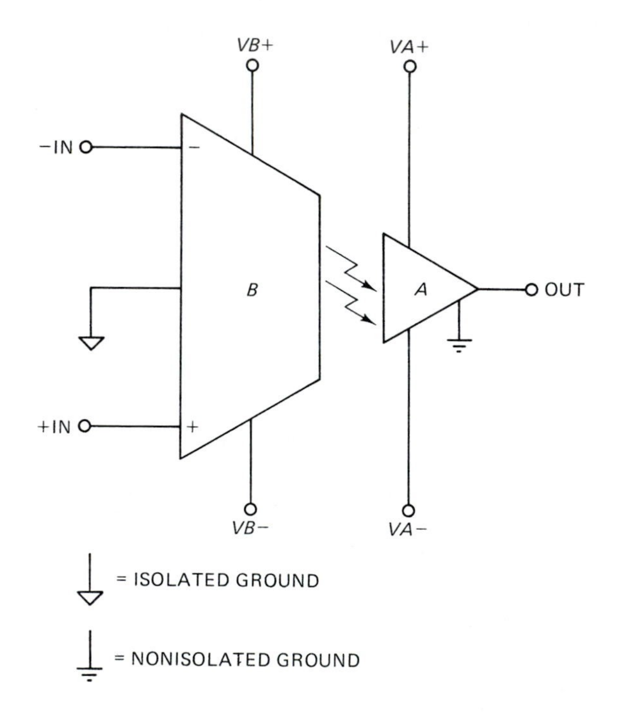


FIGURE 7-2

medicine. Many of the commercial products on the market are not, strictly speaking, integrated circuits. They are hybrids. Nonetheless, it is important to examine these devices.

An *isolation amplifier* (Fig. 7-1) has a very high impedance between the signal inputs and the power supply terminals connected to a DC power supply which are, in turn, connected to the AC power mains. Thus, in isolation amplifiers there is an extremely high resistance (on the order of $10^{12}$ ohms) between the amplifier input terminals and the AC power line. In the case of medical equipment, the goal is to prevent minute leakage from the 60-Hz AC power lines being applied to the patient.

Current levels that are normally negligible to humans can theoretically be fatal to a hospital patient in situations where the body is invaded by devices which are electrical conductors. In other cases, the high impedance is used to prevent high voltages at the signal inputs from adversely affecting the rest of the circuitry. Modern isolation amplifiers can provide more than $10^{12}$ ohms of isolation between the AC power lines and the signal inputs.

Several different circuit symbols are used to denote the isolation amplifier in schematic diagrams, but the most common one is shown in Fig. 7-2. It consists of the regular triangular amplifier symbol broken in the middle to indicate isolation between the A and B sections. The following connections are usually found on the isolation amplifier

*Nonisolated A Side.* $V+$ and $V-$ DC power supply lines (to be connected to a DC supply powered by the AC lines), output to the rest of the (nonisolated) circuitry, and (in some designs) a nonisolated ground or common. This ground is connected to the chassis or main system ground also served by the main DC power supplies.

*Isolated B Side.* Isolated $V+$ and $V-$, isolated ground or common and the signal inputs. The isolated power supply and ground are not connected to the main power supply or ground systems. Batteries are sometimes used for the isolated side. In other cases special isolated DC power supplies derived from the main supplies are used.

## 7-3 APPROACHES TO ISOLATION AMPLIFIER DESIGN

Different manufacturers use different approaches to design isolation amplifiers. Common circuit approaches to isolation include, *battery power*, *carrier operated*, *optically coupled*, and *current loading*.

### 7-3.1 Battery Powered Isolation Amplifiers

The battery approach to isolation amplifier design is perhaps the simplest to implement, but it is not always the most suitable because of problems in battery upkeep. A few products exist, however, that use a battery-powered frontend amplifier, even though the remainder of the equipment is powered from the 110 volt AC power line. Other products are entirely battery-powered. A battery-powered amplifier or instrument is isolated from the AC power mains only if the battery is disconnected from the charging circuit during use. Some battery-powered instruments used in medicine, have mechanical interlocks and electrical logic circuitry to prevent the instrument from being turned on if the AC power cord is still attached. Later in this chapter we will study a battery-powered cardiac output computer (Section 7-4.1) as a design example of this type of isolation.

### 7-3.2 Carrier Operated Isolation Amplifiers

Figure 7-3A shows an isolation amplifier which uses the carrier signal technique to provide isolation. The circuitry inside the dashed line is isolated from the AC power lines (in other words, the B side of Fig. 7-2). The voltage gain of the isolated section is typically in the range $\times 10$ to $\times 500$.

The isolation is provided by separation of the ground, power supply, and signal

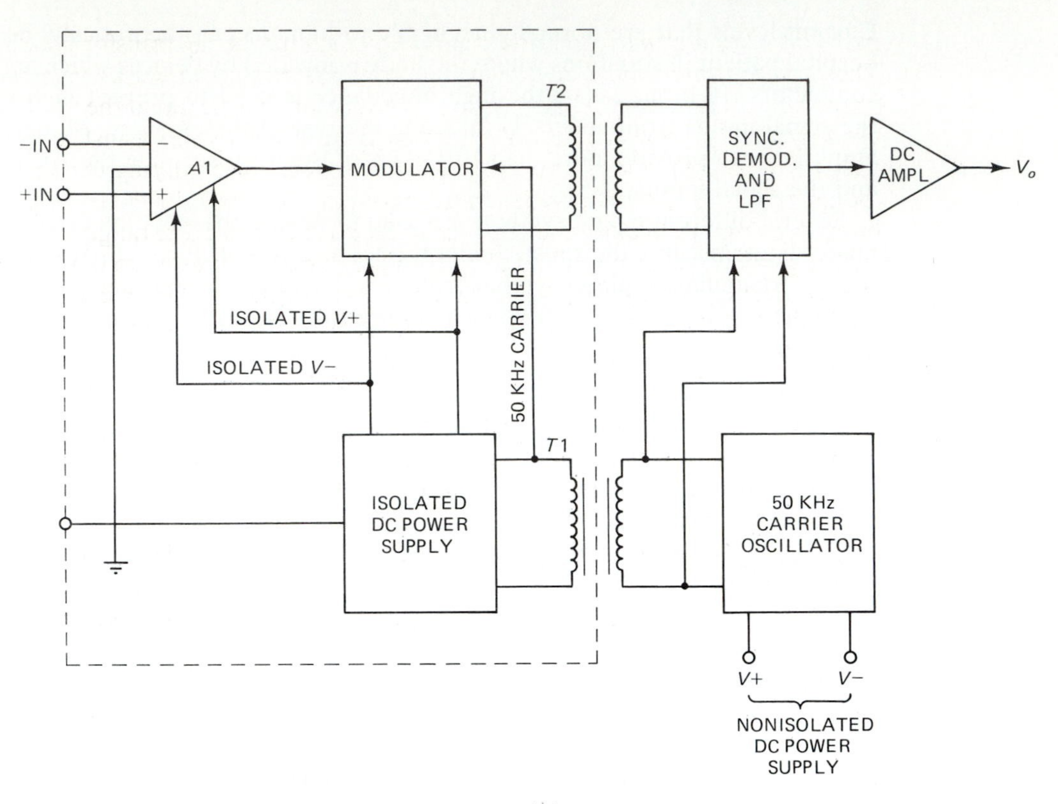
−IN
+IN
A1
MODULATOR
T2
SYNC.
DEMOD.
AND
LPF
DC
AMPL
$V_o$
ISOLATED V+
ISOLATED V−
50 KHz CARRIER
T1
ISOLATED
DC POWER
SUPPLY
50 KHz
CARRIER
OSCILLATOR
V+
V−
NONISOLATED
DC POWER
SUPPLY

(A)

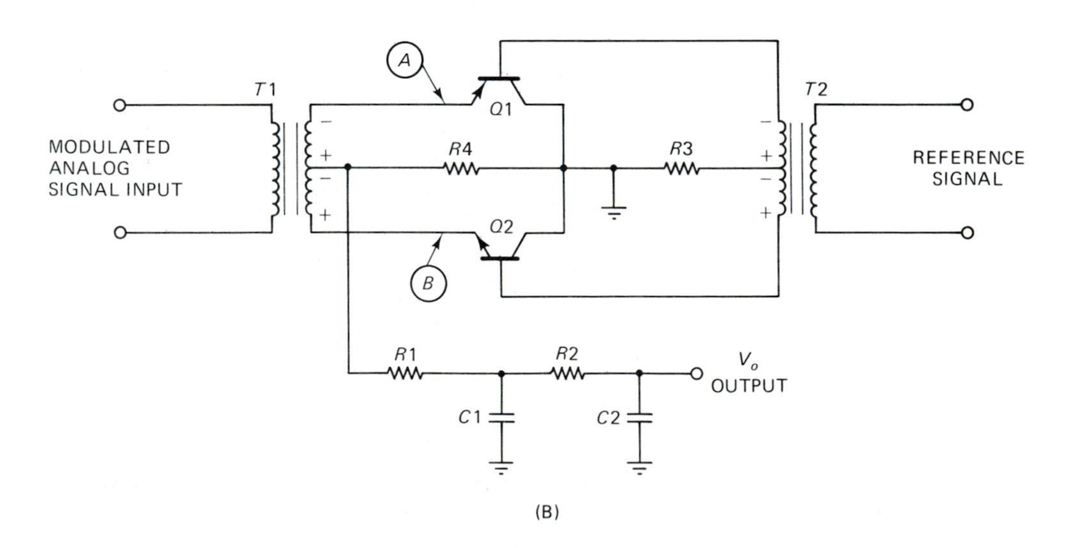
A
T1
Q1
T2
MODULATED
ANALOG
SIGNAL INPUT
R4
R3
REFERENCE
SIGNAL
Q2
B
R1
R2
$V_o$
OUTPUT
C1
C2

(B)

FIGURE 7-3

paths into two mutually exclusive sections by high frequency transformers $T1$ and $T2$. These transformers have a design and core material that works very well in the ultrasonic (20 KHz to 500 KHz) region. But it is very inefficient at the 60 Hz frequency used by the AC power lines. This design feature allows the transformers to easily pass the high frequency carrier signal, while severely attenuating 60 Hz energy. Although most models use a carrier frequency in the 50 KHz to 60 KHz range, examples of carrier amplifiers exist over the entire 20 KHz to 500 KHz range.

The carrier oscillator signal is coupled through transformer $T1$ to the isolated stages. Part of the energy from the secondary of $T1$ is directed to the modulator stage. The remaining energy is rectified and filtered, and then used as an isolated DC power supply. The DC output of this power supply is used to power the input B amplifiers and the modulator stage.

An analog signal applied to the input is amplified by $A1$. It is then applied to one input of the modulator stage. This stage amplitude modulates the signal onto the carrier. Transformer $T2$ couples the signal to the input of the demodulator stage on the nonisolated side of the circuit. Either envelope or synchronous demodulation may be used, although the latter is considered superior. Part of the demodulator stage is a low-pass filter that removes any residual carrier signal from the output signal. Ordinary DC amplifiers following the demodulator complete the signal processing chain.

An example of a synchronous demodulator circuit is shown in Fig. 7-3B. These circuits are based on switching action. Although the example shown uses bipolar PNP transistors as the electronic switches, other circuits use NPN transistors, FETs, or CMOS electronic switches (for example, 4066 device).

The signal from the modulator has a fixed frequency in the range from 20 KHz to 500 KHz, and is amplitude-modulated with the input signal from the isolated amplifier. This signal is applied to the emitters of transistors $Q1$ and $Q2$ (via $T1$) in push-pull. On one half of the cycle, therefore, the emitter of $Q1$ will be positive with respect to the emitter of $Q2$. On alternate half-cycles, the opposite situation occurs, $Q2$ is positive with respect to $Q1$. The bases of $Q1$ and $Q2$ are also driven in push-pull, but by the carrier signal (called here the reference signal). This action causes transistors $Q1$ and $Q2$ to switch on and off out of phase with each other.

On one half of the cycle, the polarities are those shown in Fig. 7-3B; transistor $Q1$ is turned on. In this condition point A on $T1$ is grounded. The voltage developed across load resistor $R4$ is positive with respect to ground.

On the alternate half-cycle, $Q2$ is turned on, so point B is grounded. But the polarities have reversed, so the polarity of the voltage developed across $R4$ is still positive. This causes a full-wave output waveform across $R4$, which when low-pass filtered, becomes a DC voltage level proportional to the amplitude of the input signal. This same description of synchronous demodulators also applies to the circuits used in some carrier amplifiers (a specialized laboratory amplifier used for low-level signals).

A variation on this circuit replaces the modulator with a *voltage controlled oscillator* (VCO). This allows the analog signal to *frequency modulate* (FM) a carrier signal generated by the VCO. The power supply carrier signal is still required, however. A phase detector, phase-locked loop (PLL), or pulse-counting FM detector on the non-isolated side recovers the signal.

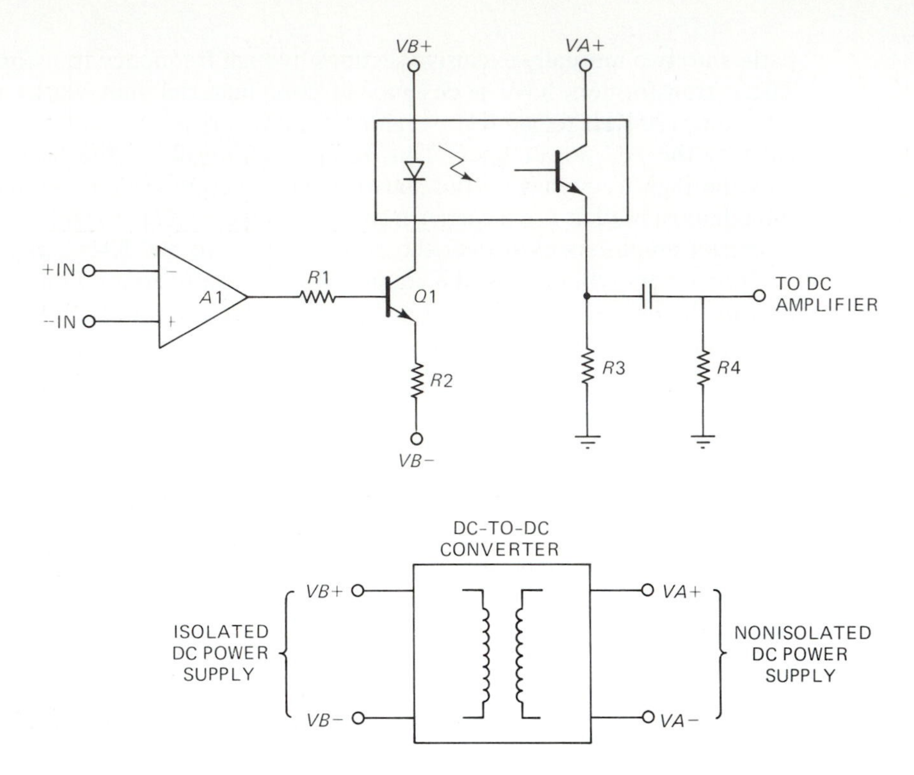


FIGURE 7-4

### 7-3.3 Optically Coupled Isolation Amplifier Circuits

Electronic optocouplers (also called "optoisolators") are sometimes used to provide isolation. In early designs of this class, a light emitting diode (LED) was mounted together with a photoresistor or phototransistor. Modern designs, however, use integrated circuit (IC) optoisolators that contain an LED and phototransistor inside of a single DIP IC package.

There are several approaches to optical coupling. Two common methods are the *carrier* and *direct* methods. The carrier method is the same as discussed in the previous section, except an optoisolator replaces transformer $T2$. The carrier method is not common in optically coupled isolation amplifiers because of frequency response limitations in some IC optoisolators. Only recently have these problems been resolved.

The more common direct approach is shown in Fig. 7-4. This circuit uses the same DC-to-DC converter to power the isolated stages as was used in other designs. It keeps $A1$ isolated from the AC power mains but it is not used in the signal coupling process. In some designs, the high frequency carrier power supply is actually a separate block from the isolation amplifier.

The LED in the optoisolator is driven by the output of isolated amplifier $A1$. Transistor $Q1$ serves as a series switch to vary the light output of the LED proportional to the analog signal from $A1$. Transistor $Q1$ normally passes sufficient collector current to bias the LED into a linear portion of its operating curve. The output of the

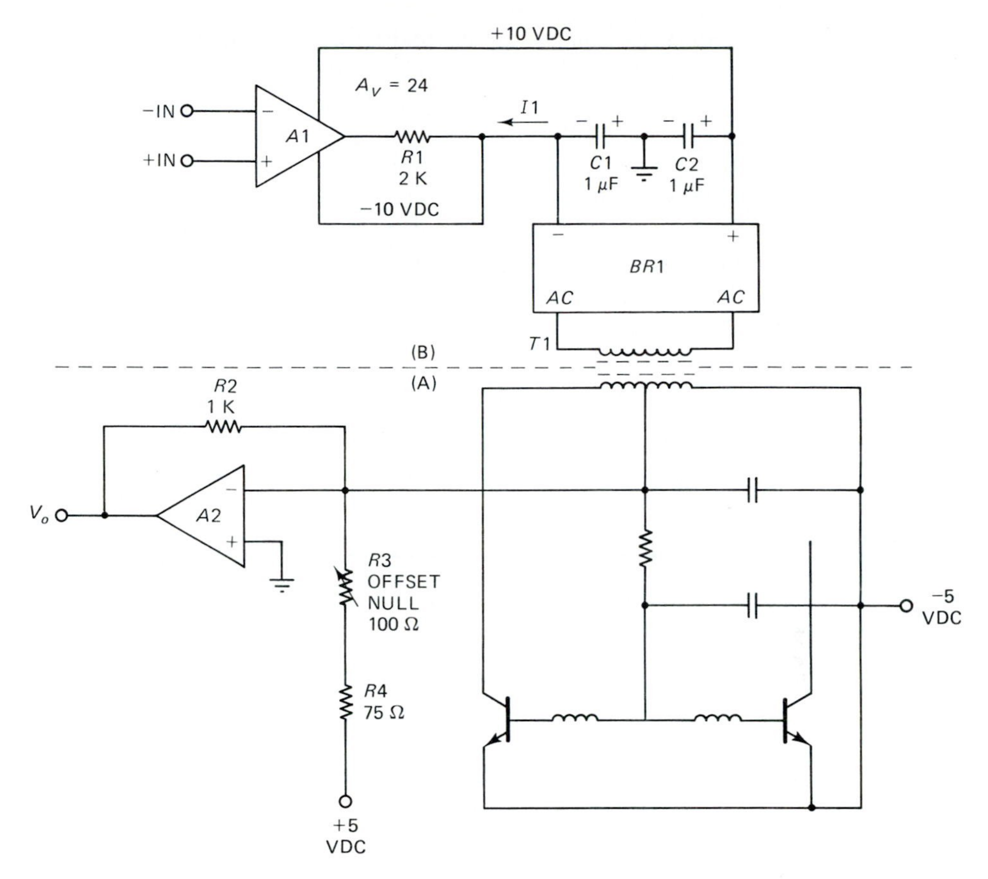


FIGURE 7-5

phototransistor is AC coupled to the remaining amplifiers on the nonisolated side of the circuit, so the offset condition created by the LED bias is eliminated.

Although it is not an actual isolated amplifier by our definition there is another category of optical isolation which is especially attractive for applications where the environment is too hostile for ordinary electronics. It is possible to use LED and phototransistor transmitter and receiver modules in a fiber optic system to provide isolation. A battery-powered (or otherwise isolated) amplifier will sense the desired signal, convert it to an AM or FM light signal, and transmit it down a length of fiber optic cable to a phototransistor receiver module. At that point the signal will be recovered and processed by the nonisolated electronics.

### 7-3.4 Current Loading Isolation Amplifier Methods

A unique current loading isolation amplifier was used in the frontend of a popular electrocardiograph (ECG) medical monitor. A simplified schematic of this device is shown in Fig. 7-5. Notice there is no obvious coupling path for the signal between the isolated and nonisolated sides of the circuit.

The gain-of-24 isolated input preamplifier ($A1$) in Fig. 7-5 consists of a high input impedance operational amplifier. This amplifier is needed to interface with the very high source impedance normal to electrodes in ECG systems. The output of $A1$ is connected to the isolated $-10$ volt DC power supply through load resistor $R1$. This power supply is a DC-to-DC converter operating at 250 KHz. Transformer $T1$ provides isolation between the floating power supplies on the isolated B side of the circuit and the nonisolated A side of the circuit (which are AC line powered).

An input signal causes the output of $A1$ to vary the current loading of the floating $-10$ volt DC power supply. Changing the current loading proportional to the analog input signal causes variation of the $T1$ primary current which is also proportional to the analog signal. This current variation is converted to a voltage variation by amplifier $A2$. An offset null control ($R3$) is provided in the $A1$ circuit to eliminate the offset at the output due to the quiescent current flowing when the analog input signal is zero. In that case, the current loading of $T1$ is constant, but it still provides an offset to the $A2$ amplifier.

## 7-4 DESIGN EXAMPLES

In the sections below you will find several design examples which were selected to illustrate the applications of isolation amplifiers. These applications are drawn from medical, scientific, and process engineering fields.

### 7-4.1 Cardiac Output Computer

The problems presented in electronic signals acquisition are simple compared to the problem of measuring human cardiac output. This is because cardiac output is usually measured using an invasive surgical technique on living humans. This measurement is presented here in order to demonstrate a data acquisition technique which for the sake of absolute safety requires an isolation amplifier to interface with the signal source.

Cardiac output (CO) is defined as the rate of blood volume pumped by the heart. The question being asked of the CO measurement is, "how much blood is this person pumping per unit of time?" Cardiac output is measured in units of liters of blood per minute of time (l/min). In healthy adults CO typically reaches a value between 3 and 5 l/min.

A quantitative measure of cardiac output is the product of the stroke volume and the heart rate. The stroke volume is merely the volume of blood expelled from the heart ventricle (lower chamber) during a single contraction of the heart. Cardiac output is calculated from

$$CO = V \times R \tag{7-1}$$

where

$CO$ = the cardiac output in liters per minute (l/min)

$V$ = the stroke volume in liters per beat (l/beat)

$R$ = the heart rate in beats per minute (beat/min)

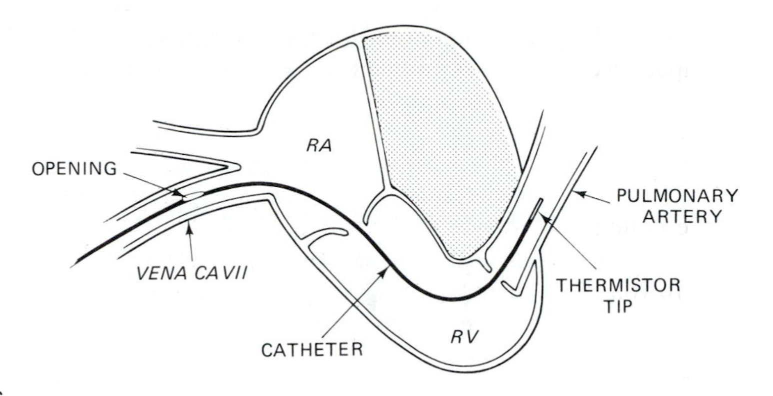


FIGURE 7-6

It is difficult, and usually impossible (except on animals in laboratory settings), to directly measure cardiac output using any technique based on the above equation. The problem is in obtaining good stroke volume data without excessive risk to the patient.

The *thermodilution method* of cardiac output measurement has become the standard indirect method for measuring cardiac output in clinical settings. It is also popular among laboratory scientists. Thermodilution technique forms the basis for most clinical and research cardiac output computers now on the market. One reason why thermodilution is preferred is no poisonous injectates are used as they are in radio-opaque or optical dye dilution methods. Only ordinary medical intravenous (IV) solutions such as normal saline or 5% dextrose in water ($D_5W$) are used.

The thermodilution measurement of cardiac output is made using a special hollow catheter that is inserted into one of the patient's veins, usually on the right arm (the brachial vein is popular). The catheter is multilumened. One of the lumens has its output hole several centimeters from the catheter tip. This proximal lumen is situated so it is outside the heart (close to the input valve on the right atrium) when the tip is all the way through the heart, resting in the pulmonary artery (Fig. 7-6). Other lumens in the catheter output at the tip, so they are used to measure pressures in the pulmonary artery in other procedures. A thermistor in the tip registers a resistance change with changes in blood temperature.

Most thermodilution cardiac output computers operate on a version of the following equation

$$CO = \frac{(64.8) \times C(t) \times V(i) \times [T(b) - T(i)]}{\int T(b)'\,dt} \qquad (7\text{-}2)^*$$

where

$CO$ = the cardiac output in liters per minute (l/min).

64.8 = a collection of other constants and the conversion factor from seconds to minutes.

---

*The mathematical integral symbol and the *dt* in the denominator indicate "integration" takes place and the result is the time-average of temperature $T(b)'$.

$C(t)$ = a constant supplied with the injectate catheter which accounts for the temperature rise in the outside of the patient's body.

$T(b)$ = the blood temperature in °C.

$T(i)$ = the temperature of the injectate in °C.

$T(b)'$ = the temperature of the blood as it changes due to mixing with the injectate.

The mathematical symbol in the denominator of Eq. (7-2) tells us the temperature of the blood at the output side of the heart is integrated, that is, the computer finds the time-average of the temperature as it changes.

## EXAMPLE 7-1

A special cardiac output computer dummy catheter test fixture enters a temperature signal which simulates a temperature change of 10°C for a period of 10 seconds. Find the expected reading during a test of the instrument if the following front panel settings are entered: $C(t) = 49.6$, $T(b) = 37°$, injectate temperature $T(i)$ is 25°, and injectate volume $V(i)$ is 10 ml.

### Solution

$$CO = \frac{(64.8) \times C(t) \times V(i) \times [T(b) - T(i)]}{\int T(b)'\,dt} \text{ (l/min)}$$

$$CO = \frac{(64.8) \times (49.6) \times (10 \text{ ml} \times 1 \text{ l}/1000 \text{ ml}) \times [(37)-(25)]}{(10°)(10 \text{ sec})}$$

$$CO = \frac{3214 \times 0.01 \times 12}{100}$$

$$CO = 385.68/100 = 3.9 \text{ l/min}$$

■

The thermistor in the end of the catheter is usually connected in a Wheatstone bridge circuit (see Fig. 7-7). The DC excitation of the bridge is critical. Either the short-term stability of this voltage must be very high or a ratiometric method must be used to cancel excitation potential drift. It is also necessary to limit the bridge excitation potential to about 200 mV for reasons of safety to the patient (electrical leakage is especially dangerous because the thermistor is *inside the heart or pulmonary artery*). This low value of excitation voltage promotes both patient safety and thermistor stability through freedom from self-heating induced drift, even though it imposes a greater burden on the amplifier design.

Figure 7-7 shows a simplified schematic of a typical cardiac output computer front-end circuit. The thermistor is in a Wheatstone bridge circuit which also consists of $R1$ through $R3$, with potentiometer $R5$ serving to balance the bridge. An autobalancing or zeroing method is sometimes used for this function. Those circuits use a digital-to-analog converter (DAC) to inject a current into one node of the bridge. That current nulls the bridge circuit to zero. The physician waits a few minutes for the thermistor to equilibrate with blood temperature (usually it is in this condition by the time it is threaded through the venous system to the pulmonary artery). The output

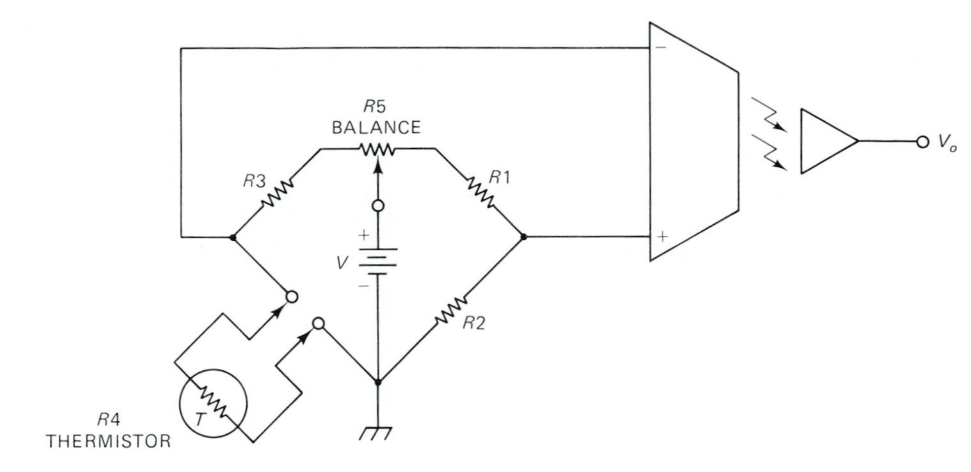

FIGURE 7-7

of the bridge, depending on the design, is typically 1.2 to 2.5 millivolts per degree Celsius, with 1.8 mV/°C being common. This signal is amplified approximately 1000 times to 1 volt/°C by the preamplifier. This preamplifier is an isolated amplifier for reasons of patient safety. The output of this circuit ($V_o$) is used in the denominator of an equation like the one above.

The block diagram for a sample analog cardiac output computer is shown in Fig. 7-8. The frontend circuitry from Fig. 7-7 is in the blocks marked "Bridge" and "Preamp." The isolator circuit is merely a buffer amplifier that permits $V_o$ to be output

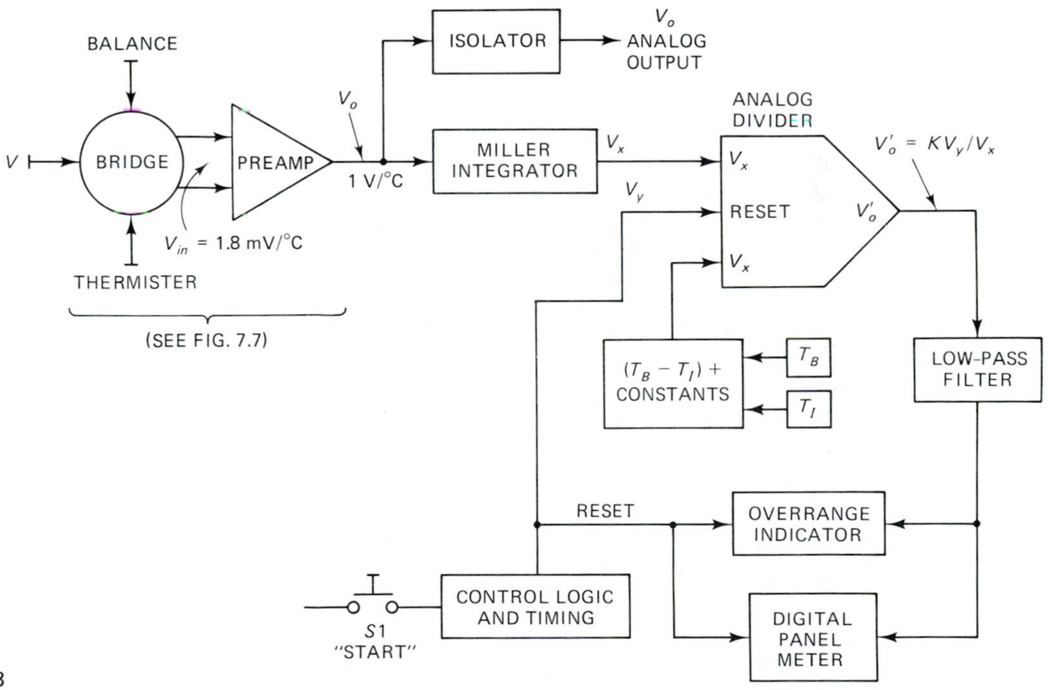

FIGURE 7-8

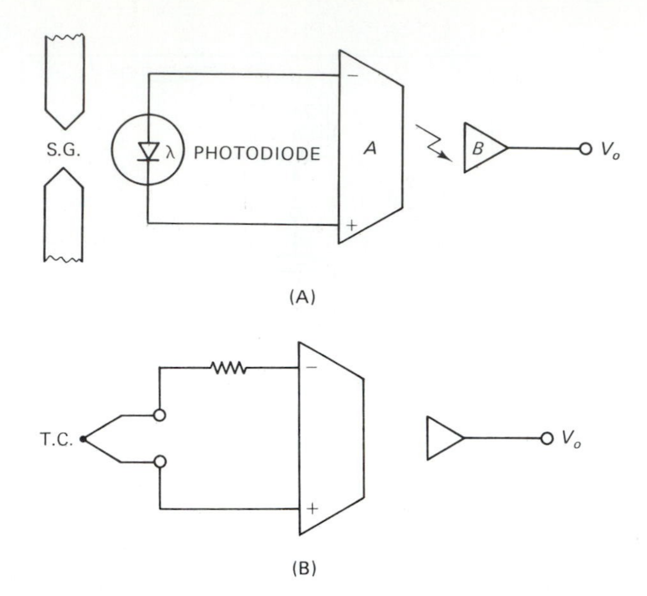


FIGURE 7-9

to an analog paper-chart recorder. Analysis of the waveshape reveals errors of technique, and thus explains odd readings not supported by other clinical facts. For this reason the physician demands an analog output. The temperature signal ($V_o$) is integrated and then sent to an analog divider where it is combined with the temperature difference signal $[T(b)-T(i)]$ and the constants (all represented by a single voltage). The low-pass filtered output of the analog multiplier is a measure of the cardiac output, and is displayed on a digital voltmeter.

## 7-4.2 Sensor Isolation Applications

Medical applications of isolation amplifiers are perhaps the best known because they greatly contribute to patient safety. But there is also a substantial body of nonmedical applications for isolation amplifiers. These amplifiers are especially useful whenever the environment (high voltages) can adversely affect the device or a conventional amplifier can perturb the signal source (certain transducers fall into this category). Figure 7-9 shows two situations where an isolation amplifier is interfaced to different forms of sensor. In Fig. 7-9A the sensor is a photodiode used to detect the existence of an electrical arc across a spark gap (SG). In some such systems strong electric fields can destroy conventional ground-referenced systems.

Another sensor application is shown in Fig. 7-9B. A *thermocouple* (TC) is a temperature transducer consisting of two dissimilar metals forming a junction. If the metals have different work functions, an electric potential is generated across the ends of the wires which is proportional to both the work function difference and the junction temperature. Ordinarily, the TC can be connected to any amplifier. But in certain cases the local electrical environment is too adverse for the electronic components. An example is a high voltage melting vessel for smelting metals. It is necessary to know and control the temperature, but the vessel floats electrically at a high AC potential.

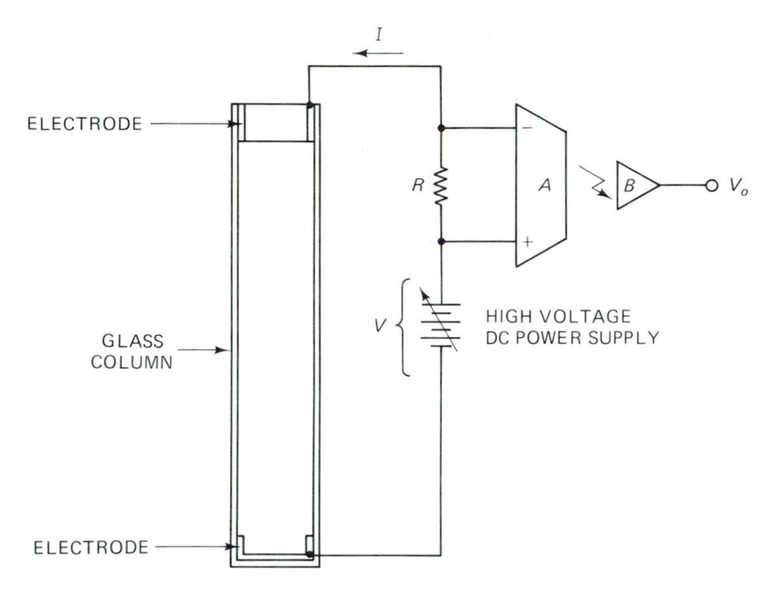


FIGURE 7-10

### 7-4.3 Electrophoresis Column

Electrophoresis is a process used in biochemistry laboratories to separate proteins from biological fluids. In an electrophoresis column (Fig. 7-10), a pair of platinum electrodes connected to a constant-current, high voltage DC power supply creates an electrical field between the top and bottom of the column. Various protein molecules within the fluid react specifically to the field and will migrate to and settle at a specific level in the field (thus height within the column). The chemist can then extract a particular protein if the height of the column and the potential gradient are known.

It is necessary to continuously monitor the current flowing in the column. If computer data logging is used, an isolation amplifier may be required to keep the high voltage of the circuit from the computer. A series resistor is inserted into the column circuit and the voltage drop across it is proportional to the current flowing (Ohm's law). The differential inputs of the isolated side of the amplifier are bridged across this resistor to indirectly measure the current drain.

### 7-4.4 A Commercial Product

Figure 7-11 shows the circuit of an isolation amplifier based on the Burr-Brown 3652 device.

The DC power for both the isolated and nonisolated sections of the 3652 is provided by the 722 dual DC-to-DC converter. This device produces two independent ± 15 Vdc supplies which are isolated from the 60-Hz AC power mains and from each other. The 722 device is powered from a +12 Vdc source derived from the AC power mains. In some cases, the nonisolated section (which is connected to the output terminal) is powered from a bipolar DC power supply derived from the 60-Hz AC mains, such

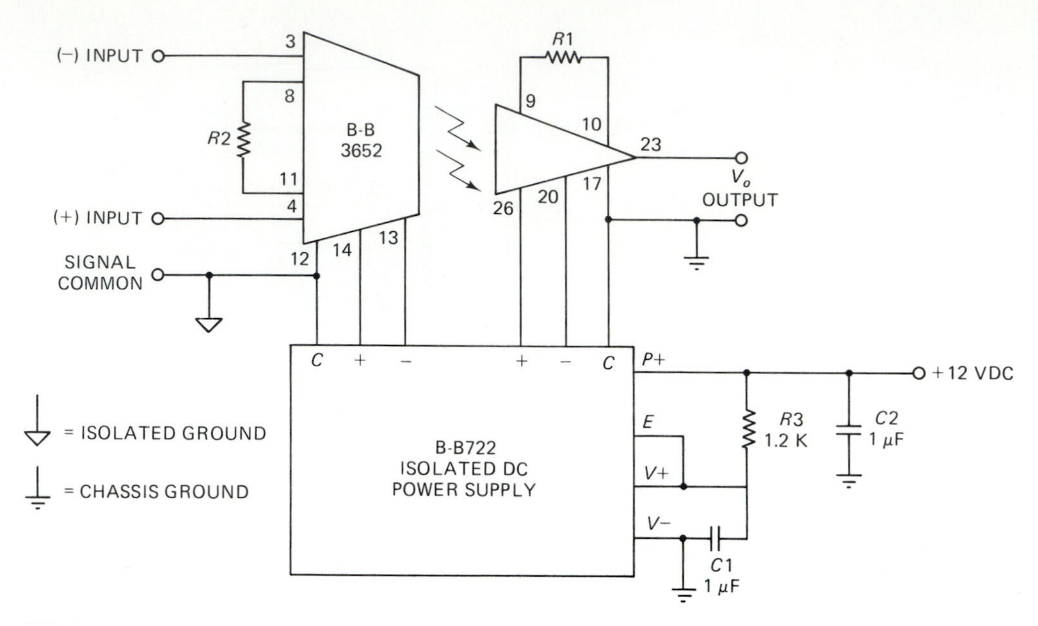


FIGURE 7-11

as a ± 12 Vdc or ± 15 Vdc supply. In no instance, however, should the isolated DC power supplies be derived from the AC power mains.

There are two separate ground systems in this circuit, symbolized by the small triangle and the regular three-bar chassis ground symbol. The isolated ground is not connected to either the DC power supply ground/common, or the chassis ground. It is kept floating at all times and becomes the signal common for the input signal source.

The gain of the circuit is approximately

$$\text{GAIN} = \frac{1{,}000{,}000}{R1 + R2 + 115} \tag{7-3}$$

In most design cases, the issue is the unknown values of the gain setting resistors. We can rearrange the equation above to solve for $(R1 + R2)$

$$(R1 + R2) = \frac{1{,}000{,}000 - (115 \times \text{GAIN})}{\text{GAIN}} \tag{7-4}$$

where

$R1$ and $R2$ are in ohms

GAIN is the voltage gain desired

### EXAMPLE 7-2

An amplifier requires a differential voltage gain of 1,000. What combination of $R1$ and $R2$ will provide that gain figure?

**Solution**

If GAIN = 1000

$$(R1 + R2) = \frac{1{,}000{,}000 - (115 \times 1000)}{1000}$$

$$(R1 + R2) = \frac{1{,}000{,}000 - (115{,}000)}{1000}$$

$$(R1 + R2) = \frac{885{,}000}{1000}$$

$$(R1 + R2) = 885 \text{ ohms}$$

In this case, we need some combination of $R1$ and $R2$ which adds up to 885 ohms. The value 440 ohms is standard and will result in only a tiny gain error if used. ■

## 7-5 CONCLUSION

Although the isolation amplifier is considerably more expensive than a common IC linear amplifier, there are applications where these amplifiers are absolutely critical. Wherever the instrument could cause injury to a human, or wherever electronics must be isolated as much as possible, the isolation amplifier is the device of choice, at least in the frontend.

## 7-6 SUMMARY

1. The purpose of isolation amplifiers is to increase the resistance between the inputs and either the output or the AC power lines to as high a number as possible. Isolation impedances on the order of $10^{12}$ ohms are possible.
2. There are four basic approaches to the design of isolation amplifiers, *battery-powered*, *carrier-operated*, *optically coupled* and *current-loading*.
3. Typical examples of isolation amplifier applications include: medical biopotentials amplifiers (ECG, EEG, and EMG), medical cardiac output computers, intra-aortic pressure meters, sensor or transducer isolation, and isolation of process control electronics from harsh or hostile electrical environments.

## 7-7 RECAPITULATION

Now return to the objectives and Prequiz questions at the beginning of the chapter and see how well you can answer them. If you cannot answer certain questions, mark them and review the appropriate parts of the text. Next, try to answer the questions and work the problems below, using the same procedure.

## 7-8 STUDENT EXERCISES

[NOTE: *In the exercises below limit the −3 dB frequency response of the analog amplifier to a range of 1 to 100 Hz.*]

1. Design, build, and test an isolation amplifier based on optical coupling. Assume the analog input signal will directly modulate the phototransmitter.
2. Design an isolation amplifier based on fiber optic coupling. State in your laboratory report how this design might be superior to others in certain applications.
3. Design, build, and test an isolation amplifier based on optical coupling. In this exercise use a voltage controlled oscillator at some convenient frequency in order to frequency modulate the light beam.
4. Design, build, and test carrier-operated isolation amplifier. In order to use ordinary audio components, limit the carrier frequency to 20 KHz.

## 7-9 QUESTIONS AND PROBLEMS

1. Define "isolation amplifier."
2. List several applications for isolation amplifiers and explain why each requires the isolation factor.
3. What is the goal of isolation amplifier design?
4. List the connections typically found on an isolation amplifier which uses differential inputs.
5. List four methods or approaches for designing an isolation amplifier.
6. An amplifier has isolated input and output sections, but uses the same common or ground connection for both sections. Is this amplifier isolated when powered from 110 volt AC power lines?
7. A synchronous detector is used on a ____________ -operated isolation amplifier.
8. A Burr-Brown 3652 is used in an application requiring an isolation amplifier. What is the gain if both gain-setting resistors are 600 ohms?
9. Use a Burr-Brown 3652 to design an isolation amplifier frontend for a cardiac output computer. Select resistor values that will yield a gain of 400 for the input amplifier.
10. In a carrier-type isolation amplifier, a voltage controlled oscillator is ____________ modulated by the input signal before being applied to an optocoupler.
11. Why are isolation amplifiers used in the "frontends" of ECG preamplifiers and cardiac output computers?
12. A special cardiac output computer test fixture simulates the thermistor catheter resistance change representing a 20°C temperature change. Calculate the cardiac output reading that would be expected if the temperature change is in the circuit for 7 seconds, and the following constants are entered into the computer: $C(t) = 64.5$, $T(b) = 37$, $T(i) = 20$, and $V(i) = 20$ ml.
13. Why are isolation amplifiers sometimes used in nonmedical instrumentation applications?

CHAPTER 8

# Nonoperational IC Linear Amplifiers

## OBJECTIVES

1. Learn basic theory for the operational transconductance amplifier (OTA).
2. Learn typical applications for the OTA.
3. Learn basic theory for the current difference (Norton) amplifier (CDA).
4. Learn typical applications for the CDA.

## 8-1 PREQUIZ

These questions test your prior knowledge of the material in this chapter. Try answering them before you read the chapter. Look for the answers (especially those you answered incorrectly) as you read the text. After you have finished studying the chapter, try answering these questions again and those at the end of the chapter (see Section 8-8).

1. Write the transfer equation for an operational transconductance amplifier.
2. Write the transfer equation for a current difference amplifier.
3. How can an OTA be used as a voltage amplifier?
4. Design a low output impedance OTA voltage amplifier circuit.

## 8-2 INTRODUCTION

In this chapter we will discuss two popular IC linear amplifiers, the *operational transconductance amplifier* (OTA) and the *current difference amplifier* (CDA), also called the *Norton amplifier.* These devices are not likely to replace the operational amplifier, but they serve their purpose in the integrated electronics marketplace.

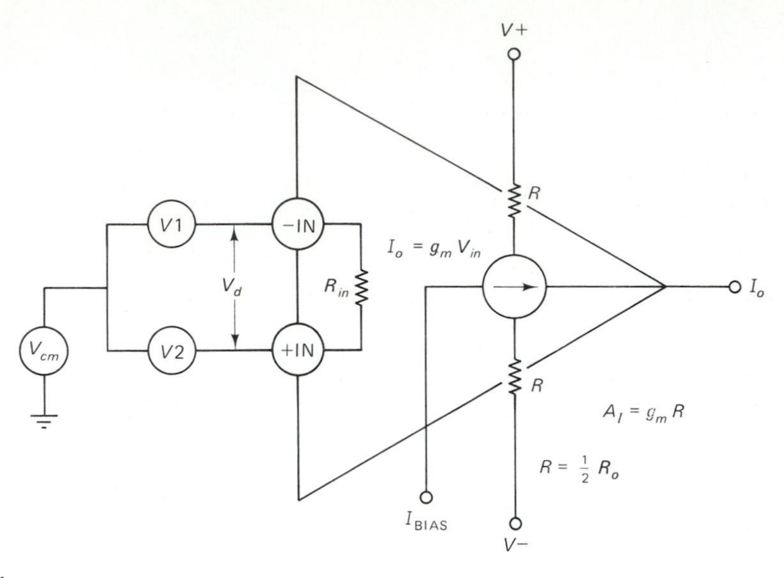


FIGURE 8-1

## 8-3 OPERATIONAL TRANSCONDUCTANCE IC AMPLIFIERS (OTA)

The OTA's transfer function *relates an output current to an input voltage.* In other words

$$G_m = \frac{I_o}{V_{\text{in}}} \tag{8-1}$$

where

$G_m$ = the transconductance in mhos or micromhos

$I_o$ = the output current

$V_{\text{in}}$ = the input voltage

The operational transconductance amplifier equivalent circuit is shown in Fig. 8-1. The differential input circuit is similar to the input circuit of the operational amplifier because each are differential voltage inputs (−IN and +IN). The input voltages are differential signals ($V_d$) ($V2 - V1$) and common mode signals ($V_{cm}$). The output side of the amplifier, however, is a *current source* which produces an output current ($I_o$). This is proportional to the gain and the input voltage. The current gain ($A_{gm}$) of this circuit is a function of the transconductance ($I_o/V_{\text{in}}$) and the load resistance ($R$)

$$A_{gm} = G_m \times R \tag{8-2}$$

where

$A_{gm}$ = the gain

$G_m$ = the transconductance ($I_o/V_{in}$)

$R$ = the load resistance (one half the output resistance $R_o$)

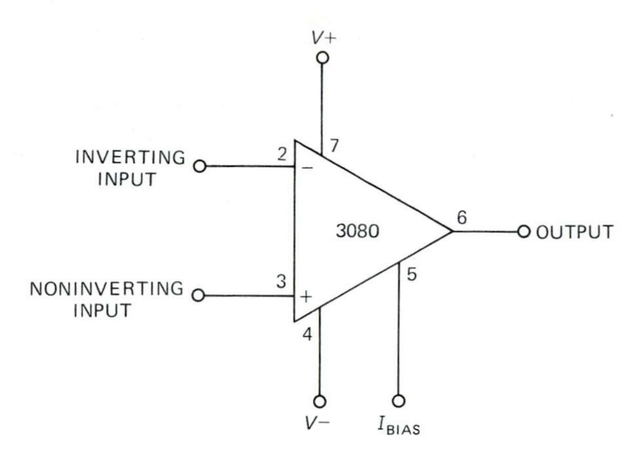


FIGURE 8-2

In this equation, $G_m$ and $R$ must be expressed in equivalent reciprocal units. In other words, when $G_m$ is in mhos or Siemens, (where 1 S = 1 Mho), $R$ is in ohms. Likewise, millimhos and milliohms, micromhos and microohms are also paired.

The most common commercial version of OTAs are the RCA CA-3080, CA-3080A, and CA-3060 devices. The CA-3080 device is available in the eight-pin metal IC package using the pinouts shown in Fig. 8-2. It will operate over DC power supply voltages from ± 2 volts to ± 15 volts, with adjustable power consumption of 10 microwatts to 30 milliwatts. The gain is 0 to the product $G_mR$. The input voltage spread is ± 5 volts. The bias current can be set as high as 2 mA.

The pinouts for the CA-3080 device are industry standard operational amplifier pinouts, except for the bias current applied to pin number 5:

$V-$ on pin 4

$V+$ on pin 7

Inverting input (−IN) on pin 2

Noninverting input (+IN) on pin 3

Output on pin 6

The operating parameters of the operational transconductance amplifier are set by the bias current ($I_{\text{bias}}$). For example, on the CA-3080 the transconductance is 19.2 times higher than the bias current:

$$G_m = 19.2 \times I_{\text{bias}} \tag{8-3}$$

where

$$G_m = \text{in millimhos}$$

$$I_{\text{bias}} = \text{in milliamperes}$$

In many actual design cases you will know the required value of $G_m$ from $I_o/V_{\text{in}}$, so it can be set by adjusting the bias current. In those cases, the $I_{\text{bias}}$ is found by rewriting the above expression

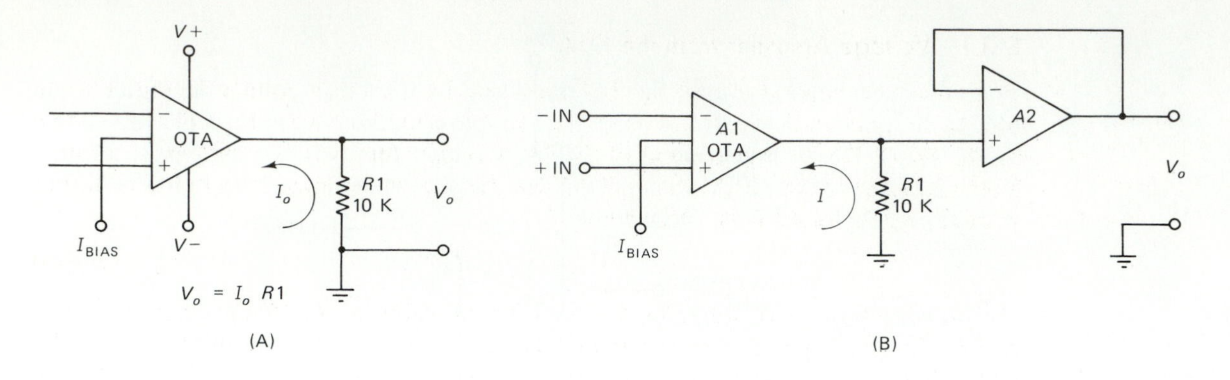


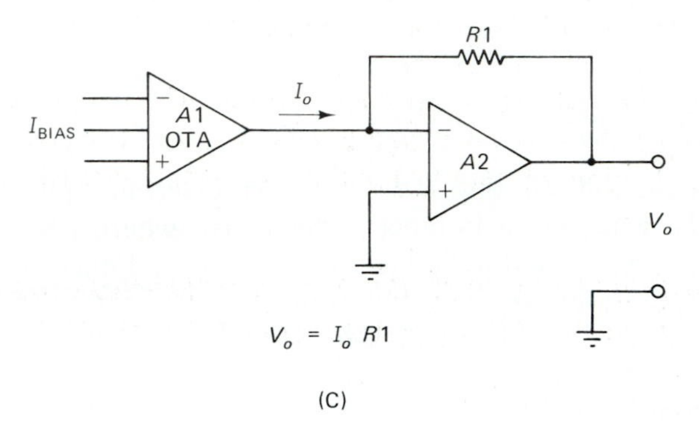


FIGURE 8-3

$$I_{\text{bias}} = \frac{G_m}{19.2} \tag{8-4}$$

The output resistance of the CA-3080 is also a function of the bias current

$$R_o = \frac{7.5}{I_{\text{bias}}} \tag{8-5}$$

where

$R_o$ = the output resistance in megohms

$I_{\text{bias}}$ = the bias current in milliamperes (mA)

### EXAMPLE 8-1

What is the output resistance ($R_o$) when the bias current is 500 μA (0.5 mA)?

### Solution

$$R_o = 7.5/I_{\text{bias}}$$

$$R_o = 7.5/0.5 = 15 \text{ megohms}$$

■

### 8-3.1 Voltage Amplifier from the OTA

The OTA is a current-output device, but it can be used as a voltage amplifier when one of the circuits in Fig. 8-3 is used. The simplest method is the resistor load shown in Fig. 8-3A. Because the output of the OTA is a current ($I_o$), we can pass it through a resistor ($R1$) to create a voltage drop. The value of the voltage drop (and the output voltage, $V_o$) is found from Ohm's law

$$V_o = I_o \times R1 \tag{8-6}$$

The very high source impedance, equal to the value of $R1$, is a problem with this circuit. In Fig. 8-3A, the output impedance is 10 kohms. We can overcome this by adding a unity gain noninverting operational amplifier such as $A2$ in Fig. 8-3B. The output voltage in this case is the same as that in the nonamplifier version ($I_o \times R1$), but the output impedance is very low.

Another form of a low-impedance output circuit is shown in Fig. 8-3C. This uses an inverting follower operational amplifier ($A2$). The output voltage is the product of the OTA output current ($I_o$) and the op-amp feedback resistor ($R1$)

$$V_o = I_o \times R1 \tag{8-7}$$

Both circuits in Figs. 8-3B and 8-3C have an output impedance equal to the operational amplifier output impedance, which is typically less than 100 ohms.

### 8-3.2 OTA Applications

#### Wideband Unity Gain Buffer

Figure 8-4 shows the circuit for a unity gain voltage follower based on the CA-3080 OTA device. The circuit is designed for operation with the standard 50-ohm input/output impedances common to "RF" amplifiers. This circuit is typical of those used for buffering and isolation in applications operating to about 300 KHz.

The OTA is prevented from oscillation by the use of a shunt capacitor ($C2$) across the feedback resistor (which also sets the gain), and by the input lag network ($R3/C1$) between the inverting and noninverting inputs of the CA-3080.

#### Analog X-Y Multiplier

An analog multiplier circuit produces a voltage which is the product of two input voltages

$$V_o = r\, V_x\, V_y \tag{8-8}$$

where

$V_o$ = the output voltage

$V_x$ = the voltage applied to the X input

$V_y$ = the voltage applied to the Y input

$r$ = a proportionality constant

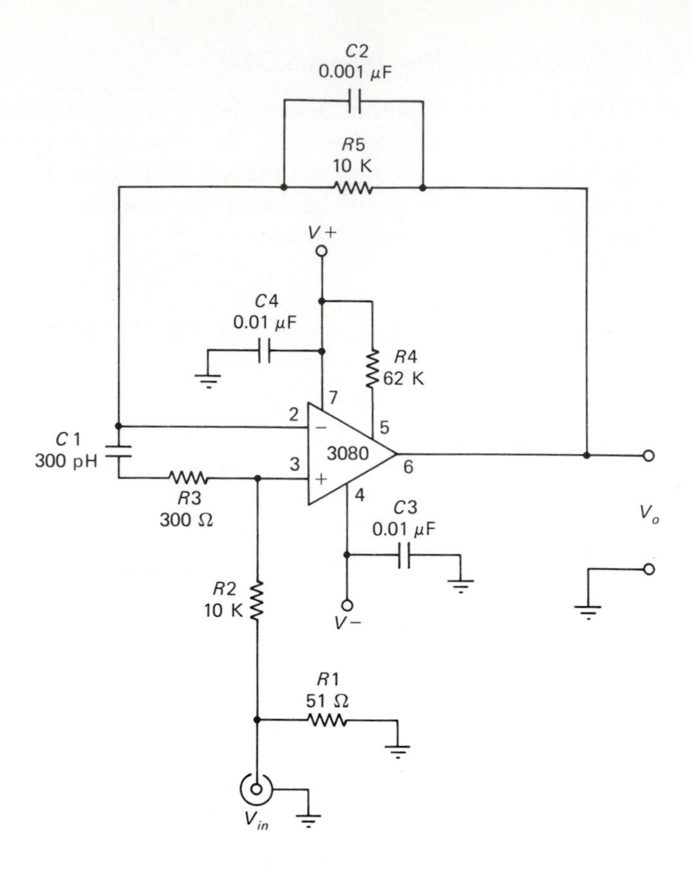


FIGURE 8-4

There are many applications for the multiplier circuit, even though some of its functions are now performed by digital computers or processors. These are instrumentation applications, however, even in this era of computers. Also analog multipliers can perform amplitude modulation and demodulation tasks. Operational transconductance amplifiers can be used to make two-quadrant and four-quadrant analog multipliers. Consider Fig. 8-5, which is an analog X-Y multiplier based on the CA-3060 quad OTA. Recall from Eq. (8-1) that

$$G_m = \frac{I_o}{V_{\text{in}}}$$

therefore

$$I_{O1} = (-V_x)(G_{M1}) \tag{8-9}$$

and

$$I_{O2} = (+V_x)(G_{M2}) \tag{8-10}$$

Because the value of $R_o$ for each OTA is very large compared to the load, we can simply sum the two output currents

$$I_o = I_{O2} + I_{O1} \tag{8-11}$$

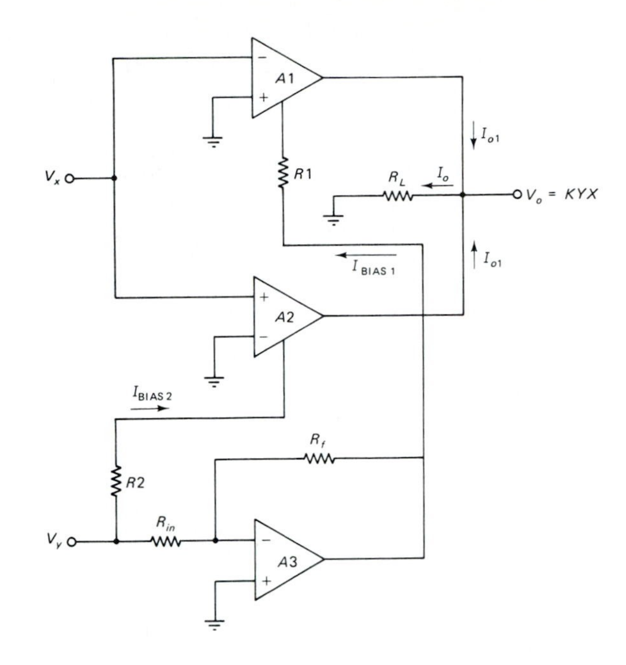


FIGURE 8-5

by $V_o = I_o R_L$

$$V_o = [I_{O2} + I_{O1}]R_L \tag{8-12}$$

$$V_o = [(+V_x)(G_{M2}) + (-V_x)(G_{M1})]R_L \tag{8-13}$$

$$V_o = (G_{M2} - G_{M1})V_x R_L \tag{8-14}$$

Recall $G_m = kI_{\text{bias}}$. For amplifier $A2$ we know that

$$I_{\text{bias}} = \frac{(V-) + (V_x)}{R1} \tag{8-15}$$

therefore

$$G_{M2} = K[(V-) + (V_x)] \tag{8-16}$$

Using a similar line of reasoning we arrive at

$$G_{M1} = K[(V-) + (V_y)] \tag{8-17}$$

Combining Eqs. (8-15), (8-16), and (8-17) yields

$$V_o = V_x K R_L[(V-) + (V_x)] - [(V-) - (V_y)] \tag{8-18}$$

or, after simplifying terms

$$V_o = 2\,K\,R_L\,V_x\,V_y \tag{8-19}$$

Equation (8-19) is the transfer equation for Fig. 8-5. It has the same form as Eq. (8-8) in which $r = 2KR_L$.

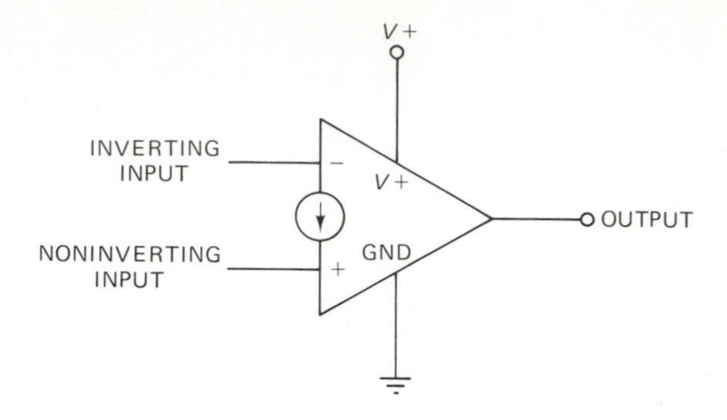


FIGURE 8-6

## 8-4 CURRENT DIFFERENCE AMPLIFIERS

The *current difference amplifier* (CDA), also called the *Norton amplifier*, is another nonoperational linear IC amplifier that performs similarly to the op-amp. The CDA is uniquely useful for certain applications. One place where the CDA is more useful than the operational amplifier is in processing AC signals where there is only a single polarity DC power supply. An example is automotive electronics equipment which is limited to a single 12 to 14.4 volt DC battery power supply and uses the car chassis for negative common return.

There are other cases where the linear IC amplifier is only a minor feature of the circuit, most of which operates from a single DC power supply. It might be wasteful in such circuits to use the operational amplifier. We would either have to bias the operational amplifier with an external resistor network or provide a second DC power supply.

The normal circuit symbol for the CDA is shown in Fig. 8-6. It looks like the regular op-amp symbol, except a "current source" is placed along the side opposite the apex. This symbol is used for several products such as the National Semiconductor LM-3900 which is a quad Norton amplifier. You may sometimes find schematics where the op-amp symbol is used for the CDA. But this is technically incorrect.

### 8-4.1 CDA Circuit Configuration

The input circuit of the CDA differs radically from the operational amplifier. Recall the op-amp used a differential input amplifier driven from a constant current source supplying the collector-emitter current. The CDA is quite different, however, as can be seen in Fig. 8-7.

The overall circuit of a typical CDA is shown in Fig. 8-7A and an alternate form of the input circuit is shown in Fig. 8-7B. Transistor *Q*7 in Fig. 8-7A forms the output transistor and *Q*5 is the driver. Both the NPN output transistor and the PNP driver transistor operate in the emitter follower configuration. Transistors *Q*4, *Q*5, *Q*6, and *Q*8 are connected to serve as current sources. The input transistor is *Q*3. It operates in the common emitter configuration. The base of *Q*3 forms the inverting (−IN) input for the CDA.

The noninverting input of the CDA is formed with a current mirror transistor, *Q*1

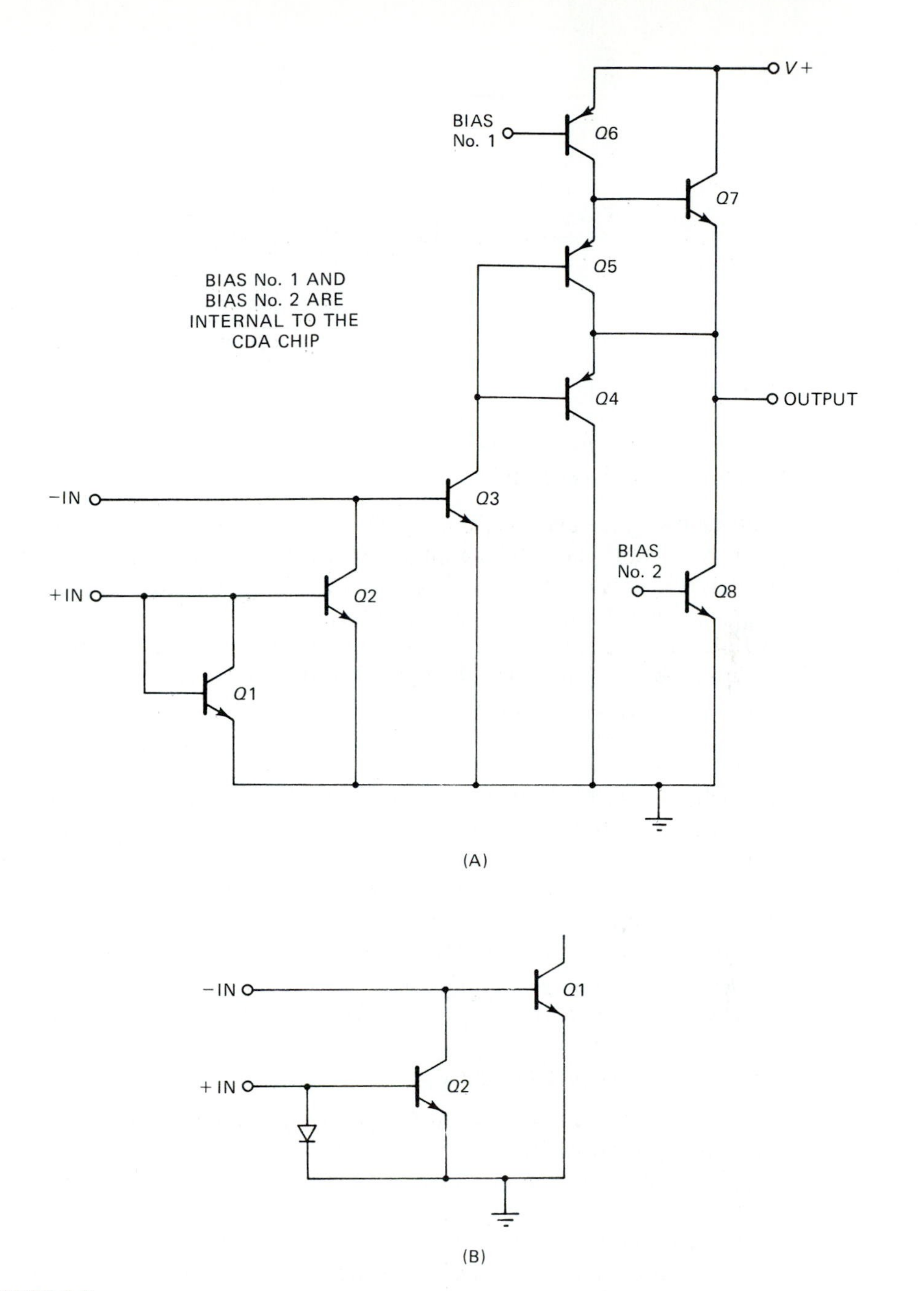


FIGURE 8-7

(transistor $Q1$ in Fig. 8-7A is diode-connected and serves the same function as diode $D1$ in Fig. 8-7B). The dynamic resistance offered by the current mirror transistor ($Q2$) is given by

$$r = 26/I_b \tag{8-20}$$

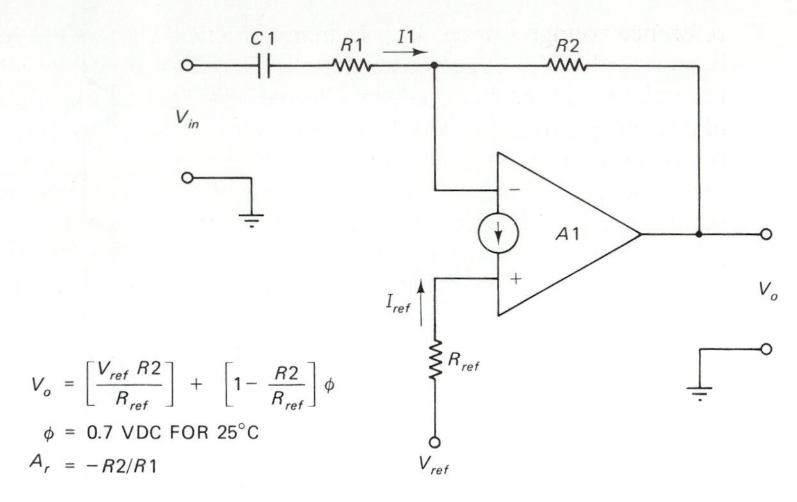


FIGURE 8-8

where

$$r = \text{the dynamic resistance of } Q2 \text{ in ohms}$$

$$I_b = \text{the base bias current of } Q3 \text{ in milliamperes}$$

Equation (8-20) is only used at normal room temperature because $I_b$ will vary with wide temperature excursions. For most common applications, however, the room temperature version of the equation will suffice. Data sheets for specific current difference amplifiers give additional details for amplifiers that must operate outside the relatively narrow temperature range specified for the simplified equation.

### 8-4.2 CDA Inverting Follower Circuits

Like the operational amplifier, the CDA can be configured in either inverting or noninverting follower configurations. The inverting follower is shown in Fig. 8-8. In many respects this circuit is very similar to operational amplifiers. The voltage gain of the circuit is set approximately by the ratio of the feedback to the input resistor:

$$A_v = -R2/R1 \tag{8-21}$$

where

$$A_v = \text{the voltage gain}$$

$$R2 = \text{the feedback resistance}$$

$$R1 = \text{the input resistance}$$

$R1$ and $R2$ are in the same units. The minus sign indicates a 180° phase reversal occurs between input and output signals.

We must provide a bias to the current mirror transistor ($Q2$ in Fig. 8-7A), so resistor $R_{ref}$ is connected in series with the noninverting input of the CDA and a reference voltage source, $V_{ref}$. In many practical circuits the reference voltage source is merely the $V+$ supply used for the CDA. In other cases, however, some other potential might be required, or alternatively the reference current needs to be regulated more tightly (or with less noise) than the supply voltage. Ordinarily we set the reference current to some convenient value between 5 μA and 100 μA. For $V+$ power supply values of +12 VDC, for example, it is common to use a 1 megohm resistor for $R_{ref}$. In that case, the input reference current is $I_{ref} = 12/1{,}000{,}000 = 12$ μA.

A limitation of CDAs is that the input resistor ($R1$) used to set gain must be high compared to the value of the current mirror dynamic resistance [Eq. (8-21)]. The CDA becomes nonlinear (distorts the input signal) if the input resistor value approaches the current mirror resistance, $r$. In that case, the voltage gain is not $-R2/R1$, but rather

$$A_v = \frac{R2}{R1 + r} \tag{8-22}$$

where

$A_v$ = the voltage gain

$R2$ = the feedback resistance

$R1$ = the input resistance

$r$ = the current mirror resistance

Equation (8-22) essentially reduces to Eq. (8-21) when we force $R1$ to be much larger than $r$. This is easily achieved in most circuits because $r$ is small. To test this, work a few examples with normal bias currents in Eq. (8-20).

The output voltage of the CDA will exhibit an offset potential even when the AC input signal is zero. This potential is given by

$$V_o = \left[\frac{V_{ref}\, R2}{R_{ref}}\right] + \left[1 - \frac{R2}{R_{ref}}\right] \Psi \tag{8-23}$$

where

$V_o$ = the output potential in volts

$V_{ref}$ = the reference potential in volts (usually $V+$)

$R2$ = the feedback resistance in ohms

$R_{ref}$ = the current mirror bias resistance in ohms

$\Psi$ = a temperature dependent factor (0.70 volt for room temperature)

The capacitor in series with the input circuitry limits the low-end frequency response. The −3 dB cutoff frequency is a function of the value of this capacitor and

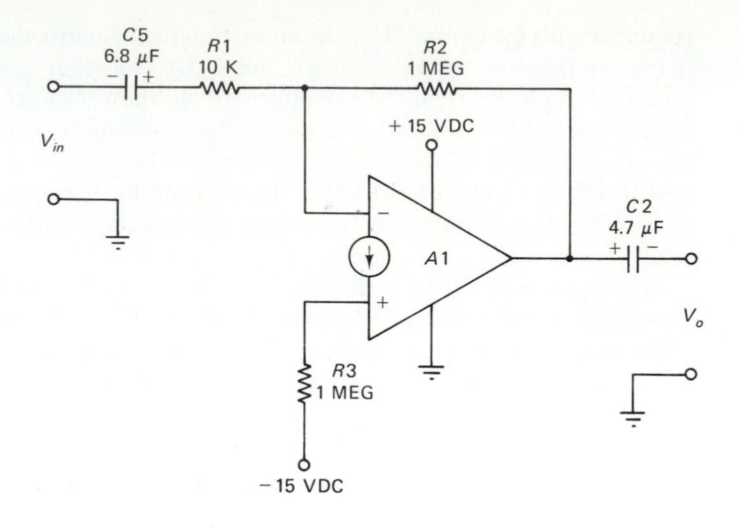


FIGURE 8-9

the input resistance, $R1$. This frequency, $F$, is given by

$$F = \frac{1{,}000{,}000}{2\,\pi\,R1\,C1} \tag{8-24}$$

where

$F$ = the lower end $-3$ dB frequency in hertz (Hz)

$R1$ = in ohms

$C1$ = in microfarads

In some CDA circuits there is also a capacitor in series with the output terminal. This prevents the DC offset, inherent in this type of circuit, from affecting following circuits. The output capacitor will also limit the low-end frequency response. Equation (8-24) is used to determine this frequency, but the input resistance of the load is the $R$ term.

In most cases, you will know input resistance ($R1$) from the application. It is typically not less than ten times the source impedance, and is part of the gain equation. The driving source impedance and voltage gain will determine the value of $R1$. The required low-end frequency response is usually determined from the application. You will generally know or can find out the frequency spectrum of input signals. From the lower limit of the frequency spectrum you can determine the value of $F$.

Thus, you determine $F$ and $R1$ from other sources than the circuit. You therefore need a version of Eq. (8-24) that will provide the value of capacitor $C1$

$$C1 = \frac{1{,}000{,}000}{2\,\pi\,R1\,F} \tag{8-25}$$

## DESIGN EXAMPLE

Design a gain of 100 AC amplifier based on a CDA in which the input impedance is at least 10 kohms and the $-3$ dB frequency response is 3 Hz or lower. Assume DC power supplies of $\mp$ 15 volts DC (see Fig. 8-9 for the final circuit).

1. Set the reference current to the noninverting input to between 5 and 100 μA. Select 15 μA.

$$R3 = V/I_{\text{ref}}$$

$$R3 = (15 \text{ Vdc})/(0.000015) = 1 \text{ megohm}$$

2. Set the gain resistors. $R1$ can be 10,000 ohms to meet the input impedance requirement. From $A_v = R2/R1$, we know that $R2 = A_v \times R1$.

$$R2 = A_v \times R1$$

$$R2 = (100) \times (10{,}000) = 1 \text{ megohm}$$

3. Find the value of input capacitor $C1$ when $F_{-3\text{ dB}}$ is 3 Hz.

$$C1 = \frac{1{,}000{,}000}{2\ \pi\ R1\ F}$$

$$C1 = \frac{1{,}000{,}000}{(2)(3.14)(10{,}000)(3)}$$

$$C1 = \frac{1{,}000{,}000}{188{,}400} = 5.3\ \mu\text{F}$$

■

Because 5.3 μF is a nonstandard value, we select the next higher standard value (6.8 μF).

The value of output capacitor $C2$ can be arbitrarily set to 6.8 μF if the load impedance is 10,000 ohms. If the load impedance is higher, use 4.7 μF or 6.8 μF. If the load impedance is much higher or lower than 10,000 ohms, calculate the value using the same equation for $C1$ but with the load resistance substituted for the input impedance.

### 8-4.3 Noninverting Amplifier Circuits

The noninverting amplifier CDA configuration is shown in Fig. 8-10. This circuit retains the reference current bias applied to the noninverting input, but some of the other components are rearranged. As in the case of the inverting amplifier, the noninverting amplifier uses $R2$ to provide negative feedback between the output terminal and the inverting input. Unlike the inverting CDA circuit, however, input resistor $R1$ is connected in series with the noninverting input. The gain of the noninverting CDA amplifier is given by

$$A_v = \frac{R2}{\left[\dfrac{26\ R1}{I_{\text{ref}}\ (\text{mA})}\right]} \tag{8-26}$$

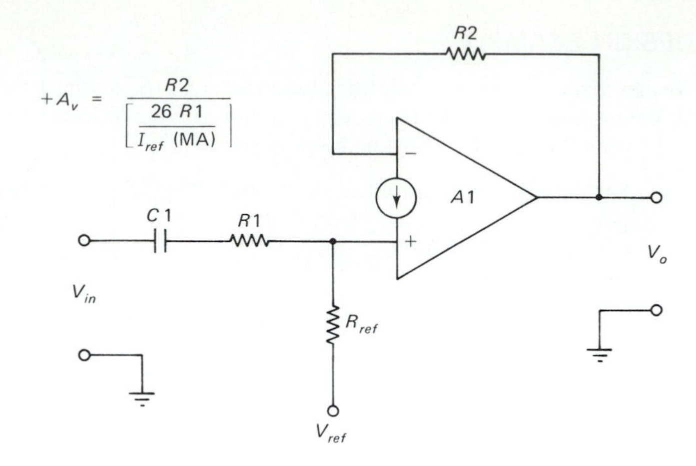


FIGURE 8-10

where

$A_v$ = the voltage gain

$R1$ and $R2$ are in ohms

$I_{\text{ref}}$ = the bias current in milliamperes (mA)

The reference current ($I_{\text{ref}}$) is set to a value between 5 μA and 100 μA (0.005 to 0.1 mA). Unlike the inverting amplifier, the value of this current is partially responsible for setting the gain of the circuit. Some clever designers have even used this current as a limited gain control for some CDA stages. The value of the resistor which provides the reference current ($R_{\text{ref}}$) is set by Ohms law, considering the required value of reference current and the reference voltage ($V_{\text{ref}}$). In most common applications, the reference voltage is merely one of the supply voltages. The value of $R_{\text{ref}}$ is determined from

$$R_{\text{ref}} = \frac{V_{\text{ref}}}{I_{\text{ref}}} \tag{8-27}$$

where

$R_{\text{ref}}$ = the reference resistor in ohms

$V_{\text{ref}}$ = the reference potential in volts

$I_{\text{ref}}$ = the reference current in amperes (1 μA = 0.000001 ampere)

The input impedance is approximately equal to $R1$, provided $R1$ is much higher than the dynamic resistance of the current mirror inside the CDA. This is typically the case.

As was true in the inverting follower, the input capacitor ($C1$) sets the low-end frequency response of the amplifier. The −3 dB frequency is given by the same equation used for the inverting case [see Eqs. (8-24) and (8-25)].

Figure 8-11 shows a modification of the noninverting follower circuit which limits noise in the reference source. This might be used where the DC power supplies for

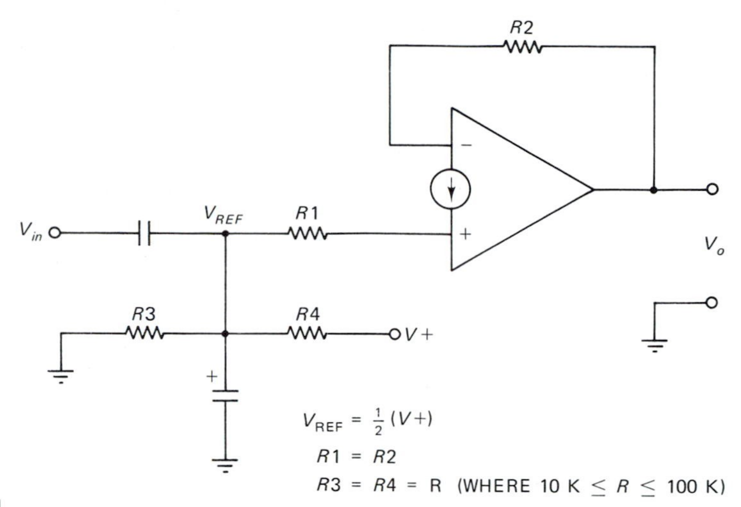


FIGURE 8-11

reference voltage are electrically noisy. Such noise could come from other stages in the circuit or from outside sources.

The circuit in Fig. 8-11 forms a reference voltage from a resistor voltage divider circuit consisting of $R3$ and $R4$. The value of $V_{\text{ref}}$ will be

$$V_{\text{ref}} = \frac{(V+)\,R3}{R3 + R4} \tag{8-28}$$

where

$V_{\text{ref}}$ = the reference potential in volts

$V+$ = the supply potential in volts

$R3$ and $R4$ are in ohms

In Eq. (8-28) $V_{\text{ref}} = (V+)/2$ when $R3 = R4$, which is usually the case in practical circuits. The reference current is

$$I_{\text{ref}} = \frac{V_{\text{ref}}}{R1} \tag{8-29}$$

### 8-4.4 Supergain Amplifier

There is a practical limit to voltage gain using standard resistor values and standard circuit configurations (this is also a problem in operational amplifiers). In Fig. 8-12 we see a method to overcome this limitation. This supergain amplifier circuit forms a noninverting follower similar to the earlier circuit, except feedback resistor $R2$ is driven from an output voltage divider network rather than directly from the output terminal of the CDA. The voltage gain of the circuit of Fig. 8-12 is given by

$$A_v = \left[\frac{R2}{R1}\right]\left[\frac{R3 + R4}{R3}\right] \tag{8-30}$$

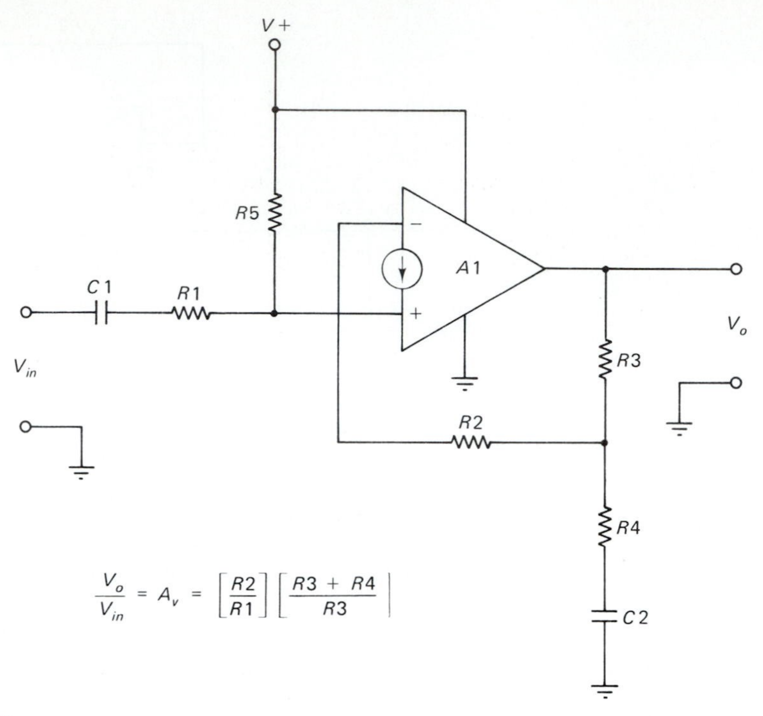


FIGURE 8-12

## EXAMPLE 8-2

Find the gain of the circuit shown in Fig. 8-7 if $R1 = 10$ kohms, $R2 = 100$ kohms, $R3 = 10$ kohms and $R4 = 100$ kohms.

### Solution

$$A_v = \left[\frac{R2}{R1}\right]\left[\frac{R3 + R4}{R3}\right]$$

$$A_v = \left[\frac{(100\text{k})}{10\text{k}}\right]\left[\frac{(100\text{k}) + (10\text{k})}{(10\text{k})}\right]$$

$$A_v = (10) \times (110/10)$$

$$A_v = (10) \times (11) = 110$$

■

Capacitor $C1$ is set using Eq. (8-25) and $C2$ is set to have a capacitive reactance of $R4/10$ at the lowest frequency of operation (in other words, the low end $-3$ dB point).

### 8-4.5 CDA Differential Amplifiers

A differential amplifier will produce an output proportional to the gain with the difference between potentials applied to the inverting and noninverting inputs. Figure 8-13 shows the circuit of a CDA differential amplifier. It is similar to the operational

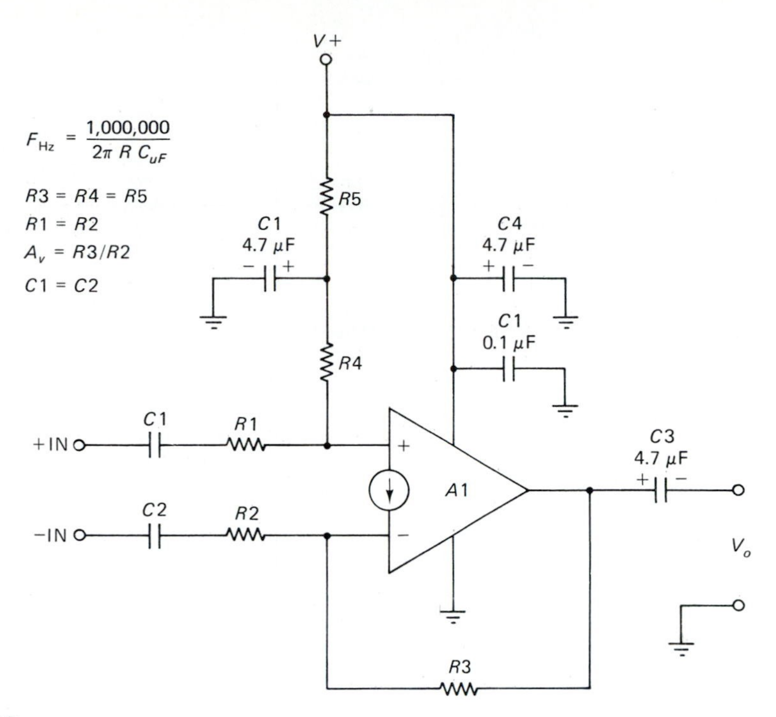


FIGURE 8-13

amplifier version of this simple circuit. For example, the two input resistors are equal. The differential voltage gain is the ratio of the negative feedback resistor ($R3$) and the input resistor

$$A_v = \frac{R3}{R2} \tag{8-31}$$

If

$$R1 = R2$$

$$R3 = R4 = R5$$

The input impedance (differential) of this circuit is twice the value of the input resistances ($R_{in} = R1 + R2$). The bias current is provided through the two series resistors, $R4$ and $R5$.

Assuming that $R1 = R2 = R$ and $C1 = C2 = C$, we can calculate the low-end $-3$ dB frequency from the equation

$$F = \frac{1{,}000{,}000}{2\ \pi\ R\ C} \tag{8-32}$$

where

$F$ = the $-3$ dB frequency in hertz (Hz)

$R$ = resistance in ohms

$C$ = capacitance in microfarads (μF)

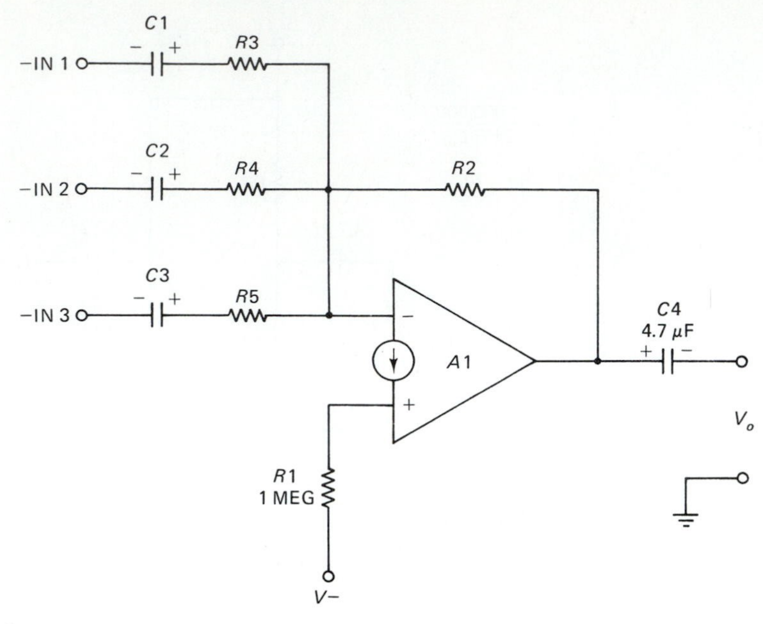


FIGURE 8-14

When this circuit is used as a 600-ohm line receiver, we can make the two input resistors 330 ohms each (or 270 ohms) if a small mismatch can be tolerated. Ideally, the input resistors will be 300 ohms each, which may require two resistors for each input resistor.

### 8-4.6 AC Mixer/Summer Circuits

The CDA mixer or summer circuit is shown in Fig. 8-14. It combines two or more inputs into one channel. The basic circuit is an inverting follower. Each input has a gain which is the quotient of the feedback resistor to its input resistor

$$A_{V1} = -R2/R3 \tag{8-33}$$

$$A_{V2} = -R2/R4 \tag{8-34}$$

$$A_{V3} = -R2/R5 \tag{8-35}$$

From Eqs. (8-33) through (8-35) we can deduce the output voltage is found from

$$V_o = R2\left[\frac{V1}{R3} + \frac{V2}{R4} + \frac{V3}{R5}\right] \tag{8-36}$$

The frequency response of each channel is found from the usual equation for −3 dB frequency

$$F = \frac{1{,}000{,}000}{2\,\pi\,R\,C}$$

where

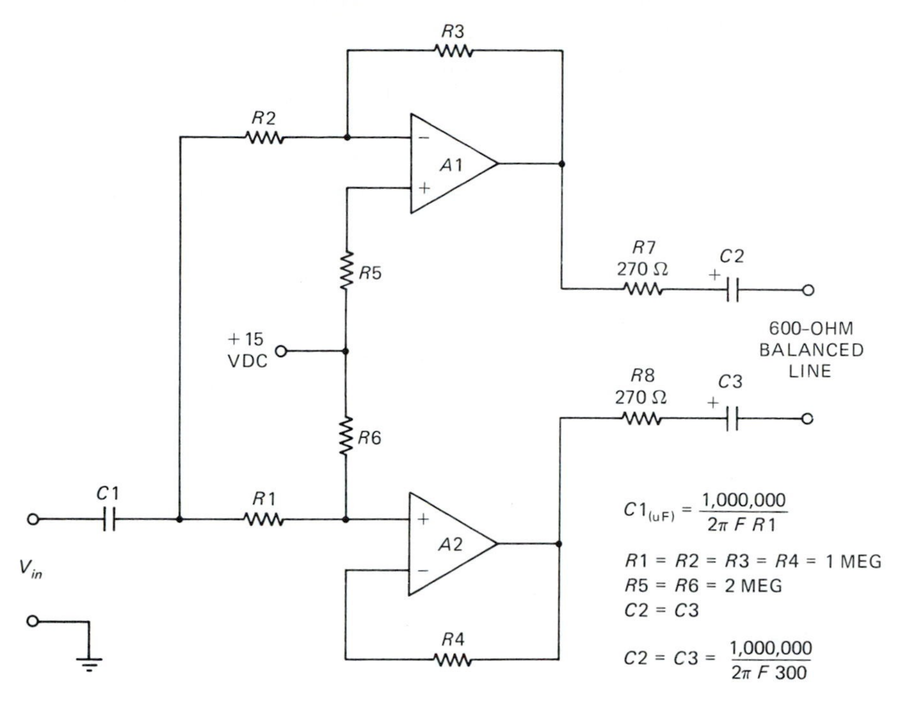


FIGURE 8-15

$F$ is the $-3$ dB frequency in hertz (Hz)

$R$ is the input resistance (R3, R4 or R5)

$C$ is the input capacitance (C1, C2 or C3)

### 8-4.7 Differential Output 600-OHM Line Driver Amplifier: A Circuit Example

The 600-ohm line used in broadcasting electronics and professional audio recording equipment requires either a center-tapped output transformer, or a linear amplifier with a push-pull output to drive the line. Figure 8-15 shows the circuit of a CDA 600-ohm line driver amplifier. This circuit consists of two separate amplifiers, an inverting follower and a noninverting follower.

The bias resistors ($R5$ and $R6$) are set to provide a small bias current from 5 to 100 μA. This current is found from Ohm's law, $I_{ref} = (V+)/R$. In the example shown in Fig. 8-15 the resistors are set to 2 megohms for a supply voltage of +15 volts DC.

The capacitors in the circuit set the low-end $-3$ dB point in the frequency response curve. These capacitor values are set from the following

Assuming that $R1 = R2 = R3 = R4$, and $R5 = R6$

$$C1 = \frac{1{,}000{,}000}{2\,\pi\,F\,R1} \tag{8-37}$$

Assuming $C2 = C3$

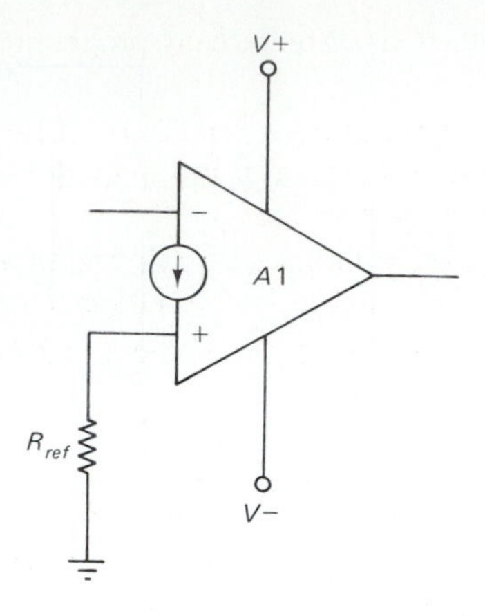


FIGURE 8-16

$$C2 = \frac{1{,}000{,}000}{2\ \pi\ F\ 300} \tag{8-38}$$

where

$C1$ and $C2$ are in microfarads

$F$ is in hertz (Hz)

$R1$ is in ohms

For best results a multiple CDA integrated circuit like the LM-3900 should be used for this application. Such a circuit control the drift because the two halves would both drift at the same rate (they share a common thermal environment).

### 8-4.8 Using Bipolar DC Power Supplies

The current difference amplifier is designed primarily for single-polarity power supply circuits. In most cases, the CDA will operate with a $V+$ DC power supply in which one side is grounded. We can, however, operate the CDA in a circuit with a bipolar DC power supply using a circuit such as Fig. 8-16. The reference resistor ($R_{ref}$) is connected from the noninverting input to ground. The $V-$ and $V+$ power supplies are each ground referenced and of equal potential. Thus, the 5 to 100 $\mu$A bias current is found from $(V+)/R_{ref}$.

## 8-5 SUMMARY

1. Operational transconductance amplifiers (OTA) have a transfer function which relates an output current to an input voltage. Thus, the units of gain in the OTA are units of transconductance (mhos, Siemens, and the subunits). Examples of OTA devices include the CA-3080 and CA-3060 (which is a quad OTA).
2. An OTA can be used as a voltage amplifier by virtue of the voltage drop caused by the output current in an external load resistance.

3. Gain on the OTA is externally programmed with a bias current applied to a special pin on the IC.
4. Current difference amplifiers (CDA), also known as Norton amplifiers, use current inputs but nonetheless have a transfer equation approximately equal to the ratio of input and feedback resistances.
5. The CDA is used especially where only one power supply is available, although bipolar DC power supplies can be accommodated.
6. An example of the CDA is the LM-3900 quad Norton amplifier offered by National Semiconductor.

## 8-6 RECAPITULATION

Now return to the objectives and Prequiz questions at the beginning of the chapter and see how well you can answer them. If you cannot answer certain questions, mark them and review the appropriate parts of the text. Next, try to answer the questions and work the problems below, using the same procedure.

## 8-7 STUDENT EXERCISES

1. Design a gain of 200 AC-coupled amplifier based on the LM-3900 CDA. Make the input impedance at least 10 kohms, and the $-3$ dB low-end frequency response equal to 10 Hz or less. Assume a DC power supply of $\mp$ 12 VDC.
2. Design a gain of 10 AC differential amplifier based on the CDA. Assume an input impedance of 10 kohms or more, and DC power supplies of $\mp$ 15 VDC.
3. Design a gain-of-50 voltage amplifier based on the CA-3080 operational transconductance amplifier.
4. Design an analog voltage multiplier based on an OTA.

## 8-8 QUESTIONS AND PROBLEMS

1. Write the generalized transfer equation for an operational transconductance amplifier (OTA).
2. What bias current is required on a CA-3080 OTA to create a transconductance of 30 mmhos?
3. Calculate the output resistance ($R_o$) of a CA-3080 if the bias current is 0.75 mA.
4. An OTA has an output load resistance of 10 kohms. When connected as a voltage amplifier an output potential of 5 volts is found across this resistance. Calculate the output current.
5. Draw a circuit diagram that will convert a current-output OTA to a voltage amplifier configuration with a low output impedance.
6. In a current difference amplifier (CDA) the base bias current of the input transistors is 0.25 mA. Calculate the approximate dynamic resistance ($r$) at room temperature.
7. Using the dynamic resistance calculated in the above problem, calculate the gain of a CDA voltage amplifier in which the feedback resistance is 100 kohms and the input resistance is 12 kohms.

8. What value of reference resistor ($R_{ref}$) is needed to create a 50 μA reference current from a 12 Vdc power supply?
9. A CDA is connected to ∓ 15 Vdc power supplies. Assuming a 1.5 megohm resistor ($R_{ref}$) between the 15 Vdc supply and the input bias terminal, and a feedback resistance of 120 kohms, calculate the DC offset potential at room temperature when no input signal is present.
10. In a noninverting follower CDA circuit the feedback resistance is 220 kohms, an input resistance of 15 kohms, and a reference current of 100 μA. Calculate the voltage gain.
11. Find the gain of a supergain CDA circuit (Fig. 8-7) if $R1$ = 12 kohms, $R2$ = 120 kohms, $R3$ = 10 kohms, and $R4$ = 120 kohms.
12. Calculate the transconductance of an amplifier in which an output current change of 2 mA is created by an input voltage change of 2 volts.
13. Calculate the output resistance of an OTA in which a bias current of 200 μA is flowing.
14. Draw the circuit for an inverting follower OTA.
15. What bias current for a CA-3080 OTA will create a transconductance of 40 mmhos?
16. What bias current is required in a CA-3080 if the transconductance is 25 mmhos?
17. Design a CDA circuit in which the voltage gain is 230.
18. A CDA inverting follower has a feedback resistance of 12 kohms, an input resistance of 1 kohm and a current mirror resistance of 100 ohms.
19. A CDA reference resistor has a value of 1 megohm, and is connected to the +15 Vdc power supply; the feedback resistance is 220 kohms. Calculate the zero-signal DC output potential that will be produced.
20. Draw the circuit for a CDA amplifier that has a gain of 200 and a lower −3 dB frequency response of 10 Hz. Select the DC power supply potentials and label the components for value.
21. Find the gain of a supergain CDA circuit (Fig. 8-7) if $R1$ = 15 kohms, $R2$ = 150 kohms, and $R4$ = 100 kohms.
22. Find the gain of a supergain CDA circuit (Fig. 8-7) if $R1$ = 22 kohms, $R2$ = 120 kohms and $R4$ = 120 kohms.
23. Draw the circuit for a CDA line driver amplifier with a differential 600 ohm output.

CHAPTER 9

# High Frequency, VHF, UHF, and Microwave Linear IC Devices

## OBJECTIVES

1. Understand the problems inherent in high frequency IC applications.
2. Learn the properties of high frequency linear devices.
3. Learn the properties of broadband devices.
4. Know the properties of microwave linear IC devices.

## 9-1 PREQUIZ

These questions test your prior knowledge of the material in this chapter. Try answering them before you read the chapter. Look for the answers (especially those you answered incorrectly) as you read the text. After you have finished studying the chapter, try answering these questions again and those at the end of the chapter (see Section 9-10).

1. A Darlington amplifier is constructed with a single collector-to-base feedback resistor and an unbypassed emitter resistor. Calculate the value of the emitter resistor if the feedback resistor is 470 ohms and the input/output impedances must be 50 ohms.
2. Two microwave NPN transistors are connected in a Darlington configuration. Calculate the overall *beta* if both are rated at $B = 120$.
3. Calculate the width of a stripline to match an impedance of 50 ohms if the stripline is 0.125 inch above a groundplane on a printed wiring board with a dielectric constant of 3.45.
4. Draw a simple schematic of two MIC amplifiers in parallel. Describe the effect on (a) driving power requirements, (b) output power delivered, (c) overall power and voltage gain, and (d) 1 dB compression point.

## 9-2 HIGH FREQUENCY LINEAR INTEGRATED CIRCUITS

High frequency linear solid-state amplifiers (those operating from near-DC to VHF, UHF or microwave frequencies), with consistent performance across a wide passband are difficult to design and build. These amplifiers often have gain irregularities such as "suck-outs" and peaks. Others have large variations of input and output impedance over the frequency range. Still others have spurious oscillation at certain frequencies within the passband.

Barkhausen's criteria for oscillation require (a) loop gain of unity or more, and (b) 360° (in-phase) feedback at the frequency of oscillation. At some frequency, the second criterion may be met by adding the normal 180° phase shift inherent in an inverting amplifier. The result is oscillation at the frequency where the propagation phase shift is 180°. Until recently only a few applications required such amplifiers. Consequently, they were either very expensive or didn't work well. Today, one can buy linear IC devices that work well in the microwave region.

## 9-3 WHAT ARE HMICs AND MMICs?

MMICs are small gain block monolithic integrated circuits which operate from DC or near DC to a frequency in the microwave region. HMICs, on the other hand, are hybrid devices which combine discrete and monolithic technology. One product (Signetics NE-5205) offers up to +20 dB of gain from DC to 0.6 GHz, while another low-cost device (Minicircuits Laboratories, Inc. MAR-x) offers +20 dB of gain over the range DC to 2 GHz depending on the model. Other devices can produce gains to +30 dB and frequencies to 18 GHz. These devices are unique because they present input and output impedances which are a good match to the 50 or 75 ohms normally used as system impedances in RF circuits.

Because the material in this chapter could apply to either HMIC or MMIC devices, we shall refer to all devices in either sub-family as "microwave integrated circuits" (MIC) unless otherwise specified.

**Monolithic Integrated Circuits.** These devices are formed through photoetching and diffusion on a substrate of silicon or other semiconductor material. Both active devices (such as transistors and diodes) and some passive devices can be formed in this manner. Passive components such as on-chip capacitors and resistors can be formed using various thin and thick film technologies. In the MMIC device, interconnections are made on the chip via built-in *planar transmission lines.*

**Hybrids.** Even though they may physically resemble larger monolithic integrated circuits from the outside, hybrids are actually a level closer to regular discrete circuit construction than ICs. Passive components and planar transmission lines are laid down on a glass, ceramic, or other insulating substrate by vacuum deposition or some other method. Transistors and unpackaged monolithic chip dies are cemented to the substrate and then connected to the substrate circuitry via mil-sized gold or aluminum bonding wires.

**Characteristics.** Three things specifically characterize the MIC device. Simplicity is the first characteristic. As you will see in the circuits discussed below, the MIC device usually has only input, output, ground, and power supply connections. Other wideband IC devices often have up to 16 pins, most of which must be either biased or capacitor bypassed. The second feature of the MIC is the very wide frequency range (DC-GHz) of the devices. The third is its constant input and output impedance over several octaves of frequency.

Although not always the case, MICs tend to be unconditionally stable because of a combination of series and shunt negative feedback internal to the device. The input and output impedances of the typical MIC device are close to either 50 or 75 ohms, so it is possible for a MIC amplifier to operate without any external impedance matching schemes. This makes it easy to broadband.

A typical MIC device generally produces a standing wave ratio (SWR) of less than 2:1 at all frequencies within the passband, provided it is connected to the design system impedance (50 ohms). The MIC is not usually regarded as a low-noise amplifier (LNA), but it can produce noise figures (NF) in the 5 to 8 dB range for frequencies up to several gigahertz. Some MICs are LNAs, however, and produce noise figures from 2.5 to 4 dB.

Narrowband and passband amplifiers can be built using wideband MICs. A narrowband amplifier is a version of the passband amplifier and it is typically tuned to a single frequency. An example is the 70-MHz IF amplifier used in microwave receivers. Because of input and output tuning, such an amplifier will only respond to signals near the 70-MHz center frequency.

## 9-4 VERY WIDEBAND AMPLIFIERS

Engineering wideband amplifiers like those used in MIC devices seem simple but they have traditionally caused difficulty for designers. Figure 9-1A shows the most fundamental form of a MIC amplifier circuit. It is a common emitter NPN bipolar transistor amplifier. Because of the high frequency operation of these devices, the amplifier in MICs is usually made of a material such as gallium arsenide (GaAs).

In Fig. 9-1A, the emitter resistor ($R_e$) is unbypassed so it introduces a small amount of negative feedback into the circuit. Resistor $R_e$ forms *series feedback* for transistor $Q1$. The *parallel feedback* is provided by collector-base bias resistor $R_f$. Typical values for $R_f$ are in the 500-ohm range and for $R_e$ in the 4 to 6 ohms range. In general, the designer tries to keep the ratio $R_f/R_e$ high in order to obtain higher gain, higher output power compression points, and lower noise figures. The input and output impedances ($R_o$) are equal and defined by the patented equation

$$R_o = \sqrt{R_f \times R_e} \tag{9-1}$$

where

$R_o$ = the output impedance in ohms

$R_f$ = the shunt feedback resistance in ohms

$R_e$ = the series feedback resistance in ohms

A more common MIC amplifier circuit is shown in Fig. 9-1B. Although based on the Darlington amplifier circuit, this amplifier still has the same sort of series and

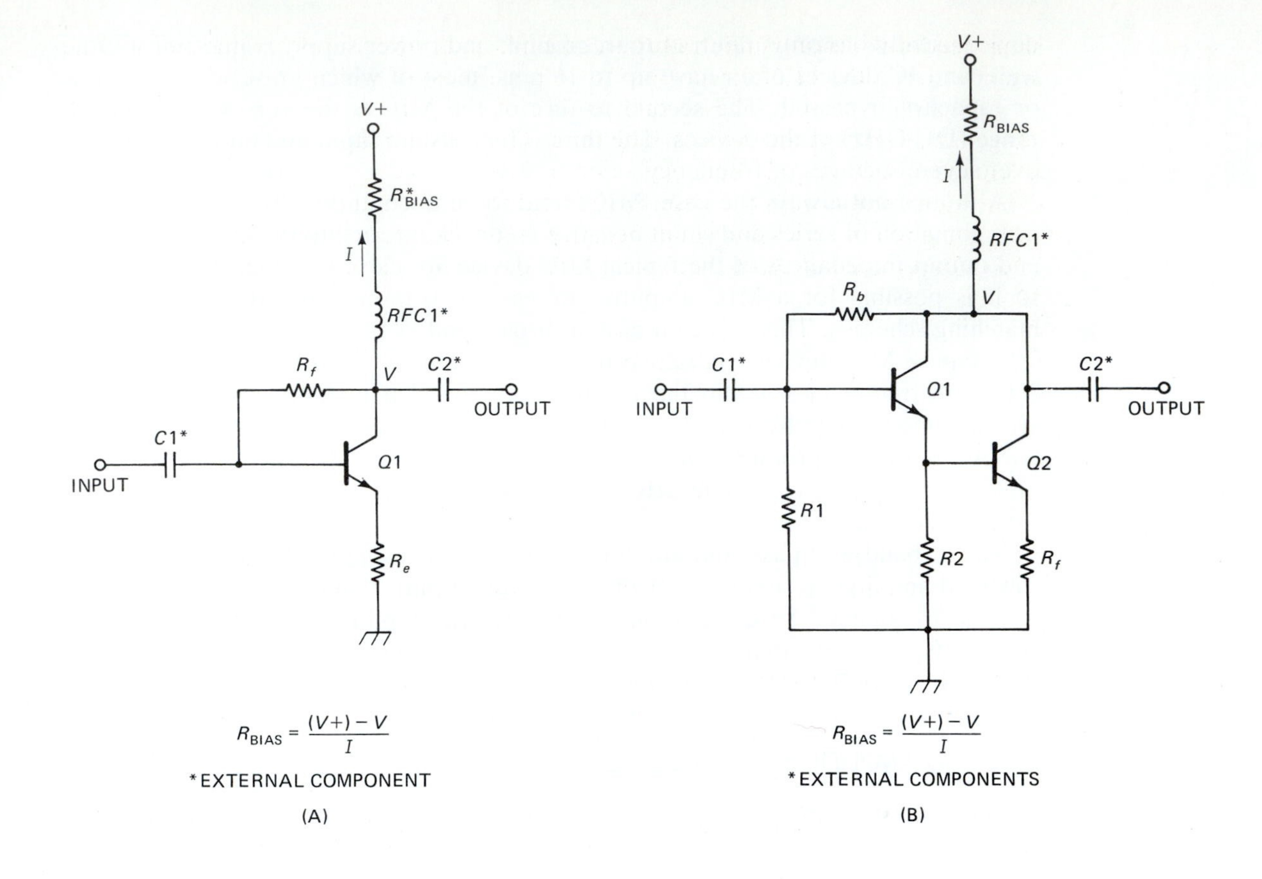


V+

$V_d$ R

I

IN C1 A1 C2 OUT

$$V = (V+) - (V_d)$$

$$R = \frac{(V+) - (V)}{I}$$

(C)

FIGURE 9-1

shunt feedback resistors ($R_e$ and $R_f$) as the previous circuit. All resistors except $R_{\text{bias}}$ are internal to the MIC device.

A Darlington amplifier, also called a Darlington pair or superbeta transistor, consists of a pair of bipolar transistors ($Q1$ and $Q2$) connected so $Q1$ is an emitter follower driving the base of $Q2$. Both collectors are connected in parallel. The Darlington connection permits both transistors to be treated as though they were a single transistor with higher than normal input impedance and a *beta* gain ($B$) equal to the product of the individual *beta* gains. For the Darlington amplifier therefore, the *beta* ($B$, or $H_{fe}$) is

$$B_o = B_{Q1} \times B_{Q2} \tag{9-2}$$

where

$B_o$ = the total beta gain of the $Q1/Q2$ Darlington pair

$B1$ = the beta gain of $Q1$

$B2$ = the beta gain of $Q2$

Two facts are apparent. First, the *beta* gain is very high for a Darlington amplifier made with transistors of relatively modest individual *beta* figures. Second, the *beta* of a Darlington amplifier made with identical transistors is the square of the common *beta* rating.

**External Components.** Figures 9-1A and 9-1B show several components which are usually external to the MIC device. The bias resistor ($R_{\text{bias}}$) is sometimes internal, although on most MIC devices it is external. RF choke $RFC1$ is in series with the bias resistor and is used to enhance operation at the higher frequencies. $RFC1$ is considered optional by some MIC manufacturers. The reactance of the RF choke is in series with the bias resistance and increases with frequency according to the $2\pi FL$ rule. Thus, the transistor experiences a higher impedance load at the upper end of the passband than at the lower end. Use of $RFC1$ as a peaking coil helps overcome the adverse effect of stray circuit capacitance. A general rule is to make the combination of $R_{\text{bias}}$ and $X_{RFC1}$ form an impedance of at least 500 ohms at the lowest frequency of operation. The gain of the amplifier may drop about 1 dB if $RFC1$ is deleted. This effect is caused by the bias resistance shunting the output impedance of the amplifier.

The capacitors are used to block DC potentials in the circuit. They prevent intra-circuit potentials from affecting other circuits, and prevent potentials in other circuits from affecting MIC operation. More will be said about these capacitors later, but for now understand that practical capacitors are not ideal; real capacitors are actually complex RLC circuits. While the $L$ and $R$ components are negligible at low frequencies, they are substantial in the VHF through microwave region. In addition, the LC characteristic forms a self-resonance that can either reduce (suck out) or enhance (peak up) gain at specific frequencies. The result is either uneven frequency response or spurious oscillations.

## 9-5 GENERIC MMIC AMPLIFIER

Figure 9-1C shows a generic MIC amplifier circuit. This circuit is nearly complete. The MIC device usually has only input, output, ground, and power connections. Some models do not have separate DC power input. There is no DC biasing, no bypassing (except at the DC power line) and no unnecessary pins on the package. MICs tend to use either microstrip packages like UHF/microwave small-signal transistor packages or small versions of the miniDIP or metallic IC packages. Some HMICs are packaged in larger transistor-like cases and others are packaged in special hybrid packages.

The bias resistor ($R_{bias}$) is connected to either the DC power supply terminal (if any) or the output terminal. It must be set to a value which limits the current to the device and drops the supply voltage to a safe value. MIC devices typically require a low DC voltage (4–7 Vdc), and a maximum current of about 15 to 25 milliamperes (mA). There may also be an optimum current for a specific device. For example, one device advertises it will operate over a range of 2 to 22 mA, but the optimum design current is 15 mA. The value of resistor needed for $R_{bias}$ is found from Ohm's law

$$R_{bias} = \frac{(V+) - V}{I_{bias}} \tag{9-3}$$

where

$R_{bias}$ = in ohms

$V+$ = the DC power supply potential in volts

$V$ = the rated MIC device operating potential in volts

$I_{bias}$ = the operating current in amperes

The construction of amplifiers based on MIC devices must follow microwave practices. This requirement means short, wide, low-inductance leads made of printed circuit foil and stripline construction. Interconnection conductors tend to behave like transmission lines at microwave frequencies, so they must be treated as such. In addition, capacitors should be capable of passing the frequencies involved, but have as little inductance as possible. In some cases, the series inductance of common capacitors forms a resonance at some frequency within the passband of the MIC device. These resonant circuits can sometimes be detuned by placing a small ferrite bead on the capacitor lead. Microwave chip capacitors are used for ordinary bypassing.

MIC technology is currently able to provide low-cost microwave amplifiers with moderate gain and noise. Better performance is available at a higher cost. Manufacturers have extended MIC operation to 18 GHz and dropped noise figures substantially. In addition, it is possible to build the entire frontend of a microwave receiver into a single HMIC or MMIC, including the RF amplifier, mixer and local oscillator stages.

Although MIC devices are available in a variety of packages, those shown in Fig. 9-2 are the most typical. Because of the very high frequency operation of these devices, MICs are packaged in stripline transistor-like cases. The low-inductance leads for these packages are essential in UHF and microwave applications.

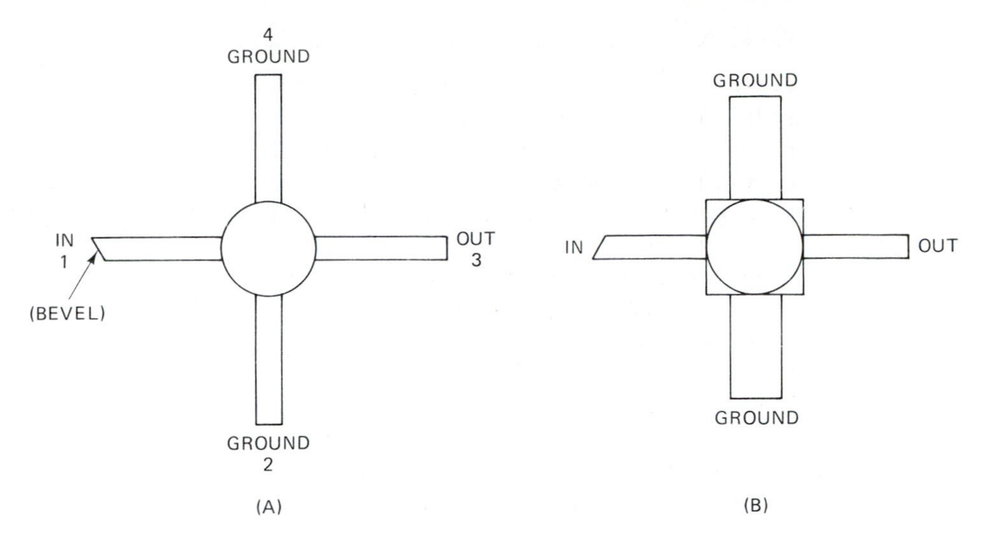


FIGURE 9-2

### 9-5.1 Attenuators in Amplifier Circuits

It is common practice to place attenuator pads in series with the input and output signal paths of microwave circuits in order to swamp out impedance variations where impedance matching or a constant impedance is required. Especially when dealing with LC filters (low-pass, high-pass, bandpass), VHF/UHF amplifiers, matching networks, and MIC devices, it is useful to insert 1 dB, 2 dB, or 3 dB resistor attenuator pads in the input and output lines. Many RF circuits depend on having the design impedance at input and output terminals. With the attenuator pad (see Fig. 9-3) in the line, source and load impedance changes do not affect the circuit nearly as much.

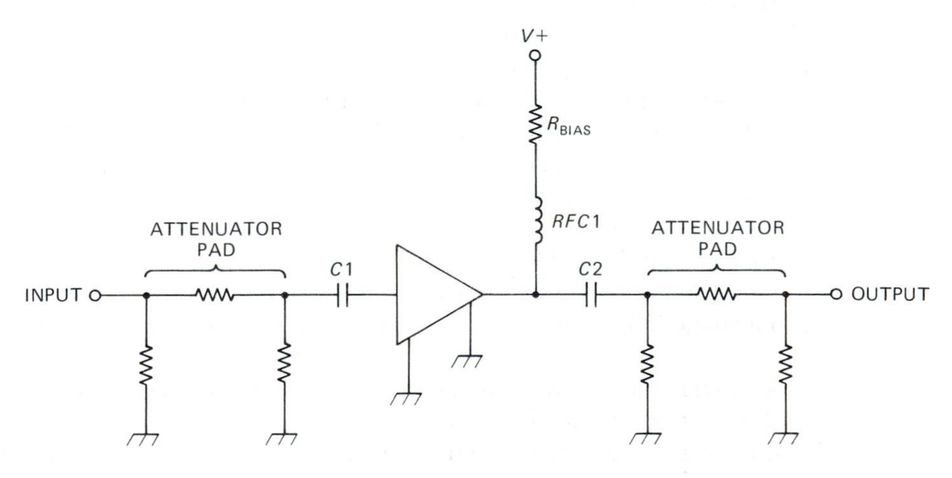


FIGURE 9-3

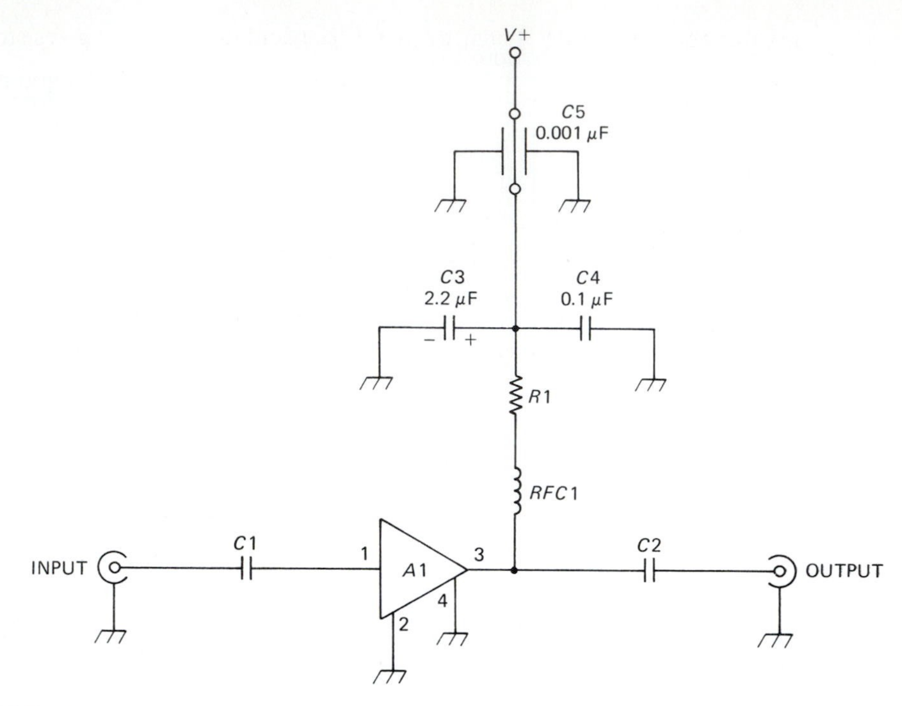


FIGURE 9-4

This is also sometimes useful with unstable very wideband amplifiers. Insert a 1 dB pad in series with both input and output lines of the unstable amplifier. This will cost about 2 dB of voltage gain, but it often cures instabilities from frequency dependent load or source impedance changes.

### 9-5.2 MiniCircuits® MAR-x Series Devices: A MIC Example

The MiniCircuits Laboratories, Inc. MAR-x series of MIC devices offers gains from +13 dB to +20 dB, and top-end frequency response of either 1 GHz or 2 GHz depending upon the type. The package used for the MAR-x device (Fig. 9-2A) is similar to that used for UHF and microwave transistors. Pin 1 (RF input) is marked by a color dot and a bevel. The usual circuit for the MAR-x series devices in shown in Fig. 9-4. The MAR-x device requires a voltage of +5 Vdc on the output terminal. It must derive this potential from a DC supply of greater than + 7 Vdc.

The RF choke (*RFC*1) is called optional in the engineering literature on the MAR-x, but it is recommended for applications where a substantial portion of the total bandpass capability of the device is used. The choke tends to pre-emphasize the higher frequencies, and thereby overcomes de-emphasis caused by circuit capacitance. In traditional video amplifier terminology the coil is called a "peaking coil" because it peaks up the higher frequencies.

It is necessary to select a resistor for the DC power supply connection. The MAR-x device needs +5 Vdc at a current not to exceed 20 mA. In addition, *V*+ must be greater than +7 volts. Thus, we need to calculate a dropping resistor ($R_d$) of

$$R_d = \frac{(V+) - 5 \text{ Vdc}}{I} \quad (9\text{-}4)$$

where

$$R_d = \text{in ohms}$$

$$I = \text{in amperes}$$

In an amplifier designed for +12 Vdc operation select a trial bias current of 15 mA (0.015 amperes). The calculated resistor value is 467 ohms (a 470 ohm, 5% film resistor should prove satisfactory). As recommended, a 1 dB attenuator pad is inserted in the input and output lines. The $V+$ is supplied to this chip through the output terminal.

## 9-6 CASCADE MIC AMPLIFIERS

MIC devices can be connected in cascade (Fig. 9-5) to provide greater gain than that available from a single device. However a few precautions must be observed. For example, MICs have a substantial amount of gain from frequencies near DC to well into the microwave region. In all cascade amplifiers feedback from stage to stage has to be prevented. Two factors must be addressed. First, as always, is component layout. The output and input circuitry external to the MIC must be physically separated to prevent coupling feedback. Second, it is necessary to decouple the DC power supply lines that feed two or more stages. Signals carried on the DC power line can easily couple into one or more stages, resulting in unwanted feedback.

Figure 9-5 shows a method for decoupling the DC power line of a two-stage MIC amplifier. In lower frequency cascade amplifiers the $V+$ ends of resistors $R1$ and $R2$ would normally be joined together and connected to the DC power supply. Only a single capacitor would be needed at that junction to ensure adequate decoupling between stages. But as the operating frequency increases, the situation becomes more complex, in part due to the nonideal nature of practical components. For example, in an audio amplifier the same aluminum electrolytic capacitors used in the power supply ripple filter may provide sufficient decoupling. At RF frequencies, however, electrolytic capacitors are almost useless as capacitors because they act more like resistors at those frequencies.

The decoupling system in Fig. 9-5 consists of RF chokes $RFC3$ and $RFC4$, and capacitors $C4$ through $C9$. The RF chokes help prevent high frequency AC signals from traveling along the DC power line. These chokes are selected for a high reactance at VHF through microwave frequencies, while having a low DC resistance. For example, a 1-microhenry (1-μH) RF choke might have only a few milliohms of DC resistance, but (by $2\pi FL$) it has a reactance of more than 3000 ohms at 500 MHz. $RFC3$ and $RFC4$ must be mounted to minimize mutual inductance due to interaction of their respective magnetic fields.

Capacitors $C4$ through $C9$ are used for bypassing signals to ground. A wide range of values and several types of capacitors are used in this circuit. Capacitor $C8$, an electrolytic type, is used to decouple very low frequency signals (those up to several

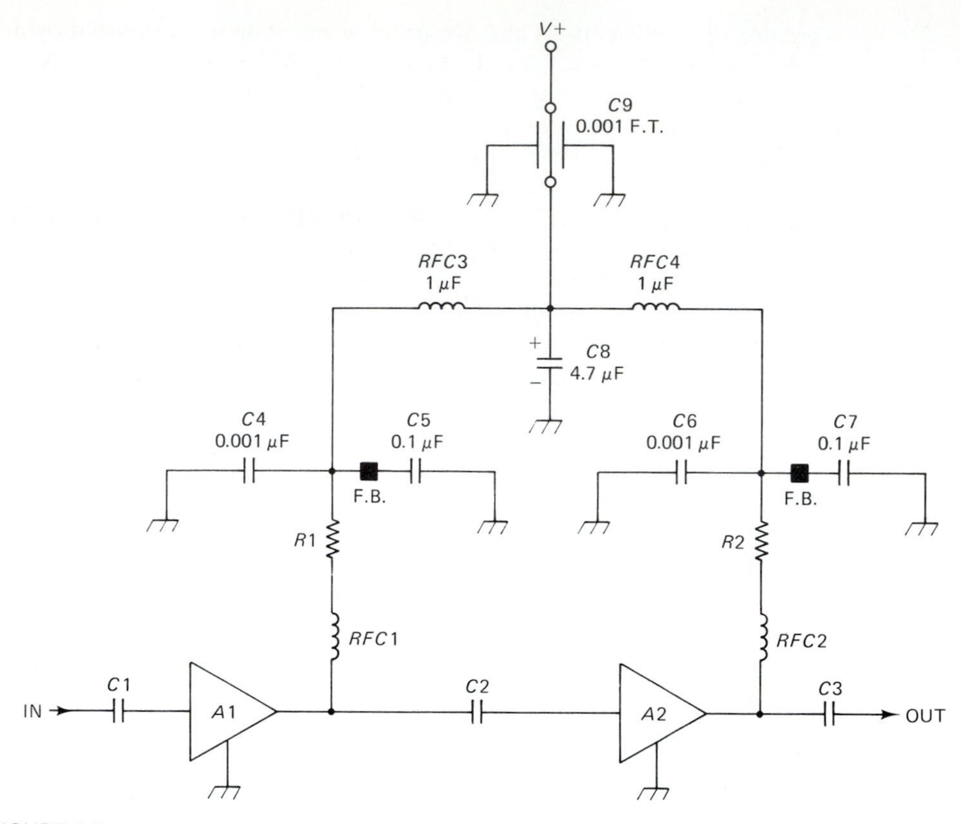


FIGURE 9-5

hundred kilohertz). Because *C*8 is ineffective at higher frequencies, it is shunted by capacitor *C*9, shown in Fig. 9-5 as a feedthrough capacitor. Such a capacitor is usually mounted on the shielded enclosure housing the amplifier. Capacitors *C*5 and *C*7 are used to bypass signals in the HF region. Because these capacitors are likely to exhibit substantial series inductance, they may form undesirable resonances within the amplifier passband. Ferrite beads (FB) are sometimes installed on each capacitor lead in order to detune capacitor self-resonances. Like *C*9, capacitors *C*4 and *C*6 are used to decouple signals in the VHF-and-up region. These capacitors must be of microwave chip construction, or they may prove ineffective above 200 MHz.

### 9-6.1 Gain in Cascade Amplifiers

In low frequency amplifiers we might expect the composite gain of a cascade amplifier to be a product of the individual stage gains

$$G = G1 \times G2 \times G3 \times \ldots \times G_n \tag{9-5}$$

This may be valid for low-frequency voltage amplifiers but it is not true for RF amplifiers (especially in the microwave region) where input-output standing wave ratio (SWR) becomes significant. In fact, the gain of any RF amplifier cannot be accurately measured if the SWR is greater than about 1.15:1.

SWR is a measure of the RF signal power which is absorbed by a load versus the power reflected back towards the source. SWR will not be discussed in detail here. It is sufficient to note it can be calculated by taking the ratio of source impedance ($R_o$) and load impedance ($Z_L$).

There are several ways in which SWR can be greater than 1:1 and all of them involve an impedance mismatch. For example, the amplifier may have an input or output resistance other than the specified value. This situation can arise due to design errors or manufacturing tolerances. Another source of mismatch are the source and load impedances. If these impedances are not exactly the same as the amplifier input or output impedance, a mismatch will occur.

An impedance mismatch at either input or output of the single-stage amplifier will result in a gain mismatch loss (ML) of

$$ML = -10 \text{ LOG} \left[ 1 - \left[ \frac{SWR - 1}{SWR + 1} \right]^2 \right] \tag{9-6}$$

where

$ML$ = the mismatch loss in decibels (dB)

$SWR$ = the standing wave ratio (dimensionless)

In a cascade amplifier we have the distinct possibility of an impedance mismatch and thus an SWR, at more than one point in the circuit. An example (refer again to Fig. 9-5) might be where neither the output impedance ($R_o$) of the driving amplifier ($A1$) nor the input impedance ($R_i$) of the driven amplifier ($A2$) is matched to the 50-ohm ($Z_o$) microstrip transmission line which interconnects the two stages. Thus, $R_o/Z_o$ or its inverse forms one SWR, while $R_i/Z_o$ or its inverse forms the other. For a two-stage cascade amplifier the mismatch loss is

$$ML = -20 \text{ LOG} \left[ 1 \pm \left[ \frac{SWR1 - 1}{SWR1 + 1} \right] \left[ \frac{SWR2 - 1}{SWR2 + 1} \right] \right] \tag{9-7}$$

## EXAMPLE 9-1

An amplifier ($A1$) with a 25-ohm output resistance drives a 50-ohm stripline transmission line (Fig. 9-6). The other end of the stripline is connected to the input of another amplifier ($A2$) in which $R_i$ = 100 ohms. (A) calculate the maximum and minimum gain loss for this system and (b) calculate the range of system gain if $G1$ = 6 dB and $G2$ = 10 dB.

### Solution

$$SWR1 = Z_o/R_o = 50/25 = 2{:}1$$

$$SWR2 = R_i/Z_o = 100/50 = 2{:}1$$

$$ML1 = 20 \text{ LOG} \left[ 1 + \left[ \frac{SWR1 - 1}{SWR1 + 1} \right] \left[ \frac{SWR2 - 1}{SWR2 + 1} \right] \right]$$

$$ML1 = 20 \text{ LOG} \left[ 1 + \left[ \frac{2 - 1}{2 + 1} \right] \left[ \frac{2 - 1}{2 + 1} \right] \right] \text{ dB}$$

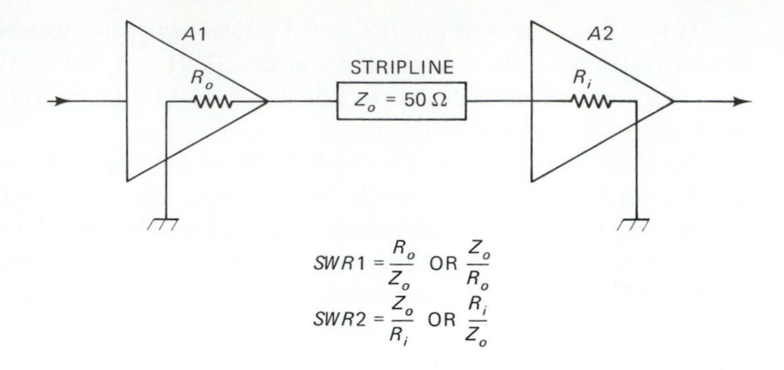


$$SWR1 = \frac{R_o}{Z_o} \text{ OR } \frac{Z_o}{R_o}$$

$$SWR2 = \frac{Z_o}{R_i} \text{ OR } \frac{R_i}{Z_o}$$

FIGURE 9-6

$$ML1 = 20 \text{ LOG } [1 + (1/3)(1/3)] \text{ dB}$$
$$ML1 = 20 \text{ LOG } [1 + 0.11] \text{ dB}$$
$$ML1 = 20 \text{ LOG } [1.11] \text{ dB}$$
$$ML1 = (20)(0.045) = 0.91 \text{ decibels}$$
$$ML2 = 20 \text{ LOG } (1 - 0.11) \text{ dB}$$
$$ML2 = 20 \text{ LOG } (0.89) \text{ dB}$$
$$ML2 = (20)\,(-0.051) \text{ dB}$$
$$ML2 = -1.02 \text{ decibels}$$

Thus

$$ML1 = +1.11 \text{ dB}$$
$$ML2 = -1.02 \text{ dB}$$

Without the SWR, gain in decibels is $G = G1 + G2 = (6 \text{ dB}) + (10 \text{ dB}) = 16$ dB. With SWR considered we find that $G = G1 + G2 \mp ML$. So

$$G_a = G1 + G2 + ML1$$
$$G_a = (6 \text{ dB}) + (10 \text{ dB}) + (1.11 \text{ dB})$$
$$G_a = 17.11 \text{ decibels}$$

and

$$G_b = G1 + G2 + ML2$$
$$G_b = (6 \text{ dB}) + (10 \text{ dB}) + (-1.02 \text{ dB})$$
$$G_b = 14.98 \text{ decibels}$$

■

The mismatch loss can vary from a negative loss resulting in less system gain ($G_b$), to a positive loss (which is actually a gain) resulting in greater system gain ($G_a$). The reason for this apparent paradox is it is possible for a mismatched impedance to be connected to its complex conjugate impedance.

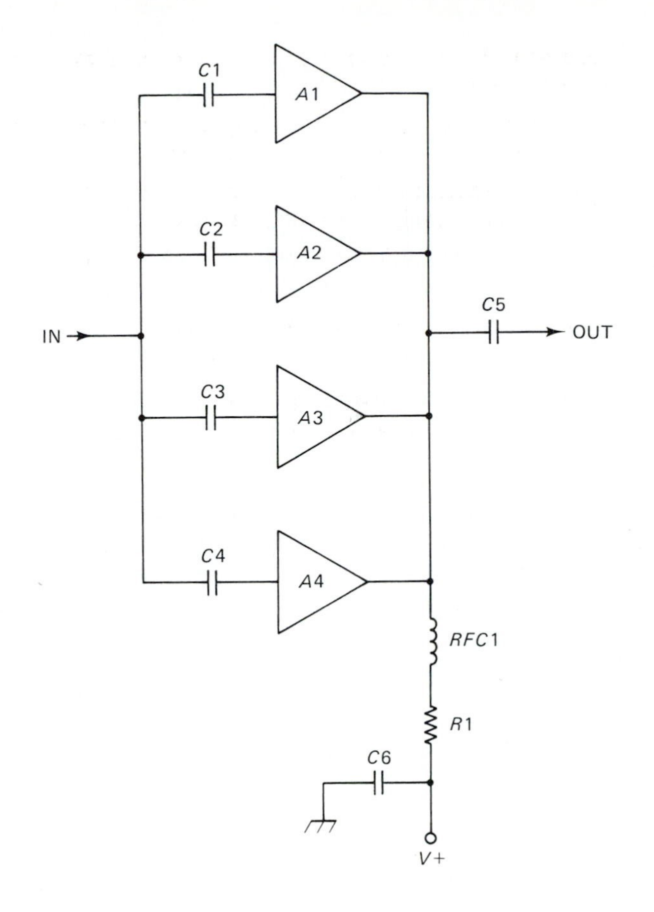


FIGURE 9-7

## 9-7 COMBINING MIC AMPLIFIERS IN PARALLEL

Figures 9-7 through 9-10 show several MIC applications involving combinations of two or more MIC devices. The simplest of these is Fig. 9-7. Although each input must have its own DC blocking capacitor to protect the MIC device's internal DC bias network, the outputs of two or more MICs may be connected in parallel and share a common power supply connection and output coupling capacitor.

Several advantages are realized with the circuit of Fig. 9-7. First, the power output increases even though total system gain ($P_o/P_m$) remains the same. As a consequence, however, drive power requirements also increase. The output power increases 3 dB when two MICs are connected in parallel, and 6 dB when four are connected as shown. The 1 dB output power compression point also increases in parallel amplifiers in the same manner, 3 dB for two amplifiers, and 6 dB for four amplifiers in parallel.

The input impedance of parallel MIC devices reduces to $R_i/N$, where $N$ is the number of MIC devices in parallel. In Fig. 9-7, the input impedance would be $R_i/4$, or 12.5 ohms if the MICs are designed for 50-ohm service. Because 50/12.5 represents a 4:1 SWR, some form of input impedance matching must be used. Such a matching

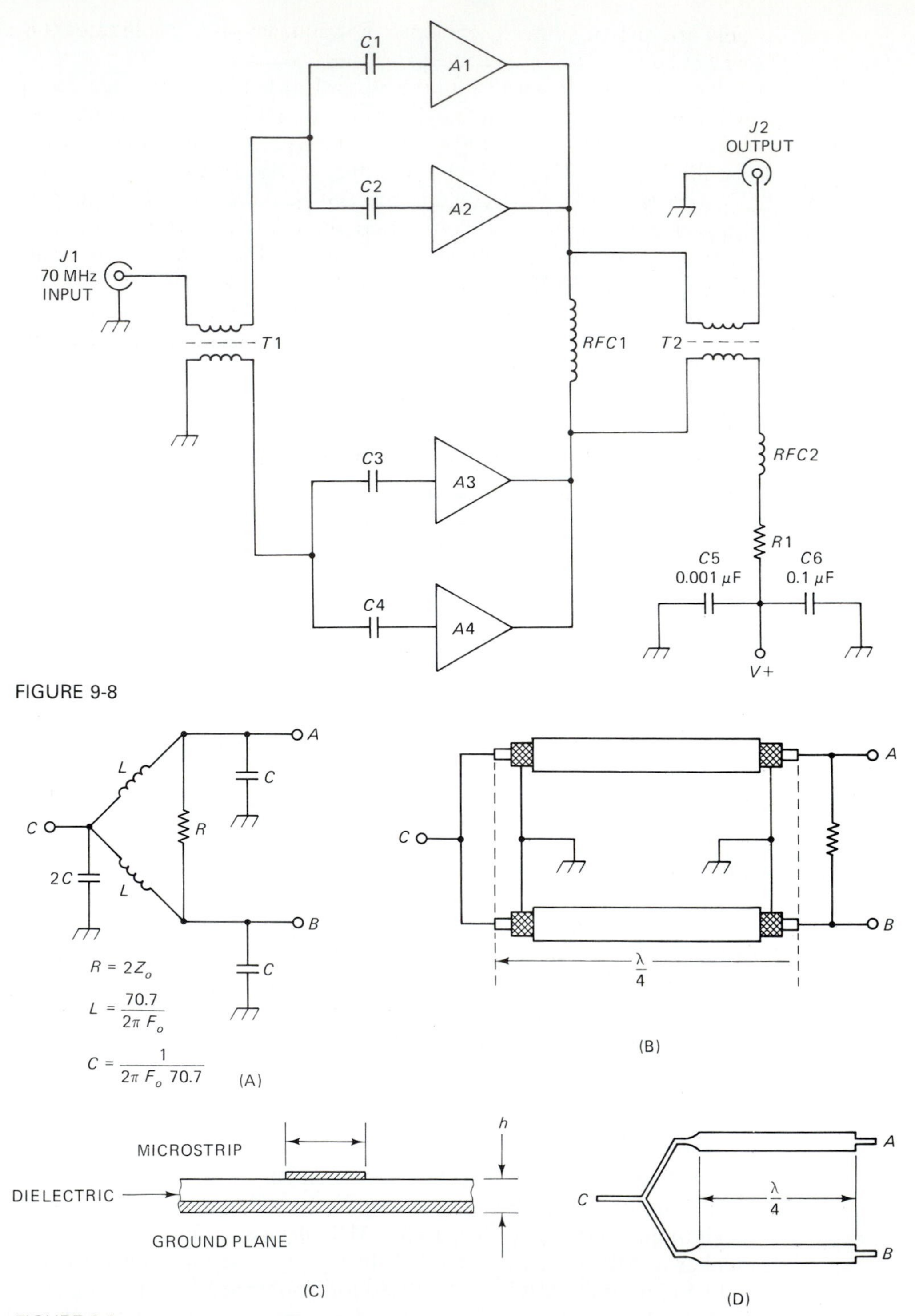


FIGURE 9-8

FIGURE 9-9

network can be either broadband or frequency-specific. Figures 9-8 and 9-9 show ways to accomplish impedance matching.

The method shown in Fig. 9-8 is used at operating frequencies up to about 100 MHz and is based on broadband ferrite toroidal RF transformers. These transformers dominate the frequency response of the system because they are less broadbanded than the usual MIC device. This type of circuit can be used as a gain block in microwave receiver IF amplifiers (which are frequently in the 70 MHz region), or in the exciter section of master oscillator power amplifier (MOPA) transmitters.

Another method is more useful in the UHF and microwave regions. In Fig. 9-9 we see several forms of the Wilkinson power divider circuit. An LC network version is shown in Fig. 9-9A. Coaxial (Fig. 9-9B) or stripline (Figs. 9-9C and 9-9D) transmission line versions are used in microwave applications. The LC version is used up to frequencies of 150 MHz. The circuit is bidirectional, so it can be used as either a *power splitter* or *power divider*. RF power applied to port C is divided equally between port A and port B. Alternatively, power applied to ports A/B are added together and appear at port C. The component values are found from the following relationships

$$R = 2Z_o \tag{9-8}$$

$$L = \frac{70.7}{2\pi F_o} \tag{9-9}$$

$$C = \frac{1}{2\pi 70.7 F_o} \tag{9-10}$$

where

$R$ is in ohms

$L$ is in henrys (H)

$C$ is in farads (F)

$F_o$ is in hertz (Hz)

Figure 9-9B shows a coaxial cable version of the Wilkinson divider which can be used at frequencies up to about 2 GHz. The lower frequency limit is set by practicality because the transmission line segments become too long to be handled easily. The upper frequency limit is set by the handling of very short lines, and by the dielectric losses which are frequency dependent. The transmission line segments are each quarter wavelength. Their length is found from

$$L = \frac{2952\ V}{F} \tag{9-11}$$

where

$L$ = the physical length of the line in inches (in.)

$F$ = the frequency in megahertz (MHz)

$V$ = the *velocity factor* of the transmission line (0-1)

An impedance transformation can take place across a quarter wavelength transmission line if the line has a different impedance than the source or load impedances being matched. Such an impedance matching system is often called a "Q-section." The required characteristic impedance for the transmission line is found from

$$Z'_o = \sqrt{Z_L Z_o} \tag{9-12}$$

where

$Z'_o$ = the characteristic impedance of the quarter wavelength section

$Z_L$ = the load impedance

$Z_o$ = the system impedance (e.g. 50 ohms)

In parallel MIC devices, the nominal impedance at port C of the Wilkinson divider is one half the reflected impedance of the two transmission lines. For example, if the two lines are each 50-ohm transmission lines, the impedance at port C is 50/2 ohms, or 25 ohms. Similarly, if the impedance of the load (the reflected impedance) is transformed to some other value, port C receives the parallel combination of the two transformed impedances. In a parallel MIC amplifier we might have two devices with 50 ohms input impedance each. Placing these devices in parallel halves the impedance to 25 ohms, which forms a 2:1 SWR with a 50-ohm system impedance. But if the quarter wavelength transmission line transforms the 50-ohm input impedance of each device to 100 ohms, then the port C impedance is 100/2, or 50 ohms.

At the upper end of the UHF spectrum and in the microwave spectrum, it may be better to use a stripline transmission line instead of coaxial cable. A stripline (see Fig. 9-9C) is formed on a printed circuit board. The board must be double-sided so one side can be used as a ground plane while the stripline is etched into the other side. The length of the stripline depends on the frequency of operation; either halfwave or quarterwave lines are normally used. The impedance of the stripline is a function of stripline width ($w$), height of the stripline above the ground plane ($h$), and dielectric constant ($\varepsilon$) of the printed circuit material

$$Z_o = 377 \frac{h}{w\sqrt{\varepsilon}} \tag{9-13}$$

where

$h$ = the height of the stripline above the groundplane

$w$ = the width of the stripline ($h$ and $w$ in same units)

$Z_o$ = the characteristic impedance in ohms

$\varepsilon$ = the dielectric constant

The stripline transmission line is etched into the printed circuit board (Fig. 9-9D). Stripline methods use the printed wiring board to form conductors and tuned circuits. In general, for operation at VHF and above, the conductors must be very wide (relative to their simple DC and RF power requirements) and very short to reduce lead inductance.

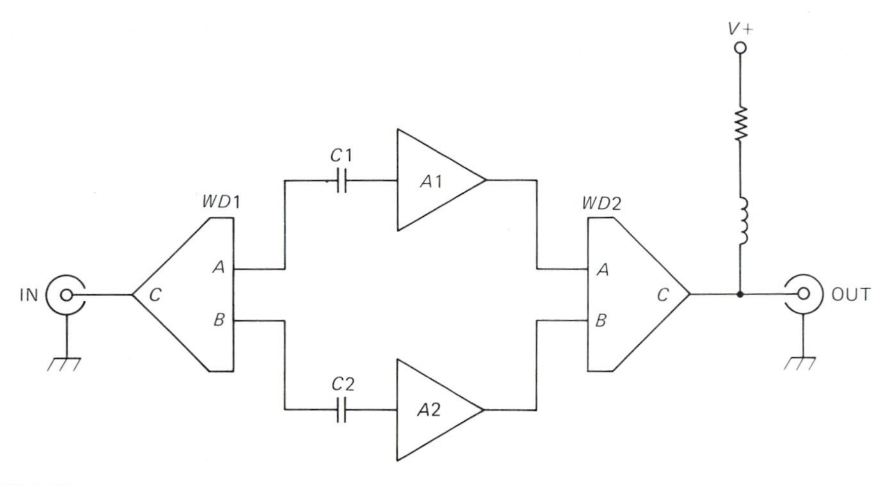


FIGURE 9-10

The actual striplines are transmission line segments and follow transmission line rules. The printed circuit material must have a *large permittivity* and a *low loss tangent*. For frequencies up to about 3 GHz it is permissible to use ordinary glass-epoxy double-sided board ($\varepsilon = 5$), but for higher frequencies a low-loss material such as Rogers Duroid ($\varepsilon = 2.17$) must be used.

When soldering connections in these amplifiers, it is important to use as little solder as possible and keep the soldered surface smooth and flat. Otherwise, the surface wave on the stripline will be interrupted.

Figure 9-10 shows a general circuit for a multiMIC amplifier based on a Wilkinson power divider. Because the divider can be used as either a splitter or combiner, the same type can be used as input (WD1) and output (WD2) terminations. In the case of the input circuit, port C is connected to the amplifier main input, while ports A/B are connected to the individual inputs of the MIC devices. At the output circuit another divider (WD2) is used to combine power output from the two MIC amplifiers and direct it to a common output connection. Bias is supplied to the MIC amplifiers through a common DC path at port C of the Wilkinson dividers.

## 9-8 SUMMARY

1. Microwave integrated circuits (MICs) are monolithic or hybrid gain blocks which have bandwidth from DC or near DC to the microwave range and constant input and output impedance over several octaves of frequency, only the bare essentials for connections (input, output, ground, and sometimes DC power). Some MIC devices are unconditionally stable over the entire frequency range.
2. MICs have all built-in components except for input and output coupling capacitors and an external bias resistor. For some types an external peaking inductor or RF choke is recommended. Narrowband or bandpass amplifiers can be built by using turning devices or frequency-selective filtering in the input and output circuits.
3. MIC devices can be connected in cascade in order to increase overall system gain. Care must be taken to ensure the SWR is low, that is, the impedances are matched. Otherwise, matching must be provided or gain errors tolerated.

4. The noise figure in a cascade MIC amplifier is dominated by the noise figure of the input amplifier stage.
5. MICs can be connected in parallel. The output terminals may be directly connected in parallel, but the input terminals of each device must be provided with its own DC blocking capacitor. MICs can also be combined using broadband RF transformers, Wilkinson power dividers/splitters or other methods.
6. Printed circuit boards used for MIC devices must have a high permittivity and a low-loss tangent. For frequencies up to 3 GHz glass-epoxy is permissable, but for higher frequencies a low-loss product is required.

## 9-9 RECAPITULATION

Now return to the objectives and Prequiz questions at the beginning of the chapter and see how well you can answer them. If you cannot answer certain questions, mark them and review the appropriate parts of the text. Next, try to answer the questions and work the problems below, using the same procedure.

## 9-10 QUESTIONS AND PROBLEMS

1. What condition may result in a microwave amplifier if the loop gain is unity or greater and the feedback in inadvertently in phase with the input signal?
2. On-chip interconnections in MIC devices are usually made with ____________ transmission lines.
3. What type of semiconductor material is usually used to make the active devices in MIC amplifiers?
4. A MIC amplifier contains a common emitter Darlington amplifier transistor and internal ____________ and ____________ negative feedback elements.
5. In a common emitter MIC amplifier the ____________ ratio must be kept high in order to obtain higher gain, higher output power compression points and lower noise figure.
6. A common emitter amplifier inside a MIC device has a collector-base resistor of 450 ohms and an emitter resistor of 12.5 ohms. Is this amplifier a good impedance match for 75-ohm cable television amplications?
7. Calculate the input and output impedances of a MIC amplifier in which the series feedback resistor is 5 ohms and the shunt feedback resistor is 550 ohms.
8. Calculate the gain of the amplifiers in the previous two questions.
9. Two transistors are connected in a Darlington amplifier configuration. $Q1$ has a *beta* of 100, and $Q2$ has a *beta* of 75. What is the *beta* gain of the Darlington pair?
10. An MIC amplifier must be operated over a frequency range of 0.5–2.0 GHz. The DC power circuit consists of an RF choke and a bias resistor in series between $V+$ and the MIC output terminal. If the resistor has a value of 220 ohms, what is the minimum appropriate value of the RF choke required to achieve an impedance of at least 500 ohms?
11. A *MCL* MAR-1 must operate from a +12 Vdc power supply. This device requires a + 5 Vdc supply and has an optimum current of 17 mA. Calculate the value of bias resistor.
12. A MIC amplifier has an optimum current of 22 mA and requires +7 Vdc at the output terminal. Calculate the bias resistor required to operate the device from (a) 12 Vdc and (b) 9 Vdc.

13. Because they operate into the microwave region, MIC devices are packaged in ________ cases.
14. In order to overcome problems associated with changes of source and load impedance it is sometimes the practice to insert resistive ____________ in series with the input and output signal paths of a MIC device.
15. The SWR in a microwave amplifier should be less than ____________ in order to make accurate gain measurements.
16. Calculate the mismatch loss in a single-stage MIC amplifier if the output impedance is 50 ohms and the load impedance is 150 ohms.
17. Calculate the mismatch loss in a single stage MIC amplifier if the input SWR is 1.75:1.
18. Calculate the mismatch loss in a cascade MIC amplifier if the output impedance of the driver amplifier and the input impedance of the final amplifier are both 50 ohms, but are interconnected by a 90-ohm transmission line. Remember that there are two values (a) ____________ dB and (b) ____________ .
19. A cascade MIC amplifier has two SWR mismatches, SWR1 = 2:1 and SWR2 = 2.25:1. Calculate both values of mismatch loss.
20. The amplifiers in the previous question have the following gains, $A1$ = 16 dB and $A2$ = 12 dB. Calculate the range of gain that can be expected from the cascade combination $A1 \times A2$ considering the possible mismatch losses.
21. A MIC amplifier has two stages $A1$ and $A2$. $A1$ has a gain of 20 dB and a noise figure of 3 dB; $A2$ has a gain of 10 dB and a noise figure of 4.5 dB. Calculate the noise figure of the system (a) when $A1$ is the input amplifier and $A2$ is the output amplifier and (b) the reverse situation, when $A2$ is the input amplifier and $A1$ is the output amplifier. What practical conclusions do you draw from comparing these results?
22. Two MIC amplifiers are connected in parallel. What is the total input impedance if each amplifier has an input impedance of 75 ohms?
23. Four MIC amplifiers are connected in parallel. The output compression point is now *raised* or *lowered* ____________ dB.
24. An LC Wilkinson power divider must be used to connect two MIC devices in parallel. The system impedance ($Z_o$) is 50 ohms and each MIC amplifier has an impedance of 50 ohms. Calculate the values of $R$, $L$, and $C$ required for this circuit.
25. A quarter wavelength transmission line for 1.296 GHz must be made from Teflon® coaxial cable which has a dielectric constant of 0.77. What is the physical length in inches?
26. A 75 ohm load must be transformed to 125 ohms in a quarter-wave $Q$-section transmission line. Calculate the required characteristic impedance.
27. A stripline transmission line is being built on a double-sided printed circuit board made with low-loss ($e$ = 2.5) material. If the board is 0.125 inch thick, what width stripline must be used to match 70.7 ohms?

CHAPTER 10

# Nonlinear (Diode) Applications of Linear IC Devices

## OBJECTIVES

1. Learn the function of and applications for the precise diode circuit.
2. Understand how zero-bound and dead-band circuits work.
3. Be able to describe the operation of active clipper and clamper circuits.
4. Understand the peak follower and sample-and-hold circuits.

## 10-1 PREQUIZ

These questions test your prior knowledge of the material in this chapter. Try answering them before you read the chapter. Look for the answers (especially those you answered incorrectly) as you read the text. After you have finished studying the chapter, try answering these questions again and those at the end of the chapter (see Section 10-10).

1. Draw the $V_o$ against $V_{in}$ curves for both ideal and practical rectifiers.
2. Give two reasons why an active precise rectifier might be preferred over a single-diode rectifier.
3. Design a zero-bound amplifier which has a threshold point of +3 volts.
4. Draw the schematic for a peak follower circuit.

## 10-2 INTRODUCTION

Although the operational amplifier is well known for its capabilities as a linear amplifier, there are also numerous nonlinear applications for which the op-amp is also well suited. In this chapter you will study circuit techniques commonly employed in fields as diverse as instrumentation, control, and communications. Of particular interest here are circuits in which PN junction diodes are used, *precise rectifiers, bounded circuits*, and *clippers/clampers*.

## 10-3 REVIEW OF THE PN JUNCTION DIODE

The PN junction diode is the oldest solid-state electronic component available. Indeed, naturally occurring diodes of galena crystals (lead sulphide, or PbS) were used prior to World War I as the demodulator (detector) in crystal set radio receivers. During World War II, radar research led to the development of the 1N34 and 1N60 germanium video detector diodes, and the 1N21 and 1N23 microwave diodes.

The PN junction diode ideally has a transfer characteristic like the one shown in Fig. 10-1A. When the anode is positive with respect to the cathode (Fig. 10-1B), the diode is forward-biased so it conducts current. Alternatively, when the anode is negative with respect to the cathode (Fig. 10-1C), the diode is reverse-biased and no current flows. Figure 10-1D shows the effect of this unidirectional current flow on a sinewave input signal. Notice in the *halfwave rectified* output only the positive peaks are present.

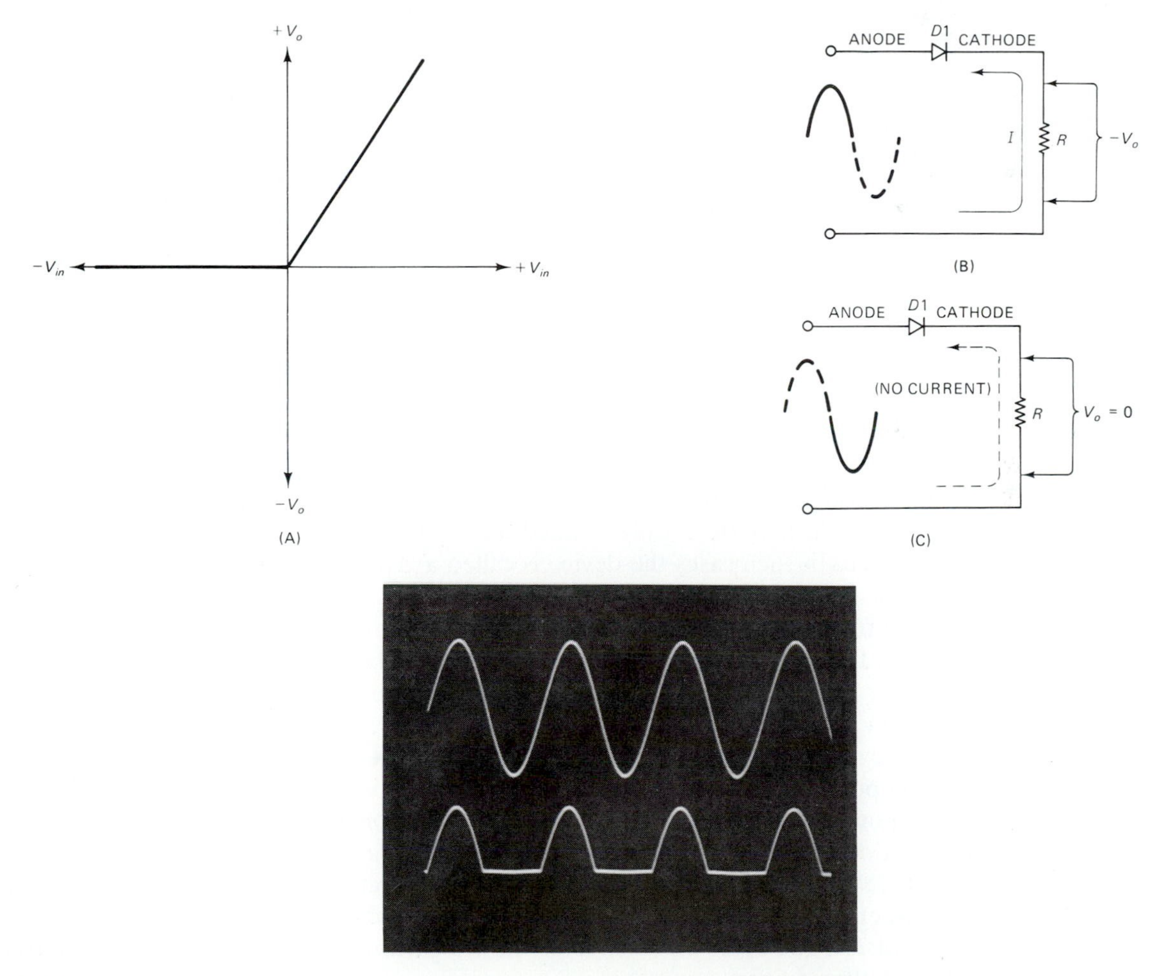


FIGURE 10-1

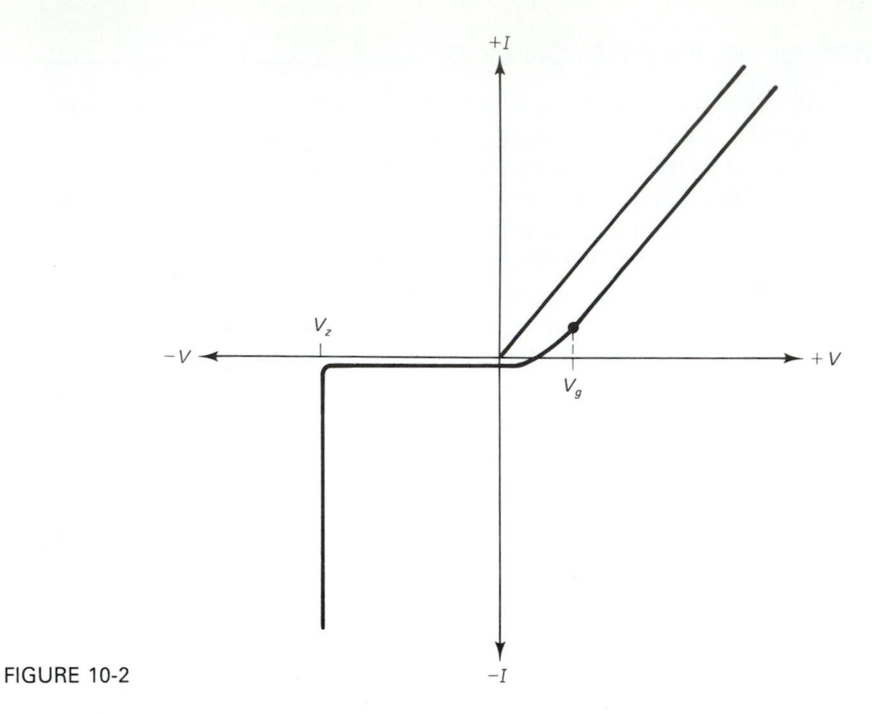


FIGURE 10-2

Real diodes fail to meet the ideal in several important respects. Figure 10-2 shows the transfer characteristic for a practical, nonideal diode. For the ideal diode the reverse current is always zero, while in real diodes there is a minute current leakage ($I_L$) flowing backwards across the junction. This current can be detected by measuring the forward and reverse resistances of a PN junction diode. The forward resistance is very low, while the reverse resistance is very high, but not infinite as one might expect from a supposed open circuit.

Another departure from the ideal in the reverse-bias region is the avalanche point ($V_Z$) where reverse-current flow increases sharply. At this point, the reverse-bias voltage is too great and causes breakthrough. When carefully regulated, the breakdown potential is both sharply defined and stable, except for a slight temperature dependence. In such cases the device is called a *zener diode* and is used as a voltage regulator.

In the forward-biased region there are other anomalies which are departures from the ideal. In the ideal there is an ohmic relationship between current flow and applied forward voltage. Similarly, there is a linear relationship between applied forward voltage ($V_f$) and output voltage $V_o$. In real diodes, however, there is a significant departure from the ideal transfer characteristic. Between zero volts and a critical junction potential ($V_g$) the characteristic curves are very nonlinear. The actual value of this potential is a function of both the type of semiconductor material used and the junction temperature. In general, $V_g$ will be 0.2 and 0.3 volts for germanium (Ge) diodes (1N34, 1N60, etc.), and 0.6 to 0.7 volts for silicon (Si) diodes (1N400x, 1N914, 1N4148, etc.). In the 0 to $V_g$ region the diode forward resistance is a variable function of $V_f$ and $T$. The $I$ against $V_f$ characteristic is logarithmic. Above $V_g$ the characteristic becomes almost linear.

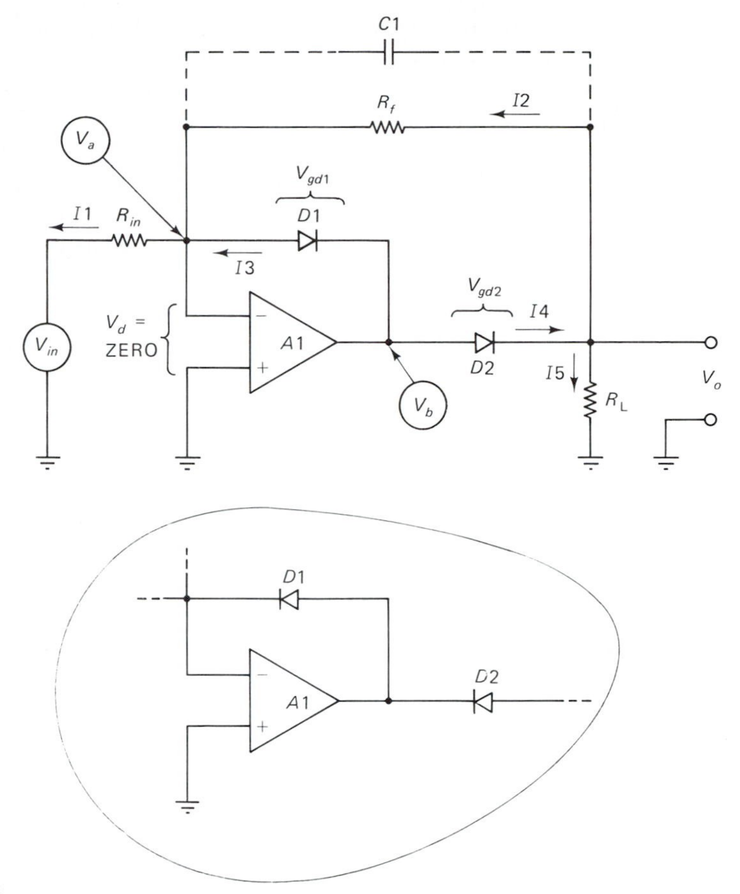


FIGURE 10-3

## 10-4 PRECISE DIODE CIRCUITRY

A *precise diode circuit* combines an active device like an operational amplifier with a pair of diodes to servo out the errors of the nonideal diode. There are two advantages to this arrangement. First, the circuit will rectify very small AC signals between zero volts and $V_g$ ($0 < V < V_g$, about 0.65 volts). Second, the rectification will be almost linear, which is not the case with just the diode alone, even in the diode's ohmic range.

Figure 10-3 shows the circuit for the inverting halfwave precise rectifier. Here the load impedance $R_L$ is purely resistive and therefore contains no energy production or storage elements. The circuit is essentially an inverting follower amplifier with two PN junction diodes ($D1$ and $D2$) added. Halfwave rectification occurs because the circuits offers two different gains which are dependent on the polarity of the input signal. For positive values of $V_{in}$ the gain ($V_o/V_{in}$) is zero. For negative values of $V_{in}$ the voltage gain is $R_f/R_{in}$.

Consider operation of the circuit for positive values of $V_{in}$. The noninverting input (+IN) is grounded, so it is zero volts. By the properties of the ideal op-amp we must consider the inverting input (−IN) as if it is also grounded ($V_a = 0$). Recall this concept is called a *virtual ground.* Thus, differential voltage $V_d$ is zero.

When $V_{in} > 0$ (when it is positive), current $I1 = +V_{in}/R_{in}$. In order to maintain the equality $I1 + I3 = 0$, conserving Kirchoff's Current Law (KCL), the op-amp output voltage $V_b$ swings negative. However, it is limited by the $D1$ junction potential to $V_g$ (about 0.6 to 0.7 volts). With $V_b < 0$ (even by only 0.6 to 0.7 volts), diode $D2$ is reverse-biased and therefore cannot conduct. Currents $I2$, $I4$ and $I5$ are zero. Thus for positive values of $V_{in}$, the output voltage $V_o$ is zero.

Now consider operation for $V_{in} < 0$. Under this input condition, the op-amp output voltage $V_b$ swings positive, forcing diode $D1$ to become reverse-biased and $D2$ to conduct. To conserve KCL in this case ($I1 + I2 = 0$), current $I2$ will have the same magnitude but opposite direction relative to $I1$. Because $V_{in}/R_{in} = -V_o/R_f$, the voltage gain ($A_v = V_o/V_{in}$) reduces to $-R_f/R_{in}$ as is appropriate for an inverting amplifier. Thus, the gain for negative input voltages ($V_{in} < 0$) is $-R_f/R_{in}$, while for positive input voltage ($V_{in} > 0$) it is zero. Halfwave rectification comes from this difference.

The voltage drop across diode $D2$ is about +0.6 to +0.7 volts. It is servoed out because $D2$ is in the negative feedback loop around $A1$. Voltage $V_b$ is correspondingly higher than $V_o$ to null the effects of $V_{gd2}$.

The precise rectifier is capable of halfwave rectifying very low input signals. The minimum signal allowed is

$$V_{in} > \frac{V_g}{A_{vol}} \tag{10-1}$$

where

$$\begin{aligned} V_{in} &= \text{the input signal voltage} \\ V_g &= \text{the diode junction potential (0.6 to 0.7 volts)} \\ A_{vol} &= \text{the open-loop gain of the amplifier} \end{aligned}$$

The open-loop gain ($A_{vol}$) of DC and low-frequency AC signals is extremely high. But at some of the frequencies at which precise diodes operate, the input frequency is a substantial fraction of the gain-bandwidth product, so $A_{vol}$ will be less than might be expected. For example, if the gain-bandwidth product is 1.2 MHz, the gain at 100 Hz is 12,000. But at 1000 Hz, a typical frequency for precise rectifier operation, the gain is only 1200.

### EXAMPLE 10-1

Calculate the minimum signal that can be accurately rectified in a precise rectifier circuit where the open-loop gain is 3000. Assume that silicon diodes are used, so $V_g = 0.7$ volts.

#### Solution

$$V_{in} > V_g/A_{vol}$$

$$V_{in} > (0.7\ \text{volts})/(3000) = 0.00023\ \text{volts}$$

■

Circuit operation of the precise rectifier is shown by the waveforms in Fig. 10-4. If a sinewave is applied (Fig. 10-4A), the output voltage $V_o$ will be zero from time $T1$ to $T2$ (positive input voltage), while $V_b$ will rest at $-V_g$ (about $-0.6$ to $-0.7$ volts). The input is negative between $T2$ and $T3$ so $V_o$ will be positive with a halfwave sine shape (Fig. 10-4B). But note the behavior of $V_b$, the op-amp output (Fig. 10-4C). From $T1$ to $T2$ the output rests at $-V_g$, but at $T2$ it snaps to a positive value of $2 \times V_g$. The halfwave sine shape rests on top of the $+V_g$ offset caused by $V_{gd2}$. Figure 10-4D shows the same situation in the form of the transfer characteristic ($V_o$ against $V_{in}$).

The circuit in Fig. 10-3 will rectify and invert negative peaks of the input signal. To accommodate the positive peaks, the polarity of diodes $D1$ and $D2$ can be reversed.

### 10-4.1 Precise Diodes as AM Demodulators

Demodulation or detection of amplitude modulated (AM) carrier signals is one of several applications for the precise diode circuit. An AM signal (Fig. 10-5) is a sinusoidal carrier signal of frequency $F_c$ varied in amplitude by a lower frequency modulating frequency $F_m$. The modulating signal might be either sinusoidal or non-sinusoidal.

The most familiar use of AM is in radio broadcasting and communications. Not so familiar, perhaps, is the use of AM in nonradio applications such as instrumentation. One common use is in certain AC-excited Wheatstone bridge transducers. In one popular pressure transducer the excitation signal is a 10 volt peak-to-peak 400 Hz sinewave. The output of the transducer is an AC signal which is proportional to the excitation voltage and the applied pressure. Thus, the pressure signal is used to modulate the 400 Hz carrier signal.

Demodulation of an AM signal is usually done by an *envelope detector*. This is a halfwave rectifier and a low-pass filter. The precise diode circuit of Fig. 10-5 can be used to demodulate AM signals and offers an advantage because of its ability to accommodate weak signals. Low-pass filtering is obtained by shunting capacitor $C1$ across feedback resistor $R_f$.

For any given type of operational amplifier, there is a maximum carrier frequency which can be accepted for AM demodulation. This frequency is related to the gain bandwidth product of the particular device selected

$$F_{c(\max)} = \frac{F_t}{100\ \mathrm{ABS}(A_v)} \tag{10-2}$$

where

$F_{c(\max)}$ = the maximum allowable carrier frequency.

$F_t$ = the device gain-bandwidth product (the frequency at which $A_{vol} = 1$).

$\mathrm{ABS}(A_v)$ = the absolute value of the closed-loop voltage gain.

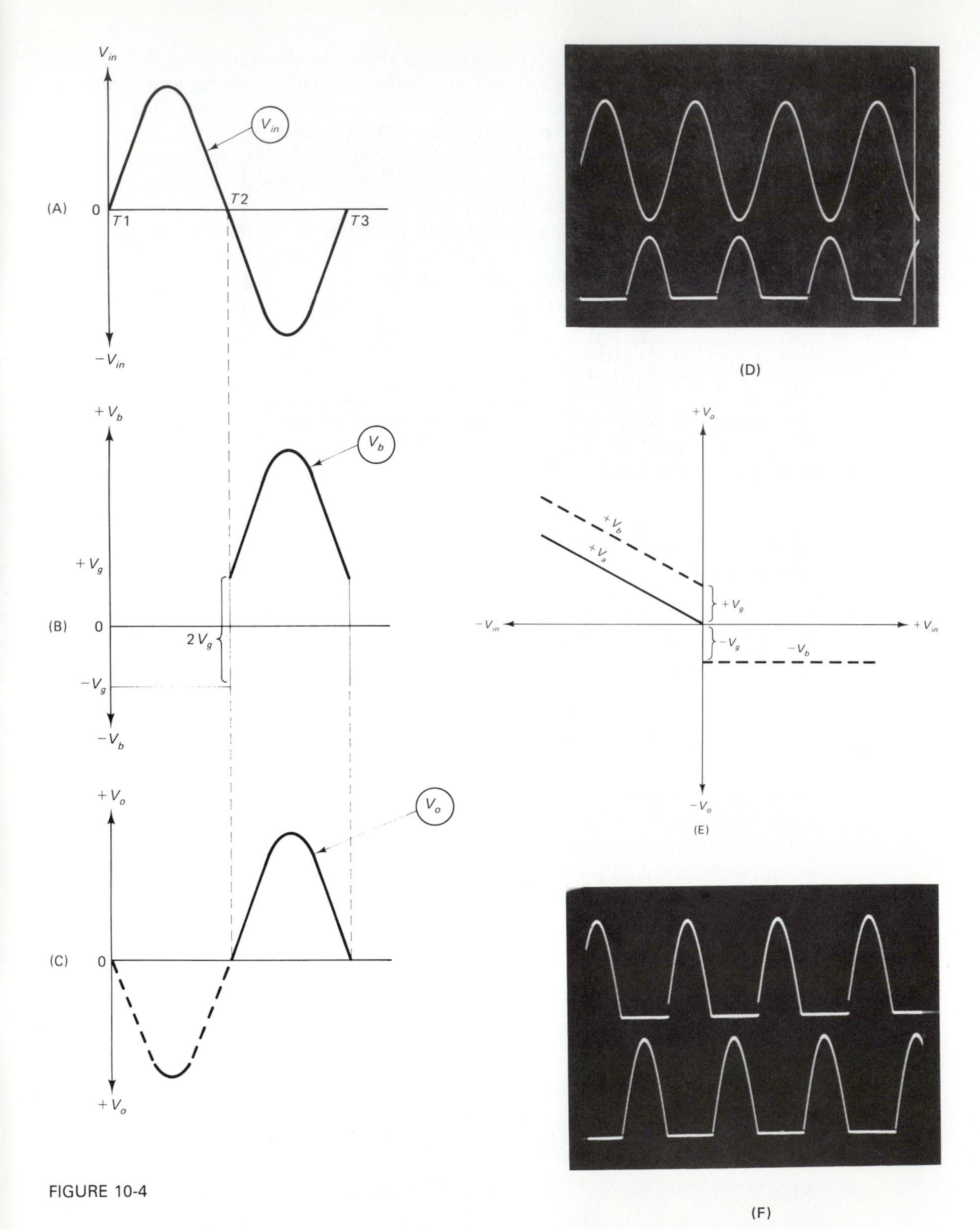

$V_{in}$
$V_{in}$
(A)
0
$T1$
$T2$
$T3$
$-V_{in}$
$+V_b$
$V_b$
$+V_g$
(B)
0
$2V_g$
$-V_g$
$-V_b$
$+V_o$
$V_o$
(C)
0
$+V_o$
(D)
$+V_o$
$+V_b$
$+V_a$
$+V_g$
$-V_{in}$
$+V_{in}$
$-V_g$
$-V_b$
$-V_o$
(E)
(F)


FIGURE 10-4

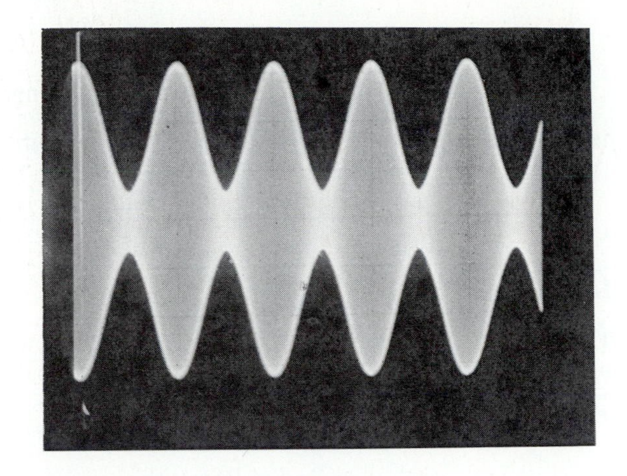

FIGURE 10-5

### EXAMPLE 10-2

A BiMOS operational amplifier with a gain-bandwidth product of 12 MHz is used as a precise rectifier AM demodulator. Calculate the maximum carrier frequency the circuit can handle in a gain-of-100 circuit.

### Solution

$$F_{c(max)} = \frac{F_t}{100\ \mathrm{ABS}(A_v)}$$

$$F_{c(max)} = \frac{1.2 \times 10^7\ \mathrm{Hz}}{(100)\ (\mathrm{ABS}(-100))}$$

$$F_{c(max)} = \frac{1.2 \times 10^7\ \mathrm{Hz}}{(100)\ (100)}$$

$$F_{c(max)} = 1.2 \times 10^3\ \mathrm{Hz} = 1{,}200\ \mathrm{Hz}$$

■

The operational amplifier slew rate ($S$) must be sufficient to handle the input signal carrier frequency. In general:

$$S > \frac{\Delta V_{in}}{\Delta t} \tag{10-3}$$

or, for the specific case of sinusoidal carriers

$$S > 2\pi F_c\ \mathrm{ABS}(A_v) V_{\mathrm{in(peak)}}$$

where

$$S = \text{the slew rate in volts per microsecond } (\mu S)$$

$$\Delta V_{in} = \text{the change in input voltage over time } \Delta t$$

$$\Delta t = \text{the time over which } V_{in} \text{ changes}$$

The value of capacitor $C1$ (Fig. 10-3) is found from

$$C1 = \frac{[F_c/F_m]^{1/2}}{2 \pi F_c R_f} \tag{10-4}$$

### EXAMPLE 10-3

A 1000 Hz sinusoidal carrier is modulated by a 15 Hz sinewave. Calculate $C1$ when 10 kohm resistors are used for both $R_f$ and $R_{in}$. Assume a peak signal voltage of 2 volts.

### Solution

(a) $S > 2\pi F_c \text{ABS}(A_v) V_{\text{in(peak)}}$

$$S > (2)(3.14)(100 \text{ Hz})(10\text{k}/10\text{k}) \ (2 \text{ volts})$$

$$S > 12{,}560\text{V/S} = 0.013 \text{ uV/S}$$

(b) $C1 = \dfrac{[F_c/F_m]^{1/2}}{2 \pi F_c R_f}$

$$C1 = \frac{[(1000/15)]^{1/2}}{(2)(3.14)(1000 \text{ Hz})(10{,}000 \text{ ohms})}$$

$$C1 = \frac{[66.7]^{1/2}}{6.3 \times 10^7}$$

$$C1 = \frac{8.2}{6.3 \times 10^7} \text{ farads} = 0.13 \text{ uF}$$

■

### 10-4.2 Polarity Discriminator Circuits

A *polarity discriminator* will produce outputs which indicate whether the input voltage is zero, positive, or negative. Applications for this circuit include alarms, controls, and instrumentation.

Figure 10-6A is a typical polarity discriminator circuit. The basic configuration is an inverting follower op-amp circuit, but with two negative feedback circuits. Each feedback path contains a diode. But the diodes are connected in the opposite polarity sense, so the polarity of the output potential will determine which one conducts and which is reverse-biased.

First consider the case where the input signal $V_{in}$ is positive. In this case, current $I_{in}$ flows away from the summing junction toward the source and has a magnitude of $+V_{in}/R_{in}$. The output terminal of the op-amp will swing negative, causing diode $D1$ to be reverse-biased and $D2$ to be forward-biased. Current $I1$ is zero and $I2$ is equal

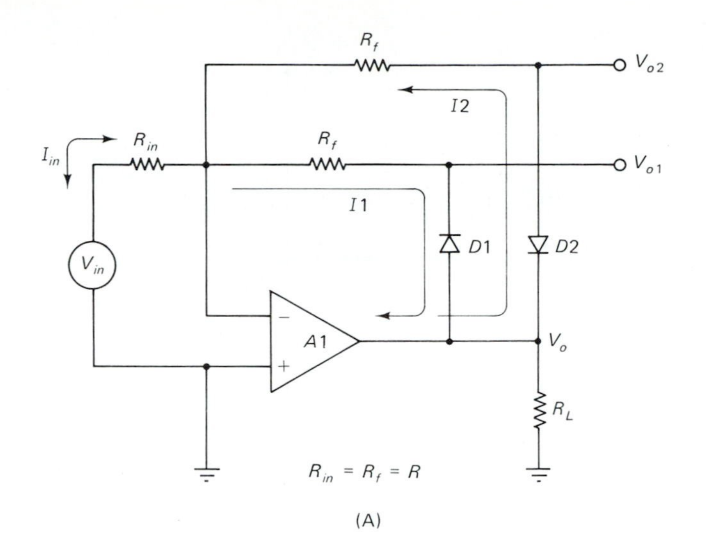


(A)

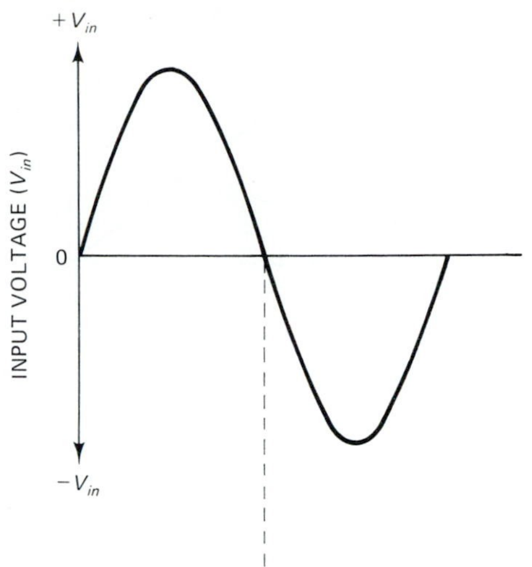


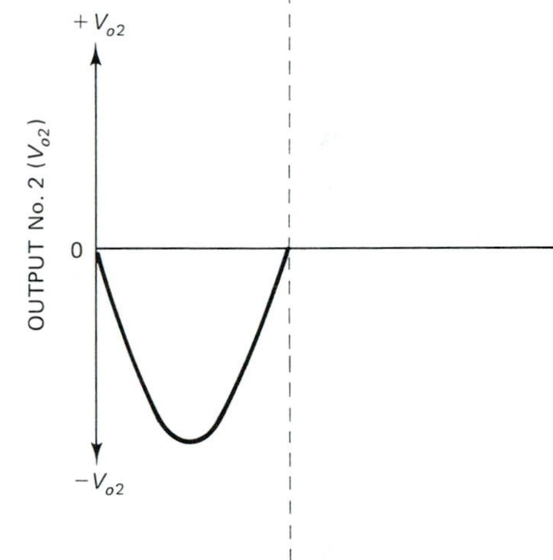


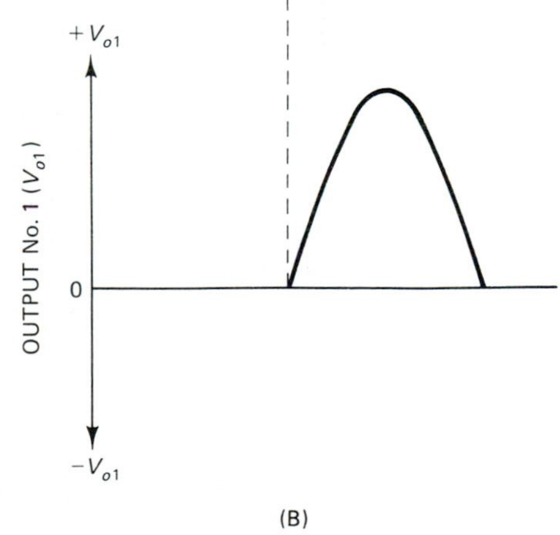


(B)

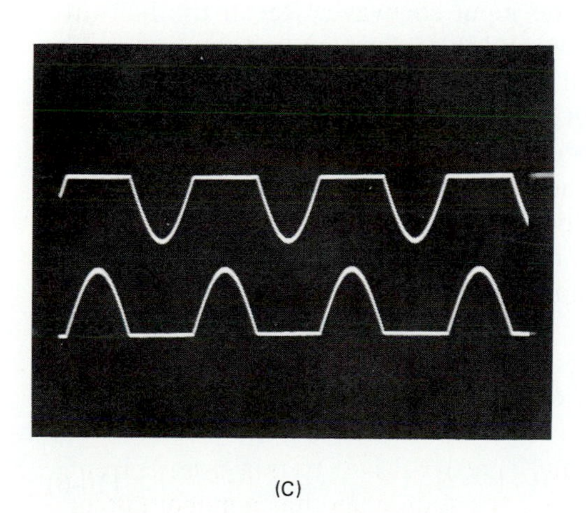

(C)

FIGURE 10-6

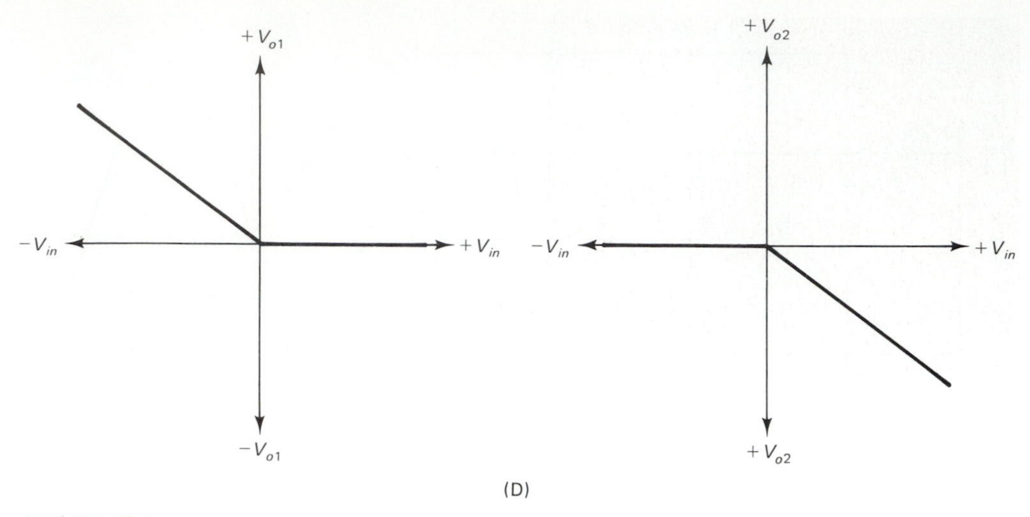


FIGURE 10-6

to $[V_{o2}]/R_f$. Output voltage $V_{o2}$ is negative and has a value of $V_{o2} = V_o - 0.6$ volts. The output voltage $V_{o1}$ is zero.

Now consider the opposite case where $V_{in}$ is negative. The current flows away from the source toward the summing junction. The output terminal of the op-amp swings positive, causing diode $D1$ to become forward-biased, while $D2$ is reverse-biased. Current $I2$ is zero, while $I1$ is $V_{o1}/R_f$. In this case $V_{o1}$ is positive, while $V_o$ is zero.

The operation of this circuit can be seen in the waveforms shown in Fig. 10-6B, and in the transfer characteristics shown in Fig. 10-6C.

### 10-4.3 Fullwave Precise Rectifier

The *fullwave rectifier* uses both halves of the input sinewave. Recall the halfwave rectifier removes one polarity of the sinewave and the fullwave rectifier preserves it. Figure 10-7 shows the relationships in a fullwave rectifier circuit. In Fig. 10-7A we see the input sinewave and the pulsating DC output of a fullwave rectifier. Note the negative halves of the sinewaves are flipped over and appear in the positive-going direction. The characteristic function for the fullwave rectifier is shown in Fig. 10-7B. Because the output voltage is always positive, regardless of whether the input signal is positive or negative, the fullwave rectifier can be called an *absolute value circuit*. Depending on the direction of the diodes within the circuit, the output voltage will be either

$$V_o = k|V_{in}| \tag{10-5}$$

or

$$V_o = -k|V_{in}| \tag{10-6}$$

The fullwave rectifier is important in DC power supplies. But its absolute value feature makes it useful for instrumentation and related applications.

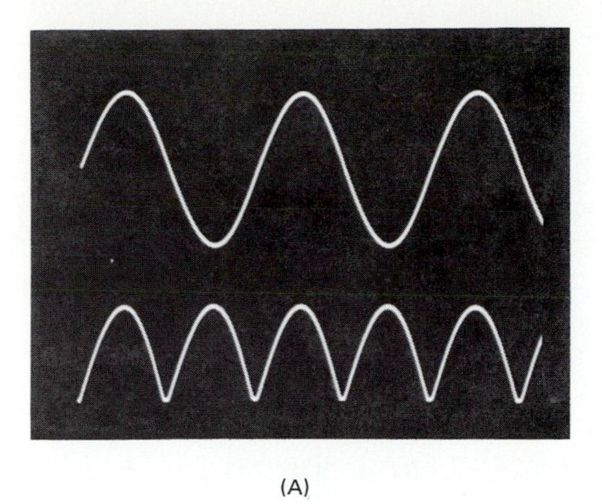

(A)

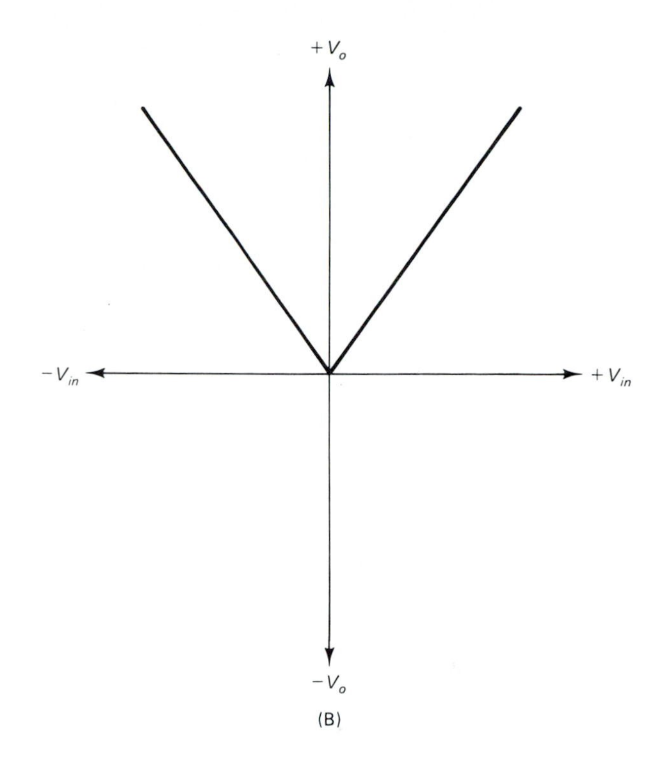


(B)

FIGURE 10-7

Some methods for creating a precise fullwave rectifier (absolute value amplifier) are shown in Fig. 10-8. The first circuit (Fig. 10-8A) is based on the polarity discriminator circuit of Fig. 10-6. The two outputs ($V_{o1}$ and $V_{o2}$) are applied to the inputs of a DC differential amplifier.

A second approach is shown in Fig. 10-7B. Here a pair of oppositely connected diodes are applied to the inputs of a simple DC differential amplifier. This approach is not highly regarded because the diodes in the input stage are not in the feedback loop, so their voltage drops ($V_g$) are not servoed out.

Figure 10-8C shows the usual absolute value amplifier circuit. It consists of a pair of precise halfwave rectifier circuits connected so their respective outputs are summed at the input of an output buffer amplifier. Amplifier $A1$ is connected as an inverting precise rectifier and amplifier $A2$ is connected as a noninverting precise rectifier. The waveform of the signal at the summing point (the input of amplifier $A3$) is the absolute value of the input waveform.

There are two common variations to this circuit. As shown, the input circuit of $A3$ uses a resistor to ground which is much larger than the summing resistors ($R$). This prevents loading of the signal by voltage divider action between $R$ and $R_i$. In some modern BiFET and BiMOS amplifiers $R_i$ is not needed because the bias current is almost zero.

The other alternative is to make $R_i = R$, causing the input voltage to $A3$ to be reduced to one half by $V(R)/(R + R)$. In this case, it is necessary to make $A3$ a gain of two noninverting follower amplifiers (see inset to Fig. 10-8C).

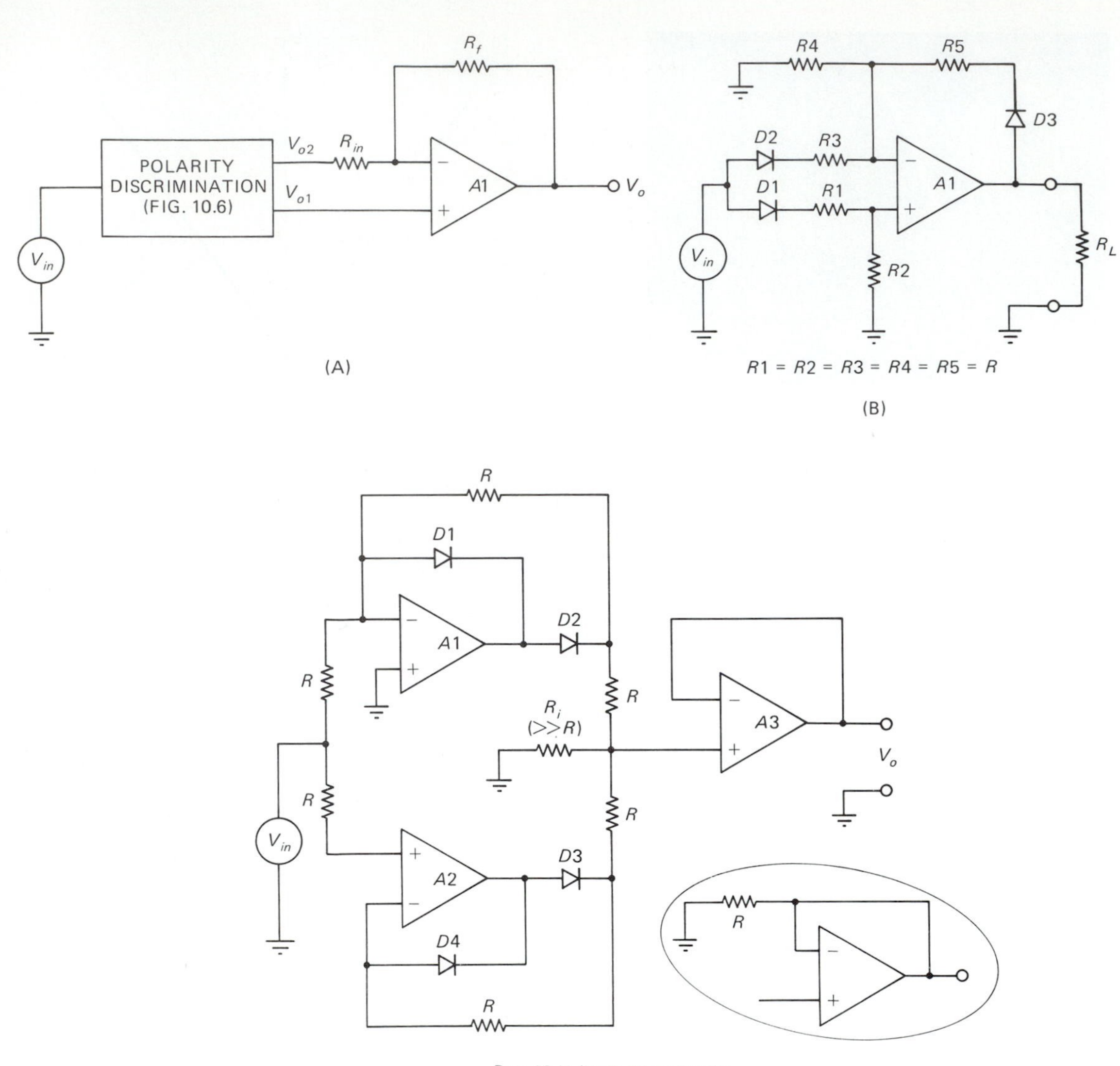


FIGURE 10-8

### 10-4.4 Zero-Bound and Dead-Band Circuits

**Zero-Bound Circuits.** A zero-bound circuit is one where the output voltage is limited so it will be nonzero for certain values of input voltage and zero for all other input voltages. The term does not mean the values of $V_{in}$ are constrained. It means there are constraints on allowable output voltages. The output of a zero-bound circuit indicates when the input signal exceeds a certain threshold and by how much.

Figure 10-9 shows a zero-bound amplifier circuit. It is based on the halfwave precise rectifier circuit of Fig. 10-3 and functions in the same way except for the extra input

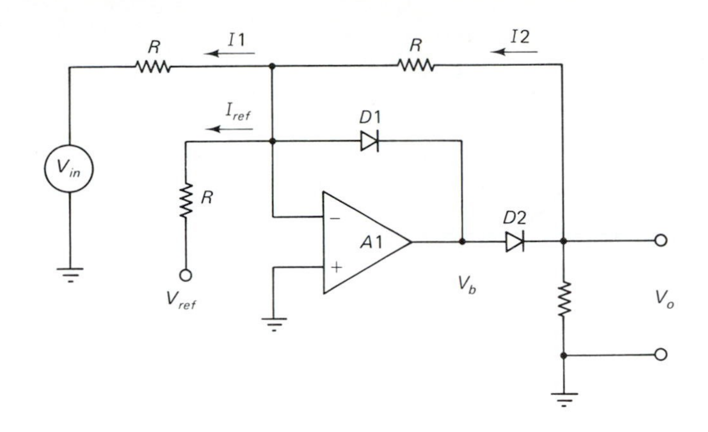


FIGURE 10-9

reference current ($I_{ref}$). $I_{ref}$ offsets the trip point where the input voltage takes effect.

We know from Kirchoff's Current Law (KCL) and the fact op-amp inputs neither sink nor source current, that the following relationship is true

$$I1 + I_{ref} + I2 = 0 \tag{10-7}$$

or

$$I1 + I_{ref} = -I2 \tag{10-8}$$

We also know that

$$I1 = \frac{V_{in}}{R} \tag{10-9}$$

$$I_{ref} = \frac{V_{ref}}{R} \tag{10-10}$$

$$I2 = \frac{V_o}{R} \tag{10-11}$$

thus

$$\frac{V_{in}}{R} + \frac{V_{ref}}{R} = -\frac{V_o}{R} \tag{10-12}$$

and after multiplying both sides by $R$

$$V_{in} + V_{ref} = -V_o \tag{10-13}$$

Thus, the output voltage is still proportional to the input voltage, but it is offset by the value of $V_{ref}$. The transfer characteristics for this circuit are shown in Fig. 10-10. In Fig. 10-10A the value of $V_{ref}$ is negative. In Fig. 10-10B the value of $V_{ref}$ is positive. In both cases, however, the transfer curve is offset by the reference signal potential.

Consider the operation of the circuit of Fig. 10-9 under two conditions, $V_{in} > 0$ and $V_{in} < 0$. First assume that $V_{ref} = 0$. For the positive input ($V_{in} > 0$), the output

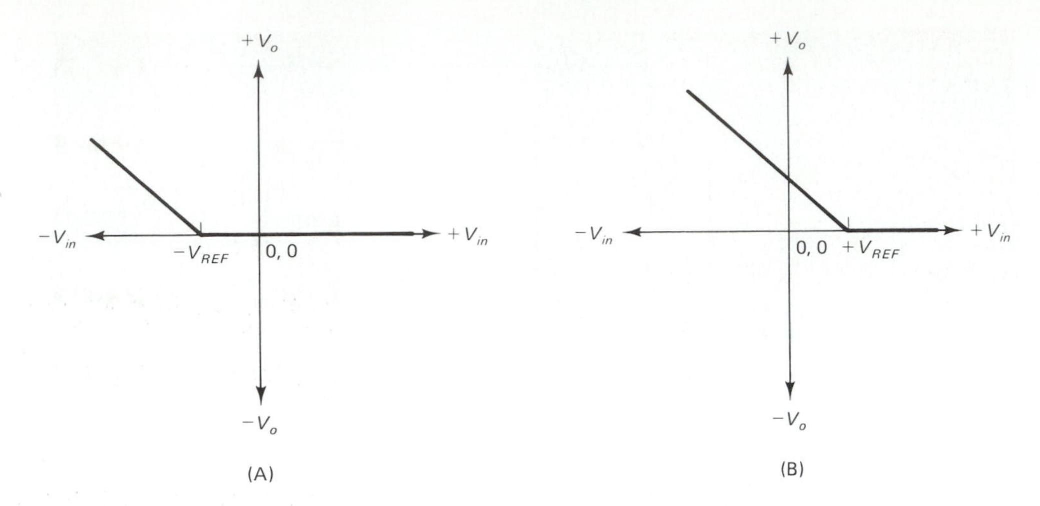


FIGURE 10-10

of the operational amplifier ($A1$) swings negative (the circuit is an inverter). This causes diode $D2$ to be reverse-biased and $D1$ to be forward-biased. Output voltage $V_o$ is zero. The output voltage is bound to zero for all values of $V_{in} > 0$.

Now consider the cases where $V_{ref}$ is not zero. There are two cases, $V_{ref} > 0$ and $V_{ref} < 0$. Figure 10-11 shows several cases of a zero-bound circuit such as Fig. 10-9. In all of these examples, a 741 operational amplifier was connected with a pair of 1N4148 silicon signal diodes. The value of $R$ was selected to be 22 kohms. The excitation signal was an 8-volt p-p sinewave at a frequency of 700 Hz. Figure 10-11A shows the action of the circuit without the reference voltage applied ($V_{ref} = 0$). The circuit operates as a normal precise rectifier. As is true in all the Fig. 10-11 examples, the input sinewave is shown in the upper trace and $V_o$ is shown in the lower trace on the dual-beam oscilloscope.

In Fig. 10-11B $V_{ref} = -1.2$ volts. Notice the clipping action. The amplitude of the waveform is 5.2 volts base-to-peak. Because the input signal is 8 volts p-p, the positive peak is 4 volts peak. Thus, the baseline of the output signal voltage is at a level of $[(+4V) - (5.2\ V)]$, or $-1.2$ volts, which is the value of $V_{ref}$. In this circuit, the zero-bounding occurs at all negative potentials greater than $-1.2$ volts. Only those signals which are more positive than this can pass to the output of the circuit.

The waveforms in Fig. 10-11C are similar to those in Fig. 10-11B, but the reference voltage has been increased to $-3.4$ volts. In this case only 0.6 volts (4 − 3.4 volts), of the negative peak is unable to pass to the output. Exactly the opposite situation is shown in Fig. 10-11D. Here the reference voltage has the same magnitude, but it is reversed in polarity ($+3.4$ volts DC). Note that only the top 0.6 volts of the positive peak shows. All lower voltages are zero-bounded.

**Dead-Band Circuits.** In a dead-band circuit two zero-bound circuits work together to produce a summed output. Figure 10-12A shows the transfer characteristic of such a circuit. There are two differential threshold values shown in this curve. The circuit outputs signals only when the input signal is less than the lower threshold ($V_{in} <$

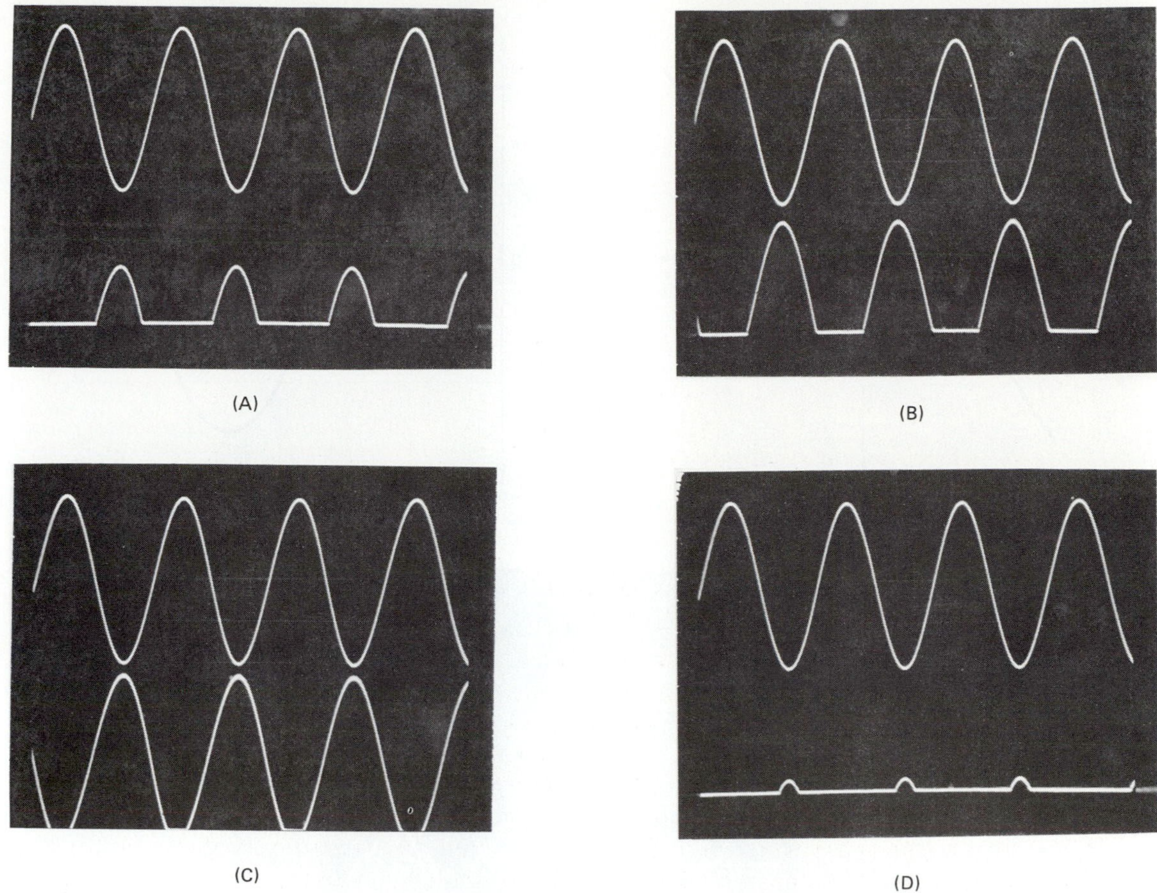

FIGURE 10-11

$-V_{th}$) or greater than the upper threshold ($V_{in} > +V_{th}$). This is shown relative to a sinewave input signal. The output will be zero for all values of the input signal within the shaded zone.

Keep in mind the output voltage will not suddenly snap to a high value above the threshold potential, but rather it will be equal to the difference between the peak voltage and the threshold voltage. Assuming unity gain for both reference voltage and input signal voltage, the output peaks will be $[(+V_p - (+V_{th})]$ and $[(-V_p - (-V_{th})]$. This type of waveform is shown in Fig. 10-12C. Note here the two threshold voltages are not equal so they produce different peak values.

Both zero-bound circuits of a dead-band amplifier are similar to Fig. 10-9. Zero-bound circuit 1 uses diodes in the same polarity as Fig. 10-9, while zero-bound circuit 2 uses reverse polarity diodes. Both are shown in the insets to Fig. 10-13.

In the first circuit, the $V+$ DC power supply is used as $V_{ref}$. In the second, the $V-$ DC power supply is used as $V_{ref}$. In both cases the magnitude is the same, but the polarities are reversed. The difference in the threshold levels in this case is set by using different values of reference resistor, $3R$ in the first zero-bound circuit and $5R$ in the second zero-bound circuit. The result is the waveform of Fig. 10-12C.

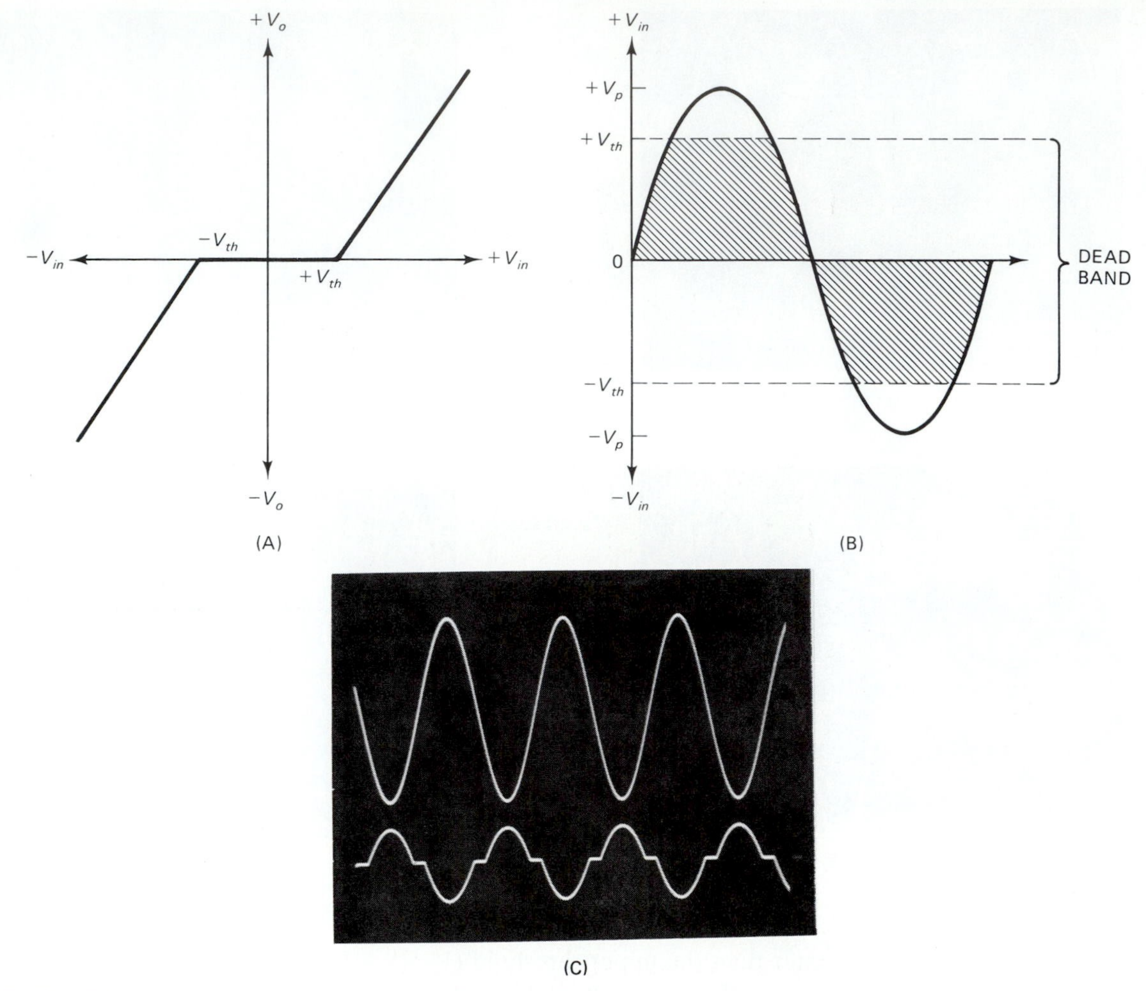


FIGURE 10-12

## 10-5 PEAK FOLLOWER CIRCUITS

The peak follower will output the highest value input voltage applied to it, regardless of what happens to the input voltage after that point. Figure 10-14 shows the action of a typical peak follower. Input voltage $V_{in}$ varies over a wide range. The output voltage (shown by the heavy line profile) however, always remains at the highest value previously reached and only increases if a new peak is encountered.

A typical peak follower is shown in Fig. 10-15. This circuit is a noninverting halfwave precise rectifier in which a unity gain, noninverting buffer amplitude is inserted into the feedback loop between diode *D*2 and the feedback resistor. A capacitor to hold the charge (*C*1) and a reset switch to discharge the capacitor are also added. The value of the capacitor is small enough to allow it to rapidly charge when an input signal is applied, but large enough to not saturate too quickly.

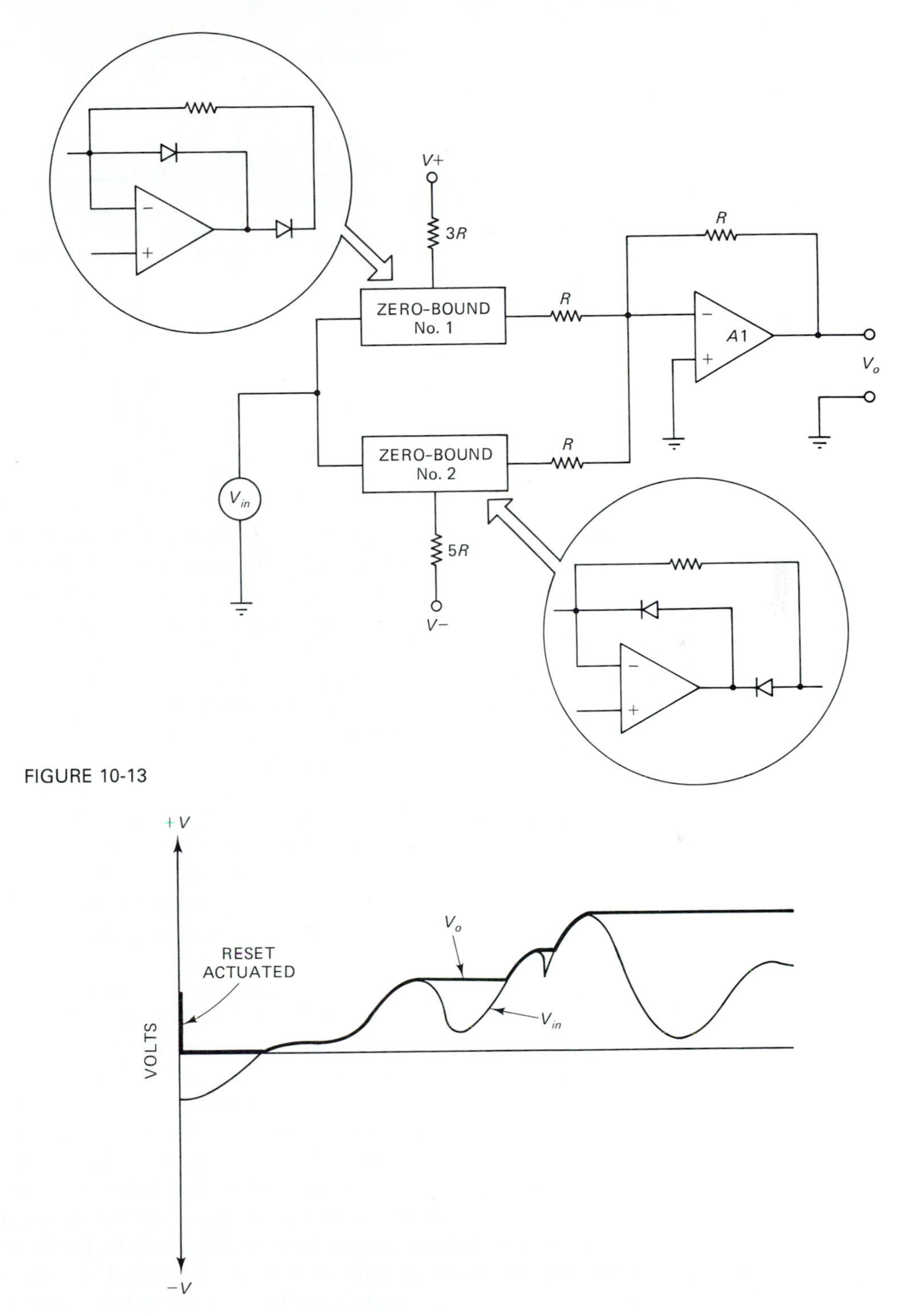


FIGURE 10-13

FIGURE 10-14

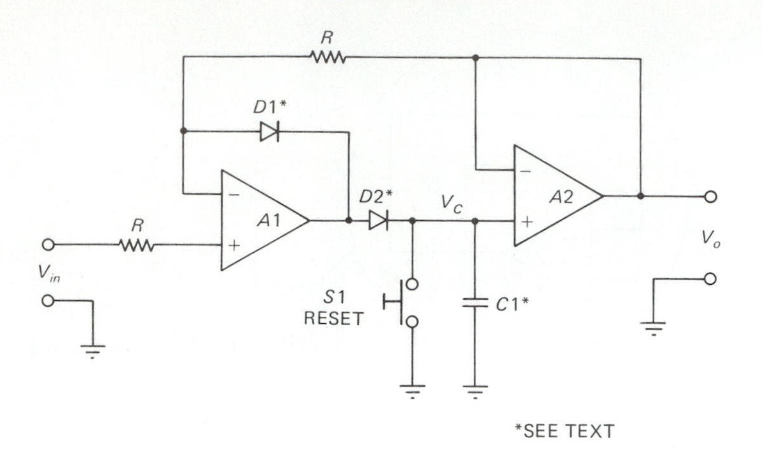


FIGURE 10-15

When the input voltage is positive ($V_{in} > 0$), the output of $A1$ is also positive. This forces diode $D1$ to be reverse-biased and $D2$ to be forward-biased. At initial turn on, or immediately after the reset switch is closed and then reopened, the capacitor is discharged. In this case $V_c = 0$. If a positive input voltage is applied to the input, this potential begins to charge $C1$, causing $V_c$ to increase to $V_{in}$. When the voltage across $C1$ is equal to the output voltage of the amplifier, current flow into the capacitor ceases and the value of $V_c$ is at the maximum value reached by $V_{in}$. Because amplifier $A2$ is a noninverting, unity gain follower, the output voltage is equal to the voltage across the capacitor ($V_o = V_c$).

As long as $V_{in}$ is equal to or less than the capacitor voltage, the capacitor voltage remains unchanged. In other words, the capacitor voltage (and the output voltage) remains at the previous high value. But if the input voltage rises to a point greater than the previous peak, $V_{in} > V_c$. Then a current will flow into $C1$ and cause its voltage to reach the new level. Again, the output voltage will track the previous high value of $V_{in(\max)}$. Only after reset switch $S1$ is momentarily closed will the output return to zero.

For negative input voltages, diode $D2$ is reverse-biased, so no current will pass either to or from the capacitor. This circuit ignores negative input potentials.

There are special precautions to take with some components in this circuit. Capacitor $C1$, for example, must be a very low leakage type. If there is leakage current, which implies a shunt resistance across the capacitance, the charge on the capacitor will bleed off with time. For the same reason, the input impedance of amplifier $A2$ must have an extremely high input impedance. For this reason, special premium operational amplifiers are selected. BiMOS, BiFET, and other very low input bias current models are preferred for this application. Also, the diode selected for $D2$ must have an extremely high reverse resistance. In other words, the leakage current which passes through $D2$ must be kept as low as possible. The reason for these precautions is to prevent the charge on $C1$ from bleeding off prematurely. The outward result of this circuit action is apparent "droop" of the output voltage, $V_o$.

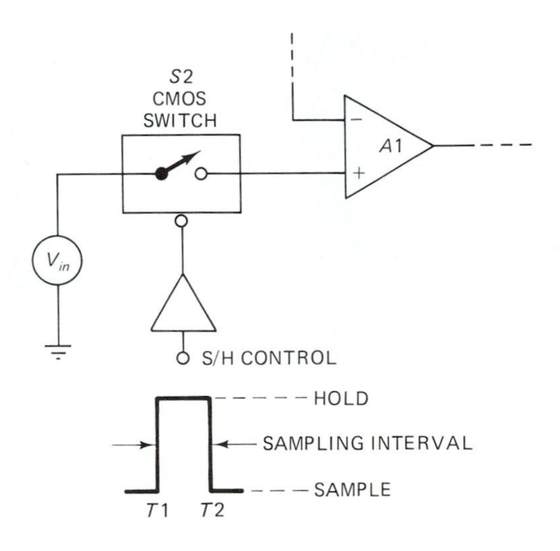


FIGURE 10-16

### 10-5.1 Sample-and-Hold Circuit

The peak follower circuit of Fig. 10-15 can be modified to form a sample-and-hold circuit. By adding a series switch ($S2$) at the input (Fig. 10-16) the peak follower will admit signals only at a discrete time determined by the S/H control signal. The switch is a CMOS electronic switch. It is used to allow a logic signal (which might be provided by a signal) to produce the S/H action. A similar switch can be used for $S1$ (Fig. 10-15). The correct action will first drive $S1$ closed to discharge $C1$, and then open $S1$ and close $S2$ in order to charge $C1$ with the maximum value of $V_{in}$ reached during the $T2-T1$ sampling interval.

## 10-6 CLIPPER CIRCUITS

Clipper (or clamp) circuits are the opposite of the dead-band circuit. The output voltage in these circuits will swing at will around zero if the input signal does not exceed a certain predetermined threshold. Either the positive peak, the negative peak, or both, will be limited to a certain clamped value. Figure 10-17 shows a typical example. These waveforms were taken in an inverting follower circuit with a gain of ($-100$ kohm/22 kohm), or about 4.6. In Fig. 10-17A the input signal (upper trace) is close to, but below the critical value. The output voltage is therefore unclipped on the positive peak and only moderately clipped on the negative peak. In Fig. 10-17B, however, the input signal is considerably increased, but the amplifier has no further output voltage. As a result, the peaks of the amplifier output are clamped to a value determined by the DC power supply potential applied to the amplifier.

Figure 10-17C shows a transfer characteristic for an inverting clipper. Output voltage $V_o$ is allowed to swing only between the lower and upper limits ($-V_{\text{LIM}}$ and $+V_{\text{LIM}}$ respectively). The dotted lines represent the output voltage which would exist in the absence of limiting.

Generally, it is not satisfactory to limit the DC power supply potentials to achieve clipping. The usual procedure is to use the full power supply potential, but to limit

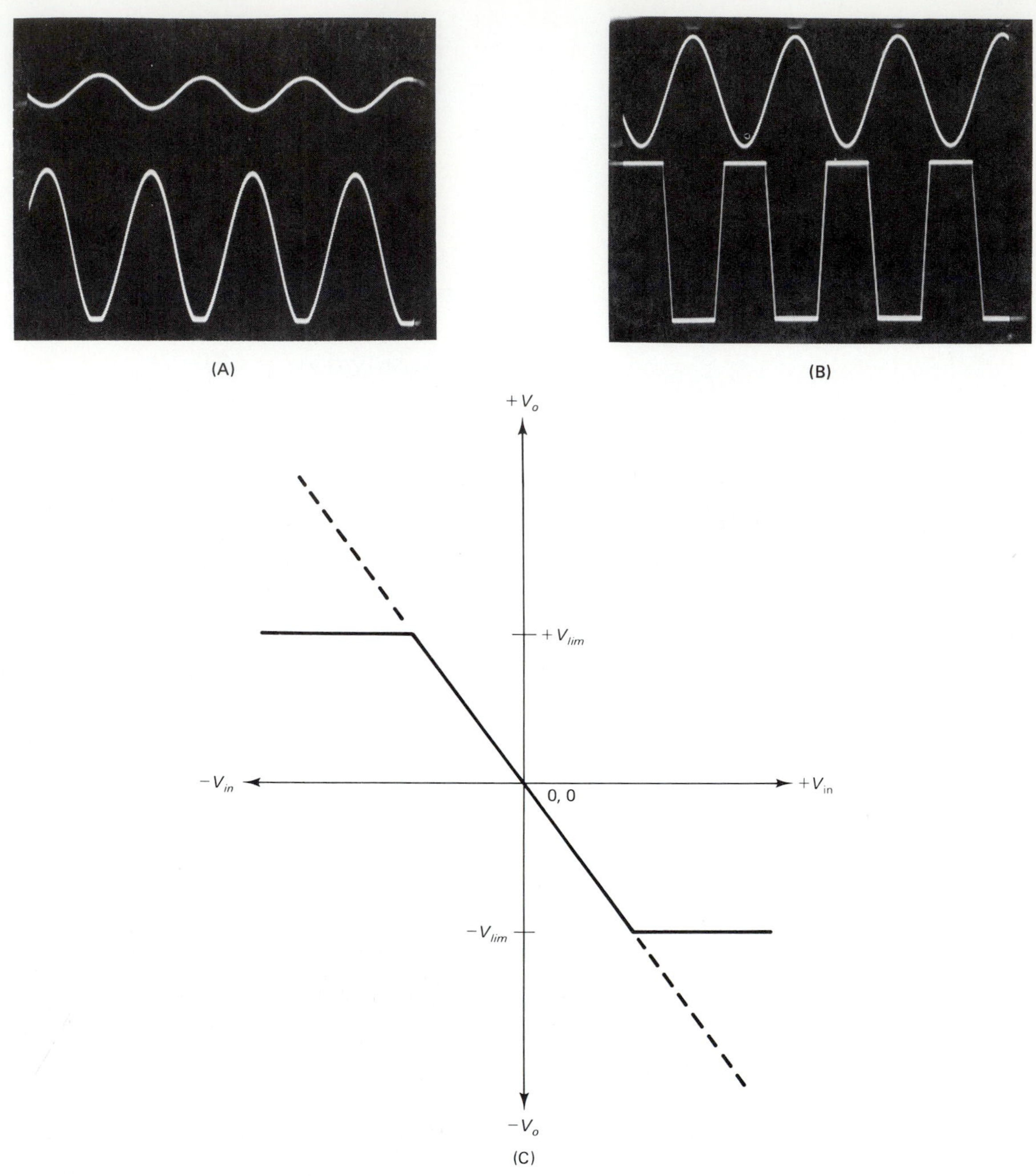


FIGURE 10-17

amplifier output voltage by certain circuit methods. Figure 10-18 shows one popular, but largely unsatisfactory, method for limiting the output swing. Here, the feedback resistor is shunted by a pair of back-to-back zener diodes. On positive output voltages, $D2$ is forward-biased and $D1$ is reverse-biased. As long as $+V_o$ is less than $V_{Z1}$ plus

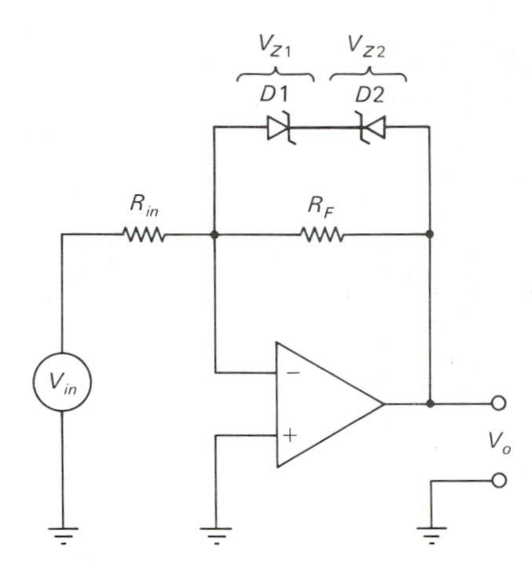


FIGURE 10-18

the forward drop across $D2$ (about 0.6 volts), it follows the requirements of the usual transfer equation ($V_o = (V_{in}(R_f)/R_{in})$. At values greater than $V_{Z1} + 0.6$ volt, however, the output is clamped. This is also true with negative output potentials. On negative swings of the output voltage the clamp occurs at $-V_{Z2} - 0.6$ volts.

The circuit in Fig. 10-18 is not highly regarded because it requires a relatively heavy signal level to sustain the zener diodes in the avalanche condition. A different approach is shown in Fig. 10-19A. In this case the amplifier drives a diode ($D1$). In the absence of $V_{ref}$, the diode will be forward-biased for positive values of $V_{in}$. But when $V_{ref}$ is applied, it will reverse bias $D1$ and prevent an output voltage until $V_a$ overcomes the reference potential. The results of this circuit are also not satisfactory (see Fig. 10-19B).

A more satisfactory approach is shown in Fig. 10-20. A diode bridge ($D1$–$D4$) is inserted into the feedback loop of amplifier $A1$. As a result, some of the nonlinearities that afflict the other circuits are servoed out of this circuit by the $1/BA_v$ factor (in which $B$ is the transfer equation of the feedback network). There are two limiting conditions for this circuit

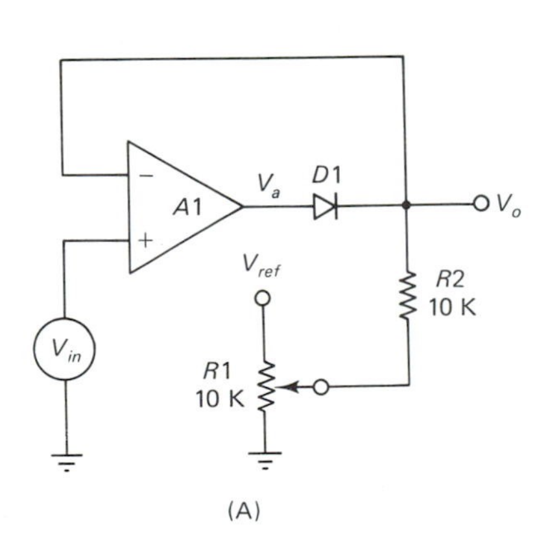


(A)

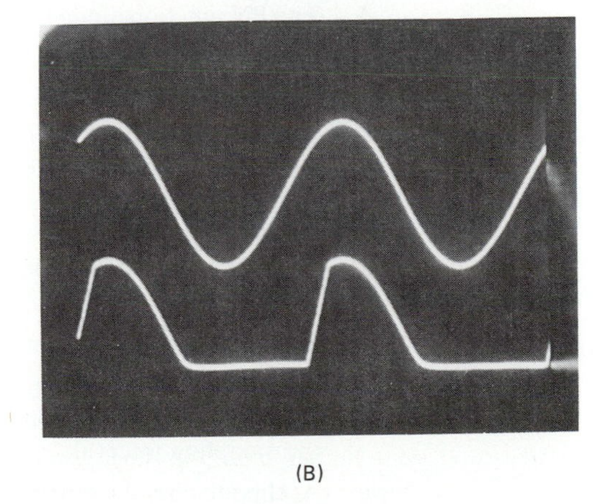

(B)

FIGURE 10-19

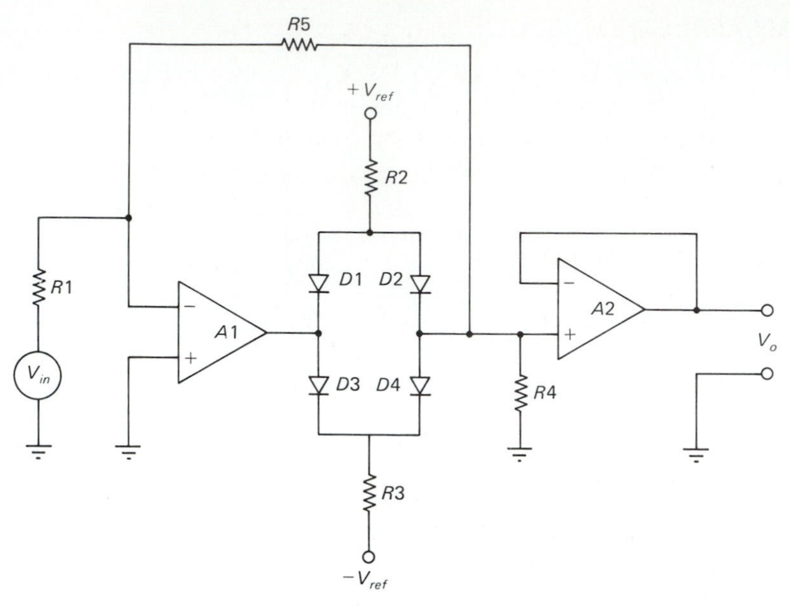


FIGURE 10-20

1. To set $+V_{\text{LIM}}$

$$+V_{\text{LIM}} = \frac{(+V_{ref})R5R4}{R5R4 + R5R2 + R4R2} \tag{10-14}$$

2. To set $-V_{\text{LIM}}$

$$-V_{\text{LIM}} = \frac{(-V_{ref})R5R4}{R5R4 + R5R3 + R4R3} \tag{10-15}$$

## 10-7 SUMMARY

1. The PN junction diode is imperfect. It behaves in a nonlinear logarithmic manner between zero and a certain critical voltage ($V_g$) which is 0.2 to 0.3 for Ge diodes and 0.6 to 0.7 volts for Si diodes.
2. A precise rectifier is an operational amplifier circuit which produces a nearly ideal transfer characteristic by servoing out the errors in the feedback loop.
3. A zero-bound circuit has a reference voltage which limits a portion of the transfer characteristic to zero. A dead-band circuit combines two zero-bound circuits which limit output voltage to zero or a zone around zero.
4. A peak follower circuit stores the input voltage in a capacitor and always outputs the highest value of that input voltage even if $V_{in}$ subsequently drops.
5. A sample-and-hold circuit is a peak follower circuit which is switched on and off by a control signal. The output voltage is the highest level attained by the input signal during the sampling interval.
6. A clipper circuit limits the maximum attainable value of the output voltage. It is the opposite of a dead-band circuit.

## 10-8 RECAPITULATION

Now return to the objectives and Prequiz questions at the beginning of the chapter and see how well you can answer them. If you cannot answer certain questions, mark them and review the appropriate parts of the text. Next, try to answer the questions and work the problems below using the same procedure.

## 10-9 STUDENT EXERCISES

1. Design and build an inverting halfwave precise rectifier using an operational amplifier. Examine the input and output signals on an oscilloscope. Alternatively, plot the transfer characteristic for data sampled with a voltmeter.
2. Design and build an inverting zero-bound circuit. Select different values of $V_{ref}$, both positive and negative.
3. Design and build a dead-band circuit (a) with equal thresholds for positive and negative peaks (b) with asymmetrical peaks.
4. Design and build a clipper based on Fig. 10-19.
5. Design and build a clipper based on Fig. 10-20.

## 10-10 QUESTIONS AND PROBLEMS

1. Draw the transfer characteristic of ideal and practical PN junction diodes.
2. Draw the circuit for the inverting halfwave precise diode circuit. Label all components and describe circuit action in your own words.
3. Sketch the output waveforms for a sinewave input for a precise diode circuit such as Fig. 10-3.
4. A 741 operational amplifier has an open-loop gain of 250,000 and a gain-bandwidth product of 1.25 MHz. Calculate the smallest input signal that a precise diode based on this amplifier will accept. Assume that silicon diodes are used and that $V_g = 0.65$ volts.
5. A precise diode is used as the demodulator in an amplitude modulated carrier amplifier. Calculate the maximum frequency of the carrier frequency which will still result in low-error demodulation. Assume a closed-loop voltage gain of 10 and a gain-bandwidth product of 1.25 MHz.
6. What would happen in the problem above if the gain was increased to 1000?
7. What is the minimum slew rate required for a precise diode used as an AM detector if the carrier frequency is 1000 Hz, the gain is 10, and the input voltage is 4 volts p-p?
8. In Fig. 10-20 all resistors are equal. $+V_{ref} = 12$ volts, and $-V_{ref} = -12$ volts. Calculate the values of the limit potentials.
9. Why are ideal diode rectifier circuits preferred in some applications than actual PN junction diode rectifiers?
10. What are the nominal junction voltages of (a) germanium PN junction diodes, and (b) silicon PN junction diodes?
11. How does the "ideal" or "precise" rectifier circuit remove the normal junction voltage drop inherent in PN junction diodes.

12. The minimum signal allowed in a silicon-diode based precise rectifier when the open-loop voltage gain is 300,000 is: ____________ .
13. Calculate the minimum signal that can be accurately rectified in a precise rectifier circuit in which the open-loop gain is 2000. Assume that a silicon diode is used.
14. What is the maximum allowable carrier frequency for a precise rectifier AM demodulator if the gain-bandwidth product of the op-amp is 4.5 MHz and the voltage gain is 15?
15. Calculate the values of $C1$ in Fig. 10-3 if a 2 KHz carrier is used and both resistors are 12 kohms. Assume a peak signal voltage of 1.25 volts.
16. Sketch the circuit of a polarity discriminator circuit.
17. Sketch the circuit of a *fullwave* precise rectifier.
18. Sketch the circuit for a zero-bound circuit.
19. Sketch the circuit for a dead-band circuit.
20. Sketch the circuit for a peak follower circuit.
21. Sketch the circuit for a sample and hold circuit.
22. Calculate the boundary limits for a circuit such as Fig. 10-20 if $-V_{ref} = -6$ Vdc, $+V_{ref} = +9$ Vdc and all resistors are 10 kohms.

CHAPTER 11

# Signal Processing Circuits

## OBJECTIVES

1. Learn how active differentiator and integrator circuits work.
2. Learn how active logarithmic and antilog amplifiers work.
3. Understand the practical applications of analog signals processing circuits.

## 11-1 PREQUIZ

These questions test your prior knowledge of the material in this chapter. Try answering them before you read the chapter. Look for the answers (especially those you answered incorrectly) as you read the text. After you have finished studying the chapter, try answering these questions again and those at the end of the chapter (see Section 11-11).

1. An *integrator* circuit finds the area under a curve over an interval, so for a time-varying analog signal it finds the ____________ ____________ of the signal.
2. Draw the simplified circuit for a logarithmic amplifier.
3. An operational amplifier Miller integrator circuit is connected to a DC input signal of 10 μV DC. If the resistor is 1 megohm and the capacitor is 1 μF, how long does it take the output voltage to reach 1 Vdc?
4. What is the gain of a Miller integrator that uses a feedback capacitance of 100 pF and an input resistance of 100 kohms?

## 11-2 INTRODUCTION

Electronic instrumentation usually requires the amplification or processing of analog electrical signals. Even when many instruments have been computerized, there is still need for the analog subsystem. For example, it might be necessary to boost an analog signal to where it can be input to the A/D converter connected to a computer. In addition to this simple scaling function, it may be desirable to do some of the signal processing in the analog subsystem. This may seem unrealistic to computer-oriented people, but it is often a reasonable tradeoff.

There may be a situation where computer hardware or time constraints make it less costly to use a simple analog circuit. It is often asserted that a computer solution is "better" than an analog circuit solution. While this claim may be true some of the time, it is not universally true. As device manufacturer catalogs attest, analog signal processing is far from extinct. It is alive, healthy, and larger than ever.

In this chapter we will examine standard laboratory amplifiers which are used for certain signal acquisition tasks in electronic instrumentation. In addition, we will examine certain linear IC circuits used for analog signals processing.

The term "laboratory amplifiers" describes a wide range of instruments. Although some are quite complex, many of them can be designed using simple linear IC devices.

These instruments can be categorized by several schemes. For example, we can divide them by input coupling methods, *DC or AC*. In the case of AC amplifiers there is often a frequency response characteristic which will take some of the burden of filtering in the system.

We can also categorize amplifiers according to gain

| | |
|---|---|
| Low Gain | 1 to 100 |
| Medium Gain | 100 to 1000 |
| High Gain | >1000 |

These ranges are common, but because they are popularly established rather than established through formal industry standards, they may vary somewhat from one manufacturer to another.

Some amplifiers have names which represent certain special applications. For example, the *biopotentials amplifier* gathers natural electrical signals from living things. They tend to have very high input impedances and certain other characteristics which define them as a separate class.

Laboratory amplifiers can be either free-standing models, or part of plug-in mainframe data logging or instrumentation systems. In the following sections we will discuss some of the special forms of laboratory amplifiers.

## 11-3 CHOPPER AMPLIFIERS

Simple DC amplifiers may be noisy and possess an inherent *thermal drift* of both gain and DC offset (especially the latter). In low- and medium-gain applications these problems are less important than in high-gain amplifiers, especially in the low-gain ranges. As gain increases, however, these problems grow. For example, a drift of 50 uV/°C in an ×100 medium-gain amplifier produces an output voltage change of

$$(50 \text{ uV/°C}) \times 100 = 5 \text{ mV/°C}$$

A drift of 5 mV/°C is tolerable in most low-gain circuits. But in an $\times 20{,}000$ high-gain amplifier the output voltage would escalate to

$$(50\ \text{uV/°C}) \times 20{,}000 = 1\ \text{V/°C}$$

This level of drift will obscure any real signals in a short period of time.

Similarly, noise can be a problem in high-gain applications, where it had been negligible in most low- to medium-gain applications. Operational amplifier noise is usually specified in terms of nanovolts of noise per square root hertz (NOISE(rms) $= \text{nV}/(\text{Hz})^{1/2}$). A typical low-cost operational amplifier has a noise specification of $100\ \text{nV}/(\text{Hz})^{1/2}$, so at a bandwidth of 10 KHz the noise amplitude will be

$$\text{NOISE(rms)} = 100\ \text{nV} \times (10{,}000)^{1/2}$$

$$\text{NOISE(rms)} = 100\ \text{nV} \times 100$$

$$\text{NOISE(rms)} = 10{,}000\ \text{nV} = 0.00001\ \text{volts}$$

In an $\times 100$ amplifier without low-pass filtering, the output amplitude will only be 1 mV, but in an $\times 100{,}000$ amplifier it will be 1 volt.

A *chopper amplifier* can solve both problems because it makes use of a relatively narrowband AC-coupled amplifier in which the advantages of feedback can be optimized. The drift problem is reduced significantly by two properties of AC amplifiers. One property is the inability to pass low-frequency (near-DC) changes such as those caused by drift. To the amplifier, drift is a valid low-frequency (sub-hertzian) signal. The other property is the ability to regulate the amplifier through the use of negative feedback.

Many low-level analog signals are very low frequency, in the DC to 30 Hz range (for example, human electrocardiogram signals have frequency components down to 0.05 Hz). So they will not pass through a narrowband AC amplifier. The solution is to chop the signal so it passes through the AC amplifier, and then to demodulate the amplifier output signal to recover the original waveshape, but at a higher amplitude.

Figure 11-1 shows a block diagram of the basic chopper amplifier circuit. The traditional chopper mechanism is a vibrator-driven SPDT switch ($S1$) connected so that it alternately grounds first the input and then the output of the AC amplifier. An example of a chopped waveform is shown in Fig. 11-2. A low-pass filter following the amplifier will filter out any residual chopper hash and miscellaneous noise signals.

Most older mechanical choppers used a chop rate of either 60 Hz or 400 Hz, although 100 Hz, 200 Hz, and 500 Hz choppers are also found. The main criterion for the chop rate is that it be at least twice the highest component frequency present in the input waveform (Nyquist's sampling criterion).

A differential chopper amplifier is shown in Fig. 11-3. In this circuit an input transformer with a center-tapped primary is used. One input terminal is connected to the transformer primary center-tap, while the other input terminal is switched back and forth between the respective ends of the transformer primary winding. A synchronous demodulator following the AC amplifier detects the signal and restores the original but now amplified waveshape. Again, a low-pass filter smoothes out the signal.

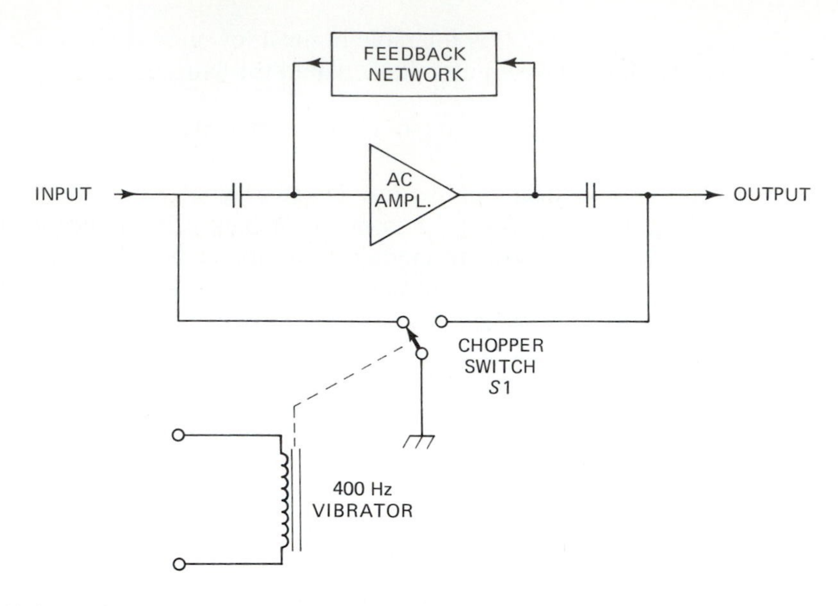


FIGURE 11-1

The modern chopper amplifier may not use mechanical vibrator switches as the chopper. A pair of CMOS or JFET electronic switches driven out of phase will perform the same function. Other electronic switches used in commercial chopper amplifiers include PIN diodes, varactors, and optoisolators. Figure 11-4 shows a modern electronically chopped amplifer which can be obtained in either IC or hybrid form.

Chopper amplifiers limit the noise because of both the low-pass filtering required and because the AC amplifier frequency response can be set to a narrow passband around the chopper frequency.

At one time the chopper amplifier was the only practical way to obtain low drift in high-gain situations. However, modern IC and hybrid amplifiers have improved

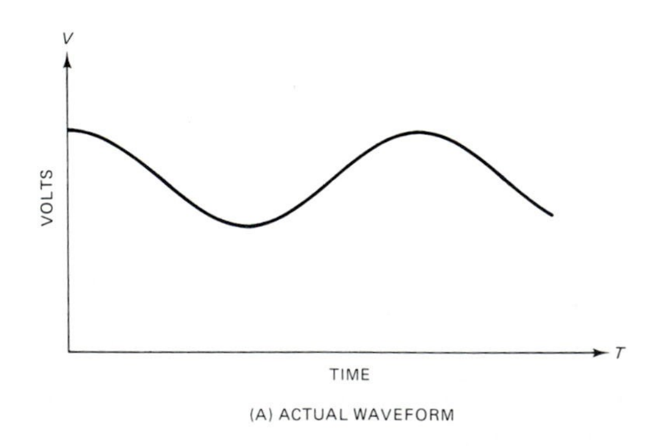


(A) ACTUAL WAVEFORM

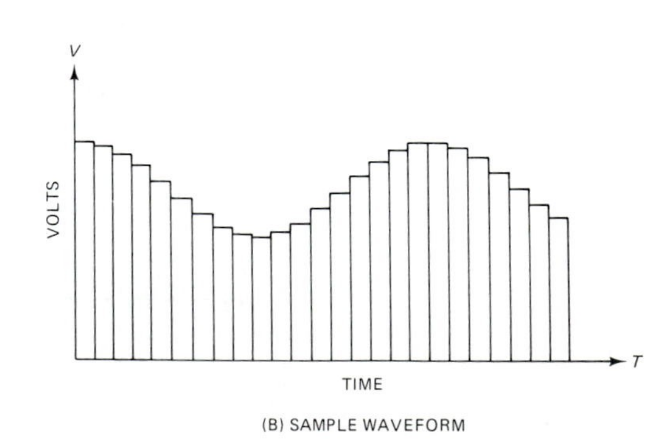


(B) SAMPLE WAVEFORM

FIGURE 11-2

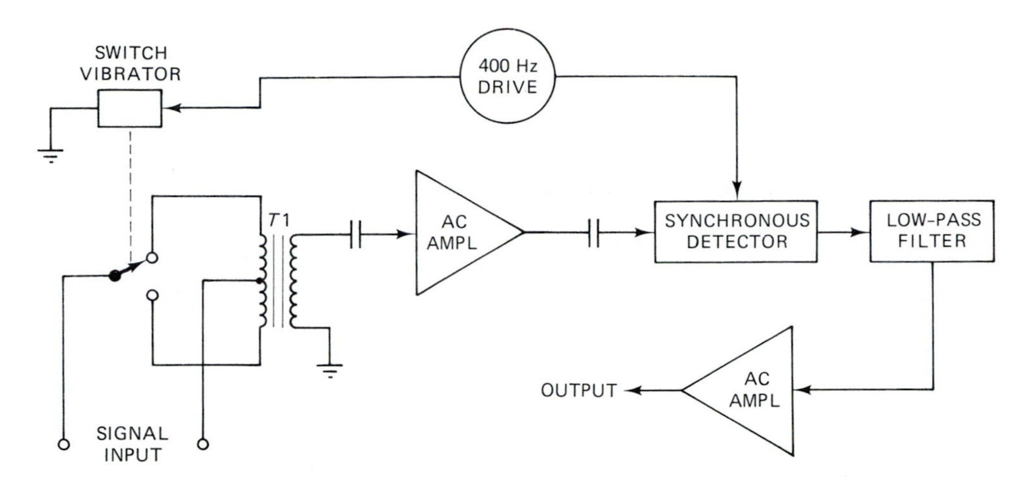


FIGURE 11-3

drift properties so no chopper is needed, especially in the lower end of the high gain range. The Burr-Brown OPA-103 is a monolithic IC operational amplifier in a TO-99 8-pin metal can package. This low-cost device has a drift characteristic of 2 uV/°C. Amplifiers are also available (especially in hybrid form) which are actually electronic chopper stabilized, but to the outside world the device simply looks like a low-drift amplifier. The Burr-Brown 3271/25 device is an example that exhibits a drift characteristic of 0.1 uV/°C.

## 11-4 CARRIER AMPLIFIERS

A *carrier amplifier* is any signal processing amplifier in which the signal is used to modulate another (higher frequency) signal, a "carrier" signal. The chopper amplifier fits this definition, but it is usually regarded as a unique type of carrier amplifier. Two principal types of carrier amplifiers are DC-excited and AC-excited.

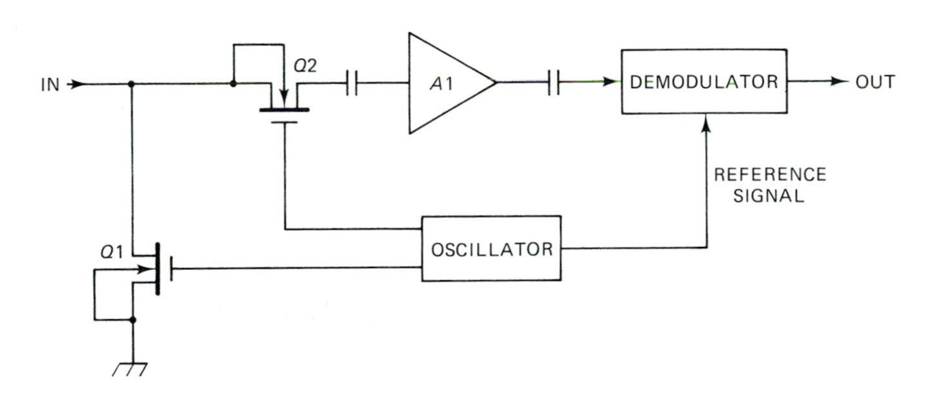


FIGURE 11-4

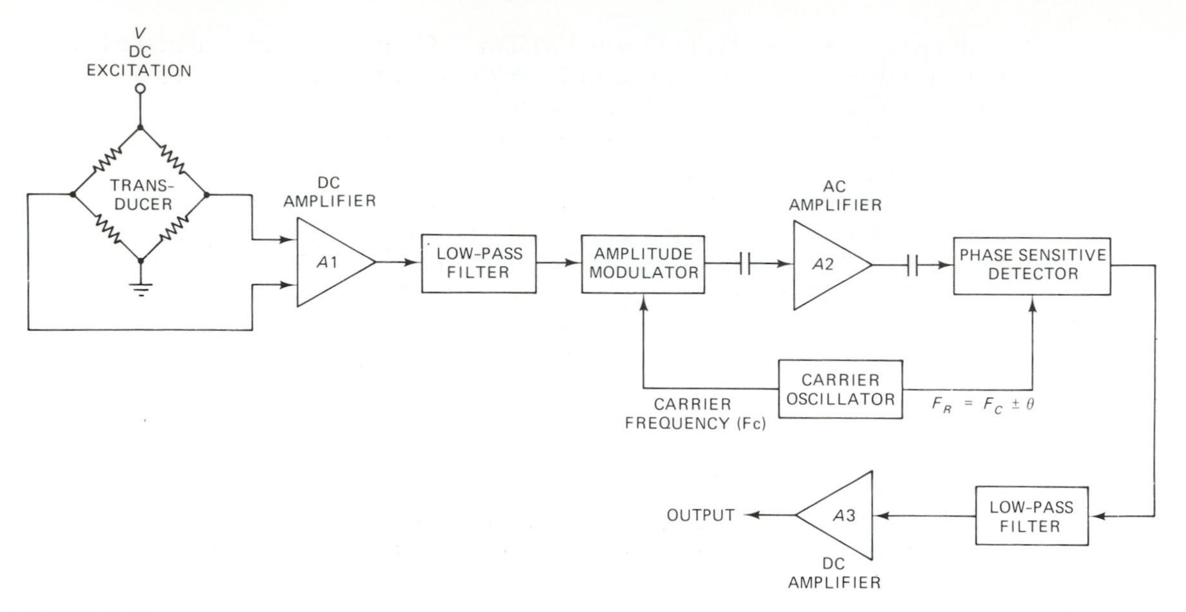


FIGURE 11-5

Figure 11-5 shows a DC-excited carrier amplifier. A Wheatstone bridge transducer provides the input signal. This is excited by a DC potential ($V$). The output of the transducer is a low-level DC voltage which varies with the value of the stimulating parameter. The transducer signal is usually of very low amplitude, and may be noisy. An amplifier increases the signal amplitude. A low-pass filter removes much of the noise. In some models the first stage is actually a composite of these two functions, being essentially a filter with gain.

The signal at the output of the amplifier-filter section is used to amplitude modulate a carrier signal. Typical carrier frequencies range from 400 Hz to 25 KHz, with 1 KHz and 2.4 KHz being common. The signal frequency response of a carrier amplifier is a function of the carrier frequency. At maximum, it is usually one-fourth of the carrier frequency. A carrier frequency of 400 Hz, then, is capable of signal frequency response of 100 Hz, while the 25-KHz carrier will support a frequency response of 6.25 KHz. Further amplification of the signal is provided by an AC amplifier.

A key to the performance of any carrier amplifier is the *phase sensitive detector* (PSD) which demodulates the amplified AC signal. Envelope detectors, while very simple and low cost, cannot discriminate between the real signals and certain spurious signals.

One advantage of the PSD is its ability to reject signals which are not part of the carrier frequency. It will also reject certain signals which are part of the carrier frequency. For example, the PSD will reject even harmonics of the carrier frequency and any components which are out of phase with the reference signal. It will, however, respond to odd harmonics of the carrier frequency. Some carrier amplifiers cannot handle this problem at all.

In some cases, manufacturers will design the AC amplifier section to be a bandpass amplifier with a response limited to $F_c \pm (F_c/4)$. This response will eliminate any third or higher order odd harmonics of the carrier frequency before they reach the PSD. It is then only necessary to ensure the reference signal has acceptable purity (harmonic distortion and phase noise).

Another common form of carrier amplifier is the AC-excited circuit shown in Fig. 11-6A. In this circuit the transducer is AC-excited by the carrier signal, eliminating the need for the amplitude modulator. The small AC signal from the transducer is amplified and filtered before being applied to the PSD circuit. Again, some designs use a bandpass AC amplifier to eliminate odd harmonic response. This circuit allows adjustment of transducer offset errors in the PSD circuit instead of in the transducer, by varying the phase of the reference signal.

Figure 11-6B shows the block diagram to a transducer amplifier which uses pulsed excitation for the Wheatstone bridge. A short duty cycle pulse generator produces either monopolar or as in the case shown here, bipolar pulses used to provide the excitation potential. In the pulsed method the short duty cycle limits the amount of power dissipated in the transducer resistance elements, and therefore reduces the effects of transducer self-heating on the thermal drift. This is an advantage of the pulsed method. Several methods are used to demodulate the pulse waveform at the output of amplifier $A1$. A Miller integrator, which finds the time average of the signal, will create a DC potential proportional to the amplified transducer output signal. Alternative schemes use CMOS or other electronic switches to demodulate the signal.

Another advantage of the pulsed scheme is that an amplifier drift cancellation circuit can be implemented. Switching is provided which shorts together the input of $A1$ during the off time of the pulse, and connects a capacitor to the output of $A1$. When the capacitor is charged, it can be connected as an offset null potential to amplifier $A2$ during the on time of the pulse. Tektronix, Inc. once used this scheme as a baseline stabilization method in a medical patient monitor oscilloscope.

## 11-5 LOCK-IN AMPLIFIERS

The amplifiers discussed so far produce relatively large amounts of noise and will respond to any noise present in the input signal. They suffer from the usual shot noise, thermal noise, H-field noise, E-field noise, and ground loop noise which affects all amplifiers. The noise at the output is directly proportional to the square root of the circuit bandwidth. The *lock-in amplifier* is a special case of the carrier amplifier where the bandwidth is very narrow. Some lock-in amplifiers use the carrier amplifier circuit of Fig. 11-6, but have an input amplifier with a very high $Q$ bandpass characteristic. The carrier frequency will be between 1 KHz and 200 KHz.

The lock-in principle works because the information signal is made to contain the carrier frequency in a way which is easy to demodulate and interpret. The AC amplifier accepts only a narrow band of frequencies centered about the carrier frequency. The narrowness of the amplifier bandwidth, which makes possible the improved signal-to-noise ratio, also limits the lock-in amplifier to very low frequency input signals. Even then, it is sometimes necessary to integrate (time-average) the signal in order to obtain the needed data.

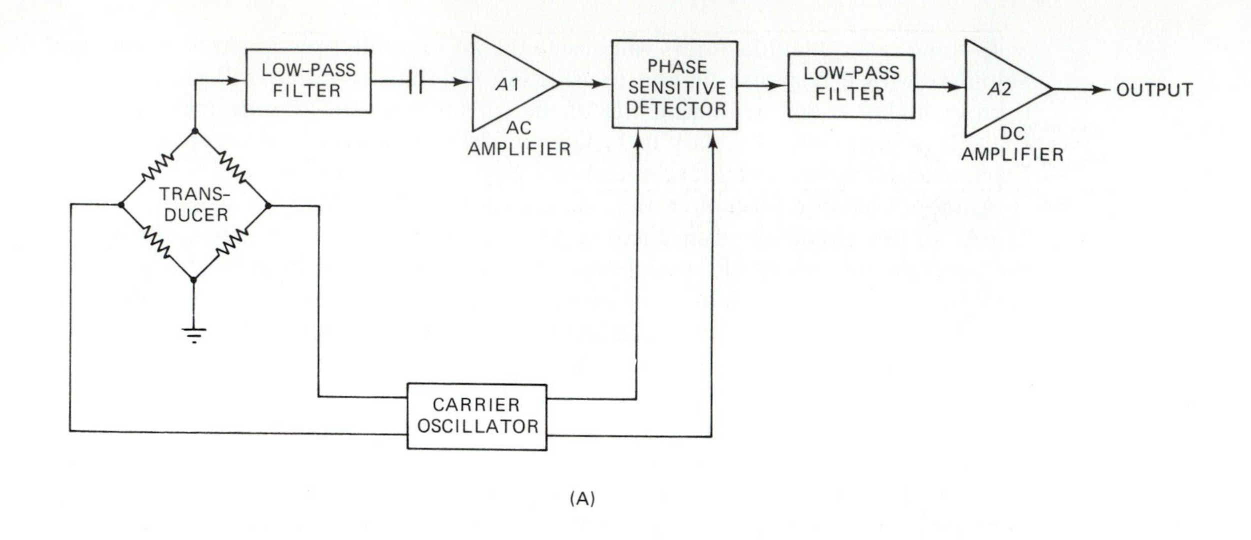


(A)

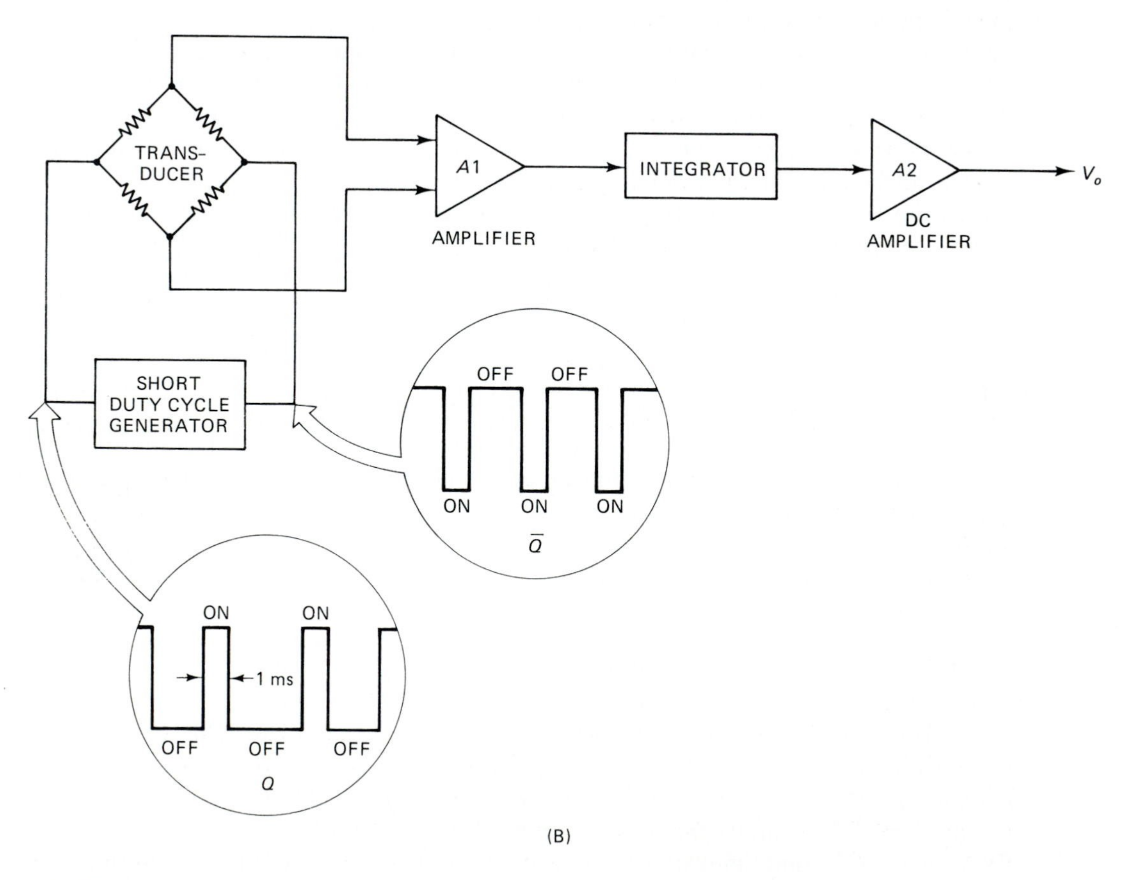


(B)

FIGURE 11-6

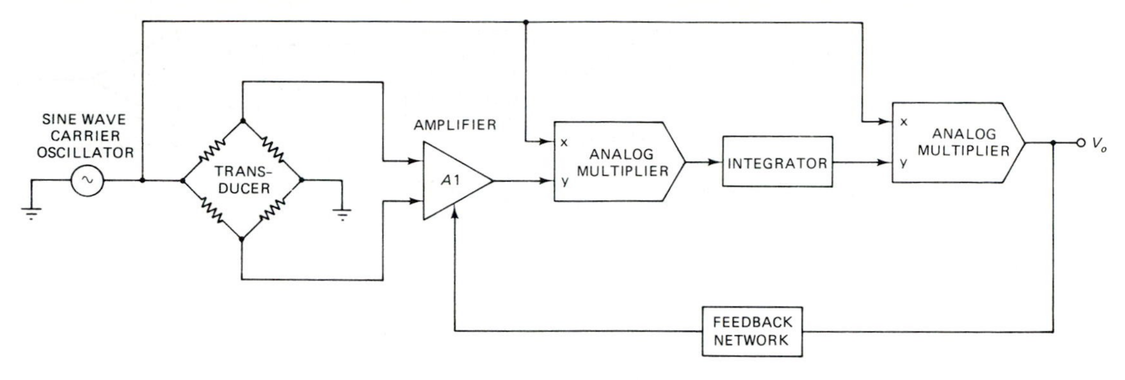


FIGURE 11-7

Lock-in amplifiers are capable of reducing noise and retrieving signals which are otherwise buried below the noise level. Improvements of up to 85 dB are fairly easily obtained and up to 100 dB reduction is possible if cost is not important.

There are several different forms of lock-in amplifiers. The type discussed above is the simplest. It is a narrowband version of the AC-excited carrier amplifier. The lock-in amplifier of Fig. 11-7, however, uses a slightly different technique. It is called an *autocorrelation amplifier.* The carrier is modulated by the input signal and then integrated. The output of the integrator is demodulated in a product detector circuit. The circuit of Fig. 11-7 produces very low output voltages for input signals which are not in phase with the reference signal. But it produces relatively high output voltages at the proper input frequency and phase.

## 11-6 ELECTRONIC INTEGRATORS AND DIFFERENTIATORS

There are two mathematical processes which are very important to electronic instrumentation and signals processing, *integration* and *differentiation*. These processes are inverses of each other. In other words, a function which is first integrated and then differentiated, returns to the original function. A similar relationship occurs when a function is differentiated and is then integrated. This is the normal nature of mathematically inverse processes.

These processes are used elsewhere in electronics, but sometimes under different names. The differentiator is sometimes called a *rate-of-change circuit*, or if the time constant is correct, a *high-pass filter*. Similarly, the integrator might be called a *time averager circuit* or *low-pass filter*. This section will provide an introduction to both of these circuits.

**Integration.** Figure 11-8 shows an analog voltage waveform which varies as a function of time ($V = F[t]$). How do you find the area under the curve between $T1$ to $T2$? If the voltage is constant over the range $T1$ to $T2$ (as in the case of DC), then you simply take the product $V \times (T2 - T1)$. But $V$ is not always constant over the interval of interest. However, if you break the curve into tiny intervals ($T_b - T_a$ in the inset to Fig. 11-8), the voltage change over short interval is small enough to consider it constant. We can then approximate the area under the curve during this short interval as $V1 \times (T_b - T_a)$, or, simply, $V1\Delta T$. Assuming the time width of all

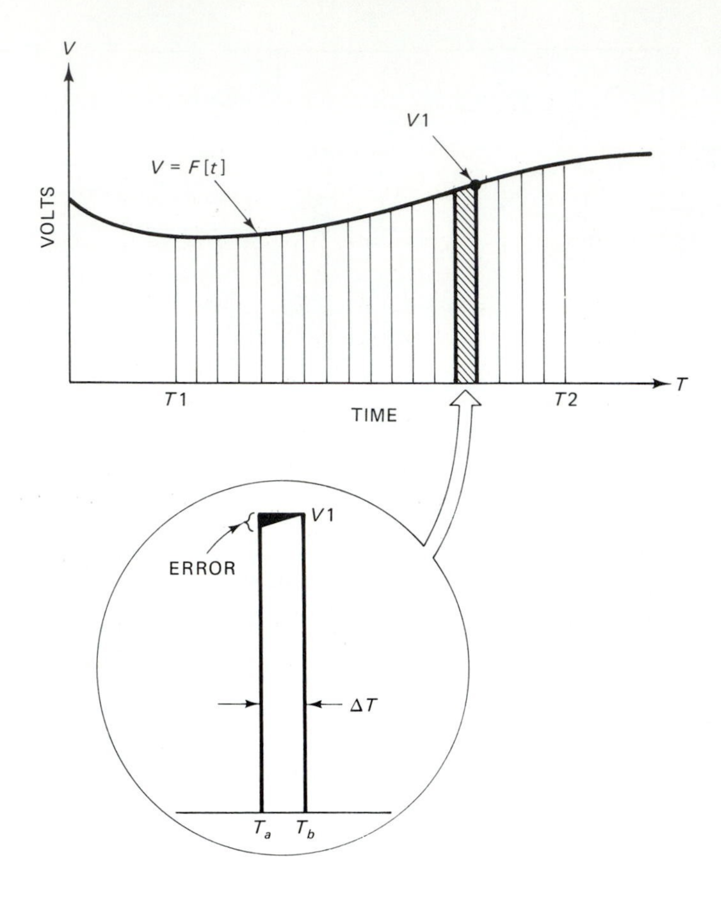


FIGURE 11-8

intervals is the same, the total area under the curve $V = F[t]$ over the interval $T1$ to $T2$ is approximated by summing the individual interval areas

$$AREA = \sum_{i=1}^{n} V_i \Delta T \tag{11-1}$$

where

$$V_i = \text{the voltage at the } ith \text{ interval}$$

$$\Delta T = \text{the time interval common to all samples}$$

The validity of this approximation increases as the time interval decreases. When $\Delta T$ becomes very small, the approximation becomes exact, and the notation used for Eq. (11-1) changes to

$$\text{AREA} = \frac{1}{T} \int V \, dt \tag{11-2}$$

where

$V$ = the voltage as a function of $T$

$T = T2 - T1$

$dt$ is used to indicate that each time interval is infinitesimal

Multiplying by the factor $1/T$ reduces the result to the *time average* of the function.

In Fig. 11-9 a voltage represents a pressure transducer output. In this case it is the output of a human blood pressure transducer. Such transducers are common in hospital intensive care units. Notice the pressure voltage varies with time from a low (diastolic) to a high (systolic) between $T1$ and $T2$. This interval represents one complete cardiac cycle. If you want to know the *mean arterial blood pressure* (MAP), you would want to find the *area under the pressure vs. time curve* over one cardiac cycle.

An electronic integrator circuit serves to compute the time-average of the analog voltage waveform which represents the time-varying arterial blood pressure. In an electronic blood pressure monitoring instrument, a voltage serves to represent the pressure. If, for example, a scaling factor of 10 mV/mmHg is used (common in medical devices), a pressure of 100 mmHg is represented by a potential of 1000 mV or 1.000 volt. This voltage will vary over the range 800 mV to 1200 mV for the case shown in Fig. 11-9 (pressure varies from 80 mmHg to 120 mmHg).

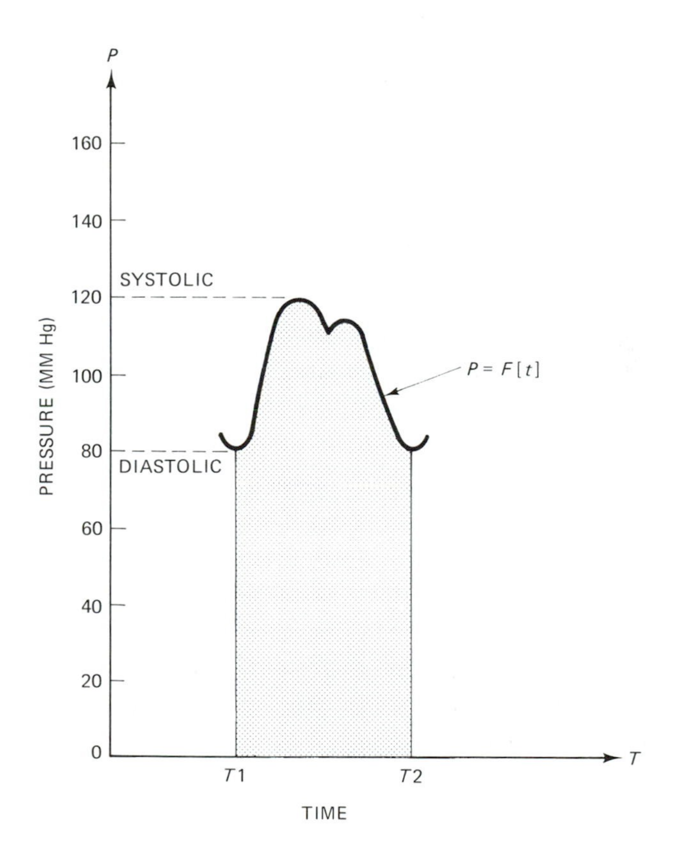


FIGURE 11-9

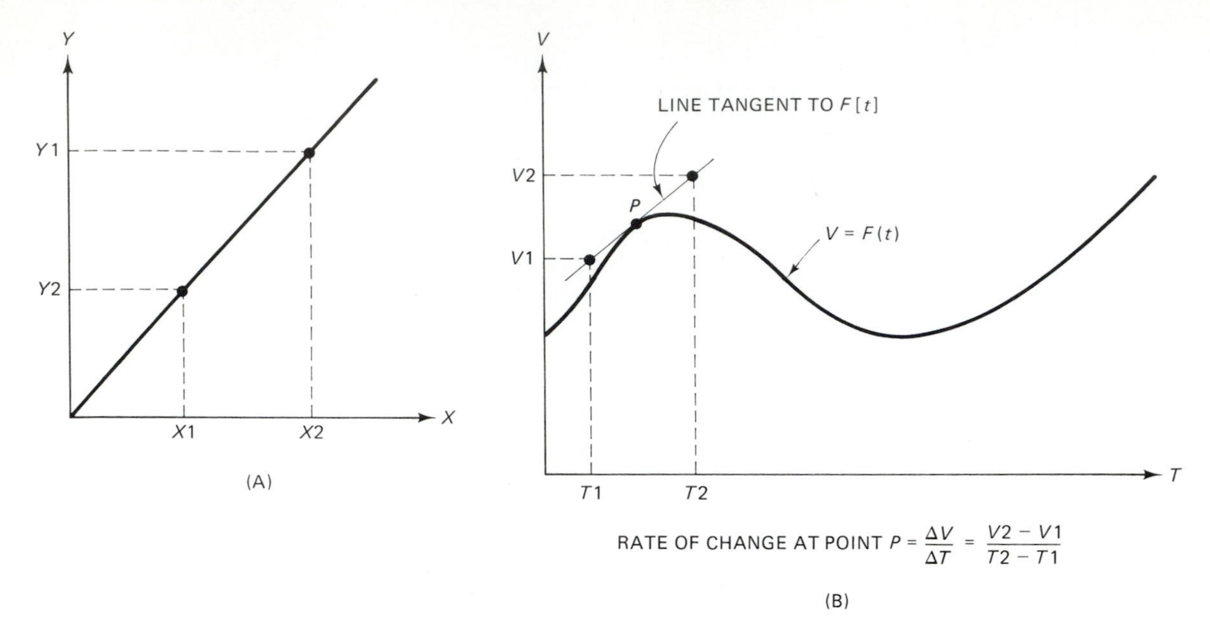


FIGURE 11-10

**Differentiation.** In mathematics, differentiation is finding the *derivative* of a curve, which is merely its *time rate of change.* In the simplest case, a straight line ($Y = mX + b$) as shown in Fig. 11-10A, the *derivative* is simply the *slope* of the line, or $(Y2 - Y1)/(X2 - X1)$. Here we usually write the expression for the slope with the Greek letter delta ($\Delta Y/\Delta X$) to indicate a small change in $X$ and a small change in $Y$.

The derivative of a straight line is simple to calculate. But in electronics one frequently encounters situations when a line is not so straight, as in Fig. 11-10B where a voltage varies with time. If we want to know the *instantaneous rate of change* (the derivative) at a specific point, we can find the *slope of a line tangent to that point.*

Electronic integrators and differentiators affect signals in different ways. Figure 11-11A shows the example of a squarewave applied to the inputs of an integrator and differentiator. The integrator output is shown at Fig. 11-11B. The differentiator output is shown at Fig. 11-11C.

First consider the operation of the integrator circuit. The integrator output waveform in Fig. 11-11B shows a constant positive-going slope between $T1$ and $T2$. The steepness of the slope is dependent on the amplitude of the input squarewave, but the line is linear. You can see from curve B in Fig. 11-11 the squarewave into the integrator produces a triangle waveform. This phenomenon will be seen again in Chapter 12 when we discuss signal generator circuits.

Now consider the operation of the differentiator circuit (see the output waveform Fig. 11-11C). At time $T1$ the squarewave makes a positive-going transition to maximum amplitude. At this instant it has a very high rate of change, so the output of the differentiator is very high (see waveform in Fig. 11-11C at $T1$). But then the amplitude of the input signal reaches a maximum and remains constant until $T2$, when it drops back to its previous value. Thus, the differentiator will produce a sharp

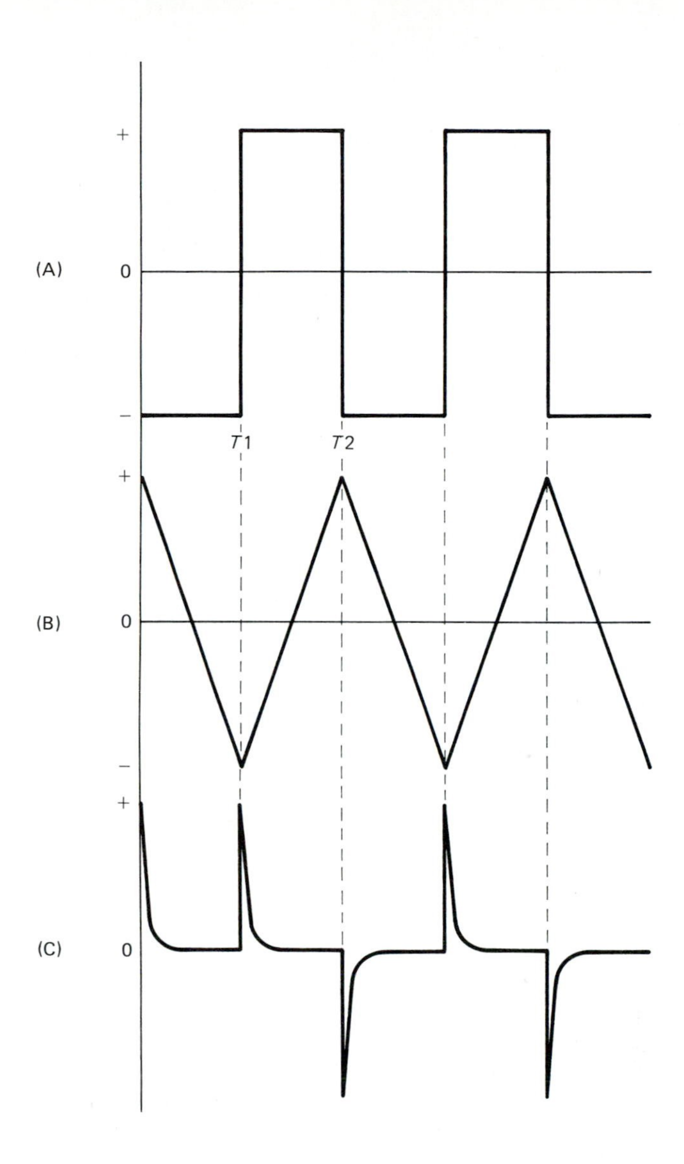


FIGURE 11-11

positive-going spike at $T1$ and a sharp negative-going spike at $T2$. In an ideal circuit there is no transition between these states, but in real circuits there is an exponential transition proportional to the RC time constant of the circuit and the rise time of the waveform. Differentiator output spikes are frequently used in circuits such as timers and zero-crossing detectors.

If a sinewave is applied to the inputs of either integrators or differentiators, the result is a sinewave output which is shifted in phase 90 degrees. The principal difference between the two forms of circuit is in the direction of the phase shift. Such circuits are frequently used to provide quadrature or sine-cosine outputs from a sinewave oscillator.

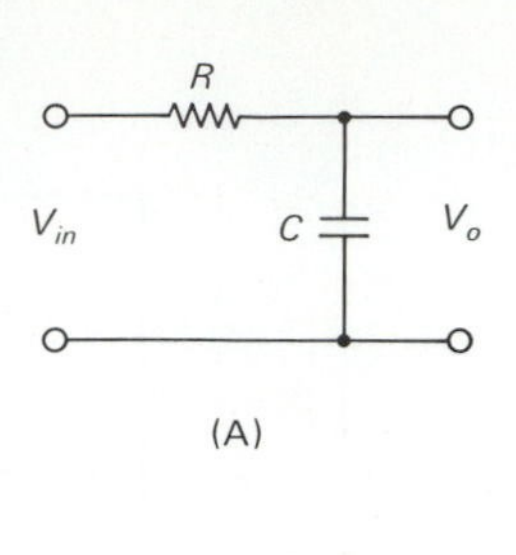


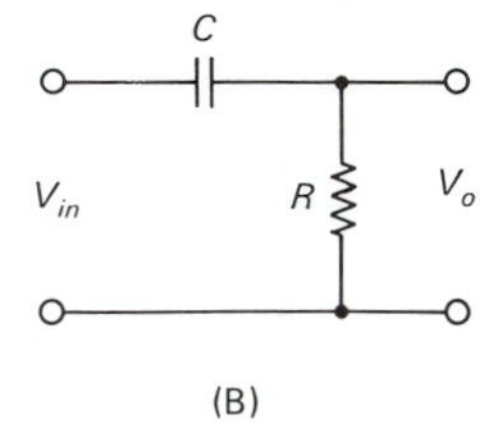


FIGURE 11-12

### 11-6.1 RC Integrator Circuits

The simplest forms of an integrator and differentiator are simple resistor and capacitor circuits, such as shown in Fig. 11-12. The integrator is shown in Fig. 11-12A and the differentiator is shown in Fig. 11-12B. The integrator consists of a resistor element in series with the signal line and a capacitor across the signal line. The differentiator is just the opposite, the capacitor is in series with the signal line and the resistor is in parallel with the line. These circuits are also known as low-pass and high-pass RC filters, respectively. The low-pass filter (integrator) has a $-6$ dB/octave falling characteristic frequency response. The high-pass filter (differentiator) has a $+6$ dB/octave rising frequency response.

The operation of the integrator and differentiator is dependent on the time constant of the RC network ($R \times C$). The integrator time constant is set long ($>10\times$) compared with the period of the signal being integrated. In the differentiator the RC time constant is short ($<1/10\times$) compared to the period of the signal. Several integrators can be connected in cascade in order to increase the time-averaging effect or increase the slope of the frequency response fall-off.

### 11-6.2 Active Differentiator and Integrator Circuits

The operational amplifier makes it relatively easy to build high quality active integrator and differentiator circuits. Previously, one had to construct a stable, drift-free, high-gain transistor amplifer for this purpose. Figure 11-13 shows the basic circuit of the operational amplifier differentiator. Again RC elements are used, but in a slightly different manner. The capacitor is in series with the op-amp's inverting input, while the resistor is the op-amp feedback resistor.

To derive the transfer function we follow a procedure similar to that followed for inverting and noninverting followers earlier in this book.

From Kirchoff's Current Law (KCL)

$$I2 = -I1 \tag{11-3}$$

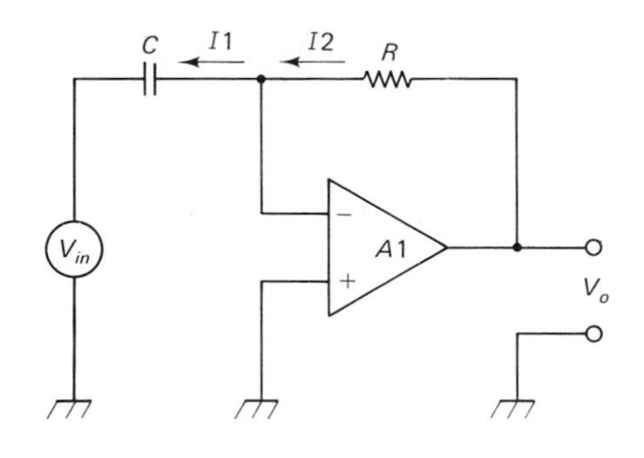


$$I1 = \frac{C\,\Delta V_{in}}{\Delta T}$$

$$I2 = \frac{V_o}{R}$$

FIGURE 11-13

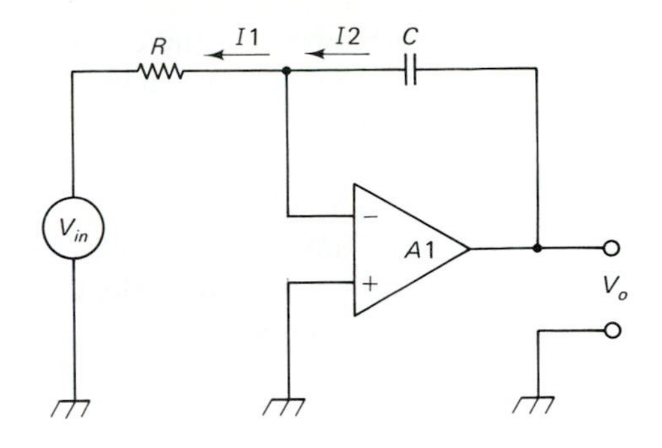


FIGURE 11-14

From basic passive circuit theory, including Ohm's law

$$I1 = \frac{C\Delta V_{in}}{\Delta t} \tag{11-4}$$

$$I2 = \frac{V_o}{R} \tag{11-5}$$

Substituting (11-4) and (11-5) into (11-3)

$$\frac{V_o}{R} = \frac{-C\Delta V_{in}}{\Delta t} \tag{11-6}$$

or, with the terms rearranged

$$V_o = -RC\frac{\Delta V_{in}}{\Delta t} \tag{11-7}$$

where

$V_o$ and $V_{in}$ are in the same units (volts, millivolts, etc.)

$R$ = in ohms

$C$ = in farads

$t$ = in seconds.

Equation (11-7) is a mathematical way of saying output voltage $V_o$ is equal to the product of the $RC$ time constant, the derivative of input voltage $V_{in}$ with respect to time (the $\Delta V_{in}/\Delta t$). Since the circuit is essentially a special case of the familiar inverting follower circuit, the output is inverted, thus the negative sign.

Figure 11-14 shows a classical operational amplifier version of the *Miller integrator circuit*. Again, an operational amplifier is the active element, while a resistor is in series with the inverting input and a capacitor is in the feedback loop. Notice the placement of the capacitor and resistor elements are exactly opposite in both the RC and operational amplifier versions of integrator and differentiator circuits. In other words, the RC elements reverse roles between Figs. 11-13 and 11-14. This will tell the astute student quite a bit about the nature of integration and differentiation.

The output of the integrator is dependent on the input signal amplitude and the RC time constant. The transfer function for the Miller integrator is derived in a manner similar to that of the differentiator:

From KCL

$$I2 = -I1 \tag{11-8}$$

from Ohm's law

$$I1 = V_{in}/R \tag{11-9}$$

and

$$I2 = C(\Delta V_o/\Delta t) \tag{11-10}$$

Substituting Eq. (11-9) and Eq. (11-10) into Eq. (11-8)

$$\frac{C\Delta V_o}{\Delta t} = \frac{-V_{in}}{R} \tag{11-11}$$

Integrating both sides

$$\int \frac{C\Delta V_o}{\Delta t}\,dt = \int \frac{-V_{in}}{R}\,dt \tag{11-12}$$

$$CV_o = \int \frac{-V_{in}}{R}\,dt \tag{11-13}$$

Collecting and rearranging terms

$$CV_o = \frac{-1}{R}\int V_{in}\,dt \tag{11-14}$$

$$V_o = \frac{-1}{RC}\int V_{in}\,dt \tag{11-15}$$

And accounting for initial conditions

$$V_o = \frac{-1}{RC}\int V_{in}\,dt + K \tag{11-16}$$

where

$V_o$ and $V_{in}$ are in the same units (volts, millivolts, etc.)

$R$ = in ohms

$C$ = in farads

$t$ = in seconds.

This expression is a way of saying the output voltage is equal to the time-average of the input signal plus some constant ($K$) which is the voltage that may have been stored in the capacitor from some previous operation (often zero in electronic applications).

### 11-7.3 Practical Circuits

The circuits shown in Fig. 11-13 and Fig. 11-14 are classic textbook circuits. Unfortunately, they do not work well in some practical cases because these circuits are too simplistic in that they depend on the properties of ideal operational amplifiers. In real circuits, differentiators may ring or oscillate and integrators may saturate because of their tendency to integrate bias currents and other inherent DC offsets shortly after turn-on.

Another problem with this kind of circuit magnifies the problem of saturation. Namely, the integrator circuit in Fig. 11-14 has a very high gain with some values of $R$ and $C$. The *voltage gain* of this circuit is given by the term $-1/RC$ which can be quite high depending on the values selected for $R$ and $C$.

### EXAMPLE 11-1

Calculate the gain of a Miller integrator circuit which uses a 0.01 μF capacitor and a 10,000 ohm resistor.

**Solution**

$$A_v = -1/RC$$

$$A_v = -1/(10{,}000 \text{ ohms})\ (0.00000001 \text{ farads})$$

$$A_v = -1/0.0001 = -10{,}000$$

■

In other words, with a gain of $-10{,}000$, a $+1$ volt applied to the input will produce a $-10{,}000$ volt output. The operational amplifier output is limited to an approximate range of $-10$ volts to $-20$ volts, depending on the device and the applied Vdc power supply voltage. In the above case, the operational amplifier will saturate very rapidly. In order to prevent this, it is necessary to keep the input signal from rising too high. If the maximum output voltage allowable is 10 volts, the maximum input signal is 10 volts/10,000 or 1 millivolt. Obviously, it is necessary to keep the RC time constant within certain bounds.

Fortunately, there are some design solutions which will keep the integration aspects of the circuit and remove the problems. A practical integrator is shown in Fig. 11-15. The heart of this circuit is an RCA BiMOS operational amplifier, type CA-3140, or an equivalent BiFET device. This works well because it has a low input bias current (being MOSFET input).

Capacitor $C1$ and resistor $R1$ in Fig. 11-15 form the integration elements. They are used in the transfer equation to calculate performance. Resistor $R2$ is used to discharge $C1$ to prevent DC offsets on the input signal. It also prevents the op-amp itself from saturating the circuit. The RESET switch is used to set the capacitor voltage back to zero (to prevent a $K$ factor offset) before the circuit is used. In some meas-

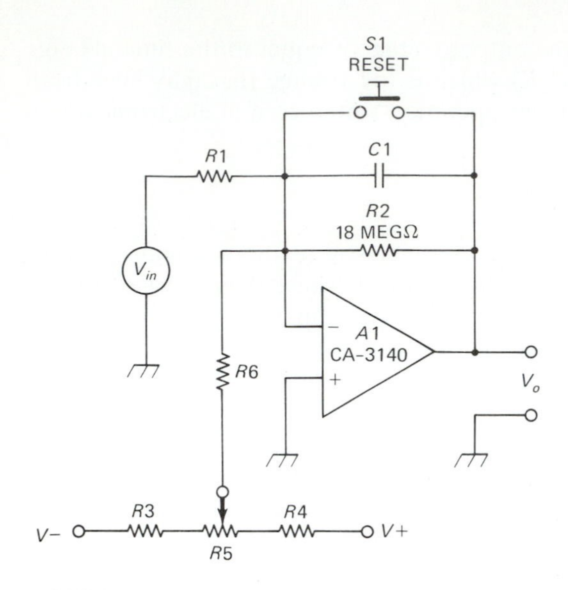


FIGURE 11-15

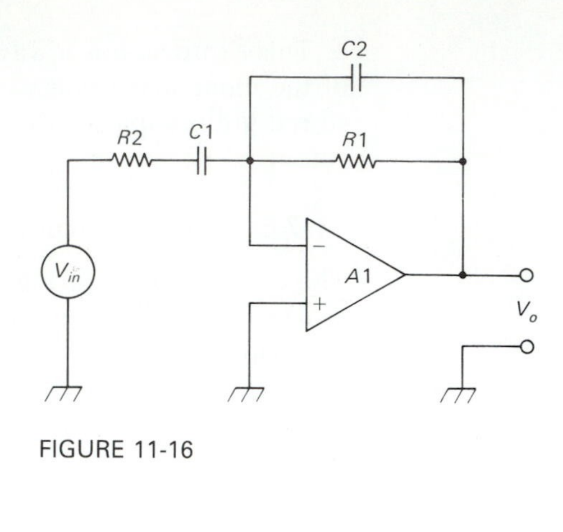


FIGURE 11-16

urement applications the circuit is initialized by closing $S1$ momentarily. In actual circuits, $S1$ may be a mechanical switch, an electromechanical relay, a solid-state relay, or a CMOS electronic switch.

If there is still a minor drift problem in the circuit, potentiometer $R5$ can be added to the circuit to cancel it. This component adds a slight counter current to the inverting input through resistor $R6$. To adjust this circuit, initially set $R5$ to midrange. The potentiometer is adjusted by shorting the $V_{in}$ input to ground (setting $V_{in} = 0$) and then measuring the output voltage. Press $S1$ to discharge $C1$, and note the output voltage (it should go to zero). If $V_o$ does not go to zero, turn $R5$ in the direction which counters the change of $V_o$. This change can be observed after each time RESET switch $S1$ is pressed. Keep pressing $S1$. Then make small changes in $R5$ until a setting is found where the output voltage stays very nearly zero and constant, after $S1$ is pressed (there may be some very long-term drift).

Figure 11-16 shows a practical version of the differentiator circuit. The differentiation elements are $R1$ and $C1$. The previous equation for the output voltage is used. Capacitor $C2$ has a small value (1 pF to 100 pF). It is used to alter the frequency response of the circuit in order to prevent oscillation or ringing on fast rise-time input signals. Similarly, a snubber resistor ($R2$) in the input also limits this problem. The operational amplifier can be almost any type with a fast enough slew rate. The CA-3140 is often recommended. The values of $R2$ and $C2$ are often determined by rule-of-thumb, but their justification is taken from the Bode plot of the circuit.

## 11-7 LOGARITHMIC AND ANTILOG CIRCUITS

*Logarithmic amplifiers* are often used in instrumentation circuits, especially where data compression is required. The overall transfer equation for an operational amplifier circuit is determined by the transfer equation of the feedback network. A

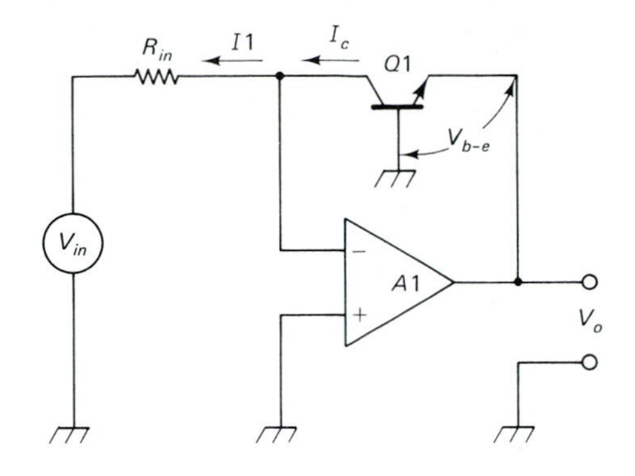


FIGURE 11-17

logarithmic transfer equation can be created in an operational amplifier circuit by placing a nonlinear element in the negative feedback loop. An ordinary PN junction transistor meets the requirement. A logarithmic amplifier has a transfer equation of either

$$V_o = k \text{ LN } (V_{in}) \tag{11-17}$$

or

$$V_o = k \text{ LOG } (V_{in}) \tag{11-18}$$

Figure 11-17 shows the basic circuit for an inverting logarithmic amplifier. As with any inverting amplifier, we can assume the summing junction potential is zero because the noninverting input is grounded.

In basic transistor theory the base-emitter voltage of the transistor is given by

$$V_{b-e} = \frac{KT}{q} \text{ LN} \left[\frac{I_c}{I_s}\right] \tag{11-19}$$

where

$V_{b-e}$ = the base-emitter potential is volts

$K$ = Boltzmann's constant ($1.38 \times 10^{-23}$ J/K)

$T$ = the temperature in degrees Kelvin (°K)

$q$ = the electronic charge ($1.6 \times 10^{-19}$ coulombs)

LN indicates the natural (or "base-e") logarithms

$I_c$ = the collector current of the transistor in amperes

$I_s$ = the reverse saturation current of the transistor (approximately $10^{-13}$ amperes at 300 °K)

Because the configuration of Fig. 11-17 makes $V_{b-e} = V_o$

$$V_o = \frac{KT}{q} \text{ LN} \left[\frac{I_c}{I_s}\right] \tag{11-20}$$

or, for those who prefer base-10 logarithms

$$\frac{\text{LOG } X}{\text{LN } X} = 0.4343 \tag{11-21}$$

and

$$\frac{\text{LN } X}{\text{LOG } X} = 2.3 \tag{11-22}$$

so

$$V_o = \frac{2.3\ KT}{q} \text{ LOG} \left[\frac{I_c}{I_s}\right] \tag{11-23}$$

When the constants $KT/q$ are accounted for

$$V_o = 60 \text{ mV LOG} \left[\frac{I_c}{I_s}\right] \tag{11-24}$$

The equation above demonstrates the output voltage $V_o$ is a logarithmic function. From KCL it is known that

$$I1 = -I_c \tag{11-25}$$

And from Ohm's law

$$I1 = \frac{V_{in}}{R_{in}} \tag{11-26}$$

Substituting Eq. (11-25) and Eq. (11-26) into Eq. (11-24) yields

$$V_o = 60 \text{ mV LOG} \left[\frac{I1}{I_s}\right] \tag{11-27}$$

or

$$V_o = 60 \text{ mV LOG} \left[\frac{V_{in}/R_{in}}{I_s}\right] \tag{11-28}$$

Thus, the output voltage $V_o$ is proportional to the logarithm of the input voltage $V_{in}$.

The simple circuit in Fig. 11-17 is usually published in textbooks, but in a practical sense it only works some of the time. This is because of the realities of nonideal operational amplifiers. Figure 11-18 shows one common problem: *oscillation.*

The oscilloscope waveform shown in Fig. 11-18 was taken from a 741 op-amp connected in the simple logarithmic amplifier configuration of Fig. 11-17. The input signal was a linear ramp sawtooth signal, so we would expect a logarithmically decreasing output voltage. Unfortunately, the amplifier oscillates in the basic configuration.

A modified version of the circuit is shown in Fig. 11-19. In this circuit a compensation network ($R3/R4/C1$) is added to prevent the oscillation. The values of the

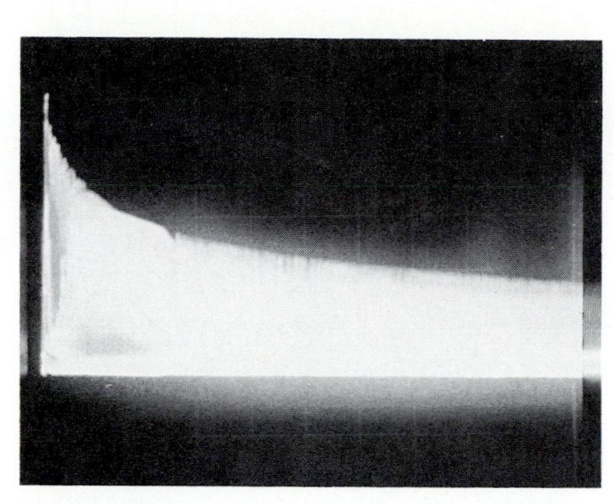

FIGURE 11-18

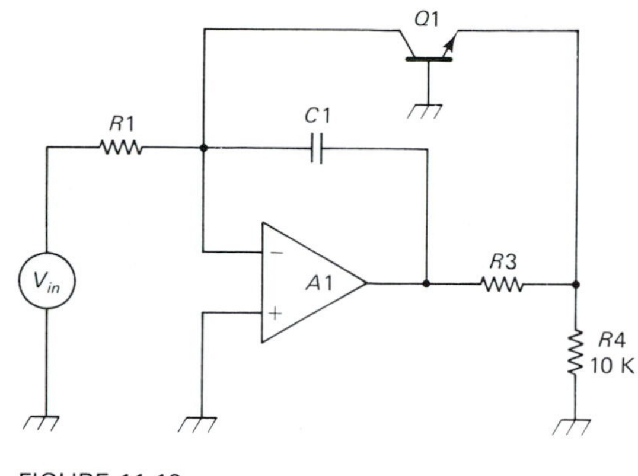


FIGURE 11-19

network components (except $R4$) are found from empirical data based on the following

$$R3 = \frac{V_{o(\max)} - 0.7}{(V_{in(\max)}/R1) + (V_{o(\max)}/R4)} \tag{11-29}$$

and

$$C1 = \frac{1}{\pi F\ R3} \tag{11-30}$$

Figure 11-20 shows the output signal of Fig. 11-19 with the same linear sawtooth ramp input signal as before. Note the output signal is clean and free of oscillation. It also has the characteristic decreasing logarithmic shape expected of an inverting log amplifier.

Another problem of the logarithmic amplifer is *temperature sensitivity*. Recall the operant equation for the logarithmic amplifier is

$$V_o = \frac{KT}{q} \text{LN} \left[ \frac{(V_{in}/R_{in})}{I_s} \right] \tag{11-31}$$

The $T$ term in the equation is temperature in degrees Kelvin (room temperature is approximately 300 °K). Temperature is a variable, so we can expect the output voltage to be a function of both the applied input signal voltage and the temperature of the *b-e* junction in the transistor. This temperature is, in turn, a function of ambient temperature. In order to prevent pollution of the output signal data it is necessary to *temperature compensate* the logarithmic amplifier circuit. Figure 11-21 shows two approaches to temperature compensation. The value of $R1$ in Fig. 11-21A is approximately 15.7 times the temperature of the thermistor ($R_t$) at room temperature. The circuit of Fig. 11-21B is more complex, but it offers a greater dynamic range than the previous circuit.

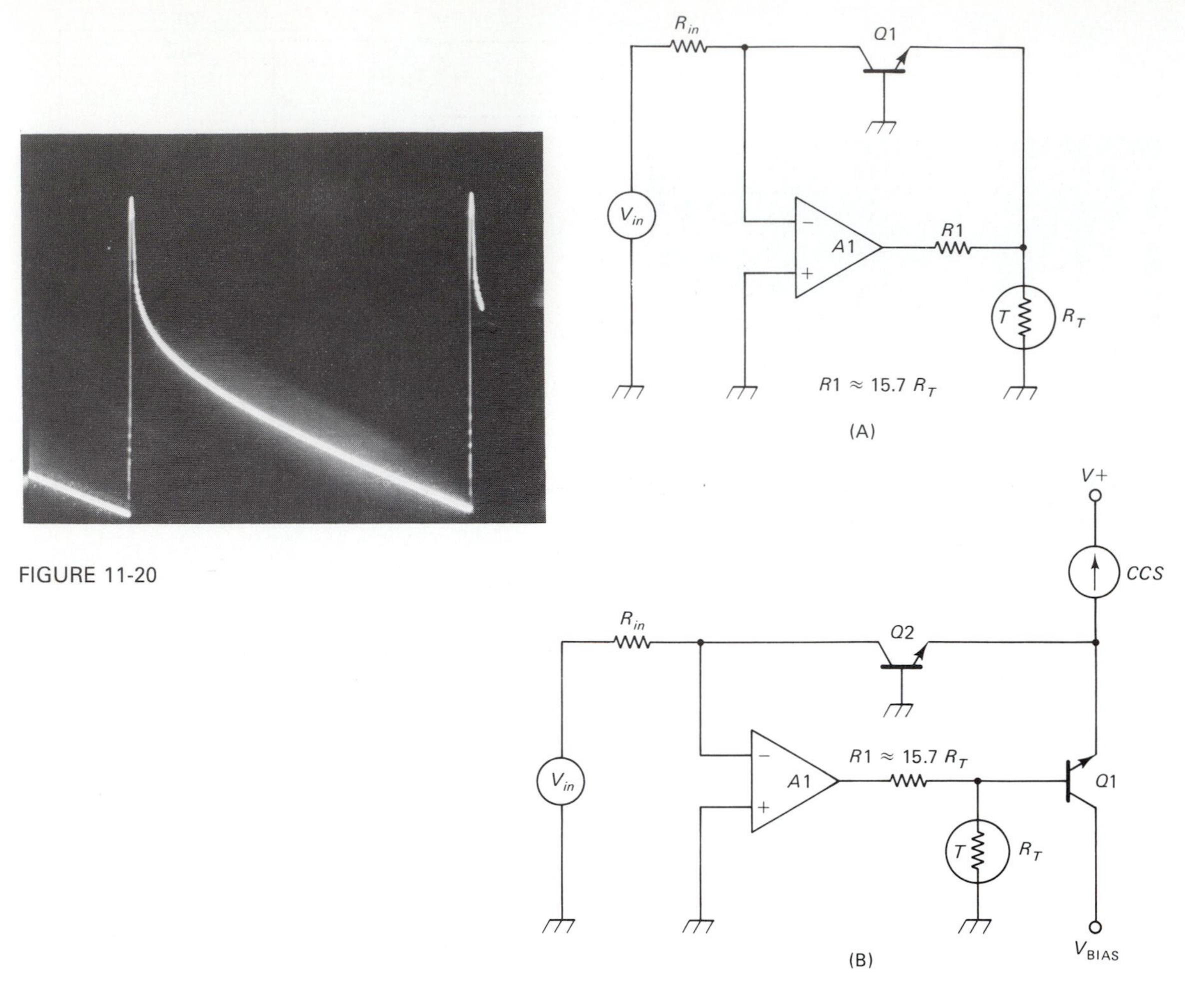


FIGURE 11-20

FIGURE 11-21

### 11.7-1 Antilog Amplifiers

The *antilog amplifier* performs the inverse function of the logarithmic amplifier. The output voltage from the antilog amplifier is

$$V_o = K \text{ ANTILOG } (V_{in}) \tag{11-32}$$

The simplified circuit for a conventional antilog amplifier is shown in Fig. 11-22. Again, a PN junction from a bipolar transistor is used because of its logarithmic transfer function. But note here the respective positions of the transistor and resistor are reversed from the logarithmic amplifier.

From basic operational amplifier theory and Kirchoff's Current Law (KCL)

$$|I1| = |I2| \tag{11-33}$$

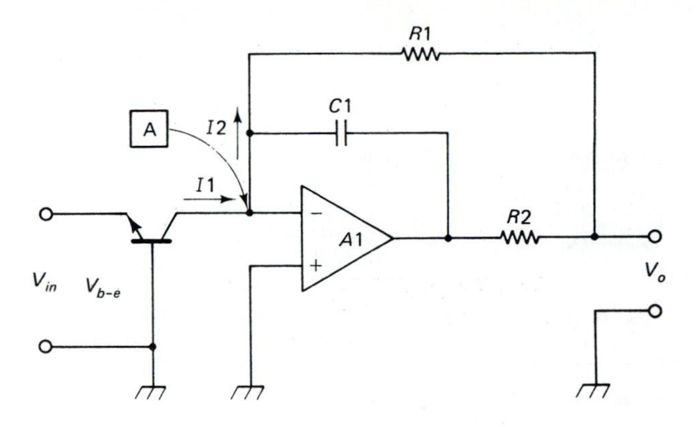


FIGURE 11-22

Further, by Ohm's law and the fact that the summing junction (point A) is at virtual ground (zero) potential

$$I2 = \frac{V_o}{R1} \tag{11-34}$$

In Fig. 11-22 voltage $V_{in}$ is applied across the *b-e* junction of $Q1$, so $V_{b-e} = V_{in}$. Current $I1$ is the collector current of $Q1$, so $I_c = I1$. Recall Eq. (11-19)

$$V_{b-e} = \frac{K\,T}{q}\,\text{LN}\left[\frac{I_c}{I_s}\right] \tag{11-35}$$

Which can be rewritten to the form

$$V_{in} = \frac{K\,T}{q}\,\text{LN}\left[\frac{I1}{I_s}\right] \tag{11-36}$$

Because $I1 = I2$

$$V_{b-e} = \frac{K\,T}{q}\,\text{LN}\left[\frac{(V_o/R1)}{I_s}\right] \tag{11-37}$$

Collecting terms

$$\frac{q\,V_{in}}{K\,T} = \text{LN}\left[\frac{(V_o/R1)}{I_s}\right] \tag{11-38}$$

Taking the antilog of Eq. (11-38)

$$\text{EXP}[(qV_{in}/KT)] = \frac{V_oR1}{I_s} \tag{11-39}$$

$$\frac{I_s\,\text{EXP}[(qV_{in}/KT)]}{R1} = V_o \tag{11-40}$$

Equation (11-40) is the transfer equation for the antilog amplifier.

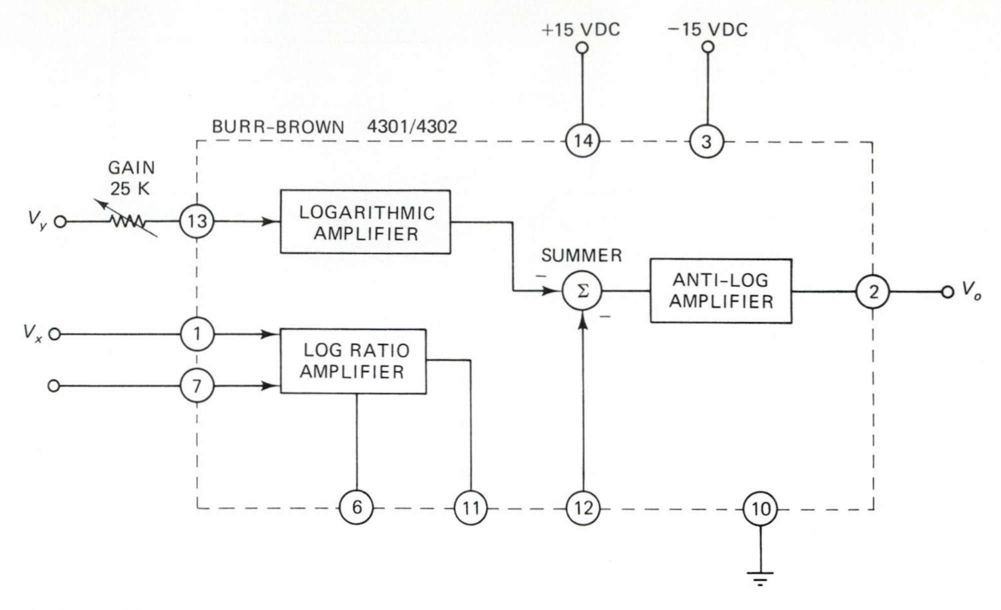


FIGURE 11-23

### 11-7.2 Special Function Circuits

Circuit elements can be combined into a single circuit to form a special function circuit. The combined circuit can be fabricated either in the monolithic IC form or as a hybrid. Although there is a wide range of possible special function circuits, a few examples will illustrate the concept.

**Multifunction Converter.** Figure 11-23 shows the block diagram for a Burr-Brown 4301 or 4302 Multi-Function Converter. This circuit can perform analog multiplication or division, squaring, square rooting, taking other roots, exponentiation, sine, cosine, arctan, and root sum squares (RSS).

The 4301 and 4302 devices consist of a logarithmic amplifier, a log-ratio amplifier, a summer, and an antilog amplifier (see Fig. 11-23). The transfer function for this circuit is

$$V_o = V_y \left[\frac{V_z}{V_x}\right]^m \tag{11-41}$$

The output signal can swing to +10 volts output and will supply 5 mA of output current. The three input signal voltages ($V_x$, $V_y$, and $V_z$) must be ±10 volts.

The function performed by these devices depends on the nature of the resistor network connected between pins 6, 11, and 12. Figure 11-24 shows three types of resistor network used to program the 4301 and 4302. The exponent $m$ (see Eq. (11-41)) determines the function performed. The value of $m$ is set by the external resistors. It can be any value between 0.2 and 5. For root functions, $m < 1$. For example, the square root function is $m = 0.5$. The resistor network which yields this function is shown in Fig. 11-24A. The values of the resistors for this type of function are set by $m < 1$.

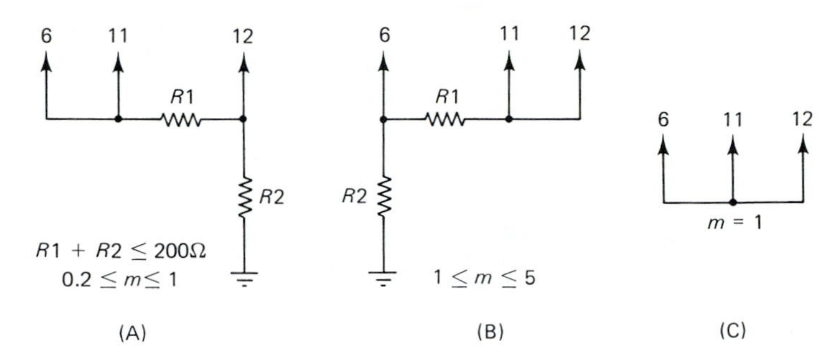


FIGURE 11-24

$$m = \frac{R2}{R1 + R2} \tag{11-42}$$

Values greater than one ($1 < m < 5$) allow the 4301 and 4302 to be used as squarers and exponentiators. For these functions use the resistor network in Fig. 11-24B. The values of the resistors are found from

$$m = \frac{R1 + R2}{R2} \tag{11-43}$$

Using the 4301 and 4302 as analog multipliers or dividers requires no resistor networks. In these circuits, $m = 1$. All three terminals are strapped together (Fig. 11-24C). Here, two inputs will represent signals, while the third represents a constant and is set to a fixed voltage. For example, to build an analog multiplier, $V_y$ and $V_z$ are used as signal inputs and $V_x$ is connected through a resistor to a fixed reference potential. The transfer function in this case is

$$V_o = KV_yV_z \tag{11-44}$$

where

$K$ is proportional to $V_x$

Similarly, to make an analog divider, $V_y$ becomes the constant ($K$) and the input signals are $V_x$ and $V_z$. The transfer equation is

$$V_o = K\frac{V_z}{V_x} \tag{11-45}$$

**Programmable Gain Amplifiers.** Amplifier gains must sometimes be changed for different applications. In other situations the gain must be changed to calibrate a system. A programmable gain amplifier allows the gain to be set (or programmed) from an external source. In some simple cases the programming is done by setting a bias current. A single resistor is connected between the programming terminal and a reference DC power supply. This type of amplifier is not amenable to either computer control or control by other digital circuits.

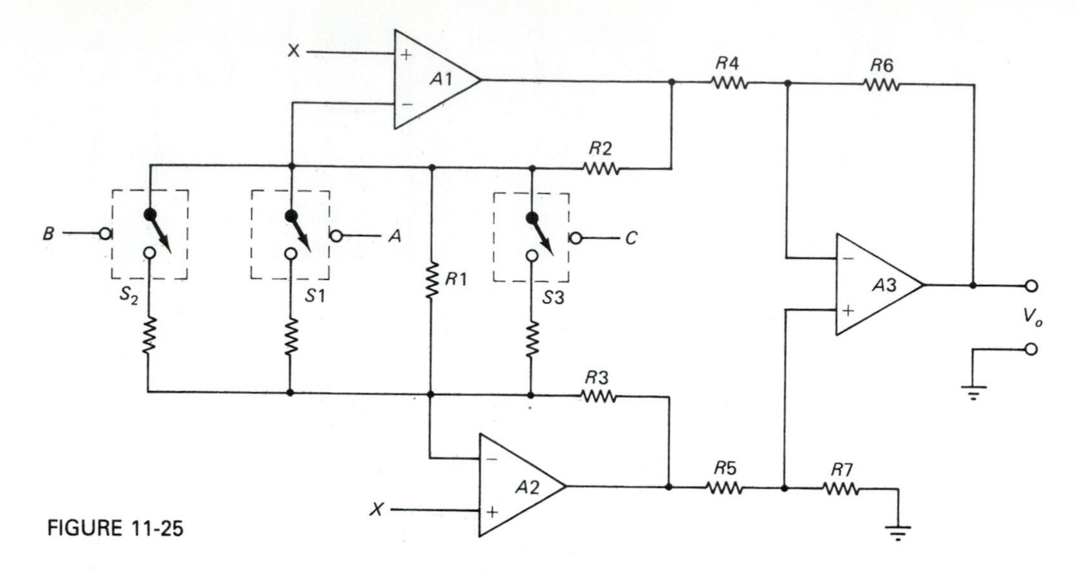

FIGURE 11-25

There are two methods for digitally programming an amplifier. One is to use a multiplying digital-to-analog converter (MDAC). All DACs provide an output which is proportional to a binary word and a DC reference input. In a multiplying DAC or MDAC, the DC reference is supplied from an external source. It is possible on some MDACs to use either a fixed DC source, a slowly varying nearly DC signal, an AC signal with a DC offset, or a symmetrical AC signal (in which case the offset is provided by the designer) as the reference. If the reference is the signal to be amplified, the output will be a function of the reference signal input and the binary word applied to the digital inputs.

Another approach to designing the programmable amplifier is shown in Fig. 11-25. The amplifier circuit is the three-device instrumentation amplifier circuit. Gain is set by

$$A_v = \left[\frac{2\,R2}{R1} + 1\right]\left[\frac{R6}{R4}\right] \tag{11-46}$$

Assuming

$$R2 = R3$$

$$R4 = R5$$

$$R6 = R7$$

By making all resistors fixed, the programmable amplifier gain can be controlled solely by varying $R1$. In Fig. 11-25 this function is performed by shunting additional resistors across a relatively high value of fixed resistor which is always in the circuit. The switches ($S1$ through $S3$) are CMOS electronic switches with active-LOW digital control terminals. When the control terminals ($A$, $B$, and $C$) are HIGH, then the switch is open and when the control terminal is LOW the switch is closed. Gain is set by selecting any (or several) switches to be closed in order to alter the resistance of $R1$.

The circuit of Fig. 11-25 is actually quite crude compared to programmable amplifiers offered by some manufacturers. In these products more than one resistance parameter can be varied, resulting in gain settings of small increments from unity to 1024.

**Thermocouple Amplifiers.** A *thermocouple* is a junction of two dissimilar metals. These devices are used to measure temperature over a very wide range (>2700°C) and can be very precise when properly built and calibrated.

The thermocouple works because of the *Seebeck effect.* All metals have a certain work function which defines how much energy is needed to loosen electrons from the atom. If two metals with different work functions are joined together in a fused junction, the difference between work functions (a voltage) is generated across the two ends of the wires. This voltage is proportional to the temperature of the junction and the work function differential. Thermocouples are given letter designations (*K* and *T*) to denote both a certain characteristic and a pair of specific metals used to form the junction.

The inverse function, called the *Peltier effect*, is responsible for the so-called solid-state refrigerators. Applying a voltage across the loose ends of the thermocouple causes one wire to loose heat and the other to absorb heat.

It is common practice in thermocouple temperature measurements to use two thermocouples. One is placed in a zero-degree Celsius ice bath and becomes a reference pole. Iced water is exactly 0°C when there is equilibrium between ice and liquid water in the bath. The other thermocouple is the measurement pole. It is connected in series with the ice point thermocouple. A differential amplifier measures the difference potential between the two thermocouples. The resulting output voltage is (within the linearity of the system) proportional to the temperature difference between 0°C and the measured temperature. This arrangement is called "ice point compensation."

Analog Devices, Inc. makes a special function IC device which is used as a transducer signal conditioner. A DC differential input amplifier converts the thermocouple signal into a single-ended output signal ($V_o$). The AD-594 and AD-595 devices (Fig. 11-26) contain the circuitry necessary to properly operate the thermocouple. These devices also contain electronic circuitry which performs the ice point compensation function mentioned above. Both Type T and Type K thermocouples can be accommodated. The output of the IC will produce a scaling function of 10 mV/°C.

## 11-8 SUMMARY

1. Analog circuits are often used for signals processing functions including *amplification*, *integration*, *differentiation*, *exponentiation*, and *logarithmic amplification.*
2. *Chopper amplifiers* use electronic switching at the input and output terminals of an AC amplifier to slowly process changing or DC signals. The principal advantages of the chopper amplifier are drift and noise reduction.
3. *Carrier amplifiers* are used to process transducer signals. The excitation signal is an AC carrier between 1 KHz and 25 KHz. The signal is processed in an AC amplifier and the analog data is recovered in a phase sensitive detector.
4. *Lock-in amplifiers* are used to process very low level signals. There are two forms. One is a very narrow band, high-Q, AC carrier amplifier. The other is an autocorrelation amplifier.

TEMPERATURE MEASUREMENT COMPONENTS

Temperature Transducer Signal Conditioners

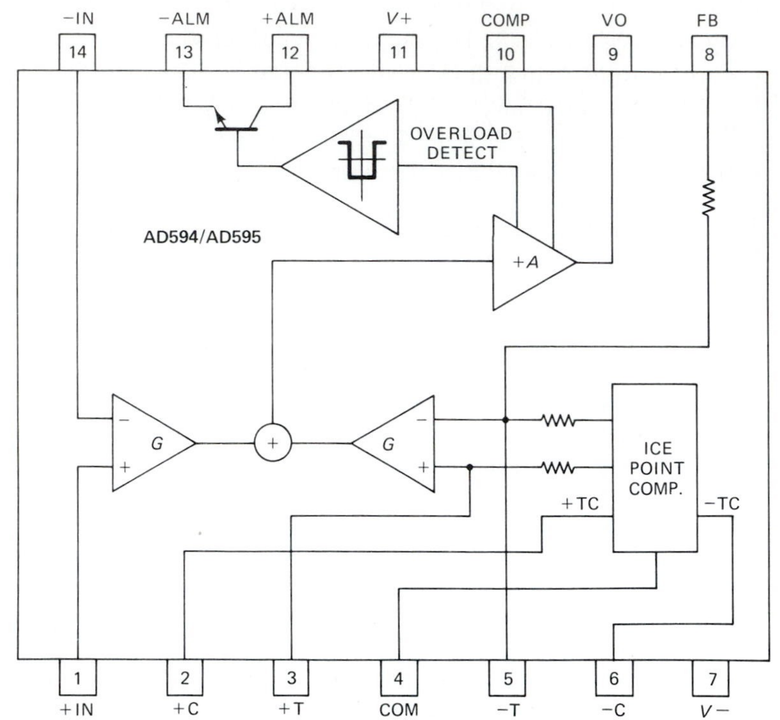


**AD594/AD595**

- Pretrimmed for Type J (AD594) or Type K (AD595) Thermocouples
- Can Be Used with Type T Thermocouple Inputs
- Low Impedance Voltage Output: 10mV/°C
- Built-In Ice Point Compensation
- Wide Power Supply Range: +5V to ±15V
- Low Power: < 1mW typical
- Thermocouple Failure Alarm
- Laser Wafer-Trimmed to 1°C Calibration Accuracy
- Set-Point Mode Operation
- Self-Contained Celsius Thermometer Operation
- High Impedance Differential Input
- Side-Brazed DIP or Low Cost CERDIP

FIGURE 11-26

5. *Integrators* find the area underneath a voltage against time function over a specified interval. *Differentiators* find the instantaneous rate of change of an analog signal. Both circuits can be made using the inverting follower configuration. In the case of the integrator an input resistor and a feedback capacitor are used. In the case of the differentiator an input capacitor and a feedback resistor are used.
6. In a *logarithmic amplifier* a nonlinear PN junction is used in the feedback network of an inverting follower to provide a logarithmic transfer function. In the *antilog amplifier* the transistor is connected in series with the input signal of the inverting follower.
7. The various forms of analog signals processing circuit are often combined into a single package to form *special function circuits.*

## 11-9 STUDENT EXERCISES

1. Select several resistors in the 10 kohm to 1 megohm range and several capacitors in the 0.001 μF to 10 μF range. Fashion an RC integrator (Fig. 11-12A) from one resistor and one capacitor. Apply the output of a function generator to the RC integrator, and observe

both input and output waveforms on a dual-beam oscilloscope. Vary both the frequency of the signal source and the output waveform to see how the RC integrator affects the signal.

2. Perform the same exercise but connect the components as a differentiator (Fig. 11-12B).
3. Build a Miller integrator circuit such as Fig. 11-14 using an input resistor of 100 kohms and a feedback capacitor of 0.1 μF. Apply a squarewave of 1/RC Hz to the input and observe the output signal on an oscilloscope. Vary the frequency of the signal to see how the circuit affects the waveform. Explain any anomalies observed.
4. Build a Miller integrator circuit such as Fig. 11-15 and compare the operation to that of the previous exercise.
5. Design and build a logarithmic amplifier such as Fig. 11-19. First use a 1-volt sawtooth to observe the operation. Then use other waveforms.

## 11-10 RECAPITULATION

Now return to the objectives and Prequiz questions at the beginning of the chapter and see how well you can answer them. If you cannot answer certain questions, mark them and review the appropriate parts of the text. Next, try to answer the questions and work the problems below using the same procedure.

## 11-11 QUESTIONS AND PROBLEMS

1. Draw the schematic for a simple chopper amplifier using a mechanical SPDT switch. Similarly, draw a circuit replacing the mechanical switch with CMOS electronic switches.
2. An operational amplifier in a noninverting configuration has a noise figure of 75 $nV/[Hz]^{1/2}$. Calculate the output RMS noise when the bandwidth is 2000 Hz.
3. Draw the schematic for the input circuit of a typical differential chopper amplifier.
4. Draw the block diagram for a carrier amplifier and explain the function of the various stages.
5. What is the advantage of a phase-sensitive detector (PSD)?
6. Draw the block diagram for a lock-in amplifier.
7. (a) Draw the circuit diagram for a Miller integrator; (b) write the transfer function for the Miller integrator; (c) what are the alternate names for the Miller integrator?
8. Draw the circuit diagram for an operational amplifier differentiator.
9. The time constant for a differentiator should be __________ relative to the period of the input signal; the time constant for an integrator circuit should be __________ relative to the period of the input signal.
10. How is an integrator used to make a quadrature circuit (a circuit that has both sine and cosine outputs)?
11. Calculate the gains for Miller integrator circuits in which the following *RC* time constants are used, 1 megohms and 0.01 μF, 10 kohms and 0.1 μF, 100 kohms and 0.001 μF.
12. A Miller integrator consists of a 100 kohm input resistor and a 1 μF feedback capacitor. The maximum allowable output potential is +11 volts. Calculate the time required to rise from an initial condition of 0 volts to saturation when a 0.1 Vdc signal is applied to the input.
13. Draw the circuit for (a) logarithmic amplifier; (b) antilog amplifier.

14. An electronic integrator is used on a medical blood pressure monitor to find the ______ ______________ of the applied analog pressure waveform. This signal is essentially the area under the ______________ curve.
15. What are the advantages of a carrier amplifier? A chopper amplifier?
16. A pH (acid-base measurement) transducer has a drift factor of 1.4 mV/°C, and the gain-of-10 voltage amplifier processing the transducer signal has a drift of 0.5 mV/°C. Calculate the total drift component if the amplifier ambient temperature rises from 30°C to 45°C, and the solution in which the transducer is immersed rises from 25°C to +38°C.
17. A 400-Hz sinewave is applied to the input of a Miller integrator circuit. Sketch the output waveform expected from this circuit.
18. Derive the transfer equation of the operational amplifier Miller integrator using the Kirchoff's law method.
19. Derive the transfer equation of an operational amplifier differentiator using the Kirchoff's law method.
20. Derive the transfer equation of the simple op-amp logarithmic amplifier.
21. Describe some of the problems that might be encountered in designing and building practical operational amplifier Miller integrators, and possible solutions to those problems.
22. Describe some of the problems that might be encountered in designing and building practical operational amplifier differentiators and possible solutions to those problems.
23. Design a practical Miller integrator in which the output voltage will rise at a rate of 1 volt per second when a 100 mV input signal is applied.
24. Describe why the logarithmic amplifier must be temperature compensated in order to avoid a large temperature sensitive error. How can the amplifier be temperature compensated (draw a sketch)?
25. A digital-to-analog converter that can be used as a programmable gain amplifier is also called a ______________ DAC.
26. Describe the properties of, and differences between, analog signals, sampled analog signals, and digital signals.
27. Describe how a sampled analog signal might be integrated. Use a sketch if appropriate.
28. What is the purpose of resistor $R2$ in Fig. 11-15? And the purpose of $S1$?
29. Why is the resistor network $R3$ through $R6$ required on practical integrators (Fig. 11-15), even though it is not always shown in textbook integrator circuits?
30. Describe the functions of resistor $R2$ and capacitor $C2$ in Fig. 11-16.
31. Describe the functions of $C1/R3$ in Fig. 11-19.
32. Calculate the values for $R3$ and $C1$ in Fig. 11-19 if the maximum operating frequency is 500 Hz, $R1$ is 12 kohms, the maximum input signal voltage is +5 volts, and the maximum output voltage is +16 volts.
33. Design a programmable gain amplifier (Fig. 11-25) with the following differential voltage gains: ×100, ×500, ×1000.

CHAPTER 12

# IC Waveform Generator and Waveshaping Circuits

## OBJECTIVES

1. Understand the difference between relaxation and feedback oscillators.
2. Learn the operation of monostable multivibrators.
3. Learn the operation of astable squarewave, triangle, and sawtooth multivibrators.
4. Learn the operation of sinewave oscillators.

## 12-1 PREQUIZ

These questions test your prior knowledge of the material in this chapter. Try answering them before you read the chapter. Look for the answers (especially those you answered incorrectly) as you read the text. After you have finished studying the chapter, try answering these questions again and those at the end of the chapter (see Section 12-11).

1. Calculate an $RC$ time constant which will allow a capacitor to charge from $-10$ Vdc to $+10$ Vdc in 250 mS.
2. A monostable multivibrator (MMV) must produce a 20 mS pulse. Select an $RC$ time constant.
3. A monostable multivibrator has a timing resistor of 100 kohms, and a timing capacitor of 0.001 μF. What is the duration of the output pulse if the positive feedback resistors are equal?
4. Calculate the frequency of an astable multivibrator in which the positive feedback resistors are equal and the $RC$ time constant is 0.030 seconds.

## 12-2 INTRODUCTION TO WAVEFORM GENERATORS

Waveform generators are used to produce a large variety of electronic waveforms. Some are sinewave oscillators. Others like the *astable multivibrator* (AMV) may produce square waves, triangle waves or other nonsinusoidal waveforms. A digital clock is a special case of the astable multivibrator often used in digital logic and computer circuits.

In general, the term oscillator may be used to describe the three cases above, sinewave oscillators, astable multivibrators, and digital clocks. An oscillator is a circuit which produces a *periodic waveform* (one that repeats itself). The output waveform can be a sine wave, square wave, triangle wave, sawtooth wave, pulses, or any of several other waveshapes. The important thing is that the waveform is *periodic*.

A class of waveform generator which is not an oscillator is the *monostable multivibrator*, or *one-shot*, circuit. This circuit produces only a single pulse when triggered, so it is not periodic.

There are two types of oscillator circuits, *relaxation oscillators* and *feedback oscillators*. Feedback oscillators use an active device such as an amplifier and provide feedback which produces regeneration instead of degeneration. These circuits account for a large number of the oscillators used in practical electronic circuits.

Relaxation oscillators use any of several breakdown devices (neon lamps) or negative resistance devices (tunnel diodes). Negative resistance devices operate according to Ohm's law under certain conditions and opposite Ohm's law under other conditions. Breakdown relaxation oscillators pass little or no current at voltages below some threshold and pass a large current at voltages above the threshold. Examples of these devices are neon glow lamps and unijunction transistors (UJT).

There is also a subclass of oscillators based on IC devices such as voltage comparators, operational amplifiers, and integrators. These circuits are discussed in this chapter. Because they are based on the charge and discharge properties of resistor-capacitor networks it is prudent to review the operation of simple *RC* networks.

## 12-3 REVIEW OF RC NETWORKS

Many of the waveform generators discussed in this chapter depend on *RC* network characteristics for their operation. Here, we will briefly review *RC* network DC theory. Consider Fig. 12-1A. In the initial condition switch *S*1 is in position A and is thus open-circuited. Initially there is no charge stored in capacitor $C$ ($V_c = 0$). If switch *S*1 is moved to position B, however, voltage $V$ is applied to the *RC* network. The capacitor begins to charge with current from the battery and $V_c$ begins to rise towards $V$ (see curve $V_{cb}$ in Fig. 12-1B). The instantaneous capacitor voltage is found from

$$V_c = V[1 - e^{(-T/RC)}] \tag{12-1}$$

where

$V_c$ = the capacitor voltage

$V$ = the applied voltage from the source

$T$ = the elapsed time (in seconds) after charging begins

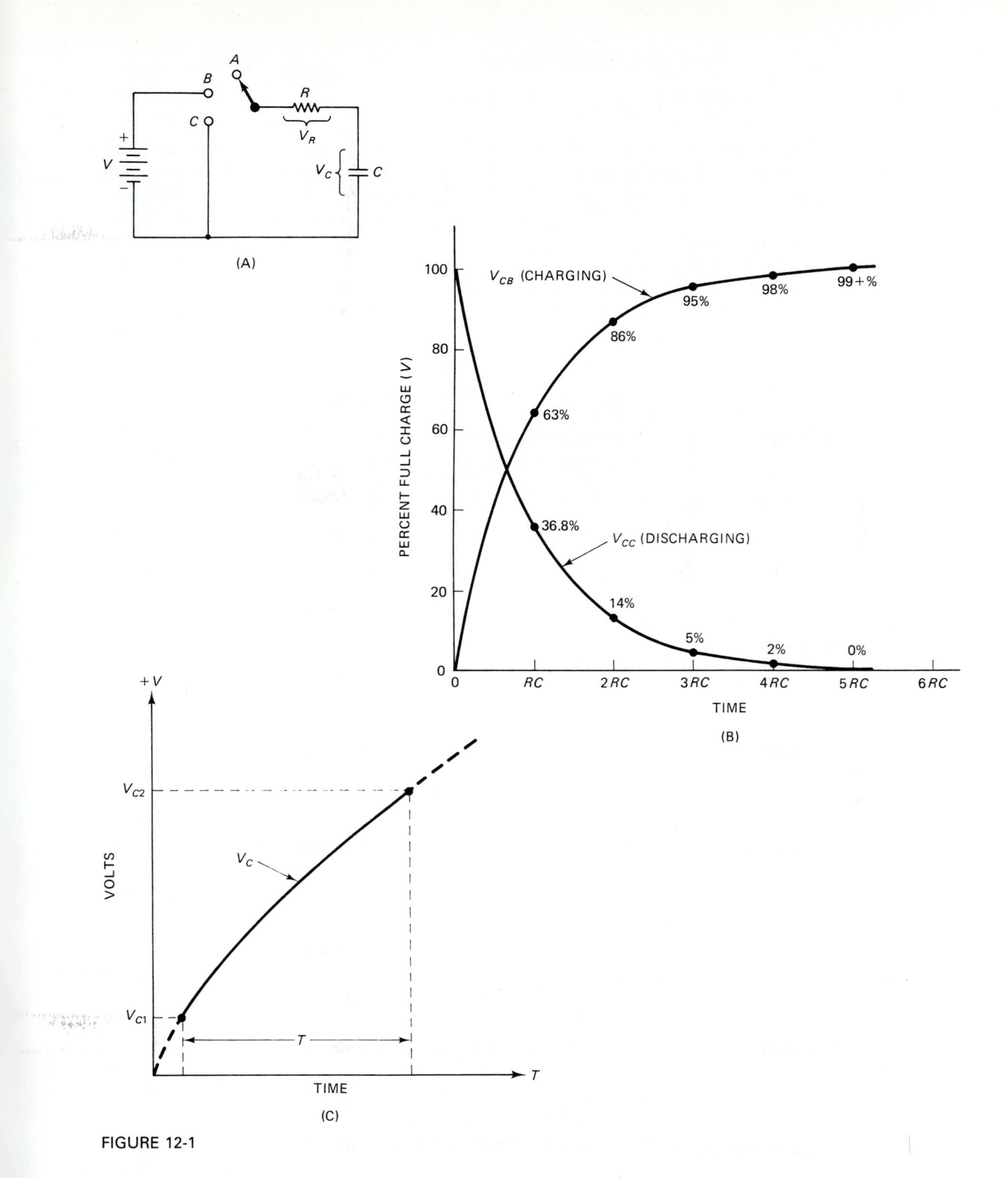
A
B
C
R
$V_R$
V
$V_C$
C
+
−
(A)
100
80
60
40
20
0
PERCENT FULL CHARGE (V)
$V_{CB}$ (CHARGING)
$V_{CC}$ (DISCHARGING)
99+%
98%
95%
86%
63%
36.8%
14%
5%
2%
0%
0
RC
2RC
3RC
4RC
5RC
6RC
TIME
(B)
+V
VOLTS
$V_{C2}$
$V_C$
$V_{C1}$
T
TIME
(C)

FIGURE 12-1

$R$ = the resistance in ohms

$C$ = the capacitance in farads

The product $RC$ is called the *RC time constant* of the network. It is sometimes abbreviated T. If $R$ is in *ohms* and $C$ is in *farads*, then the product $RC$ is specified in *seconds*. The capacitor voltage rises to approximately 63.2% of the final value after $1RC$, 86% after $2RC$ and >99% after $5RC$. A capacitor in an $RC$ network is considered fully charged after five time constants.

If switch $S1$ in Fig. 12-1A is next set to position C, the capacitor will begin to discharge through the resistor. In the discharge condition

$$V_c = V[e^{(-T/RC)}] \tag{12-2}$$

Voltage $V_c$ drops to 36.8% of the full charge level after one time constant ($1RC$) and to nearly zero after $5RC$.

Figure 12-1C depicts a situation commonly encountered in waveform generator circuits. In this graph the capacitor is required to charge from some initial condition ($V_{C1}$), which may or may not be 0 volts, to a final condition ($V_{C2}$), which may or may not be the fully charged $5RC$ point, in a specified time interval ($T$). What $RC$ time constant will force $V_{C1}$ to rise to $V_{C2}$ in time $T$? Assuming $V_{C1} < V_{C2} < V$

$$V - V_{C2} = (V - V_{C1})\,[e^{(-T/RC)}] \tag{12-3}$$

$$\frac{V - V_{C2}}{V - V_{C1}} = e^{(-T/RC)} \tag{12-4}$$

$$\text{LN}\left[\frac{V - V_{C2}}{V - V_{C1}}\right] = \frac{-T}{RC} \tag{12-5}$$

or, rearranging terms

$$RC = \frac{-T}{\text{LN}\left[\dfrac{V - V_{C2}}{V - V_{C1}}\right]} \tag{12-6}$$

### EXAMPLE 12-1

An $RC$ network is connected to a +12 Vdc source. What $RC$ product will permit voltage $V_c$ to rise from +1 Vdc to +4 Vdc in 200 mS when $V$ = +12 Vdc, $V_{C2}$ = +4 Vdc, and $V_{C1}$ = +1 Vdc?

### Solution

$$RC = \frac{-T}{\text{LN}\left[\dfrac{V - V_{C2}}{V - V_{C1}}\right]}$$

$$RC = \frac{-\left[200 \text{ mS} \times \frac{1 \text{ S}}{1000 \text{ mS}}\right]}{\text{LN}\left[\frac{12 - 4}{12 - 1}\right]}$$

$$RC = \frac{-0.200 \text{ Sec.}}{\text{LN}\left[\frac{8}{11}\right]}$$

$$RC = \frac{-0.200 \text{ Sec.}}{\text{LN } [0.727]}$$

$$RC = (-0.200 \text{ Sec.})/(-0.319) = 0.627 \text{ sec.}$$

■

Equation (12-3) can be used to derive the timing or frequency setting equations of many different $RC$-based waveform generator circuits. The key voltage levels frequently will be device trip points or critical values set by the circuit design.

## 12-4 MONOSTABLE MULTIVIBRATOR CIRCUITS

The *monostable multivibrator* (MMV) has two permissible output states (HIGH and LOW), but only one of them is stable. The MMV produces *one output pulse in response to an input trigger signal* (Fig. 12-2). The output pulse ($V_o$) has a duration ($T$) in which the output is in the *quasistable state.* The MMV is also called *one-shot*, *pulse generator*, and *pulse stretcher*. The latter name derives from the fact that the output duration $T$ is longer than the trigger pulse ($T > T_c$).

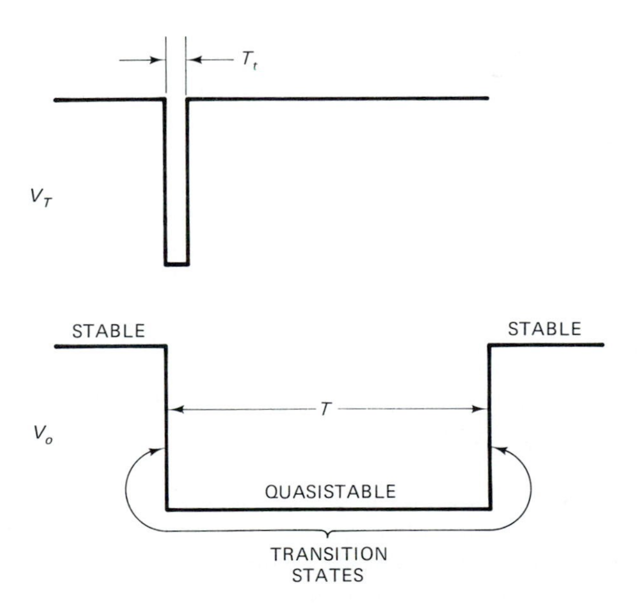


FIGURE 12-2

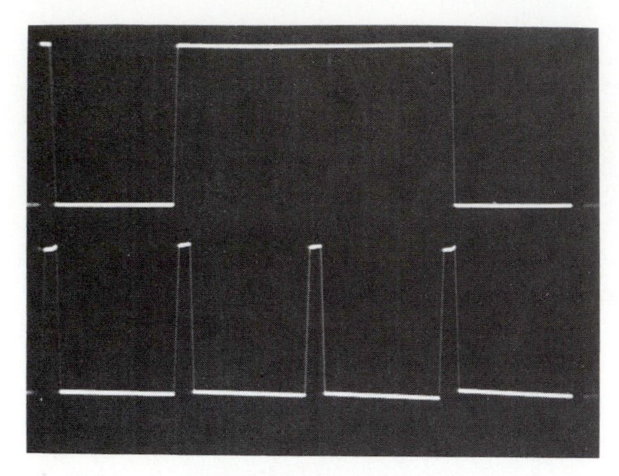

FIGURE 12-3

Monostable multivibrators have a wide variety of applications. Besides the pulse stretcher mentioned above, the MMV also serves to lock out unwanted pulses. Figure 12-3 shows how the output responds to only the first trigger pulse. The next two pulses occur during the active time ($T$) and are ignored. Such an MMV is said to be *nonretriggerable*.

A common application of this feature is in switch contact debouncing. All mechanical switch contacts bounce a few times on closure, creating a short run of exponentially decaying pulses. If an MMV is triggered by the first pulse from the switch, and if the MMV remains quasi-active long enough for the bouncing pulses to die out, the MMV output signal becomes the debounced switch closure. The main requirement is the MMV duration be longer than the switch contact bounce pulse train (5 mS is generally considered adequate for most switch types).

The range of possible MMV applications is so broad we will only discuss a general set of categories here. These include, *pulse generation*, *pulse stretching*, *contact debouncing*, *pulse signal cleanup*, *switching*, and *synchronization* of circuit functions (especially digital).

Figure 12-4A is a circuit for a nonretriggerable monostable multivibrator based on the operational amplifier using a voltage comparator circuit (Chapter 1). When there is no feedback, the effective voltage gain of an op-amp is its open-loop gain ($A_{vol}$). When both $-$IN and $+$IN are at the same potential, the differential input voltage ($V_{id}$) is zero, so the output is also zero. But if $V_{(-\text{IN})}$ does not equal $V_{(+\text{IN})}$, the high gain of the amplifier forces the output to either its positive or negative saturation values. If $V_{(-\text{IN})} > V_{(+\text{IN})}$, the op-amp reacts to a positive differential input signal, so the output saturates at $-V_{sat}$. However, if $V_{(-\text{IN})} < V_{(+\text{IN})}$, the amplifier reacts to a negative differential input signal and the output saturates to $+V_{sat}$. The operation of the MMV depends on the relationship of $V_{(-\text{IN})}$ and $V_{(+\text{IN})}$.

Four states of the monostable multivibrator must be considered, *stable state*, *transition state*, *quasistable state*, and *refractory state*.

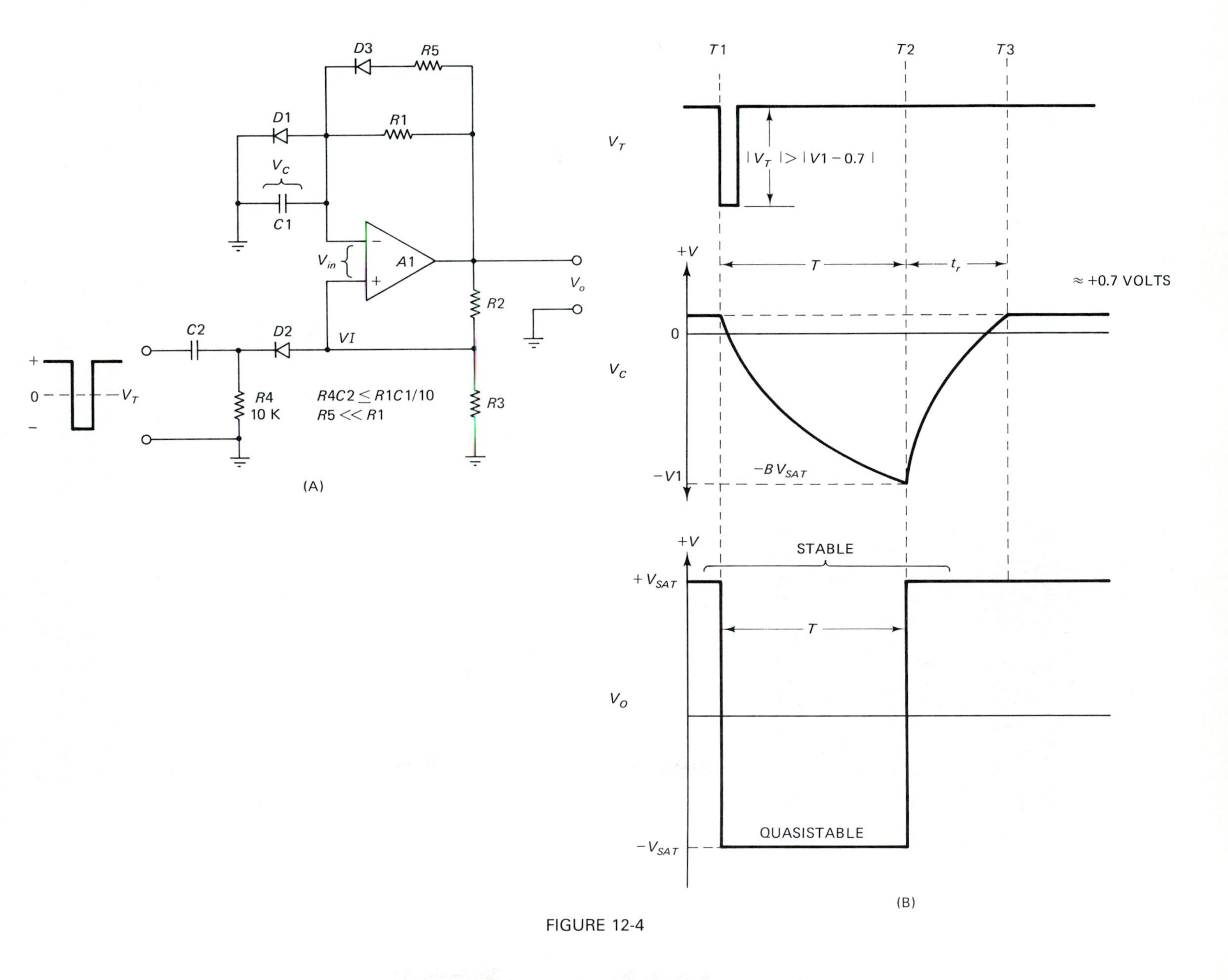
D3
R5
D1
R1
$V_C$
C1
$V_{in}$
A1
$V_o$
R2
C2
D2
VI
+
0
−
$V_T$
R4
10 K
R4C2 ≤ R1C1/10
R5 << R1
R3
(A)
T1
T2
T3
$V_T$
$|V_T| > |V1 - 0.7|$
+V
T
$t_r$
≈ +0.7 VOLTS
0
$V_C$
−V1
$-BV_{SAT}$
+V
STABLE
$+V_{SAT}$
T
$V_O$
$-V_{SAT}$
QUASISTABLE
(B)
FIGURE 12-4

### 12-4.1 Stable State

The output voltage $V_o$ is initially at $+V_{sat}$. Capacitor $C1$ will attempt to charge in the positive-going direction because $+V_{sat}$ is applied to the $R1C1$ network. But, because of diode $D1$ shunted across $C1$, the voltage across $C1$ is clamped to $+V_{D1}$. For a silicon diode such as the 1N914 or 1N4148, $+V_{D1}$ is about +0.7 Vdc. Thus, the inverting input (−IN) is held to +0.7 Vdc during the stable state. The noninverting input (+IN) is biased to a level $V1$, which is

$$V1 = \frac{R3(+V_{sat})}{R2 + R3} \quad (12\text{-}7)$$

or in the special case of $R2 = R3$

$$V1 = \frac{+V_{sat}}{2} \quad (12\text{-}8)$$

The factor $R/3/(R2 + R3)$ is often designated by the Greek letter *beta* (β), so

$$\beta = \frac{R3}{R2 + R3} \quad (12\text{-}9)$$

Therefore

$$V1 = \beta(+V_{sat}) \quad (12\text{-}10)$$

The amplifier ($A1$) sees a differential input voltage ($V_{id}$) of ($V1 - V_{D1}$), or ($V1 - 0.7$) volts. Using the previous notation

$$V_{id} = \frac{R3(+V_{sat})}{R2 + R3} - 0.7 \quad (12\text{-}11)$$

As long as $V1 > V_{D1}$, the amplifier effectively reacts to a negative DC differential voltage at the inverting input and, with its high open-loop again ($A_{vol}$), it will remain saturated at $+V_{sat}$. Here, the amplifier is a 741 operated at DC power supply potentials of ±12 Vdc, so $V_{sat}$ typically will be ±10 volts.

### 12-4.2 Transition State

The input trigger signal ($V_t$) is applied to the MMV in Fig. 12-4A through $RC$ network $R4C2$. The general design rule for this network is its time constant should be not more than one-tenth the time constant of the timing network

$$R4\ C2 < \frac{R1\ C1}{10} \quad (12\text{-}12)$$

At time $T1$ (see Fig. 12-4B) trigger signal $V_t$ makes an abrupt HIGH-to-LOW transition to a peak value less than ($V1 - 0.7$) volts. Under this condition the polarity of $V_{id}$ is now reversed and the inverting input now reacts to a positive voltage, ($V1 + V_t - 0.7$) is less than $V_{D1}$. The output voltage $V_o$ now rapidly snaps to $-V_{sat}$. The fall time of the output signal is dependent on the slew rate and the open-loop gain of the operational amplifier ($A1$).

### 12-4.3 Quasi-Stable State

The output signal from the MMV is quasi-stable between $T1$ and $T2$ in Fig. 12-4B. It is called quasi-stable because it does not change over $T = T2 - T1$. But when $T$ expires, the MMV times out and $V_o$ reverts to the stable state ($+V_{sat}$).

During the quasi-stable time $D1$ is reverse-biased. Capacitor $C1$ discharges from +0.7 Vdc to zero and then recharges toward $-V_{sat}$. When $-V_o$ reaches $-V1$, however, the value of $V_{id}$ crosses zero which forces $V_o$ to snap once again to $+V_{sat}$.

Equation (12-6) makes it possible to derive the *timing equation* for the MMV. The timing capacitor must charge from an initial value ($V_{C1}$) to a final value ($V_{C2}$) in time $T$. Then what value of $R1C1$ will cause the required transitions? Consider the case $R2 = R3$ ($V1 = 0.5V_{sat}$)

$$R1C1 = \frac{-T}{\text{LN}\left[\dfrac{V_{sat} - V_{C2}}{V_{sat} - V_{C1}}\right]} \qquad (12\text{-}13)$$

$$R1C1 = \frac{-T}{\text{LN}\left[\dfrac{V_{sat} - ((0.5)\,(V_{sat} + 0.7))}{V_{sat} - 0.7}\right]} \qquad (12\text{-}14)$$

and for the case where $V_{sat} = 10$ Vdc

$$R1C1 = \frac{-T}{\text{LN}\left[\dfrac{10\text{ Vdc} - ((0.5)\,(10 + 0.7))}{10\text{ Vdc} - 0.7\text{ volts}}\right]} \qquad (12\text{-}15)$$

$$R1C1 = \frac{-T}{\text{LN}\left[\dfrac{10\text{ Vdc} - 5.35\text{ volts}}{10\text{ Vdc} - 0.7\text{ volts}}\right]} \qquad (12\text{-}16)$$

$$R1C1 = \frac{-T}{\text{LN}\left[\dfrac{4.65}{9.3}\right]} \qquad (12\text{-}17)$$

$$R1C1 = \frac{-T}{\text{LN}\,[0.5]} \qquad (12\text{-}18)$$

$$R1C1 = \frac{-T}{-0.69} \qquad (12\text{-}19)$$

Thus

$$T = 0.69R1C1 \qquad (12\text{-}20)$$

**EXAMPLE 12-2**

A monostable multivibrator circuit is based on a 741 operational amplifier with an output saturation voltage of ±10 volts. Calculate the $RC$ time constant needed to produce a 100 mS output pulse when feedback resistors $R2$ and $R3$ are each 10 kohms.

**Solution**

$$R1C1 = T/0.69$$

$$R1C1 = \frac{\left[100 \text{ mS} \times \frac{1 \text{ S}}{1000 \text{ mS}}\right]}{0.69}$$

$$R1C1 = 0.1/0.69 = 0.145 \text{ S}$$

■

Equation (12-20) represents the special case in which $B = 1/2$ ($R2 = R3$). Although $R2 = R3$ may be the usual case for this class of circuit, $R2$ and $R3$ may not be equal in some cases. A more generalized expression is

$$RC = \frac{T}{\text{LN}\left[\frac{1 + 0.7\ V/V_{sat}}{1 - B}\right]} \tag{12-21}$$

In which

$$\beta = \frac{R3}{R2 + R3} \tag{12-22}$$

When the quasistable state times out, the circuit returns to a stable state where it remains dormant until triggered again.

## EXAMPLE 12-3

A monostable multivibrator (Fig. 12-4A) is constructed with $R2 = 10$ kohms, and $R3 = 3.3$ kohms. The device is a 741 operational amplifier operated such that $|V_{sat}| = 10$ volts. Calculate the $RC$ time constant required to produce a 5 mS output pulse.

**Solution**

(a) *First calculate B*

$$\beta = \frac{R3}{R2 + R3}$$

$$\beta = \frac{3.3 \text{ kohms}}{(10 \text{ kohm} + 3.3 \text{ kohms})}$$

$$\beta = 3.3/13.3 = 0.248$$

(b) *Calculate RC time constant*

$$RC = \frac{T}{\text{LN}\left[\frac{1 + 0.7\ V/V_{sat}}{1 - \beta}\right]}$$

$$R = \frac{\left[5 \text{ mS} \times \frac{1 \text{ S}}{1000 \text{ mS}}\right]}{\text{LN}\left[\frac{1 + (0.7\ V/10\ V)}{1 - 0.248}\right]}$$

$$RC = \frac{0.005 \text{ S}}{\text{LN}\ [1.07/0.752]} = 0.0142 \text{ sec.}$$

■

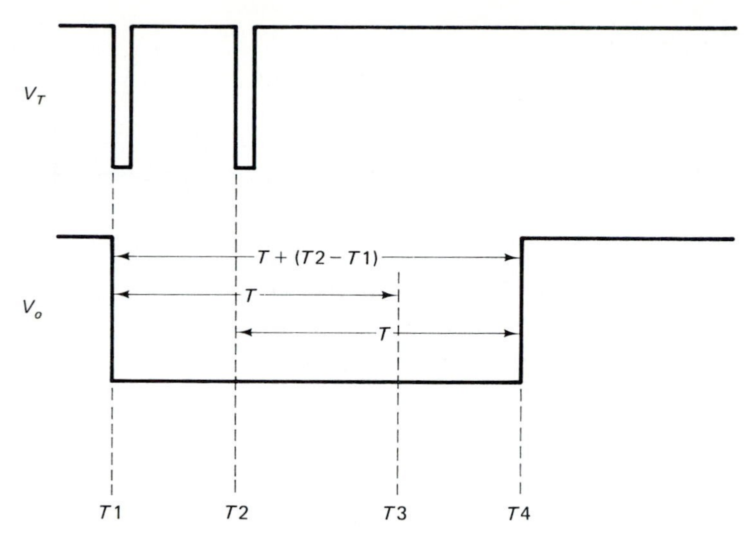


FIGURE 12-5

### 12-4.4 Refractory Period

At time $t2$ the output signal voltage $V_o$ switches from $-V_{sat}$ to $+V_{sat}$. Although the output has timed out, the MMV is not yet ready to accept another trigger pulse. The *refractory state* between $t2$ and $t3$ is characterized by the output being in a stable state, but the input is unable to accept a new trigger input stimulus. The refractory period must wait for the discharge of $C1$ under the influence of the output voltage to satisfy $V1 < (V1 - 0.7)$ volts.

### 12-4.5 Retriggerable Monostable Multivibrators

The circuit of Fig. 12-4A is a nonretriggerable MMV. Once it is triggered, the circuit will not respond to further trigger inputs until after both the quasistable and refractory states are completed. This characteristic can be used to advantage in some applications. But in other cases the MMV may have to be retriggered. A *retriggerable monostable multivibrator* (RMMV) one-shot will respond to further trigger signals.

Figure 12-5 shows the retriggerable MMV response. An initial trigger signal ($V_t$) is received at time $t1$. The output snaps LOW and under normal circumstances, it would remain in this quasi-stable state until time $t3$ when the duration $T$ expires. But at time $t2$ a second trigger pulse is received. The circuit is now retriggered for another duration $T$, so will not time out until $t4$. The total time that the RMMV is in the quasistable state is $[T + (t2 - t1)]$. In other words, the RMMV output is active for the entire duration $T$ plus the portion of the previous active time which expired when the next trigger pulse was received.

Figure 12-6A shows the circuit for a simple RMMV based on an operational amplifier. The two inputs are biased from a reference voltage source, $+V_{ref}$. The potential applied to +IN is a fraction of $+V_{ref}$. That is, $[(R3)(+V_{ref})/(R2 + R3)]$. The potential applied to −IN is a function of $+V_{ref}$ and time-constant $R1C1$. If the circuit is not triggered at turn on, the capacitor ($C1$) charges up to $+V_{ref}$, so −IN is more positive

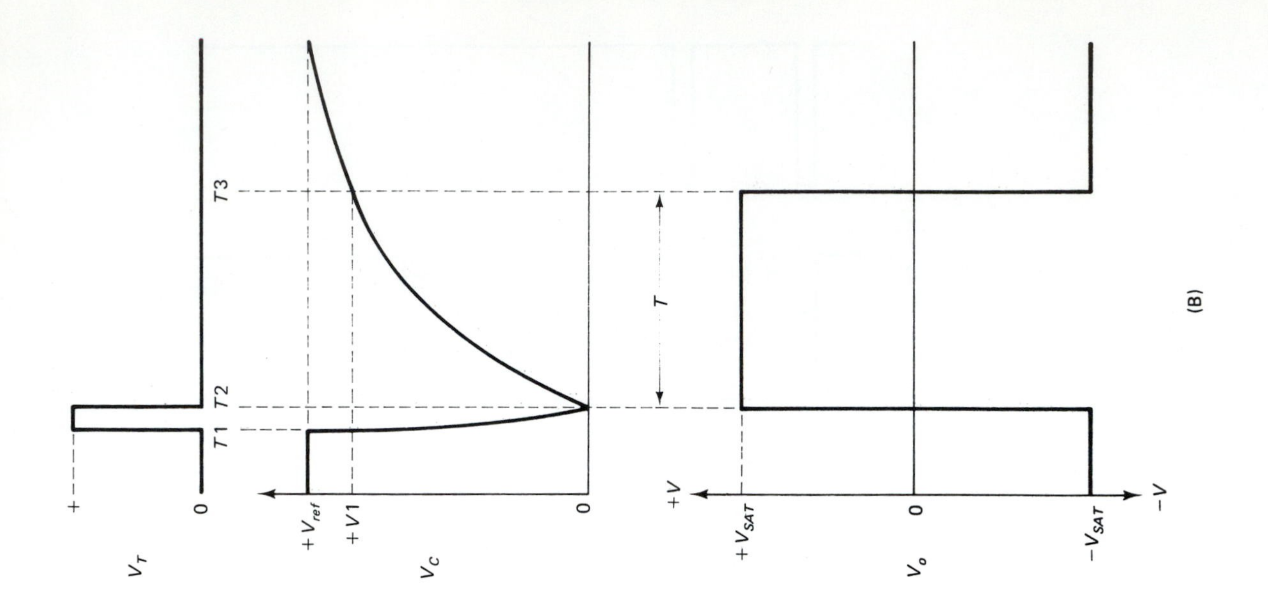

$V_T$
+
0
T1
T2
T3
$V_C$
$+V_{ref}$
$+V1$
0
T
+V
$+V_{SAT}$
0
$V_o$
$-V_{SAT}$
−V


(B)

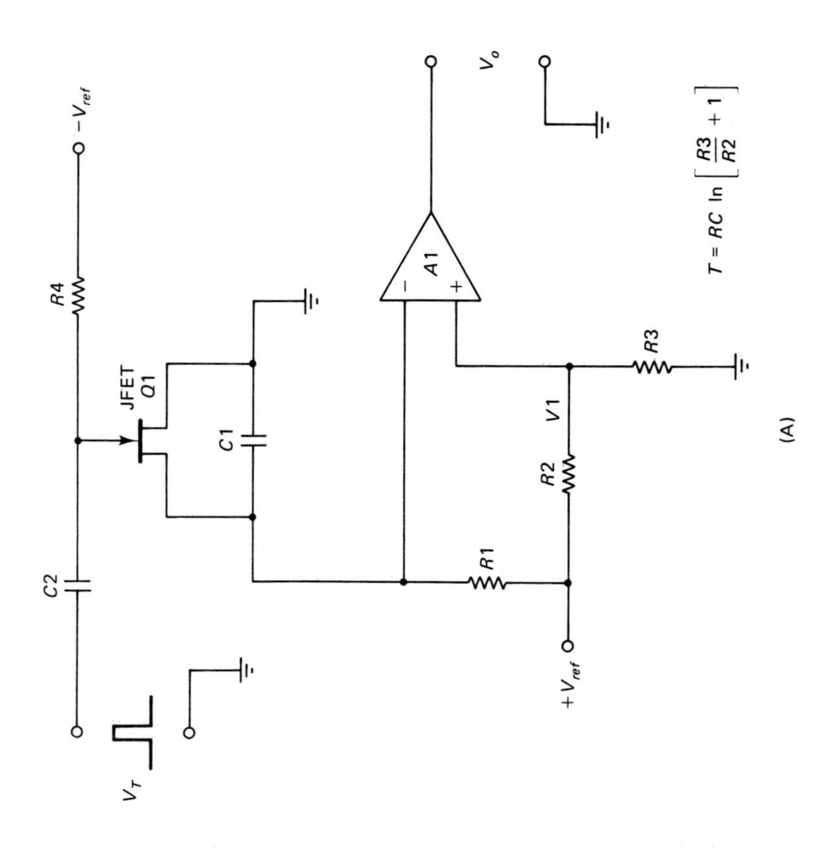

R4
$-V_{ref}$
JFET
Q1
C2
C1
R1
R2
R3
V1
A1
$+V_{ref}$
$V_o$
$V_T$
$T = RC \ln \left[ \frac{R3}{R2} + 1 \right]$


(A)

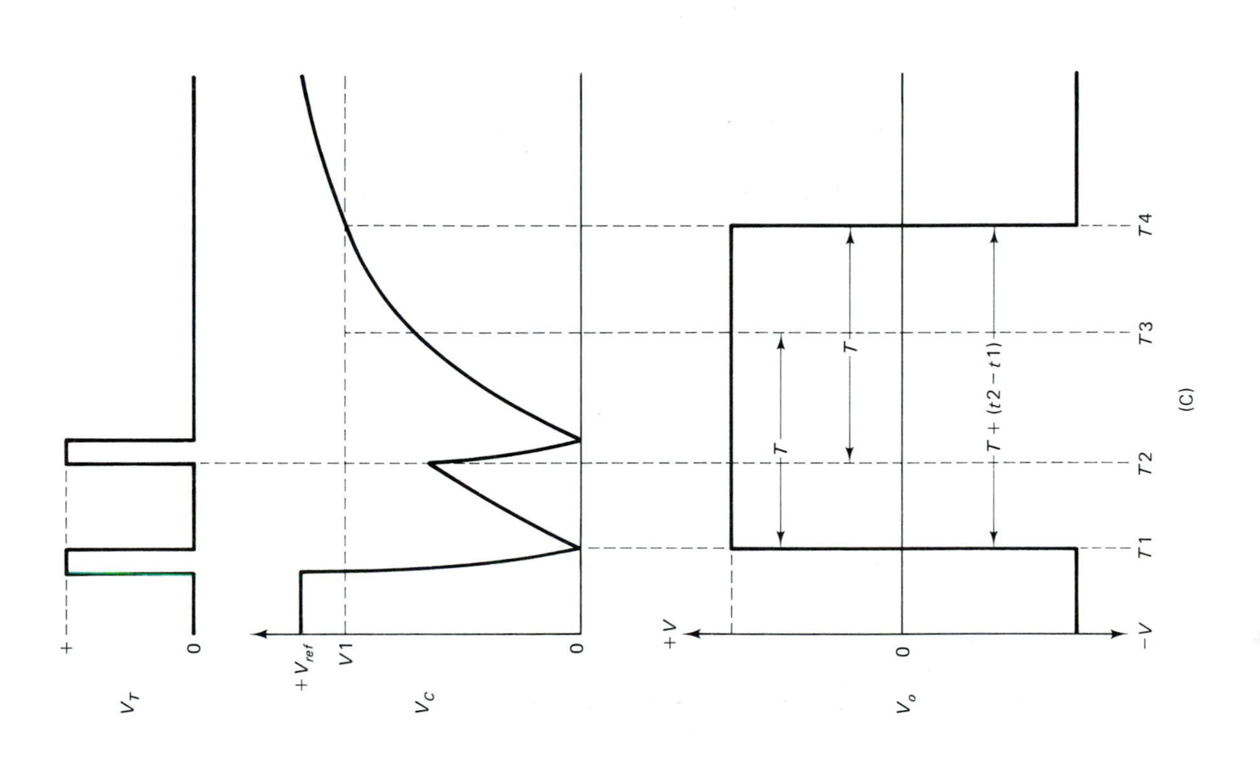


(C)

FIGURE 12-6

than $+$IN. This situation forces $V_o$ to $-V_{sat}$, which is the stable state. When a positive-going trigger pulse ($V_t$) is received (see Fig. 12-6B), it biases the junction field effect transistor (JFET), $Q1$, hard on. The JFET drain-source channel resistance drops very low, causing $C1$ to rapidly discharge between $t1$ and $t2$. With $V_c$ close to 0 Vdc, $+$IN is more positive than $-$IN, so the output snaps abruptly to $+V_{sat}$ at time $t1$. During the interval $t2$ to $t3$ capacitor $C1$ begins charging towards $+V_{ref}$. $V_o$ remains at $+V_{sat}$. Once $V_c$ reaches $+V1$, however, the output of $A1$ snaps back to $-V_{sat}$.

The duration ($T$) is found from

$$T = R1C1 \text{ LN}\left[\frac{R3}{R2} + 1\right] \tag{12-23}$$

### EXAMPLE 12-4

Calculate the duration of a retriggerable monostable multivibrator (Fig. 12-6A) in which $R2 = R3 = 10$ kohms, $R1 = 15$ kohms and $C1 = 0.1$ uF.

### Solution

$$T = R1C1 \text{ LN}\left[\frac{R3}{R2} + 1\right]$$

$$T = (15 \text{ kohms}\left[0.01 \text{ uF} \times \frac{1 \text{ F}}{10^6 \text{ uF}}\right] \text{LN } [(10\text{K}/10\text{K}) + 1]$$

$$T = 1.5 \times 10^{-4} \text{ LN}(2) = 1.04 \times 10^{-4} \text{ sec.}$$

■

The operation discussed above, and depicted in Fig. 12-6B, is for normal nonretriggered operations. Figure 12-6C shows the retriggered case. Here the RMMV receives a second trigger pulse at time $t2$, which forces the JFET ($Q1$) to turn on again and rapidly discharge $C1$. The charging process then starts over again and will continue until the circuit times out, unless a further trigger pulse is received.

The RMMV is triggered by an external event and will continually retrigger as long as the external event keeps occurring. A common use for the RMMV is in alarm or sensing circuits. If no event is sensed prior to time out, the RMMV returns to the stable state and the following circuitry will be triggered to alarm status. An example is a medical respirator alarm. A sensor in the respirator line senses variations in either pressure or air temperature caused by breathing. Each time a breath is sensed it retriggers the RMMV. But if the patient ceases breathing, the RMMV will time out and cause an alarm to nearby medical personnel.

## 12-5 ASTABLE (FREE-RUNNING) CIRCUITS

The circuits discussed in the previous section were aperiodic, that is an output pulse occurs only once in response to a stimulus or trigger. Such circuits are said to be *monostable* because they possess only one stable state. An *astable multivibrator* (AMV) is free-running. The output of the AMV is a pulse or wave train which is *periodic*. In a periodic signal the wave repeats itself indefinitely until the circuit is either turned off or otherwise inhibited.

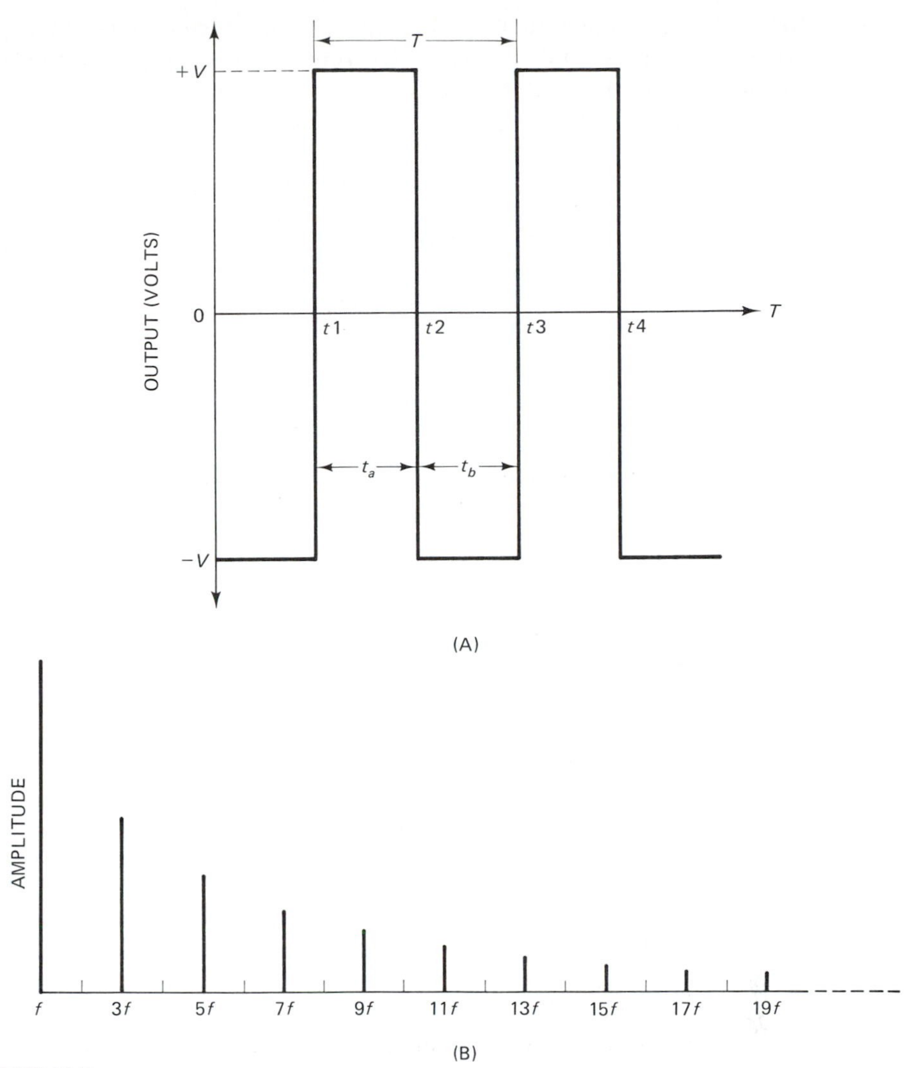


FIGURE 12-7

Astable multivibrators are oscillators. The AMV produces square, triangle, and sawtooth waves. Sine waves are also produced by oscillator circuits, but these circuits operate differently (see Sec. 12-6).

### 12-5.1 Nonsinusoidal Waveform Generators

The nonsinusoidal AMV circuits generate square, triangle, or sawtooth waves. A pulse generator results when it is combined with a monostable multivibrator (MMV).

### 12-5.2 Squarewave Generators

The square wave generator is the most basic of the three wave generators we will discuss in this section. Figure 12-7 shows the classical square wave. Each time interval of the wave is quasistable, so we may conclude the square wave generator has no stable states (it is *astable*). The waveform snaps back and forth between $-V$ and $+V$, dwelling on each level of a duration of time ($t_a$ or $t_b$). The period, $T$, is

$$T = t_a + t_b \tag{12-24}$$

where

$T$ = the period of the square wave ($t1$ to $t3$)

$t_a$ = the interval $t1$ to $t2$

$t_b$ = the interval $t2$ to $t3$

The frequency of oscillation ($F$) is the reciprocal of $T$

$$F = \frac{1}{T} \tag{12-25}$$

The ideal square wave is both baseline and time line symmetrical. That means that $|+V| = |-V|$ and $t_a = t_b$. Under time line symmetry $t_a = t_b = t$, so $T = 2t$ and $f = 1/2t$.

All continuous mathematical functions can be constructed from a series of fundamental frequency sine waves ($f$) added to a series of sine and cosine harmonics ($2f$, $3f$, $4f$, . . ., $nf$). The *Fourier series* of the waveform is a mathematical expression which lists the harmonics present, their respective amplitudes, and phase relationships. The Fourier series is usually depicted as a bar graph spectrum (Fig. 12-7B).

In the ideal symmetrical square wave, the Fourier spectrum (Fig. 12-7B) consists of the fundamental frequency ($f$) plus the *odd order* harmonics ($3f$, $5f$, $7f$). Furthermore, the harmonics are in phase with the fundamental. Theoretically, an infinite number of odd number harmonics are present in the ideal square wave. However, in practical square waves the ideal is considered satisfied with harmonics to about $999f$. This ideal is almost never reached, however, due to the normal bandwidth limitations of the circuit. An indicator of harmonic content is the rise time of the square waves, the faster the rise time, the higher the number of harmonics.

The circuit for an operational amplifier square wave generator is shown in Fig. 12-8A. The basic circuit is similar to the simple voltage comparator and the MMV (see Section 12-4). Like the MMV, the AMV depends on the relationship between $V_{(-IN)}$ and $V_{(+IN)}$. In Fig. 12-8A the voltage applied to the noninverting input ($V_{(+IN)}$) is determined by a resistor voltage divider, $R2$ and $R3$. This voltage is called $V1$ in Fig. 12-8A and is

$$V1 = \frac{V_o R3}{R2 + R3} \tag{12-26}$$

or when $V_o$ is saturated

$$V1 = \frac{V_{sat} R3}{R2 + R3} \tag{12-27}$$

Once again, the factor $R3/(R2 + R3)$ is often designated $\beta$

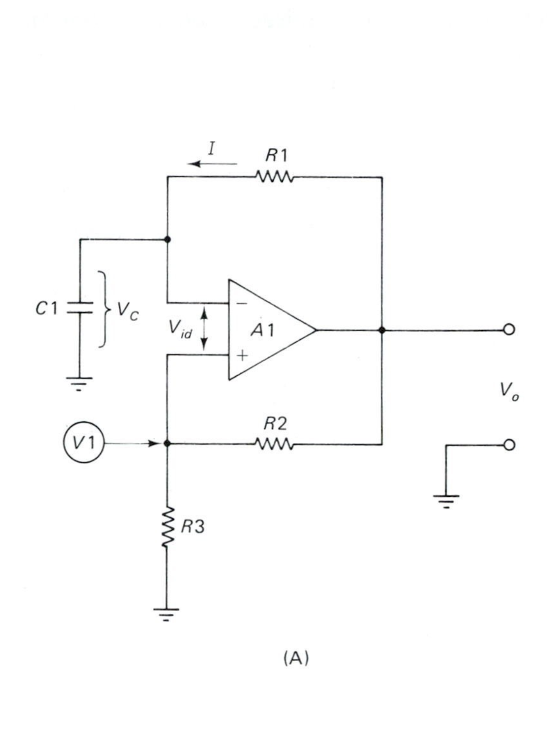


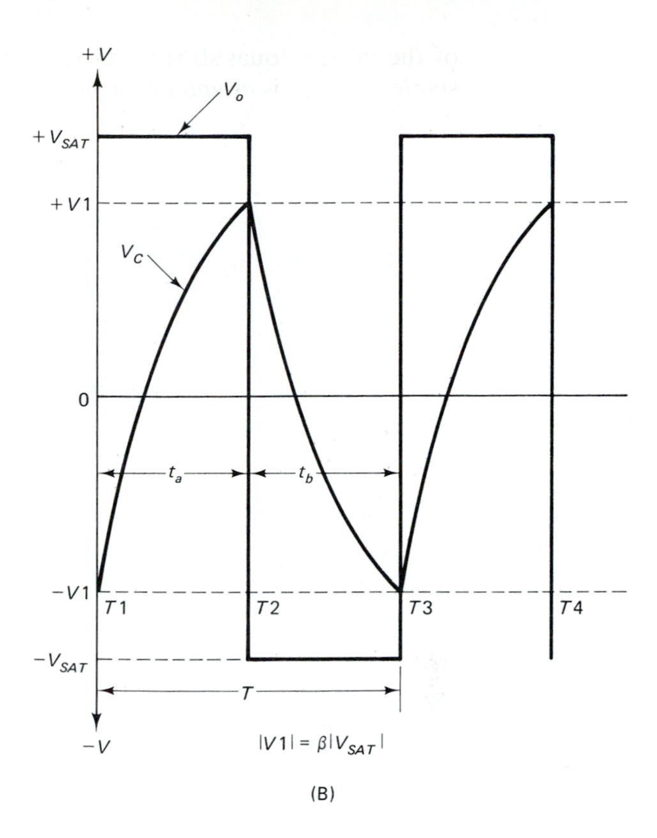


FIGURE 12-8

$$\beta = \frac{R3}{R2 + R3} \tag{12-28}$$

Because Eq. (12-28) is always a fraction, $V1 < V_{sat}$, $V1$ has the same polarity as $V_{sat}$.

The voltage applied to the inverting input ($V_{(-IN)}$) is the voltage across capacitor $C1$ ($V_{C1}$). This voltage is created when $C1$ charges under the influence of current $I$, which in turn is a function of $V_o$ and the time constant of $R1C1$. Timing of the circuit is shown in Fig. 12-8B.

At turn on $V_{C1} = 0$ volts and $V_o = +V_{sat}$, so $V1 = +V1 = \beta(+V_{sat})$. Because $V_{C1} < V1$, the op-amp sees a negative differential input voltage so the output remains at $+V_{sat}$. During this time, however, $V_{C1}$ is charging towards $+V_{sat}$ at a rate of

$$V_{C1} = V_{sat}[1 - e^{(-t_a/R1C1)}] \tag{12-29}$$

When $V_{C1}$ reaches $+V1$, however, the op-amp interprets $V_{C1} = V1$, so $V_{id} = 0$. The output snaps from $+V_{sat}$ to $-V_{sat}$ (time $t2$ in Fig. 12-8B). The capacitor begins to discharge from $+V1$ towards zero and then recharges towards $-V_{sat}$. When it reaches $-V1$, the inputs are once again zero so the output again snaps to $+V_{sat}$. The output continues to snap back and forth between $-V_{sat}$ and $+V_{sat}$ and thereby produces a square wave output signal.

Again with Eq. (12-6) we can find the time constant required to charge from an initial voltage $V_{C1}$ to an end voltage $V_{C2}$ in time $t$

$$RC = \frac{-T}{\text{LN}\left[\frac{V - V_{C2}}{V - V_{C1}}\right]} \quad (12\text{-}30)$$

In Fig. 12-8A the $RC$ time constant is $R1C1$. From Fig. 12-8B it is apparent that, for interval $t_a$, $V_{C1} = -\beta V_{sat}$, $V_{C2} = +\beta V_{sat}$ and $V = V_{sat}$. To calculate the period $T$, then:

$$2\ R1C1 = \frac{-T}{\text{LN}\left[\frac{V_{sat} - \beta V_{sat}}{V_{sat} - (-\beta V_{sat})}\right]} \quad (12\text{-}31)$$

or, rearranging Eq. (12-31)

$$-T = 2R1C1\ \text{LN}\left[\frac{V_{sat} - \beta V_{sat}}{V_{sat} - (-\beta_{sat})}\right] \quad (12\text{-}32)$$

$$-T = 2R1C1\ \text{LN}\left[\frac{1 - \beta}{1 + \beta}\right] \quad (12\text{-}33)$$

$$T = 2R1C1\ \text{LN}\left[\frac{1 + \beta}{1 - \beta}\right] \quad (12\text{-}34)$$

Because $\beta = R3/(R2 + R3)$

$$T = 2R1C1\ \text{LN}\left[\frac{1 + (R3/(R2 + R3))}{1 - (R3/(R2 + R3))}\right] \quad (12\text{-}35)$$

which reduces to

$$T = 2R1C1\ \text{LN}\left[\frac{2\ R2}{R3}\right] \quad (12\text{-}36)$$

### EXAMPLE 12-5

Calculate the oscillating frequency for an astable multivibrator in which $R1$ = 10 kohms, $R2$ = 10 kohms, $R3$ = 4.7 kohms, and $C1$ = 0.005 uF.

#### Solution

$$T = 2R1C1\ \text{LN}\left[\frac{2\ R2}{R3}\right]$$

$$T = (2)\ (10^4\ \text{ohms})\left(0.005\ \text{uF} \times \frac{1\ F}{10^6\ \text{uF}}\right)\ \text{LN}\left[\frac{(2)\ (10\ \text{kohms})}{4.7\ \text{kohms}}\right]$$

$$T = (0.0001)\ \text{LN}\ [4.26] = 0.000145\ \text{sec.}$$

■

Equation (12-36) defines the frequency of oscillation for any combination of $R1$, $R2$, $R3$, and $C1$. In the special case $R2 = R3$, $\beta = 0.5$

$$T = 2R1C1 \text{ LN} \left[\frac{1 + 0.5}{1 - 0.5}\right] \tag{12-37}$$

$$T = 2R1C1 \text{ LN} \left[\frac{1.5}{0.5}\right] \tag{12-38}$$

$$T = 2R1C1 \text{ LN} [1.09] = 2.2 \; R1C1 \tag{12-39}$$

### EXAMPLE 12-6

A square wave oscillator is constructed so $R2 = R3$. Calculate the time constant required for a 1000-Hz symmetrical square wave.

### Solution

$$R1C1 = \frac{1}{2.2 \text{ f}}$$

$$R1C1 = \frac{1}{(2.2)\,(1000 \text{ Hz})} = 0.00046 \text{ sec.}$$

■

The circuit of Fig. 12-8A produces time line symmetrical square waves ($t_a = t_b$). If time line asymmetrical square waves are required, a circuit like Fig. 12-9 or 12-11A is required. The circuit in Fig. 12-9 uses a potentiometer ($R4$) and a fixed resistor ($R5$) to establish a variable duty cycle asymmetry. The circuit is similar to Fig. 12-8A, but with an offset circuit ($R4/R5$) added. The assumptions are $R5 = R1$ and $R4 << R1$. If $V_a$ is the potentiometer output voltage, $C1$ charges at a rate of $(R1/2)C1$ towards a potential of $(V_a + V_{sat})$. After output transition, however, the capacitor discharges at the same $(R1/2)C1$ rate toward $(V_a - V_{sat})$. Therefore, the two interval times are different ($t_a$ and $t_b$ are no longer equal). Figure 12-10 shows three extremes of $V_a$ where $V_a = +V$ (Fig. 12-10A), $V_a = 0$ (Fig. 12-10B), and $V_a = -V$ (Fig. 12-10C). These traces represent very long, equal, and very short duty cycles, respectively.

The circuit of Fig. 12-11A also produces asymmetrical square waves, but the duty cycle is fixed instead of variable. Once again the basic circuit is like Fig. 12-8A, but with additional components. In Fig. 12-11A the $RC$ timing network is altered so the resistors are different on each swing of the output signal. During $t_a$, $V_a = +V_{sat}$ so diode $D1$ is forward-biased and $D2$ is reverse-biased. For this interval

$$t_a = (R1A)\,(C1) \text{ LN} \left[1 + \frac{2\,R2}{R3}\right] \tag{12-40}$$

During the alternate half-cycle ($t_b$), the output voltage $V_o$ is at $-V_{sat}$, so $D1$ is reverse-biased and $D2$ is forward-biased. During this interval, $R1B$ is the timing resistor, while $R1A$ is effectively out of the circuit. The timing equation is

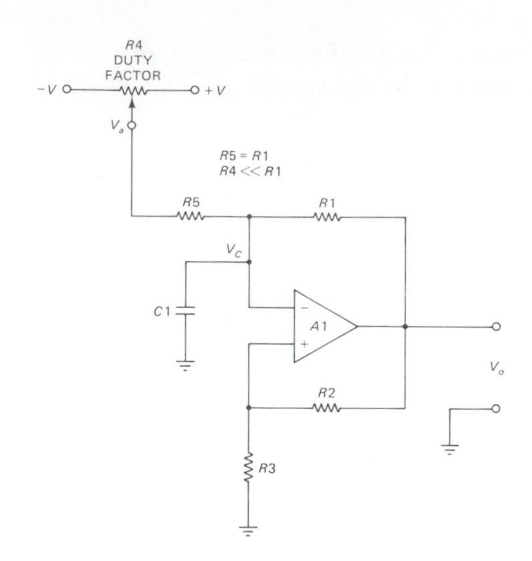


FIGURE 12-9

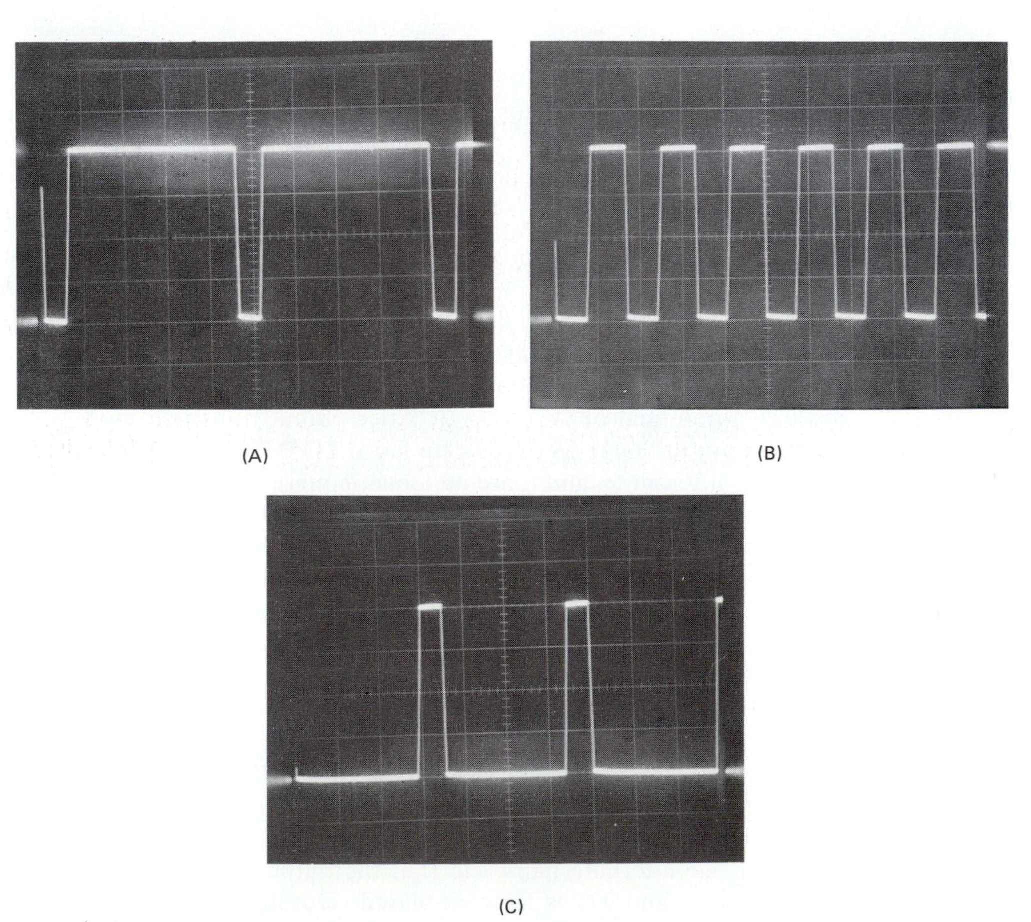


FIGURE 12-10

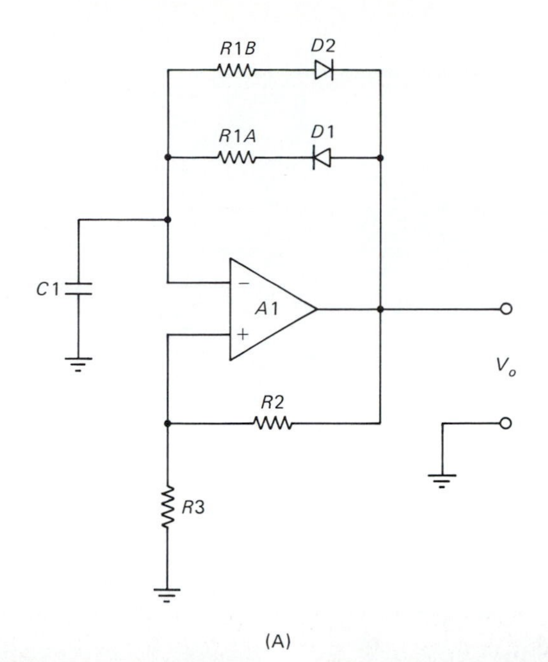


(A)

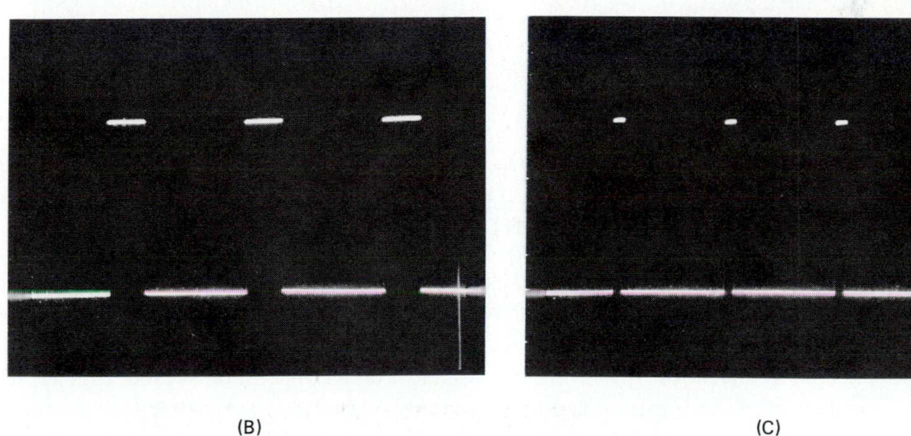

(B)

(C)

FIGURE 12-11

$$t_b = (R1B)\,(C1)\,\mathrm{LN}\left[1 + \frac{2\,R2}{R3}\right] \tag{12-41}$$

The total period ($T$) is $t_a + t_b$ so

$$T = (R1A)\,(C1)\,\mathrm{LN}\left[1 + \frac{2R2}{R3}\right] + (R1B)\,(C1)\,\mathrm{LN}\left[1 + \frac{2\,R2}{R3}\right] \tag{12-42}$$

Collecting terms

$$T = (R1A + R1B)\,(C1)\,\mathrm{LN}\left[1 + \frac{2R2}{R3}\right] \tag{12-43}$$

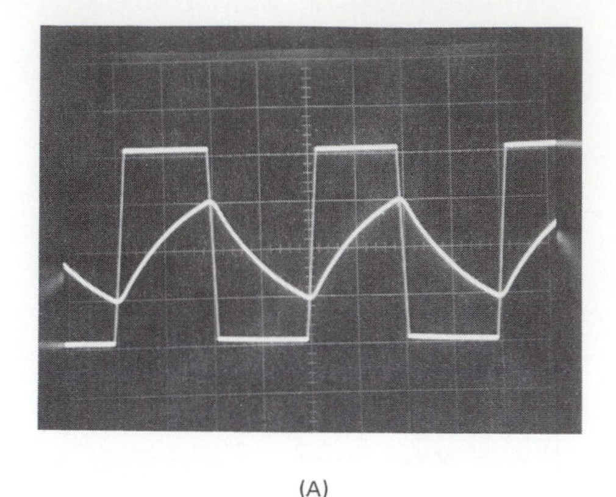

(A)

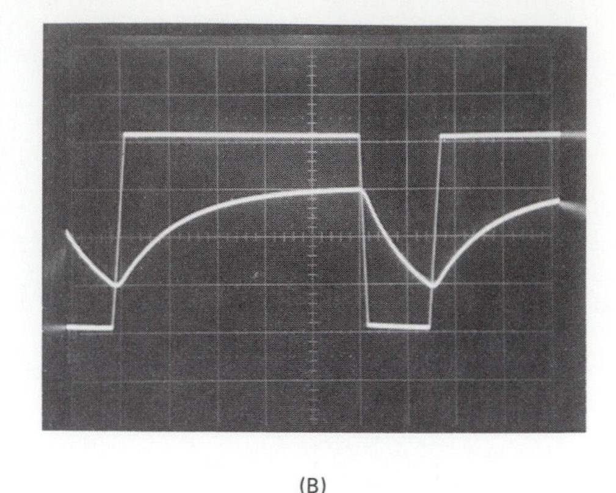

(B)

FIGURE 12-12

## EXAMPLE 12-7

An asymmetrical astable multivibrator is designed so $R1A$ = 100 kohms, $R1B$ = 10 kohms, $C1$ = 0.1 μF, and $R2 = R3$. Calculate the period of oscillation.

### Solution

Let $R2 = R3 = R$.

$$T = (R1A + R1B)\,(C1)\ \mathrm{LN}\left[1 + \frac{2\,R2}{R3}\right]$$

$$T = (100\ \mathrm{k} + 10\ \mathrm{k})\left(0.1 \times \frac{1\ F}{10^6\ \mathrm{uF}}\right)\ \mathrm{LN}\left[1 + \frac{2R}{R}\right]$$

$$T = (0.011)\ \mathrm{LN}\ (3) = 0.012\ \mathrm{sec.}$$

■

Equation (12-43) defines the oscillation frequency of the circuit in Fig. 12-11A. Figures 12-11B and 12-11C show the effects of two values of the $R1A/R1B$ ratio. In Fig. 12-11B the ratio $R1A/R1B$ is 3:1, in Fig. 12-11C it is 10:1.

The effect of this circuit on capacitor charging can be seen in Fig. 12-12. A relatively low $R1A/R1B$ ratio is seen in Fig. 12-12A. Notice in the lower lower trace the capacitor charge time is long compared with the discharge time. The effect is even more pronounced in the case of a high $R1A/R1B$ ratio (Fig. 12-12B).

### 12-5.2.1 Output Voltage Limiting

The standard op-amp MMV or AMV circuit sometimes produces a relatively sloppy square output wave. By adding a pair of back-to-back zener diodes (Fig. 12-13A) across the output, however, the signal can be cleaned up, but at the expense of amplitude. For each polarity, the output signal reacts to one forward-biased and one reverse-biased zener diode. On the positive swing, the output voltage is clamped at $[V_{Z1} + 0.7]$ volts. The 0.7 volts represents the normal junction potential across the forward-biased diode ($D2$). On negative swings of the output signal, the situation reverses. The output signal is clamped to $[-(V_{Z2} + 0.7)]$ volts. Figure 12-13B shows the unclamped output signal of a square wave generator. The signal swings ±10 volts.

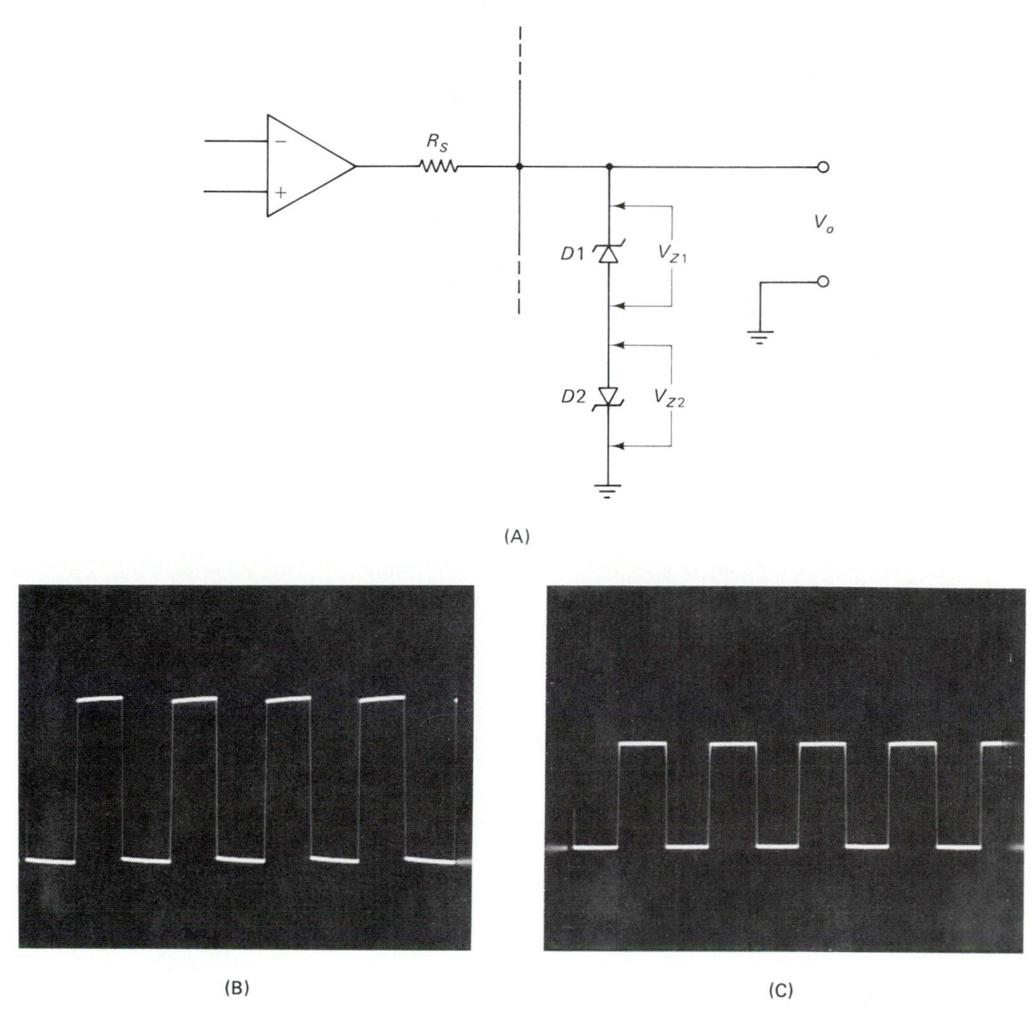


FIGURE 12-13

Figure 12-13C shows the output of the same circuit when a pair of 5.6-volt zener diodes are connected across the output. The output is reduced to $\pm[5.6 + 0.7]$ volts, but the corners are sharper.

### 12-5.2.2 Squarewaves from Sinewaves

Figure 12-14 shows a method for converting sinewaves to squarewaves. The circuit is shown in Fig. 12-14A and the waveforms are shown in Fig. 12-14B. The circuit is an operational amplifier connected as a comparator. Because the op-amp has no negative feedback path, the gain is very high ($A_{vol}$). Thus, a voltage difference across the input terminals of only a few millivolts will saturate the output. From this, the operation of the circuit and the waveform in Fig. 12-14B can be understood.

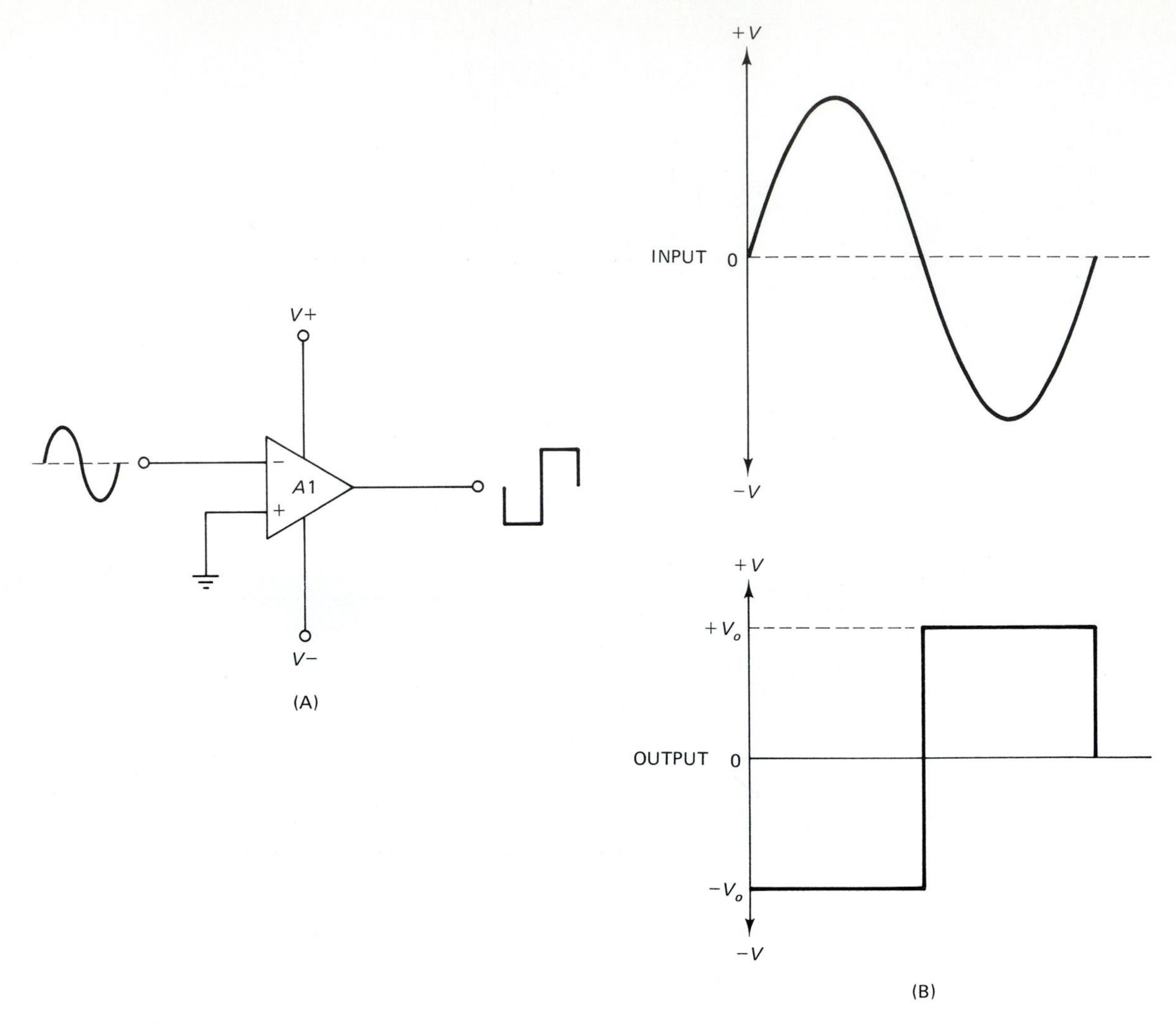


FIGURE 12-14

The input waveform is a sinewave. Because the noninverting input is grounded (Fig. 12-14A), the output of the op-amp is zero only when the input signal voltage is also zero. When the sinewave is positive, the output signal will be at $-V_o$. When the sinewave is negative, the output signal will be at $+V_o$. The output signal will be a squarewave at the sinewave frequency, with a peak-to-peak amplitude of $[(+V_o) - (-V_o)]$.

### 12-5.3 Triangle and Sawtooth Waveform Generators

*Triangle* and *sawtooth waveforms* (Fig. 12-15) are examples of *periodic ramp* functions. The sawtooth (Fig. 12-15A) is a single ramp waveform. The voltage begins to rise linearly at time *t*1. At time *t*2 the waveform abruptly drops back to zero, where it

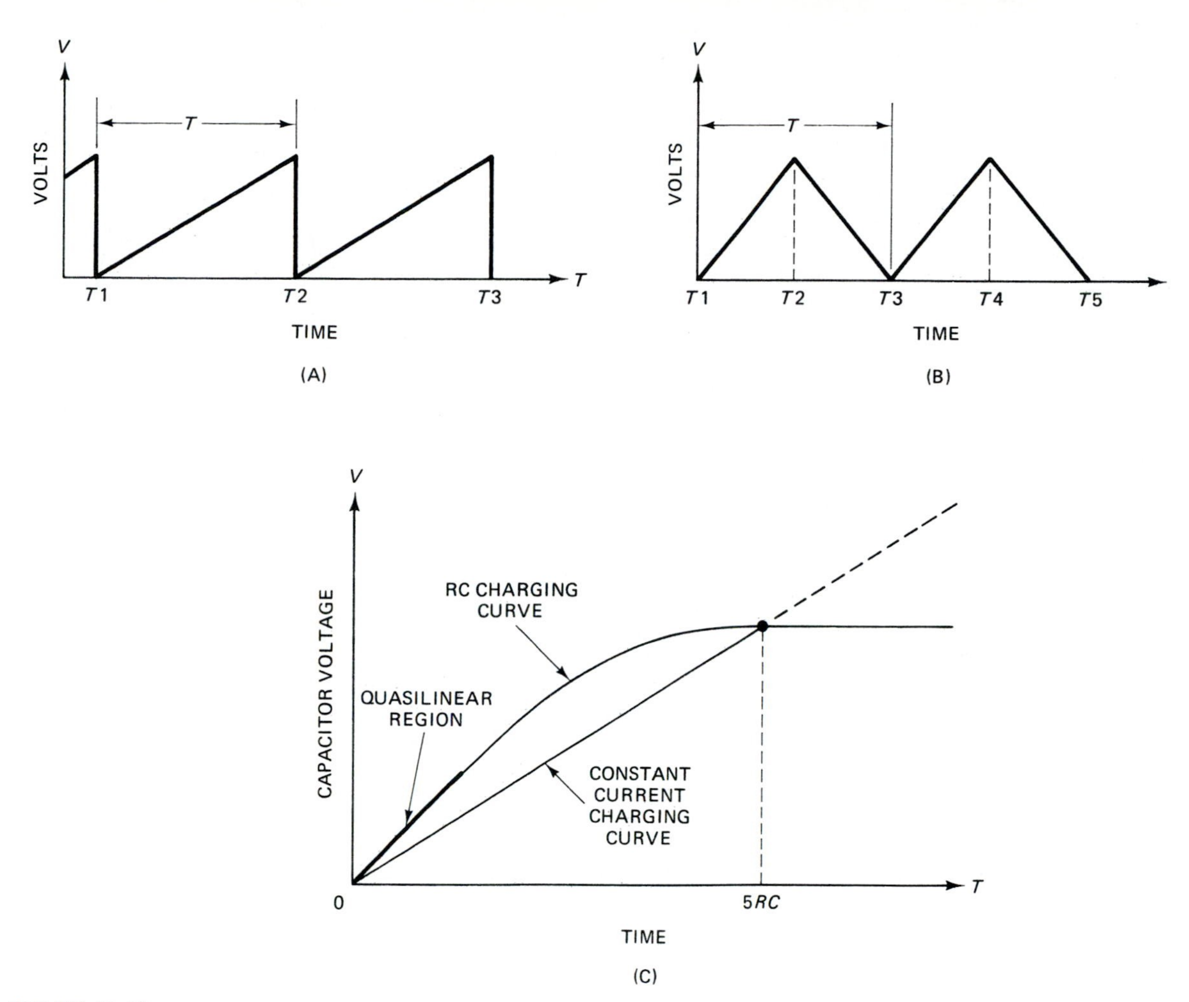


FIGURE 12-15

again starts to linearly ramp up. The sawtooth is usually periodic (although single sweep variants are sometimes used) and the period is designated $T$ (see Fig. 12-15A). The frequency is therefore $1/T$.

The triangle waveform (Fig. 12-15B) is a double ramp. The waveform begins to linearly ramp up at time $t1$. It reverses direction at time $t2$ and then downward linearly ramps until time $t3$. At time $t3$ the waveform again reverses direction and begins ramping upwards. The period of the triangle waveform ($T$) is $T1 - T3$.

*Ramp generators* are derived from capacitor charging circuits. The familiar $RC$ charging curve we discussed earlier in this chapter is reproduced in simplified form in Fig. 12-15C. The $RC$ charging waveform has an exponential shape, so is not well-suited to generating a linear ramp function.

There are two ways to force the capacitor charging waveform to be more linear. The first is to limit the charging time to the short quasilinear segment shown in Fig. 12-15C. The ramp thus obtained is not very linear, is limited in amplitude to a small

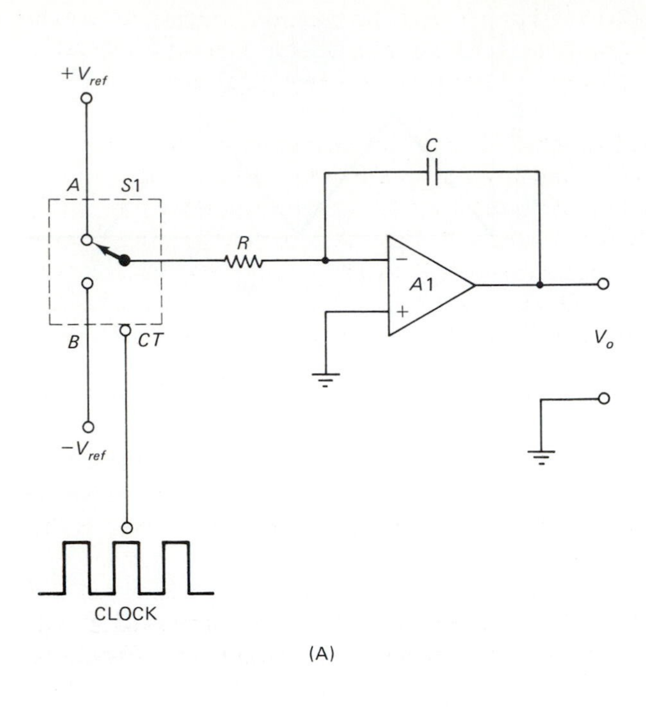


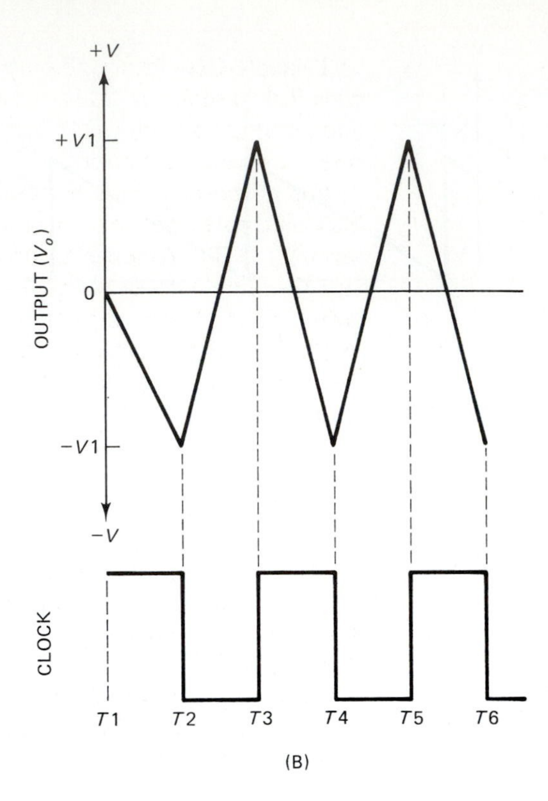


FIGURE 12-16

fraction of $V1$, and has a relatively steep slope that may or not be useful. A better method is to charge the capacitor through a *constant current source* (CCS). Using the CCS to charge the capacitor results in the linear ramp shown in Fig. 12-15C.

Triangle and sawtooth waveform oscillators create the constant current form of ramp generator by using a Miller integrator (see Chapter 11) to charge the capacitor (Fig. 12-16A). When a Miller integrator is driven by a *stable reference voltage*, the output is a linearly rising ramp. The ramp voltage ($V_o$) is

$$V_o = \frac{V_{ref}}{T} \tag{12-44}$$

or, because $T = RC$

$$V_o = \frac{V_{ref}}{RC} \tag{12-45}$$

If $V_{ref} = +10$ Vdc and the $RC$ time-constant is $T = RC = 0.001$ seconds, the ramp slope is

$$V_o = \frac{10 \text{ volts}}{0.001 \text{ sec.}} \tag{12-46}$$

$$V_o = 0.10 \text{ volts/sec.} \tag{12-47}$$

**Triangle Generators.** Figure 12-16A shows a simplified circuit model of a triangle waveform generator. This circuit consists of a Miller integrator as the ramp generator and an SPDT switch ($S1$) which can select either positive ($+V_{ref}$) or negative ($-V_{ref}$) reference voltage sources.

For purposes of this discussion switch $S1$ is an electronic switch which is toggled back and forth between positions A and B by a squarewave applied to the control terminal (CT). Assume an initial condition at time $t2$ (at which point $V_o = -V1$) and the input of the integrator is connected to $-V_{ref}$. At time $t2$ the square wave switch driver changes to the opposite state, so $S1$ toggles to connect $+V_{ref}$ to the integrator input. The ramp output will rise linearly at a rate of $+V_{ref}/RC$ until the switch again toggles at time $t3$. At this point, the ramp is under the influence of $-V_{ref}$, so it drops linearly from $+V1$ to $-V1$. The switch continuously toggles back and forth between $-V_{ref}$ and $+V_{ref}$. Thus, the output ($V_o$) continuously ramps back and forth between $-V1$ and $+V1$.

The circuit of Fig. 12-16A is not practical, but it serves as an analogy for the actual circuit. Figure 12-17A shows the circuit for a triangle waveform generator in which a Miller integrator forms the ramp generator and a voltage comparator serves as the switch. The comparator uses the positive feedback configuration, so it operates as a noninverting Schmitt trigger. Such a circuit snaps HIGH ($V_B = +V_{sat}$) when the input signal crosses a certain threshold voltage in the positive-going direction. It will snap LOW again ($V_B = -V_{sat}$) when the input signal crosses a second threshold in a negative going direction. The two thresholds are not always the same potential.

Because zener diodes $D1$ and $D2$ are in the circuit, the maximum allowable value of $+V_B$ is $[V_{ZD1} + 0.7]$ volts. The limit for $-V_B$ is $-[V_{ZD2} + 0.7]$ volts. If $V_{ZD1} = V_{ZD2}$, then $|+V_B| = |-V_B|$. These potentials represent $\pm V_{ref}$ discussed in the analogy presented above, so they are the potentials which affect the ramp generator input.

Consider an initial state in which $V_B$ is at the negative limit $-V_B$. The output $V_o$ will begin to ramp upwards from a minimum voltage of

$$V1 = \frac{V_A(R2 + R4)}{R4} - \frac{V_B R2}{R4} \tag{12-48}$$

The output will contine to ramp upwards toward a maximum value of

$$V3 = \frac{V_A(R2 + R4)}{R4} + \frac{V_B R2}{R4} \tag{12-49}$$

Causing a peak swing voltage of

$$V_p = V3 - V1 \tag{12-50}$$

$$V_p = \left[\frac{V_A(R2 + R4)}{R4} + \frac{V_B R2}{R4}\right] - \left[\frac{V_A(R2 + R4)}{R4} - \frac{V_B R2}{R4}\right] \tag{12-51}$$

$$V_p = \frac{V_B R2}{R4} + \frac{V_B R2}{R4} = \frac{2V_B R2}{R4} \tag{12-52}$$

Switching of the comparator occurs when the differential input voltage $V_{id}$ is zero. The inverting input ($-$IN) voltage is $V_A$. This is a fixed reference potential. The

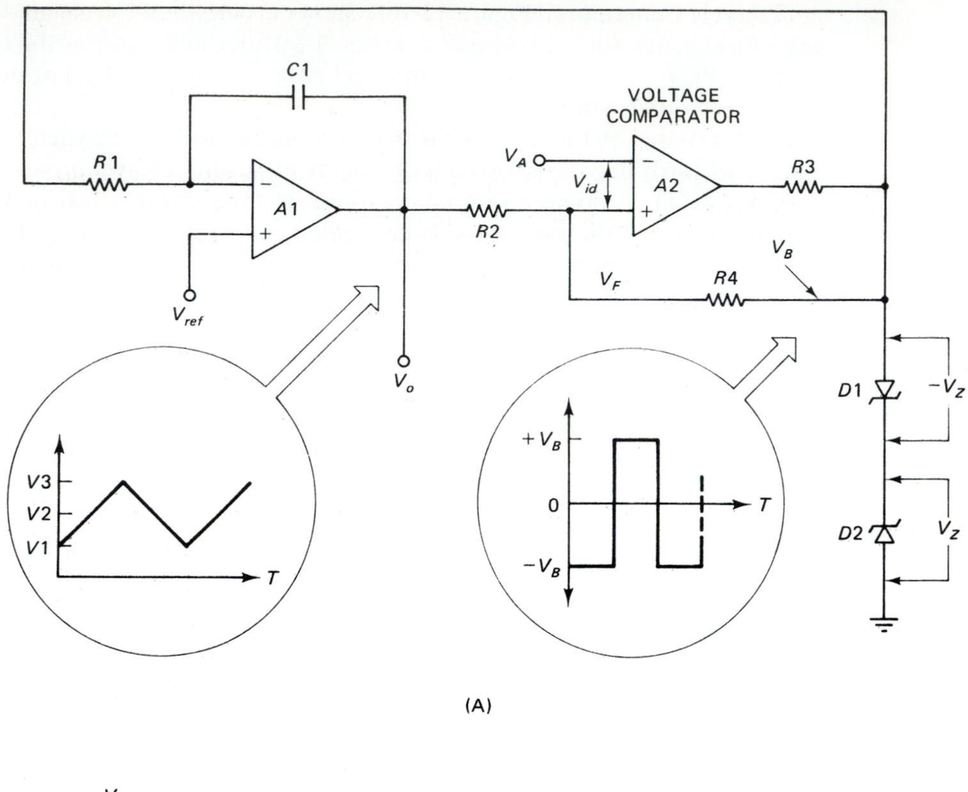


(A)

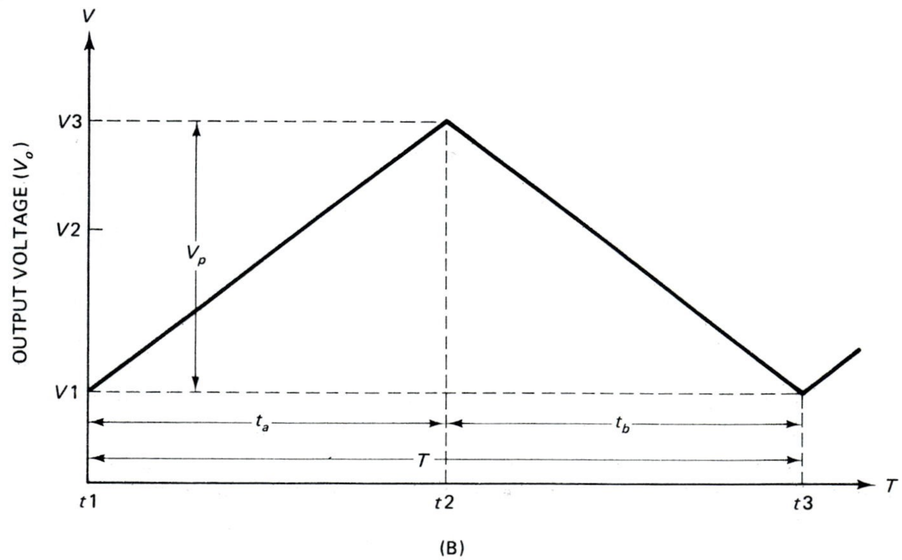


(B)

FIGURE 12-17

noninverting input (+IN) is at a voltage ($V_F$) which is the superposition of two voltages, $V_o$ and $V_B$

$$V_F = \frac{V_o R4}{R2 + R4} \pm \frac{\pm V_B R2}{R2 + R3} \qquad (12\text{-}53)$$

If $+V_B = -V_B$, the positive and negative thresholds are equal.
The duration of each ramp ($t_a$ and $t_b$) can be found from

$$t_{a,b} = \frac{V_p}{\left[\dfrac{V_B}{R1C1}\right]} \qquad (12\text{-}54)$$

The value of $V_B$ is selected from $-V_B$ or $+V_B$ as needed. In Eq. (12.52) it was found that $V_p = 2V_B R2/R4$, so

$$t_{a,b} = \frac{\left[\dfrac{2V_B R2}{R4}\right]}{\left[\dfrac{V_B}{R1C1}\right]} \qquad (12\text{-}55)$$

$$t_{a,b} = \left[\frac{R1C1}{V_B}\right]\left[\frac{2V_B R2}{R4}\right] \qquad (12\text{-}56)$$

$$t_{a,b} = R1C1\left[\frac{2V_B R2}{R4}\right] \qquad (12\text{-}57)$$

or, in the less general, but more common case of $t_a = t_b$

$$T = 2R1C1\left[\frac{2\ R2}{R4}\right] \qquad (12\text{-}58)$$

The frequency of the triangle wave is the reciprocal of the period ($1/T$), so

$$F = \frac{1}{T} \qquad (12\text{-}59)$$

$$F = \frac{1}{\dfrac{4R1C1R2}{R4}} \qquad (12\text{-}60)$$

$$F = \frac{R4}{4R1C1R2} \qquad (12\text{-}61)$$

**Sawtooth Generators.** The *sawtooth wave* (Fig. 12-15A) is a single slope ramp function. The wave ramps linearly upwards or downwards, and then abruptly snaps back to the initial baseline condition. Figure 12-18A shows a simple sawtooth generator circuit. A constant current source charges a capacitor which generates the linear ramp function (Fig. 12-18B). When the ramp voltage ($V_c$) reaches the maximum point ($V_p$)

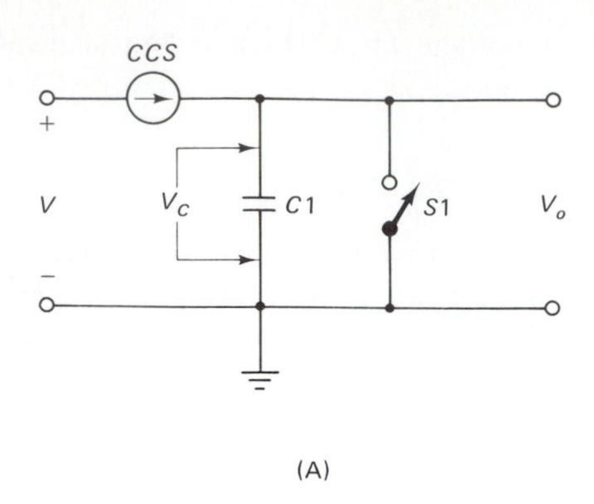


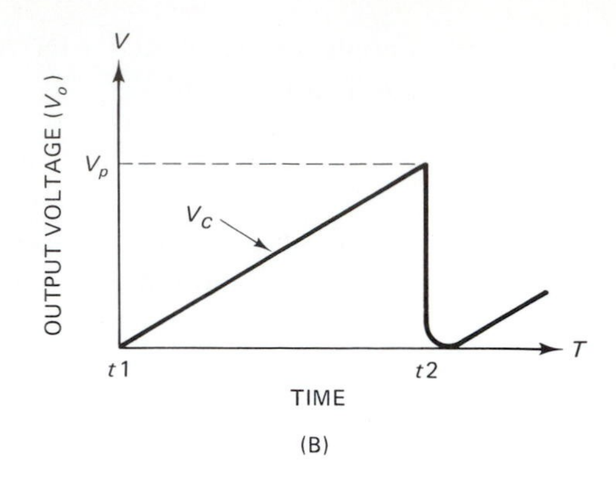


FIGURE 12-18

switch $S1$ is closed, forcing $V_c$ back to zero by discharging the capacitor. If switch $S1$ remains closed, the sawtooth is terminated. But if $S1$ reopens, however, a second sawtooth is created as the capacitor recharges.

Figure 12-19A shows the circuit for a periodic sawtooth oscillator. It is similar to Fig. 12-18A except a junction field effect transistor (JFET), $Q1$, is used as the discharge switch. When $Q1$ is turned off, the output voltage ramps upwards (see Fig. 12-19B). When the gate is pulsed hard on, the drain-source channel resistance drops from a very high value to a very low value, forcing $C1$ to rapidly discharge. In the absence of a gate pulse, however, the channel resistance remains very high. At time $t1$ the gate is turned off, so $V_c$ begins ramping upwards. At $t2$ the JFET gate is pulsed, so $C1$ rapidly discharges back to zero. When the pulse ($t2$–$t3$) ends, however, $Q1$ turns off again and the ramp starts over. The same circuit can be used for single sweep operation by replacing the pulse train applied to the gate of $Q1$ with the output of a monostable multivibrator.

The circuit of Fig. 12-20A shows a sawtooth generator which uses a Miller integrator ($A1$) as a ramp generator and replaces the discharge switch with an electronic switch

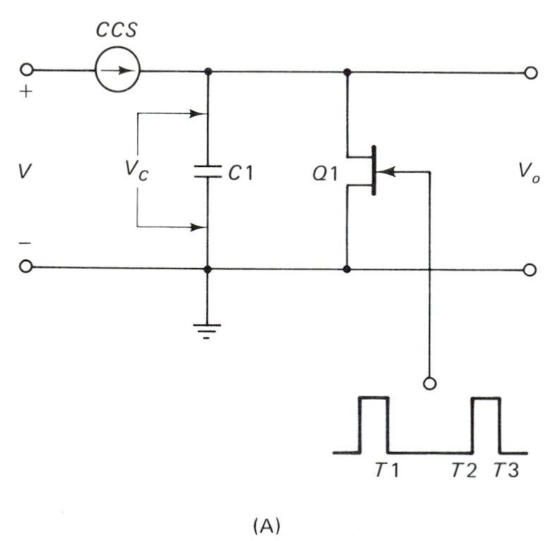


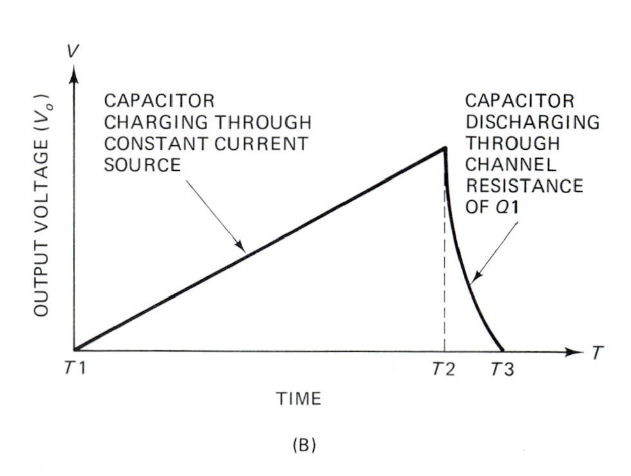


FIGURE 12-19

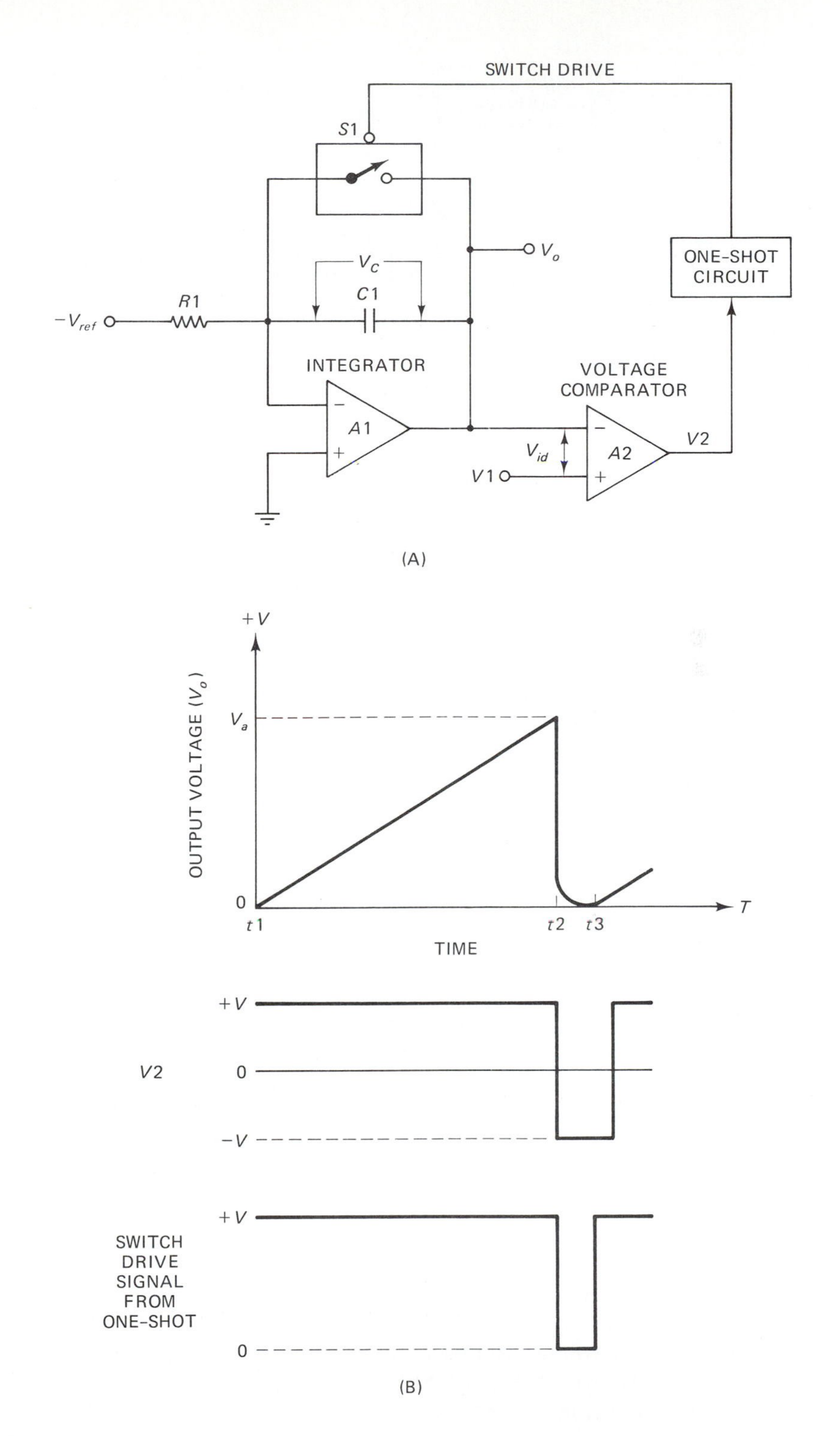

SWITCH DRIVE
S1
$V_o$
$V_C$
C1
R1
$-V_{ref}$
ONE-SHOT CIRCUIT
INTEGRATOR
VOLTAGE COMPARATOR
A1
A2
$V_{id}$
V1
V2
(A)
+V
OUTPUT VOLTAGE ($V_o$)
$V_a$
0
t1
t2
t3
T
TIME
+V
V2
0
−V
+V
SWITCH DRIVE SIGNAL FROM ONE-SHOT
0
(B)


FIGURE 12-20

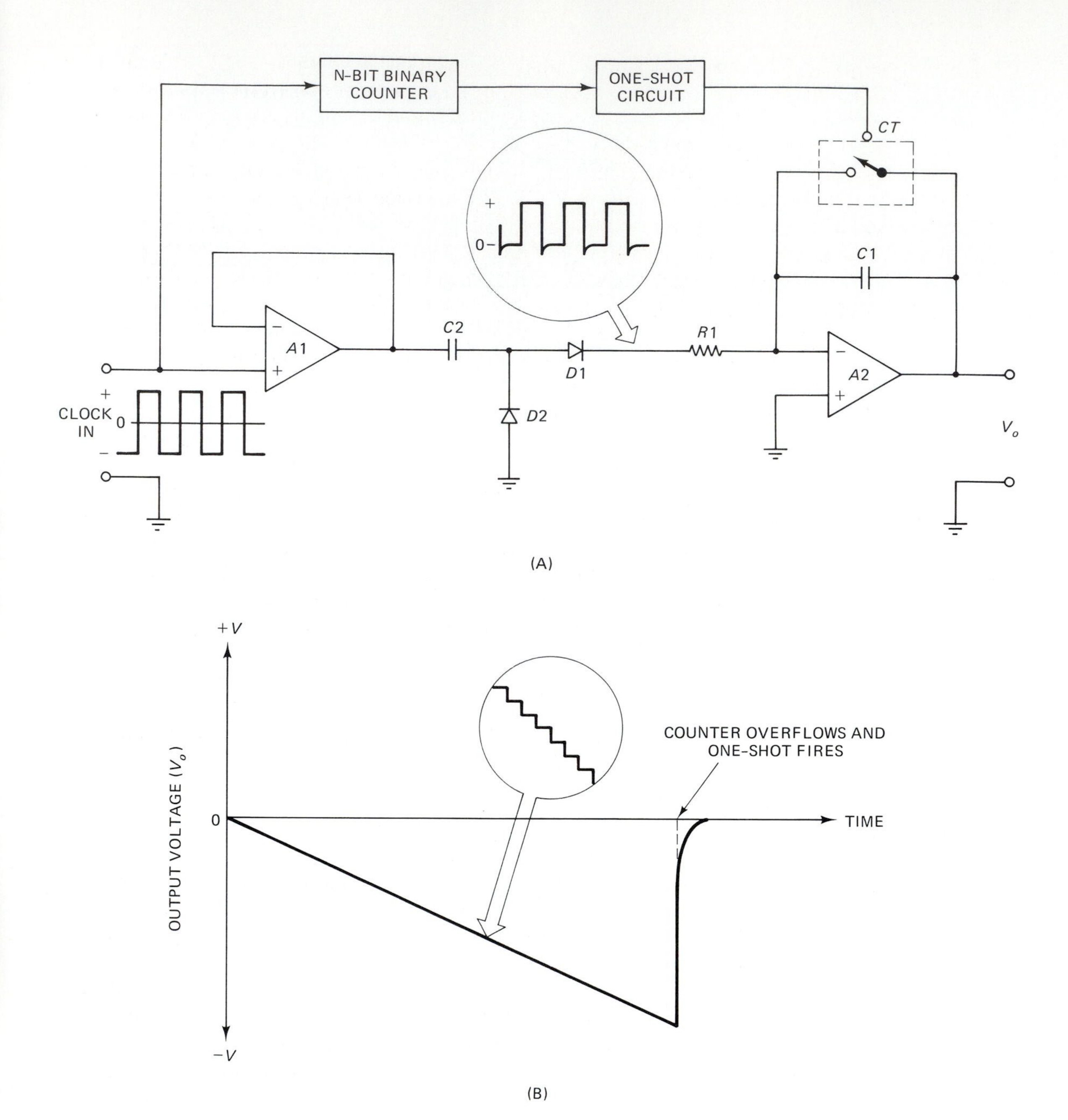


FIGURE 12-21

driven by a voltage comparator and one-shot circuit. The timing diagram for this circuit is shown in Fig. 12-20B. Under the initial condition at time $t1$, the output voltage ($V_o$) ramps upwards at a rate of $[-(-V_{ref})/R1C1]$. The voltage comparator ($A2$) is biased with the noninverting input (+IN) set to $V1$ and the inverting input set to $V_o$. The comparator differential input voltage $V_{id} = (V1 - V_o)$. As long as $V1 > V_o$ the comparator sees a negative input and produces a HIGH output of $+V_{sat}$. At the point where $V1 = V_o$ the differential input voltage is zero, so the output of $A2$ (voltage $V2$) drops LOW ($-V_{sat}$). The negative going edge of $V2$ at time $t2$ triggers the one-shot circuit. The output of the one-shot briefly closes electronic switch $S1$, causing the capacitor to discharge. The one-shot pulse ends at time $t3$, so $S1$ reopens and allows $V_o$ to again ramp upwards.

The *staircase generator* (Fig. 12-21A) is a variant of the sawtooth generator circuit. The input amplifier ($A1$) provides buffering. A square wave clock signal applied to the input of $A1$ is passed through capacitor $C2$ to a diode clipping network ($D1$, $D2$). The clipping circuit removes the negative excursions of the square wave (see inset to Fig. 12-21A). The remaining positive polarity pulses are applied to the input of the inverting Miller integrator ramp generator circuit. Each pulse adds a slight step increase to the capacitor charge voltage, and unless there is significant droop between pulses, the output will ramp up to a negative potential in the staircase fashion shown in the inset to Fig. 12-21B.

The reset circuitry here is a little different. Although the comparator method of Fig. 12-20A would also work, this circuit takes advantage of the input square wave to provide the period timing of the sawtooth. The square waves are applied to the input of an N-bit binary digital counter circuit. When $2^N$ pulses have passed, the counter overflows on $2^N + 1$ and triggers a one-shot circuit. As in the previous case, the one-shot output pulse momentarily closes the electronic reset switch shunted across capacitor $C1$.

## 12.6 FEEDBACK OSCILLATORS

A feedback oscillator (Fig. 12-22) consists of an amplifier with an open-loop gain of $A_{vol}$ and a feedback network with a gain or transfer function $B$. It is called a feedback oscillator because the output signal of the amplifier is fed back to the amplifier's own input by way of the feedback network. Figure 12-22 is a block diagram model of the feedback oscillator. It is no coincidence that it bears more than a superficial resemblance to a feedback amplifier. As anyone who has misdesigned or misconstructed an amplifier knows, a feedback oscillator is an amplifier in which special conditions prevail. Barkhausen's criteria for oscillation must be observed.

The amplifier can be any of many different devices. In some circuits it will be a common-emitter bipolar transistor (NPN or PNP devices). In others it will be a junction field effect transistor (JFET) or metal oxide semiconductor field effect transistor (MOSFET). In older equipment it was a vacuum tube. In modern circuits, it will probably be either an integrated circuit operational amplifier, or some other form of linear IC amplifier.

The amplifier is most frequently an inverting type, so the output is out of phase with the input by 180°. As a result, in order to obtain the required 360° phase shift, an additional phase shift of 180° must be provided in the feedback network *at the frequency of oscillation only*. If the network is designed to produce this phase shift at only one frequency, then the oscillator will produce a sinewave output on that frequency.

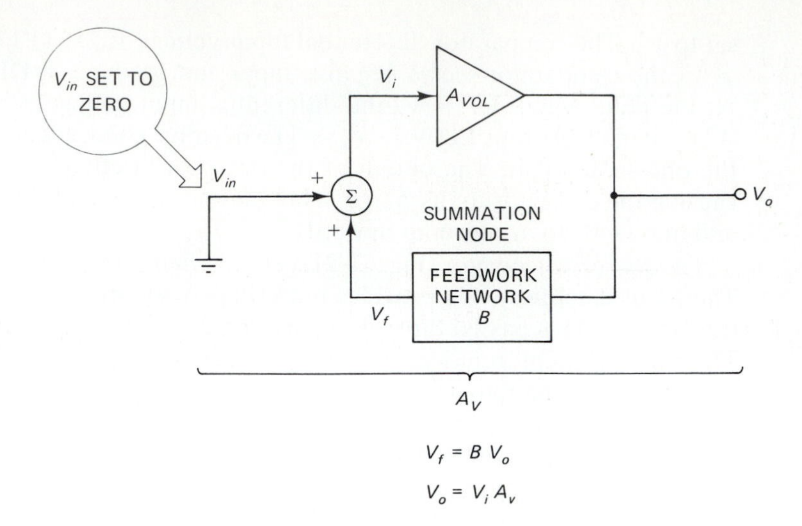


$V_f = B\,V_o$

$V_o = V_i\,A_v$

FIGURE 12-22

Before considering specific sinewave oscillator circuits let's examine Fig. 12-22 more closely. Several things can be determined about the circuit

$$V_i = V_{in} + V_F \tag{12-62}$$

so

$$V_{in} = V_{in} - V_F \tag{12-63}$$

and also

$$V_F = BV_o \tag{12-64}$$

$$V_o = V_i A_{vol} \tag{12-65}$$

The transfer function (gain) $A_v$ is

$$A_v = \frac{V_o}{V_{in}} \tag{12-66}$$

Substituting Eq. (12-63) and Eq. (12-65) into Eq. (12-66)

$$A_v = \frac{V_i A_{vol}}{V_i - V_F} \tag{12-67}$$

From Eq. (12-64), $V_F = BV_o$, so

$$A_v = \frac{V_i A_{vol}}{V_i - BV_o} \tag{12-68}$$

But Eq. (12-65) shows $V_o = V_i A_{vol}$, so Eq. (12-68) can be written

$$A_v = \frac{V_i A_{vol}}{V_i - \beta V_i A_{vol}} \tag{12-69}$$

and dividing both numerator and denominator by $V_i$

$$A_v = \frac{A_{vol}}{1 - \beta A_{vol}} \tag{12-70}$$

Equation (12-70) serves for both feedback amplifiers and oscillators. But in the special case of an oscillator $V_{in} = 0$, so $V_o \rightarrow \infty$. Implied, therefore, is the denominator of Eq. (12-70) also must be zero

$$1 - \beta A_{vol} = 0 \tag{12-71}$$

Therefore, for the case of the feedback oscillator

$$\beta A_{vol} = 1 \tag{12-71}$$

$BA_{vol}$ is the loop gain of the amplifier and feedback network, so Eq. (12-72) meets Barkhausen's second criterion.

## 12-7 IC SINEWAVE OSCILLATORS

Sinewave oscillators produce a sinusoidal output signal. Such a signal is ideally very pure. If it is perfect, its Fourier spectrum will contain only the fundamental frequency and no harmonics. It is the harmonics in a nonsinusoidal waveform which give it a characteristic shape. The operational amplifier is the active element in the circuits described here. However, any linear amplifier will also work in place of the operational amplifier.

*Stability* in oscillator circuits refers to several different phenomena. First is *frequency stability*, which refers to the ability of the oscillator to remain on the design frequency over time. Several different factors affect frequency stability, but the most important are *temperature* and *power supply voltage variations*.

*Amplitude stability* is another form of stability. Because sinewave oscillators do not operate in the saturated mode, it is possible for minor variations in circuit gain to affect the amplitude of the output signal. Again the factors most often cited for this problem include temperature and DC power supply variations. The latter is solved by using regulated DC power supplies for the oscillator. The former is solved by either a temperature-compensated design or maintaining a constant operating temperature. Some variable sinewave oscillators will exhibit amplitude variation of the output signal when the operating frequency is changed. In these circuits either a self-compensation element or an *automatic level control* amplifier stage is used.

Still another form of stability regards the purity of the output signal. If the circuit exhibits spurious oscillations, these will be superimposed on the output signal. As with any circuit containing an op-amp or any other high gain linear amplifier, it is necessary to properly decouple the DC power supply lines. It may also be necessary to frequency-compensate the circuit.

### 12-7.1 RC Phase Shift Oscillator Circuits

The RC phase shift oscillator is based on a three-stage cascade resistor capacitor network such as shown in Fig. 12-23A. An *RC* network will exhibit a phase shift $\phi$ (Fig. 12-23B) which is a function of resistance ($R$) and capacitive reactance ($X_c$). Because $X_c$ is inversely proportional to frequency ($1/2\pi fC$), the phase angle is a

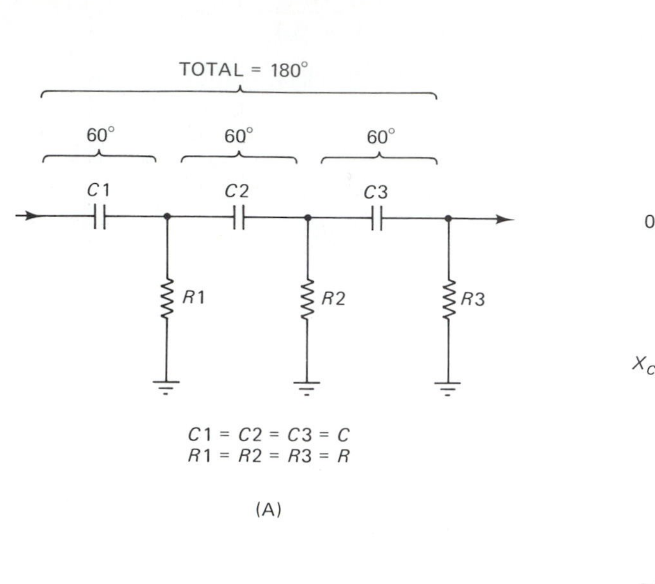


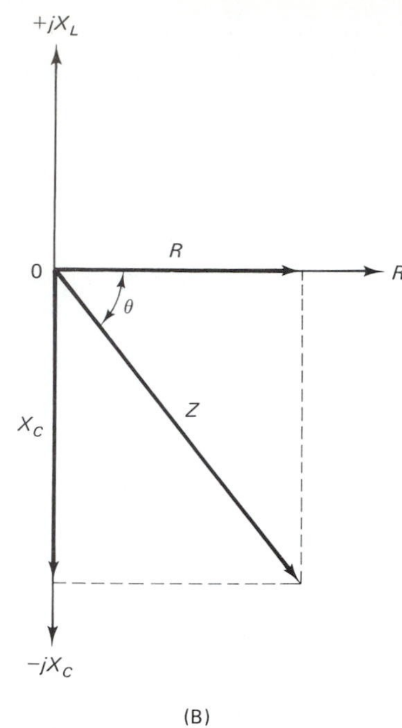


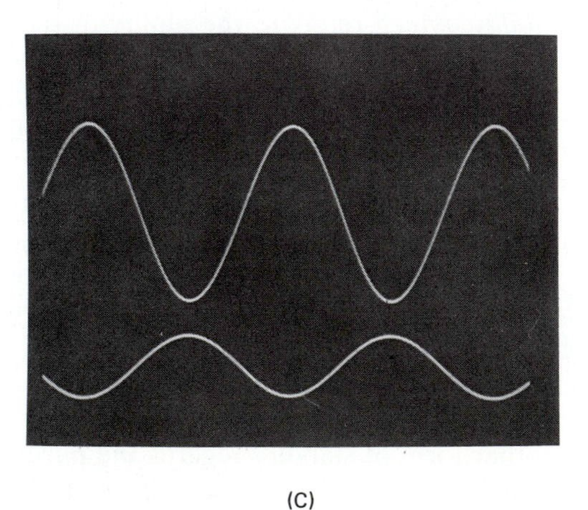

(C)

FIGURE 12-23

function of frequency. The goal in designing the *RC* phase shift oscillator is to create a phase shift of 180° between the input and output of the network at the desired frequency of oscillation. It is conventional practice to make the three stages of the network identical, so each provides a 60° phase shift. This is not necessary, however, provided the total phase shift is 180°. But one reason for using identical stages is that it is possible for nonidentical designs to find more than one frequency for which the total phase shift is 180°. This problem leads to undesirable multimodal oscillation.

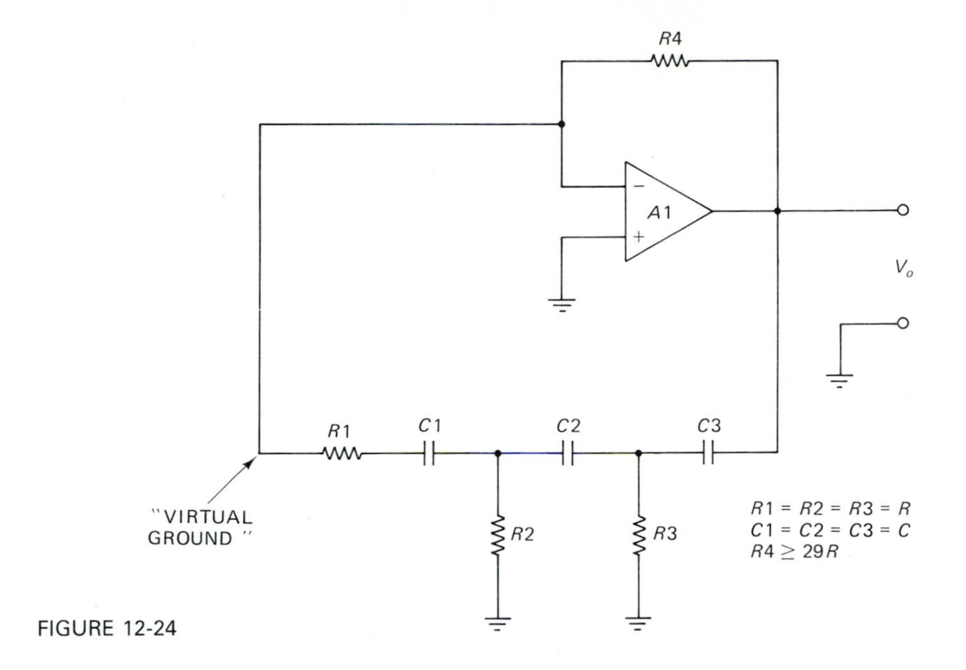


FIGURE 12-24

Figure 12-23C shows the input and output waveforms of an *RC* network where three stages of 10 kohms/0.01 uF each were used with an input frequency of 650 Hz. Note the 180° phase shift in the lower trace. Also, be aware that the vertical input scale factors for these two traces are different. The peak-to-peak amplitude of the upper trace is 7.8 volts, while that of the lower trace is 0.268 volts, or 1/29 of the input amplitude. This attenuation factor is important because it establishes the gain requirement in the amplifier.

Figure 12-24 shows the circuit for an operational amplifier *RC* phase shift oscillator. The cascade phase shift network *R*1*R*2*R*3/*C*1*C*2*C*3 provides 180° of phase shift at a specific frequency. The amplifier provides another 180° because it is an inverting follower. The total phase shift is therefore 360° at the frequency for which the *RC* network provides a 180° phase shift. The frequency of oscillation ($f$) for this circuit is given by

$$f = \frac{1}{2\pi RC[6]^{1/2}} \tag{12-72}$$

where

$f$ is in hertz (Hz)

$R$ is in ohms

$C$ is in farads

## EXAMPLE 12-8

Find the frequency of oscillation for an *RC* phase shift oscillator (Fig. 12-24) if the three-stage cascade feedback network uses $R$ = 10 kohms and $C$ = 0.005 uF.

**Solution**

$$f = \frac{1}{2\pi RC[6]^{1/2}} \text{ Hz}$$

$$f = \frac{1}{(2)(3.14)(10000 \text{ ohms})\left[0.005\ \mu\text{F} \times \frac{1 \text{ Farad}}{10^6\ \mu\text{F}}\right][6]^{1/2}} \text{ Hz}$$

$$f = \frac{1}{7.7 \times 10^{-4}} \text{ Hz} = 1300 \text{ Hz}$$

■

It is common practice to combine the constants in Eq. (12-72) to arrive at a simplified expression

$$f = \frac{1}{15.39RC} \tag{12-73}$$

Because the required frequency of oscillation is usually determined from the application, it is necessary to select an $RC$ time constant to force the oscillator to operate as needed. Also because capacitors come in fewer standard values, it is best to select an arbitrary trial value of capacitance and then select the resistance which will cause the oscillator to produce the correct frequency. To make the calculations simpler, it is prudent to express the equation in a way that permits specifying the capacitance ($C$) in *microfarads*. As a result, Eq. (12-73) is sometimes rewritten as

$$R = \frac{1{,}000{,}000}{15.39C_{\mu F}f} \tag{12-74}$$

The attenuation through the feedback network must be compensated by the amplifier if loop gain is to be unity or greater. At the frequency of oscillation the attenuation is 1/29. The loop gain must be unity, so the gain of amplifier $A1$ must be at least 29 in order to satisfy $AB = 1$. For the inverting follower (as shown), $R1 = R$, and $A_v = R4/R1$. Therefore, we can conclude $R4 \geq 29R$ in order to meet Barkhausen's criterion for loop gain.

## EXAMPLE 12-9

Calculate (a) the resistance needed to make an $RC$ phase shift oscillator operate on a frequency of 1000 Hz when a capacitance of 0.01 μF is used for $R$ in the three-stage cascade phase shift network; (b) the minimum resistance of $R4$.

**Solution**

(a) Resistance $R$ required:

$$R = \frac{1{,}000{,}000}{15.39C_{\mu F}f}$$

$$R = \frac{1{,}000{,}000}{(15.39)(0.01\ \mu\text{F})(1000 \text{ Hz})}$$

$$R = \frac{1{,}000{,}000}{153.9} = 6498 \text{ ohms}$$

(b) Minimum allowable value of $R4$

$$R4 \geqslant 29R$$

$$R4 \geqslant (29)(6498 \text{ ohms}) = 188{,}442 \text{ ohms}$$

■

### 12-7.2 Wein Bridge Oscillator Circuits

The *Wein bridge* circuit is shown in Fig. 12-25A. Like several other bridge circuits, the Wein bridge consists of four impedance arms. Two of the arms ($R1$, $R2$) form a resistive voltage divider which produces a voltage $V1$ of

$$V1 = \frac{V_{ac}R2}{R1 + R2} \tag{12-75}$$

The remaining two arms ($Z1$, $Z2$) are complex $RC$ networks. Each consists of one capacitor and one resistor. Impedance $Z1$ is a series $RC$ network, while $Z2$ is a parallel $RC$ network. The voltage and phase shift produced by the $Z1/Z2$ voltage divider are

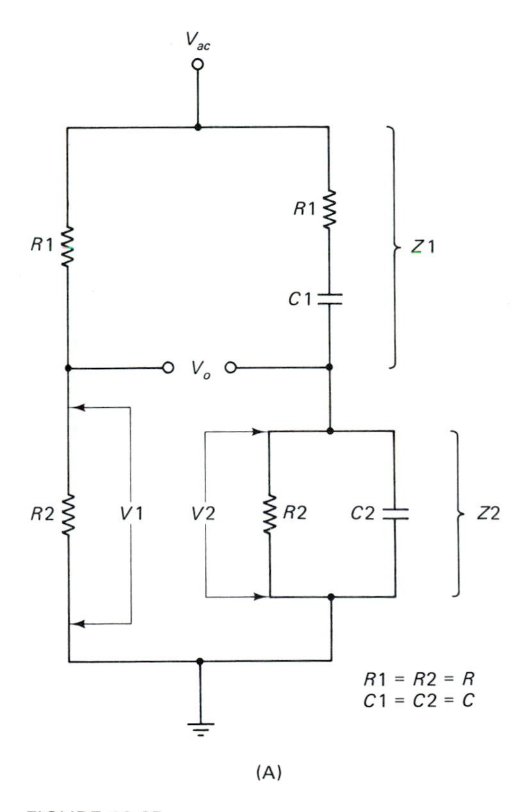


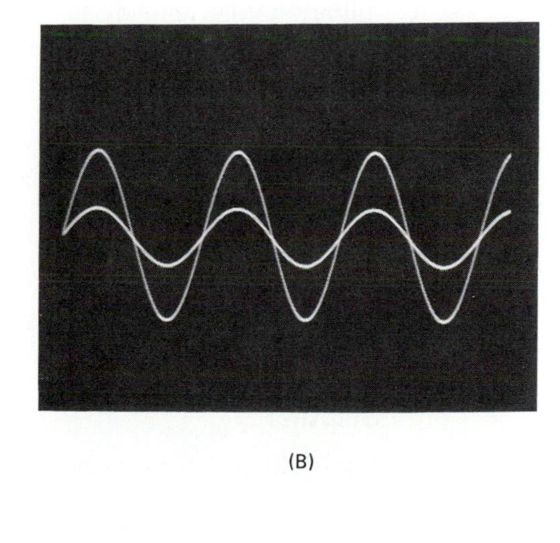


FIGURE 12-25

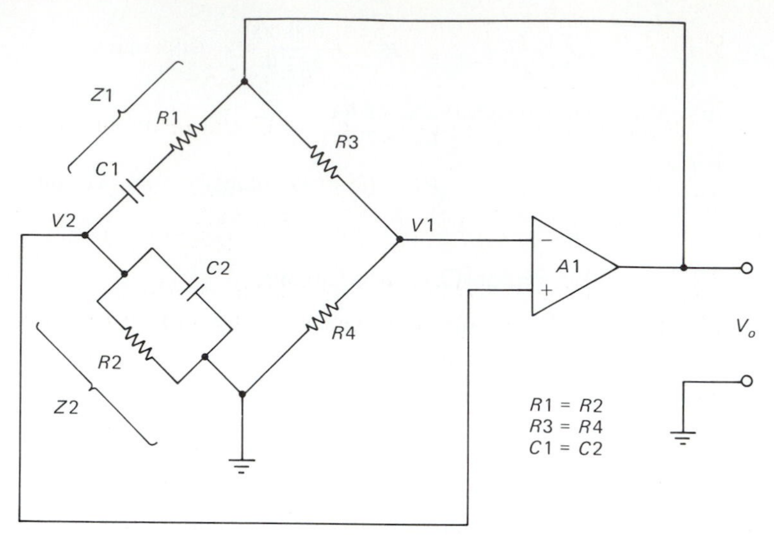


FIGURE 12-26

functions of the $RC$ values and the applied frequency. Figure 12-25B shows $V_{ac}$ superimposed on $V2$. Note that $V2 = V_{ac}/3$ and that $V2$ and $V_{ac}$ are in phase with each other.

Figure 12-26 shows the circuit for a *Wein bridge oscillator.* The resistive voltage divider supplies $V1$ to the inverting input (−IN), while $V2$ is applied to the noninverting input (+IN). In Fig. 12-26 the bridge signal source is the output of the amplifier ($A1$). The AC signal is applied to +IN, so the gain is found from

$$A_v = \frac{R3}{R4} + 1 \tag{12-76}$$

The AC feedback applied to +IN is

$$\beta = \frac{Z2}{Z1 + Z2} \tag{12-77}$$

At resonance $B = 1/3$, so (as shown in Fig. 12-25B)

$$V2 = \frac{V_o}{3} \tag{12-78}$$

Because $A_v = V_o/V2$ by definition, satisfying Barkhausen's loop-gain criterion $(-A_vB = 1)$ requires that $A_v = V_o/V2 = 3$. Using this result

$$A_v = \frac{R3}{R4} + 1 \tag{12-79}$$

$$3 = \frac{R3}{R4} + 1 \tag{12-80}$$

$$2 = \frac{R3}{R4} \tag{12-81}$$

or

$$R3 = 2\ R4 \tag{12-82}$$

IF $R1 = R2 = R$ and $C1 = C2 = C$, the resonant frequency of the Wein bridge is

$$f = \frac{1}{2\pi RC} \tag{12-83}$$

For the standard Wein bridge oscillator, in which $R1 = R2 = R$ and $C1 = C2 = C$, and $R3 = 2R4$, a sinewave output will result on frequency $f$.

### EXAMPLE 12-10

Calculate the resistor values for a Wein bridge oscillator which produces a frequency of 1000 Hz.

### Solution

(a) Let

$$C = 0.01\ \mu\text{F}$$

$$R4 = 10\ \text{kohms}$$

(b) Set gain resistor for $A_v = 3$

$$R3 = 2\ R4$$

$$R3 = (2)(10\ \text{kohms}) = 20\ \text{kohms}$$

(c) Set $R = R1 = R2$ value

$$R = \frac{1}{2\pi RC}$$

$$R = \frac{1}{(2)(3.14)(1000\ \text{Hz})\left[0.01\text{-}\mu\text{F} \times \dfrac{1F}{10^6\ \mu\text{F}}\right]}$$

$$R = \frac{1}{6.28 \times 10^{-5}} = 15{,}924\ \text{ohms}$$

(d) The design values are, therefore

$R1$, $R2$: 15,924 ohms

$R3$: 20 kohms

$R4$: 10 kohms

$C1$, $C2$: 0.01 $\mu$F

$f$: 1000 Hz

■

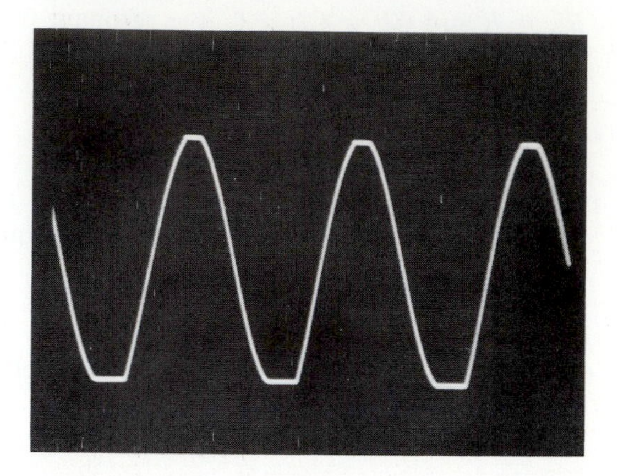

FIGURE 12-27

**Amplitude Stability.** The oscillations in the Wein bridge oscillator circuit will build up without limit when the gain of the amplifier is high. Figure 12-27 shows the result of the gain being only slightly above that required for stable oscillation. Note some clipping is beginning to appear on the sinewave peaks. At even higher gains the clipping becomes more severe, and will eventually look like a square wave. Figure 12-28 shows several methods for stabilizing the waveform amplitude.

Figure 12-28A shows the use of small signal diodes such as the 1N914 and 1N4148 devices. At low signal amplitudes the diodes are not sufficiently biased, so the gain of the circuit is

$$A_v = \frac{R1 + R3}{R2} + 1 \qquad (12\text{-}84)$$

As the output signal voltage increases, however, the diodes become forward-biased. $D1$ is forward-biased on negative peaks of the signal, while $D2$ is forward-biased on positive peaks. Because $D1$ and $D2$ are shunted across $R3$, the total resistance $R3'$ is less than $R3$. Equation (12-84) shows one can determine that reducing $R3$ to $R3'$ reduces the gain of the circuit. The circuit is thus self-limiting.

Another variant of the gain-stabilized Wein bridge oscillator is shown in Fig. 12-28B. In this circuit a pair of back-to-back zener diodes provides the gain limitation function. With the resistor ratios shown, the overall gain is limited to slightly more than unity, so the circuit will oscillate. The output peak voltage of this circuit is set by the zener voltages of $D1$ and $D2$ which should be equal for low distortion.

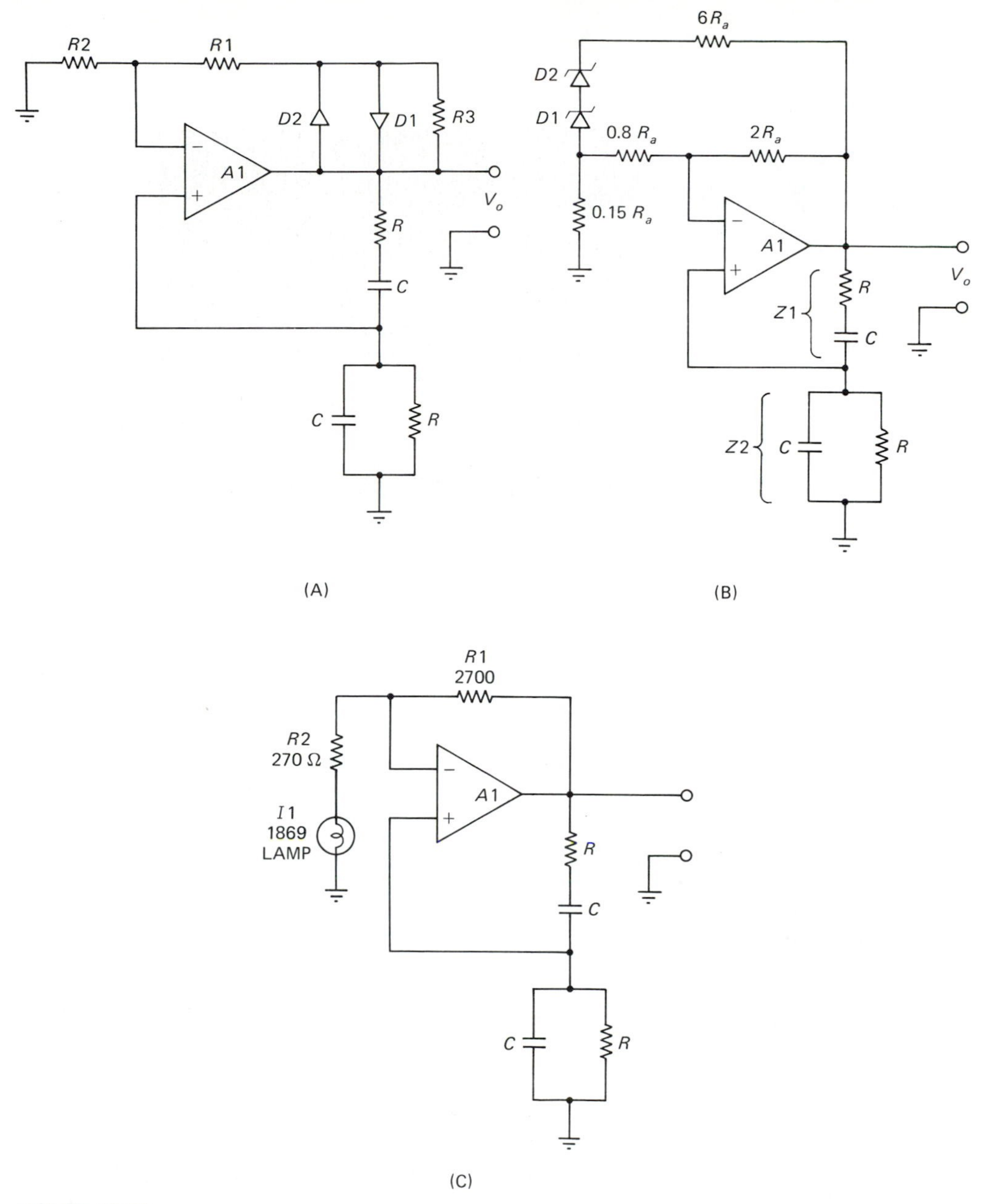


FIGURE 12-28

One final version of the gain-stabilized oscillator is shown in Fig. 12-28C. Here a small incandescent lamp is connected in series with resistor *R*2. When the amplitude of the output signal tries to increase above a certain level, the lamp will draw more current causing the gain to reduce. The lamp-stabilized circuit is probably the most popular where stable outputs are required. A thermistor is sometimes substituted for the lamp.

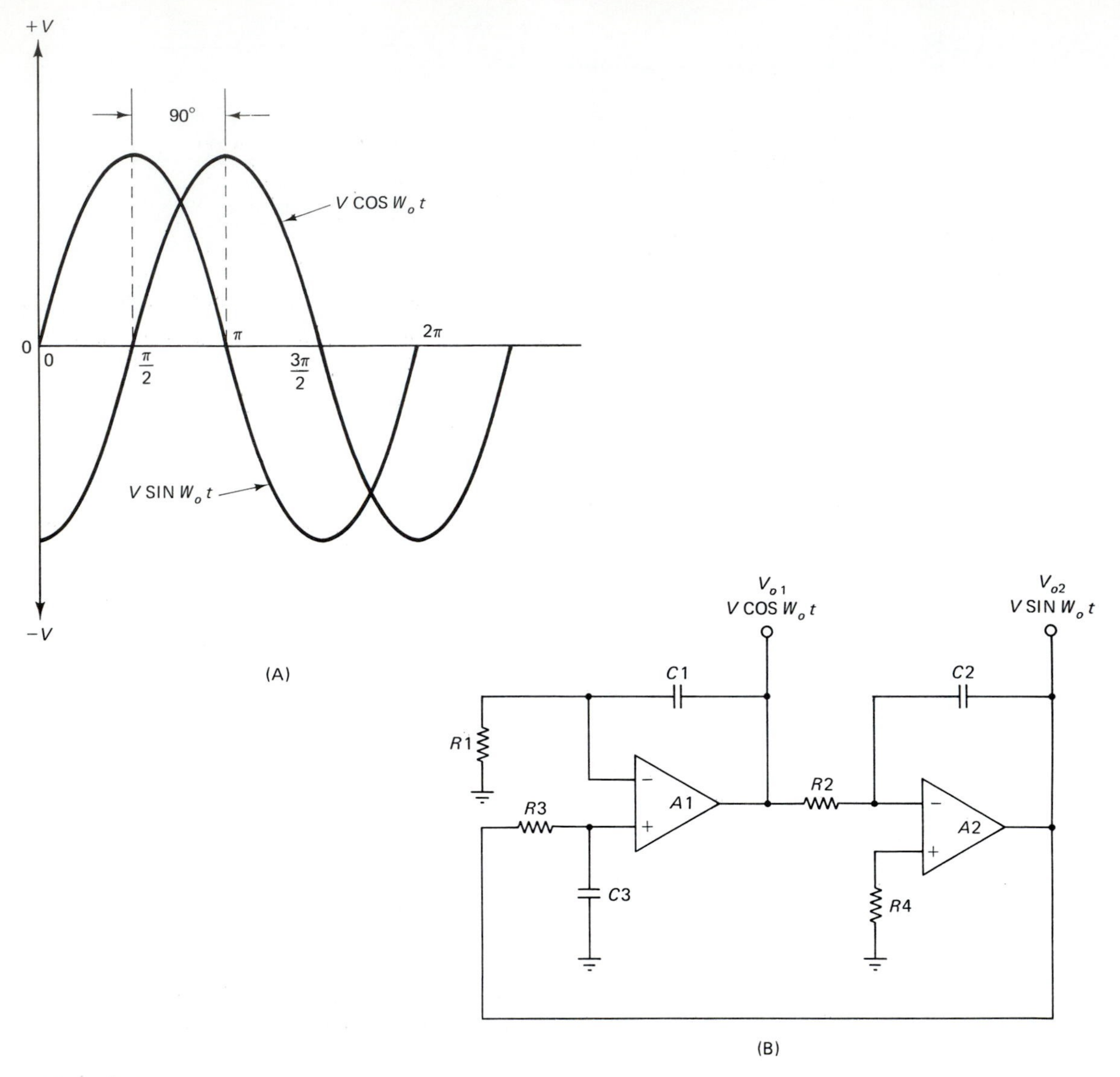


FIGURE 12-29

### 12-7.3 Quadrature and Biphasic Oscillators

Signals are said to be "in quadrature" when they are the same frequency but are phase shifted 90°. Sine and cosine waves (Fig. 12-29A) are quadrature signals. Applications for the quadrature oscillator include demodulation of phase sensitive detector signals in data acquisition systems (Chapter 13). The sine wave has an instantaneous voltage $v = V\text{Sin}(\omega_o t)$. The cosine wave is defined by $v = V\text{Cos}(\omega_o t)$.

The distinction between sine and cosine waves is meaningless unless either both are present, or some other timing method is used to establish when "zero degrees"

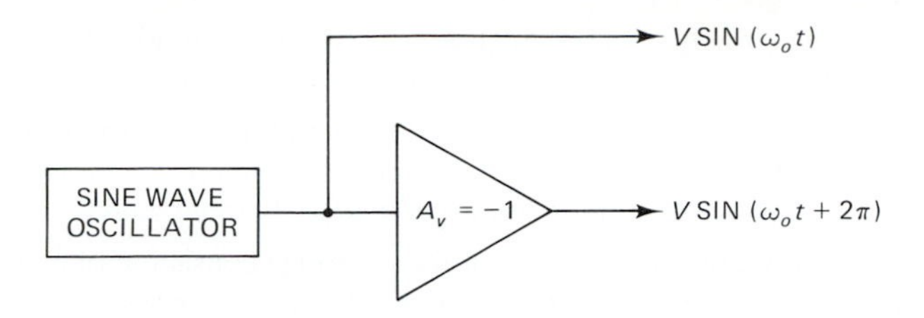


FIGURE 12-30

is supposed to occur. Thus, when sine and cosine waves are called for it is in the context of both being present, and a phase shift of 90° is present between them.

The circuit for the quadrature oscillator is shown in Fig. 12-29B. It consists of two operational amplifiers, $A1$ and $A2$. Both amplifiers are connected as Miller integrators, although $A1$ is noninverting and $A2$ is an inverting integrator. The output of $A1$ ($V_{o1}$) is assumed to be the sine wave output.

To make this circuit operate, a total of 360° of phase shift is required between the output of $A1$, around the loop and back to the input of $A1$. Of the required 360° phase shift, 180° are provided by the inversion inherent in the inverting design of $A2$. Another 90° come from $A2$ being an integrator. This inherently causes a 90° phase shift (Chapter 11). An additional 90° phase shift is provided by $RC$ network $R3C3$. If $R1 = R2 = R3 = R$ and $C1 = C2 = C3 = C$, the frequency of oscillation is given by

$$f = \frac{1}{2\pi RC} \tag{12-85}$$

The cosine output ($V_{o2}$) is taken from the output of amplifier $A2$. The relative amplitudes are approximately equal, but the phase is shifted 90° between the two stages.

A *biphasic oscillator* is a sinewave oscillator which outputs two identical sine wave signals 180° out of phase with each other. The basic circuit is simple. It is shown in block diagram form in Fig. 12-30. The biphasic oscillator consists of a sine wave oscillator followed by an inverting amplifier which has a gain of one. The output of the sine wave oscillator is $V\text{Sin}(\omega_o t)$. The output of the inverter is $V\text{Sin}(\omega_o t + 2\pi)$. Biphasic oscillators are sometimes used for transducer excitation in carrier amplifiers (Chapter 13).

## 12-8 SUMMARY

1. Monostable and astable multivibrators are based on the properties of the operational amplifier connected as a voltage comparator. In both cases timing of the output waveform is based on the properties of the simple series $RC$ network.
2. A nonretriggerable monostable multivibrator cannot be retriggered until both the output pulse duration and any refractory period expires. A retriggerable monostable multivibrator will respond to additional trigger pulses regardless of time out or refractory period.
3. The sawtooth and triangle generator circuits are designed using the Miller integrator as a ramp generator. When driven from a constant voltage the integrator outputs a linear ramp.

4. The principle causes of frequency and amplitude stability problems in fixed frequency sine wave oscillators are *temperature* and *power supply voltage variation*.
5. *RC* phase shift oscillators use the amplifier to provide 180° of phase shift, and a three-section cascade *RC* network to provide the remaining 180° required to meet Barkhausen's criteria. The amplifier must provide a voltage gain of 29 or greater.
6. The Wein bridge oscillator uses a bridge consisting of a parallel and a series *RC* network, along with a resistor voltage divider in an operational amplifier circuit.
7. Amplitude stability is provided in the Wein bridge oscillator by one of several means, a diode AGC, a zener diode AGC, and an incandescent lamp in series with one of the arms.
8. The quadrature oscillator is built using an inverting integrator, a noninverting integrator, and an *RC* phase shift network to create the required 360° phase shift.

## 12-9 RECAPITULATION

Now return to the objectives and Prequiz questions at the beginning of the chapter and see how well you can answer them. If you cannot answer certain questions, mark them and review the appropriate parts of the text. Next, try to answer the questions and work the problems below using the same procedure.

## 12-10 STUDENT EXERCISES

1. Design and build a one-shot multivibrator which has an output duration of 5 mS. Examine the output voltage and capacitor voltage with an oscilloscope. Drive the one-shot with a pulse generator or square wave generator and then compare the output pulse and trigger signal.
2. Design and build a symmetrical squarewave generator operating at a frequency of 1000 Hz using equal positive feedback resistors. Examine the output and capacitor voltages on an oscilloscope. Vary the values of *R*1 and *C*1 and observe what happens to the circuit. Replace the positive feedback resistors with a single potentiometer. Initially set the potentiometer to midscale. Observe what happens to the operating frequency and waveform when the ratio of resistances is changed by varying the potentiometer.
3. Build an asymmetrical square wave oscillator using (a) potentiometer method, (b) diode method. In both cases examine operation of the circuit on an oscilloscope.
4. Design and build a triangle waveform oscillator operating on a frequency of 500 Hz.
5. Design and build a sawtooth oscillator operating on a frequency of 100 Hz.
6. Design and build an *RC* phase shift oscillator operating on a frequency of 1000 Hz. Use a 741 or other available operational amplifier. Use an oscilloscope to check the output voltage and the feedback voltage. Confirm the feedback attenuation is 1/29. Vary the *RC* network values and see what happens to the circuit.
7. Design and build a Wein bridge oscillator operating on a frequency of 600 Hz. Use an oscilloscope to examine the output signal. Confirm the feedback attenuation factor is 1/3. Vary the component values and observe what happens in the circuit. Redesign the circuit using one or more of the amplitude stability techniques.

## 12-11 QUESTIONS AND PROBLEMS

1. Calculate the *RC* time constant that will allow a capacitor to charge from −12 Vdc to +10 Vdc in 350 mS.
2. An op-amp monostable multivibrator (MMV) must produce a 70-mS pulse. Select an *RC* time-constant.
3. A monostable multivibrator (Fig. 12-4A) is constructed so $R2 = 12$ kohms and $R3 = 4.7$ kohms. The device is a 741 op-amp operated so $|V_{sat}| = 12$ Vdc. Calculate the *RC* time constant required to produce an 8-mS output pulse.
4. A monostable multivibrator has a timing resistor of 9.1 kohms, and a timing capacitor of 0.01 μF. What is the duration of the output pulse if the positive feedback resistors are equal?
5. Calculate the duration of a retriggerable monostable multivibrator circuit in which $R2 =$ 10 kohms, $R3 = 8.2$ kohms, $R1 = 68$ kohms and $C1 = 0.068$ μF.
6. Calculate the oscillating frequency for an astable multivibrator in which $R1 = 10$ kohms, $R2 = 10$ kohms, $R3 = 5.6$ kohms, and $C1 = 0.01$ μF.
7. An asymmetrical astable multivibrator is designed such that $R1A = 82$ kohms, $R1B = 22$ kohms, $C1 = 0.08$ μF and $R2 = R3$.
8. Calculate the frequency of an astable multivibrator in which the positive feedback resistors are equal and the *RC* time constant is 0.010 seconds.
9. State Barkhausen's criteria for oscillation.
10. Calculate the oscillating frequency when an *RC* phase shift oscillator has the following component values: $R1 = R2 = R3 = 15$ kohms, $C1 = C2 = C3 = 0.1$ μF. Calculate the minimum value of gain setting feedback resistor $R4$.
11. Calculate the resonant frequency of a Wein bridge oscillator in which $R = 100$ kohms and $C = 0.001$ μF.
12. Find the frequency of oscillation for an *RC* phase shift oscillator (Fig. 12-24) if the three-stage cascade feedback network uses $R = 22$ kohms and $C = 0.002$ μF.
13. Describe the operation of a relaxation oscillator. Give two examples of electronic devices that can be used in making relaxation oscillators.
14. A series *RC* network consists of a 1.5 μF capacitor and a 2.2 megohm resistor. The network is connected to a 20 volt DC power supply for 20 minutes. The capacitor is charged to a potential of ____________ volts DC. Assume that the DC power supply is replaced with a short circuit, and calculate: (a) the length of time for the capacitor to discharge to 7.34 volts; (b) the potential across the capacitor after 3.3 seconds; and the time required for the potential across the capacitor to reach zero.
15. A series *RC* network consists of a 0.47 uF capacitor and an 82-kohm resistor. How much time is required for the capacitor to charge from +10 Vdc to −2 Vdc?
16. What is the difference between a monostable and an astable multivibrator?
17. What is the principal difference between a retriggerable one-shot and a nonretriggerable one-shot? Use waveform drawings if necessary to make your point.
18. What are the four states of a monostable multivibrator?
19. Draw the block diagram for a feedback oscillator.
20. What is the rule-of-thumb maximum value of the time-constant $R4C2$ in Fig. 12-4A if $R1C1 = 10$ mS?

21. In Fig. 12-4A, what is the value of $V1$ when $V_o = -12$ Vdc and $R2 = R3$?
22. The circuit of Fig. 12-4A is stable. Calculate differential input voltage $V_{id}$ if $R2 = R3$ and the maximum allowable output voltage is +9 Vdc.
23. Describe the *refractory period* of a monostable multivibrator.
24. What is the function of $D3/R5$ in Fig. 12-4A? What is the effect of these parts on the $V_c$ waveform in Fig. 12-4B?
25. Draw the circuit for a retriggerable multivibrator.
26. Draw the circuit for an operational amplifier astable multivibrator. Calculate values for the timing components for a 1000 Hz version of this circuit when the feedback factor (B) is 0.5.
27. Draw the timing diagram for the circuit of question 26. Label the time durations of each section.
28. Draw the schematic diagram for an astable multivibrator that has a continuously adjustable duty cycle.
29. Why does the circuit of Fig. 12-11A use two timing resistors ($R1A$ and $R1B$)? What is the function of $D1$ and $D2$?
30. In Fig. 12-13A the zener potential of $D1$ is 5.6 Vdc, and of $D2$ is 6.8 Vdc. Calculate the maximum output voltages of both positive and negative excursions of the output waveform.
31. Discuss a simple means for making a ramp generator circuit from a Miller integrator.
32. Sketch triangle and sawtooth waveforms and describe the difference between them.
33. Draw the constant current and normal $RC$ charging curves for a Miller integrator.
34. Draw the circuit for a simple $RC$ sawtooth generator. Also for an op-amp sawtooth generator based on the Miller integrator.
35. What is a staircase generator and how does it differ from a sawtooth generator?
36. Draw the frequency determining network for an $RC$ phase shift oscillator. What is the total phase shift across this network at the resonant frequency? What is the approximate voltage gain/loss of the network?
37. What is the minimum value of $R4$ in Fig. 12-24 if $R1 = R2 = R3 = 2.7$ kohms?
38. What is the purpose of $D1$ and $D2$ in Fig. 12-28A?
39. Why is the incandescent lamp used in Fig. 12-28C?
40. The input signal to an op-amp circuit is $v = V\text{Sin}(\omega_o t)$. Write the output signal expressions if the circuit is (a) an inverting follower with a gain of one, (b) an inverting follower with a gain of ten, (c) a noninverting follower with a gain of two, (d) an integrator, and (e) a differentiator.

CHAPTER 13

# Measurement and Instrumentation Circuits

## OBJECTIVES

1. Learn the basics of instrumentation circuits using linear integrated circuit devices.
2. Understand instrument signal sources such as transducers and electrodes.
3. Know the parameters pertinent to instrument circuit designs.
4. Be able to design simple electronic instruments.

## 13-1 PREQUIZ

These questions test your prior knowledge of the material in this chapter. Try answering them before you read the chapter. Look for the answers (especially those you answered incorrectly) as you read the text. After you have finished studying the chapter try answering these questions again and those at the end of the chapter (see Section 13-7).

1. A strain gage Wheatstone bridge transducer has a sensitivity factor of 35 uV/V/Torr of pressure and is excited by a 10 volt DC source. Calculate the output voltage if a 420-Torr pressure is applied.
2. Using the parameters of the problem above, calculate the gain of a DC differential amplifier required to produce an output sensitivity of 10 mV/Torr (an output potential of 4.200 volts when the applied pressure is 420 Torr).
3. Define "transduction."
4. ____________ metric measurement is often used when the excitation potential or another parameter of the measurement system is likely to independently change.

One of the functions of this book is to give the student familiarity with linear integrated circuits *and their applications*. Electronic instrumentation for control and measurement systems is one of these applications. In this section we will discuss broad categories of signals sources (transducers and electrodes) and the circuitry needed to process these signals. Our goal will be to introduce generic classes of transducers rather than provide a catalog of all available types.

## 13-2 TRANSDUCTION AND TRANSDUCERS

*Transduction* is the process of *changing energy from one form to another for purposes of measurement, tabulation, or control.* Transducers are sensors. They are the eyes and ears of an electronic instrument. These devices convert assorted forms of energy from physical systems (temperature or pressure for example) into electrical energy. For example, a transducer might convert fluid pressure to an analogous voltage or current.

There are many different forms of transducer which use resistive strain gage elements. Most of them are based on the Wheatstone bridge circuit. Various physical parameters can be measured with strain gage transducers, including force, displacement, vibration, liquid pressure, and gas pressure. For example, if a hospital patient requires intensive care, the physician may order continuous electronic blood pressure monitoring through an indwelling fluid-filled catheter inserted into an artery. The transducer used to measure blood pressure will probably be a Wheatstone bridge strain gage. First we will discuss how strain gages work and how to make them operate properly in practical cases.

### 13-2.1 Piezoresistivity

All electrical conductors possess the electrical property of resistance. The resistance of any specific conductor is directly proportional to the length (see Fig. 13-1) and inversely proportional to the cross-sectional area. Resistance is also directly proportional to a physical property of the conductor material called *resistivity*. The relationship between *length* ($L$), *crossectional area* ($A$), and resistivity ($\rho$) is

$$R = \frac{\rho L}{A} \tag{13-1}$$

*Piezoresistivity* occurs when the resistance changes because either length or cross-sectional area is changed. That is, electrical resistance changes in response to mechanical deformation. Figure 13-1A shows a cylindrical conductor with a resting length $L_o$ and a cross-sectional area $A_o$. When a *compression force* is applied, as in Fig. 13-1B, the length decreases and the cross-sectional area increases. This situation results in a *decrease in the electrical resistance.*

Similarly, when a *tension force* is applied (Fig. 13-1C), the length increases and the cross-sectional area decreases, so the *electrical resistance increases.* If the physical change is small and the conductor's elastic limit is not exceeded, the change of electrical resistance is a nearly linear function of the applied force, so it can be used to make measurements of that force.

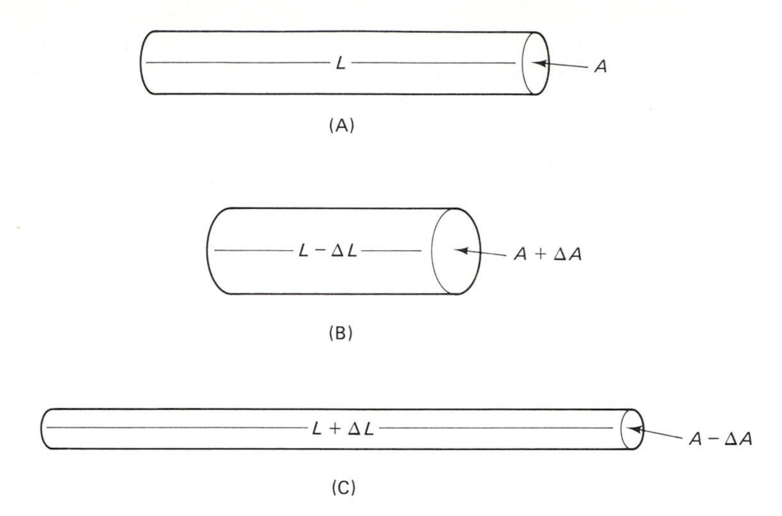


FIGURE 13-1

**Strain Gages.** A *strain gage* is a piezoresistive element made of either wire, metal foil or a semiconductor material. It is designed to create a *resistance change* when a force is applied. Strain gages can be classified as either *bonded* or *unbonded.* Figure 13-2 shows both methods of construction.

The *unbonded strain gage* is shown in Fig. 13-2A. It consists of a wire resistance element stretched taut between two flexible supports. These supports are configured so either a tension or compression force is placed on the taut wire when an external force is applied. In the example shown, the supports are mounted on a thin metal

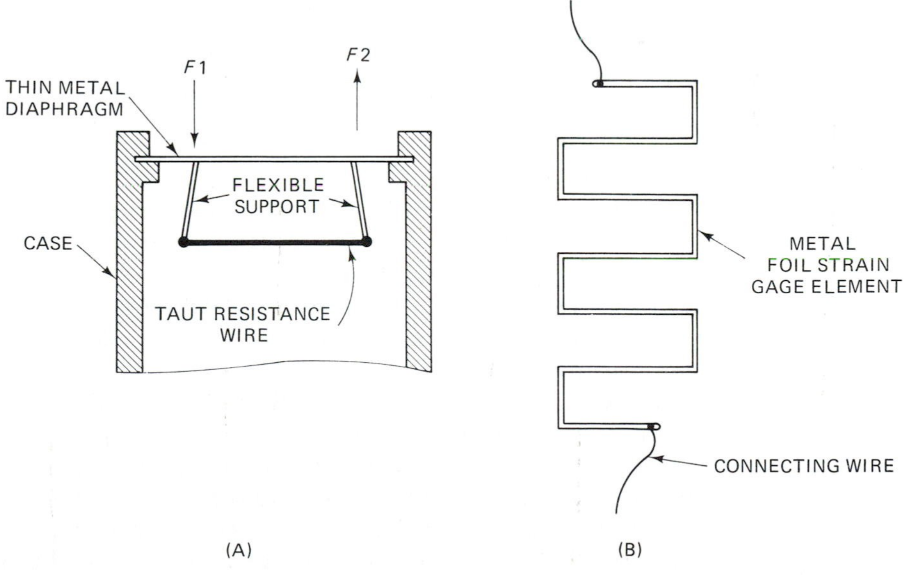


FIGURE 13-2

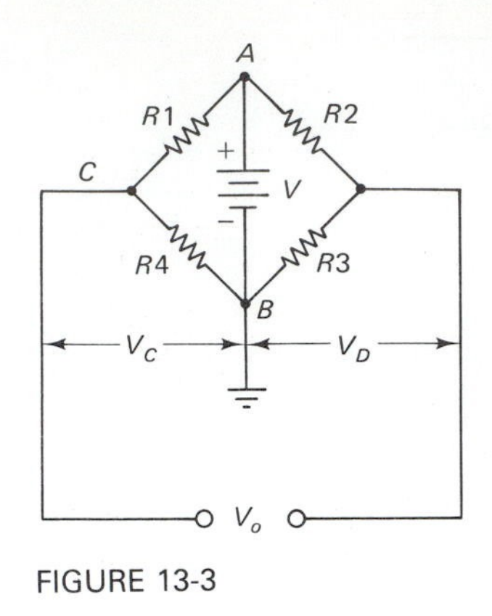


FIGURE 13-3

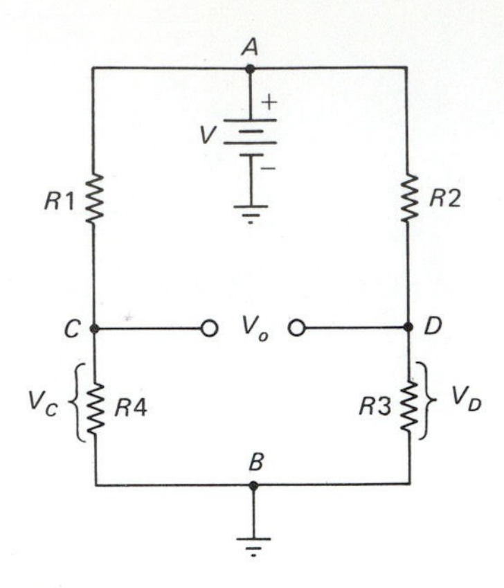


FIGURE 13-4

diaphragm that flexes when a force is applied. Force *F*1 will cause the flexible supports to spread apart, placing a tension force on the wire and increasing its resistance. Alternatively, when force *F*2 is applied, the ends of the supports tend to move closer together, effectively placing a compression force on the wire element and thereby reducing its resistance. The resting condition is slight tautness, which implies a small normal tension. Force *F*1 increases the normal tension and force *F*2 decreases the normal tension.

The *bonded strain gage* is shown in Figure 13-2B. Here, a wire, foil, or semiconductor element is cemented directly to a thin metal diaphragm. When the diaphragm is flexed, the element deforms to produce a resistance change.

The linearity of both types of strain gage can be quite good, provided the elastic limits of the diaphragm and element are not exceeded. A metallic diaphragm can be distended linearly and will regain its shape only as long as the applied force is less than the force required to exceed the limit of elasticity. It is therefore necessary to ensure the change of length is only a small percentage of the resting length.

In the past, bonded strain gages were more rugged, but less linear than unbonded models. But now recent technology has produced highly linear, reliable units of both types of construction.

**The Wheatstone Bridge.** The Wheatstone bridge is a 19th-century circuit which still has widespread application in many modern electronic instrument circuits. The classic form of Wheatstone bridge is shown in Fig. 13-3. There are four resistive arms in the bridge labeled *R*1, *R*2, *R*3 and *R*4. The excitation voltage (*V*) is applied across two of the nodes, while the output signal is taken from the alternate two nodes (labeled C and D). This circuit can be modeled as two series voltage dividers in parallel, one consisting of *R*1 and *R*4 and the other by *R*2 and *R*3 (see Fig. 13-4).

The output voltage from a Wheatstone bridge is the difference between the voltages at points C and D. The mathematics of the circuit reveals the output voltage will be zero when the ratio *R*4/*R*1 is equal to the ratio *R*3/*R*2. If these ratios are not kept equal, as is the case when one or more elements in a strain gage is not at rest, an output voltage is produced which is proportional to both the applied excitation voltage and the change of resistance.

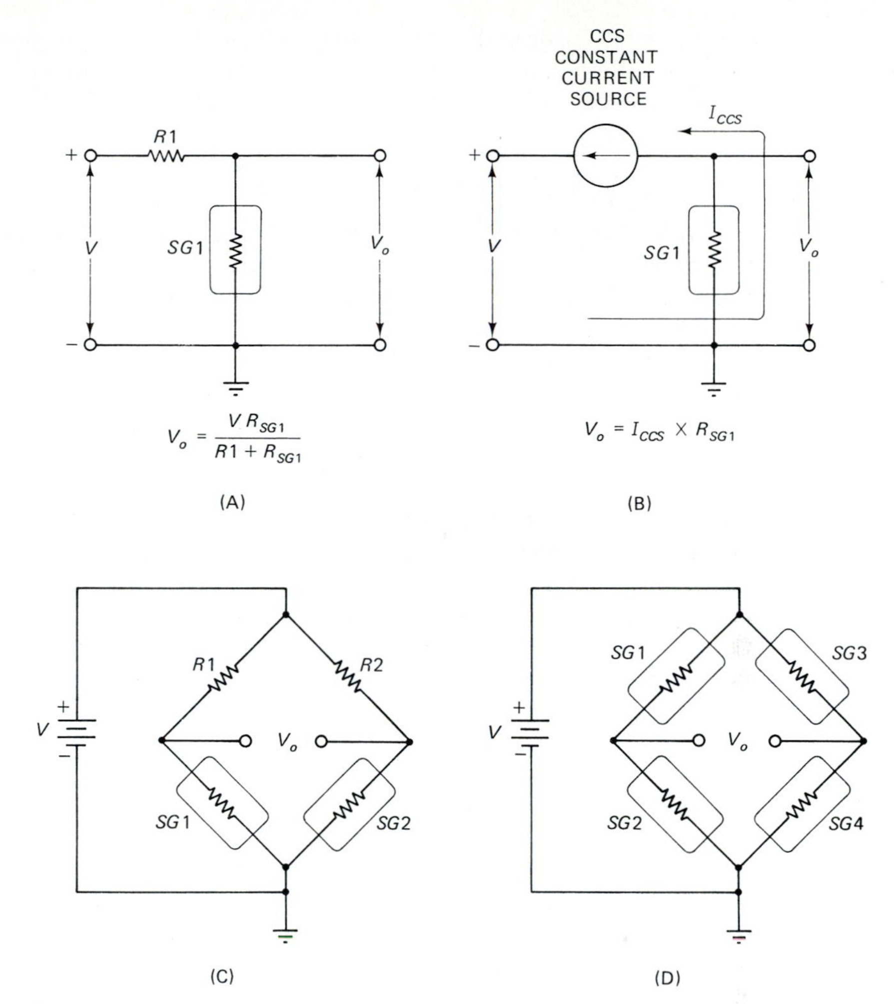


FIGURE 13-5

**Strain Gage Circuitry.** Before the resistive strain gage can be useful it must be part of a circuit which will convert its resistance changes into a current or voltage output proportional to the applied force. Most strain gage applications have voltage-output circuits.

Figure 13-5 shows several popular forms of the circuit. The circuit in Fig. 13-5A is both the simplest and least useful. It is sometimes called a half-bridge circuit, or voltage divider circuit. The strain gage element of resistance $R$ is placed in series with a fixed resistor ($R1$) across a stable DC voltage ($V$). The output voltage $V_o$ is found from the voltage divider equation

$$V_o = \frac{VR}{R + R1} \tag{13-2}$$

Equation (13-2) describes the output voltage $V_o$ when the transducer is at rest (nothing is stimulating the resistive element). When the element is stimulated, however, its resistance changes a small amount ($h$). The output voltage in that case is

$$V_o = \frac{V(R + h)}{(R \pm h) + R1} \tag{13-3}$$

Another form of half-bridge circuit is shown in Fig. 13-5B. The strain gage is connected in series with a constant current source (CCS), which will maintain current $I$ at a constant level regardless of changes in the strain gage resistance. In this case, $V_o = I(R \pm h)$.

Both half-bridge circuits suffer from one major defect, an output voltage $V_o$ will always be present regardless of the value of the stimulus applied to the transducer. Ideally, in any transducer system the output voltage should be zero when the applied stimulus is zero. For example, when a gas pressure transducer is open to atmosphere, the gage pressure is zero so the transducer output voltage ideally should also be zero. Secondly, the output voltage should be proportional to the value of the stimulus when the stimulus is not zero. A Wheatstone bridge circuit can have these properties if properly designed. Strain gage elements can be used for one, two, three, or all four arms of the Wheatstone bridge.

Figure 13-5C shows a circuit in which two strain gages ($SG1$ and $SG2$) are used in two arms of a Wheatstone bridge. Fixed resistors $R1$ and $R2$ form the alternate arms of the bridge. Usually $SG1$ and $SG2$ are configured so their actions oppose each other. That is, under any given stimulus, element $SG1$ will assume a resistance $R + h$ and $SG2$ will assume a resistance $R - h$, or vice versa. If both resistances change in the same direction, bridge balance is maintained and no output voltage is generated.

One of the most linear forms of transducer bridge is the circuit of Fig. 13-5D in which all four bridge arms contain strain gage elements. In most of these transducers, all four strain gage elements have the same resistance ($R$) which will usually be a value between 50 and 1000 ohms.

Recall the output from a Wheatstone bridge is the difference between the voltages across the two half-bridges. It is possible to calculate the output voltage for any of the standard configurations from the equations given below. Assuming all four bridge arms nominally have the same resistance, the output is

### One Active Element

$$V_o = \frac{V\,h}{4\,R} \tag{13-4}$$

This is accurate to $\pm 5\%$ provided $h < 0.1$.

### Two Active Elements

$$V_o = \frac{V\,h}{2\,R} \tag{13-5}$$

**Four Active Elements**

$$V_o = \frac{V\,h}{R} \tag{13-6}$$

where

$V_o$ = the output potential

$V$ = the excitation potential

$R$ = the bridge arm nominal resting resistance

$h$ = the change of bridge arm resistance ($\Delta R$) under the applied stimulus

The example below is for a bridge with all four arms active (similar to Fig. 13-5D).

## EXAMPLE 13-1

A force transducer is used to measure the weight of small objects. It has a resting resistance of 200 ohms and an excitation potential of +5 volts DC is applied. When a 1 gram weight is placed on the transducer diaphragm, the resistance of the bridge arms changes by 4.1 ohms. What is the output voltage?

$$V_o = Vh/R$$

$$V_o = [(5 \text{ volts})\ (4.1 \text{ ohms})/(200 \text{ ohms})]$$

$$V_o = [20.5/200]$$

$$V_o = 0.103 \text{ volts, or } 103 \text{ millivolts}$$

■

**Transducer Sensitivity.** There is some practical use for the example worked above, but most readers will probably work with a known transducer for which the sensitivity is specified. The *sensitivity factor* ($P$) relates the *transducer output voltage* ($V_o$) to the *applied stimulus value* ($Q$) and *excitation voltage* ($V$). In most cases, the transducer manufacturer will specify a number of *microvolts (or millivolts) output potential per volt of excitation potential per unit of applied stimulus.* In other words

$$P = V_o/V/Q_o \tag{13-7}$$

or, written another way

$$P = \frac{V_o}{V \times Q} \tag{13-8}$$

where

$V_o$ = the output potential

$V$ = the excitation potential

$Q$ = one unit of applied stimulus.

If the sensitivity factor is known, it is possible to calculate the output potential

$$V_o = PVQ \tag{13-9}$$

Equation (13-9) is most often used in circuit design.

### EXAMPLE 13-2

A fluid pressure transducer is used for measuring human and animal blood pressures through an indwelling fluid catheter. It has a sensitivity factor ($P$) of 5 uV/V/Torr. (1 Torr = 1 mmHg). Find the output potential when the excitation potential is +7.5 volts DC and the pressure is 400 Torr.

### Solution

$$V_o = PVQ$$

$$V_o = \frac{5 \text{ uV}}{V \times T} \times (7.5 \text{ V}) \times (400 \text{ T})$$

$$V_o = (5 \times 7.5 \times 400) \text{ uV}$$

$$V_o = 15{,}000 \text{ uV} = 15 \text{ millivolts} = 0.015 \text{ volts}$$

■

**Balancing and Calibrating a Bridge Transducer.** Few Wheatstone bridge transducers meet the ideal condition in which all four bridge arms have exactly equal resistances and exactly equal resistance changes per unit of stimulus. In fact, the bridge resistance specified by the manufacturer is only a nominal value. The actual value may vary quite a bit from the specified value. There will inevitably be an offset voltage ($V_o$ is not zero when $Q$ is zero). Figure 13-6 shows circuits that will balance the bridge when the stimulus is zero.

The method shown in Fig. 13-6A depends on the ratio of the two bridge halves being equal for null to occur $[(R_{SG1}/R_{SG2}) = (R_{SG3}/R_{SG4})]$. If the bridge is unbalanced, potentiometer $R1$ can be used to trim out the differences by adding a little more resistance to one-half the bridge. The excitation potential $V$ is applied to the wiper of the potentiometer.

The method shown in Fig. 13-6B depends on current injection into one node of the bridge circuit. Potentiometer $R1$ in Fig. 13-6B is usually a precision type with 5 to 15 turns required to cover the entire range. The purpose of the potentiometer is to inject a balancing current ($I$) into the bridge circuit at one of the nodes adjacent to an excitation node. Potentiometer $R1$ is adjusted, with the stimulus at zero, for zero output voltage.

Another application for this type of circuit is injecting an intentional offset potential. For example, on a digital weighing scale such a circuit may be used to adjust for the Tare weight of the scale. The Tare weight is the sum of the platform and all other weights acting on the transducer when no sample is being measured on the scale. This is also sometimes called empty weight compensation.

Calibration can be accomplished using either of two methods. The most accurate method is to set the transducer up in a system and apply the stimulus. The stimulus is measured by other means and the result is compared with the transducer output.

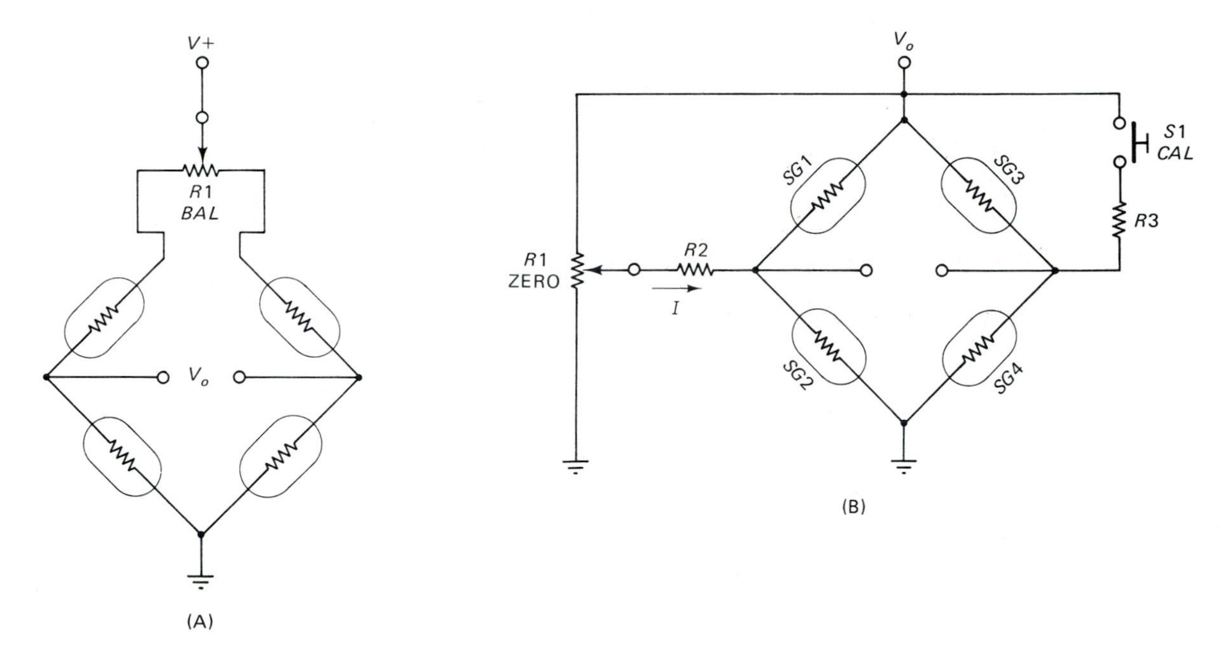


FIGURE 13-6

For example, if you are testing a pressure transducer, connect a manometer (pressure measuring device containing a column of mercury) and measure the pressure directly. The result is compared with the transducer output. All transducers should be tested in this manner when placed in service and periodically thereafter.

In less critical applications, however, an alternate (but less accurate) method may be used. It is possible to use a *calibration resistor* to synthesize the offset and thereby allow the electronics to be calibrated. The resistor and CAL switch ($S1$) in Fig. 13-6B is used for this purpose. Resistor $R3$ should have a value of

$$R3 = [(R/4QP) - (R/2)] \tag{13-10}$$

In this equation we express the output voltage from the sensitivity factor ($P$) as volts instead of microvolts.

**Auto-Zero Circuitry.** Even when the arm elements have the same nominal at-rest resistance ($R$), an offset error may exist because the bridge arm resistances each have a certain tolerance. This causes a difference in the actual values of resistance. Because of this problem, the output voltage of the Wheatstone bridge transducer will be nonzero when there is no stimulus applied. We have previously discussed a manual means for zeroing or balancing the transducer, and now we will examine an *autozero circuit*.

Figure 13-7 shows the block diagram of a typical autozero circuit. The bridge and bridge amplifier ($A1$) are the same as in the other circuits. The offset cancellation current ($I$) is generated by applying either the voltage output of a digital-to-analog converter (DAC) to resistor $R1$, or the output of a current-producing DAC to the same point on the bridge as $R1$.

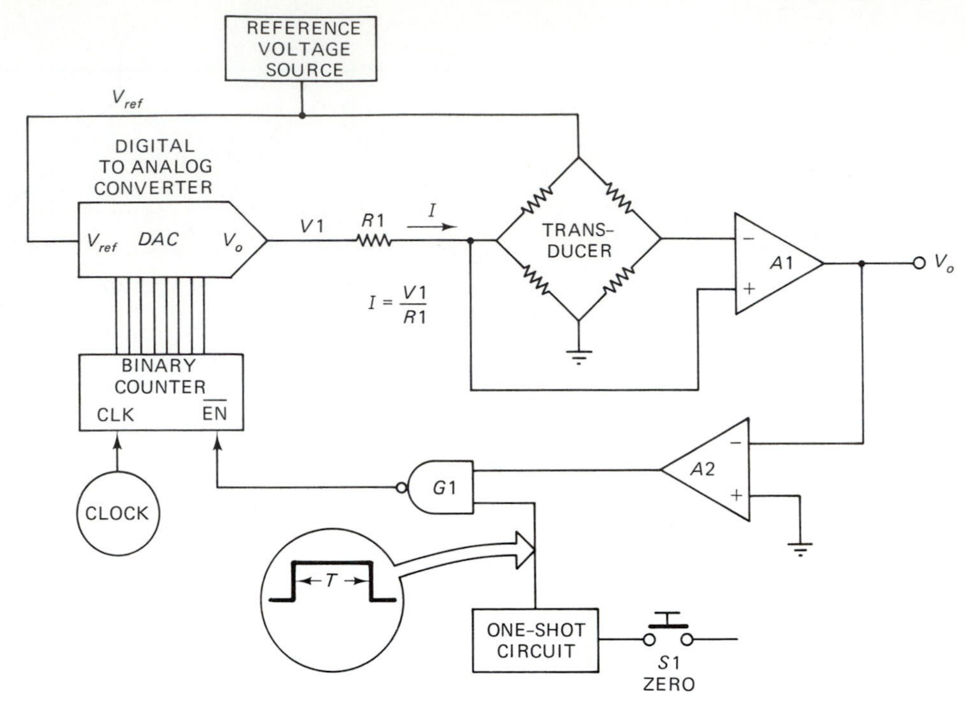


FIGURE 13-7

A voltage comparator ($A2$) monitors the output of the bridge amplifier in Fig. 13-7. When $V_o$ is zero, the $A2$ input is zero, so the output is LOW. Alternatively, when the amplifier ($A1$) output voltage is nonzero, the $A2$ output is HIGH. The DAC binary inputs are connected to the digital outputs of a binary counter which is turned on when the *enable* (EN) line is HIGH. Circuit operation is as follows

1. The operator sets the transducer to zero stimulus (for pressure transducers the valve to atmosphere is opened).
2. The ZERO button is pressed, thereby triggering the one-shot to produce an output pulse of time $T$, which makes one input of NAND gate $G1$ HIGH.
3. If voltage $V_o$ is nonzero, the $A2$ output is also HIGH, so both inputs of $G1$ are HIGH, making the $G1$ output LOW and thereby turning on the binary counter.
4. The binary counter continues to increment in step with the clock pulses, thereby causing the DAC output to rise continuously with each increment. This action forces the bridge output towards null.
5. When $V_o$ reaches zero, the $A2$ output turns off, stemming the flow of clock pulses to the binary counter and stopping the action. The DAC output voltage will remain at this voltage level.

**Transducer Linearization.** Transducers are not perfect devices. Although the ideal output function is perfectly linear, real transducers are sometimes highly nonlinear. For Wheatstone bridge strain gages the constraints on linearity include making $\Delta R$ (called $h$ in some equations) very small (less than 5%) compared to the at-rest resistance of the bridge arms.

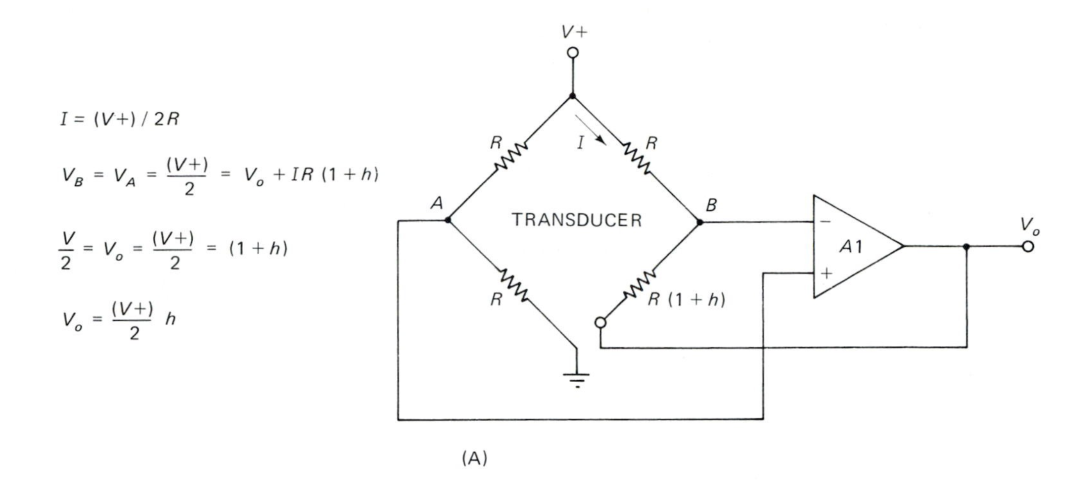


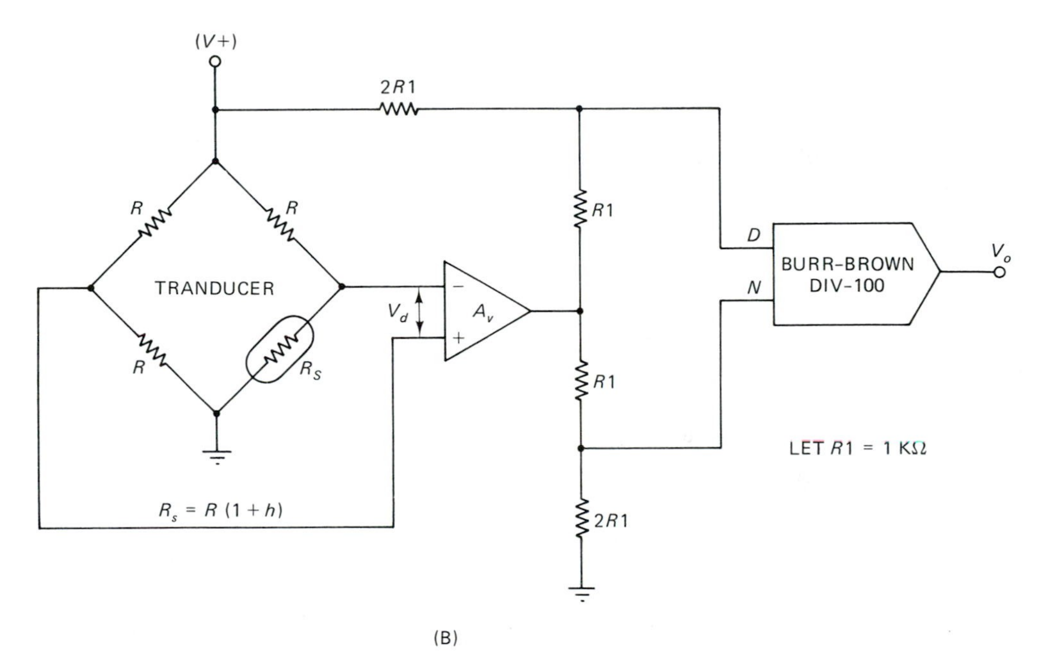


FIGURE 13-8

There are several methods for linearization. An analog method is shown in Fig. 13-8A. Here the circuit of the usual single strain gage Wheatstone bridge is modified. The ground end of one bridge arm resistor is lifted and applied to the output of a null-forcing amplifier ($A1$). In this case, the resistor element $R(1 + h)$ is in the feedback network of operational amplifier $A1$. Small amounts of nonlinearity are cancelled with this circuit.

Another analog method is shown in Fig. 13-8B. This circuit is based on the Burr-Brown DIV-100 analog divider. Two inputs are provided, $D$ and $N$. The transfer function of the DIV-100 is

$$V_o = \frac{10\,N}{D} \tag{13-11}$$

where

$V_o$ = the output signal voltage

$N$ = the signal voltage applied to the $N$ input

$D$ = the signal voltage applied to the $D$ input

It can be shown that the bridge output voltage is given by

$$V_d = \frac{-(V+)\,h}{2(2+h)} \tag{13-12}$$

If $R_n$ and $R_d$ are the input resistances of the $N$ and $D$ inputs on the DIV-100, the signal voltages applied to the inputs are found from applying the voltage divider equation

$$N = \frac{-(V+)h\,R_n}{(2R1 + 3R_n)\,(2+h)} \tag{13-13}$$

and

$$D = \frac{2(V+)R_d}{(2R1 + 3R_d)\,(2+h)} \tag{13-14}$$

Substituting Eq. (13-13) and Eq. (13-14) into Eq. (13-15), the DIV-100 transfer equation yields

$$V_o = \frac{(2R1 + 3R_d)\,(R_n h)\,(10)}{(2R1 + 3R_n)\,(2R_d)} \tag{13-15}$$

Collecting terms

$$V_o = -5h \tag{13-16}$$

This circuit is a *ratiometric* linearization method because the excitation voltage, $(V+)$, is applied to the DIV-100 inputs through the voltage divider. Thus, the output voltage taken from the DIV-100 is free of variations in the transducer excitation potential which is a common form of artifact in other circuits.

For larger degrees of nonlinearity, or for nonlinearity over a large range, other methods must be employed. Figure 13-9 shows a hypothetical transducer function in which a voltage $V_o$ is a function of an applied pressure, $P$. The ideal transducer will output a linear signal of the form $V_o = mP + b$, where $m$ is the slope of the line and $b$ is the offset.

Before digital computer chips were routinely used in electronic instrumentation applications, special function circuits or *diode breakpoint generators* were used for linearization. In cases where a special function circuit is used, it was assumed the equation of the actual curve was known. The special function circuit generated the inverse of that function and summed it with the input voltage in a difference amplifier. In the case of the diode breakpoint generator an offset voltage is added to or subtracted from the actual input signal to normalize it to the ideal. A reverse-biased diode is

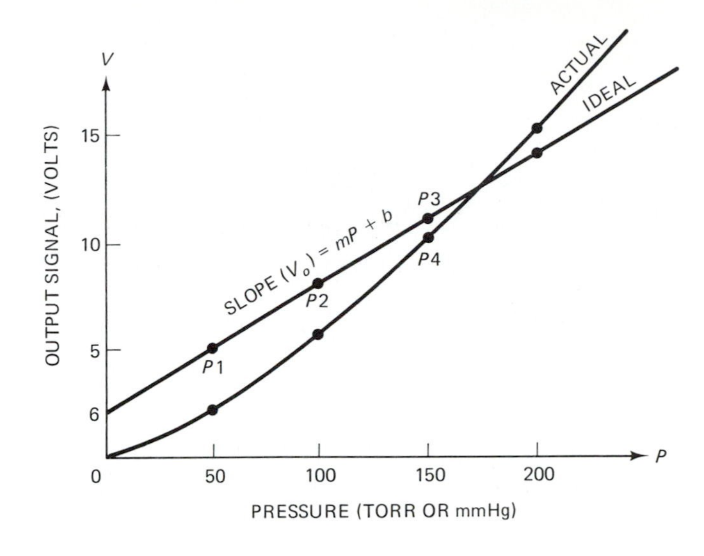


| POINT | PRESSURE (TORR) | IDEAL | ACTUAL |
|---|---|---|---|
| *P*1 | 50 | 5 | 2 |
| *P*2 | 100 | 8 | 6 |
| *P*3 | 150 | 11 | 10 |
| *P*4 | 200 | 14 | 16 |

FIGURE 13-9

used to switch on when the input signal voltage exceeds a certain value, causing an offset correction bias voltage to be added to the output voltage. This method is rare today. It was limited because of the temperature sensitivity of the diode switches, and the fact that only a piece-wise linear approximation was possible unless a large number of diode breakpoints was used.

Now that microprocessors are used in instruments, software can be used to correct transducer error. If the equation defining the actual curve is known, we could write a software program which algebraically cancels the error. There should exist a polynomial which allows errors to be mathematically smoothed out. Alternatively, the lookup table method of Fig. 13-10 may be used. This example shows only a limited number of data points for simplicity's sake. The actual number would depend on the bit length of the A/D converter. An eight-bit A/D converter can represent $2^8$ (256) different values.

The values for the ideal transfer function are stored in a lookup table in computer memory which begins at address location HF000 (the H indicates that hexadecimal or base-16 notation is used). The value HF000 is stored in the X register. When the A/D binary word is input to the computer it is added to the contents of the X register. This value becomes the indexed address in the lookup table where the correct value is found. Although a pressure transducer example is shown here, it is useful for almost any form of transducer.

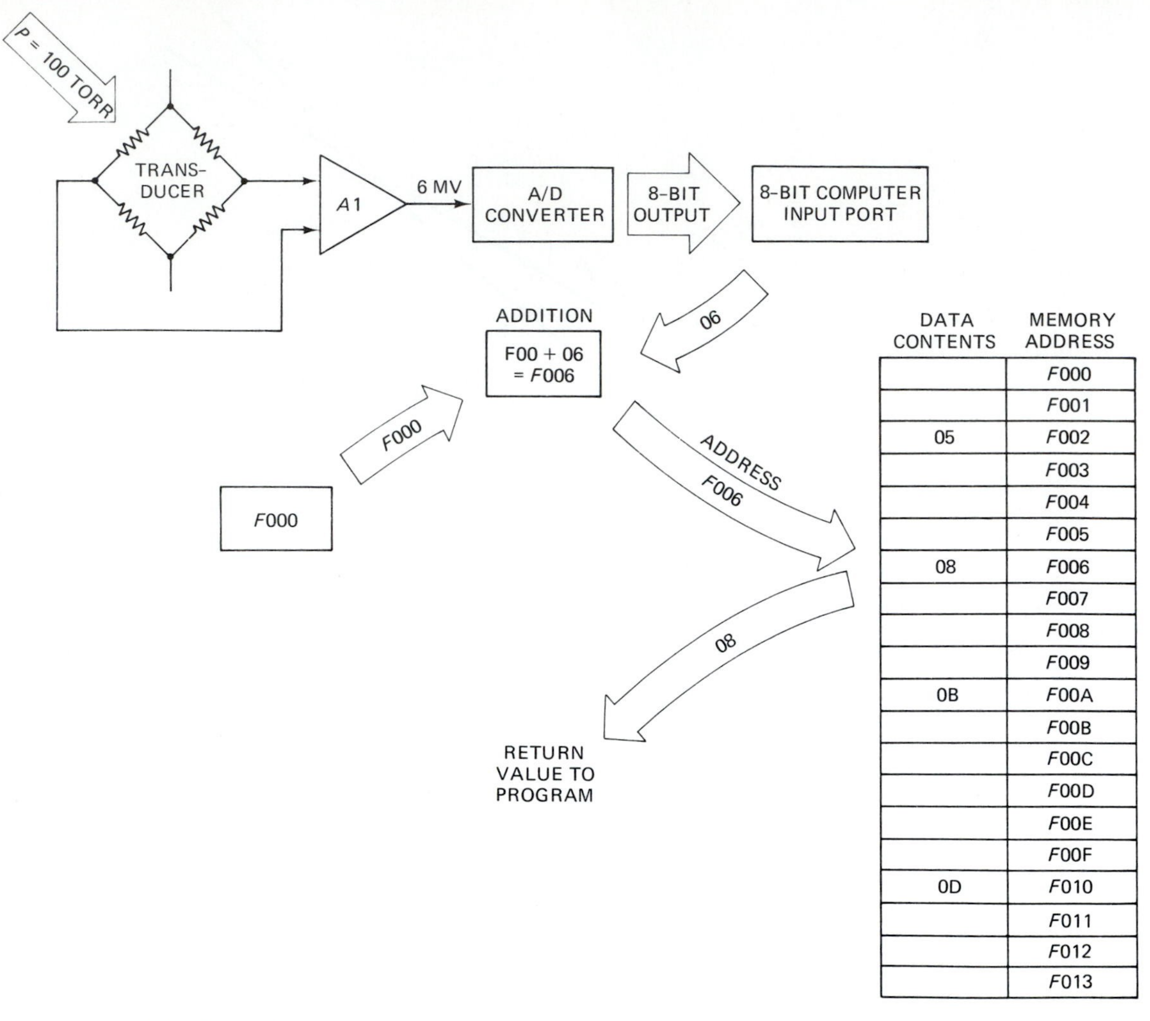


| DATA CONTENTS | MEMORY ADDRESS |
| --- | --- |
| | F000 |
| | F001 |
| 05 | F002 |
| | F003 |
| | F004 |
| | F005 |
| 08 | F006 |
| | F007 |
| | F008 |
| | F009 |
| 0B | F00A |
| | F00B |
| | F00C |
| | F00D |
| | F00E |
| | F00F |
| 0D | F010 |
| | F011 |
| | F012 |
| | F013 |

FIGURE 13-10

**Transducer Excitation Sources.** The DC Wheatstone bridge transducer requires a source of either AC or DC voltage. DC is the most common. Most transducers require an excitation voltage of 10 volts DC or less. The maximum allowable voltage is critical. Exceeding it will shorten the life expectancy of the transducer. An optimum voltage may be specified for the transducer. This potential is the highest excitation potential that will not cause thermal drift due to self-heating of the transducer resistance elements. A typical fluid pressure transducer requires +7.5 volts DC and operates best (with the least thermal drift) at +5 volts DC. A source of DC must be provided which is stable, within specifications, and precise.

The simplest form of transducer excitation is the zener diode circuit in Fig. 13-11A. A zener diode sufficiently regulates the voltage for many applications. There are two problems with this circuit, however. First, the zener potential is typically not a reasonable value like 5.0 volts, but will have a value such as 4.7, 5.6, 6.2, or 6.8 volts.

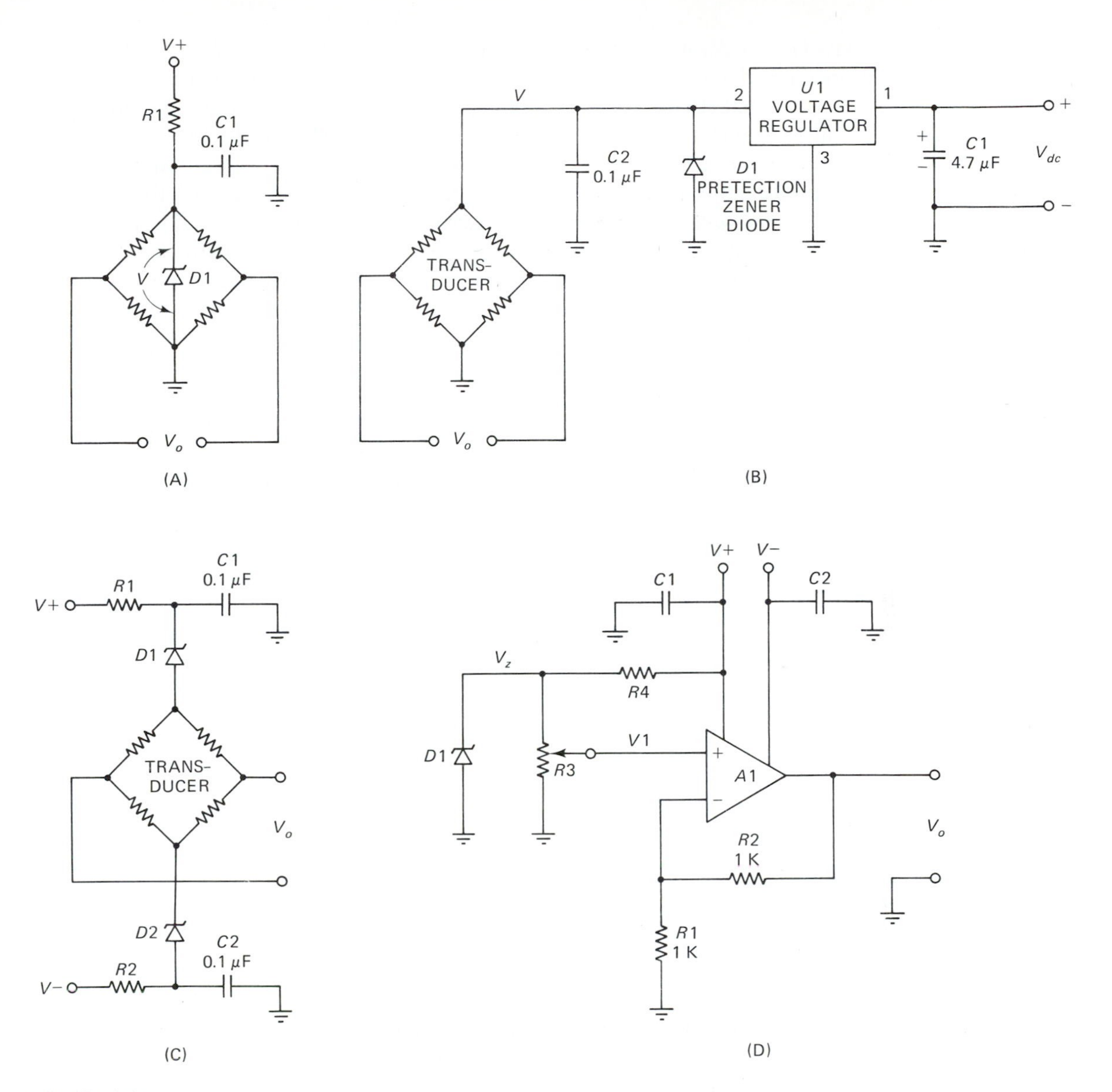


FIGURE 13-11

The second and most serious defect is *thermal drift* of the zener point. The zener voltage will vary somewhat with temperature in all but certain reference grade zener diodes. Unless the application is not critical, or the diode can be kept at a constant temperature, the method of Fig. 13-11A is not generally suitable.

Figure 13-11B shows a second type of DC excitation circuit. In this circuit the voltage regulator is a three-terminal IC voltage regulator (*U*1) of the LM-309, LM-340, 78xx, or similar families. Selection of the device depends on the voltage required and the current normally drawn by the transducer. The load current is found from

Ohm's law ($I = V/R$) where $V$ is the regulator output voltage and $R$ is the resistance of any one transducer element. In a typical case, the transducer will use a +5 volt excitation potential. If the resistance of the $R$ elements is >50 ohms, by Ohm's law the current will be less than 100 mA. A 100 mA LM-309H or LM-340H can be used.

The zener diode ($D1$) in Fig. 13-11B is not used for voltage regulation, but for overvoltage protection of the transducer. If the regulator ($U1$) fails, +8 to +16 volts from the $V+$ line will be applied to the transducer. The purpose of $D1$ is to clamp the voltage to a value greater than the excitation voltage, but less than the maximum allowable excitation voltage rating of the transducer. A small fuse is sometimes inserted in series with the input (pin 1) of $U1$. The value of this fuse is set to roughly twice the current requirements ($V/R$) of the transducer and will blow if the zener diode voltage is exceeded. The fuse adds a certain amount of protection. In other cases, a light emitting diode (LED) in series with diode $D1$ informs the user of a fault. Otherwise, the error would be interpreted simply as a higher transducer output signal which is indistinguishable from a higher value due to increased stimulus.

Some applications require a dual-polarity power supply. Figure 13-11C shows an excitation circuit in which two zener diodes are used, one each for positive and negative polarities. A certain amount of thermal stability is achieved by having opposite polarity power supplies because under temperature changes the voltage difference remains relatively constant.

Neither the zener circuit nor the three-terminal regulator circuit will deliver precise output voltages. The voltage will be stable but not precise. A typical three-terminal IC voltage regulator output voltage, for example, may vary several percent from sample to sample. If we need a precise excitation voltage, a circuit such as Fig. 13-11D might be used. This circuit is an operational amplifier voltage reference circuit in which the op-amp is a high current model, like the National Semiconductor LM-13080.

The output voltage from Fig. 13-11D is determined by $R1$, $R2$, the setting of $R3$, and the value of zener diode $D1$. The voltage $V$ will be

$$V = (V1) \times \frac{R2}{R1} + 1 \tag{13-17}$$

or, since $R1 = R2 = 1$ kohms

$$V = 2 \times (V1) \tag{13-18}$$

Voltage $V1$, at the noninverting input of $A1$, is a fraction of the zener voltage which depends upon the setting of potentiometer $R3$. We can adjust $V1$ from 0 Vdc to $V_z$ Vdc, so the transducer voltage can be set at any value from 0 to $2V_z$ volts DC. In most cases, $V$ will be set at 5.00 volts, 7.50 volts, or 10.00 volts depending on the nature of the transducer.

### 13-2.2 Transducer Amplifiers

The basic DC differential amplifier is the most common for amplifying Wheatstone bridge transducer signals. These amplifiers are easily constructed from simple operational amplifiers. Figure 13-12 shows such a circuit. Assuming that $R1 = R2$ and $R3$

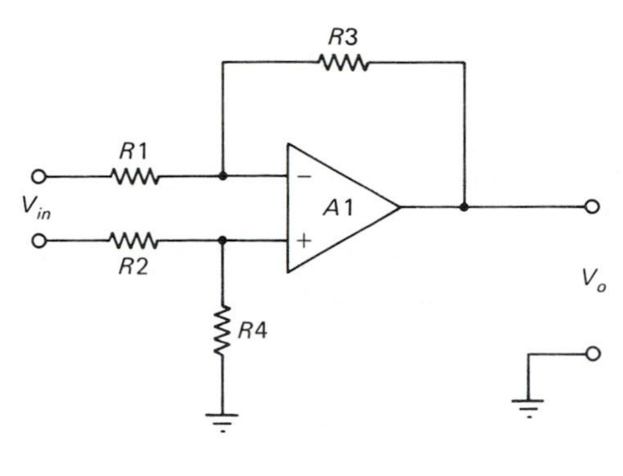


FIGURE 13-12

= $R4$, the gain of the amplifier will be $R4/R2$, or $R3/R1$. The amplifier output voltage will be found from

$$V_o = (V_{in}) \times (R3/R1) \tag{13-19}$$

where

$V_o$ = the amplifier output voltage

$V_{in}$ = the transducer output voltage

$R1$ and $R3$ are the resistors in the amplifier circuit.

The gain required from the amplifier is determined from a *Scale Factor* (*SF*) which is the ratio between the voltage representing full-scale at the output of the amplifier and the voltage representing full-scale at the output of the transducer

$$SF = \frac{V_{o(\max)}}{V_{in(\max)}} \tag{13-20}$$

where

$V_{o(\max)}$ = the voltage $V_o$ representing full scale

$V_{in(\max)}$ = the transducer output voltage $V_{in}$ which represents full scale

You may recall the transducer output voltage is found from the *excitation voltage*, the *applied stimulus*, and the *sensitivity factor* (*P*). The sensitivity factor is given in terms of millivolts (or microvolts) output per volt of excitation potential per unit of applied stimulus

$$P = \frac{V_o}{V \times Q} \tag{13-21}$$

The output voltage from the transducer is therefore found from

$$V_{in} = P \times V \times Q \tag{13-22}$$

where

$V_{in}$ = the transducer output voltage (or amplifier input voltage)

$V$ = the excitation voltage

$Q$ = the applied stimulus (force, pressure, etc.)

### EXAMPLE 13-3

A chemical requires a 0 to 1000-Torr fluid pressure transducer and amplifier. The transducer is rated at a sensitivity factor ($P$) of 50 uV/V/Torr and a range of $-200$ to $+1200$ Torr. The available excitation source was 6.95 volts. Calculate (a) the full-scale output voltage, and (b) the gain required for an output voltage scale factor of 10 mV/Torr.

(a) At the required full scale pressure (1000 Torr) the output voltage is

$$V_o = PVQ$$

$$V_o = \frac{50\ \text{uV}}{V \times T} \times (5.00\ V) \times (1000\ \text{T})$$

$$V_o = (50\ \text{uV}) \times (5.00) \times (1000)$$

$$V_o = 250{,}000\ \text{uV} \times \frac{1\ \text{mV}}{1000\ \text{uV}} = 250\ \text{mV}$$

(b) The amplifier output voltage required will depend on the desired display method. For example, a strip chart recorder might have a full-scale voltage range of 0.5 volt or 1.0 volt. Alternatively, a digital panel meter (DPM) may be used for the output display. Most low-cost DPMs have either a 0 to 1999 millivolt range or a 0 to 19.99 volt range, so a great deal of utility is gained by making the output voltage at full scale numerically the same as the DPM reading. For example, 1000 Torr would be represented by either 1000 millivolts or 10 volts. In that case, the DPM scale factor would be either 1 mV/Torr or 10 mV/Torr. The value of $V_{o(\max)}$ in the equation above is 1000 mV. The gain of the amplifier is the scale factor $SF$ described earlier

$$SF = \frac{V_{o(\max)}}{V_{in(\max)}}$$

$$SF = \frac{10{,}000\ \text{mV}}{250\ \text{mV}} = 40$$

■

### 13-2.3 A Practical Design Example

Physiologists and other life scientists sometimes use a strain gage force-displacement transducer such as the Grass FT-3 shown in Fig. 13-13A for their experiments. Small displacements of the operating arm at one end of the transducer housing produce changes in the internal Wheatstone bridge. This in turn, results in an output voltage ($V_{ot}$). The sensitivity factor ($P$) of the transducer is 35 uV/V/gram-force (the gram-force is a force equal to the force of gravity on a mass of one gram, i.e., 1 gm-F = 980 dynes). Forces to 10 gm-F can be accommodated with excitation potential up to +7.5 Vdc.

A physiologist attempted to use an FT-3 transducer without a preamplifier because the manufacturer's offering was too expensive for the scientist's research budget. The experiment was to measure the minute contractions of a guinea pig heart in vitro in response to a certain stimulus. The transducer output, therefore, was a weak pulsatile voltage. The transducer was excited from a 6 Vdc lantern battery. The balanced output of the transducer was used to drive the Channel A and Channel B inputs of an oscilloscope.

When an oscilloscope is used in the A-B input mode, it will poorly mimic a differential amplifier. Unfortunately, gain differences between the two channels resulted in a common mode rejection ratio (CMRR) problem. In addition, the lead wires were unshielded, which led to excessive 60-Hz hum pickup from surrounding AC power wiring in the laboratory. The problem was exacerbated because the transducer signals were low compared with the 60-Hz induced signal.

It was decided to design and build a transducer preamplifier for the experiment. The excitation voltage was standardized to +5 Vdc because 5-volt IC voltage regulators were easily obtained. For a maximum 4 gm-F stimulus, which the scientist believed to be the absolute maximum which would be seen, the transducer output voltage is

$$V_{ot} = PVF$$

$$V_{ot} = \frac{35 \text{ uV}}{\text{V} \times \text{gm-F}} \times 5.00 \text{ Vdc} \times 4 \text{ gm-F} \tag{13-23}$$

$$V_{ot} = 35 \text{ uV} \times 5 \times 4 = 700 \text{ uV}$$

Converting to volts

$$V_{ot} = 700 \text{ uV} \times \frac{1 \text{ volt}}{10^6 \text{ uV}} = 0.0007 \text{ volts}$$

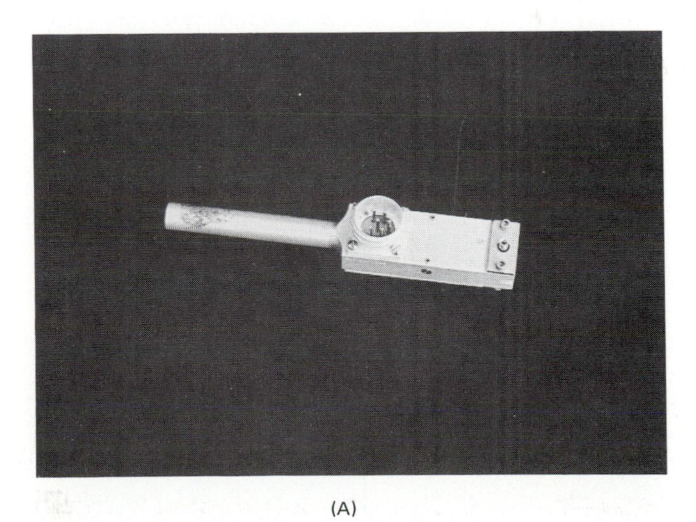

(A)

FIGURE 13-13 (*continued on next page*)

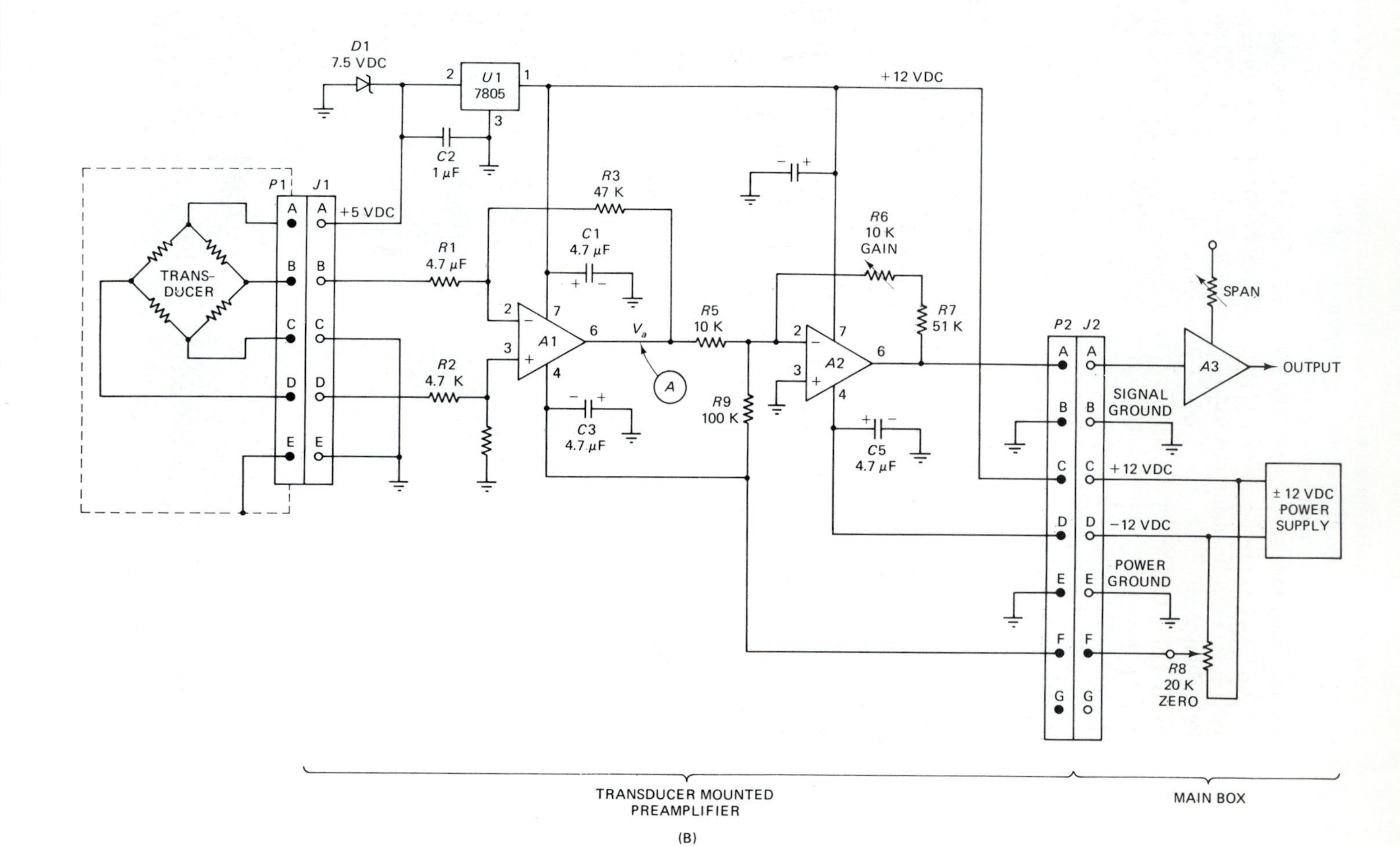


FIGURE 13-13 *(continued)*

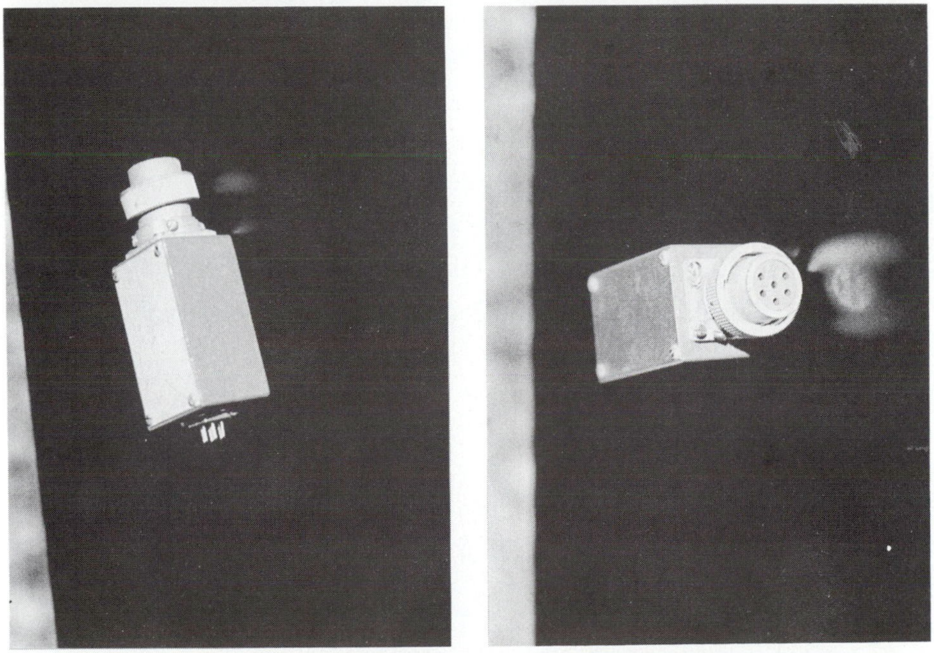

(D)

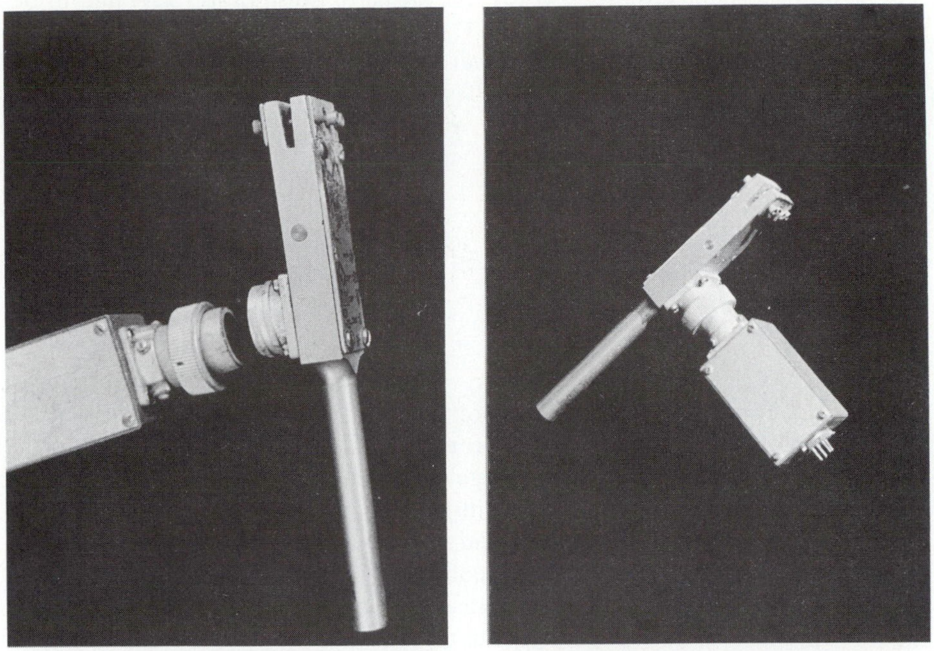

(C)

FIGURE 13-13 *(continued)*

Overcoming the 60-Hz interference problem required boosting the signal at the source to a higher level so the induced 60-Hz signal was only a tiny fraction of the desired signal. In other words, the signal to noise ratio (S/N) had to be improved. A strategy was adopted to boost $V_{ot}$ to 10 mV/gm F, or 40 mV when the maximum 4 gm-F is applied to the transducer. The required voltage gain is

$$A_{vt} = \frac{V_{ota}}{V_{ot}}$$

$$A_{vt} = \frac{\left[40 \text{ mV} \times \dfrac{1 \text{ volt}}{1000 \text{ mV}}\right]}{0.0007} \qquad (13\text{-}24)$$

$$A_{vt} = \frac{0.04 \text{ volt}}{0.0007} = 57.14$$

A two-stage preamplifier circuit (Fig. 13-13B) was designed and built. It used a gain of 10 ($A_{vt} = 10$) DC differential input stage ($A1$) to accommodate the balanced output of the Wheatstone bridge. The output signal of $A1$ ($V_a$) was boosted $10\times$ to 0.007 volts (7 mV) when a 4 gm-F was applied to the transducer.

Input resistors $R1$ and $R2$ had to be at least ten times the Thevanin's resistance of the transducer, which for a Wheatstone bridge with all elements equal is the resistance of any one element. For the Grass FT-3 transducer used here the resistance elements were 200 ohms each, so any input resistor value >2000 ohms is acceptable. A value of 4.7 kohms was selected for reasons of convenience.

The gain $A_{v1}$ of this stage is found from

$$A_{v1} = R3/R1 \qquad (13\text{-}25)$$

so

$$R3 = A_{v1}R1 \qquad (13\text{-}26)$$

Inserting the actual values

$$R3 = (10)\ (4.7 \text{ kohms}) = 47 \text{ kohms}$$

A variable gain stage ($A2$) is used to boost $V_{ota}$ to 40 mV when $V_a = 7$ mV. The gain required of $A2$ is

$$A_{v2} = \frac{V_{ota(\max)}}{V_{a(\max)}} \qquad (13\text{-}27)$$

or, with the values inserted

$$A_{v2} = \frac{40 \text{ mV}}{7 \text{ mV}} = 5.714$$

Another way to calculate $A_{v2}$ is to divide the total required gain by $A_{v1}$

$$A_{v2} = \frac{A_{vt}}{A_{v1}} \qquad (13\text{-}28)$$

$$A_{v2} = \frac{57.14}{10} = 5.714$$

A value of 10 kohms was selected for $R5$ because it is a standard value, and thus is easy to obtain. The value was therefore selected for reasons of economy and convenience. The value had to be >1000 ohms in order to accommodate the normal <100 ohms output impedance of the operational amplifier. For a gain of 5.714 the feedback resistance $R_f = (R6 + R7)$ must be

$$R6 + R7 = A_{v2}R5 \tag{13-29}$$

$$R6 + R7 = (5.714)\ (10\ \text{kohms}) = 57.14$$

Although a single precision resistor can be obtained to accommodate the required resistance, the solution adopted here is to break up the total resistance between two standard values. For example, $R6$ is set to 10 kohms (and is a potentiometer), while $R7$ is a 51 kohm fixed resistor.

The circuit of Fig. 13-13B was built inside a small aluminum instrument box (see Fig. 13-13C) which was fitted with a mate to the transducer output connector on one end. This configuration allowed the preamplifier to be mounted directly to the transducer housing (Fig. 13-13D).

A control box was connected to the preamplifier through *P2/J2*. This external box provided additional gain. A gain-of-10 amplifier will boost the signal to a healthy 400 mV. The external box also provided an offset null (or ZERO) circuit and gain control (SPAN) to calibrate the system.

### 13-2.4 Microelectrodes

The microelectrode is an ultra-fine electrical probe used to measure biopotentials at the cellular level. The microelectrode penetrates a cell immersed in an "infinite" fluid (such as physiological saline), which, in electrical terminology, serves as the ground for the measurement. The fluid is, in turn, connected to a *reference electrode* (Fig. 13-14A).

Several types of microelectrode exist, but the *metallic tip contact* is perhaps the most common because it is easily made. The other type is fluid-filled and uses the potassium chloride (KCL) column. In both types of microelectrode, an exposed contact surface of one to two micrometers (1 uM = $10^{-6}$ M) is in contact with the cell. As might be expected, this small surface area makes microelectrodes high-impedance devices.

Figure 13-14B shows the construction of a typical glass-metal microelectrode. A very fine platinum or tungsten wire is slip-fit through a 1.5 to 2 millimeter (mm) glass pipette. The tip of the pipette is acid etched and then fire formed into a shallow angle taper. The microelectrode can then be connected to one input of the differential amplifier. There are two subcategories of this electrode. In one, the metallic tip is flush with the end of the pipette taper. In the other, there is a thin layer of glass covering the metal point. This glass layer is so thin its thickness is usually measured in Angstroms (glass substantially increases the impedance of the device).

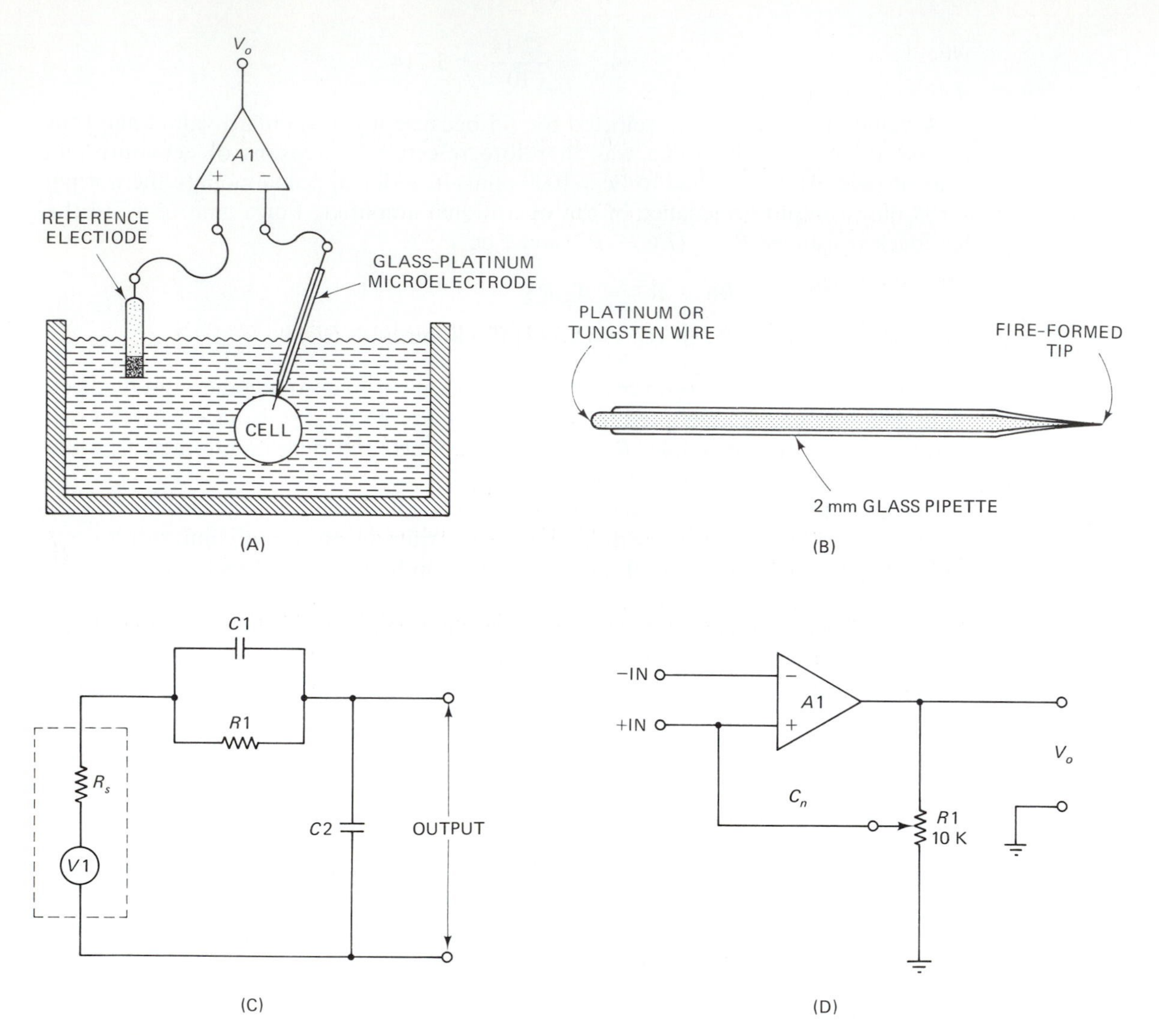


FIGURE 13-14

Figure 13-14C shows a simplified equivalent circuit for the microelectrode (disregarding the contribution of the reference electrode). Analysis of this circuit reveals a problem due to the *RC* components. Resistor *R*1 and capacitor *C*1 are used at the electrode-cell interface. They are frequency dependent. These values fall off to a negligible level at a rate of $1/(2\pi F)^2$ and are generally considerably lower than $R_s$ and *C*2. Resistance $R_s$ in Fig. 13-14C is the spreading resistance of the electrode. It is a function of the tip diameter. The value of $R_s$ in metallic microelectrodes without the glass coating is approximated by

$$R_s = \frac{P}{4\pi r} \tag{13-30}$$

where

$R_s$ = the resistance in ohms

$P$ = the resistivity of the infinite solution outside of the electrode (e.g., 70 ohm-cm for physiological saline)

$r$ = the tip radius (typically 0.5 uM for a 1 uM electrode)

### EXAMPLE 13-4

Assuming the typical values given above, calculate the tip spreading resistance of a 1 uM microelectrode.

$$Rs = P/4\pi r$$

$$Rs = \frac{70 \text{ ohm-cm}}{4\pi \; 0.5 \text{ uM} \times \dfrac{10^{-4} \text{ cm}}{1 \text{ uM}}} = 111{,}000 \text{ ohms}$$

The impedance of glass coated metallic microelectrodes is at least 1 or 2 orders of magnitude higher than this figure. ■

For fluid filled KCL microelectrodes with small taper angles ($\pi/180$ radians), the series resistance is approximated by:

$$R_s = \frac{2P}{\pi ra} \tag{13-31}$$

where

$R_s$ = the resistance in ohms

$P$ = the resistivity (typically 3.7 ohm-cm for 3$M$ KCL)

$r$ = the tip radius (typically 0.1 uM)

$a$ = the taper angle (typically $\pi/180$)

### EXAMPLE 13-5

Find the series impedance of a KCL microelectrode using the values shown above

$$R_s = \frac{2P}{\pi ra}$$

$$R_s = \frac{(2)\;(3.7 \text{ ohm-cm})}{(3.14)\; 0.1 \text{ uM} \times \dfrac{10^{-4} \text{ cm}}{1 \text{ uM}} \dfrac{3.14}{180}} = 13.5 \text{ Mohms}$$

■

The capacitance of the microelectrode is given by

$$C2 = \frac{0.55\, e}{\text{LN}(R/r)} \text{ pF/cm} \tag{13-32}$$

where

$$e = \text{the dielectric constant of glass (typically 4)}$$

$$R = \text{the outside tip radius}$$

$$r = \text{the inside tip radius } (r \text{ and } R \text{ in same units})$$

### EXAMPLE 13-6

Find the capacitance ($C2$ in Fig. 13-14C) if the pipette radius is 0.2 uM and the inside tip radius is 0.15 uM.

$$C2 = \frac{0.55\ e}{\text{LN}(R/r)}\ \text{pF/cm}$$

$$C2 = \frac{(0.55)\ (4)}{\text{LN}(0.2\ \text{uM}/0.15\ \text{uM})}\ \text{pF/cm} = 7.7\ \text{pF/cm}$$

■

How do these values affect performance of the microelectrode? Resistance $R_s$ and capacitor $C2$ operator together as an $RC$ low-pass filter. For example, a KCL microelectrode immersed in 3 cm of physiological saline has a capacitance of approximately 23 pF. Suppose it is connected to the amplifier input (15 pF) through 3 feet of small diameter coaxial cable (27 pF/ft, or 81 pF). The total capacitance is (23 + 15 + 81) pF = 119 pF. Given a 13.5-megohm resistance, the frequency response at the −3 dB point is

$$F = \frac{1}{2\pi RC} \tag{13-33}$$

where

$$F = \text{the } -3 \text{ dB point in hertz (Hz)}$$

$$R = \text{the resistance in ohms}$$

$$C = \text{the capacitance in farads}$$

### EXAMPLE 13-7

For $C = 119$ pF ($1.19 \times 10^{-10}$ farads) and $R = 1.35 \times 10^{7}$ ohms, find the frequency response upper −3 dB point.

$$F = \frac{1}{(2)\ (3.14)\ (1.35 \times 10^{7})\ (1.19 \times 10^{-10})}$$

$$F = 99\ \text{Hz} \sim 100\ \text{Hz}$$

■

Clearly, a 100-Hz frequency response, with a −6 dB/octave characteristic above 100-Hz, results in severe rounding of the fast rise-time action potentials. A strategy must be devised in the instrument design to overcome the effects of capacitance in high impedance electrodes.

**Neutralizing Microelectrode Capacitance.** Figure 13-14D shows the standard method for neutralizing the capacitance of the microelectrode and associated circuitry. A neutralization capacitance ($C_n$) is in the positive feedback path along with a potentiometer voltage divider. The value of this capacitance is

$$C_n = \frac{C}{A_v - 1} \tag{13-34}$$

where

$C_n$ = the neutralization capacitance

$C$ = the total input capacitance

$A_v$ = the gain of the amplifier

### EXAMPLE 13-8

A microelectrode and its cabling exhibit a total capacitance of 100 pF. Find the value of neutralization capacitance (Fig. 13-14D) required for a gain of 10 amplifier.

**Solution**

$$C_n = C/(A_v - 1)$$

$$C_n = (100\text{-pF})/(10 - 1)$$

$$C_n = 100 \text{ pF}/9 = 11 \text{ pF}$$

■

## 13-2.5 Light Transducers

Various classes of instruments use light as a transducible element. Colorimetry instruments are often based on transducers using this principle. Various instrumentation techniques may depend on various properties of light and light transducers. In some cases, only the existence or nonexistence of the light beam is important. An example of this is the PAPER OUT sensor on a computer printer, or a light beam that counts entrances and exits from a building. In some cases, the color of the light is important. In others, the absorption of particular colors is the important factor. In all of these cases the light beam must be sensed before it can be used.

Light is a form of electromagnetic radiation. In the ultimate sense, it is the same as radio waves, infrared (heat) waves, ultraviolet, and X-rays. The principal differences between these types of electromagnetic radiation are the frequency and wavelength. The wavelength of visible light is 400 to 800 nanometers (1 nm = $10^{-9}$ meters). Infrared has longer wavelengths than visible light. Ultraviolet has wavelengths shorter than visible light and X-radiation has wavelengths even shorter than ultraviolet. Frequency and wavelength are related in electromagnetic radiation by the equation

$$\lambda = \frac{c}{f} \tag{13-35}$$

where

$$c = \text{the velocity of light (300,000,000 meters/second)}$$

$$\lambda = \text{wavelength in meters}$$

$$f = \text{frequency in hertz}$$

From the above equation you can see light has a frequency on the order of $10^{14}$ Hz (compare with the frequencies of AM and FM broadcast bands in the radio portion of the spectrum which are $10^6$ Hz and $10^8$ Hz respectively).

Because IR, UV, and X-radiation are similar in nature and close in wavelength to visible light, many of the sensors and techniques applied to visible light also work to one extent or another in adjacent regions of the electromagnetic spectrum. Although performance may vary, and some photosensitive devices are not even useful in those areas of the spectrum, it is nonetheless true designers may find some of these devices useful.

The photosensors described in this section depend on quantum effects for their operation. Quantum mechanics arose as a new idea in physics in December 1900 with a famous paper by German physicist Max Planck. He had been working on thermodynamics problems resulting from the fact that the experimental results reported in 19th-century physics laboratories could not be explained by classical Newtonian mechanics. The solution turned out to be a simple, but terribly revolutionary idea: energy existed in discrete bundles, not as a continuum. In other words, energy comes in packets of specific energy levels; other energy levels are excluded. The name eventually given to these energy packets was *quanta* and quantum mechanics was born. The name given to energy bundles which operated in the visible light range was *photons*.

The energy level of each photon is expressed by the equation

$$E = \frac{ch}{\lambda} \tag{13-36}$$

or alternatively

$$E = h\nu \tag{13-37}$$

where

$$E = \text{the energy in electron volts (eV)}$$

$$c = \text{the velocity of light in meters per second}$$

$$\lambda = \text{the wavelength in meters}$$

$$h = \text{Planck's constant } (6.62 \times 10^{-34}\ J\text{-}s)$$

$$\nu = \text{the frequency of light waves in hertz (Hz)}$$

Sometimes the constants *ch* are combined and expressed together as 1240 electron volts per nanometer (eV/nm).

Some light sensors are constructed to allow at least one electron to be freed from its associated atom by one photon of light. Materials in which the electrons are too tightly bound for light photons to do this will not work well as light sensors.

**The Phototube.** The *phototube* is a vacuum tube which is based on the *photoelectric effect*. Physicist Albert Einstein won the Nobel Prize in physics for his explanation of the photoelectric effect, not for either the Special or General Theories of Relativity as is commonly assumed. Einstein wrote three seminal papers for the 1905 edition of *Annalen der Physik*, an explanation of Brownian motion as a molecular effect, an explanation of the photoelectric effect, and Special Relativity.

The photoelectric effect had long perplexed physicists. If you shine a light onto certain types of metallic plate in a vacuum, electrons are emitted from the surface of the plate. Oddly, increasing the brightness of the light does not increase the energy level of the emitted electrons. If this phenomenon were purely a mechanical kinetic event, one would assume that increasing the intensity of the light would increase the energy of the electrons emitted from the surface. It turned out, however, that changing the *color* of the light affected the energy level of the emitted electrons. Red light produced lower energy electrons than violet light. The *energy level of the electrons is color sensitive*, even though the *amount of current (number of electrons) emitted is intensity sensitive*. Once Planck's principle was known, however, Einstein was able to explain this effect by quantum principles. From the above equations we see the higher frequency (shorter wavelength) violet colored light is significantly more energetic than red light.

Figure 13-15A shows a phototube based on the photoelectric effect. The cathode is a metallic plate exposed to light. It is made of a material that will easily emit electrons when light is applied. The anode is positively charged (with respect to the

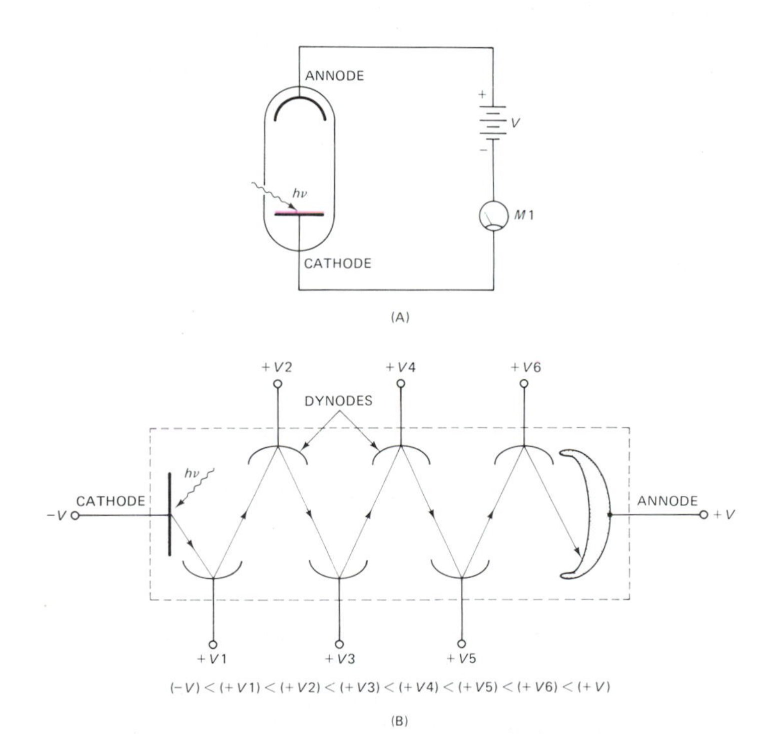


FIGURE 13-15

cathode) by an external DC power supply. Electrons emitted from the photocathode are collected by this anode, and can be read on an external meter (or other circuit).

The photoemission process is often not very efficient. We can make it more efficient by using a *photomultiplier* (PM) *tube* (Fig. 13-15B). In this photosensor there are a number of positively charged anodes, called *dynodes*, which intercept the electrons. Electrons are emitted when light impinges the cathode. They are accelerated through a high voltage potential ($V1$) to the first dynode. They acquire substantial kinetic energy during this transition, so when each electron strikes the metal it gives up its kinetic energy. Because of the conservation of energy principle, giving up kinetic energy requires some energy to be converted to heat, while some is converted to additional kinetic energy by dislodging other electrons from the dynode surface. Thus, a single electron causes two or more additional electrons to be dislodged. These electrons are accelerated by another high voltage potential ($V2$). This produces the same effect at the second dynode. The process is repeated several times, and each time several more electrons join the process for each previously accelerated electron. Finally, the electron stream is collected by the anode, and can be used in an external circuit.

**Photovoltaic Cells.** A *photovoltaic cell* generates a potential difference and thus a current flow, by shining light onto its surface. The common solar cell is an example of the photovoltaic cell. The instrumentation applications for this type of cell are limited. Figure 13-16 shows a typical circuit for instrumentation applications of the photovoltaic cell. The cell is connected across the input of a high impedance amplifier, like the noninverting operational amplifier shown. The output voltage is found from

$$V_o = V1 \times \frac{R2}{R1} + 1 \tag{13-38}$$

**Photoresistors.** A *photoresistor* changes electrical ohmic resistance when light is applied. Figure 13-17A shows the usual circuit symbol for photoresistors. It is the normal resistor symbol enclosed within a circle and represented by the greek letter *lambda* symbol to denote it is a resistor which responds to light. Figure 13-17B shows

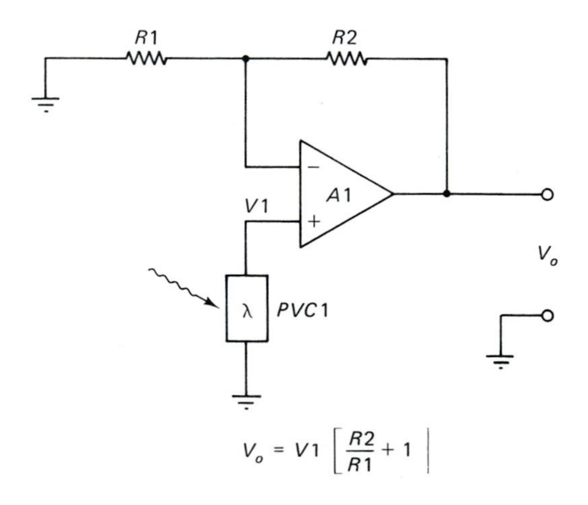


FIGURE 13-16

(A)

(B)

FIGURE 13-17

an actual photoresistor. When photoresistors are specified, both a *dark resistance* and a *light/dark resistance ratio* are listed. In most common varieties, the resistance is very high when dark, and very low under intense light. The intensity of the light affects the resistance, so these devices can be used in photographic lightmeters, densitometers, and colorimeters.

Figure 13-18 shows three circuits in which photoresistors can be used. The half-bridge circuit is shown in Fig. 13-18A. In this circuit, the photoresistor is connected across the output of a voltage divider made up of $R1$ and $PC1$. The output voltage is given by

$$V_o = \frac{V\,PC1}{R1 + PC1} \tag{13-39}$$

where

$V_o$ = the output potential

$V$ = the applied excitation potential

$R1$ and $PC1$ are in ohms

A problem with this circuit is the output potential never drops to zero. It always has an offset value.

A second way to use the photoresistor is shown in Fig. 13-18B. Here the photoresistor is the feedback resistor in an operational amplifier inverting follower circuit. The output voltage ($V_o$) is found from

$$V_o = -(-V_{\text{ref}}) \times (PC1/R1) \tag{13-40}$$

The circuit of Fig. 13-18B provides a low-impedance output. But like the half-bridge circuit (Fig. 13-18A), the output voltages does not drop to zero. In addition, with some photoresistors the dynamic range of the operational amplifier may not match the dark/light ratio of the photoresistor at practical values of $-V_{ref}$ potential.

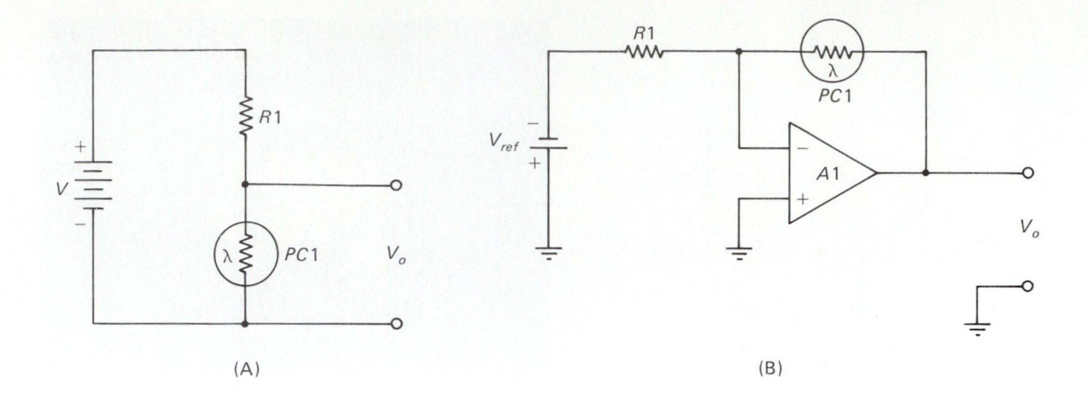


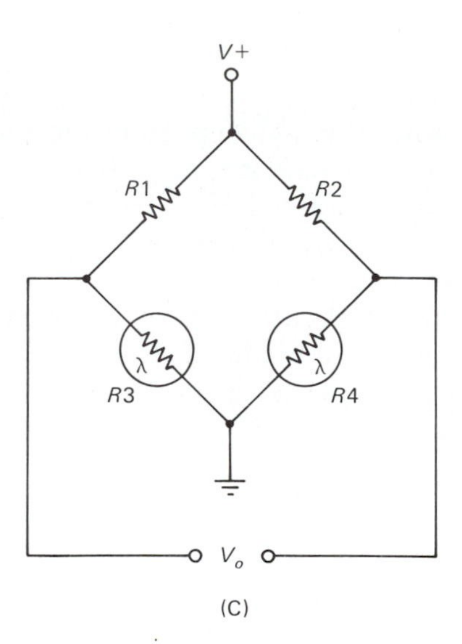


FIGURE 13-18

The last photoresistor configuration is the Wheatstone bridge shown in Fig. 13-18C. This circuit allows the output voltage to be zero under the right circumstances and it is the circuit favored by most designers. If a low impedance output is required, or additional amplification is needed, then a DC differential amplifier can be connected across output potential.

**Photodiodes and Phototransistors.** The most modern light sensor is the PN junction configured in the form of special photodiodes or phototransistors. In certain types of diodes, the level of the reverse leakage current increases when the junction is illuminated. Figure 13-19A shows the basic circuit used for these sensors. The diode is

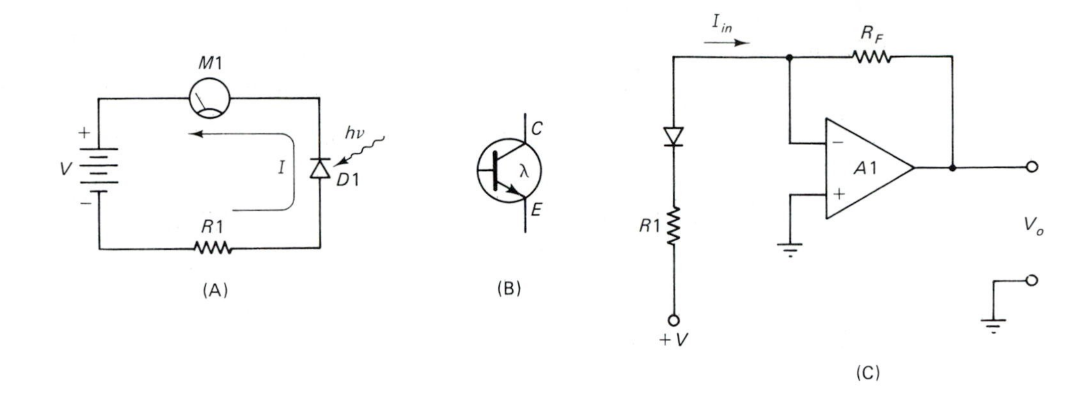


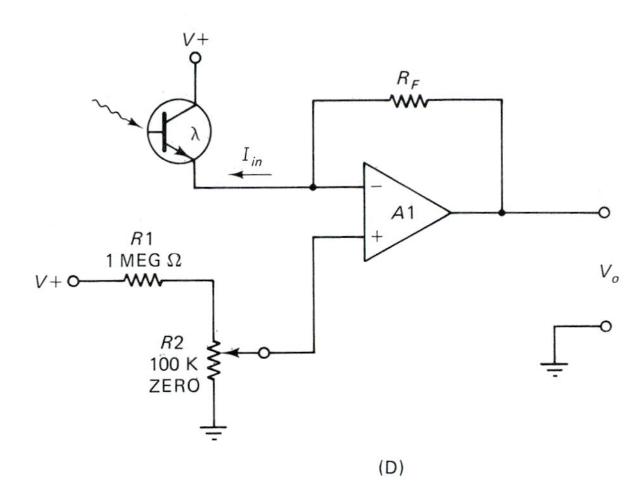


FIGURE 13-19

normally reverse-biased, with a current-limiting resistance in series. Microammeter $M1$ measures the reverse leakage current that crosses the PN junction during this type of operation. When light strikes the PN junction, the reading on $M1$ will increase.

The same principle applies to the class of NPN or PNP transistors called *phototransistors* (Fig. 13-9B). In these devices, collector to emitter current flows when the base region is illuminated. These devices are the heart of optoisolator and optocoupler integrated circuits. They are used in various instrumentation sensor applications.

Figures 13-19C and 13-19D show ways in which photodiodes and phototransistors are used. The current flow through the device under light conditions is the transducible event, so they must be connected into a circuit which makes use of that property. The inverting follower operational amplifier circuit works well. The output voltage $V_o$ is equal to the product of the input current and the feedback resistance ($I_{in} \times R_f$). In the case of Fig. 13-19D, a *zero control* is added. This can also be added to the other circuits as well.

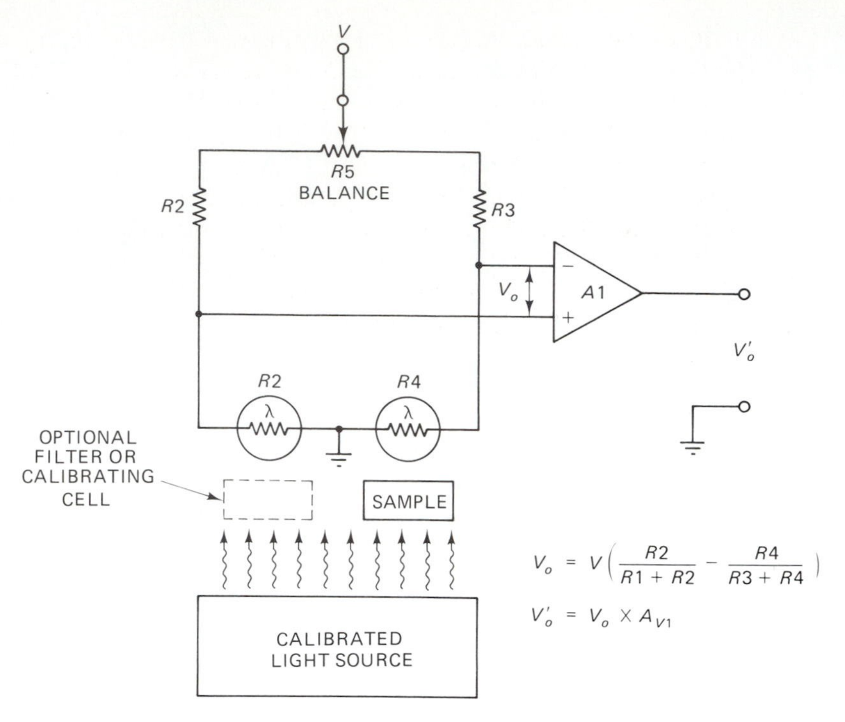


FIGURE 13-20

**Photocolorimeter.** One of the most basic instruments is also both the oldest and most commonly used, the photocolorimeter. These devices are used to measure oxygen content in blood, $CO_2$ content of air, water vapor content, blood electrolyte (Na and K) levels, and a host of other measurements in the medical laboratory.

Figure 13-20 shows the basic circuit of the most elementary form of colorimeter. It is widely used as a medical blood oxygen meter. The circuit is a Wheatstone bridge which uses a pair of photoresistors as the transducing elements. Potentiometer $R5$ is used as a balance control. It is adjusted for zero output ($V_o = 0$) when the light shines equally on both photoresistors. Recall from Chapter 10 that output voltage $V_o$ will be zero when the two legs of the bridge are balanced. In other words, $V_o$ is zero when $R1/R2 = R3/R4$. It is not necessary for the resistor elements to be equal (although it is usually the case), but their ratios be equal. Thus, a 500k/50k ratio for $R1/R2$ will produce zero output voltage when $R3/R4 = 100\text{k}/10\text{k}$.

The photoresistors are arranged so light from a calibrated source shines on both equally and fully, except when an intervening filter or sample is present. Thus, the bridge can be nulled to zero using potentiometer $R5$ under this zero condition. In most instruments, a translucent sample is placed between the light source and one of the photocells. The comparison of light transmission through air (or a vacuum or a specified wavelength filter) to light transmitted through the sample being tested is a measure of sample density, and is thus transducible. Consider some practical applications below.

**Blood O2 Level.** A classical and still used method for measuring blood oxygen level is the colorimeter in Fig. 13-20. It works on the fact that the redness of blood is a measure of its oxygenation. This instrument is nulled with neither a standard filter cell nor blood in the light path. A standard 800-nm filter cell is introduced between the light source and *R*2, and a blood sample is placed in a standardized tube between the light source and *R*4. The degree of blood saturation in the sample is thus reflected by the difference in the bridge reading. On one model, a separate resistor across the *R*1/*R*2 arm is used to bring the bridge back into null condition, and the dial for that resistor is calibrated in percent $O_2$. More modern instruments based on digital computer techniques provide a more automatic measurement.

**Respiratory $CO_2$ Level.** The exhaled air from humans is roughly 2–5% carbon dioxide, while the percentage of $CO_2$ in room air is negligible. A popular form of end tidal $CO_2$ meter is based on the fact that $CO_2$ absorbs infrared (IR) waves. The light source is actually either an IR LED or a Cal-Rod® (identical to the type used to heat electrical coffee pots). The photocells are selected for IR response. In this type of instrument, room air is passed through a *cuvette* placed between *R*2 and the heat source, while patient expiratory air is passed through the same type of *cuvette* placed between the heat source and *R*4. The difference in IR transmission is a function of the percentage of $CO_2$ in the sample circuit.

The associated electronics will allow zero and maximum span (gain) adjustment. The zero point is adjusted with room air in both cuvettes, while the maximum scale (usually 5% $CO_2$) is adjusted with the sample cuvette purged of room air and replaced with a calibration gas (usually 5% $CO_2$ and 95% nitrogen).

**Blood Electrolytes.** Blood chemistry includes levels of sodium (Na) and potassium (K). The *flame photometer* (Fig. 13-21) measures these elements. This form of colorimeter replaces the light source with a flame produced in a gas carburetor. The sample is injected into the carburetor and burned with the gas/air mixture. The colors emitted on burning are proportional to the concentrations of Na and K ions in the sample. A special gas is used to burn cleanly with a blue flame when no sample is present. In medical applications, a specified size sample of patient's blood is mixed with a premeasured indium calibrating solution. The solution is well mixed and then applied to the carburetor. By comparing the intensities of the colors generated by burning Na and K ions with the intensity of the calibration color, the instrument can infer the concentration of the two elements.

**Photospectrometer.** This class of instruments depends on the fact that certain chemical compositions absorb wavelengths of light in different degrees. If you are familiar with the spectrum of sunlight created when the light passes through a prism or is reflected from a diffraction grating, you are familiar with the physical basis for the spectrophotometer (see Fig. 13-22). Either the light beam or the photodetector must move to examine each of the light wavelengths in turn. In most cases, the sample and the photodetector are kept stable, while a diffraction grating or prism is rotated to permit all colors of the spectrum to fall on the same photodetector. By comparing the amplitudes of the light at different angles, we can infer the transmission of the respective colors, and create a chart which shows amplitude against wavelength.

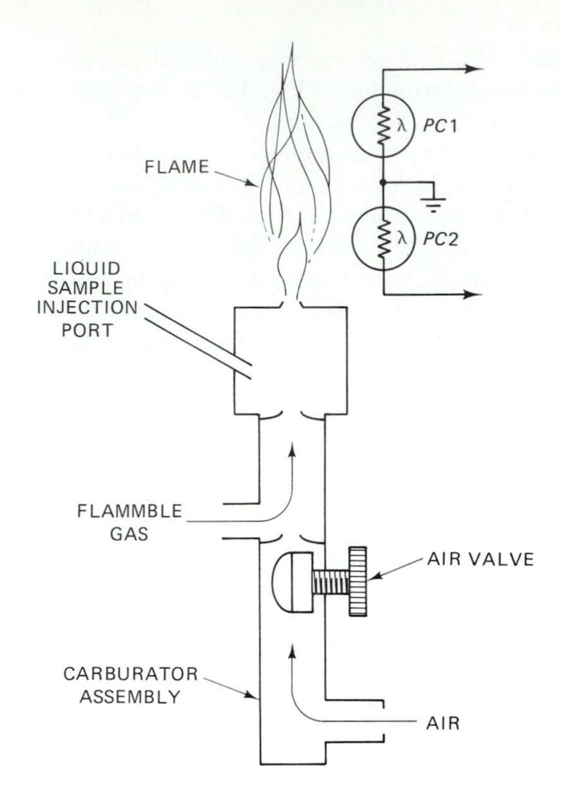


FIGURE 13-21

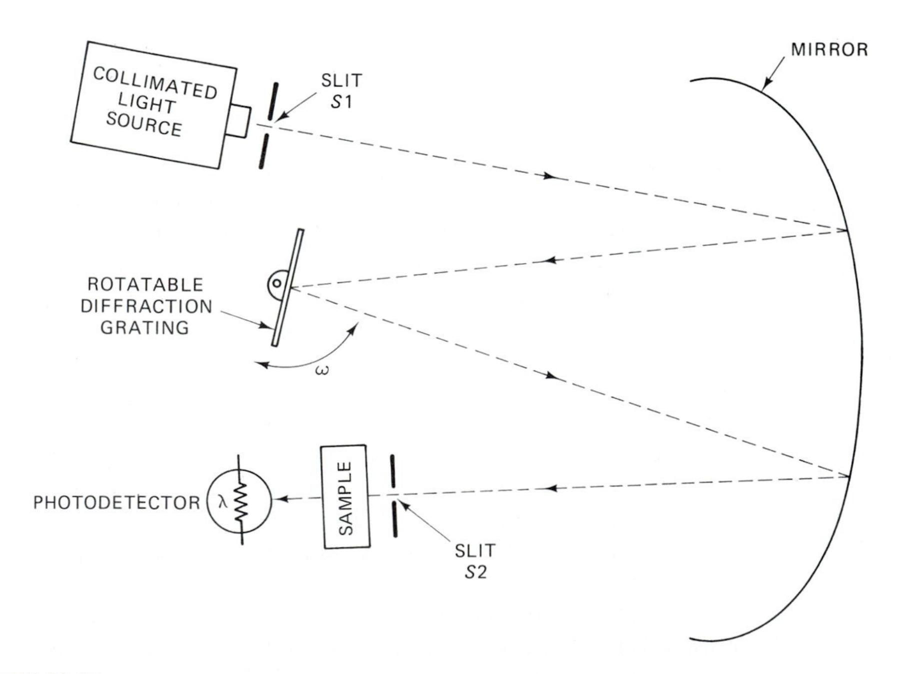


FIGURE 13-22

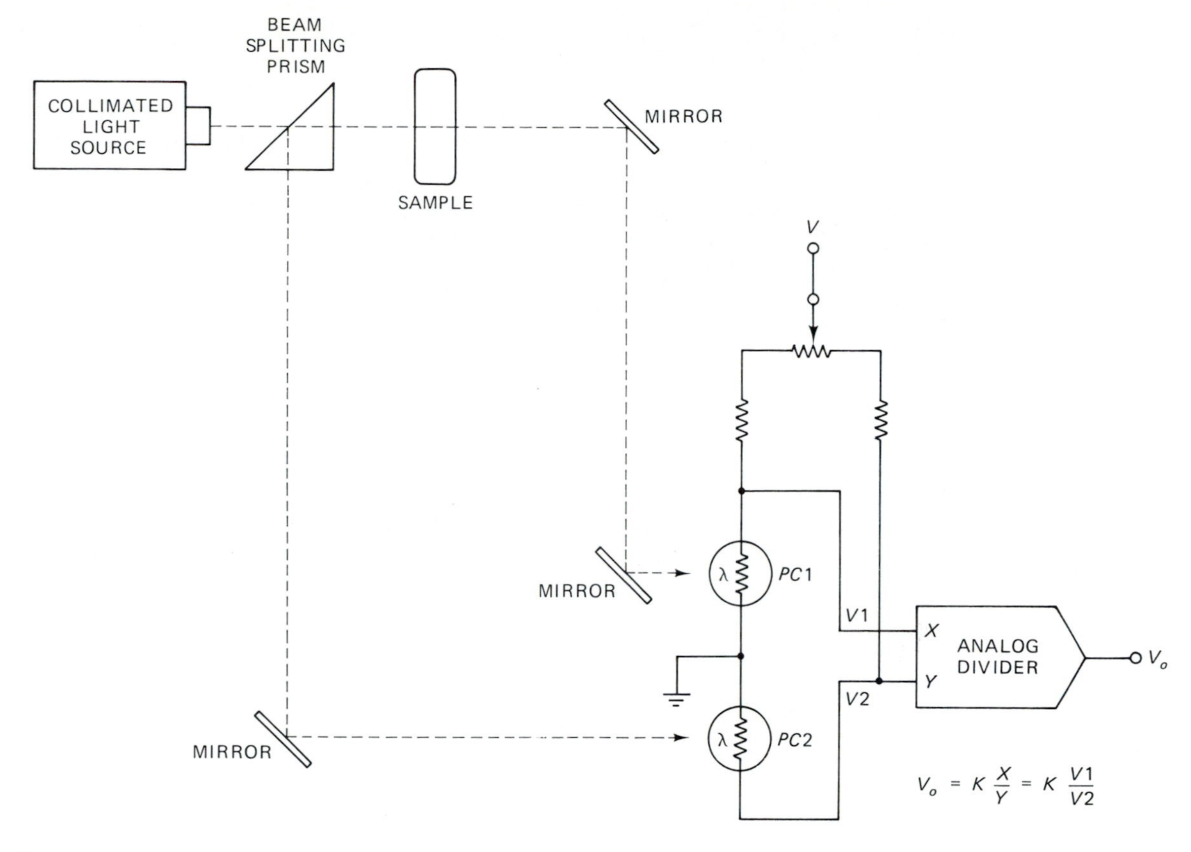


FIGURE 13-23

**Ratiometric Measurements.** A significant problem with photometric measurements is keeping the light source constant in both intensity and color. All forms of light emission change, and this introduces artifact in the data, if it occurs between calibration and measurement. The solution to many of these problems is the *ratiometric* measurement technique (Fig. 13-23).

In a ratiometric system, the collimated light beam is passed through a 50% beam-splitting prism. This optical device splits the beam into two equal amplitude beams at right angles to each other. The incident beam is passed through an optical path of length $L$ to photosensor $PC2$. The sample beam also passes through an optical path of length $L$, but also passes through the sample. If the optical properties of the two paths are the same and they are the same length, the light intensities arriving at the two colorimeter photosensors will be the same, except for any energy lost in the sample. Thus, the Wheatstone concept can be used. But first, some signal processing of the signal is needed.

The circuit must take the *ratio* of $V1$ and $V2$, the outputs of $PC1$ and $PC2$ respectively. Either in the computer, or in an analog multiplier, we must take the ratio $V1/V2$. Thus, a change in the light level affects both photosensors equally, so the only difference between the two is the properties of the sample. Analog logarithmic amplifiers can be used for this purpose.

**A Practical Design Example: The pH Meter.** Chemists use pH as a measure of the relative acidity or alkalinity (base) of a fluid. If a solution has a pH of 7.00 it is neither acid nor base and it is said to be neutral. Acidic solutions have a pH less than 7. Basic solutions have a pH more than 7. Human blood has a normal pH of 7.4 ± 0.04 (7.36 to 7.44). Values outside this range are considered pathological. A pH electrode is a special high impedance chemical glass electrode which produces an output voltage that is a function of fluid pH (Fig. 13-24A). A typical pH electrode will output 0 volts at pH = 7, plus or minus the offsets represented by a tolerance band and the electrode temperature.

In order to design a simple digital pH meter is is necessary to create a circuit that produces 1-Vdc output at pH = 1, 7 Vdc when pH = 7 and 14 Vdc when pH = 14.

The input impedance of the circuit must be more than 100 megohms because of the glass electrode source impedance. This requirement can be met by using a BiFET or BiMOS operational amplifier, or an electrometer amplifier. The unity gain configuration is used in the input stage shown in Fig. 13-24B.

The slope of the characteristic curve in Fig. 13-14A is negative, while the slope required by the design is positive (Fig. 13-24C). Therefore, an inverting stage ($A2$) is used to follow the input stage ($A1$). Accommodating electrode tolerances requires both an *offset control* and a variable voltage gain.

The offset control is adjusted to set $V_o$ to +7.00 volts when the input voltage is zero ($V_{in} = 0$). The amplifier gain ($A_{v2}$) is set so $V_o = +14.00$ volts when $V_{in} = -1.4$ volts, or

$$A_{v2} = \left.\frac{V_{o(\max)}}{V_{in(\max)}}\right|_{\mathrm{pH}=14} \tag{13-41}$$

$$A_{v2} = \frac{14 \text{ volts}}{-1.4 \text{ volts}} = -10$$

Because the tolerance band is ±80 mV, the gain has to have a range of operation. The maximum gain

$$A_{v2(\max)} = \frac{14 \text{ volts}}{-1.32 \text{ volts}} = -10.61$$

and the minimum gain

$$A_{v2(\min)} = \frac{14 \text{ volts}}{-1.48 \text{ volts}} = -9.45$$

The amplifier must have a gain that can be varied between −9.45 and −10.61. If $R1 = 10$ kohms, then ($R2 + R3$) must be variable between

$$(R2 + R3) = (10 \text{ kohms})\,(9.45) = 94.5 \text{ kohms}$$

and

$$(R2 + R3) = (10 \text{ kohms})\,(10.61) = 106.1 \text{ kohms}$$

A combination of a 91-kohms fixed resistor in series with a 20 kohms multiturn potentiometer will solve the problem.

The offset control must set voltage $V_b$ to 1.00 when $V_{in} = +1.4$ volts. Assuming a nominal gain of $-10$ for stage $A2$ and $+1$ for $A1$, voltage $V_b$ would normally be $[(V_{in})(A_v)] = [(+1.4 \text{ Vdc})(-10)] = -14.00$ Vdc. The potential at the wiper of the potentiometer ($V_c$) normally is a nominal gain of $-10$ (i.e. $(R2 + R3)/R6$). The goal is to offset $-14$ Vdc to $+1$ Vdc, a total of $+15$ Vdc change. Given the gain of $-10$

$$V_c = \frac{V_{b(\text{desired})}}{A_v} \tag{13-42}$$

$$V_c = \frac{+15 \text{ Vdc}}{-10} = -1.5 \text{ Vdc}$$

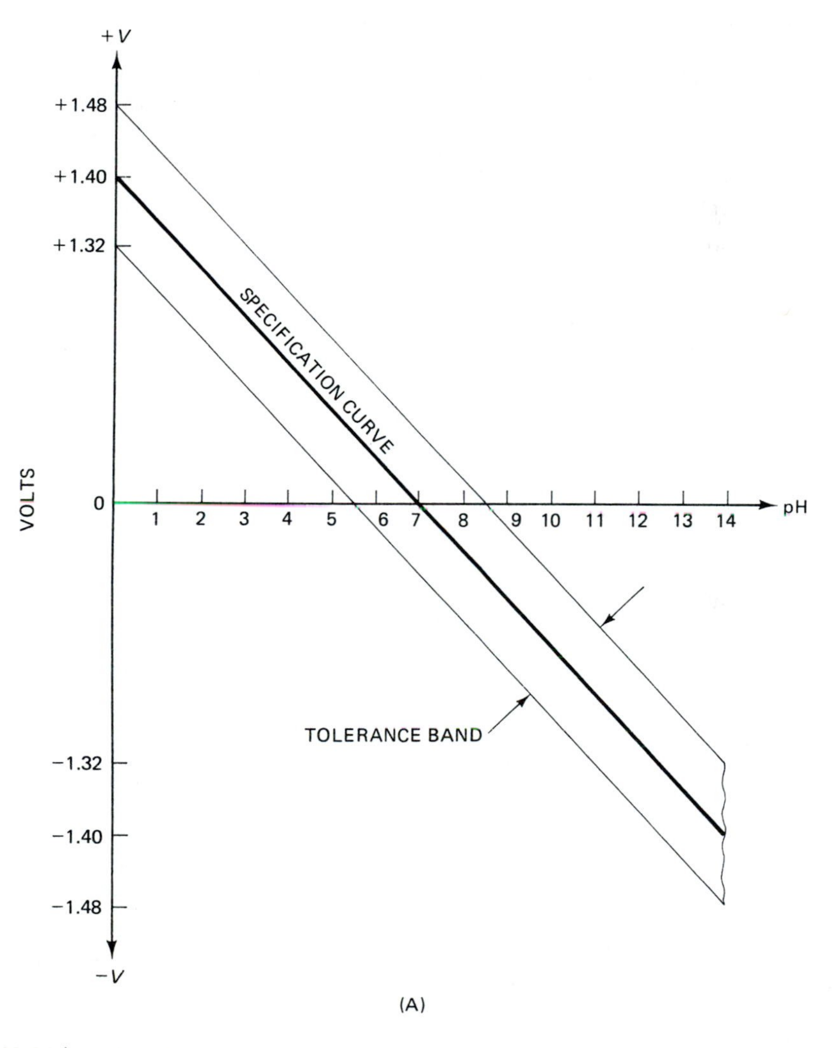


FIGURE 13-24 (*continued on next page*)

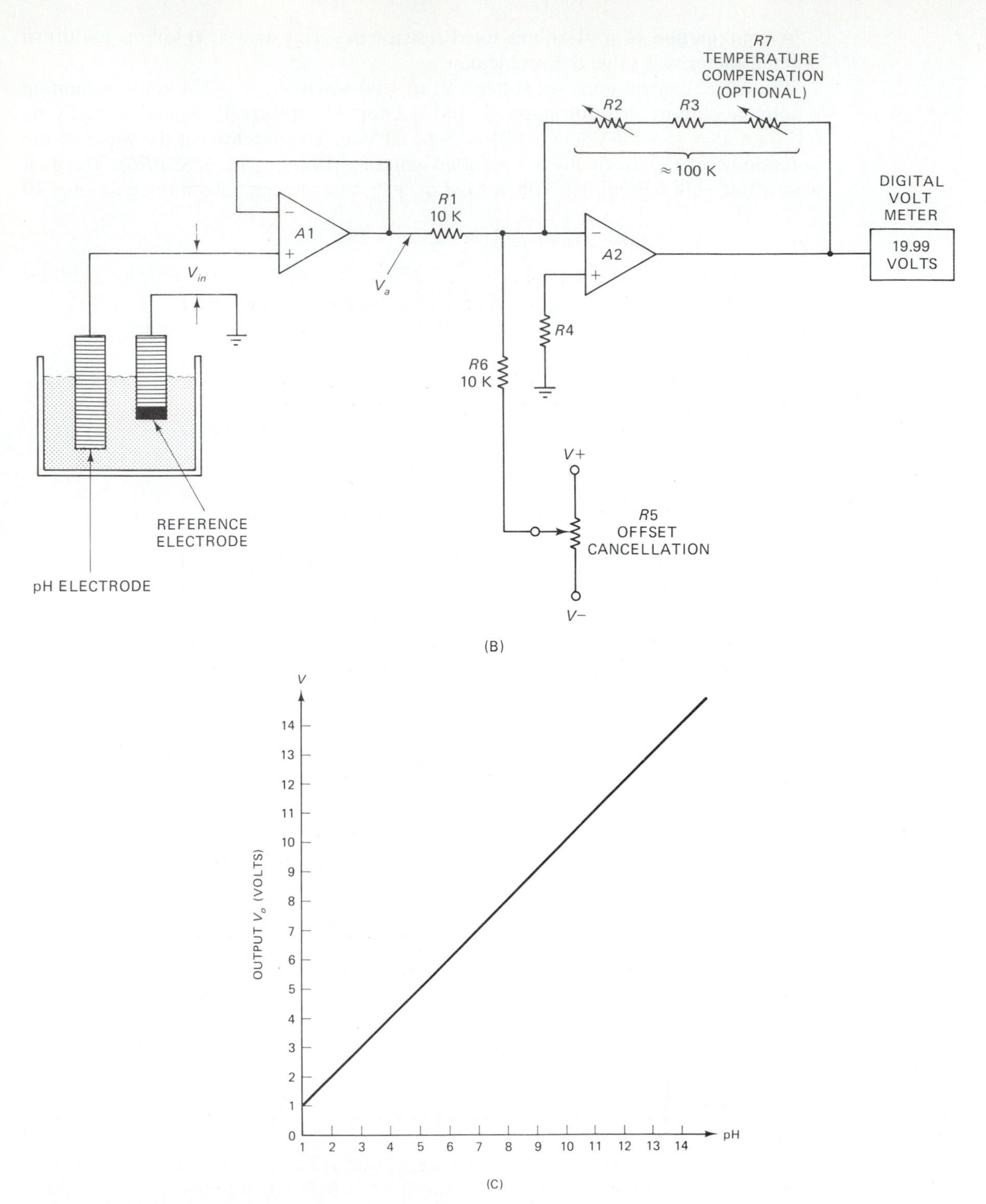


FIGURE 13-24 (*continued*)

Under the condition where $V_{in}$ is +1.4 volts (representing pH = 1) and $V_c$ = +1.3 Vdc, the total output voltage will be

$$V_b = -(R2 + R3)\left[\frac{V_a}{R1} + \frac{V_c}{R6}\right]$$

$$V_b = -\ (100\ \text{kohms})\left[\frac{(+1.4\ \text{volts})}{10\ \text{kohms}} + \frac{(-1.5\ \text{volts})}{10\ \text{kohms}}\right] \qquad (13\text{-}43)$$

$$V_b = (-100)(0.14 - 0.15) = +1.00\ \text{volts}$$

If a temperature compensation control is needed, another offset control is added to the circuit at $A2$. This control is operable from the instrument front panel. In some instruments, the temperature offset potential is provided by measuring the temperature of the solution being measured. In at least one case, a thermistor or PN junction temperature sensor is embedded in the pH electrode housing. The student is left to think about the design of an automatic temperature compensation circuit when a 10 mV/°K sensor is used and the temperature range will be +20°C to +45°C (NOTE: 0°C = 273°K).

## 13-4 SUMMARY

1. Transducers are devices which convert one form of energy (pressure, displacement force) to electrical energy for purposes of measurement, tabulation or control.
2. The phenomenon of *piezoresistivity* is often used for transducer. This phenomenon causes a change in resistance due to mechanical deformation of an electrical conductor.
3. A strain gage is a piezoresistive element which produces a resistance change in response to an applied force. Unbonded and bonded types are used.
4. Most strain gages and other resistive transducers are used in Wheatstone bridge circuits or modified bridges.
5. A DC differential amplifier can be used to process the signal output from a Wheatstone bridge strain gage.
6. Light transducers can be photovoltaic cells, photoresistors, photodiodes or phontotransistors. The latter two devices operate because PN junction leakage current is altered by light falling on the junction.

## 13-5 RECAPITULATION

Now return to the objectives and Prequiz questions at the beginning of the chapter and see how well you can answer them. If you cannot answer certain questions, mark them and review the appropriate parts of the text. Next, try to answer the questions and work the problems below, using the same procedure.

## 13-6 STUDENT EXERCISES

1. Design and build a transducer excitation source such as Fig. 13-11D to produce an output potential of 10.00 Vdc from an available zener diode (select $V_z$ between 4.2 and 6.8 Vdc).
2. Using the exercise above as a guide, produce a circuit that offers *bipolar* ±10 Vdc outputs.

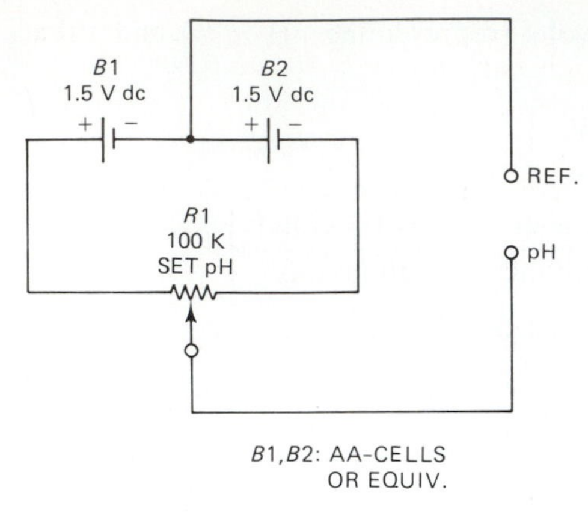

FIGURE 13-25

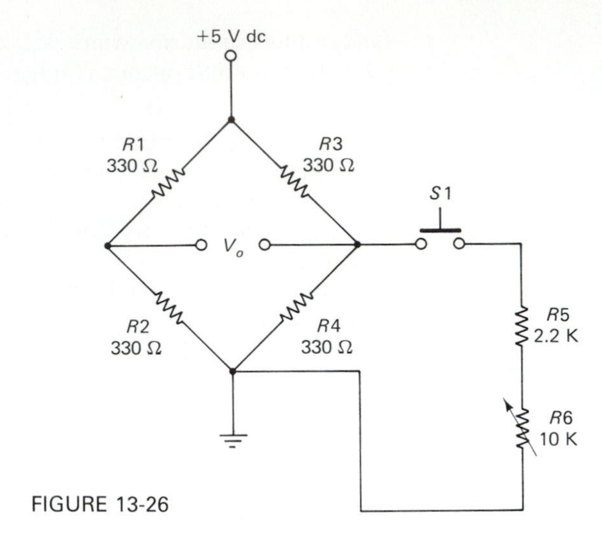


FIGURE 13-26

3. Design and build a gain-of-20 preamplifier for a microelectrode. Make provision for cancelling input capacitance (use a 20-foot length of coaxial cable between the signal generator and the amplifier to simulate electrode capacitance).
4. Design and build a photoelectric amplifier such as Fig. 13-18B. Select a photoresistor and a value of $V_{ref}$ that will produce an output signal in response to light.
5. Build a Wheatstone bridge such as Fig. 13-18C using available photoresistors. Scale *R*1 and *R*2 to the nominal photoresistor resistance in room light. Connect the outputs to a DC differential amplifier with a gain of 10. Examine what happens to the output signal of the bridge when first one and then the other photoresistor is covered.
6. Design and build a pH meter using the circuit of Fig. 13-25 as a simulated pH electrode. Measure circuit performance using various settings of the SET pH control.
7. Design and build an amplifier to process signals from the Wheatstone bridge transducer simulator shown in Fig. 13-26. The maximum output voltage when *R*6 = 0 ohms should be 2 volts. Measure the amplifier output (a) with *S*1 open and (b) with *S*1 closed (at various settings of *R*6).

## 13-7 QUESTIONS AND PROBLEMS

1. Resistance change under mechanical deformation is called ____________ .
2. A strain gage is constructed of a resistance wire held taut between two flexible supports. This is a ____________ strain gage.
3. A strain gage is constructed of semiconductor resistance material cemented to a thin metal diaphragm. This is a ____________ strain gage.
4. Draw a Wheatstone bridge circuit. If all elements are 300 ohms the output voltage will be ____________ volts.
5. Calculate the Thevanin's equivalent "looking back" resistance of a Wheatstone bridge transducer in which all four elements have a nominal resting value of 200 ohms.
6. A Wheatstone bridge consists one active strain gage element with the other three resistances being fixed and equal. The nominal resistance value of the strain gage (*R*) is equal to the

value of the three fixed resistors. If an excitation potential of +10.00 Vdc is applied, what will the output voltage be if the strain gage element changes to a value of $0.9R$?

7. Work the above problem for the cases of (a) two active elements and (b) four active elements.
8. A force transducer is used to measure the weight of small objects. The transducer elements have a resting resistance of 250 ohms. An excitation potential of +6 Vdc is applied. When an 800 mg mass is placed on the scale all four bridge arms change by 6.7 ohms. Calculate the output potential.
9. A Wheatstone bridge strain gage pressure transducer has a sensitivity factor of 48 uV/V/mmHg and an optimum excitation potential of 6 Vdc. Calculate the output voltage from this transducer at the optimum excitation when a pressure of 120 mmHg is applied.
10. Using the parameters in the question above, determine the gain required in the transducer amplifier to output 1.20 Vdc when the static pressure applied to the transducer is 120 mmHg.
11. A transducer excitation source such as Fig. 13 11D is built using a 1.36 Vdc bandgap diode as the reference source. Calculate the output voltage when $R1$ = 1000 ohms and $R2$ = 2700 ohms.
12. The full-scale output potential of a transducer is 400 uV when a force of 2600 dynes is applied. Calculate the amplifier scale factor required to produce an output voltage of 2.6 Vdc when this force is applied.
13. Calculate the source resistance of a 0.8 uM diameter glass microelectrode when immersed in normal physiological saline solution ($P$ = 71 ohm-cm).
14. A microelectrode has a source resistance of 126 kohms and a capacitance of 49 pF. Calculate the −3 dB frequency response.
15. Calculate the neutralization capacitance required to accommodate a 130-pF electrode in a gain of 10 amplifier.
16. The phototube operates on the basis of the ____________ effect.
17. A phototube that contains a number of special anodes called *dynodes* is a ____________ tube.
18. A transducer produces an output voltage when light shines on it. This is a ____________ cell.
19. Two photoresistors are connected into a Wheatstone bridge circuit. Light from a calibrated source falls on both photoresistors equally to calibrate. Then a sample is placed in the path between the light source and one photoresistor to make a measurement. This class of instrument is called a ____________ .
20. A Wheatstone bridge such as Fig. 13-3 is constructed with the following components: $R1$ = 10 kohms, $R2$ = 6.8 kohms, $R3$ = 6.8 kohms, and $R4$ = 5.6 kohms. Calculate the output voltage $V_o$ when a 1.5 Vdc battery is connected for $V$. What will the output voltage be if $R4$ is increased to 10 kohms?
21. Describe different methods of linearizing the output signal produced by a transducer.
22. Draw the circuit for a Wheatstone bridge transducer in which *balance* and *zero* controls are provided.
23. Draw the circuit for a microelectrode amplifier that provides for neutralization of electrode and cable capacitance.
24. What is the purpose of zener diode $D1$ in Fig. 13-11B?
25. What are the most desirable properties of the input amplifier ($A1$) in Fig. 13-25B?

CHAPTER 14

# Integrated Circuit Timers

## OBJECTIVES

1. Understand the internal operation of the 555 timer and related timers.
2. Be able to design monostable multivibrator circuits based on the 555 timer.
3. Be able to design astable multivibrator circuits based on the 555 timer.
4. Understand the wide range of potential applications of the 555 timer and related devices.

## 14-1 PREQUIZ

These questions test your prior knowledge of the material in this chapter. Try answering them before you read the chapter. Look for the answers (especially those you answered incorrectly) as you read the text. After you have finished studying the chapter try answering these questions again and those at the end of the chapter (see Section 14-12).

1. Calculate the duration of a 555 monostable multivibrator in which $R1$ = 56 kohms and $C1$ = 0.001 uF.
2. List the pin-outs of the 555 timer and briefly describe the operation of each.
3. An astable multivibrator made from a 555 IC timer uses the following components, $R1$ = 68 kohms, $R2$ = 33 kohms, and $C1$ = 0.01 uF. Calculate (a) the duty cycle and (b) the operating frequency.
4. A 555 timer is operated from a +12 Vdc regulated DC power supply. What is the minimum value of the load resistor that can be connected between pin 3 on the 555 and V+?

## 14-2 THE 555-FAMILY OF IC TIMERS

The integrated circuit (IC) timer is a class of chips which are extraordinarily efficient and easy to apply. These timers are based on the properties of the series *RC* timing network (see Chapter 12) and the voltage comparator. In some ways these devices are similar to circuits discussed in Chapter 12, but operational amplifiers are not explicitly used. A combination of voltage comparator circuits and digital circuits is used inside these chips. Although several devices are on the market, the most common and best known is the type 555 device. The 555 is made by a number of different semiconductor manufacturers. Today it remains one of the most widespread IC devices on the market.

The original 555s, first made by Signetics in 1970, included the SE-555, which operated a temperature range of $-55$ to $+125$°C, and the NE-555, which operated over the range 0 to $+70$°C. Several different designations are now commonly used for the 555 made by other manufacturers, including simply 555 and LM-555. A dual 555-class timer is also marketed under the number 556. There is also a low-power CMOS version of the 555 marketed as the LMC-555.

The 555 is a multipurpose chip which will operate at DC power supply potentials from $+5$ Vdc to $+18$ Vdc. The temperature stability of these devices is on the order of 50 PPM/°C (0.005%/°C). The output of the 555 can either sink or source up to 200 mA of current. It is compatible with TTL devices (when the 555 is operated from $+5$ Vdc power supply), CMOS devices, operational amplifiers, other linear IC devices, transistors, and most classes of solid-state devices. The 555 will also operate with most passive electronic components.

Several factors contribute to the popularity of the 555 device. Besides the versatile nature of the device, it is efficient because its operation is straightforward and the circuit designs are simple. Like the general purpose operational amplifier, the 555 usually works in a predictable manner, according to standard published equations.

The 555 operates in two different modes, *monostable* (one-shot) and *astable* (free-running). Figure 14-1A shows the astable mode output from pin 3 of the 555. The waveform is a series of squarewaves that can be varied in duty cycle over the range 50 to 99.9% and in frequency from less than 0.1 Hz to more than 100 KHz. Monostable operation (Fig. 14-1B) requires a trigger pulse applied to pin 2 of the 555. The trigger must drop from a level $>2(V+)/3$ down to $<(V+)/3$. Output pulse durations from microseconds up to hours are possible. The principal constraint on longer operation is the leakage resistance of the capacitor used in the external timing circuit.

## 14-3 PIN-OUTS AND INTERNAL CIRCUITS OF THE 555 IC TIMER

The package for the 555 device is shown in Fig. 14-2. Most 555's are sold in the eight-pin miniDIP package as shown, although some are found in the eight-pin metal can IC package. The latter are mostly the military specification temperature range SE-555 series. The pin-outs are the same on both miniDIP and metal can versions. The internal circuitry is shown in block form in Fig. 14-3. The following stages are found, two voltage comparators (COMP1 and COMP2), a reset-set (RS) control flip-flop (which can be reset from outside the chip through pin 4), an inverting output amplifier ($A1$) and a discharge transistor ($Q1$). The bias levels of the two comparators are

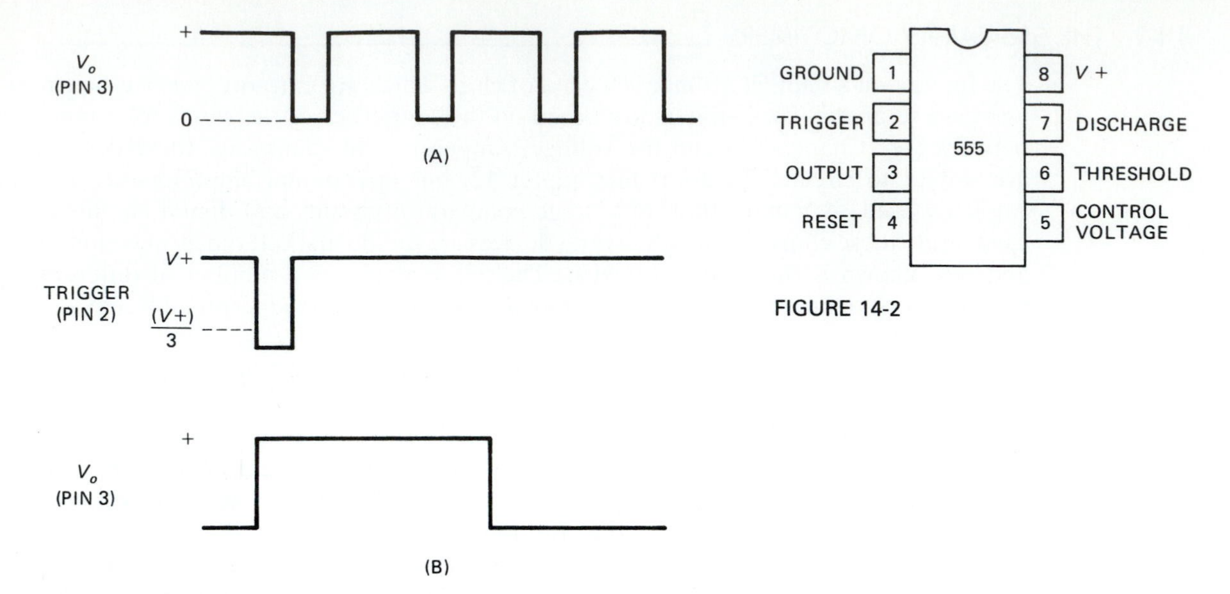


FIGURE 14-2

FIGURE 14-1

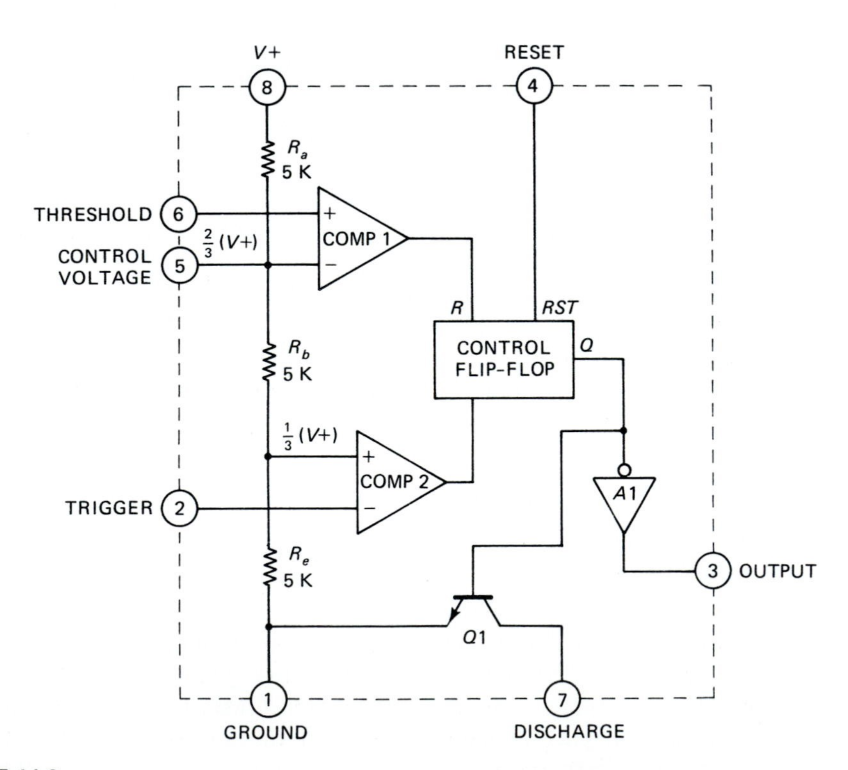


FIGURE 14-3

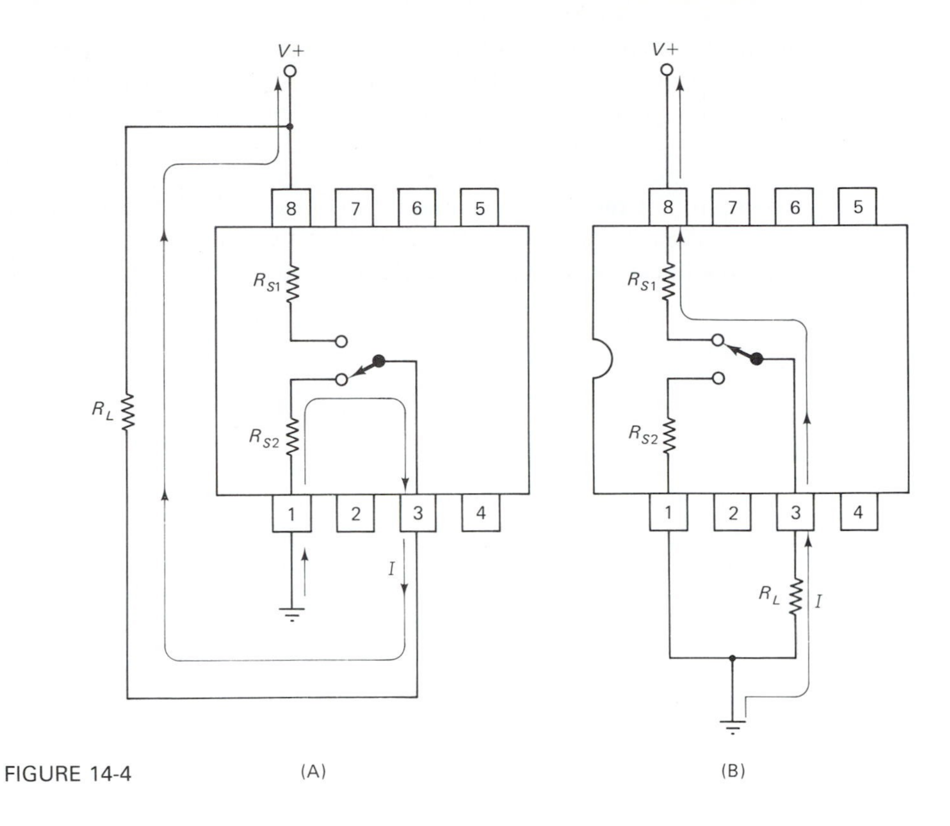


FIGURE 14-4 (A) (B)

determined by a resistor voltage divider ($R_a$, $R_b$, and $R_c$) between $V+$ and ground. The inverting input of COMP1 is set to $2(V+)/3$, and the noninverting input of COMP2 is set to $(V+)/3$.

Figures 14-3 and 14-4 show the pin-outs of the 555. In the descriptions below the term HIGH implies a level $>2(V+)/3$ and LOW implies a grounded condition ($V = 0$), unless otherwise specified in the discussion. These pins serve the following functions:

**Ground (Pin 1).** This pin serves as the common reference point for all signals and voltages in the 555 circuit both internal and external to the chip.

**Trigger (Pin 2).** The trigger pin is normally held at a potential $>2(V+)/3$. In this state the 555 output (pin 3) is LOW. If the trigger pin is brought LOW to a potential $<(V+)/3$, the output (pin 3) abruptly switches to the HIGH state. The output remains HIGH as long as pin 2 is LOW, but the output does not necessarily revert back to LOW immediately after pin 2 is brought HIGH again (see operation of the Threshold input below).

**Output (Pin 3).** The output pin of the 555 is capable of either sinking or sourcing current up to 200 mA. This is in contrast to other IC devices where the outputs will either sink or source current, but not both. Whether the 555 output operates as a sink or a source depends on the configuration of the external load. Figure 14-4 shows both types of operation.

In Fig. 14-4A the external load $R_L$ is connected between the 555 output and $V+$. Current only flows in the load when pin 3 is LOW. In that condition the external load is grounded through pin 1 and a small internal source resistance, $R_{s1}$. In this configuration the 555 output is a current sink.

The operation depicted in Fig. 14-4B is for a case where the load is connected between pin 3 of the 555 and ground. When the output is LOW the load current is zero. But when the output is HIGH, the load is connected to $V+$ through a small internal resistance $R_{s2}$ and pin 8. In this configuration the output serves as a current source.

**Reset (Pin 4).** The reset pin is connected to a preset input of the 555 internal control flip-flop. When a LOW is applied to pin 4 the output of the 555 (pin 3) switches immediately to a LOW state. In normal operation pin 4 is connected to $V+$ in order to prevent false resets from noise impulses.

**Control Voltage (Pin 5).** This pin normally rests at a potential of $2/3-(V+)$ due to an internal resistive voltage divider (see $R_a$ through $R_c$ in Fig. 14-3). Applying an external voltage to this pin or connecting a resistor to ground, will change the duty cycle of the output signal. If not used, then pin 5 should be decoupled to ground through a 0.01-uF to 0.1-uF capacitor.

**Threshold (Pin 6).** This pin is connected to the noninverting input (+IN) of comparator COMP1. It is used to monitor the voltage across the capacitor in the external *RC* timing network. If pin 6 is at a potential of $<2(V+)/3$, the output of the control flip-flop is LOW, and the output (pin 3) is HIGH. Alternatively, when the voltage on pin no. 6 is $\geqq 2(V+)/3$, the output of COMP1 is HIGH and chip output (pin 3) is LOW.

**Discharge (Pin 7).** The discharge pin is connected to the collector of NPN transistor $Q1$ and the emitter of $Q1$ is connected to the ground (pin 1). The base of $Q1$ is connected to the NOT-$Q$ output of the control flip-flop. When the 555 output is HIGH, the NOT-$Q$ output of the control flip-flop is LOW, so $Q1$ is turned off. The *c-e* resistance of $Q1$ is very high under this condition, so it does not appreciably affect the external circuitry. But when the control flip-flop NOT-$Q$ output is HIGH, however, the 555 output is LOW and $Q1$ is biased hard on. The *c-e* path is in saturation, so the *c-e* resistance is very low. Pin 7 is effectively grounded under this condition.

**V+ Power Supply (Pin 8).** The DC power supply is connected between ground (pin 1) and pin 8. Pin 8 is positive. In good practice a 0.1 uF to 10 uF decoupling capacitor will be used between pin 8 and ground.

## 14-4 MONOSTABLE OPERATION OF THE 555 IC TIMER

A monostable multivibrator (MMV), also called the one-shot, circuit produces a single output pulse of fixed duration when triggered by an input pulse. Operational amplifier versions of the monostable circuit were discussed in Chapter 12. The output of the one-shot will snap HIGH following the trigger pulse and will remain HIGH for a fixed, predetermined duration. When this time expires the one-shot is timed-out so it snaps LOW again. The output of the one-shot will remain low indefinitely unless another trigger pulse is applied to the circuit. The 555 can be operated as a monostable multivibrator by suitable connection of the external circuit.

Figure 14-5 shows the operation of the 555 as a monostable multivibrator. In order to make the operation of the circuit easier to understand, Fig. 14-5A shows the internal circuitry as well as the external circuitry, Fig. 14-5B shows the timing diagram for this circuit, and Fig. 14-5C shows the same circuit in a more conventional schematic diagram format.

The two internal comparators are biased to certain potential levels by a series voltage divider consisting of resistors $R_a$, $R_b$, and $R_c$. The inverting input of voltage comparator COMP1 is biased to $2(V+)/3$, while the noninverting input of COMP2 is biased to $(V+)/3$. These levels govern the operation of the 555 in whichever mode is selected. An external timing network ($R1C1$) is connected between $V+$ and the noninverting input of COMP1 via pin 6. Also connected to pin 6 is pin no. 7, which has the effect of connecting the transistor across capacitor $C1$. If the transistor is turned on, the capacitor is exposed to a very low resistance short circuit through the *c-e* path of the transistor.

When power is initially applied to the 555, the voltage at the inverting output of COMP1 will go to $2(V+)/3$ and the noninverting input of COMP2 will go to $(V+)/3$. The control flip-flop is in the reset condition, so the NOT-$Q$ output is HIGH. Because this flip-flop is connected to output pin 3 through an inverting amplifier ($A1$), the output is LOW at this point. Also, because NOT-$Q$ is HIGH, transistor $Q1$ is biased into saturation, creating a short circuit to ground across external timing capacitor $C1$. The capacitor remains discharged in this condition ($V_c = 0$).

If a trigger pulse is applied to pin 2 and if it drops to a voltage $<(V+)/3$ as shown in Fig. 14-5B, comparator COMP2 is exposed to an inverting input which is less positive than the noninverting input, so the output of COMP2 snaps HIGH. This action sets the control flip-flop, forcing the NOT-$Q$ output LOW, and therefore the 555 output HIGH. The LOW at the output of the control flip-flop also means transistor $Q1$ is now unbiased, so the short across the external capacitor is removed. The voltage across $C1$ begins to rise (see Figs. 14-5B and 14-5D). The voltage will continue to rise until it reaches $2(V+)/3$, at which time comparator COMP1 will snap HIGH causing the flip-flop to reset. When the flip-flop resets, its NOT-$Q$ output drops LOW again, terminating the output pulse and returning the capacitor voltage to zero. The 555 will remain in this state until another trigger pulse is received.

The timing equation for the 555 can be derived in exactly the same manner as the equations used with the operational amplifier MMV circuits. The basic equation was discussed in Chapter 12 and relates the time required for a capacitor voltage to rise from a starting point ($V_{C1}$) to an end point ($V_{C2}$) with a given $RC$ time constant

$$T = -RC \text{ LN}\left[\frac{V - V_{C2}}{V - V_{C1}}\right] \tag{14-1}$$

In the 555 timer the voltage source is $V+$, the starting voltage is zero, and the trip-point voltage for comparator COMP1 is $2(V+)/3$. Equation (14-1) can therefore be rewritten as

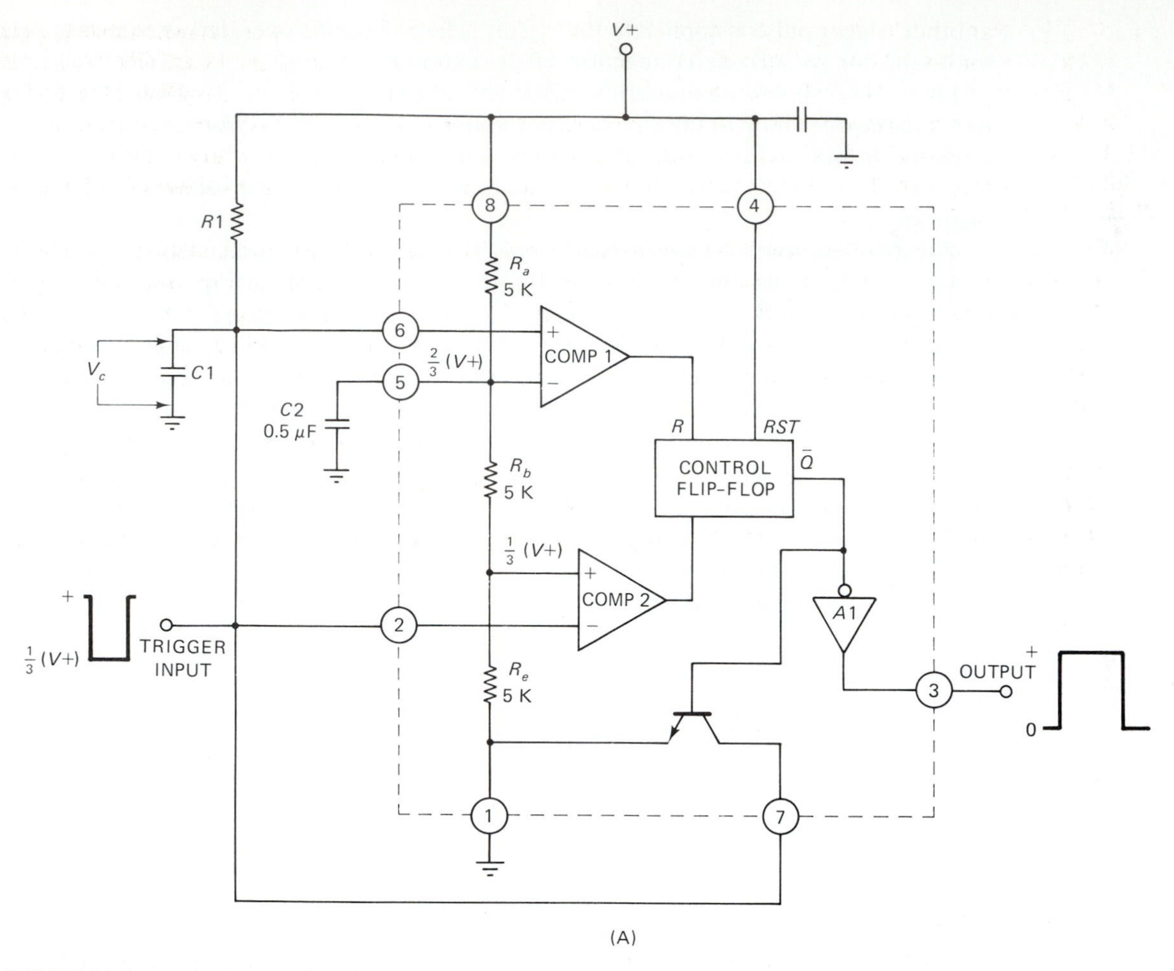


FIGURE 14-5

$$T = -R1C1 \text{ LN}\left[\frac{V - V_{C2}}{V - V_{C1}}\right] \tag{14-2}$$

$$T = -R1C \text{ LN}\left[\frac{(V+) - 2(V+)/3}{V+}\right] \tag{14-3}$$

$$T = -R1C1 \text{ LN}[1 - 0.667] \tag{14-4}$$

$$T = -R1C1 \text{ LN}[0.333] \tag{14-5}$$

$$T = 1.1R1C1 \tag{14-6}$$

### 14-4.1 Input Triggering Methods for the 555 MMV Circuit

The 555 MMV circuit triggers by bringing pin 2 from a positive voltage down to a level $<(V+)/3$. Triggering can be accomplished by applying a pulse from an external signal source or through other means. Figure 14-6 shows the circuit for a simple

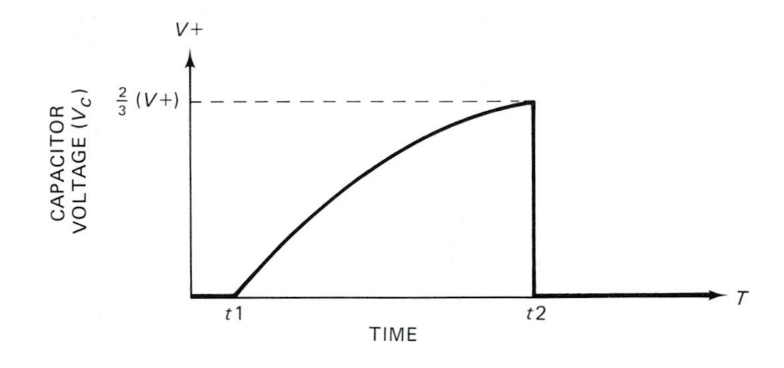


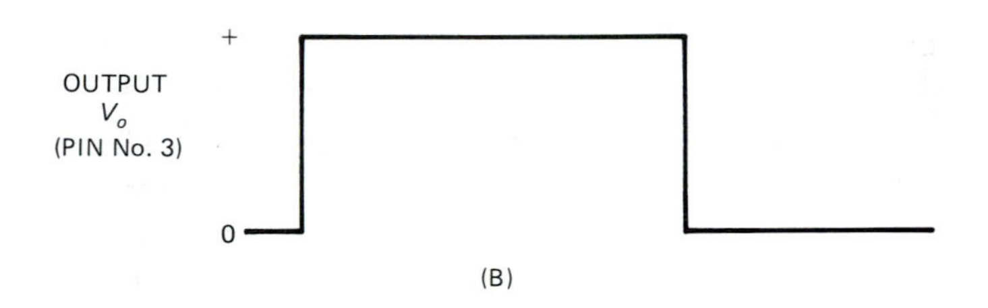


(B)

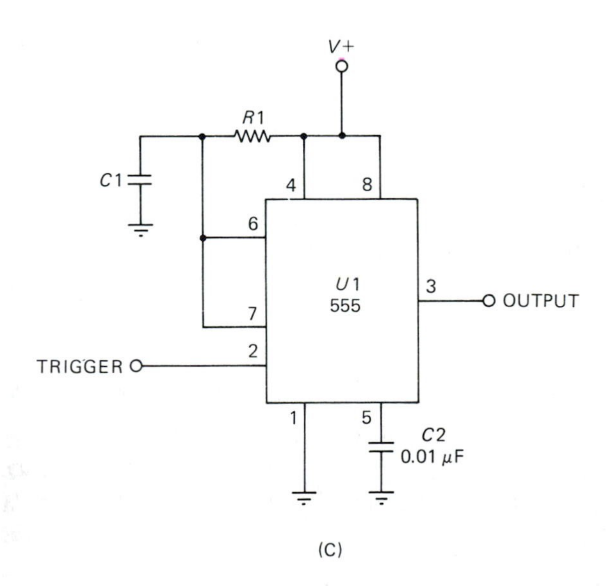


(C)

FIGURE 14-5 (*continued*)

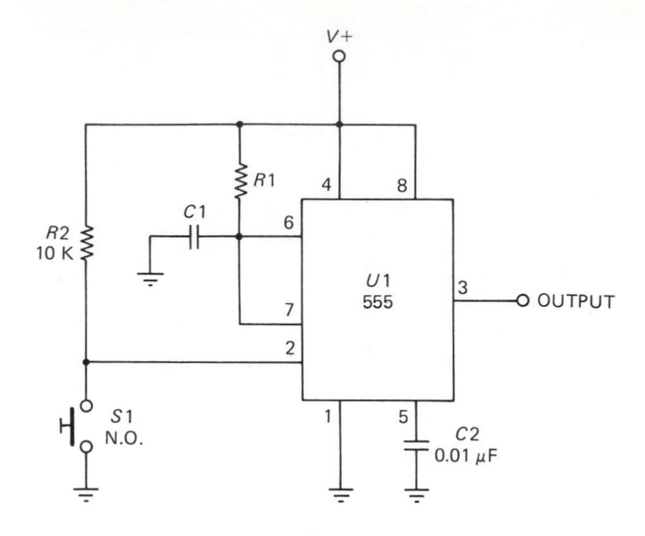


FIGURE 14-6

pushbutton switch trigger circuit. A pullup resistor ($R2$) is connected between pin 2 and $V+$. If normally open (N.O.) pushbutton switch $S1$ is open, the trigger input is held at a potential very close to $V+$. But when $S1$ is closed, pin 2 is brought LOW to ground potential. Because pin 2 is now at a potential less than $(V+)/3$ the 555 MMV will trigger. This circuit can be used for contact debouncing.

A circuit for inverting the trigger pulse applied to the 555 is shown in Fig. 14-7. In this circuit an NPN bipolar transistor is used in the common emitter mode to inverting the pulse. Again, a pullup resistor is used to keep pin 2 at $V+$ when the transistor is turned off. But when the positive polarity trigger pulse is received at the base of transistor $Q1$, the transistor saturates which forces the collector (and pin 2 of the 555) to near ground potential.

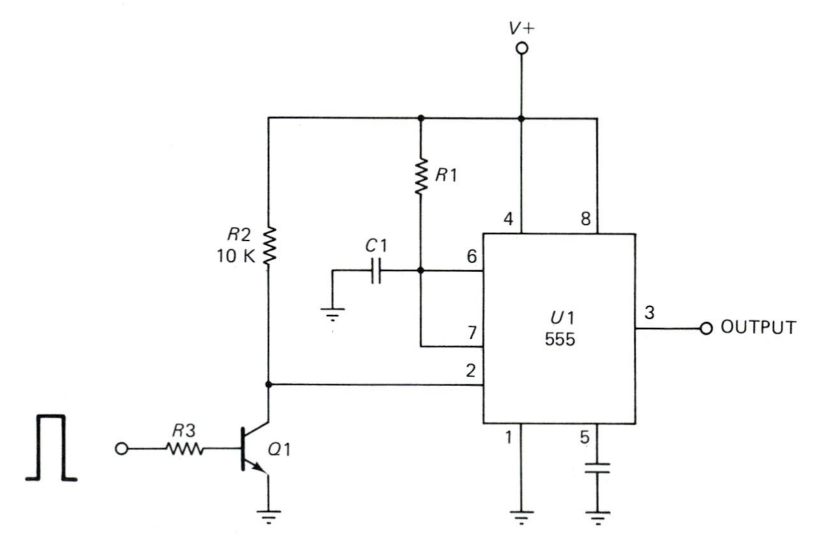


FIGURE 14-7

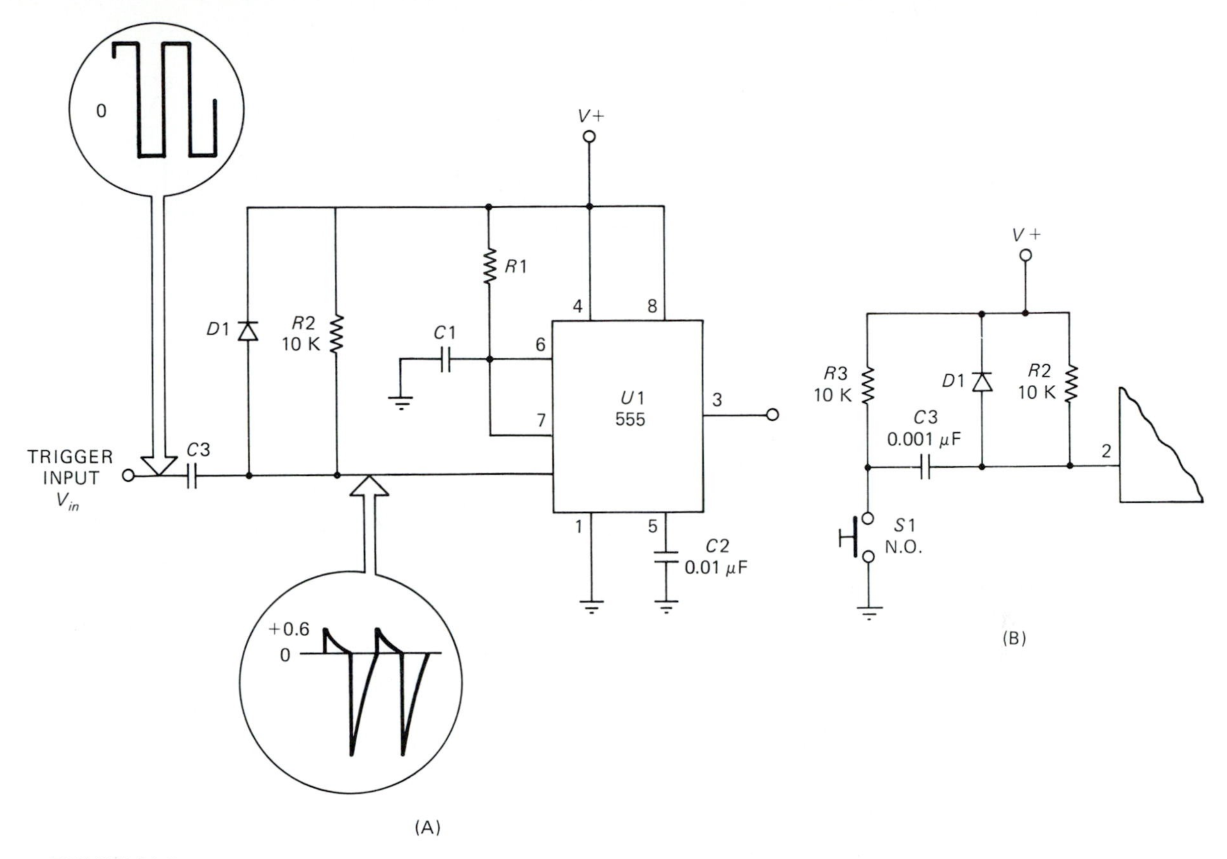


FIGURE 14-8

Figure 14-8 shows two AC-coupled versions of the trigger circuit. In these circuits a pullup resistor normally keeps pin 2 at $V+$. But when a pulse is applied to the input end of capacitor $C3$, a differentiated version of the pulse is created at the trigger input of the 555. Diode $D1$ clips the positive going spike to 0.6 or 0.7 volts, passing only the negative going pulse to the 555. If the negative going spike can sufficiently counteract the positive bias provided by $R2$ and force the voltage lower than $(V+)/3$, the 555 will trigger. A pushbutton switch version of this same circuit is shown in Fig. 14-8B.

A touchplate trigger circuit is shown in Fig. 14-9A. The pullup resistor $R2$ has a very high value (22 megohms shown here). The touchplate consists of a pair of closely spaced electrodes. As long as there is no external resistance between the two halves of the touchplate, the trigger input of the 555 remains at $V+$. But when a resistance is connected across the touchplate, the voltage ($V1$) drops to a very low value. If the average finger resistance is about 20 kohms, the voltage drops to

$$V1 = \frac{(V+)\,(20\text{K})}{R2 + 20\text{K}} \tag{14-7}$$

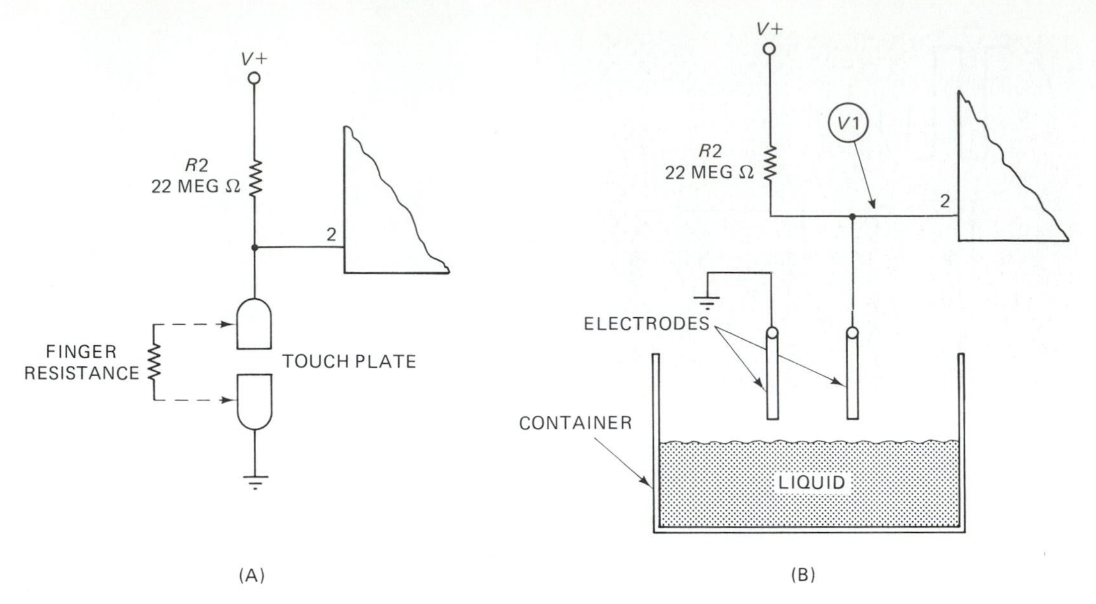


FIGURE 14-9

Which, when $R2 = 22$ megohms, to $0.0009\ (V+)$—which is certainly less than $(V+)/3$.

The same concept is used in the liquid level detector shown in Fig. 14-9B. Once again a 22-megohm pullup resistor is used to keep pin 2 at $V+$ under normal operation. When the liquid level rises enough to short out the electrodes, however, the voltage on pin 2 ($V1$) drops to a very low level, forcing the 555 to trigger.

### 14-4.2 Retriggerable Operation of the 555 MMV Circuit

The 555 is a nonretriggerable monostable multivibrator. If additional trigger pulses are received prior to the time-out of the output pulse (see Fig. 14-10A), then the additional pulses have no effect on the output. But the first pulse after time-out will cause the output to again snap HIGH. In Fig. 14-10A the trigger input signal is a square wave signal, so the output of the 555 is LOW only for the duration of one trigger pulse before snapping HIGH again.

The circuit in Fig. 14-10B will permit retriggering of the 555 device. An external NPN transistor ($Q2$ in Fig. 14-10B) is connected with its c-e path across timing capacitor $C1$. In this sense it mimics the internal discharge transistor. A second transistor ($Q1$) is connected to the trigger input of the 555 in a manner similar to Fig. 14-7. The bases of the transistors form the trigger input. When a positive pulse is applied to the combined trigger line both transistors become saturated. Any charge in $C1$ is immediately discharged and pin 2 is triggered by the collector of $Q1$ being dropped to less than $(V+)/3$.

As long as no further trigger pulses are received, this circuit behaves like any other 555 MMV circuit. But if a trigger pulse is received prior to the time-out defined by Eq. (14-6), the transistors are forward-biased once again. $Q1$ retriggers the 555, while $Q2$ dumps the charge built up in the capacitor. Thus, the 555 retriggers.

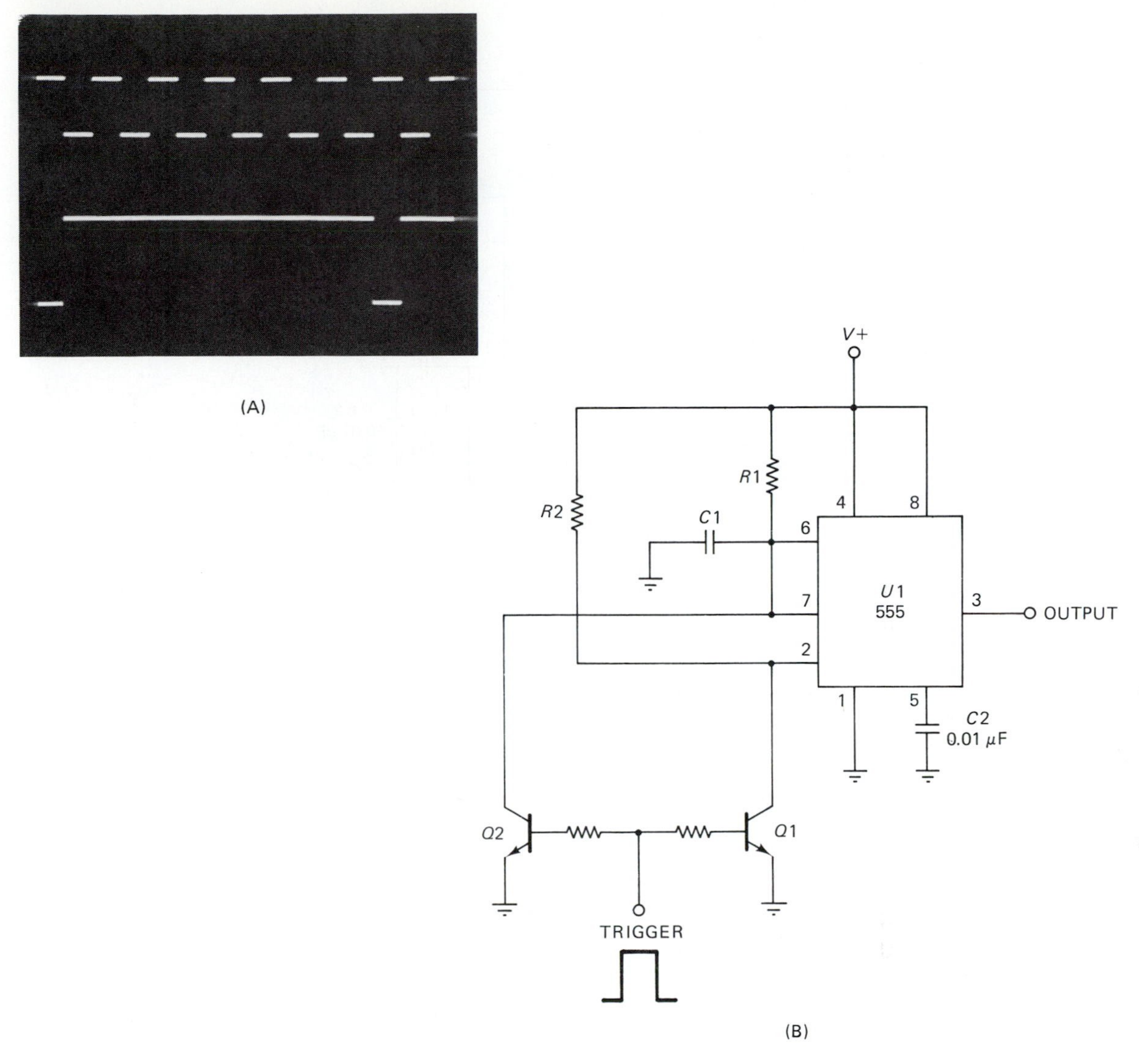


FIGURE 14-10

## 14-5 APPLICATIONS FOR THE 555 ONE-SHOT CIRCUIT

The MMV is a one-shot circuit which produces a single output pulse for every trigger input pulse, except for those that fall inside the output pulse and any associated refractory period. There are numerous applications for these circuits.

### 14-5.1 Missing Pulse Detector

A *missing pulse detector* circuit remains dormant as long as a series of trigger pulses are received, but it will produce an output pulse when an expected pulse is missing. These circuits are used in a variety of applications including alarms. For example, in

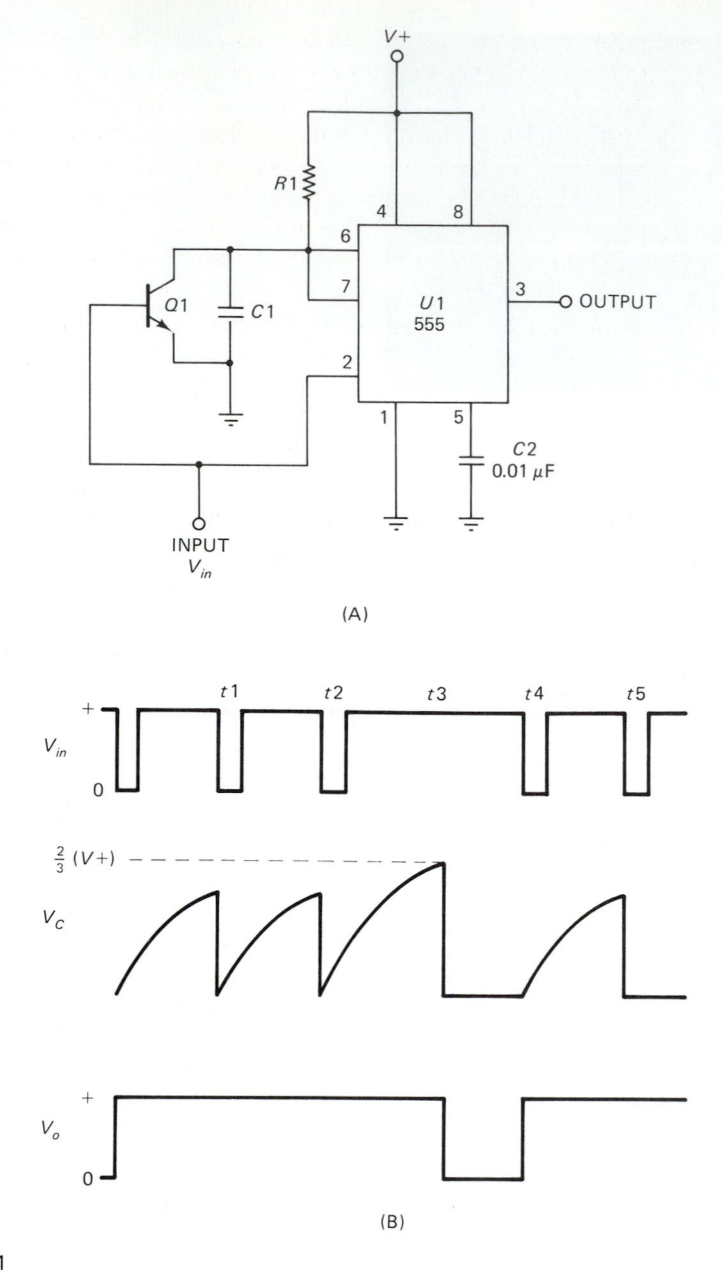


FIGURE 14-11

a bottling plant, soft drink cans are packaged into "six-packs." As each can passes a photocell, a pulse is generated to the input of a missing pulse detector. If a pulse is not received, however, the machine knows the count is one can short and issues an alarm or corrective action.

A similar circuit can be used in a wildlife photography system. An infrared light emitting diode (LED) is modulated or chopped with a pulse waveform. As long as the pulse is received at the sensor, the circuit is dormant. But if an animal passes through the IR beam even briefly, a missing pulse detector will sense its presence and issue an output that fires a camera flashgun and electrical shutter control.

Figure 14-11A shows the circuit for a missing pulse detector based on the 555 IC timer. Fig. 14-11B shows the timing waveforms. This circuit is the standard 555 MMV, except a discharge transistor is shunted across capacitor $C1$. When a pulse is applied to the input it will trigger the 555 and turn on $Q1$, causing the capacitor to discharge. After the first input pulse the output of the 555 snaps HIGH and remains HIGH until a missing pulse is detected.

Circuit action can be seen in Fig. 14-11B. At times $t1$ and $t2$ input pulses are received. As long as ($t2$-$t1$) is less than the time required for $C1$ to charge up to $2(V+)/3$ the 555 will never time-out. But if a pulse is missing, as at $t3$, the capacitor voltage continues to rise to the critical $2(V+)/3$ threshold value. When $V_c$ reaches this point the 555 will time-out, forcing its output LOW. The output remains LOW until a subsequent input pulse is received ($t4$), at which time $Q1$ turns on again and forces the capacitor to discharge. The cycle can then continue as before.

### 14-5.2 Pulse Position Circuit

A *pulse positioner* is a circuit which allows adjustment of the timing of a pulse to coincide with some external event. For example, in some instrumentation circuits a short pulse must be positioned to a certain point on a sine wave (for example, at the peak). The pulse positioner could be triggered from the zero-crossing of the sinewave, and then adjusted to place the output pulse where it is needed.

Figure 14-12A shows the concept of pulse positioning using two one-shot circuits, labelled OS1 and OS2. The circuit is shown in Fig. 14-12B. The repositioned pulse is not actually the original pulse, it is a recreated pulse with similar characteristics. The input pulse is used to trigger OS1. The duration of this one-shot circuit is fixed to the delay required of the repositioned pulse. If the delay must be variable, resistor $R1$ is made variable. When OS1 times out it will trigger OS2. The output pulse of OS2 is set to the parameters of the original input pulse. An inverter circuit is used to make the output of OS2 have the same polarity as the trigger pulse at the input of OS1. To an outside observer the pulse appears to have been repositioned, although in fact it was merely recreated at time $T$ (the delay period in Fig. 14-12B).

### 14-5.3 Tachometry

*Tachometry* is the measurement of repetition rate. In an automotive tachometer, for example, the instrument counts the pulses produced by the ignition coil to measure the engine speed in RPM. In medical instruments it is often necessary to measure heart or respiration rate electronically using tachometry circuits. A heart rate meter (*cardiotachometer*) measures the heart rate in beats per minute (BPM). The respiration meter (*pneumotachometer*) measures breathing rate in breaths per second.

There is a certain commonality among nondigital tachometer circuits. It does not matter whether the rate is audio, sub-audio, or above the audio rate, the basic circuit design is the same.

Figure 14-13 shows the basic tachometer circuit in block diagram form. Not all of these stages will be present in all circuits. For example, the AC amplifier and Schmitt

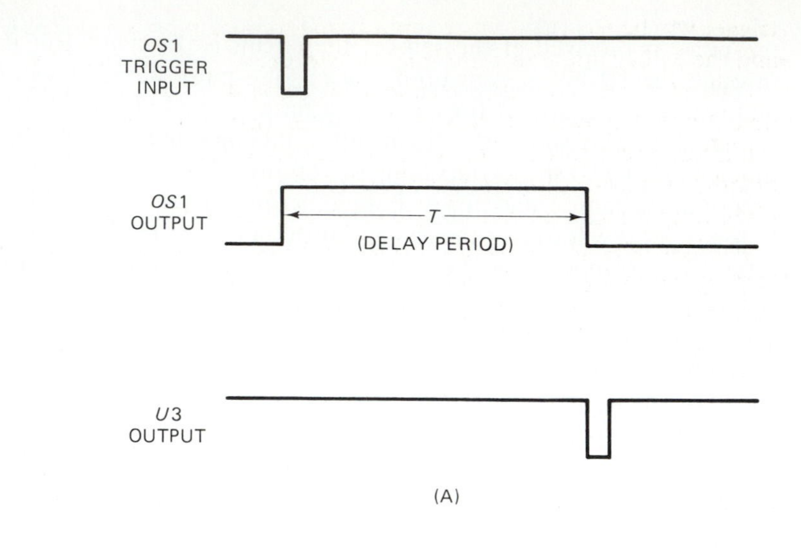


(A)

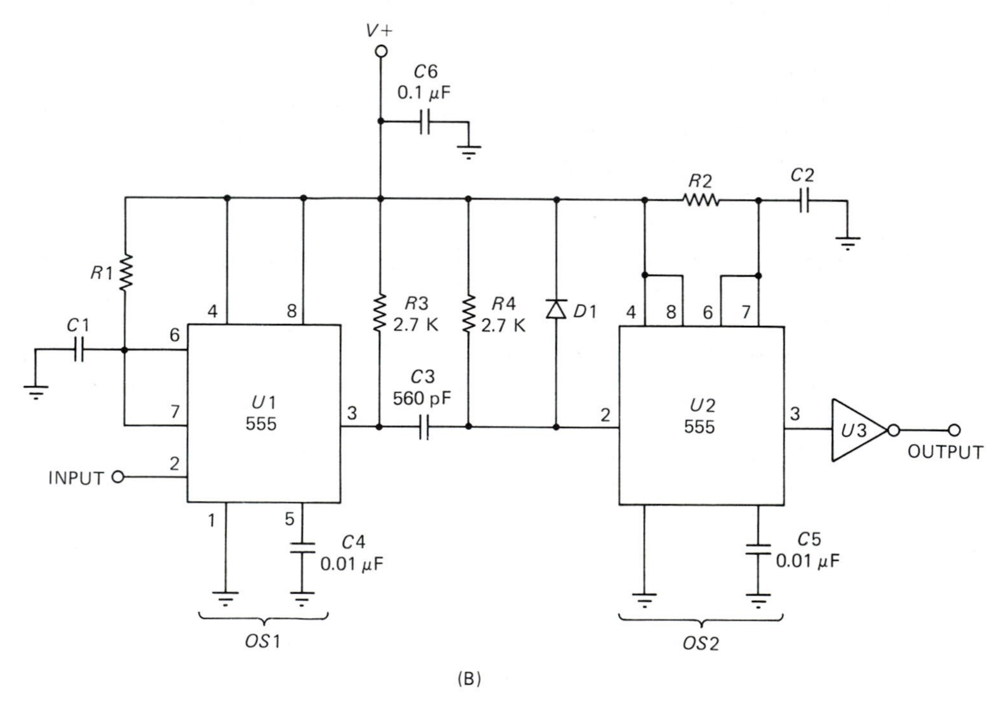


(B)

FIGURE 14-12

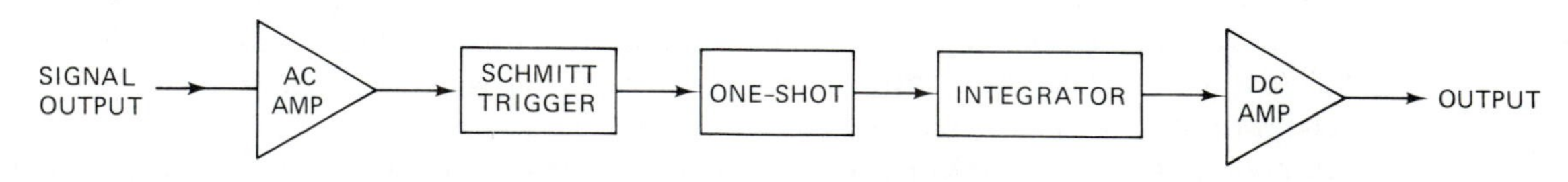


FIGURE 14-13

trigger will be used only when input signal conditioning is needed. The one-shot circuit and the Miller integrator are basic, however, so they are used for all of these circuits.

The idea is to convert a frequency or repetition rate to an analog voltage. This is done by first converting the signal to pulse form. The AC input amplifier is used only if it is necessary to scale the input signal to a level where it will drive a Schmitt trigger or other squaring circuit. The purpose of the following stage is to produce a square wave output signal at the same frequency as the input signal.

The purpose of the stages in Fig. 14-13 is to produce a DC voltage output which is proportional to the input frequency or pulse repetition rate. The integrator is designed to produce an output voltage that is the time-average of the input signal. That is, the integrator output is proportional to the area under the input signal. The job of the tachometer designer is to create a circuit in which the only variable is the frequency or repetition rate of the input signal.

The output pulse of a one-shot circuit has a constant amplitude and constant duration. The area under the pulse is the product of the amplitude and duration, so from pulse to pulse the area does not change. If the one-shot is constantly retriggered by the input signal, the total area under the resultant pulse train is a function of only the number of pulses. Therefore, the time-average of the integrator output will be a DC voltage which is proportional to the input frequency.

Figure 14-14 shows a practical application of the tachometer circuit. It was used to demodulate the audio frequency modulated signal from an instrumentation telemetry set. A similar circuit (but not based on the 555) was once popular as a *coil-less* FM detector in communications and broadcast receivers. These *pulse counting detectors* operated at 10.7 MHz (a commonly used FM IF frequency in receivers). The circuit shown in Fig. 14-14 was used to demodulate a human electrocardiograph (ECG) signal transmitted over telephone lines. The ECG is an analog voltage waveform, and was used to frequency modulate an audio voltage controlled oscillator (VCO) at the transmit end. Normally, the ECG has too low a Fourier frequency content (0.05 Hz to 100 Hz) to pass over the restricted passband of the telephone lines (300 Hz to 3000 Hz). But when used to frequency modulate a 1500 Hz carrier, however, the signal passed easily over telephone circuits.

The circuit for the demodulator circuit is shown in Fig. 14-14A. The input waveshaping function is performed by an LM-311 voltage comparator. The job of the LM-311 is to square the 200 mV peak-to-peak sine wave input signal so it is capable of triggering the 555 (*U*2). In this mode the LM-311 is operating as a zero-crossing detector circuit.

The output of the 555 is a pulse train with constant amplitude and duration. These pulses vary only in repetition rate, which is the same as the frequency of the input signal. The 555 output pulses are integrated in a passive RC integrator (*R*5-*R*7/*C*4-*C*6). The output of the integrator is a DC voltage which is a linear function of input frequency (see Fig. 14-14B). This DC voltage can be scaled to any desired level.

A related circuit is shown in Fig. 14-15. This 555-based tachometer is used to measure audio frequency over three ranges, DC to 50 Hz, DC to 500 Hz, and DC to 5000 Hz. It uses the same form of input signal conditioning as the previous circuit and uses a 555 as the one-shot circuit. The integration function is taken up by the combination of RC network *R*4/*C*4 and the mechanical inertia of the meter (*M*1) movement.

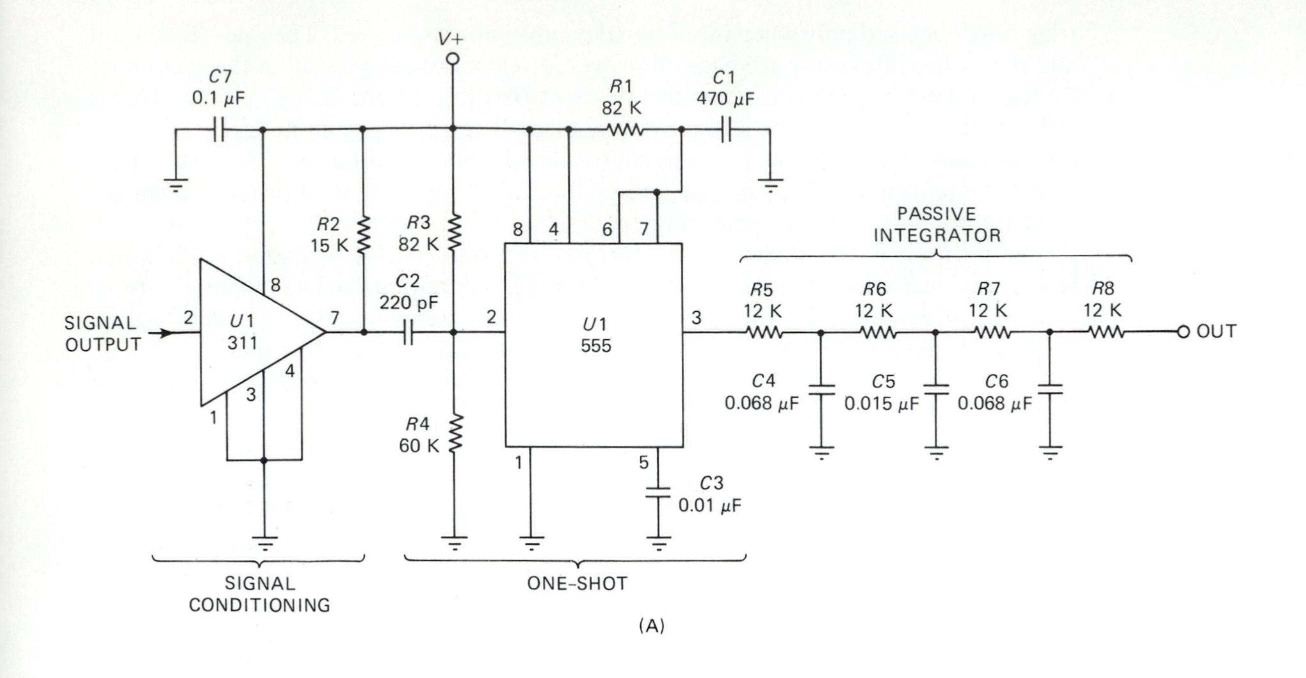


(A)

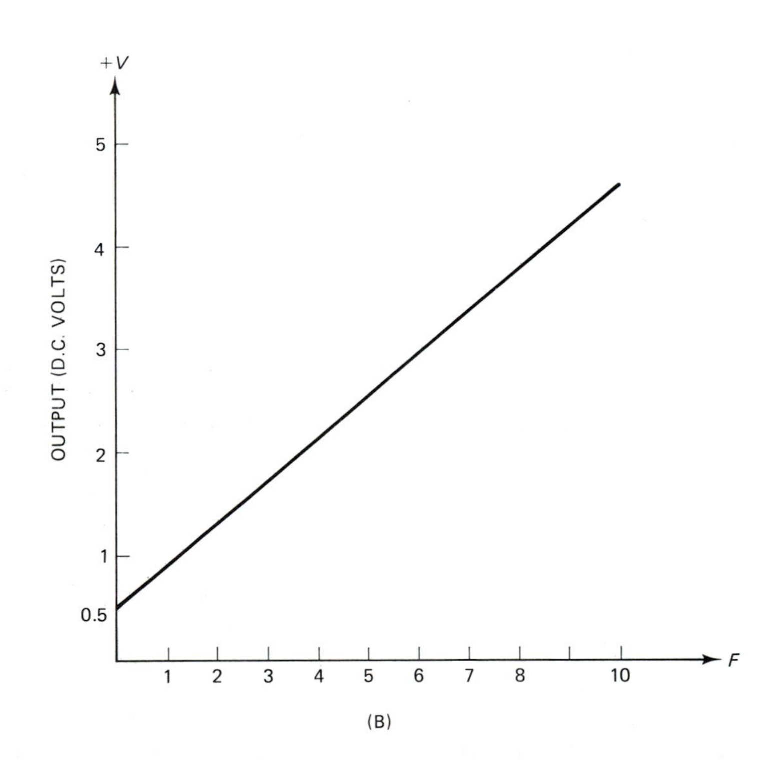


(B)

FIGURE 14-14

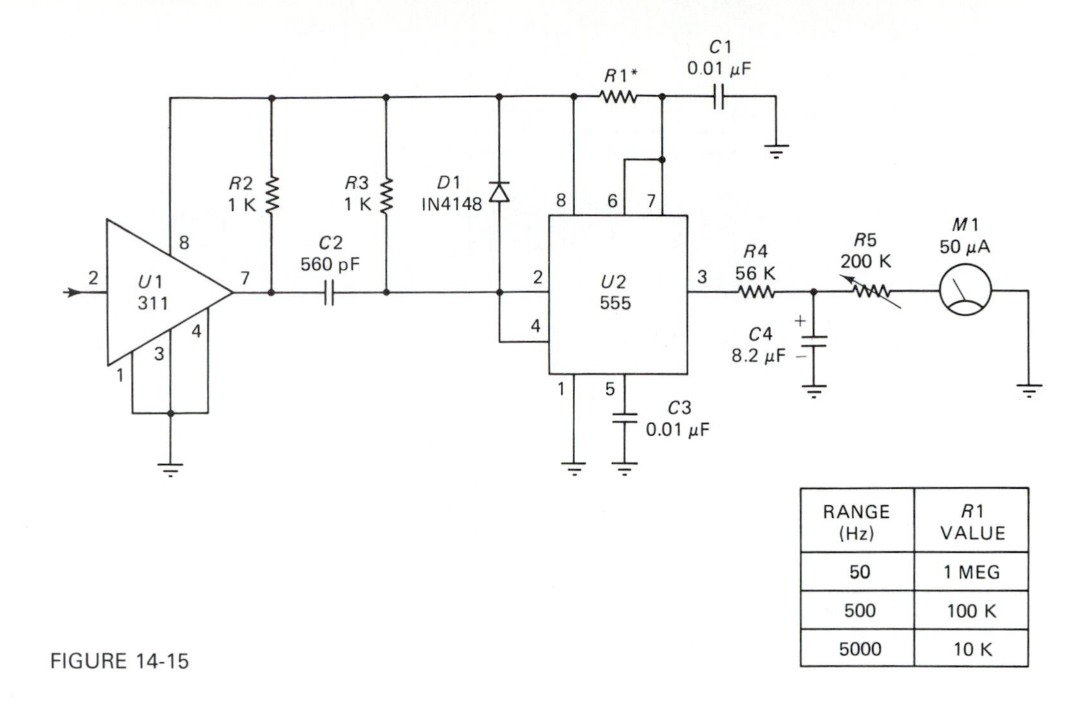


| RANGE (Hz) | R1 VALUE |
|---|---|
| 50 | 1 MEG |
| 500 | 100 K |
| 5000 | 10 K |

FIGURE 14-15

## 14-6 ASTABLE OPERATION OF THE 555 IC TIMER

An *astable multivibrator* (AMV) is a free-running circuit which produces a square wave output signal. The 555 can be connected to produce a variable duty cycle AMV circuit (Fig. 14-16). A version of the circuit showing the internal stages of the 555 is shown in Fig. 14-16A. The circuit as it normally appears in schematic drawings is shown in Fig. 14-16B. This circuit is an AMV because the threshold and trigger pins (6 and 2) are connected, forcing the circuit to be self-retriggering.

Under initial conditions at turn-on, the voltage across timing capacitor $C1$ is zero and the biases on COMP1 and COMP2 are set to $2(V+)/3$ and $(V+)/3$, respectively, by the internal resistor voltage divider ($R_a$, $R_b$, and $R_c$). The output of the 555 is HIGH under this condition, so $C1$ begins to charge through the combined resistance $[R1 + R2]$. On discharge, however, transistor $Q1$ shorts the junction of $R1$ and $R2$ to ground, so the capacitor discharges through only $R2$. The result is the waveform shown in Fig. 14-16C. The time the output is HIGH is $t1$. The LOW time is $t2$. The period ($T$) of the output square wave is the sum of these two durations: $T = (t1 = t2)$.

As with all similar $RC$-timed circuits, the oscillating frequency is determined from

$$T = -RC \text{ LN}\left[\frac{V - V_{C2}}{V - V_{C1}}\right] \tag{14-8}$$

When the output is HIGH ($t2$ in Fig. 14-16C), the resistance $R$ is $[R1 + R2]$ and the capacitance ($C$) is $C1$. Because of the internal biases of the voltage comparator stages of the 555, the capacitor will charge from $(V+)/3$ to $2(V+)/3$ and then discharge back to $(V+)/3$ on each cycle. Thus, Eq. (14-8) can be rewritten

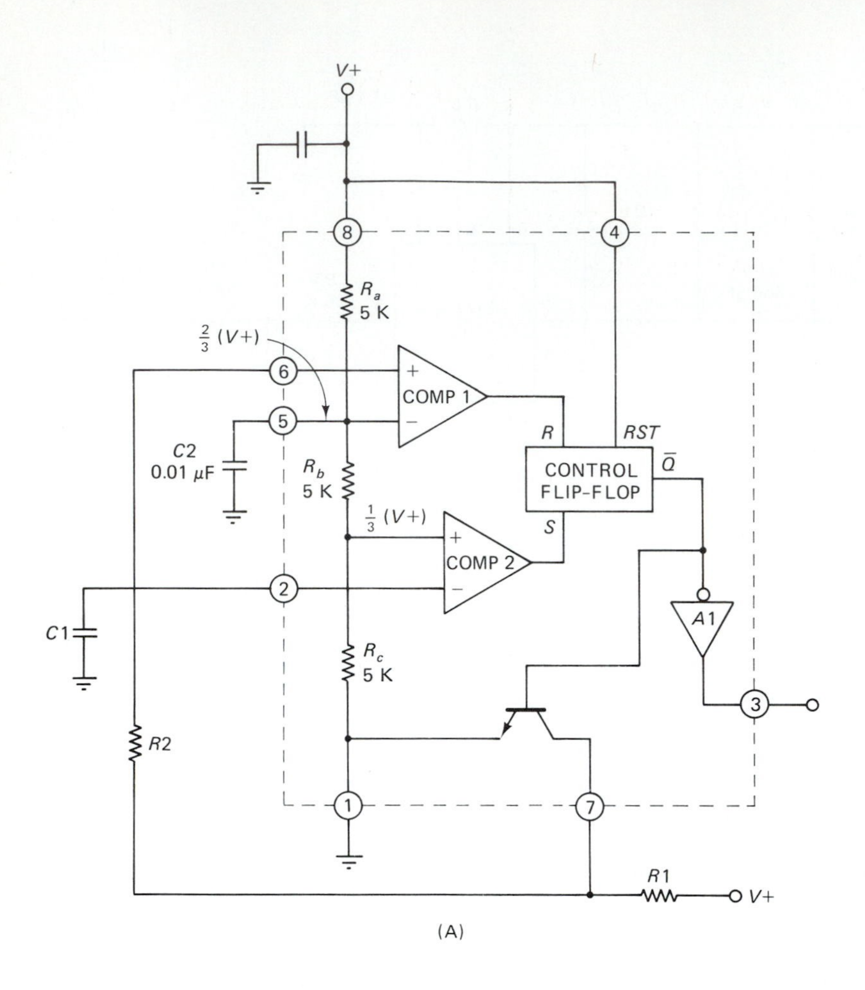
V+
8
4
$R_a$
5 K
$\frac{2}{3}$ (V+)
6
+
COMP 1
−
5
R
RST
C2
0.01 μF
$R_b$
5 K
CONTROL
FLIP-FLOP
$\bar{Q}$
$\frac{1}{3}$ (V+)
S
+
COMP 2
−
2
C1
A1
$R_c$
5 K
3
R2
1
7
R1
V+
(A)

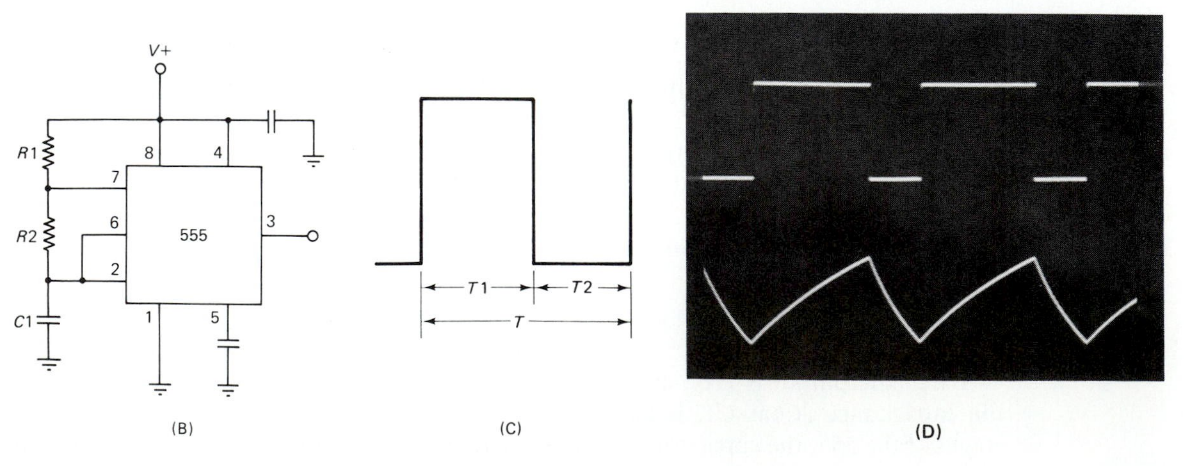
V+
R1
8
4
7
R2
6
555
3
2
C1
1
5
(B)
T1
T2
T
(C)
(D)

FIGURE 14-16

$$t1 = -(R1 + R2)\,C1\,\mathrm{LN}\left[\frac{(V+) - 2(V+)/3)}{(V+) - (V+)/3)}\right] \tag{14-9}$$

or, once the algebra is done

$$t1 = 0.695(R1 + R2)C1 \tag{14-10}$$

By similar argument it can be shown that

$$t2 = 0.695R2C1 \tag{14-11}$$

For the total period $T$

$$T = t1 + t2 \tag{14-12}$$

$$T = [0.695(R1 + R2)C1] + [0.695R2C1]$$

$$T = 0.695(R1 + 2R2)C1 \tag{14-13}$$

Equation (14-13) defines the *period* of the output square wave. In order to find the *frequency of oscillation* take the reciprocal of Eq. (14-13)

$$F = 1/T \tag{14-14}$$

$$F = \frac{1.44}{(R1 + 2R2)C1} \tag{14-15}$$

### 14-6.1 Duty Cycle of 555 Astable Multivibrator

Time segments $t1$ and $t2$ are not equal in most cases, so the charge and discharge times for capacitor $C1$ are also not equal (see Fig. 14-16D). The duty cycle of the output signal is the ratio of the HIGH period to the total period ($t1/T$). Expressed as a percent,

$$\%DC = \frac{R1 + R2}{R1 + 2R2} \tag{14-16}$$

Various methods are used for varying the duty cycle. First, a voltage can be applied to pin 5 (*Control Voltage*). Second, a resistance can be connected from pin 5 to ground. Both of these tactics alter the internal bias voltages applied to the comparator. One can also divide the external resistances $R1$ and $R2$ into three values. Figure 14-17 shows a variable duty factor 555 AMV which uses a potentiometer ($R2$) to vary the ratio of the charge and discharge resistances.

### 14-6.2 Synchronized Operation of 555 Astable Multivibrator

A synchronized AMV operates with its operating frequency locked to an external frequency. The horizontal and vertical deflection oscillators in a television receiver operate in this manner because they are locked to the sync pulses transmitted by the TV broadcast station.

A method for locking in the oscillating frequency of the 555 is shown in Fig. 14-18. In this circuit a 7400 TTL NAND gate is used to sample both the 555 AMV output

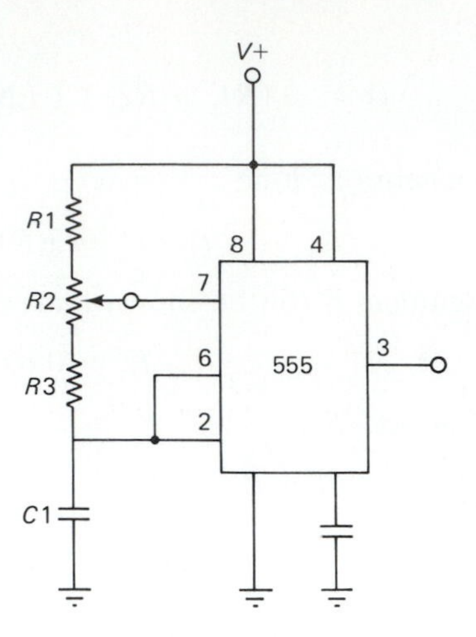


FIGURE 14-17

signal and the input sync signal. The NAND gate has the following properties

1. If either input is LOW, the output is HIGH.
2. Both inputs must be HIGH for the output to be LOW.

Because the output of the NAND gate is applied to the timing circuit of the 555 (through waveshaping network *R*3/*C*2), it will affect the relative timing of the circuit. This circuit is analogous to a mechanical pendulum oscillator which has an external nonresonant forcing frequency applied. The circuit will lock to the new frequency if it is reasonably close to the natural oscillating frequency, or an integer harmonic of the natural frequency. Students interested in modern chaos theory might want to investigate the behavior of this and similar circuits at sync frequencies away from the natural frequency or its harmonics and subharmonics.

### 14-6.3 A 555 Sawtooth Generator Circuit

A sawtooth waveform (Fig. 14-19A) rises linearly to some value, and then drops abruptly back to the initial conditions.

The circuit for a 555-based sawtooth generator (Fig. 14-19B) is simple and is based on the 555 timer IC. The circuit is the monostable multivibrator configuration of the 555 in which one of the timing resistors is replaced with a transistor operated as a current source (*Q*1). Almost any audio small signal PNP silicon replacement transistor can be used, although for this test the 2N3906 device was used here. The zener diode is a 5.6 Vdc unit. Note the output is taken from pins 6 and 7, rather than the regular chip output, pin 3, which is not used.

The circuit as shown is a one-shot multivibrator. Triggering occurs when pin 2 is brought to a potential less than 2/3 the supply potential. When a pulse is applied to pin 2 through differentiating network *R*1*C*1, the device will trigger because the negative-going slope meets the triggering criteria. To make an astable sawtooth multivibrator drive the input of this circuit we need either a square wave or pulse train which produces at least one pulse for each required sawtooth. Being a nonretriggerable monostable multivibrator the circuit of Fig. 14-19B will ignore subsequent trigger pulses during the one-shot's refractory period.

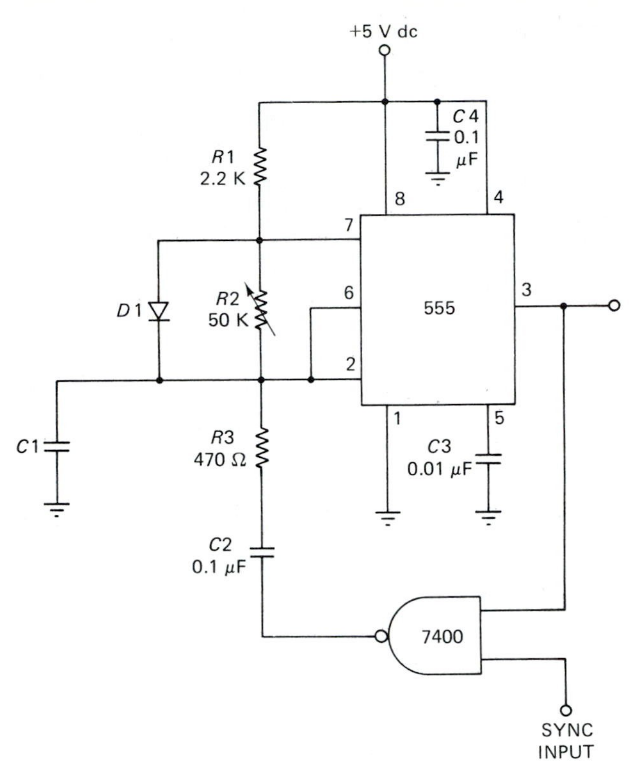


FIGURE 14-18

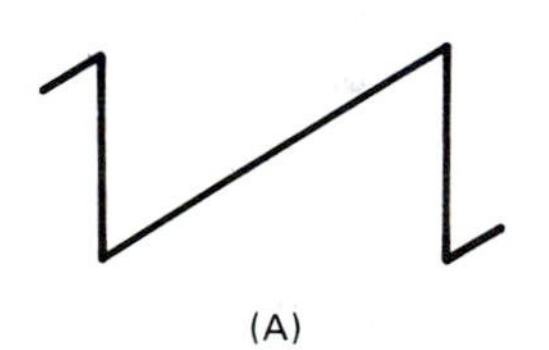

(A)

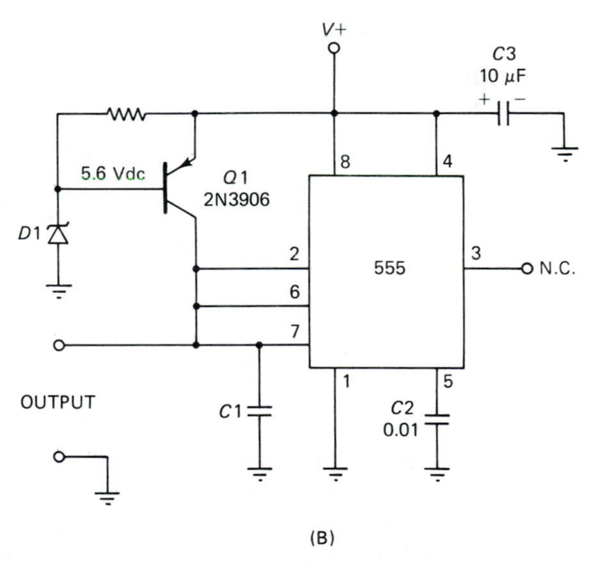


(B)

FIGURE 14-19

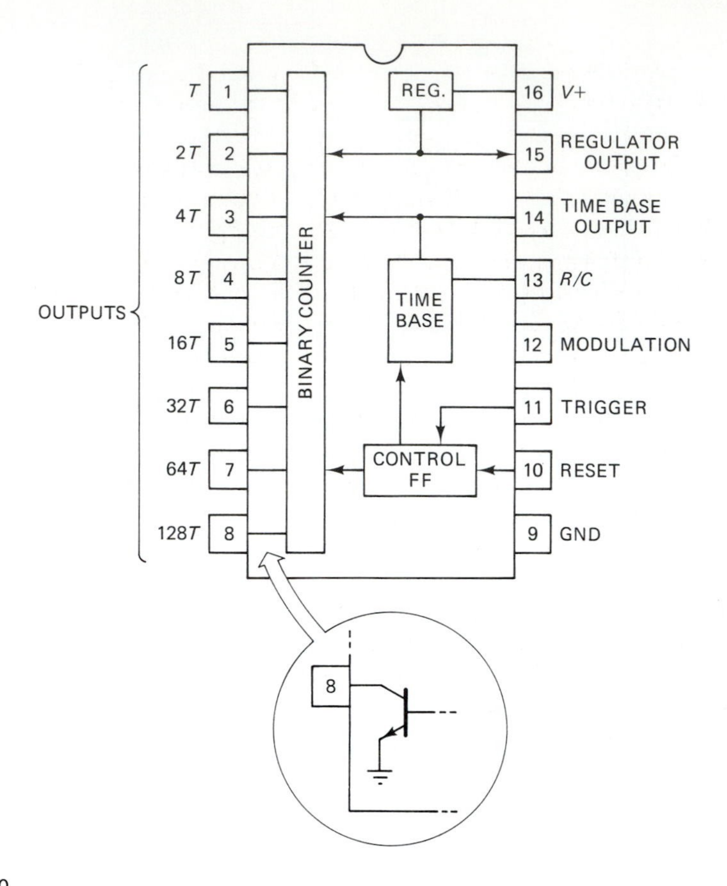


FIGURE 14-20

## 14-7 XR-2240 IC TIMER

The XR-2240 (also called 8240) IC timer (Fig. 14-20) is based on the same circuit concept as the 555 with a window comparator which sets and resets a control flip-flop. The timebase circuit must receive its power from outside the chip, so in normal operation a 20-kohm resistor is connected between the regulator output (pin 15) and timebase output (pin 14). This feature allows external timebase circuits to be used.

The timebase section of the XR-2240 is the 555-style timer modified to remove the constant 1.1 from the timing equation [see Eq. (14-6)]. The reason for the constant in the 555 is the resistors in the internal voltage divider which biases the comparators are equal. But in the XR-2240 nonequal resistances are used, making the reference levels $0.27(V+)$ and $0.73(V+)$ instead of $0.33(V+)$ and $0.67(V+)$ as used in the 555. This change was made in order to simplify the timing equation to

$$T = RC \tag{14-17}$$

The values of the timing components are 1 kohm $\leqq R \leqq$ 10 megohms, and 0.05 uF $\leqq C \leqq$ 1000 uF. The XR-2240 is packaged in a 16-pin DIP case and will operate over the range +4.5 Vdc to +18 Vdc. The XR-2240 differs from the 555 in that the trigger and reset pulses are positive-going rather than negative-going. The minimum amplitude of the trigger and reset pulses is approximately two PN-junction voltage drops, or about 1.4 Vdc. As a practical matter, however, a minimum 3 Vdc is recommended to guard against trigger failures caused by negative noise impulses.

There are other differences between the 555 device and the XR-2240. The XR-2240 constains an eight-bit binary counter with open-collector NPN transistor outputs (see inset to Fig. 14-20). If these outputs are wired in the logical-OR configuration, times from $1T$ to $255T$ can be programmed [$T$ is defined by Eq. (14-17)]. Each output is connected to $V+$ through a 10-kohms pullup resistor. The use of the open-collector configuration makes the output active-LOW. The combination of the large range of timer RC components and an eight-bit binary counter makes it possible to create very long duration timer circuits which are as stable as the RC network.

It is also possible to use the binary outputs independently if each is supplied with its own 10 kohm pullup resistor. Examples of applications include binary address sequences for digital circuits, and oscillators with base-2 related synchronized output frequencies.

Other less common versions of the timer include the XR-2250 (or 8250), which offers Binary Coded Decimal (BCD) outputs instead of binary outputs, and the XR-2260, which uses BCD outputs but limits the most significant digit to 6

Figure 14-21 shows the circuit for both the monostable and astable multivibrator configurations for the XR-2240. One interesting design feature of the XR-2240 is the sole difference between the monostable and astable modes is a single feedback resistor ($R3$) which automatically retriggers the XR-2240 at the end of each output cycle.

The timer is put into operation by applying a positive-going pulse to the trigger input (pin 11). This pulse is routed to the control logic. It performs several jobs simultaneously, resetting the binary counter to $00000000_2$, driving all outputs HIGH, and enabling the timebase circuit. As was true with the 555 timer, the XR-2240 works by charting capacitor $C1$ through resistor $R1$ from positive voltage source $V+$.

One purpose for the open-collector configuration is so different multiples of the basic timer duration can be programmed by connecting the required outputs together in a wire-OR configuration. For example, if a 57 second timer is needed it is possible to use a 1 megohm resistor and a 1 uF capacitor ($T = 1$ second) and connect the $1T$, $8T$, $16T$, and $32T$ outputs: $(1 + 8 + 16 + 32)T = 57T$. In this circuit, pins 1, 4, 5, and 6 will be connected and to $V+$ through a single 10 kohm resistor. The output will remain LOW for 57 seconds following triggering and then it will return to the HIGH state.

The XR-2240 is considered more flexible than the 555 because the duration (monostable mode) or period (astable mode) can be set by either the $RC$ timing network or through selection of which outputs are wired together at the output of the circuit.

Synchronization to an external timebase, or modulation of the pulse width, is possible by manipulation of the *modulation* input (pin 12). In normal operation the modulation input is bypassed to ground through a 0.01 uF capacitor so noise signals will not disrupt operation of the device. A voltage applied to pin 12 will modulate the pulse width of the timebase output signal. This modulating voltage should be between +2 and +5 volts for a change factor of 0.4 to 2.25.

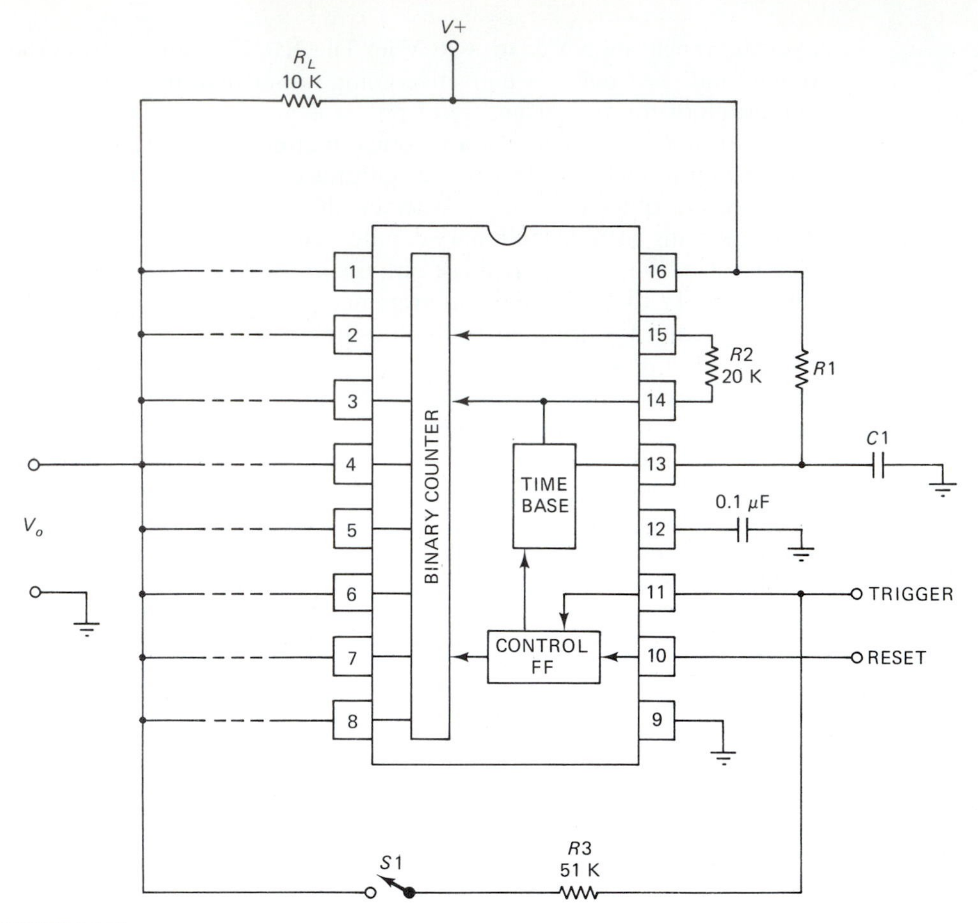

FIGURE 14-21

Synchronization of the XR-2240 to an external time base is done through a series *RC* network consisting of a 5.1 kohm resistor and a 0.1 uF capacitor (see Fig. 14-22) connected to pin 12. This network differentiates the input pulse. The synchronization pulse should have an amplitude of at least +3 volts, and a period of $0.3T$ to $0.8T$. Another method of synchronization is to use an external timebase connected directly to pin 14.

## 14-7.1 Very Long-Duration Timers

The long-duration timer presents special problems to the electronic circuit designer. The drift and other errors of some components tend to accumulate, so the total error becomes large over longer time periods. In fact, whenever the errors are a function of time the long duration time will suffer markedly. Any long duration RC network, for example, will suffer from several time related problems.

A viable alternative is to use an RC timer. But most high value capacitors are electrolytic and as a result have a tolerance of −20 to +100% of the label capacitance.

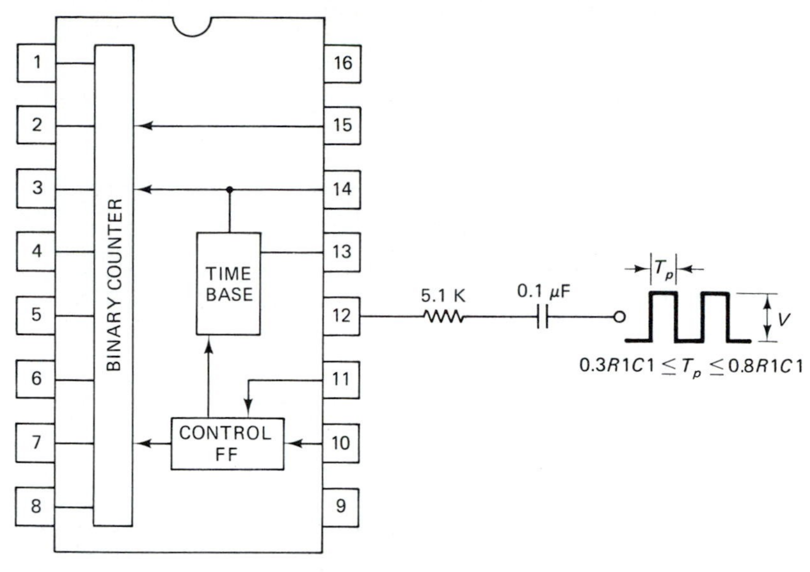


FIGURE 14-22

In addition, some capacitors show capacitance variation with both time and applied voltage. Many large value capacitors also exhibit considerable shunt resistance which must be considered in timing networks. Because of these problems the 555 is limited to practical durations of about 100 seconds if precision is a consideration.

The timebase section of the XR-2240 timer suffers from the same problems as the 555, but the *RC* values needed for very long duration monopulse operation are much lower than the 555 because of the built-in binary counter. If the outputs of the XR-2240 are wired-OR together, times from 1*RC* to 255*RC* are possible. For example, with an *RC* time constant of 10 seconds, a single XR-2240 can be used for up to 2550 seconds (42.5 minutes). Greater duration times can be accommodated by cascading two or more XR-2240 devices.

Figure 14-23 shows the use of two XR-2240 timers in cascade to produce a very long-duration timer. Unit 1 is used as the timebase. It has a frequency set by the expression 1/*RC*. Only the eighth bit of the binary counter (128*T*) is used. The output of Unit 1 becomes the timebase of Unit 2, which has all of its outputs wired-OR together. The total period is 65,536*T*. When the *RC* time constant is 1 second, the output duration is 65,536 seconds (18.2 hours!) even though the stability and accuracy attributes are those of a 1-second timer.

## 14-8 LM-122, LM-322, LM-2905, AND LM-3905 TIMERS

The LM-X22 and LM-X905 IC timers are precision devices which operate with DC power supply voltages of +4.5 Vdc to +40 Vdc to produce durations of microseconds to hours. The LM-122 and LM-322 packages are shown in Fig. 14-24A. The LM-2905 and LM-3905 are shown in Fig. 14-24B. The LM-2905/3905 devices are identical to the LM-122/322 except that the BOOST and $V_{adj}$ pins are not available. The LM-122 and LM-2905 operate over the temperature range −55°C to +125°C. The LM-322 and LM-3905 operate over 0°C to +70°C.

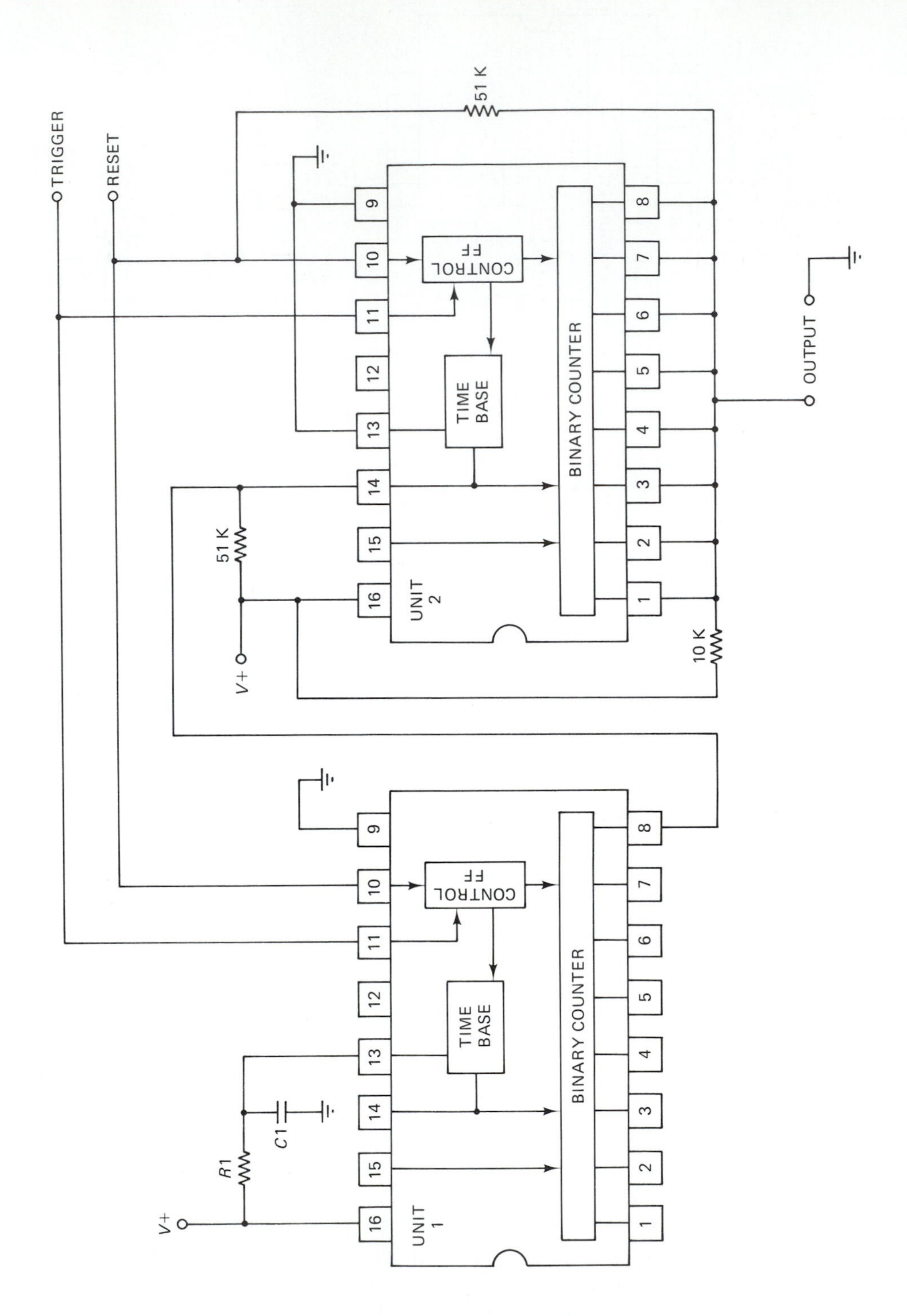
TRIGGER
RESET
51 K
CONTROL FF
TIME BASE
BINARY COUNTER
UNIT 2
V+
51 K
10 K
OUTPUT
UNIT 1
R1
C1
9
10
11
12
13
14
15
16
1
2
3
4
5
6
7
8

FIGURE 14-23

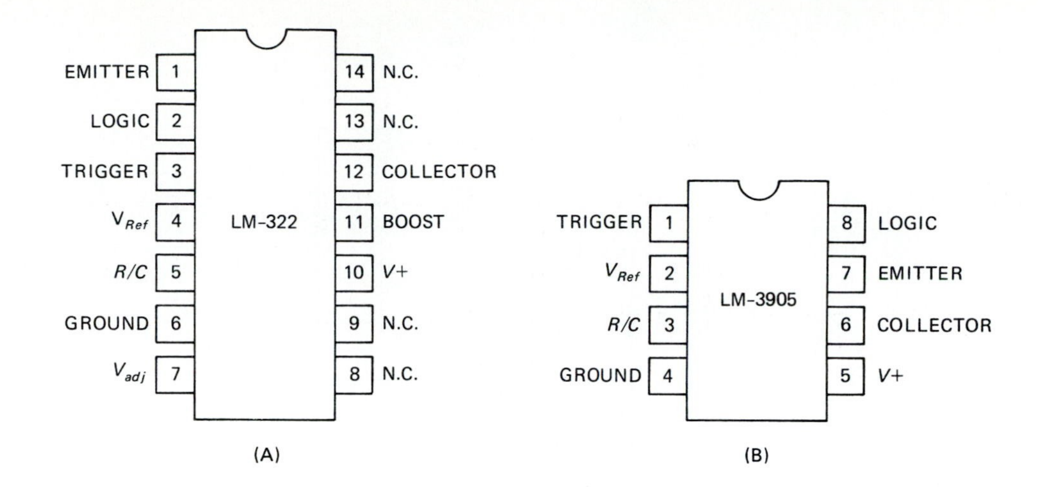
EMITTER
LOGIC
TRIGGER
$V_{Ref}$
$R/C$
GROUND
$V_{adj}$
1
2
3
4
5
6
7
14
13
12
11
10
9
8
N.C.
N.C.
COLLECTOR
BOOST
$V+$
N.C.
N.C.
LM-322
(A)
TRIGGER
$V_{Ref}$
$R/C$
GROUND
1
2
3
4
8
7
6
5
LOGIC
EMITTER
COLLECTOR
$V+$
LM-3905
(B)

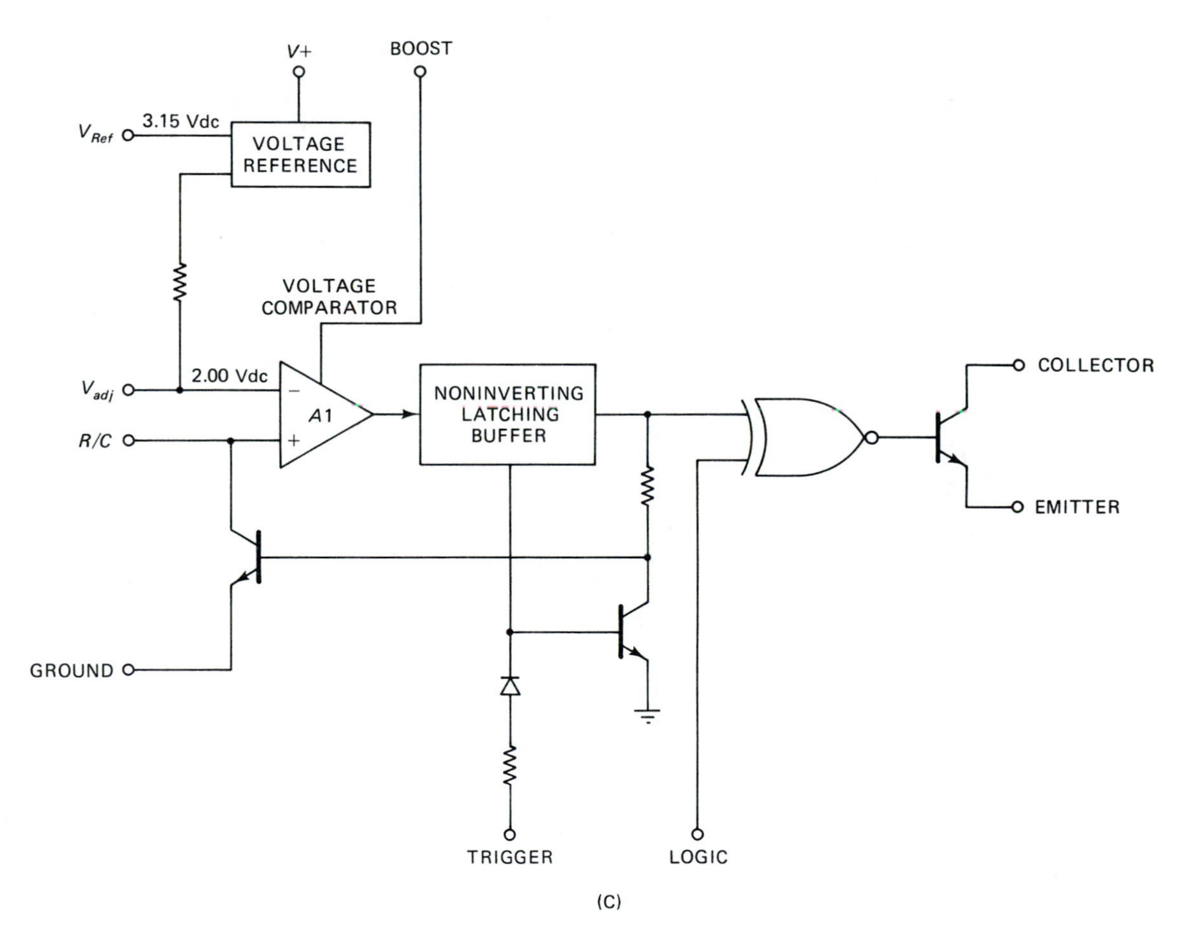
$V+$
BOOST
$V_{Ref}$
3.15 Vdc
VOLTAGE
REFERENCE
VOLTAGE
COMPARATOR
$V_{adj}$
2.00 Vdc
$R/C$
A1
NONINVERTING
LATCHING
BUFFER
COLLECTOR
EMITTER
GROUND
TRIGGER
LOGIC
(C)

FIGURE 14-24

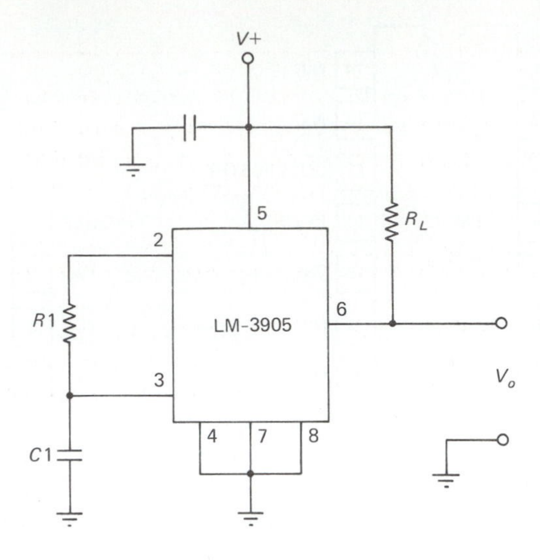


FIGURE 14-25

The internal circuitry of these timers is shown in Fig. 14-24C. The *RC* timing network is monitored by the noninverting input of a voltage comparator (*A*1). The inverting input is biased to +2.00 Vdc. The output of the comparator is routed to a noninverting latching buffer, which in turn drives an Exclusive-OR (XOR) gate. The alternate input of the XOR gate is connected to the LOGIC pin on the device. The output of these devices is a floating emitter and floating collector transistor. Both ground referenced and floating loads, at potentials up to 40 Vdc, can be accommodated with this arrangement. The $V_{adj}$ pin can be used in the LM-122 and LM-322 to vary the timing ratio up to 50:1 by using an external voltage. The circuit for the basic timer is shown in Fig. 14-25. The output duration is set by $T = R1C1$.

## 14-9 SUMMARY

1. The 555 timer is *RC*-timed and can be used as either a *monostable multivibrator* (MMV) or *astable multivibrator* (AMV). Most 555 devices are available in the 8-pin miniDIP package, although some wider temperature range devices are sold in the 8-pin metal can IC package.
2. The 555 MMV is triggered by bringing pin 2 to drop from above 2/3 − (*V*+) to less than 1/3 − (*V*+). The trigger signal can be a regular pulse, or a switch that shorts a normally HIGH level to ground potential.
3. The 555 is normally nonretriggerable. It can be made retriggerable by adding an external discharge transistor to dump the charge across the timing capacitor in response to a trigger pulse.
4. The XR-2240 timer is similar to the 555 in the time base section. The XR-2240 contains a binary counter driven by the time base. The outputs of the binary counter are open-collector NPN transistors, so external pullup resistors are needed. Times from 1*RC* to 255*RC* can be achieved by setting the time base (*RC*) and connecting the required outputs together in a wired-OR configuration. Very long duration timers (like 65,536*RC*) can be made by cascading XR-2240 devices.

## 14-10 RECAPITULATION

Now return to the objectives and Prequiz questions at the beginning of the chapter and see how well you can answer them. If you cannot answer certain questions, mark them and review the appropriate parts of the text. Next, try to answer the questions and work the problems below using the same procedure.

## 14-11 STUDENT EXERCISES

1. Design, build, and test a 555 monostable multivibrator (Fig. 14-5C) with an output duration of 10 milliseconds. Examine the output waveforms and the capacitor voltage waveforms together on a dual trace oscilloscope. Use various values of capacitance and resistance to see what happens to the duration.
2. Build the circuit of Exercise 1 above using a +5 Vdc power supply. Use a squarewave as the trigger signal, making sure it has adequate amplitude to cause triggering. Examine the output pulse and the trigger pulse on a dual trace oscilloscope and note their relationship.
3. Design, build and test a retriggerable monostable multivibrator. Test the circuit using square wave input signal to the trigger pin.
4. Design, build and test an astable multivibrator using the 555 (Fig. 14-16B). Select component values for an oscillating frequency of approximately 1000 Hz. Initially make $R1 = R2$. Examine the capacitor voltage and output waveforms on a dual trace oscilloscope. Vary the values of $R1$ and $R2$ to create different duty cycles. Also vary $R1$, $R2$, and $C1$ and examine both oscillating frequency and duty factor.

## 14-12 QUESTIONS AND PROBLEMS

1. The 555 is an example of an ____________ -timed multivibrator circuit.
2. Draw a block diagram of the 555 internal circuit and explain its operation.
3. Draw the circuit diagram for a 555 monostable multivibrator.
4. Draw the circuit diagram for a 555 astable multivibrator.
5. An $RC$ circuit is made consisting of a 0.0015 uF capacitor, and a 12 kohms resistor. What is the $RC$ time constant?
6. A 555 timer is used to produce an 8 mS output pulse. Calculate the resistance required if a 0.01 uF capacitor is used.
7. A series $RC$ network consisting of a 0.01 uF capacitor and a 100 kohms resistor is connected to a +12 Vdc voltage source. What is the time required for the capacitor voltage to rise from +2.5 Vdc to +7.5 Vdc?
8. A timer is designed like the 555, but with the internal resistor voltage divider with the following resistors: $R_a$ = 6.8 kohms, $R_b$ = 5 kohms, and $R_c$ = 10 kohms. Calculate the potentials applied to the inputs of COMP1 and COMP2. Derive the timing equation for monostable operation.
9. A 555 timer is used to make an astable multivibrator with a frequency of 500 mS. Calculate the resistances needed if the duty cycle is 75% and the timing capacitor is 0.01 uF.
10. An XR-2240 timer is used to make a 10-minute timer. What outputs must be wired-OR together if the time base uses the following timing components: $R1$ = 1,000,000 and $C1$ = 10 uF?

11. Why is the 555-class IC timer free of operating frequency drift when the $V+$ power supply potential slowly changes?
12. Define the rule for operating the *trigger* input of the 555 in the monostable multivibrator mode.
13. The output pin of the 555 is capable of sinking or sourcing up to ____________-mA of current.
14. Define the operation of the *reset* pin on the 555.
15. Draw a circuit that will allow retriggerable operation of the 555 in the monostable mode.
16. Draw the circuit and timing diagram for a missing pulse detector based on the 555.
17. Draw the block diagram for a pulse stretcher circuit that will increase a 100 uS pulse to 10 mS.
18. Draw the block diagram for a simple tachometer circuit based on the 555 that will produce a DC output that is proportional to the frequency of an input pulse signal.
19. Draw the circuit for a variable duty cycle astable multivibrator based on the 555.
20. Draw the circuit for a 555 sawtooth generator.
21. Draw the circuit for an XR-2240 monostable multivibrator that has a duration of 255 seconds.
22. All eight outputs of the XR-2240 are connected together with a 10-kohm pullup resistor. Calculate the output duration of this monostable multivibrator when the $RC$ time constant of the timing network is 0.5 mS.
23. How may very long duration timers be built using the XR-2240 circuit?
24. An XR-2240 is connected in the astable multivibrator mode. Calculate the frequency of oscillation if the RC time constant is 0.22 mS and the following outputs are connected together with a 10 kohm pullup resistor: 1, 4, 8, and 64.
25. What is the timing range of an XR-2240 if the time base is $RC$?

CHAPTER 15

# IC Data Converter Circuits and their Application

## OBJECTIVES

1. Understand the basic elements of data conversion.
2. Be able to describe the operation of the standard R-2R DAC.
3. Be able to describe the operation of the principal forms of A/D converter.
4. Be able to design circuits using common IC data converter devices.

## 15-1 PREQUIZ

These questions test your prior knowledge of the material in this chapter. Try answering them before you read the chapter. Look for the answers (especially those you answered incorrectly) as you read the text. After you have finished studying the chapter try answering these questions again and those at the end of the chapter (see Section 15-9).

1. A ____________ DAC has an external reference voltage source.
2. A DAC-08 device has a reference potential of +7.5 Vdc, and an reference input resistance of 10 kohms. Calculate the output current at full scale (binary input = $11111111_2$).
3. Describe how a DAC can be used to digitally synthesize a sawtooth waveform.
4. What is the output current of a DAC-08 when the reference current is 2 mA and the applied binary word is $10000000_2$?*

*Binary (base 2) numbers are used in digital circuits. The notation adopted in this text is to add a subscript "2" to indicate a binary number. Thus, 1000 would mean one thousand, while $1000_2$ means a binary number which evaluates to decimal eight.

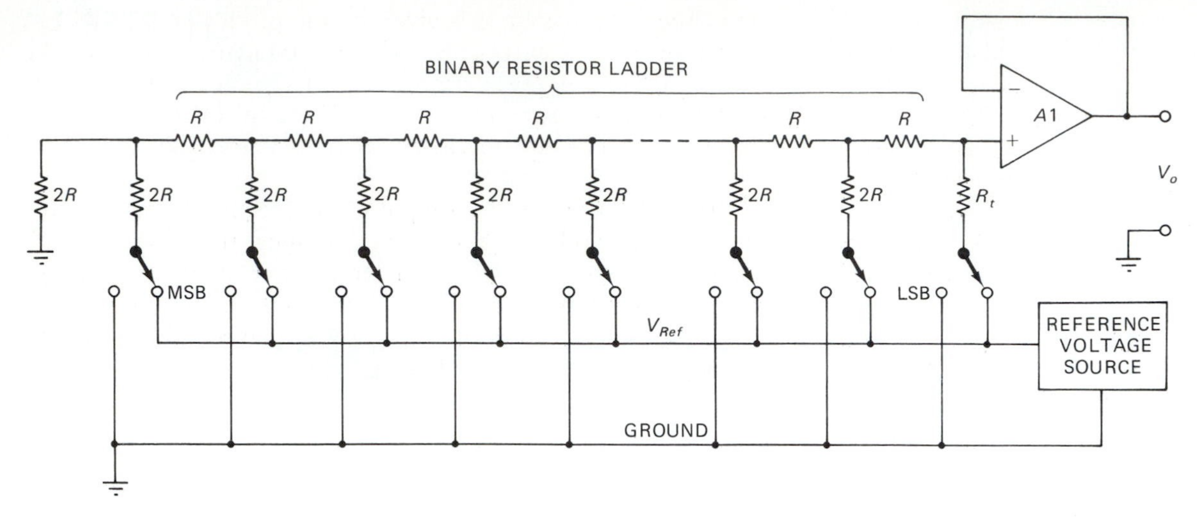


FIGURE 15-1

## 15-2 BASICS OF DATA CONVERSION

The data converter either converts a binary digital word to an equivalent analog current or voltage, or it converts an analog current or voltage to an equivalent binary word. The first is a *digital-to-analog converter* (DAC), the second is a *analog-to-digital converter* (A/D or ADC). These devices form the interface between digital computers and the analog world. This chapter will examine the basic functioning of these data conversion building blocks.

## 15-3 DIGITAL-TO-ANALOG CONVERTERS (DACs)

There are several approaches to DAC design, but all of them are varieties of a weighted current or voltage system which generates binary words by appropriate switch contacts. The most common example is the *R-2R ladder* shown in Fig. 15-1. The active element, $A1$, is an operational amplifier in a unity gain inverting follower configuration. In the circuit of Fig. 15-1 the digital inputs are shown as mechanical switches, but in a real data converter circuit the switches would be replaced by electronic switching devices (transistors). The electronic switches are driven by either a binary counter, or an N-bit parallel data line.

A *precision reference voltage* ($V_{ref}$) source is required for accurate data conversion. This voltage is most often +2.56 volts, +5.00 volts, or +10.00 volts. Other voltages can also be used, however. The accuracy of the converter depends on the precision of the reference voltage source. There are other sources of error, but if the reference voltage accuracy is poor, there is no hope for any other factors to be effective in improving the performance of the circuit. Although almost any voltage regulator can be used as the reference, it is prudent to select a precision, low-drift model.

Returning to Fig. 15-1, consider the circuit action under circumstances where various binary bits are either HIGH or LOW. If all bits are LOW, the output voltage will be zero. The value of the output voltage is given by the product $I \times R$. When all bits are LOW this current is zero. In practical circuits, though, there might be some output voltage under these circumstances due to offsets in the operational amplifier, the R-2R ladder and the electronic switches. These offsets can be nulled to zero output voltage when all bits are intentionally set to zero (or ignored, if negligible).

The unterminated R-2R ladder produces an output current. Some commercial IC DACs are current output models, and have no output amplifier. If there is a terminating resistor ($R_t$) shunting the output terminals of the DAC, the circuit produces an output voltage $I_oR_t$. The output impedance of this circuit tends to be high, so some of these DACs use an output amplifier to produce a low-impedance voltage output. The transfer function of the R-2R ladder type of DAC is

$$V_o = \frac{V_{ref}A}{2^N} \tag{15-1}$$

where

$V_o$ is the output potential

$V_{ref}$ is the reference potential

$A$ is the decimal value of the applied binary word

$N$ is the number of bits in the applied binary word.

### EXAMPLE 15-1

An eight-bit R-2R DAC has a 10 volt reference potential. What is the output voltage when the applied binary word is $11000000_2$ (decimal 192)?

#### Solution

$$V_o = \frac{V_{ref}A}{2^N}$$

$$V_o = \frac{(10\ \text{Vdc})\ (192)}{2^8}$$

$$V_o = \frac{1920}{256} = 7.5\ \text{Vdc}$$

■

If the most significant bit (MSB) is made 1 (HIGH), the output voltage will be approximately 1/2 $V_{ref}$. Similarly, if the next most significant bit is turned out (set to HIGH) and all others are LOW, the output will be 1/4 $V_{ref}$. The least significant bit (LSB) would contribute $V_{ref}/2^N$ to the total output voltage. For example, with an 8-bit DAC, the LSB changes the output $V_{ref}/2^8$, or $V_{ref}/256$. This change is called the 1-LSB value. Figure 15-2 graphs the output of a voltage DAC in response to the entire range of binary numbers applied to the digital inputs. The result is a *staircase waveform* which rises by the 1-LSB value for each 1-LSB change of the binary word. This step height represents the minimum discernible resolution of the circuit.

The reference source can be either internal or external to an integrated circuit

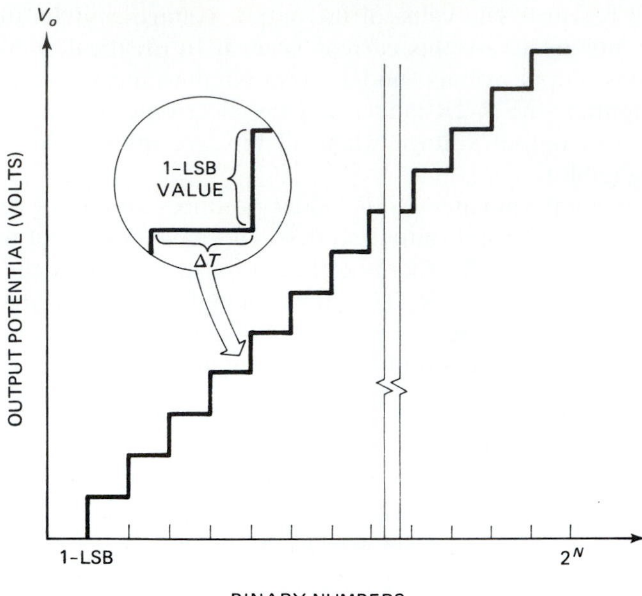


FIGURE 15-2

DAC. If the reference voltage or current source is external, the DAC is said to be a *multiplying DAC* or MDAC. The multiplication takes place between the analog reference and the fraction defined as $A/2^N$ in Eq. (15-1). If the reference source is completely internal and not adjustable (except for fine trimming), the DAC is said to be a *nonmultiplying DAC* or simply "DAC."

### 15-3.1 Coding Schemes

There are several different coding methods for defining the transfer function of the DAC. The most common are, *unipolar positive*, *unipolar negative*, *symmetrical bipolar*, and *asymmetrical bipolar*.

The unipolar coding schemes provide an output voltage of one polarity only. These circuits usually produce 0 volts for the minimum and some positive or negative value for the maximum. Because one binary number state represents 0 Vdc, there are $[2^N - 1]$ states to represent the analog voltages within the range. For example, in an 8-bit system there are 256 states, so if one state ($00000000_2$) represents 0 Vdc, there are 255 possible states for the nonzero voltages. Thus, the maximum output voltage is always 1 LSB less than the reference voltage.

For example, in a 10 Vdc system, the maximum output voltage is 10 volts/256 = 0.039 volts = 39 mV less. Thus, the maximum output voltage will be approximately 9.96 Vdc. The unipolar positive coding scheme is:

| *Unipolar Output Voltage* | *Binary Input Word* |
|---|---|
| 0.00 Vdc | $00000000_2$ |
| $V_{max}/2$ | $10000000_2$ |
| $V_{max}$ | $11111111_2$ |

The negative version of this coding scheme (unipolar negative) is identical, except the midscale voltage is $-V_{max}/2$ and the full-scale output voltage is $-V_{max}$. A variant on the theme inverts the definition so

| *Unipolar Output Voltage* | *Binary Input Word* |
|---|---|
| 0.00 Vdc | $11111111_2$ |
| $V_{max}/2$ | $10000000_2$ |
| $V_{max}$ | $00000000_2$ |

The bipolar coding scheme has a problem which requires a trade-off in the design. There is an even number of output states in a binary system. For example, in the standard 8-bit system there are 256 different output states. If one state is selected to represent 0 Vdc, then there are 255 states left to represent the voltage range. As a result, there is an even number of states to represent positive and negative states either side of 0 Vdc. For example, 127 states might be assigned to represent negative voltages, and 128 to represent positive voltages. In the asymmetrical bipolar coding, therefore, the pattern might look like

| *Bipolar Output Voltage* | *Binary Input Word* |
|---|---|
| $-V_{max}$ | $00000000_2$ |
| 0.00 Vdc | $10000000_2$ |
| $+V_{max}$ | $11111111_2$ |

A decision must be made regarding which polarity will lose a small amount of dynamic range.

The other bipolar coding system is the symmetrical bipolar scheme. The decision in the symmetrical scheme is that each polarity will be represented by the same number of binary states either side of 0 Vdc. But this scheme does not permit a dedicated state for zero.

| *Bipolar Output Voltage* | *Binary Input Word* |
|---|---|
| $-V_{max}$ | $00000000_2$ |
| −Zero (−1 LSB) | $01111111_2$ |
| 0.00Vdc | (disallowed) |
| +Zero (+1 LSB) | $10000000_2$ |
| $+V_{max}$ | $11111111_2$ |

The state plus zero is more positive than the 1-LSB value than 0 Vdc, while the "minus zero" state is more negative than 0 Vdc by the same 1-LSB value.

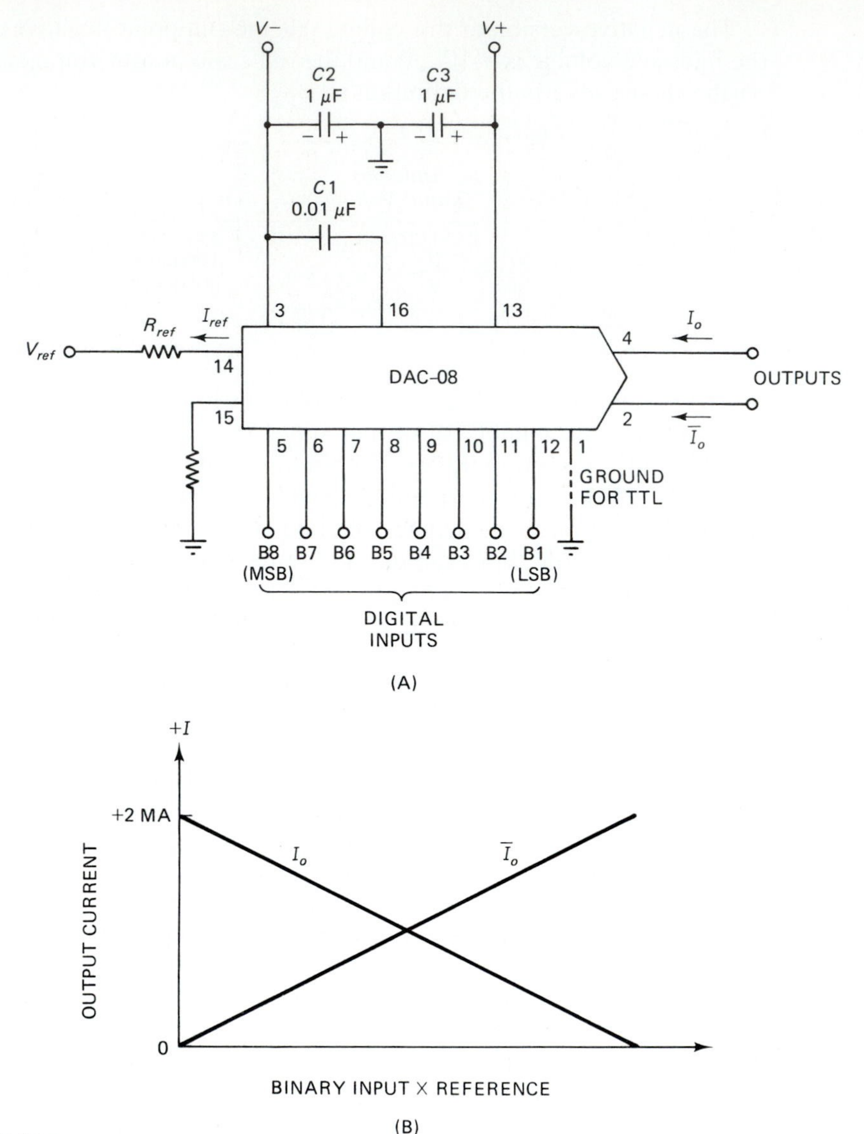


FIGURE 15-3

### 15-3.2 A Practical IC DAC Example

A number of different manufacturers offer low-cost IC DACs which contain nearly all the circuitry needed for the process, except possibly the reference source (some devices do contain the reference) and operational amplifiers for either level shifting or current to voltage conversion.

For purposes of our discussion here, the DAC-08 device is used as a practical circuit example. This eight-bit DAC is now an industry standard. The DAC-08 is a later generation version of the Motorola MC-1408. It is sometimes designated LMDAC-0800. A closely related device, the DAC-0806, is also easily available.

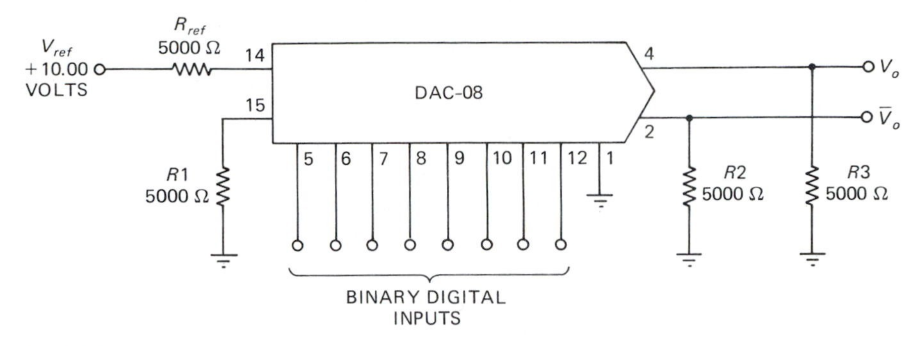


FIGURE 15-4

Figure 15-3A shows a basic circuit for the DAC-08. In subsequent circuits, the power supply terminals are deleted for simplicity. They will always be the same as shown here. The internal circuitry of the DAC-08 is the R-2R ladder shown in the previous section, but it has two outputs, $I_o$ and NOT-$I_o$. These current outputs are unipolar and complementary (Fig. 15-3B). If the full-scale output current is $I_{max}$, then $I_{max} = [I_o + \text{NOT-}I_o]$. The specified value of $I_{max}$ on the DAC-08 is 2 mA.

Two types of input signal are required to make this DAC work, an *analog reference* and an *8-bit digital signal*. The analog signal the reference current ($I_{ref}$) applied through pin 14. This current may be generated by combining a precision reference voltage source with a precision, low-temperature-coefficient resistor to convert $V_{ref}$ to $I_{ref}$. Alternatively, a constant current source may be used to provide $I_{ref}$. For TTL compatibility of the binary inputs, make $V_{ref}$ = 10.000 volts and $R_{ref}$ = 5000 ohms.

The other type of input is the eight-bit digital word which is applied to the IC at pins 5 through 12 as shown. The logic levels which operate these inputs can be preset by the voltage applied to pin 1 (for TTL operation, pin 1 is grounded). In the TTL-compatible configuration shown, LOW is 0 to 0.8 volts and HIGH is +2.4 to +5 volts.

Figure 15-4 shows the connection of the DAC-08 (less power supply and reference input) required to provide the simplest form of unipolar operation over the range of approximately 0 to 10 volts. When the input word is $00000000_2$, the DAC output is 0 volts, ± DC offset error. A half-scale voltage (−5 volts) is given when the input word is $10000000_2$. This situation occurs when the MSB is HIGH and all other digital inputs are LOW. The full-scale output will exist only when the input word is $11111111_2$ (all HIGH). The output under full-scale conditions will be −9.96 volts, rather than 10 volts as might be expected (9.96 volts is 1 LSB less than 10 volts).

The circuit in Fig. 15-4 works by using resistors *R*2 and *R*3 as current-to-voltage converters. When currents $I_o$ and NOT-$I_o$ pass through these resistors, a voltage drop of *IR*, or 5.00 × $I_o$ (mA) is created. A problem with this circuit is its high source impedance (5 kohms, with the values shown for *R*2/*R*3).

Figure 15-5 shows a simple method for converting $I_o$ to an output voltage ($V_o$) with a low output impedance (less than 100 ohms) by using an inverting follower operational amplifier. The output voltage is the product of the output current and the negative feedback resistor

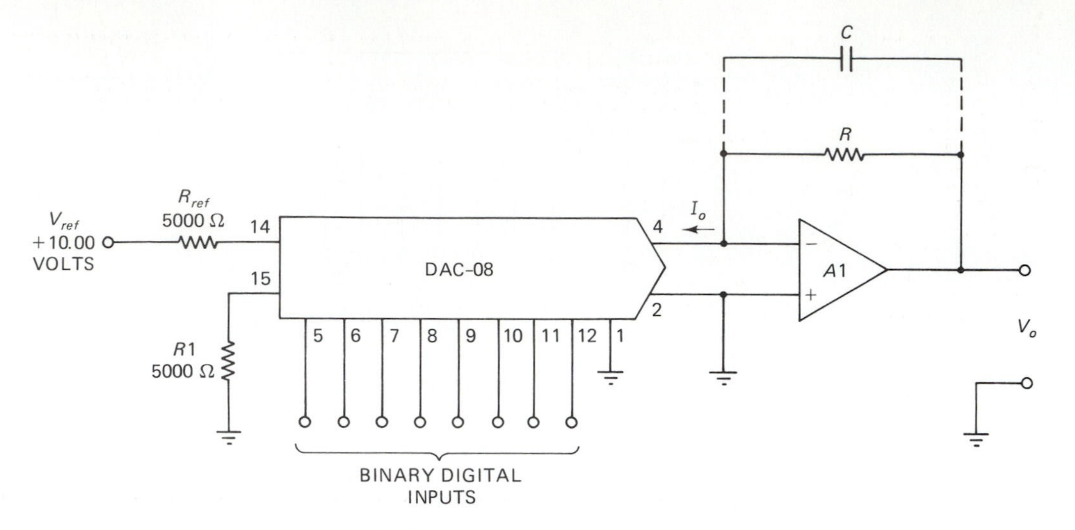


FIGURE 15-5

$$V_o = R \times I_o \tag{15-2}$$

As in the previous case previously, a 5000-ohm resistor and a 10 Vdc reference voltage will produce a 9.96 volt output voltage when the DAC-08 is set up for TTL inputs and 2.0 mA $I_{o(\max)}$.

The frequency response of the DAC circuit can be tailored to meet certain requirements. The normal output waveform of the DAC is a staircase when the digital input increments up from $00000000_2$ to $11111111_2$ in a monotonic manner. In order to make the staircase into an actual ramp function, a low-pass filter is needed at the output to remove the stepness of the normal waveform. A capacitor shunted across the feedback resistor ($R$) will offer limited filtering on the order of $-6$ dB/octave above a cutoff frequency of

$$F = \frac{1{,}000{,}000}{6.28 \; R \; C_{\mathrm{uF}}} \tag{15-3}$$

where

$F$ = the $-3$ dB frequency in hertz (Hz)

$R$ is in ohms

$C_{\mathrm{uF}}$ is in microfarads (uF)

In most practical circuits the required value of $F$ is known from the application. It is the highest frequency Fourier component in the input waveform. It is necessary to calculate the value of capacitor needed to achieve that cutoff frequency, so we would swap the $F$ and $C$ terms

$$C_{\mathrm{uF}} = \frac{1{,}000{,}000}{6.28 \; R \; F} \tag{15-4}$$

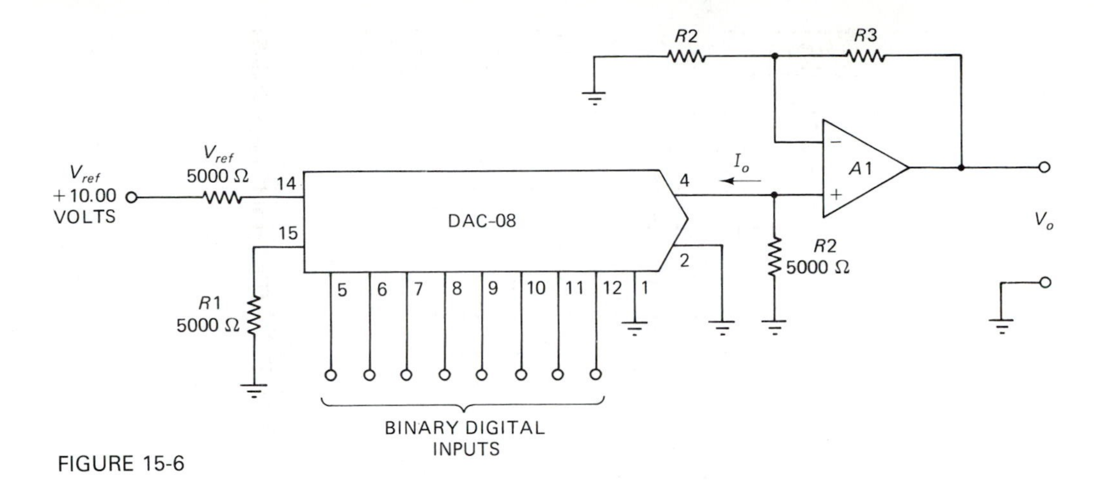


FIGURE 15-6

A related method shown in Fig. 15-6 produces an output voltage of the opposite polarity from Fig. 15-5. The circuit of Fig. 15-6 is connected to a noninverting unity-gain follower at the output. The output voltage is the product of $I_o$ and $R2$. If a higher output voltage is needed, the circuit variant shown in the inset to Fig. 15-6 can be used. In this case, the output amplifier has gain, so the output voltage will be

$$V_o = I_o \times R2 \times \left[\frac{R4}{R3} + 1\right] \tag{15-5}$$

One way to achieve bipolar binary operation is shown in Fig. 15-7. In this circuit the output amplifier is a DC differential amplifier and both current outputs of the DAC-08 are used. Note the maximum and minimum voltages are positive and negative. The zero selected can be either (+)*zero* (+1-LSB voltage), or (−)*zero* (−1-LSB voltage). It cannot be exactly zero because an even number of output codes are equally spaced around zero. In other words, the absolute value of $FS(-)$ is equal to the absolute value of $FS(+)$. There are also circuits that make zero = zero, but at the expense of uneven ranges for $FS(-)$ and $FS(+)$.

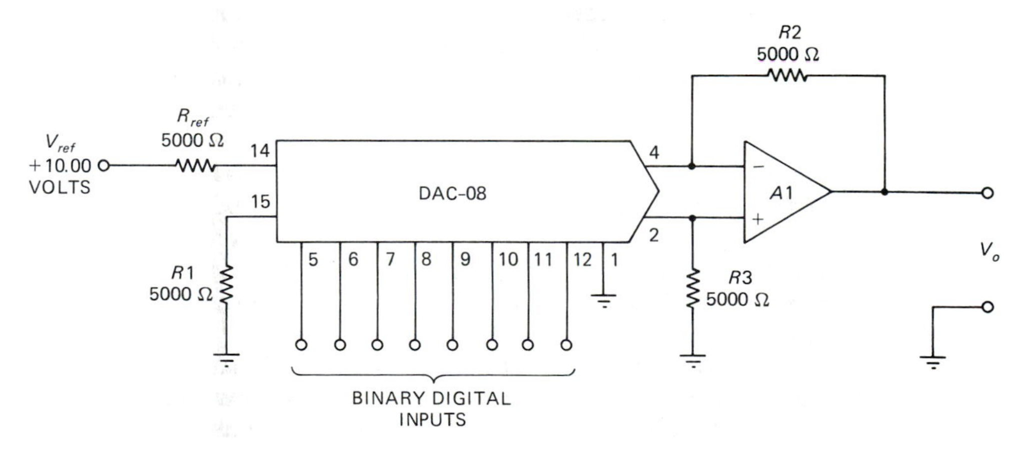


FIGURE 15-7

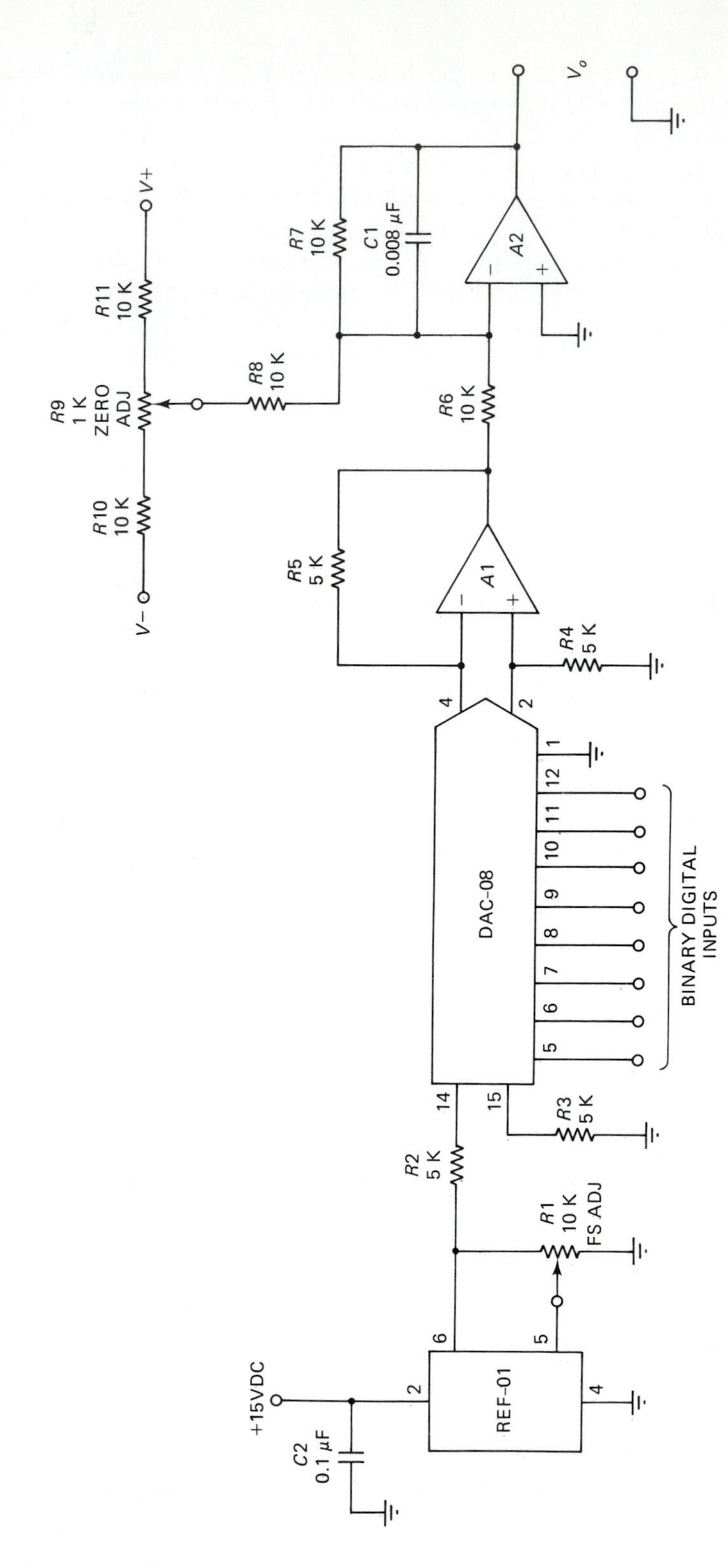

V+
V−
R11
10 K
R10
10 K
R9
1 K
ZERO
ADJ
R8
10 K
R7
10 K
C1
0.008 μF
A2
$V_o$
R6
10 K
R5
5 K
A1
R4
5 K
DAC-08
BINARY DIGITAL
INPUTS
R3
5 K
R2
5 K
R1
10 K
FS ADJ
REF-01
+15VDC
C2
0.1 μF


FIGURE 15-8

A practical DAC circuit is shown in Fig. 15-8. It combines the fragments shown earlier to make a complete circuit. The power connections are not shown. The heart of this circuit is a DAC-08 connected in the bipolar binary circuit discussed above.

The reference potential in Fig. 15-8 is a *REF-01* 10.000 volt IC reference voltage source. Potentiometer $R1$ adjusts the value of the actual voltage and also serves as a full-scale adjustment for the output voltage ($V_o$).

The output amplifier can be a 741 operational amplifier or any other type. It is not critical. Potentiometer $R9$ acts as a zero adjustment for $V_o$. The capacitor across $R7$ limits the frequency response to 200 Hz with the value shown. This frequency limit can be changed with the equation given earlier.

### Adjustment

1. Set the binary inputs all LOW ($00000000_2$).
2. Adjust $R9$ for $V_o = 0.00$ volts.
3. Set all binary inputs HIGH ($11111111_2$).
4. Adjust potentiometer $R1$ for $V_o = 9.96$ volts.

## 15-4 DIGITAL SYNTHESIS SAWTOOTH GENERATOR

Sawtooth signal generators can be used for a variety of purposes; some applications include electronic music synthesizers, sweeping RF oscillators, audio signal generators, and voltage controlled oscillators (VCO), certain laboratory bench tests, and calibrating oscilloscopes. In addition, there are many circuit applications for embedded sawtooth generators. Examples include situations where the sawtooth is used to provide precision calibration of an oscilloscope timebase. If the sawtooth used to sweep the oscilloscope horizontally is controlled from a stable crystal oscillator, a very precise sweep rate is possible. The standard solid-state sawtooth generator circuit consists of an operational amplifier Miller integrator circuit excited by a squarewave (see Chapter 12).

The circuit for a digitally synthesized sawtooth generator is shown in Fig. 15-9. The heart of this circuit is IC1, a DAC-0806 eight-bit DAC. This DAC is related to the DAC-08, and it too is based on the MC-1408 family of DACs.

A DAC-0806 produces an output current proportional to (a) the reference voltage or current and (b) the binary word applied to the digital inputs. The controlling function for the DAC is

$$I_o = I_{ref} \times \left[\frac{A}{256}\right] \tag{15-6}$$

where

$I_o$ = the output current from pin 4

$I_{ref}$ = the reference current applied to pin 14

$A$ = the decimal value of the binary word applied to the eight binary inputs (pins 5–12)

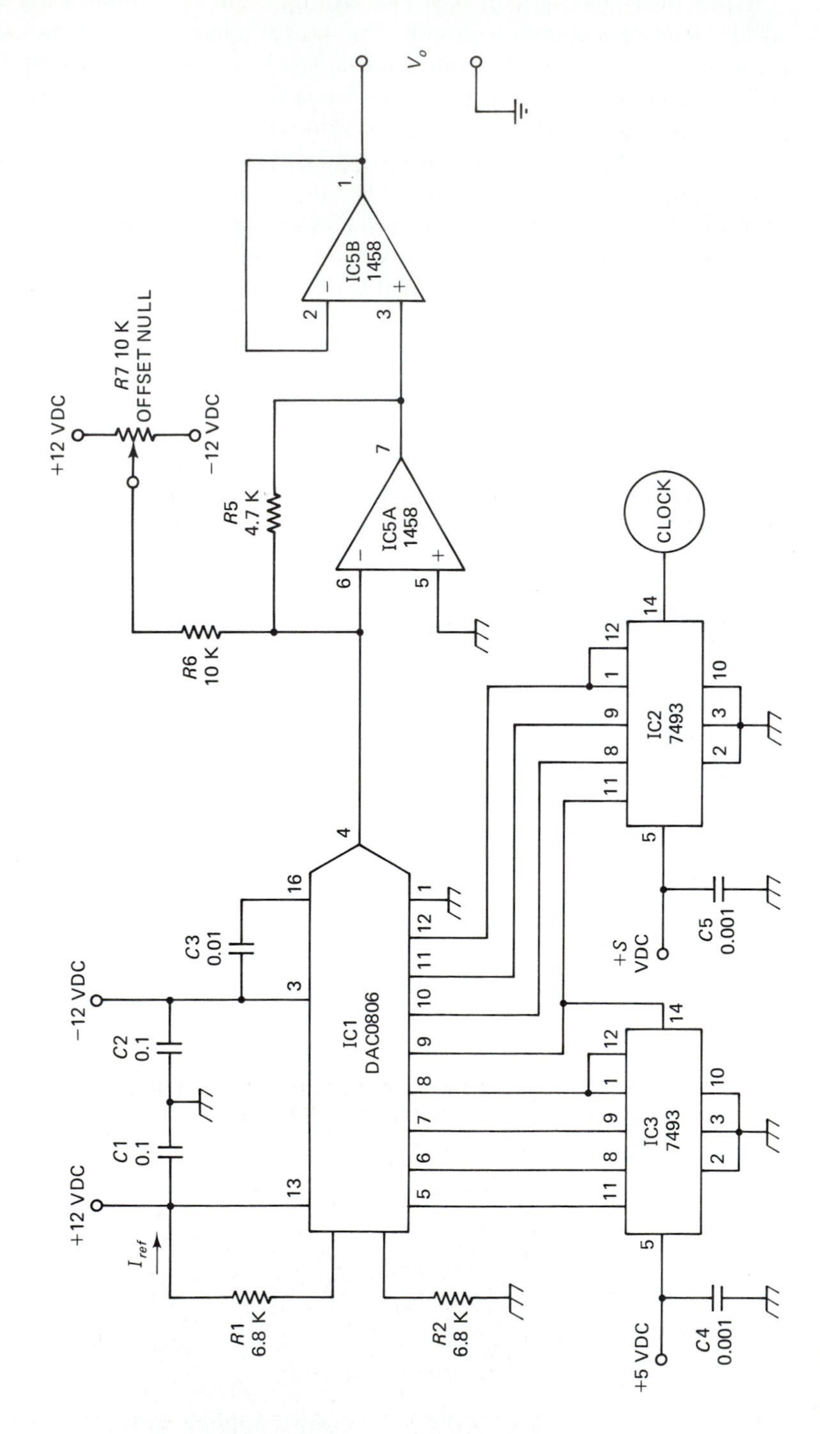

+12 VDC
−12 VDC
R7 10 K
OFFSET NULL
R5
4.7 K
R6
10 K
IC5A
1458
IC5B
1458
$V_o$
CLOCK
IC1
DAC0806
IC2
7493
IC3
7493
C1
0.1
C2
0.1
C3
0.01
C4
0.001
C5
0.001
R1
6.8 K
R2
6.8 K
$I_{ref}$
+5 VDC
+5 VDC


FIGURE 15-9

The reference current is found from Ohm's law. It is the quotient of the reference voltage and the series resistor at pin 14. In analog data acquisition and display systems the reference voltage is a precision, regulated potential. But in this case the precision may not be needed, so the $V+$ power supply is used as the reference voltage. Therefore, the reference current is (+12 Vdc)/$R1$. With the value of $R1$ shown (6800 ohms), $I_{ref}$ is 0.00018 amperes, or 1.8 mA. Values from 500 uA to 2 mA are permissible with this device.

The reference current sets the maximum value of output current ($I_o$). When a full-scale binary word ($11111111_2$) is applied to the binary inputs, the output current $I_o$ is

$$I_o = (1.8 \text{ mA}) \times \left[\frac{255}{256}\right] \tag{15-7}$$
$$I_o = (1.8 \text{ mA}) \times (0.996) = 1.78 \text{ mA}$$

The DAC-0806 is a current output DAC, so it is an op-amp current to voltage converter in order to make a sawtooth voltage output function. Such a circuit is an ordinary inverting follower without an input resistor. The output voltage ($V_o$) will rise to a value of ($I_o \times R5$).

The waveform produced by this circuit is shown in Fig. 15-10. It has a period of about 5 ms (200 Hz) and an amplitude of about 3 volts (the falling edge of the sawtooth is too fast for the oscilloscope camera to capture on film).

The actual output waveform is staircased in binary steps equal to the 1-LSB current of the DAC or the 1-LSB voltage. It is the smallest step change in output potential caused by changing the least significant bit ($B1$) either from 0 to 1 or 1 to 0. The steps in Fig. 15-10 cannot be seen because the frequency response of the 741 operational amplifier used for the current to voltage converter acts as a low-pass filter to smooth the waveform. If a higher frequency op-amp is used, then a capacitor shunting $R3$ will serve to low-pass filter the waveform. A $-3$ dB frequency ($F$) of 1 or 2 KHz will suffice to smooth the waveform. The value of the capacitor is calculated from

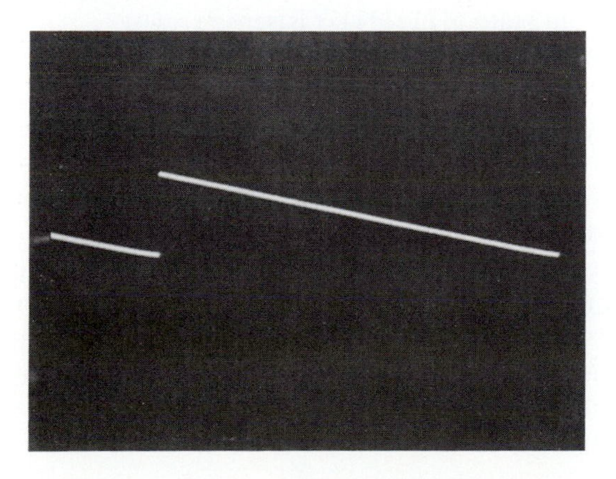

FIGURE 15-10

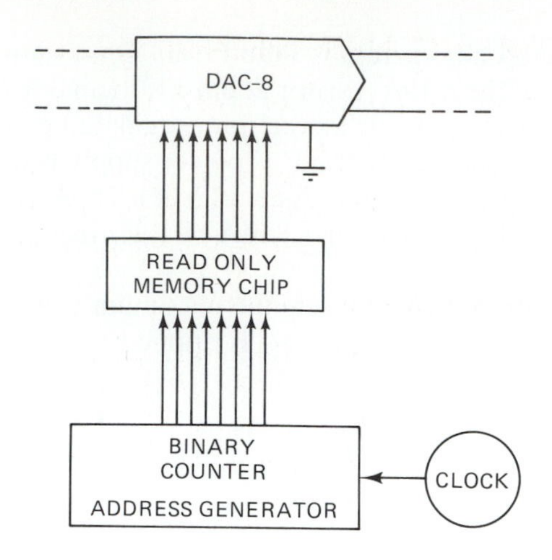


FIGURE 15-11

$$C_{\mathrm{uF}} = \frac{1{,}000{,}000}{6.28\ R3\ F} \tag{15-8}$$

where

$C_{uF}$ = the capacitance in microfarads

$F$ = the $-3$ dB cut-off frequency in hertz (Hz)

$R3$ is expressed in ohms

Select a clock frequency 256 times the desired sawtooth fundamental frequency.

## 15-4.1 Other Waveforms

The simple binary counter which allows the sawtooth to be generated also limits the circuit. It can be modified, however, to produce any other waveform. For a triangle waveform replace the straight binary counters with a base-16 up/down binary counter. Arrange the digital control logic to reverse the direction of the count when the maximum state ($11111111_2$) is sensed.

There are two ways to generate waveforms other than a sawtooth or triangle. Both of them involve using computer memory. The binary bit pattern representing the waveform is stored in memory, and then output in the right sequence. One method uses a Read Only Memory (ROM) which is preprogrammed with a bit pattern representing the waveform (Fig. 15-11). A binary counter circuit connected as an address generator selects the bit pattern sequence.

The second method is to store the bit pattern in a computer and output it under program control via an eight-bit parallel output port. This method is usable for both generating special waveforms and for linearizing the tuning characteristic of circuits such as VCOs and swept oscillators. The digital solution to the linearization problem (Chapter 13) involves storing a lookup correction table in either a ROM or computer memory.

## 15-5 ANALOG-TO-DIGITAL CONVERTERS (A/D)

The *analog-to-digital* converter (A/D) is used to convert an analog voltage or current input to an output binary word which can be used by a computer. Of the many techniques that have been published for performing an A/D conversion, only a few are of interest to us: the *voltage-to-frequency*, *single-slope integrator*, *dual-slope integrator*, *counter* (or *servo*), *successive approximation* and *flash* methods.

### 15-5.1 Integration A/D Methods

Most digital panel meters (DPM) and digital multimeters (DMM) use either the *single* or *dual-slope* integration methods for the A/D conversion process. An example of a single-slope integrator A/D converter is shown in Fig. 15-12A. The single-slope integrator is simple, but it is limited to those applications where an accuracy of 1% or 2% is acceptable.

The single-slope integrator A/D converter of Fig. 15-12A consists of five basic sections: *ramp generator*, *comparator*, *logic*, *clock*, and *output encoder*. The ramp generator is an ordinary operational amplifier Miller integrator with its input connected to a stable, fixed, reference voltage source. This makes the input current $I_{ref}$ essentially constant; so the voltage at point B will rise in a nearly linear manner, creating the voltage ramp.

The comparator is an operational amplifier with no feedback loop. The circuit gain is the open-loop gain ($A_{vol}$) of the device selected which is typically very high even in low-cost operational amplifiers. When the analog input voltage $V_x$ is greater than the ramp voltage, the output of the comparator is saturated at a logic-HIGH level.

The logic section consists of a main AND gate, a main-gate generator, and a clock. The waveforms associated with these circuits are shown in Fig. 15-12B. When the output of the main-gate generator is LOW, switch $S1$ remains closed, so the ramp voltage is zero. The main-gate signal at point A is a low frequency squarewave with a frequency equal to the desired time-sampling rate. When point A is HIGH, $S1$ is open, so the ramp will begin to rise linearly. When the ramp voltage is equal to the unknown input voltage $V_x$, the differential voltage seen by the comparator is zero; so its output drops LOW.

The AND gate requires all three inputs to be HIGH before its output can be HIGH. From times $T0$ to $T1$, the output of the AND gate will go HIGH every time the clock signal is also HIGH.

The encoder, in this case an eight-bit binary counter, will then see a pulse train with a length proportional to the amplitude of the analog input voltage. If the A/D converter is designed correctly, the maximum count of the encoder will be proportional to the maximum range (full-scale) value of $V_x$.

Several problems are found in single slope integrator A/D converters

The ramp voltage may be nonlinear.

The ramp voltage may have too steep or too shallow a slope.

The clock pulse frequency could be wrong.

It may be prone to changes in apparent value of $V_x$ caused by noise.

Many of these problems can be corrected by the dual-slope integrator of Fig. 15-13. This circuit also consists of five basic sections: *integrator*, *comparator*, *control*

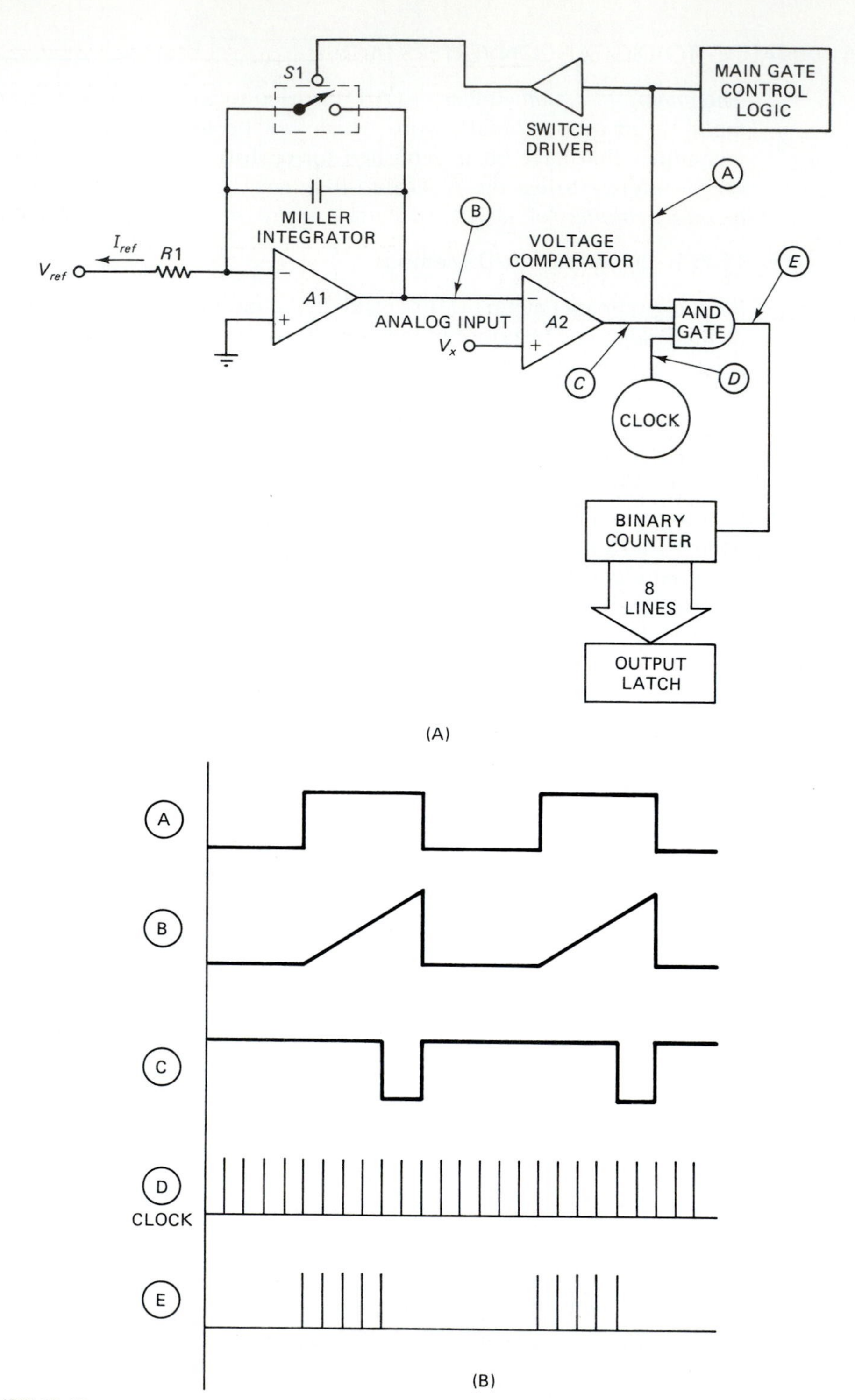
S1
MAIN GATE
CONTROL
LOGIC
SWITCH
DRIVER
MILLER
INTEGRATOR
A
B
$I_{ref}$
R1
$V_{ref}$
A1
VOLTAGE
COMPARATOR
E
ANALOG INPUT
A2
AND
GATE
$V_x$
C
D
CLOCK
BINARY
COUNTER
8
LINES
OUTPUT
LATCH
(A)
A
B
C
D
CLOCK
E
(B)

FIGURE 15-12

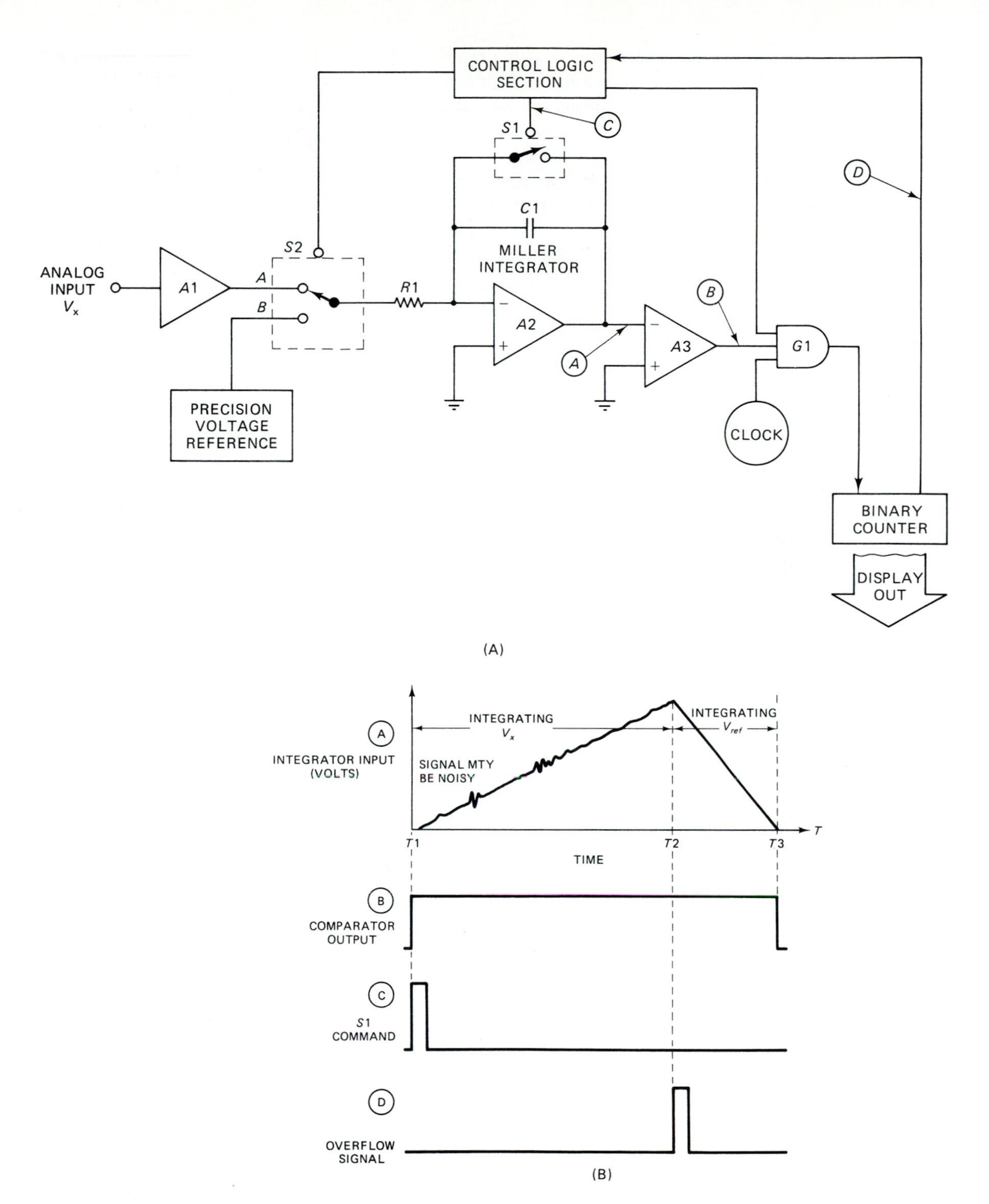
CONTROL LOGIC SECTION
S1
C
D
C1
MILLER INTEGRATOR
S2
ANALOG INPUT $V_x$
A1
A
B
R1
A2
A3
B
G1
A
PRECISION VOLTAGE REFERENCE
CLOCK
BINARY COUNTER
DISPLAY OUT
(A)
INTEGRATING $V_x$
INTEGRATING $V_{ref}$
INTEGRATOR INPUT (VOLTS)
SIGNAL MTY BE NOISY
T1
T2
T3
T
TIME
COMPARATOR OUTPUT
S1 COMMAND
OVERFLOW SIGNAL
(B)

FIGURE 15-13

*logic section*, *binary counter*, and a *reference current* or *voltage source*. An integrator is made with an operational amplifier connected to a capacitor in the negative feedback loop, as in the single-slope version. The comparator in this circuit is also the same kind of circuit as in the previous example. In this case though, the comparator is ground-referenced by connecting +IN to ground.

When a start command is received, the control circuit resets the counter to $00000000_2$, resets the integrator to 0 volts (by discharging *C*1 through switch *S*1), and sets electronic switch *S*2 to the analog input (position A). The analog voltage creates an input current to the integrator which causes the integrator output to begin charging capacitor *C*1. The output voltage of the integrator will begin to rise. As soon as this voltage rises a few millivolts above ground potential (0 Vdc) the comparator output snaps HIGH-positive. A HIGH comparator output causes the control circuit to enable the counter, which begins to count pulses.

The counter is allowed to overflow and this output bit sets switch *S*2 to the reference source (position B). The graph of Fig. 15-13B shows the integrator charging during the interval between start and the overflow of the binary counter. At time *T*2 the switch changes the integrator input from the analog signal to a precision reference source. Meanwhile, at time *T*2 the counter had overflowed, and again it has an output of $00000000_2$ (maximum counter + 1 more count is the same as the initial condition). It will, however, continue to increment as long as there is a HIGH comparator output. The charge accumulated on capacitor *C*1 during the first time interval is proportional to the average value of the analog signal that existed between *T*1 and *T*2.

Capacitor *C*1 is discharged during the next time interval (*T*2-*T*3). When *C*1 is fully discharged the comparator will see a ground condition at its active input, so it will change state and make its output LOW. Even though this causes the control logic to stop the binary counter, it does not reset the binary counter. The binary word at the counter output at the instant it is stopped is proportional to the average value of the analog waveform over the interval (*T*1-*T*2). An end-of-conversion (EOC) signal is generated to notify the computer so it knows the output data is both stable and valid (therefore ready for use).

### 15-5.2 Voltage-to-Frequency Converters

These circuits are not A/D converters in the strictest sense, but they are good for representing analog data in a form which can be tape recorded. The V/F converter output can also be used for direct input to a computer if a binary counter is used to measure the output frequency. Two forms of V/F converter are common. One is a *voltage controlled oscillator* (VCO). That is, it is a regular oscillator circuit in which the output frequency is a function of an input control voltage. If the VCO is connected to a binary or binary coded decimal (BCD) counter, the VCO becomes a V/F form of A/D converter.

The type of V/F converter shown in Fig. 15-14 is superior to the VCO method. The circuit is shown in Fig. 15-14A. The timing waveforms are shown in Fig. 15-14B. The operation of this circuit is dependent on the charging of a capacitor, although it is not an *RC* network as in the case of some other oscillator or timer circuits. The input voltage signal ($V_x$) is amplified (if necessary) by *A*1 and then converted to a

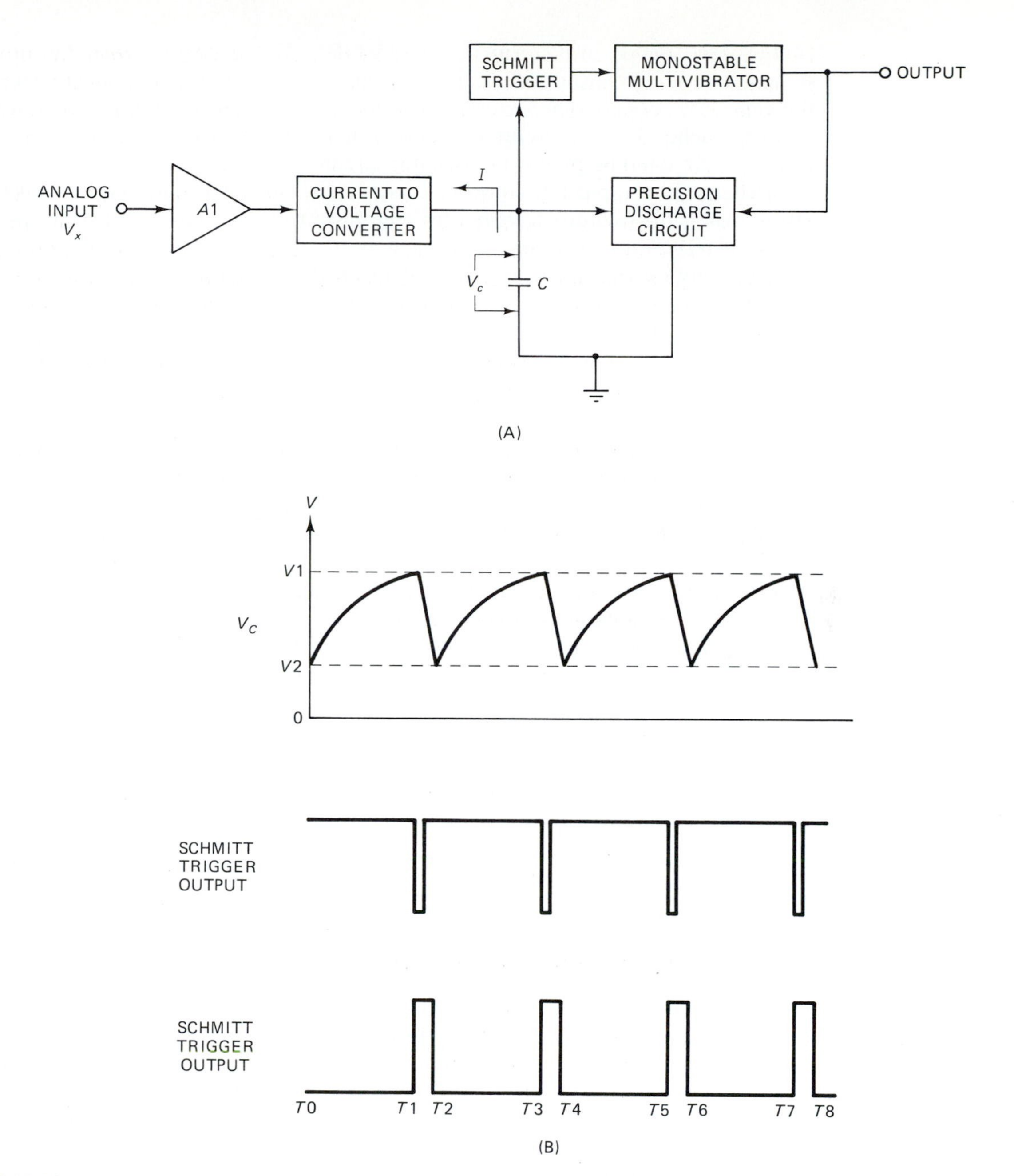


FIGURE 15-14

proportional current level in a *voltage to current converter* stage. If the voltage applied to the input remains constant, so will the current output of the $V$-to-$I$ converter ($I$).

The current from the $V$-to-$I$ converter is used to charge the timing capacitor ($C$). The voltage appearing across this capacitor ($V_c$) varies with time as the capacitor charges (see the $V_c$ waveform in Fig. 15-14B). The precision discharge circuit is designed to discharge capacitor $C$ to a certain level ($V2$) whenever the voltage across the capacitor reaches a predetermined value ($V1$). When the voltage across the capacitor reaches $V2$, a Schmitt trigger circuit is fired which turns on the precision discharge circuit. The precision discharge circuit, in its turn, will cause the capacitor to discharge in a rapid but controlled manner to value $V1$. The output pulse snaps HIGH when the Schmitt trigger fires (the instant $V_c$ reaches $V1$) and drops LOW again when the value of $V_c$ has discharged to $V2$. The result is a train of output pulses whose repetition rate is exactly dependent on the capacitor charging current which in turn is dependent upon the applied voltage. Thus, the circuit is a voltage-to-frequency converter.

Like the *VCO* circuit, the output of the *V/F* converter can be applied to the input of a binary counter. The parallel binary outputs become the data lines to the computer. Alternatively, if the frequency is relatively low, the computer can be programmed to measure the period between pulses. Also, certain interface devices such as the 6522 and Z80-CTC chips have built-in timers to measure the period.

### 15-5.3 Counter Type (Servo) A/D Converters

A counter type A/D converter (also called servo or ramp A/D converters) is shown in Fig. 15-15. It consists of a *comparator*, *voltage output DAC*, *binary counter*, and the necessary *control logic*. When the start command is received, the control logic resets the binary counter to $00000000_2$, enables the clock, and begins counting. The counter outputs control the DAC inputs so the DAC output voltage will begin to rise when the counter begins to increment. As long as analog input voltage $V_{in}$ is less than $V_{ref}$ (the DAC output), the comparator output is HIGH. When $V_{in}$ and $V_{ref}$ are equal, however, the comparator output goes LOW. This turns off the clock and stops the counter. The digital word appearing on the counter output at this time represents the value of $V_{in}$.

Both slope and counter type A/D converters take too long for many applications, on the order of $2^N$ clock cycles (where $N$ = number of bits). Conversion time becomes critical if a high frequency component of the input waveform is to be faithfully reproduced. Nyquist's criteria require the sampling rate (conversions per second) be at least twice the highest frequency to be recognized.

### 15-5.4 Successive Approximation A/D Converters

Successive approximation A/D conversion is best suited where speed is important. This type of A/D converter requires only $N + 1$ clock cycles to make the conversion. Some designs allow truncation of the conversion process after fewer cycles if the final value is found prior to $N + 1$ cycles.

The successive approximation converter operates by making several successive trials at comparing the analog input voltage with a reference generated by a DAC. An example is shown in Fig. 15-16. This circuit consists of a *comparator*, *control logic section*, a *digital shift register*, *output latches*, and a *voltage output DAC*.

When a START command is received, a binary 1 (HIGH) is loaded into the MSB of the shift register. This sets the output of the MSB latch HIGH. A HIGH in the

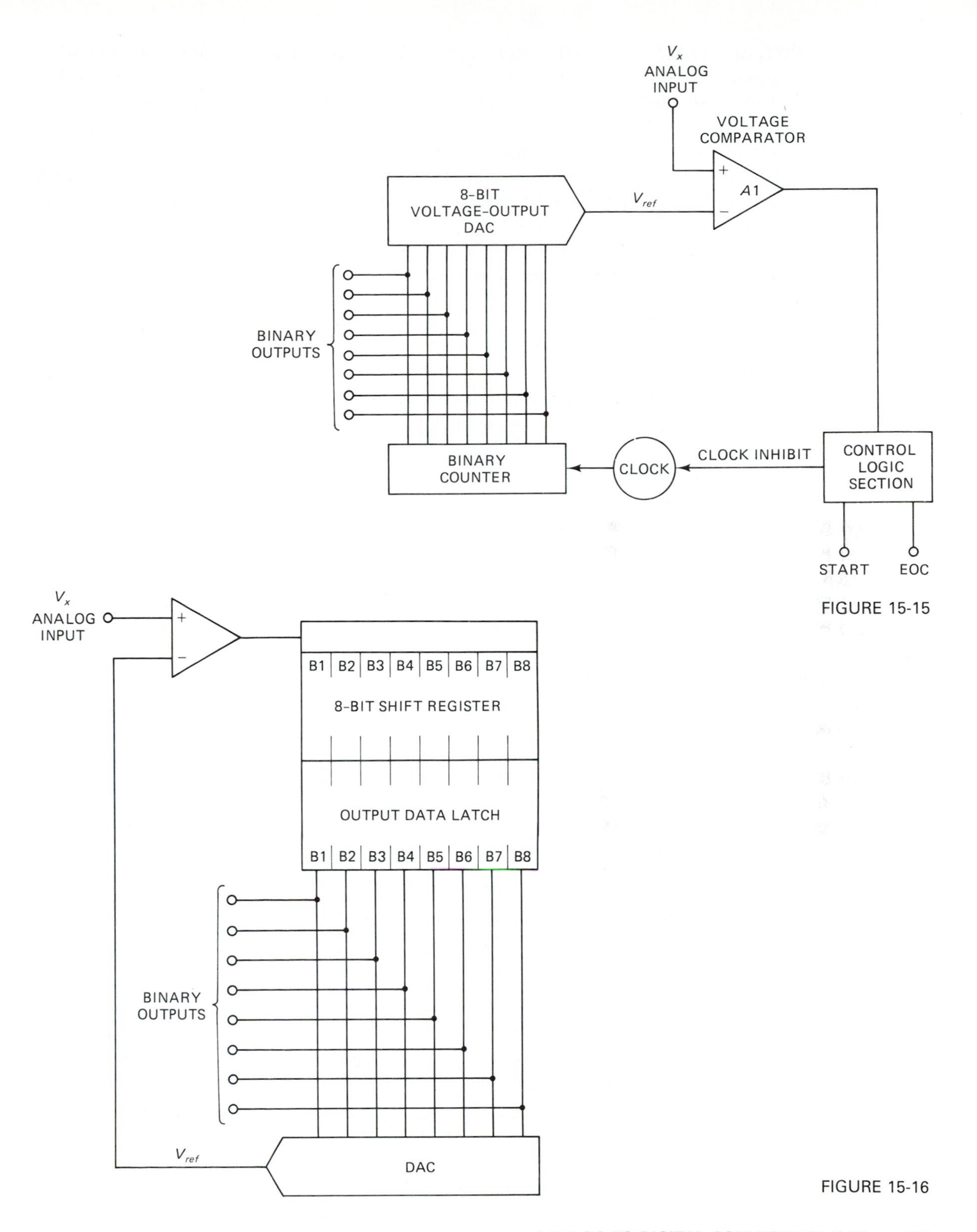


FIGURE 15-15



FIGURE 15-16

MSB of a DAC will set the output voltage $V_{ref}$ to half scale. If the input voltage $V_{in}$ is greater than $V_{ref}$, the comparator output stays HIGH and the HIGH in the shift register MSB position shifts one-bit to the right and therefore occupies the next most significant bit (bit 2). Again the comparator compares $V_{in}$ with $V_{ref}$. If the reference voltage from the DAC is still less than the analog input voltage, the process will be repeated with successively less significant bits until either a voltage is found that is equal to $V_{in}$ (in which case the comparator output drops LOW) or the shift register overflows.

On the other hand, if the first trial with the MSB indicates $V_{in}$ is less than the half-scale value of $V_{ref}$, the circuit continues making trials below $V_{ref}$. The MSB latch is reset to LOW and the HIGH in the MSB shift register position shifts one-bit to the right to the next most significant bit (bit 2). Here the trial is repeated again. This process will continue until either the correct level is found or overflow occurs. At the end of the last trial (bit 8 in this case) the shift register overflows and the overflow bit becomes an *end-of-conversion* (EOC) flag indicating the conversion is completed.

This type and most other types of A/D converters require a starting pulse and signal completion with an EOC pulse. This requires the computer or other digital instrument to engage in bookkeeping to repeatedly send the start command and look for the EOC pulse. If the start input is tied to the EOC output, conversion is continuous and the computer need only look for the periodic raising of the EOC flag to know when a new conversion is ready. Such operation is said to be *asynchronous.*

### 15-5.5 Parallel or Flash A/D Converters

The parallel A/D converter (Fig. 15-17) is probably the fastest A/D circuit known. The fastest ordinary commercial products use this method. Some sources call the parallel A/D converter the flash circuit because of its inherent high speed.

The parallel A/D converter consists of a bank of $[2^N - 1]$ voltage comparators biased by reference potential $V_{ref}$ through a resistor network which keeps the individual comparators 1 LSB apart. Since the input voltage is applied to all the comparators simultaneously, the speed of conversion is limited essentially by the slew rate of the slowest comparator in the bank and also by the decoder circuit propagation time. The decoder converts the output code to binary code needed by the computers.

## 15-6 SUMMARY

1. An analog-to-digital (A/D) converter produces a binary number output which represents the applied analog output voltage or current. A digital-to-analog converter (DAC) is exactly the opposite. It outputs an analog voltage or current proportional to an applied binary number.
2. Most DAC designs are based on the R-2R binary weighted ladder. If the output is unterminated, then the DAC is a current output type. If the output is terminated in either a load resistor or an operational amplifier, then the DAC is a voltage output type.
3. Several different coding schemes are used in data converters; *unipolar positive*, *unipolar negative*, *symmetrical bipolar*, and *asymmetrical bipolar* are the most common.
4. IC DACs can be used to generate waveforms. If the DAC digital inputs are incremented by an ordinary binary counter, a sawtooth is generated; if incremented by an up/down

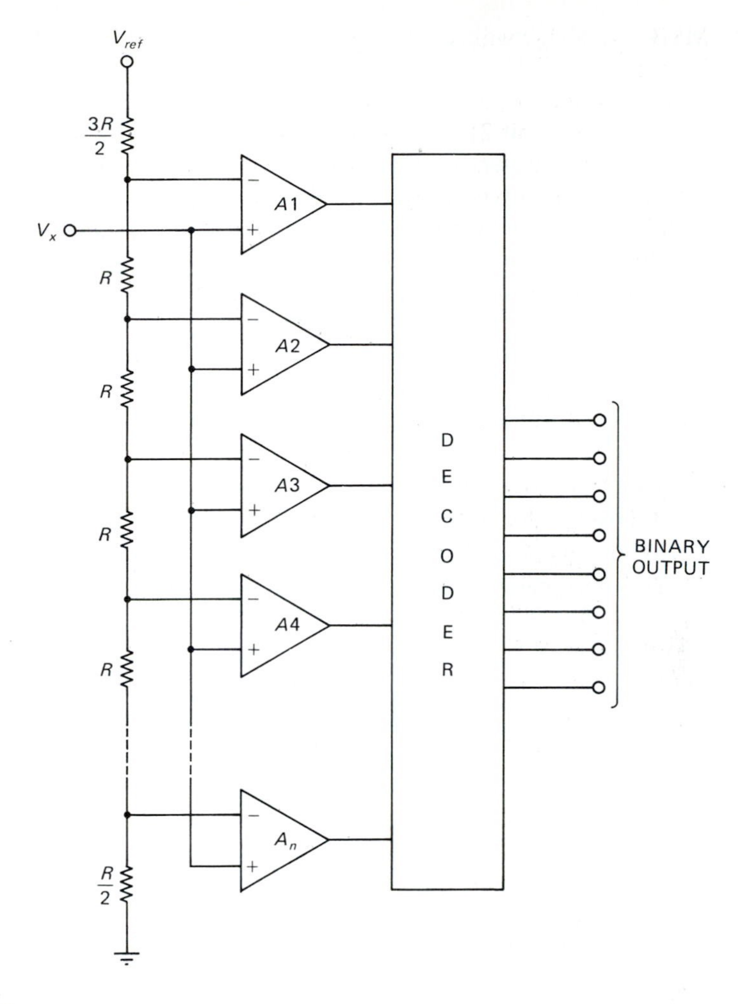


FIGURE 15-17

counter, a triangle waveform is generated; if the binary inputs are driven from a Read Only Memory (ROM) which is in turn driven by a binary counter used as an address generator, any waveform whose binary bit pattern can be stored in ROM can be created.

5. Several approaches are used for A/D conversion: *voltage-to-frequency converter*, *single-slope integrator*, *dual-slope integrator*, *counter* (or *servo*), *successive approximation*, and *flash* methods.

## 15-7 RECAPITULATION

Now return to the objectives and Prequiz questions at the beginning of the chapter and see how well you can answer them. If you cannot answer certain questions, mark them and review the appropriate parts of the text. Next, try to answer the questions and work the problems below using the same procedure.

## 15-8 STUDENT EXERCISES

1. Build a 5-bit R-2R ladder such as Fig. 15-1 from resistors in the 100 to 1000 ohms range ($R_t >> R$). Test the circuit for the output voltage when various binary numbers are applied by closing the associated switches.
2. Build a working DAC for 0 to +9.96 Vdc operation using a DAC-08, MC-1408, DAC-0806, or similar commercial IC DAC. Test the circuit for the output voltage when various binary numbers are applied to the eight binary input lines.
3. Configure the circuit in the previous exercise as an AC programmable gain amplifier.
4. Build an 8-bit A/D converter based on the DAC circuit of exercise 2 above.

## 15-9 QUESTIONS AND PROBLEMS

1. Draw the circuit for an R-2R ladder DAC with voltage output.
2. A DAC-08 device has a reference potential of +10 Vdc and an reference input resistance of 5 kohms. Calculate the output current at full scale (binary input = $10000000_2$).
3. Describe how a DAC can be used to digitally synthesize a triangle waveform. Use a circuit diagram to illustrate the case.
4. Describe a method, and provide a simplified circuit diagram, for a synthesizer based on a DAC that will output the human electrocardiogram waveform.
5. What is the output current of a DAC-08 when the reference current is 2 mA and the applied binary word is $00000001_2$?
6. What is the full-scale output of a DAC-08 IC digital to analog converter when the reference potential is +10 volts DC?
7. An eight-bit DAC is designed for symmetrical bipolar operation. If the full-scale outputs are ±5 Vdc, what are the values of +ZERO and −ZERO?
8. In addition to the binary digital inputs, the ____________ DAC has an external reference voltage.
9. The circuit of Fig. 15-4 uses a DAC-08. What is the output voltage $V_o$ if $I_o$ = 1.25 mA?
10. In the circuit of Fig. 15-5, what is the output voltage if $R$ = 4.7 kohms and $I_o$ = 2 mA?
11. List the different forms of A/D converter circuit.
12. Draw a block diagram for a servo converter circuit.
13. A ____________ A/D converter has a conversion time of $2^N$ clock cycles.
14. A ____________ A/D converter has a conversion time of $[2^N - 1]$ clock cycles.
15. A ____________ A/D converter has a conversion time of $[N + 1]$ clock cycles.
16. The ____________ A/D converter is generally regarded as the fastest form of converter, even though the output must be decoded before it can be used in binary systems.
17. Draw the output waveform of a 3-bit DAC. Label the 1-LSB value.
18. Draw the output circuit for a DAC-08 that will produce an output impedance and a maximum potential of +10 volts when $I_o$ = 0.002 amperes.
19. What is the purpose of $C$ in Fig. 15-5?
20. In Fig. 15-7, what is the output current of the DAC-08 when $V_o$ = +10 Vdc?
21. What is the purpose of REF-01 in Fig. 15-8?

22. Draw the circuit for a 3-bit flash A/D converter.
23. Discuss the terms "plus and minus zero" with respect to bipolar coded A/D converters.
24. An 8-bit A/D converter is designed to convert a range of potentials from 0 to +4.5 volts. (a) What is the 1-LSB voltage? (b) What voltage is represented by $11000010_2$?
25. Write the codes for zero, midscale, and full-scale potentials for a unipolar DAC.

CHAPTER 16

# Audio Applications of Linear IC Devices

## OBJECTIVES

1. Be able to list various audio applications for linear IC devices.
2. Understand the methods of design used for audio circuits.
3. Learn the types of linear ICs typically used in audio circuits.

## 16-1 PREQUIZ

These questions test your prior knowledge of the material in this chapter. Try answering them before you read the chapter. Look for the answers (especially those you answered incorrectly) as you read the text. After you have finished studying the chapter, try answering these questions again, and those at the end of the chapter (see Section 16-11).

1. Draw the circuit for a bridge audio power amplifier.
2. An audio preamplifier for a two-way radio transmitter must have a gain of 100 and a frequency response of 300 Hz to 3300 Hz between −3 dB points. Calculate the values for the input and feedback resistors and the value of the capacitor shunting the feedback resistor. Assume a noninverting follower.
3. In the circuit above, the input circuit contains a transformer with a 1:2 turns ratio. What is the total voltage gain of this circuit?
4. Draw the circuit for a three-channel audio mixer with an inverting characteristic.

## 16-2 AUDIO CIRCUITS

The audio frequency range is generally accepted to be the nomimal range of human hearing, 20 Hz to 20,000 Hz. Although some applications require frequency ranges lower than 20 Hz, or higher than 20 KHz, most audio circuits fall within the stated range. Some applications use a smaller range, however. For example, communications equipment sometimes uses 300 Hz to 3,000 Hz. Similarly, an AM radio station is allowed to transmit audio frequencies up to 5,000 KHz, but not higher. Thus, for audio equipment designed for use in AM radio stations, the audio frequency range may be limited to 20 Hz to 5,000 Hz. Cassette tape recorders are often limited to 16 KHz at the upper end, while other forms of tape recorder may use the entire 20-KHz range.

There are also some applications where a nominal 20-KHz frequency range actually requires a higher range of frequencies. High-fidelity enthusiasts once debated the desirability of 100 KHz or even 250 KHz as the upper end of the frequency range. The argument is based on the fact that all nonsinusoidal waveforms are composed of a fundamental sine wave frequency plus a number of harmonic sine and cosine waves. The tonal coloration of any given musical instrument is dependent on which overtones (harmonics) are present and what amplitude they exhibit. That is why a violin playing an A at 440 Hz sounds different from a piano or trumpet playing the same 440 Hz note. The difference between these instruments is the harmonic content of the note. It was argued that the brightness or coloration of the audio is believed to be compromised if the harmonics above 20 KHz are eliminated or attenuated.

Several different classes of amplifiers are used in audio circuits. The *preamplifier* is designed to accept a low-level voltage signal and boost it to a higher voltage level. For example, the output of a phonograph cartridge is on the order of 1 to 5 millivolts, while the standard line out level required to drive a power amplifier is on the order of 100 mV to 500 mV. Some tape recorder magnetic heads produce outputs in the 100 uV to 500 uV range. The preamplifier is designed to boost the signal from these levels to a level from 10 to 1000 times higher. A *control amplifier* is designed to operate at low gain ($\times 1$ to $\times 10$), but performs certain signal processing functions such as bass and treble boost, balance, or a similar function. Finally, a *power amplifier* is designed to boost a 100 mV to 500 mV voltage signal to a power level sufficient to drive a loudspeaker or other load. Audio power amplifiers tend to produce output levels from 100 milliwatts to hundreds of watts. In a typical audio system these circuits are arranged as in Fig. 16-1.

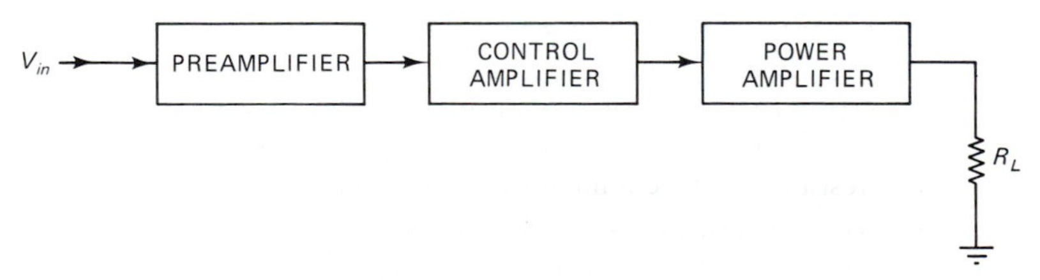


FIGURE 16-1

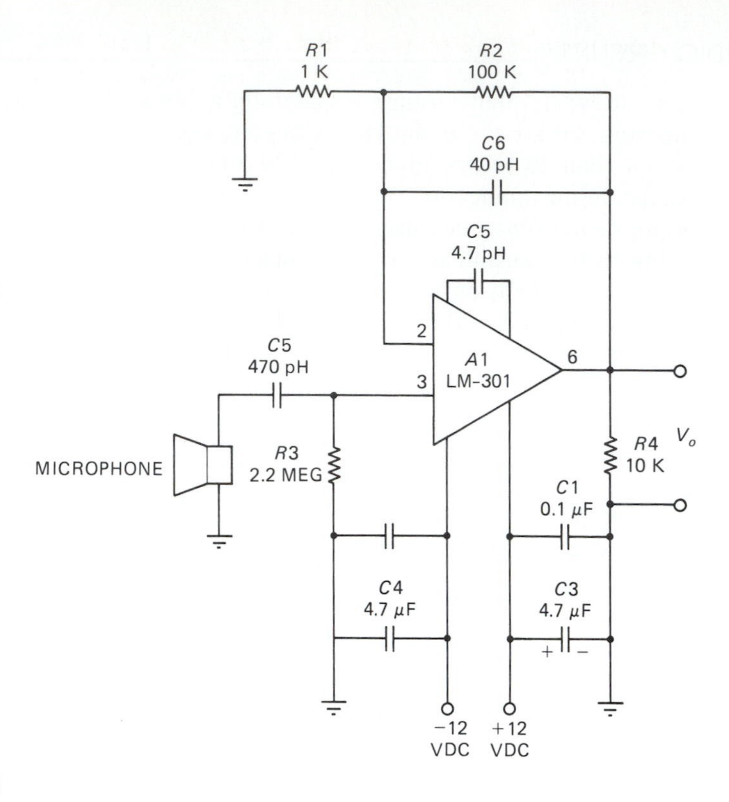


FIGURE 16-2

## 16-3 PREAMPLIFIER CIRCUITS

A preamplifier circuit gives initial amplification to a signal before passing it to another circuit for additional amplification or other processing. For example, a microphone has a low-level output (several millivolts for a dynamic type, and up to 0.2 volts for a crystal type). A preamplifier typically boosts the microphone signal to 100 to 1000 millivolts before it is applied to the input of a power amplifier (for the loudspeaker) or the input of a transmitter modulator (depending upon use).

**Microphone Preamplifier.** Figure 16-2 shows a simple gain-of-100 microphone preamplifier for communications and public address uses. This circuit uses an LM-301 operational amplifier in the noninverting follower gain configuration. With the values of feedback and input resistors (*R*1 and *R*2) shown, the gain is 101.

The input circuit of this preamplifier is capacitor coupled to the microphone. In order to keep the input bias currents of the op-amp from charging the capacitor and thereby latching up the op-amp, a 2.2 megohm resistor (*R*3) is connected from the noninverting input to ground. This circuit is general, but it can be made less complex if we only use a dynamic microphone. Those forms of mike use either a high or low impedance coil (like a loudspeaker, but reversed in function) which is permanently connected into the circuit. In this case, delete *R*3 and *C*7, and connect the microphone between ground and pin 3 of the op-amp. If the microphone is to be disconnected from time to time, however, keep *R*3 in the circuit to prevent the op-amp output from saturating at or near $V+$ when the noninverting input goes open.

The frequency response of this circuit is tailored by $C5$ and also by the capacitor shunting the feedback resistor ($C6$). With the values of capacitance shown, the upper $-3$ dB point in the response curve is a little over 3000 Hz and falls off at a rate of approximately $-6$ dB/octave above that frequency. The equation for determining the frequency response is

$$f = \frac{1}{2\pi R\ C} \tag{16-1}$$

where

$f$ = the $-3$ dB frequency of the amplifier in hertz (Hz)

$R$ = the resistance in ohms

$C$ = the capacitance in farads

### EXAMPLE 16-1

Calculate the capacitance (in uF) needed to establish a $-3$ dB frequency response of 5000 Hz when the feedback resistor is 100 kohms.

### Solution

$$C6 = \frac{1}{(2)\ (3.14)\ (100{,}000 \text{ ohms})\ (5{,}000 \text{ Hz})} \times \frac{10^6 \text{ uF}}{1\ F}$$

$$C6 = \frac{1}{3.14 \times 10^9} \times \frac{10^6 \text{ uF}}{1\ F} = 3.2 \times 10^{-4} \text{ uF}$$

■

Frequency-tailored preamplifiers are found in several varieties which are characterized by the shape of the frequency response. Figure 16-3 shows three common response characteristics. The form shown in Fig. 16-3A is the single slope version. The gain ($A_v$) is flat from DC (or some low AC frequency) to some higher frequency ($F_c$). Above this frequency the gain rolls off at a specified rate such as $-6$ dB/octave. The frequency response is determined as the point at which the gain drops off $-3$ dB from the low frequency response. The $-3$ dB point may be the natural roll-off frequency of the amplifier being used or it may be artificially tailored as in the example above.

A two-slope response characteristic is shown in Fig. 16-3B. In this case, there are two gain segments. A higher gain ($A_{V1}$) is found from the lower end of the response range to frequency $F1$, while a lower gain ($A_{V2}$) is found from $F1$ to $F_c$. This type of response is commonly found in tape recorder systems. It is used to roll off the high frequency preemphasis imparted to the tape recorded audio in order to overcome noise. If the gain of the playback preamplifier was not tailored in a manner such as Fig. 16-3B, the reproduced audio would be unbalanced toward the treble end of the range.

A three-slope version is shown in Fig. 16-3C. In this case there are three different gains, $A_{V1}$, $A_{V2}$, and $A_{V3}$. This characteristic is used to restore the audio balance of phonograph records when reproduced.

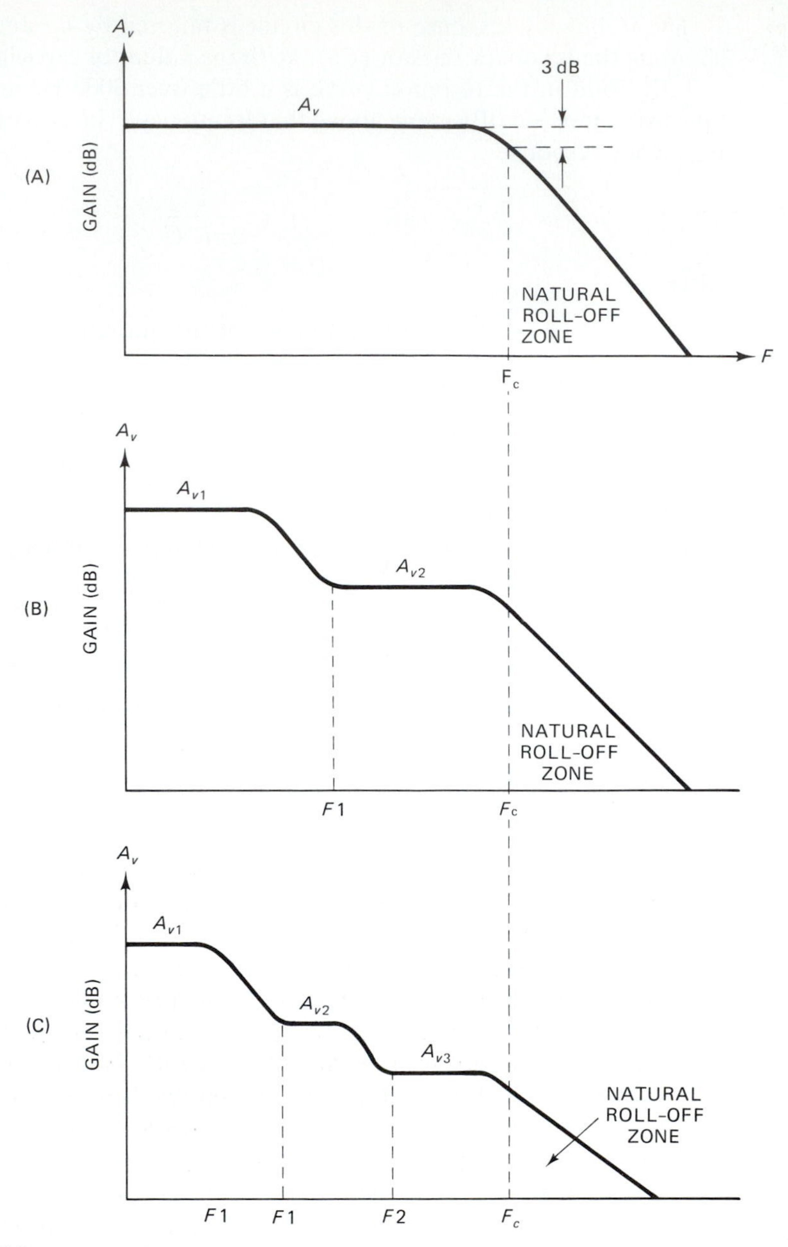


FIGURE 16-3

A typical preamplifier circuit is shown in Fig. 16-4. The basic circuit is a noninverting follower operational amplifier circuit. The input network $R1/C1$ sets the lower $-3$ dB frequency by Eq. (16-1). Resistor $R1$ is needed if there is a substantial bias current from the input of the operational amplifier. That current could charge the capacitor ($C1$), causing a DC offset which will eventually saturate the output of the amplifier. Resistor $R1$ drains the DC charge that would otherwise accumulate on $C1$. Some

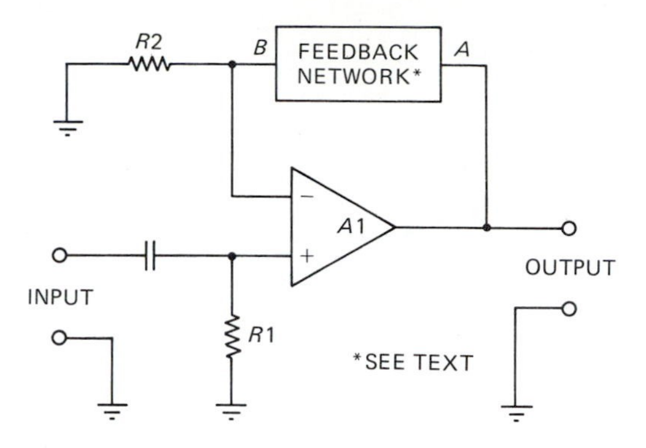


FIGURE 16-4

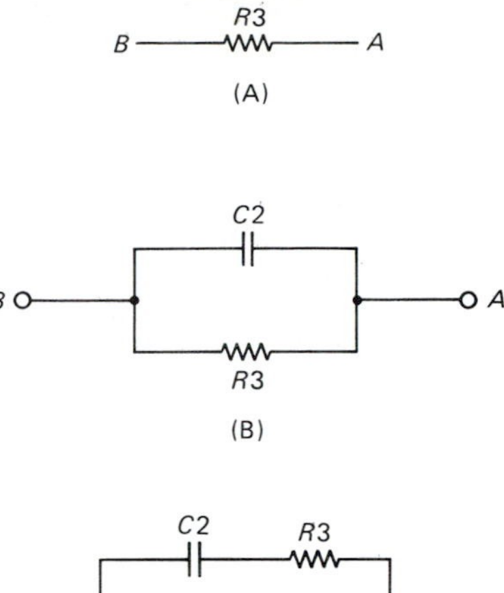


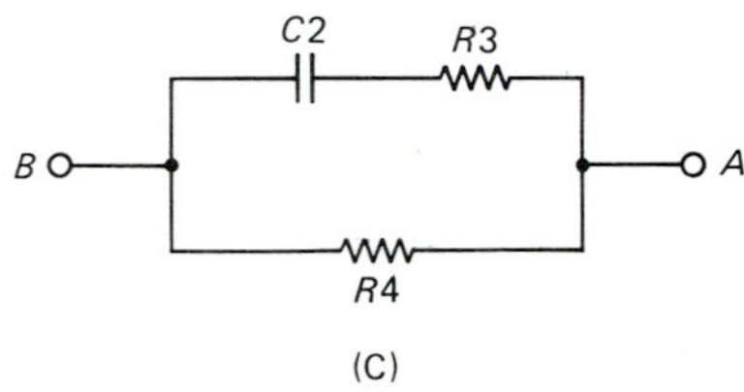


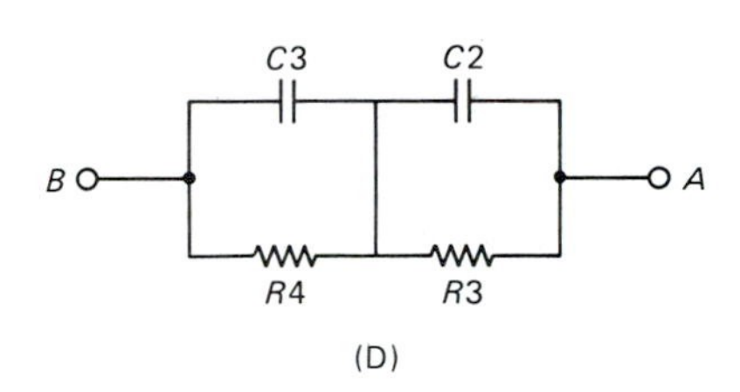


FIGURE 16-5

modern operational amplifiers have such a low value of input bias current (picoamperes) that $R1$ can sometimes be omitted.

The input impedance of the preamplifier is set by resistor $R1$, so it should have as high a value as possible. The general rule is to make the input resistor not less than ten times the source impedance of the transducer or other device used to originate the audio signal. Standard audio amplifiers use a 600-ohm output impedance (a holdover from telephone technology). In those situations almost any value of input resistor is sufficient. However, tape heads tend to have a source impedance on the order of 20 kohms to 100 kohms. Crystal microphones and phonograph cartridges are over 100 kohms. Thus, it is common to use a value of 1 megohms to 10 megohms for $R1$.

The characteristic of the upper $-3$ dB roll-off is set by the feedback network shown in Fig. 16-4 as a block. Figure 16-5 shows several forms of an $RC$ feedback network circuit. If a single resistor is used, the roll-off will be single-slope (Fig. 16-3A), but the $-3$ dB point is the natural $-3$ dB point for the amplifier. If a network such as Fig. 16-5B is used, the characteristic is still single-slope, but the $-3$ dB point is set by Eq. (16-1). A two-slope characteristic (Fig. 16-3B) is created by the network of Fig. 16-5C. In this case, the low frequency gain setting resistor ($R4$) is shunted by a roll-off network which has a $-3$ dB breakpoint determined by Eq. (16-1). The three-slope network is shown in Fig. 16-5D.

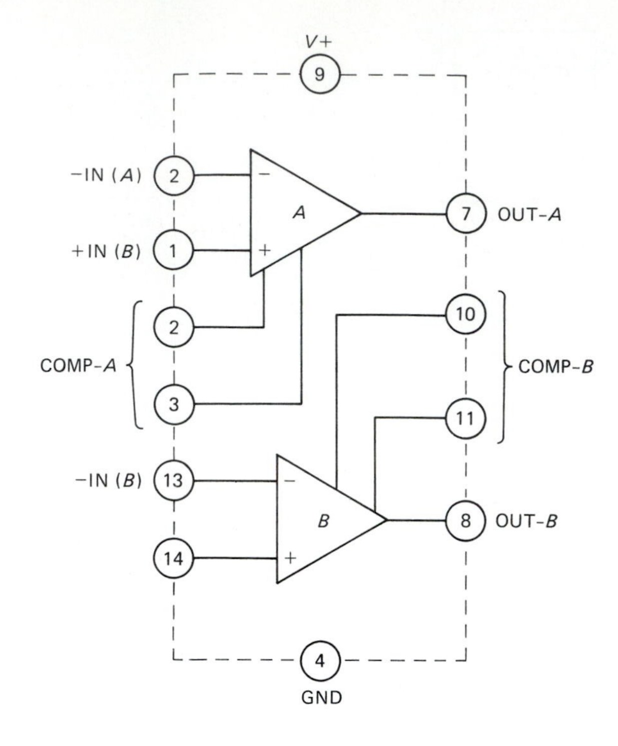


FIGURE 16-6

### 16-3.1 LM-381 IC Preamplifier

There are a number of operational and nonoperational amplifier audio preamplifiers on the market. One of the earliest was the Motorola MC-1303P, which used two gain-of-20,000 wideband operational amplifiers in a single package. Two devices are required for stereo audio applications.

Another class of nonoperational amplifiers is represented by the LM-381 shown in Fig. 16-6. This consists of a pair of differential devices similar to operational amplifiers, but it operates from a single DC power supply of +9 to +40 Vdc. The LM-381 offers voltage gains of 112 dB per channel with a unity gain bandwidth product of 15 MHz. The power bandwidth is 75 KHz at an output signal of 20 volts peak-to-peak. The device boasts a total harmonic distortion (THD) of 0.1% (1 KHz, $A_v$ = 60 dB).

The LM-381 offers a power supply rejection ratio (PSRR) of 120 dB, and a stage-to-stage isolation (at 1000 Hz) of 60 dB. The LM-381 is internally frequency compensated, yet it can also be custom-tailored through external compensation. The device also features output short circuit protection. The output terminals will sink 2 mA, or source 8 mA, of current. The total equivalent noise input is 0.5 to 1 u$V_{rms}$.

The basic circuit for the LM-381 device is shown in Fig. 16-7A. The input signal is applied directly to the noninverting input through a DC blocking capacitor. The gain and frequency response characteristic is set by the feedback network between output and inverting input. Examples of feedback networks are shown in Fig. 16-7B (NAB tape preamplifier) and Fig. 16-7C (RIAA phonograph preamplifier).

An inverting amplifier configuration of the LM-381 is shown in Fig. 16-8. In this

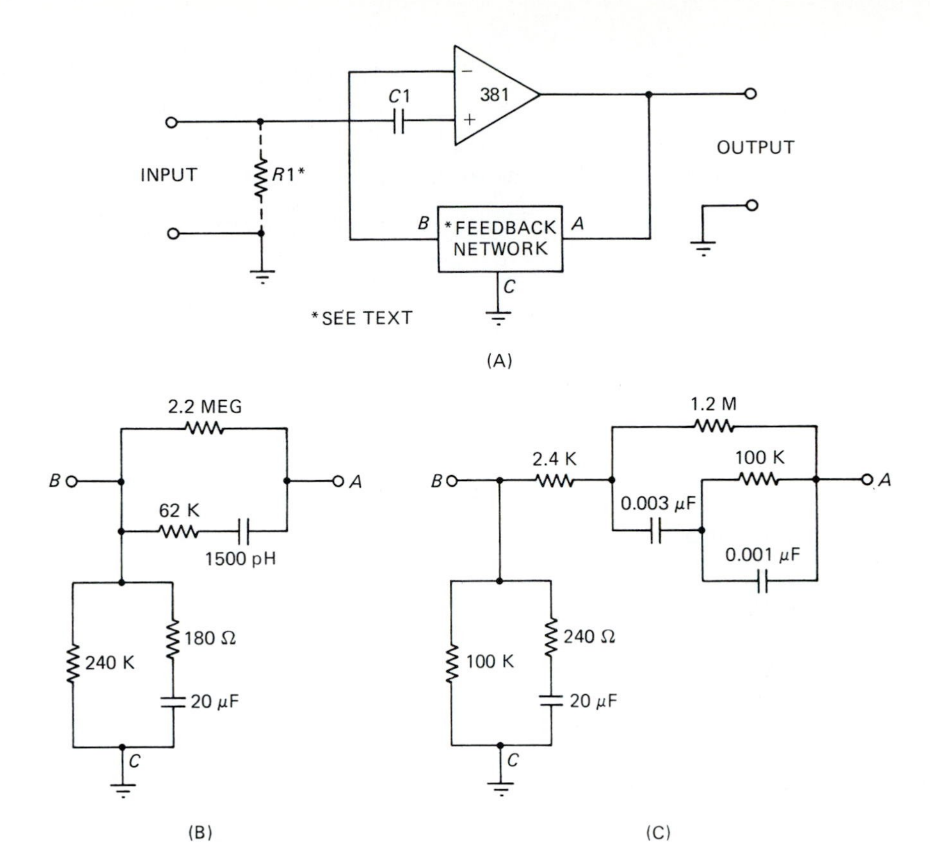


FIGURE 16-7

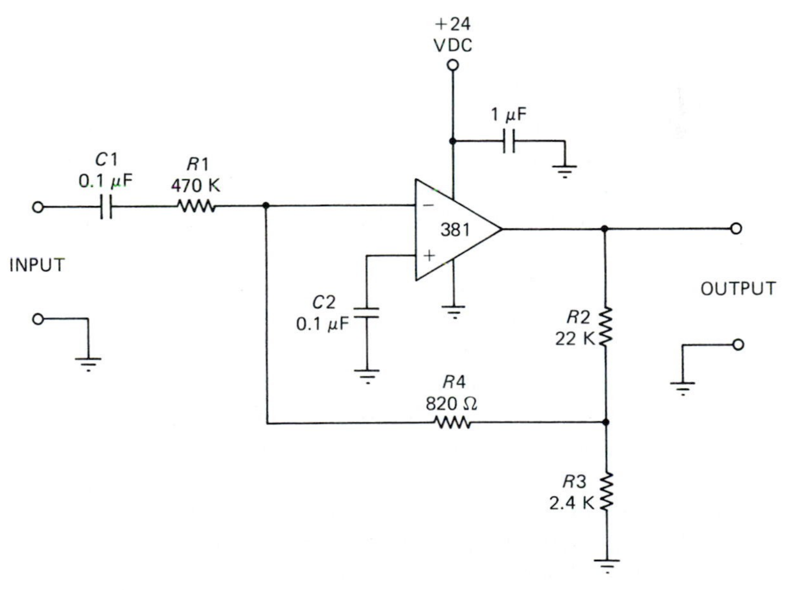


FIGURE 16-8

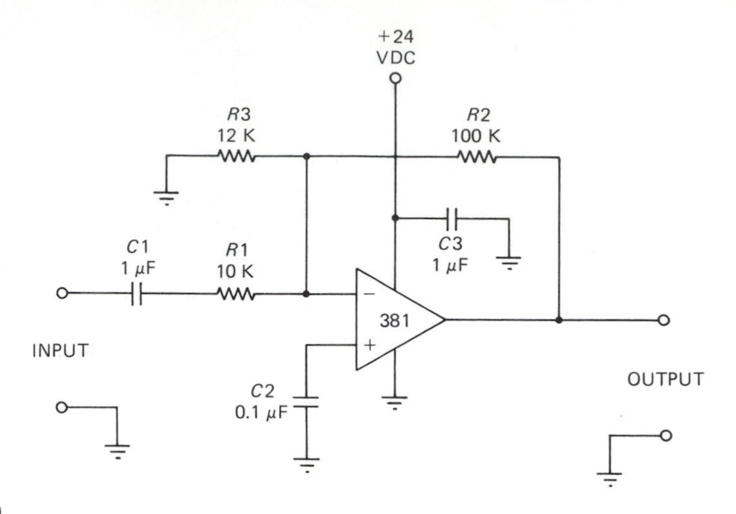


FIGURE 16-9

circuit the noninverting input is decoupled to ground through a capacitor ($C2$). The input signal is applied to the inverting input through an RC network ($C1R1$). The gain is set by a feedback network consisting of a resistor voltage divider ($R2/R3$) and a series resistor ($R4$).

A low-distortion version of this same circuit is shown in Fig. 16-9. The THD of the more general circuit (above) is on the order of 0.1%, while that of Fig. 16-9 is 0.05% or less for a gain ($A_v$) of ten. In this circuit the gain-setting circuit is the feedback network $R2/R3$ connected to the inverting input. This circuit is almost the same as the inverting operational amplifier configuration. However, the input signal is applied not through $R3$ but rather through an RC network ($R1C1$) similar to the previous circuit.

**CMOS Preamplifier.** Figure 16-10 shows two general-purpose preamplifiers based on the RCA CA-3600E device. This IC is not an operational amplifier. It is a complementary COS/MOS transistor array. A single-stage design is shown in Fig. 16-10A. A multistage design is in Fig. 16-10B. The internal transistor array equivalent circuit for one transistor pair is shown in the inset to Fig. 16-10A. The single-stage design is capable of up to 30 dB of gain at a $V+$ of 15 volts DC, and slightly more at lower potentials (but only at a sacrifice of the 1 MHz $-3$ dB point).

The multistage design shown in Fig. 16-10B is capable of voltage gains to 100 dB at frequencies up to 1 MHz (assuming a 10 volt DC supply, gain drops to 80 dB for +15 volt supplies). This gain and frequency response is quite useful in audio and other applications, but it must be approached with caution when actually built. Be sure to keep the input and output sides of the amplifier separated as much as possible. Also make sure that the power supply decoupling capacitors are mounted as close as possible to the body of the IC.

The 50 ohm input impedance of this amplifier takes it out of the audio amplifier category (which usually uses higher system impedances), unless a preamplifier stage with a higher impedance is used. A good candidate for this is the circuit of Fig. 16-10A.

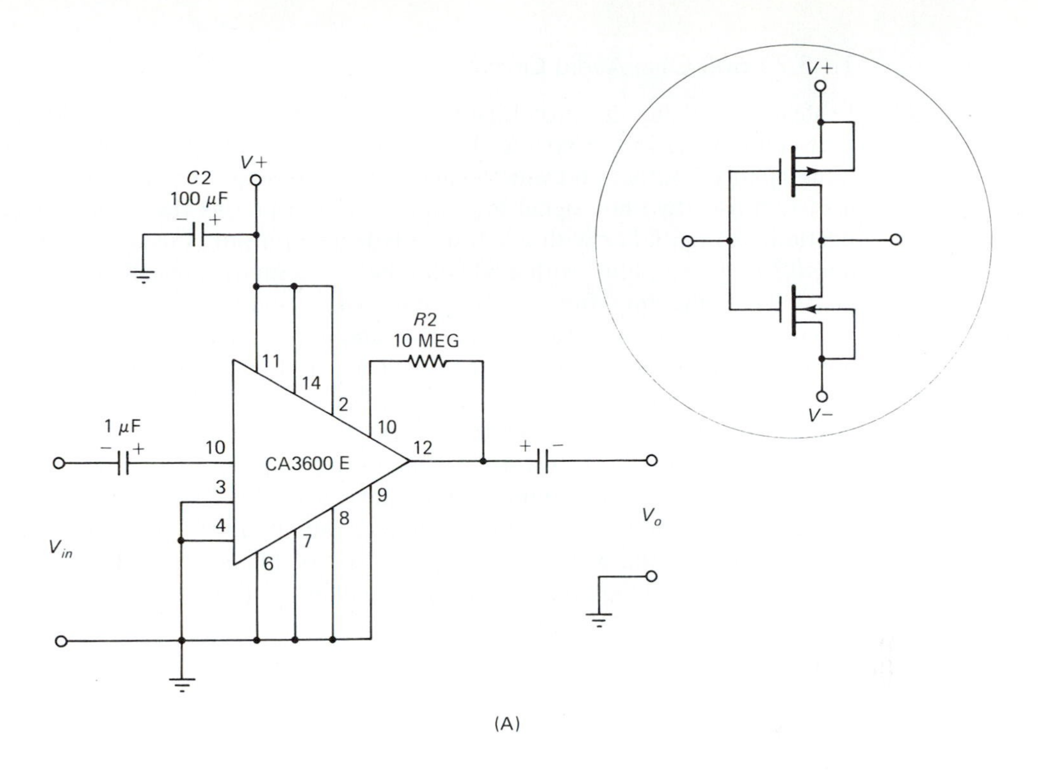


(A)

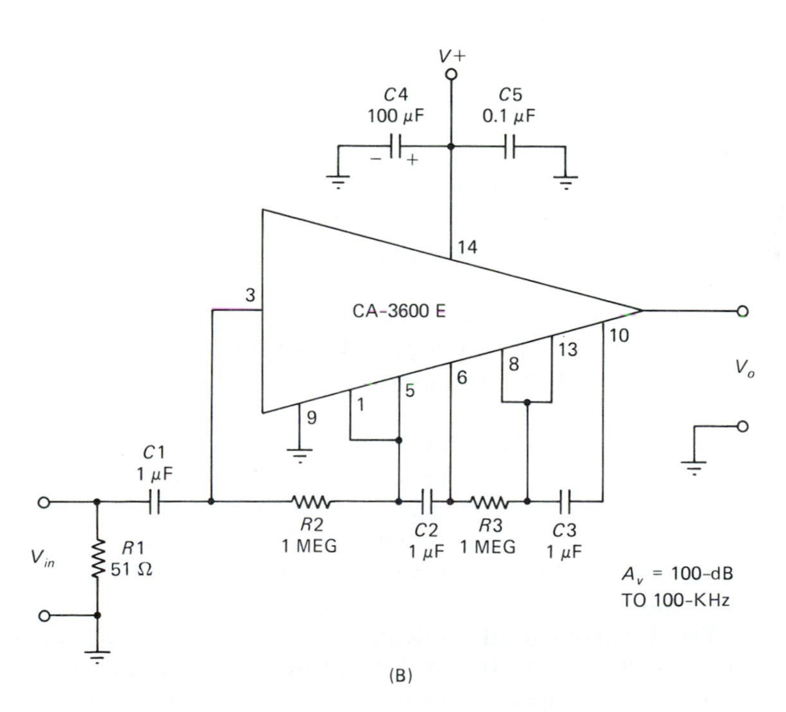


(B)

FIGURE 16-10

### 16-3.2 600 Ohm Audio Circuits

Professional audio and broadcast applications generally use a 600 ohm balanced line between devices in the system. For example, a remote preamplifier will have a 600 ohm balanced output and will connect to the next stage through three-wire line. Such a system uses two hot signal lines and a ground line as the stage-to-stage interconnection. An amplifier with a 600 ohm balanced output is usually called a *line driver amplifier*. An amplifier with a 600 ohm balanced input is usually called a *line receiver amplifier*. Some amplifiers are both line drivers and line receivers.

Figure 16-11A shows a line receiver amplifier based on the LM-301 operational amplifier used in the unity gain, noninverting configuration. The input circuit is a line transformer. If the turns ratio of transformer $T1$ is 1:1, the overall gain of the circuit is unity. But if the transformer has a turns ratio other than 1:1, the gain is essentially the turns ratio of the transformer. For example, suppose a transformer is selected with a 600 ohm balanced input winding and a 10,000 ohm secondary winding. Such a transformer will have a sec/pri impedance ratio of 10,000/600, or about 17:1, so the turns ratio is on the order of the square root of 17, or about 4.1. Thus, the voltage actually applied to the noninverting input will be 4.1 times higher than the input signal voltage from the 600 ohm line.

$$A_v = \left[\frac{N_s}{N_p}\right]\left[\frac{R_f}{R_{in}} + 1\right] \tag{16-2}$$

Like most other op-amp circuits, the one in Fig. 16-11A requires two DC power supplies ($V-$ and $V+$). Typically, these power supplies can be anywhere from $\pm 6$ to $\pm 15$ volts. As in other circuits, each DC power supply is decoupled with a pair of capacitors. The output of this circuit is an ordinary single-ended voltage amplifier (as in other op-amp circuits), so it will typically have a very low impedance. Some designers use a transformer coupled output circuit. It is possible to get away with a 600 ohm 1:1 transformer if the natural output impedance of the op-amp is on the order of 50 ohms or so. The general rule is the primary impedance of any transformer selected should be 10 times the natural output impedance of the device for best voltage transformation.

Another way to make a 600 ohm line input amplifier is to use the simple DC differential circuit (Chapter 5) and make sure the input resistors are 300 ohms each (see Fig. 16-11B). The gain of the circuit is $R3/300$, provided that $R1 = R2 = 300$ ohms and $R3 = R4$.

Figure 16-11C shows a line driver amplifier based on a pair of operational amplifiers (the DC power supply connections are deleted for simplicity). The output circuitry is balanced because it is made from two single-ended op-amps driven out of phase with each other. The low output impedance of the operational amplifier, plus the 270 ohms series resistance, makes the balanced output impedance a total of approximately 600 ohms.

The circuit of Fig. 16-11C is a good example of the clever use of one of the properties of the ideal op-amp. Recall that one of the ideal properties is inputs stick together. In other words, applying a voltage to one input causes the same voltage to appear at the other input. In this case, for example, the AF input signal voltage applied to the

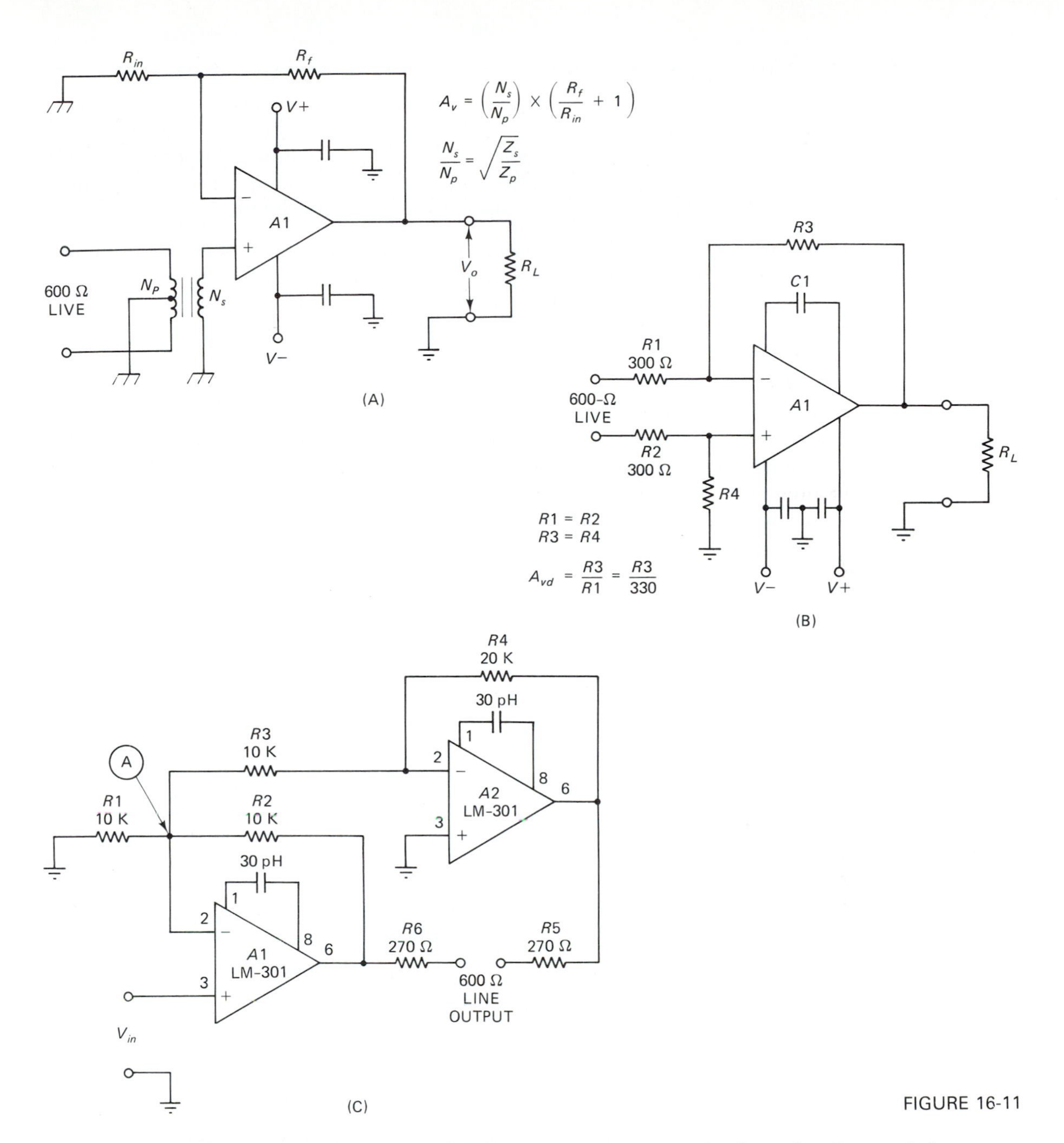


FIGURE 16-11

noninverting input of amplifier $A1$ also appears on the inverting input of that same amplifier. Thus, $V_{in}$ appears on both the noninverting input and at point A in Fig. 16-11C. Therefore, the circuit uses point A to feed the other half of the balanced circuit, amplifier $A2$. Because $A1$ is a noninverting gain of two circuit and $A2$ is a gain of two inverting circuit, the two sides are out of phase with each other, the condition required of the two balanced output lines.

## 16-4 AUDIO MIXERS

An *audio mixer* is a circuit which linearly combines audio signals from two or more inputs into a single channel. Application examples include multiple microphone public address systems, multiple guitar systems (music), or radio station audio consoles where inputs from tape players, record players, and two or more microphones are combined into a single line which goes to the transmitter's modulator input.

**Audio Mixer I.** One form of audio mixer is an operational amplifier version shown in Fig. 16-12A. This circuit is nothing more than a unity gain-inverting follower with multiple inputs. Three audio lines are identified here, $V1$, $V2$, and $V3$. Each of these are applied to the input of the operational amplifier, and are exposed to gains of $R4/R1$, $R4/R2$ and $R4/R3$, respectively. Because all resistors are 100 kohms, the gains for all three channels are unity.

Gain can be customized on a channel-by-channel basis by varying the input resistance value. The gain of any given channel will be 100 kohm/$R$, where $R$ is the input resistance ($R1$, $R2$, or $R3$) in kohms. The output voltage for multiple inputs mixers is given by

$$V_o = R4\left[\frac{V1}{R1} + \frac{V2}{R2} + \frac{V3}{R3} + \ldots + \frac{V_n}{R_n}\right] \tag{16-3}$$

Be careful not to reduce the input resistance so the source is loaded. If the source is another operational amplifier preamplifier (or other voltage amplifier), the input resistance can be reduced to several kohms without a problem. But if the source is a high-impedance phono cartridge or some similar high-impedance device, then 50 kohms is probably the minimum acceptable value.

In some cases, it is beneficial to increase the value of the feedback resistance to 1 megohm or so, in order to make the corresponding input resistances higher for any given gain. Remember, the input impedance seen by any single channel is the value of the input resistor.

A *master gain control* is provided by making feedback resistor $R4$ variable. If no gain control is needed, make this resistor fixed. An audio taper potentiometer is used for most audio applications. If the application calls for a one-time "set and forget" gain control adjustment (as might be true in radio station applications), make $R4$ a trimmer potentiometer; otherwise it should be a panel type with a shaft appropriate for a knob (quarter-inch half- or full-round).

The operational amplifier can be almost any good op-amp with a gain bandwidth product sufficient for audio applications. Because the gain is unity, any GBW over 20 KHz will suffice (which means almost all devices except the 741 family, which will work in communications applications).

The circuit of Fig. 16-12A can also be used to add a DC offset level to an audio signal. An example showing why this might be desirable is shown in Fig. 16-12B. Here the audio amplifier is used to drive an optocoupler device. These IC devices have a light emitting diode (LED) on the input side and a phototransistor on the output side. A DC bias current ($I1$) is used to raise the LED into a linear portion of its operating range and to prevent negative excursions of the input signal from turning

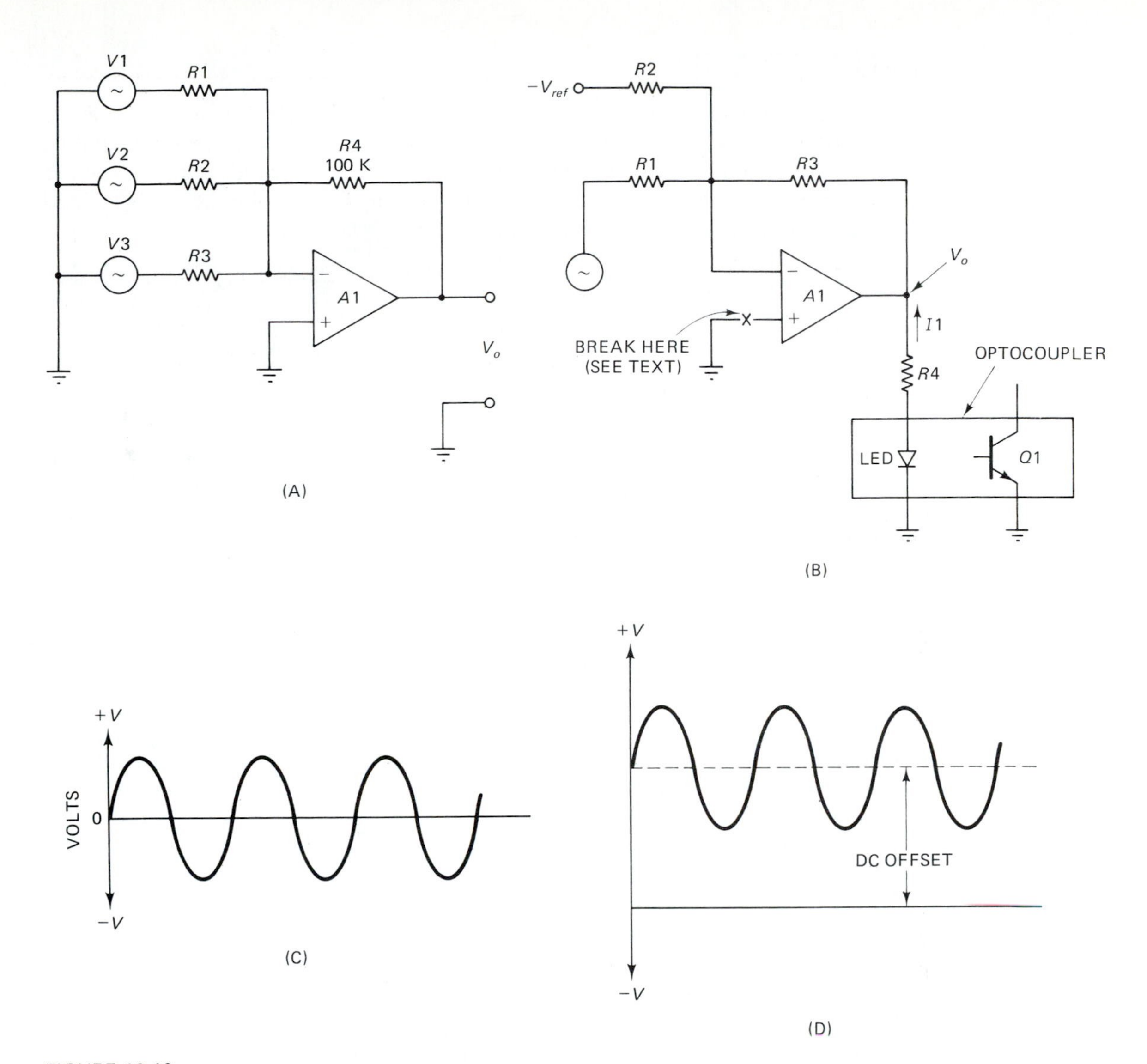


FIGURE 16-12

it off. Figure 16-12C shows the output signal without the DC offset bias. Figure 16-12D shows the output signal riding on a positive DC output level. The DC bias in Fig. 16-12B comes from a second input in which a negative reference voltage ($V_{ref}$) is used as the input signal. The DC offset level will be $[-(-V_{ref}R3)/R2]$.

Another method of providing the DC offset bias is to break the ground connection at the noninverting input and apply a DC bias of the same polarity as the required output offset. The gain of the circuit in that case will be $[(R3/R2) + 1]$.

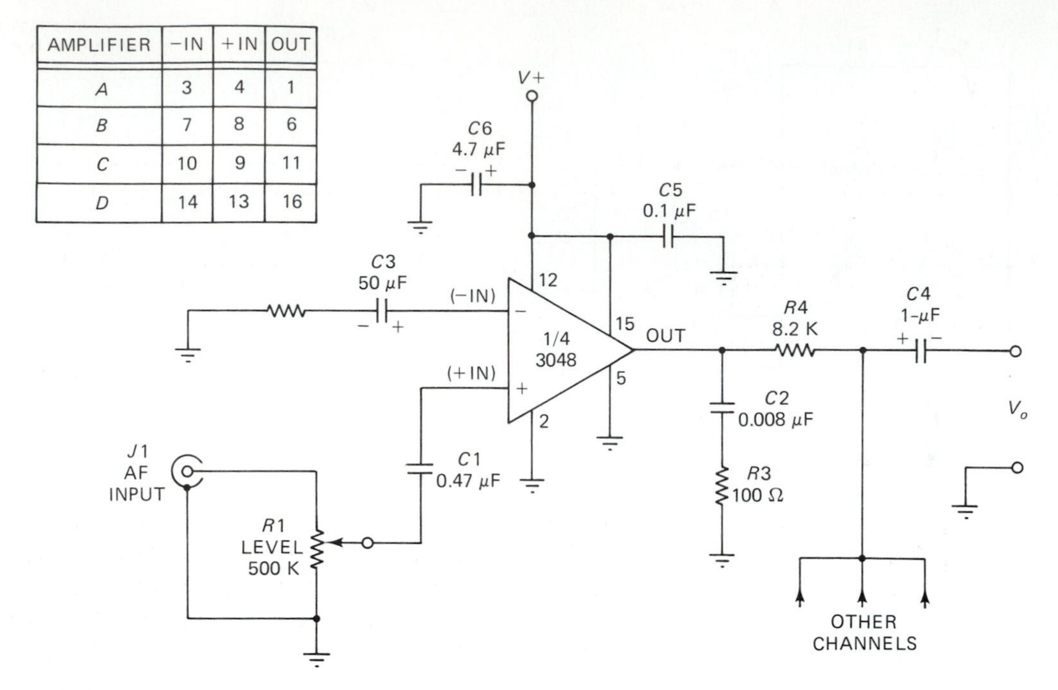

| AMPLIFIER | −IN | +IN | OUT |
|---|---|---|---|
| A | 3 | 4 | 1 |
| B | 7 | 8 | 6 |
| C | 10 | 9 | 11 |
| D | 14 | 13 | 16 |



FIGURE 16-13

**Audio Mixer II.** The circuit in Fig. 16-13 is an improved audio mixer based on the RCA CA-3048 CDA amplifier array. This circuit provides approximately 20 dB of gain at each channel. The CA-3048 is a 16-pin DIP integrated circuit which contains four independent AC amplifiers. Offering a gain of 53 dB with a GBW of 300 KHz (typical), the CA-3048 has a 90-kohm input impedance and an output impedance of less than 1 kohm. It will produce a maximum low-distortion output signal of 2 volts RMS and can accept input signals up to 0.5 volts RMS.

Each DC power supply can be up to +16 volts DC. There are two *V*+ and two ground connections. These multiple connections are used to reduce the internal coupling between amplifiers. The two *V*+ terminals are tied together externally. The two ground terminals are also tied together. The *V*+ terminals are bypassed with a dual capacitor. *C*5 in Fig. 16-13 is a 0.1 uF unit. It is used for high frequencies. *C*6 is a 4.7 uF tantalum electrolytic for low frequencies. Both capacitors must be mounted as close as possible to the CA-3048 body, with *C*5 taking precedence for closeness over *C*6 because high frequencies are more critical.

An RC network from the amplifier output to ground (*R*3/*C*2) is used to stabilize the amplifier, thus preventing oscillation. Like the power supply bypass/decoupling capacitors, these components need to be mounted as close as possible to the body of the amplifier.

Only one channel is shown in detail in Fig. 16-13 because of space limitations. But each of the other three channels is identical. They are joined with the circuitry shown at the output capacitor (*C*4) as shown. Each of the four channels has its own level control (*R*1), which also provides a high input impedance for the mixer.

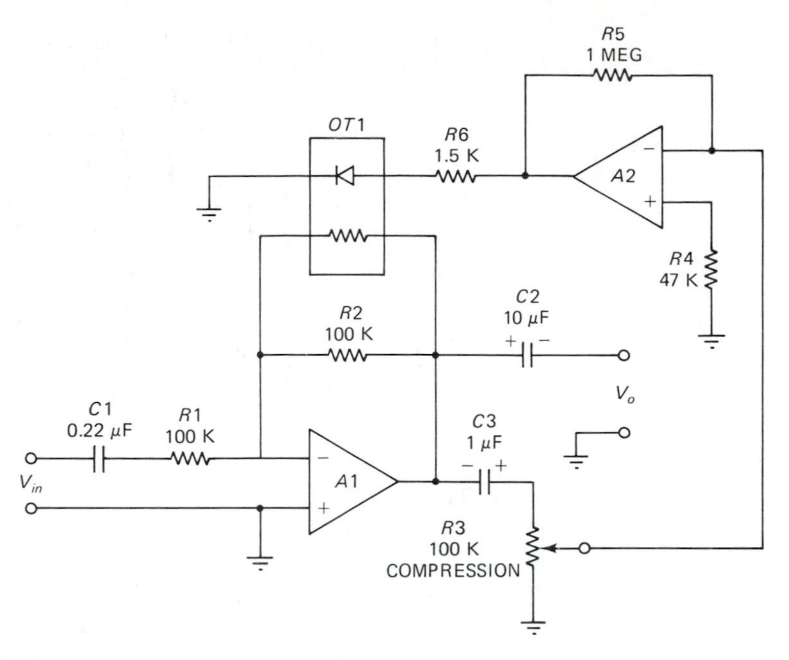


FIGURE 16-14

## 16-5 COMPRESSION AMPLIFIER

A compression amplifier reduces the gain on input signal peaks and increases the gain in the signal valleys. These circuits are used by electronic music fans and by broadcasters to raise the average power in the signal without appreciable increase in total harmonic distortion. Figure 16-14 shows a compression amplifier circuit.

The amplifier (*A*1) is an audio operational amplifier, such as the LM-301 (power supply and compensation not shown). The gain of the circuit is set by the input resistor (*R*1), and a feedback resistance which consists of the parallel combination of *R*2 and the resistance of the optocoupler (OT1) output element. The OT1 resistance is set by the intensity of the light emitting diode (LED) brightness. This in turn is set by the signal amplitude produced by *A*2. Because the *A*2 output signal is proportional to the *A*1 output signal, overall gain reduces itself (compresses). Optoisolator OT1 can be any resistance output device such as the Clairex, or a modern type (H11) which uses a JFET for the resistance element.

## 16-6 CONTROL AMPLIFIERS

Generally speaking, control amplifiers are a category of circuits which will perform some processing function other than simple gain. Control amplifiers are low gain circuits (×1 to ×10). It is common to find the output signal at the same amplitude as the input signal. In commercial and consumer audio equipment the primary functions of control amplifiers are *tone control*, *input source selection*, and *channel balance* (in stereo equipment).

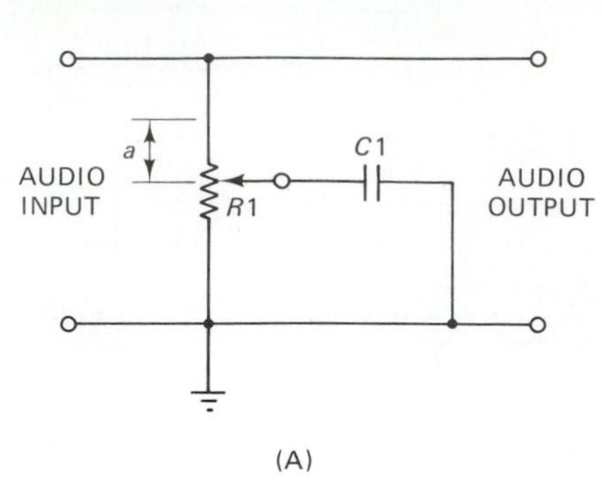


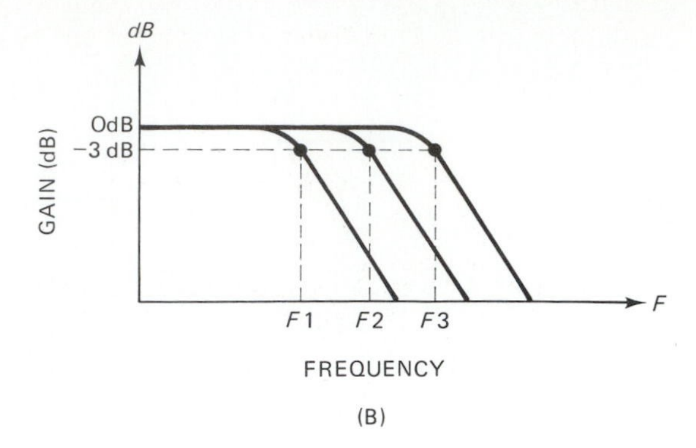


FIGURE 16-15

## 16-6.1 Tone Control Circuits

The tone control circuit either accentuates or attenuates a particular band of frequencies within the audio spectrum. The simplest tone control circuit is the *treble roll-off* circuit of Fig. 16-15A. This method is used only in the cheapest equipment. The basis for the treble roll-off control is an RC network shunted across the audio line. The RC time-constant of the active section of $R1$ (a) and $C1$ is $T = R1(a) \times C1$. The apparent tone change is dependent on this time constant. It sets the $-3$ dB frequency and the roll-off slope (see Fig. 16-15B).

The treble roll-off control functions only by attenuating the high frequencies. Thus, the bass setting does not boost the low frequencies, but rather it gives the illusion of bass by reducing the amplitude of signals in the treble range. Similarly, the treble setting of the control merely restores the original tone balance, rather than boost the high frequency ranges.

A *shelf equalizer* will boost or cut a specific range of frequencies in either the bass or treble ranges. This type of circuit produces a characteristic such as Fig. 16-16. The 0 dB line represents the *flat* condition in which the signal is neither boosted nor cut. In the *boost* condition, frequencies within the defined range (treble or bass) are amplified more than frequencies within the flat zone. In the *cut* condition the signals within the band are amplified less than those in the flat zone.

In simple equipment only two bands are affected by the tone controls, *bass* (low) and *treble* (high). In *graphic equalizers* there are more controls, each of which can either boost or cut. Each control operates over a narrower range of audio frequencies than simple bass and treble controls.

A circuit for a single band in a shelf equalizer is shown in Fig. 16-17A. It consists of a unity gain inverting follower based on an operational amplifier. In this circuit $R1 = R2 = R$. The flat zone gain is $-R2/R1 = -R/R = -1$. This gain figure is modified at certain frequency ranges by a frequency sensitive feedback network which shunts $R2/R1$.

A treble range feedback network is shown in Fig. 16-17B. The bass version of this circuit is shown in Fig. 16-17C. It is similar to the treble version except a capacitive reactance shunts the control.

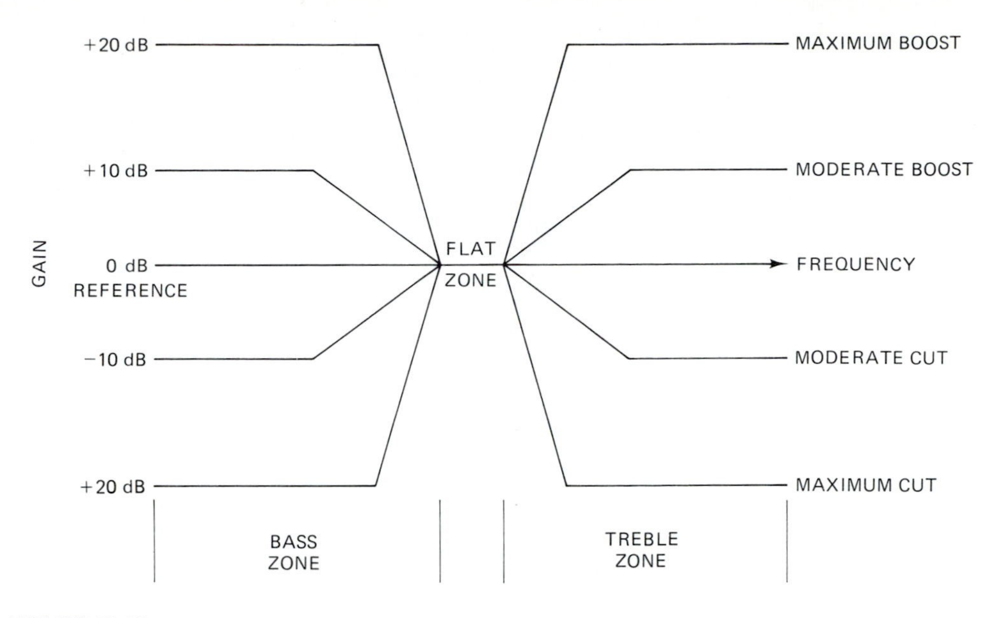


FIGURE 16-16

Let's evaluate the treble version. The network consists of two equal resistances ($R_a$) in series with a potentiometer ($R_b$). The total resistance of the network is $2R_a + R_b$. The wiper of the potentiometer is connected to the summing junction of the inverting follower through a capacitor ($C1$). The reactance of $C1$ is $X_{C1} = 1/2\pi FC1$, so the frequency selectivity is a function of a circuit resistances and $X_{C1}$.

The potentiometer ($R_b$) serves as the front panel adjustment control. The resistance of the pot is divided into two portions (labelled a and b). The relative magnitudes of $R_{ba}$ and $R_{bb}$ are determined by the setting of the pot's adjustment shaft. When the shaft is set to the midpoint, $R_{ba} = R_{bb}$.

Input current $I1$ is determined by input signal voltage $V_{in}$ and the parallel combination of $R1$ and the impedance ($Z1$) consisting of $R_a$, $R_{b1}$ and $X_{C1}$. The feedback current ($I2$) is determined by output voltage $V_o$ and the parallel combination of resistance $R2$ and the impedance ($Z2$) consisting of $R_a$, $R_{bb}$, and $X_{C1}$.

Inverting amplifiers like the one in Fig. 16-17A can be evaluated by the fact the summing junction is at ground potential because the noninverting input is grounded.

If $R_b$ is set to the mid-point, both segments have the same resistance. The current flows due to $V_{in}$ and $V_o$ are therefore equal, so they cancel each other. If the potentiometer is set so $R_{ba} < R_{bb}$, input current $I1$ is increased due to a decrease of $Z1$. For frequencies at which $Z1$ is minimum, the gain is maximum, so the action is to boost those frequencies.

When $R_b$ is rotated toward the output side of the circuit, the condition will be $R_{ba} > R_{bb}$. In this condition, impedance $Z2$ is reduced. This setting has the effect of reducing the gain at the design frequency. That frequency is the point at which the gain either increases or decreases 3 dB from the 0 dB flat zone reference gain

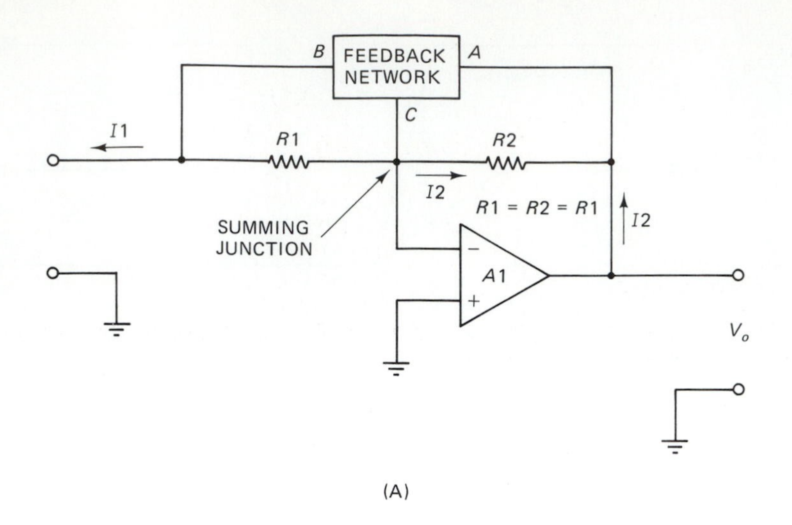


(A)

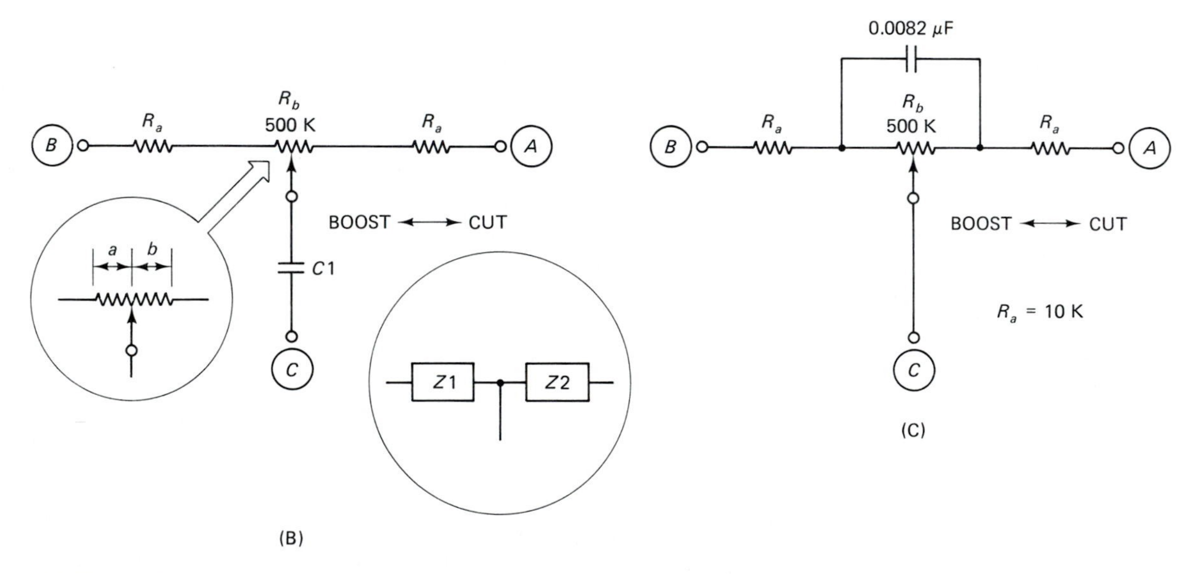


(B)

(C)

FIGURE 16-17

$$F_{3\text{ dB}} = \frac{1}{2\pi R1\ C1} \tag{16-4}$$

Because the feedback network is symmetrical, boost and cut frequencies are the same. The maximum boost or cut will be

$$B = \frac{R1 \parallel R_b}{R_a \parallel R1} \tag{16-5}$$

or

$$B = \frac{\left[\dfrac{R1R_b}{R1 + R_b}\right]}{\left[\dfrac{R1R_a}{R1 + R_a}\right]} \tag{16-6}$$

**EXAMPLE 16-2**

Calculate the maximum boost or cut in a high frequency shelf equalizer in which $R1 = 100$ kohms, $R_b = 500$ kohms, and $R_a = 10$ kohms.

**Solution**

$$B = \frac{\left[\dfrac{R1R_b}{R1 + R_b}\right]}{\left[\dfrac{R1R_a}{R1 + R_a}\right]}$$

$$B = \frac{\left[\dfrac{(100\text{-k})\ (500\text{-k})}{(100\text{-k} + 500\text{k})}\right]}{\left[\dfrac{(10\text{-K})\ (100\text{-K})}{(10\text{-k} + 100\text{-k})}\right]}$$

$$B = \frac{\left[\dfrac{50{,}000}{600}\right]}{\left[\dfrac{1{,}000}{110}\right]}$$

$$B = \frac{83.3}{9.09} = 9.2$$

A gain of 9.2 represents a gain of about 20 dB. ■

Other forms of tone control are based on *RL*, *RC*, and *RLC* networks in the feedback path. Figure 16-18 shows the block diagram for a type of tone control circuit in which a reactance is used between the wiper of the potentiometer and ground. The bass control uses a capacitive reactance to ground. The treble control uses an inductive reactance.

The resonant form of tone control is shown in Fig. 16-19. This circuit is similar to the earlier circuit, except the capacitor between the potentiometer wiper and the op-amp summing junction is a series resonant circuit. A series resonant tank circuit has a very low impedance at the frequency of resonance, while exhibiting a high impedance when frequencies are removed from resonance. The resonant frequency of this circuit is found from

$$F = \frac{1}{2\pi[LC]^{1/2}} \tag{16-7}$$

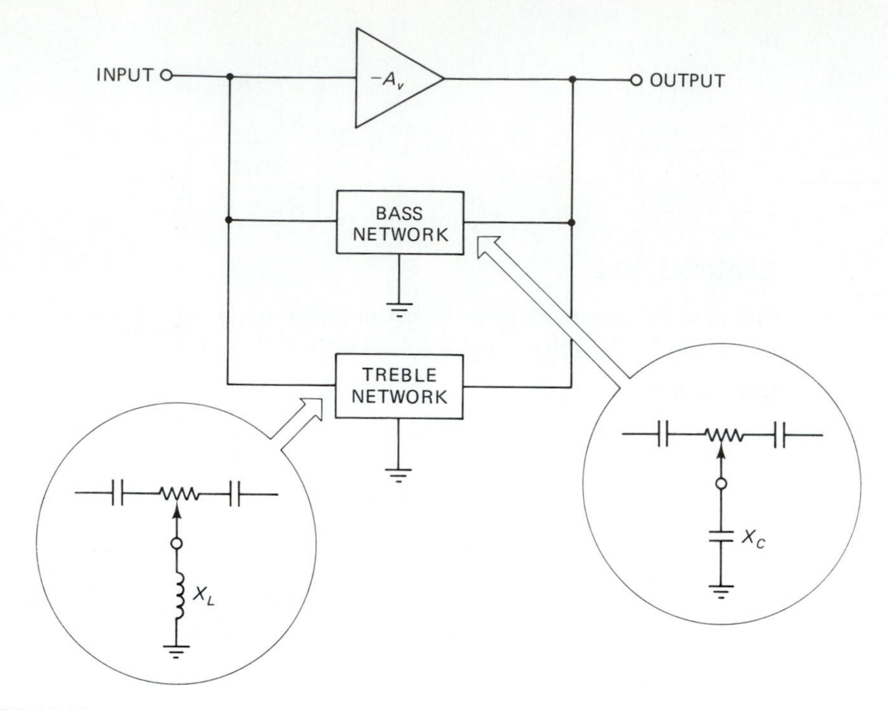


FIGURE 16-18

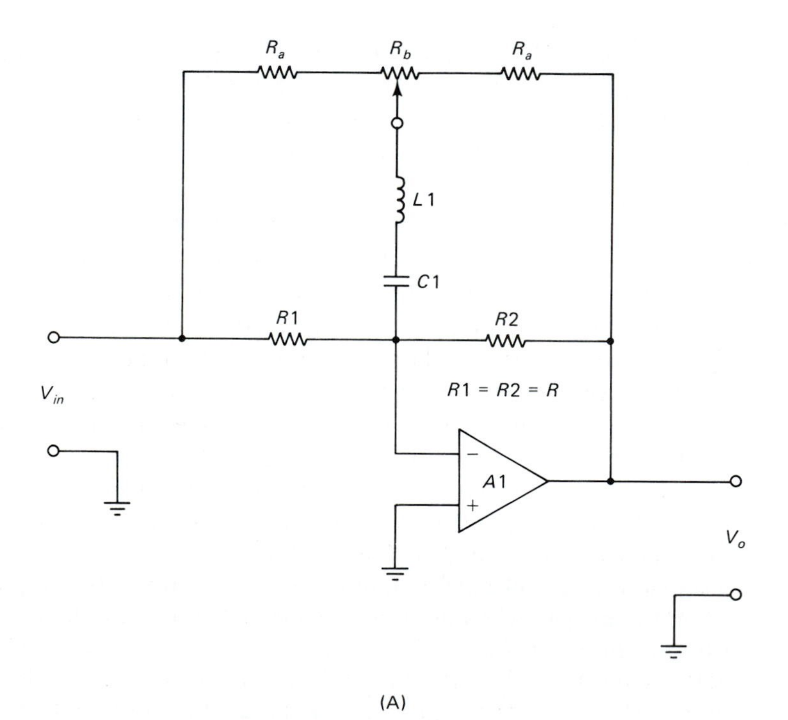


(A)

FIGURE 16-19

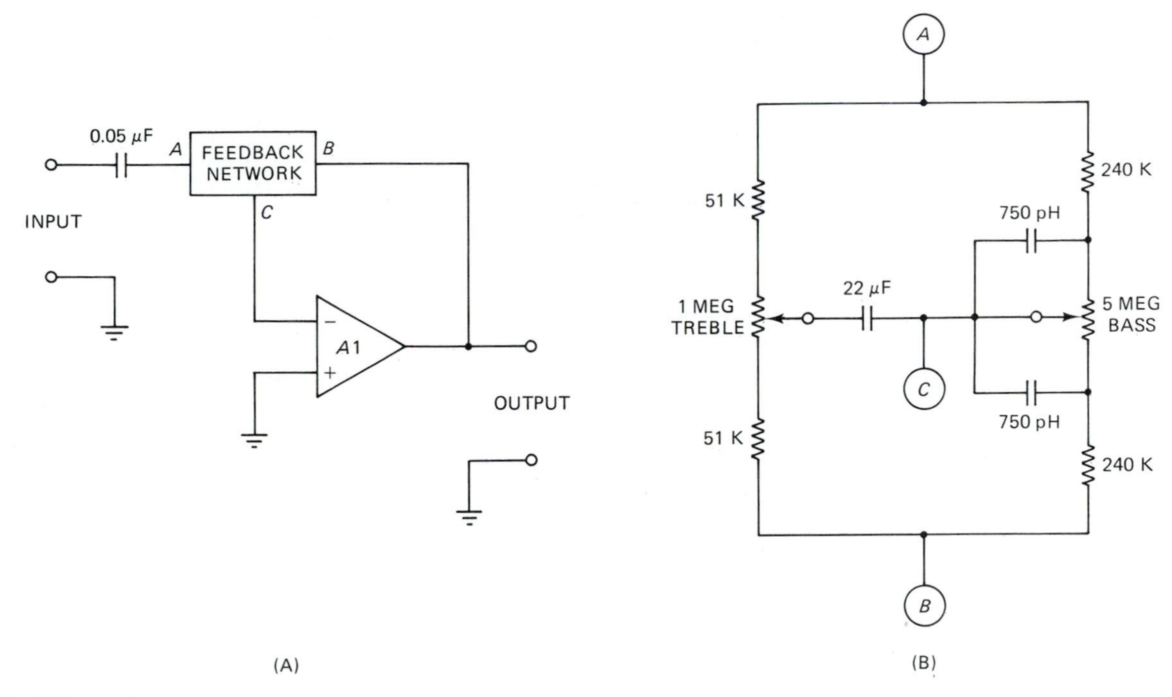


FIGURE 16-20

The *Baxandall tone control* circuit is shown in Fig. 16-20. The operational amplifier is connected in the inverting follower configuration, but with a different type of feedback network than the earlier example. This type of circuit uses a single bass and treble control. It once was the basis for most earlier high fidelity equipments.

## 16-7 POWER AMPLIFIERS

The power amplifier in an audio system is designed to boost the signal from a voltage level up to a level sufficient to drive a load such as a loudspeaker. While most preamplifiers using small-signal transistors are operated class A, power amplifiers are operated in class B, or class AB. A class A amplifier is one in which the output current flows over all 360° of the input signal cycle. The result of 360° conduction reproduces the signal in its entirety in the output circuit. Class A amplifiers also produce a large amount of heat at any given power level because they are only about 25% efficient. Thus, for a 1 watt output there will be 3 watts wasted as heat. For this reason, class A amplifiers are only rarely used as audio power amplifiers (some older car radios used class A power amplifiers).

A class B amplifier offers output current conduction only over 180° of the input signal cycle. Thus, the output will be only halfwave. In Fig. 16-21A a class B amplifier with a gain $+A$ is used. The input signal is a sinewave and the output signal is only half a sinewave. This type of amplifier, if used alone, would result in unacceptable distortion of the input signal. A correction for this problem is the *push-pull power amplifier* shown in Fig. 16-21B. Here, a pair of complementary amplifiers is used,

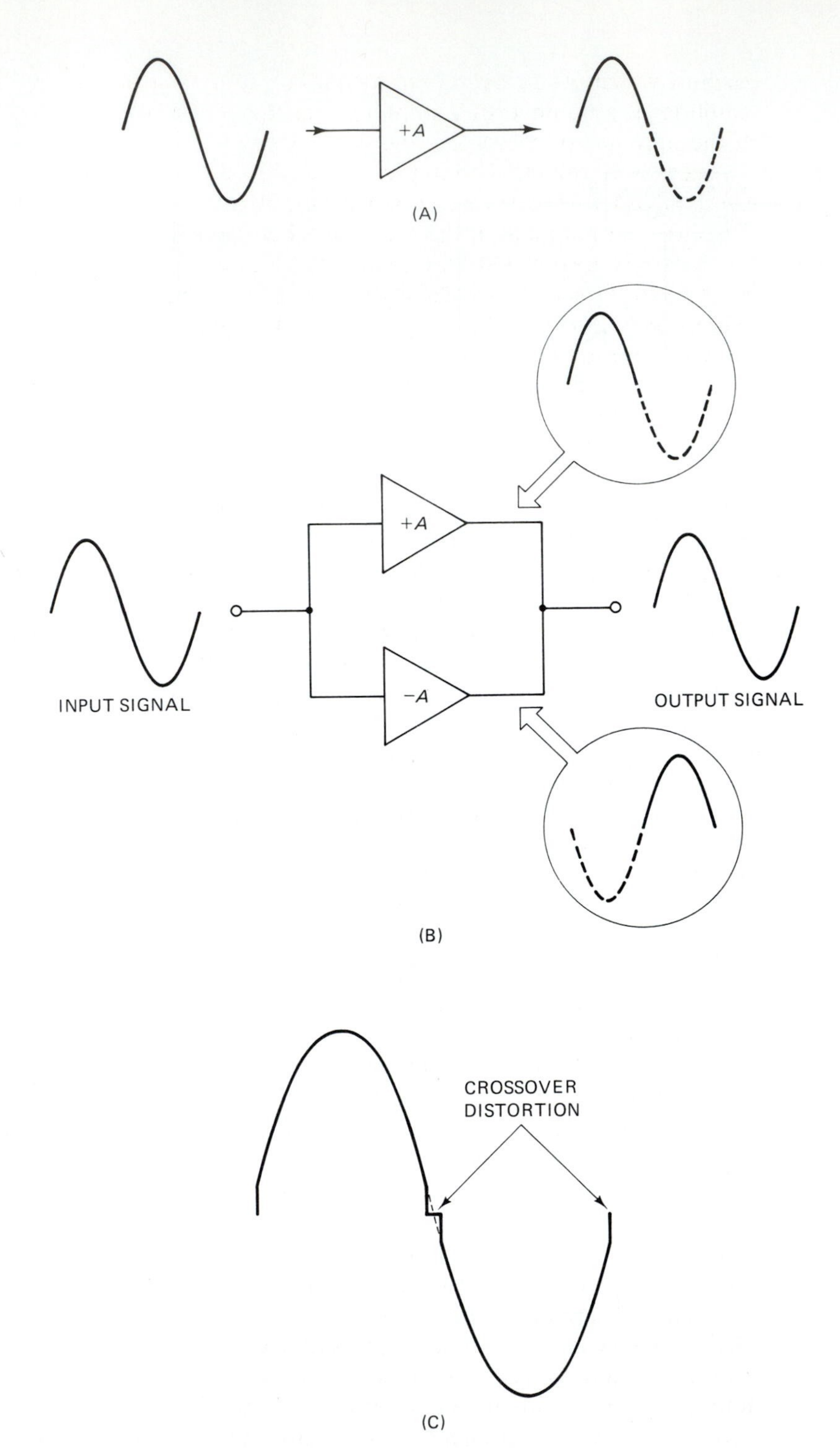


FIGURE 16-21

each of which produces the same gain but an opposite phase relationship. The $+A$ amplifier is a noninverting amplifier and the $-A$ amplifier is an inverting amplifier. If the outputs are combined, the full sinewave is reproduced.

There is a problem with improperly designed solid-state push-pull amplifiers, however. Figure 16-21C shows *crossover distortion.* This distortion is caused by the 0.7 Vdc junction potential of the transistors being nonlinear. Crossover distortion was at the root of the so-called "transistor sound" imputed to early solid-state high fidelity equipment. Bias arrangements are used to overcome crossover distortion. These amplifiers are called class AB amplifiers because the current flows somewhat more than 180° but considerably less than 360°.

Linear IC devices tend to be class AB amplifiers or even preamplifiers. The operational amplifier (which forms the basis of most preamplifier circuits) uses one of several class AB circuits in the output stage, even though only producing a few milliwatts of power.

Figure 16-22 shows several implementations of the class AB power amplifier as used in many integrated circuits. The version shown in Fig. 16-22A is called the *complementary symmetry push-pull power amplifier.* It uses a pair of bipolar transistors which are electrically identical except for polarity; one is an NPN and the other is PNP. An NPN transistor turns on harder when base-emitter (*b-e*) signal goes more positive and turns off when the *b-e* signal goes less positive. An NPN transistor, on the other hand, works exactly the opposite: it will turn on harder when the *b-e* signal goes less positive and turns off when the *b-e* signal goes more positive. Thus, NPN transistor $Q1$ will turn on when the input signal is positive. PNP transistor $Q2$ turns on when the input signal is negative. The transistors are operated as emitter followers, and the output signal is taken at the common junction of the two emitters. The load resistor (for example, a loudspeaker) forms the emitter resistor for $Q1$ and $Q2$. The voltage across the load tends to be low compared with certain other amplifier circuits.

It is difficult to build identically matched NPN and PNP power transistors, although somewhat less difficult at smaller signal levels. Because it is easier to build identical transistors of the same polarity on the same silicon substrate, many manufacturers use the *totem pole push-pull amplifier* (Fig. 16-22B). In this circuit a pair of identical NPN transistors are connected in series across the DC power supply.

The two transistors are identical so no inherent phase shift exists to accommodate the two halves of the signal. Because of this limitation, a phase inverter circuit is used between the input signal and the base terminals of the two power transistors. The phase inverter has two outputs which are 180° out of phase with each other.

If only a signal polarity DC supply is used, the voltage at the junction of the two transistors will be $(V+)/2$. Because this voltage will interfere with external loads, a large value capacitor ($C1$) is used to block DC from the load. This capacitor typically has a value of at least 1,000 uF, and may have a value as high as 10,000 uF. These capacitors are electrolytic types and are therefore quite large. They also introduce undesirable phase shift to the output signal, especially at lower frequencies. The phase shift is created by the capacitive reactance of $C1$. Because of this problem, some totem pole power amplifiers are designed to use bipolar DC power supplies. Such an arrangement is possible, as in the complementary case, because the positive and negative contributions to the DC level at the output cancel each other.

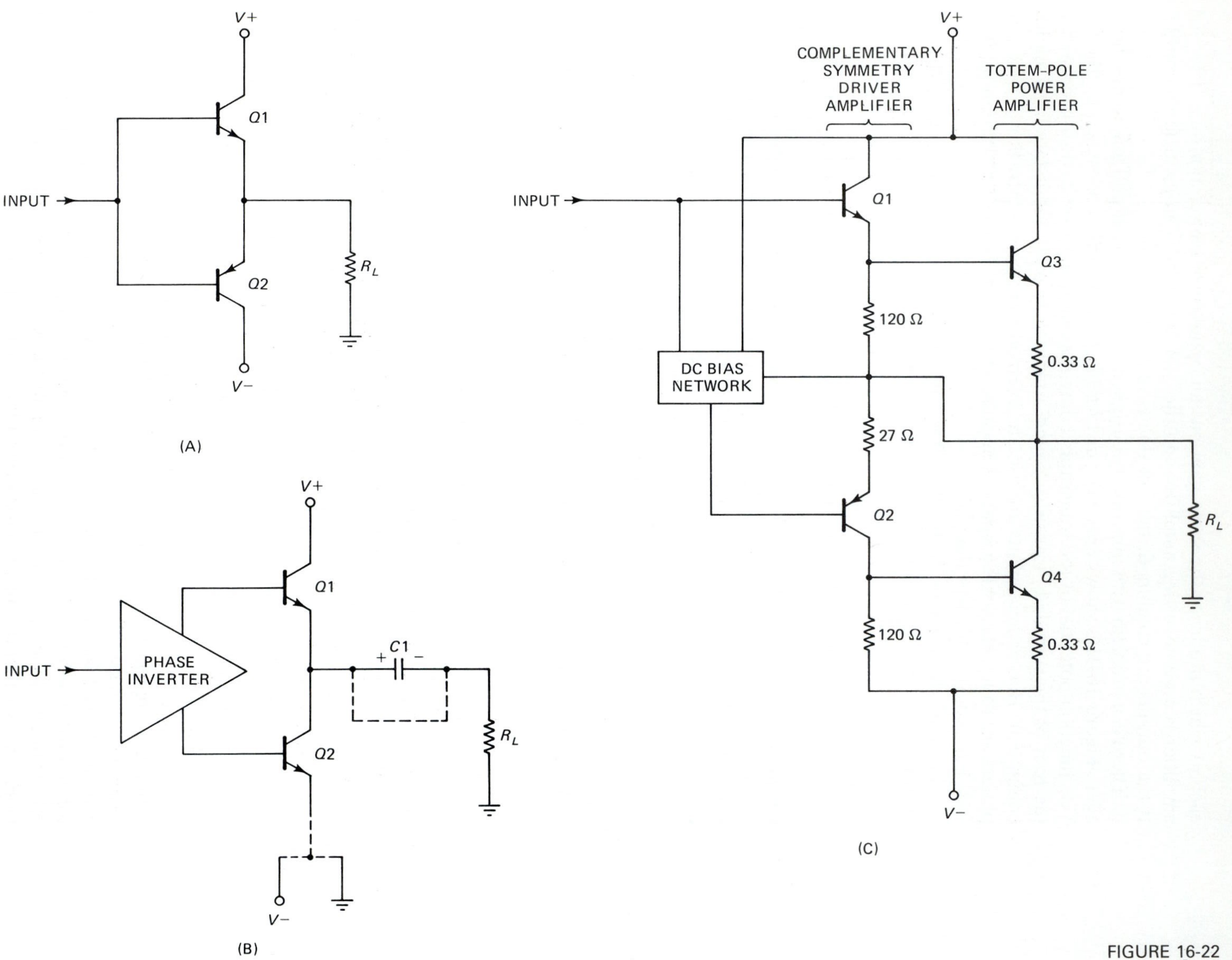

V+
Q1
INPUT
Q2
$R_L$
V−
(A)
V+
Q1
INPUT
PHASE INVERTER
C1
+
−
$R_L$
Q2
V−
(B)
V+
COMPLEMENTARY SYMMETRY DRIVER AMPLIFIER
TOTEM-POLE POWER AMPLIFIER
INPUT
Q1
Q3
120 Ω
0.33 Ω
DC BIAS NETWORK
27 Ω
Q2
$R_L$
Q4
120 Ω
0.33 Ω
V−
(C)


FIGURE 16-22

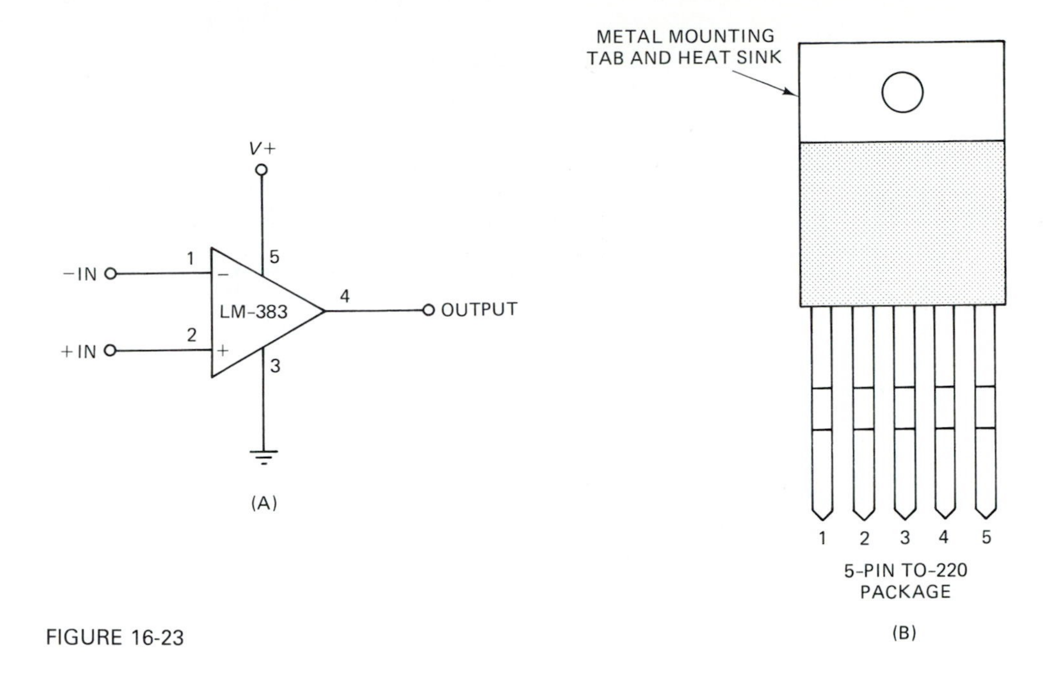


FIGURE 16-23

The circuit used in most IC devices, whether power amplifiers or preamplifiers, is the *quasicomplementary power amplifier* shown in Fig. 16-22C. The output section ($Q3/Q4$) is a totem-pole amplifier using a pair of identical NPN transistors. The driver amplifiers are medium power level transistors connected in a complementary symmetry amplifier circuit. The phase inversion function takes place in this stage.

### 16-7.1 IC Power Amplifiers

The main limitation to producing power amplifiers in IC form was the dissipation of internal heat built up in the process. Class AB amplifiers can be more than 75% efficient if carefully designed and properly built, but that still leaves considerable heat energy to be dissipated into the environment. But methods were worked out, and IC device power levels started upwards in the mid-1970s. Today, IC and hybrid audio power amplifier devices are available at power levels from 250 mW up to 250 watts. We will discuss only a few of the many devices which are available.

The LM-383 (Fig. 16-23A) is a 7-watt audio power amplifier packaged in a 5-pin version of the TO-220 plastic power transistor case (see Fig. 16-23B). It will operate at monopolar DC power supply potentials of +5 Vdc to +20 Vdc and offers a peak current capability of 3.5 amperes. Because this amplifier operates from a monopolar supply, however, there will always be a DC output level of $(V+)/2$ at the output terminals (Fig. 16-24). Similarly, the AC feedback path also requires a DC blocking capacitor.

The LM-383 offers a gain-bandwidth product of 30 KHz at a gain of 40 dB, and a total harmonic distortion (THD) level of 0.2% or less. The series RC network ($C4/R1$) between the output terminal and ground is a lag compensation circuit designed to prevent oscillation which could occur when high gains are selected.

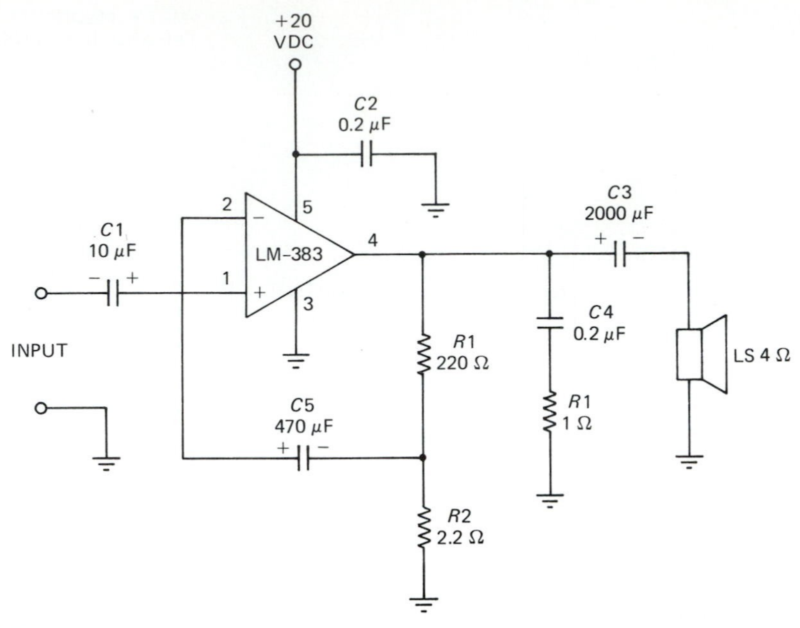


FIGURE 16-24

The circuit of Fig. 16-24 is a single power amplifier for the LM-383. This circuit would be used whenever the required power level is within the range of the device. However, another way to use these amplifiers is the *bridge amplifier* circuit of Fig. 16-25. This circuit uses a pair of LM-383s (*U*1 and *U*2) connected so the load, a loudspeaker, is connected between the two output terminals. One amplifier is connected as a noninverting amplifier (*U*1). The other is connected as an inverting amplifier (*U*2). The inverting amplifier derives its input signal from the output of *U*1.

No DC blocking capacitors are needed in the output circuits of *U*1 and *U*2 because the DC levels are equal. For example, in a 14.4-Vdc auto radio power amplifier each IC will exhibit +7.2 volts at the output terminal, so the voltage across the load will be [(+7.2 Vdc) − (+7.5 Vdc)] = 0. A DC BALANCE control is used at the noninverting input of *U*2 to maintain the zero difference potential over the range of tolerance of the LM-383.

Another example of a power IC device is the Burr-Brown OPA-501 shown in Fig. 16-26. The OPA-501 is not a completely monolithic IC, but rather an example of a hybrid. The OPA-511 and OPA-512 are similar devices. These devices can be used as audio amplifiers, servo amplifiers, motor drivers, synchro drivers, and DC power supply regulators. It will operate over a wide range of bipolar DC power supplies: ±10 Vdc to ±30 Vdc.

The OPA-501 uses an operational amplifier for the preamplifier stages. A class AB output stage (*Q*1/*Q*2) which will deliver up to 260 watts of peak power and a current of ±10 amperes peak. The OPA-501 is packaged in a small size eight-pin version of the TO-3 power transistor case.

The output current is limited by external circuitry to a level determined by external set-current resistors (see $+R_{sc}$ and $-R_{sc}$ in Fig. 16-27). The value of these two resistors

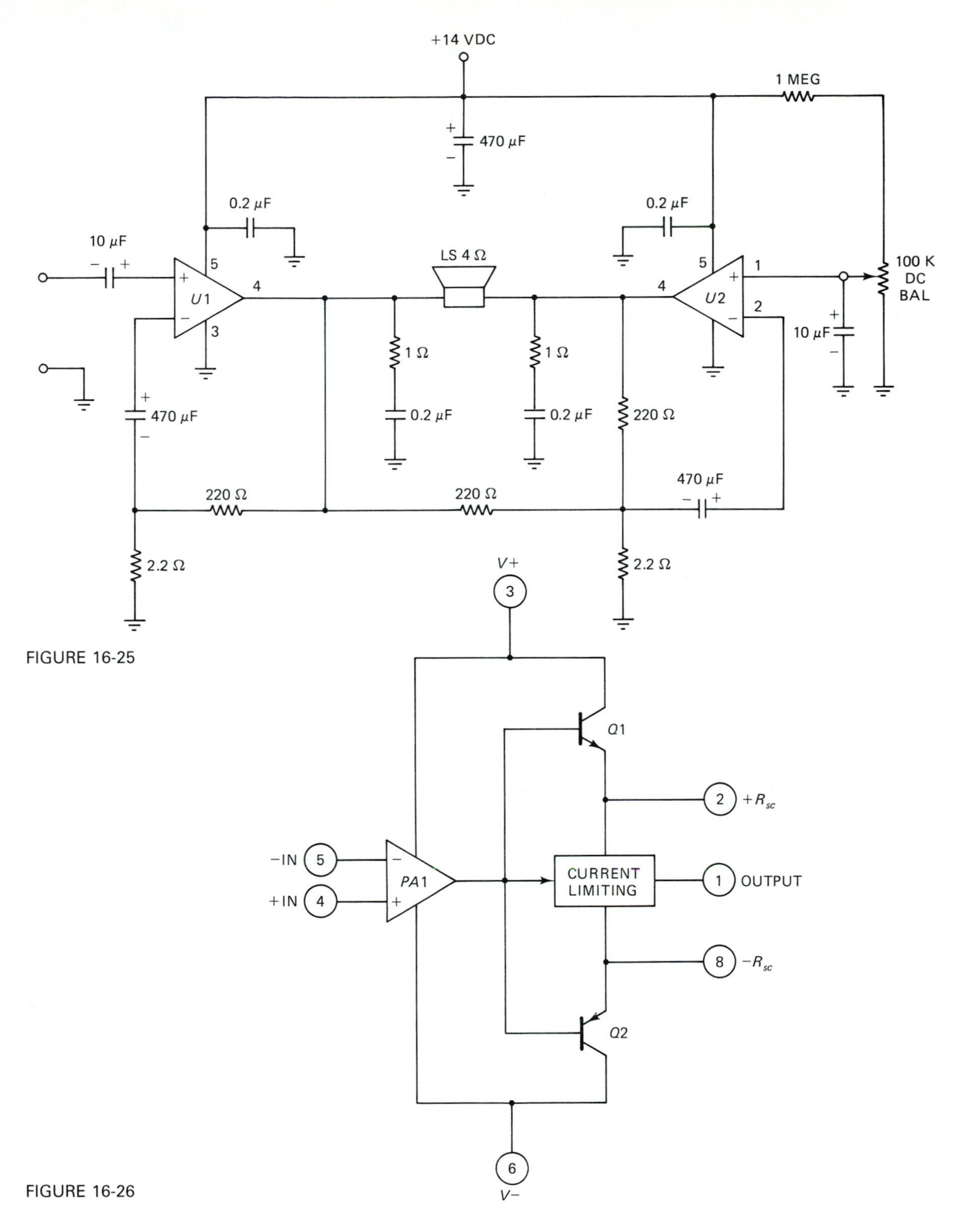


FIGURE 16-25

FIGURE 16-26

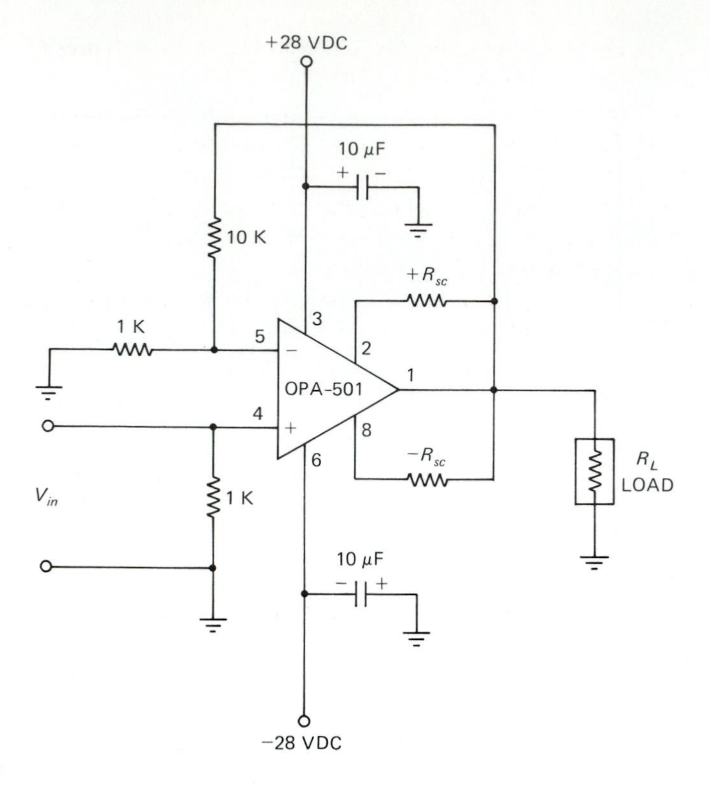


FIGURE 16-27

is

$$R_{sc} = \left[\frac{-0.65}{I_{\text{limit}}} - 0.0437\right] \text{ ohms} \quad (16\text{-}8)$$

Current $I_{\text{limit}}$ is the maximum output current the designer wishes to allow. The power dissipation of these resistors is found from

$$P_d = R_{sc}(I_{\text{limit}})^2 \text{ watts} \quad (16\text{-}9)$$

The resistors selected for the limiting function should have a power rating higher than given by Eq. (16-9) by a margin sufficient for reliable and safe operation.

## 16-8 SUMMARY

1. Audio circuits operate in the range 20-Hz to 20-Hz, although some amplifiers characterized as audio operate over a slightly wider range.
2. Three different forms of audio amplifiers are typically used. *Preamplifiers* are (generally) high gain voltage amplifiers, and are used to boost low-level transducer or other input signals to a higher level (usually 100 mV to 500 mV). *Control amplifiers* are usually low gain (×1 to ×10), but perform some signals processing function such as tone control. *Power amplifiers* create the audio power level needed to drive the external load.
3. Commercial and broadcast audio amplifiers tend to use 600 ohm input and output imped-

ances. The 600 ohm line is a balanced line which uses two out of phase signal lines and one ground or common line.

4. Audio mixers linearly combine the signals from two or more sources into one channel. For example, a broadcast studio might mix together signals from two microphones, a tape player, and a record turntable.
5. Compression amplifiers offer less gain on signal peaks than on valleys, resulting in an output signal with a higher average power. These devices are used in electronic music applications and in broadcasting.
6. Tone control amplifiers are used to boost or cut specific frequency ranges in order to custom balance the frequency response of the amplifier.

## 16-9 RECAPITULATION

Now return to the objectives and Prequiz questions at the beginning of the chapter and see how well you can answer them. If you cannot answer certain questions, mark them and review the appropriate parts of the text. Next, try to answer the questions and work the problems below, using the same procedure.

## 16-10 STUDENT EXERCISES

1. Design, build, and test a noninverting audio amplifier with a frequency response of 300 Hz to 3000 Hz using an operational amplifier.
2. Design, build, and test a shelf equalizer (Fig. 16-17A) with a center frequency of 1500 Hz.
3. Design, build, and test an audio amplifier based on an LM-381.

## 16-11 QUESTIONS AND PROBLEMS

1. What is the nominal "audio" range?
2. List four applications for audio amplifiers.
3. Draw the circuit for a bridge audio power amplifier based on the LM-383 integrated circuit power amplifier. Is it necessary to use output DC blocking capacitors?
4. An audio preamplifier for an AM broadcast radio transmitter must have a gain of 100, and a frequency response of 30 Hz to 5000 Hz between $-3$ dB points. Calculate the values for the input and feedback resistors, and the value of the capacitor shunting the feedback resistor. Assume a noninverting follower.
5. In the circuit above, the input circuit contains a transformer with a 1:2 turns ratio. What is the total voltage gain of this circuit?
6. Draw the circuit for a five-channel op-amp audio mixer with an inverting characteristic.
7. A ______________ circuit is used to boost the 5-mV output of a dynamic microphone to 1.00 volt needed to drive the input of a radio transmitter.
8. An inverting audio amplifier is built from an operational amplifier with a 150-kohm resistor in the feedback network. Calculate the value of capacitance that must be shunted across the feedback resistor in order to limit the upper $-3$ dB point in the frequency response to 4 KHz.
9. In an inverting amplifier the feedback resistor is 220 kohms, and it is shunted with a 100 pF capacitor. Calculate the frequency at which the gain drops $-3$ dB.

10. A four-input mixer circuit must be designed and built. Assume a feedback resistance of 150 kohms and that each input has a 100 kohm resistance. Calculate the total output voltage if the following DC voltages are applied, $V1 = 0.1$ Vdc, $V2 = 0.2$ Vdc, $V3 = 0.1$ Vdc, and $V4 = 0.4$ Vdc. Calculate the peak output voltage if the DC sources are replaced with the following rms signal voltages, $V1 = 0.05\ V_{rms}$, $V2 = -0.1\ V_{rms}$, $V3 = -.43\ V_{rms}$ and $V4 = 0.23\ V_{rms}$.
11. The simplest form of tone control is the ____________ ____________-off.
12. Draw the circuit diagram for a shelf equalizer stage (Fig. 16-17A) in which $R1 = R2 =$ 100 kohms, $R_a = 10$ kohms, and $R_b = 500$ kohms. Calculate the capacitance needed for $C1$ for a 1500 Hz $-3$ dB point. Calculate the maximum boost or cut for this circuit (in dB).
13. A resonant equalizer tone control (Fig. 16-19) requires a resonant frequency of 2000 Hz. Calculate the capacitance required if the inductor is a 900-mH unit.
14. What is the purpose of $C6$ in Fig. 16-2?
15. What is the input impedance of the circuit in Fig. 16-2? The output impedance?
16. Draw the frequency response curve of Fig. 16-2.
17. Sketch the frequency response curves of Fig. 16-4 when the feedback network is (a) Fig. 16-5C, and (b) Fig. 16-5D.
18. In Fig. 16-11A the transformer has a turns ratio of 1:3, and both resistors in the feedback network are equal to each other. What is the peak-to-peak output voltage when a 100 $mV_{rms}$ 1000 Hz sinewave signal is applied to the input?
19. Select a value for $R3$ and $R4$ in Fig. 16-11B when the overall differential gain must be 100.
20. Sketch the circuit diagram for a transformerless 600 ohm line driver amplifier based on op-amps.
21. Sketch the circuit for a Baxandall tone control circuit.
22. A ____________ amplifier reduces its gain on input signal peaks and increases the gain on signal valleys.
23. An audio power amplifier drives an 8 ohm resistance load. A potential of 22.4 volts peak-to-peak is measured across the load when a 1000 Hz sinewave is applied to the input of the amplifier. (a) Calculate the output power; (b) calculate the amplifier efficiency if the DC power consumption of the stage is 34 watts.
24. An audio power amplifier produces a maximum output power of 100 watts into a 4 ohm resistive load. What is the DC power drawn by this amplifier if it is 33% efficient?
25. What is the cause of crossover distortion? What is the usual circuit fix?

CHAPTER 17

# Communications Applications of Linear IC Devices

## OBJECTIVES

1. Learn the types of applications for linear ICs in communications circuits.
2. Understand how to use phase-locked loop IC devices.
3. Understand how to use the analog multiplier/divider as a modulator.
4. Learn how to use current loop systems.

## 17-1 PREQUIZ

These questions test your prior knowledge of the material in this chapter. Try answering them before you read the chapter. Look for the answers (especially those you answered incorrectly) as you read the text. After you have finished studying the chapter try answering these questions again and those at the end of the chapter (see Section 17-10).

1. List two applications for current loop communications.
2. List the voltage levels that represent logical-1 and logical-0 in the RS-232C standard.
3. Amplitude modulation is a ____________ process.
4. The ____________ signal is a form of amplitude modulation in which the carrier is suppressed, and one sideband is filtered out.

## 17-2 INTRODUCTION

The broad area of "communications" takes in a wide range of technology, only a small portion of which can be addressed in a single chapter. The areas we can touch on, however, are varied and broad because they relate to the applications of linear IC devices in communications. In this chapter a number of different areas are considered including current loop communications, serial voltage level communications, modulation and demodulation, and phase locked loops.

## 17-3 CURRENT LOOP SERIAL COMMUNICATIONS

Current loop communications is one of the oldest forms of machine communication known. Applications of the current loop include teletypewriters and process control instrumentation (a varied field in itself). The current loop offers some advantages even today, despite the fact it seems old-fashioned and has been largely replaced by other methods. First, the current loop is somewhat less sensitive to noise problems caused by voltage droops or spikes on the lines. Second, it is relatively easy to daisy chain several instruments into one system. Third, the current loop can be implemented using a simple twisted pair of wires rather than a multiwire cable or a coaxial transmission line as is required in other (faster) forms of data communications.

A disadvantage is that a single point failure can take the entire system down. Just like a string of series-connected Christmas tree lights, if a single load on the series current loop opens up, the entire chain is inoperative (although not all failures in series loop machines will cause this problem). In this text two forms of current loop will be examined, *teletypewriters* and *process instrumentation*.

### 17-3.1 Teletypewriter Current Loops

One of the earliest forms of alphanumeric data communications was the teletypewriter machine. These were typewriter-like (or printer-like) machines which used a mechanism of electrical solenoids to pull in the type bars or to position the type printing cylinder which struck an inked ribbon. The original devices used the now-obsolete 5-bit BAUDOT code and a 60 milliampere current loop. Later versions of the teletypewriter machine used the modern 7-bit ASCII code (still in common use) and a 20 milliampere current loop. These machines used the same basic technique as earlier machines, but were generally more sophisticated. A few modern teletypewriters use dot matrix printing and contain a floppy disk to store a magnetic copy of the data transmitted and received. They nonetheless operate from the 20 mA current loop.

Figure 17-1 shows the basic elements of a teletypewriter (or other printer) based on the 20 milliampere current loop. The keyboard and printer are actually separate units. They usually have to be wired together if a local loop is desired (where the keystroke on the keyboard produces a printed character on the same machine). This circuit is simplified. In a real typetypewriter there will be an encoder wheel or circuit that produces the binary coded output. The keyboard consists of a series of switches (that actuate the encoder). Since these switches and their associated encoder are in series with the line, a LOCAL switch must be provided to bypass the transmitter section on receive.

The receiver consists of a decoder and the receiver solenoids, which operate the typebar mechanism. Note in Figure 17-1 that a 1N4007 diode is shunted across the receive solenoid. This diode is used to suppress the high voltage inductive spike which is generated when the reactive solenoids are de-energized. The diode is placed in the circuit so it will be reverse-biased under normal operation. But the transient counter electromotive force (CEMF) produced as a result of inductive kick briefly forward-biases the diode. The diode therefore clamps the high voltage spike to a harmless level. In some older machines the inductive spike was safely ignored because the mass of the mechanism effectively integrated the spike to nothingness. But modern solid-

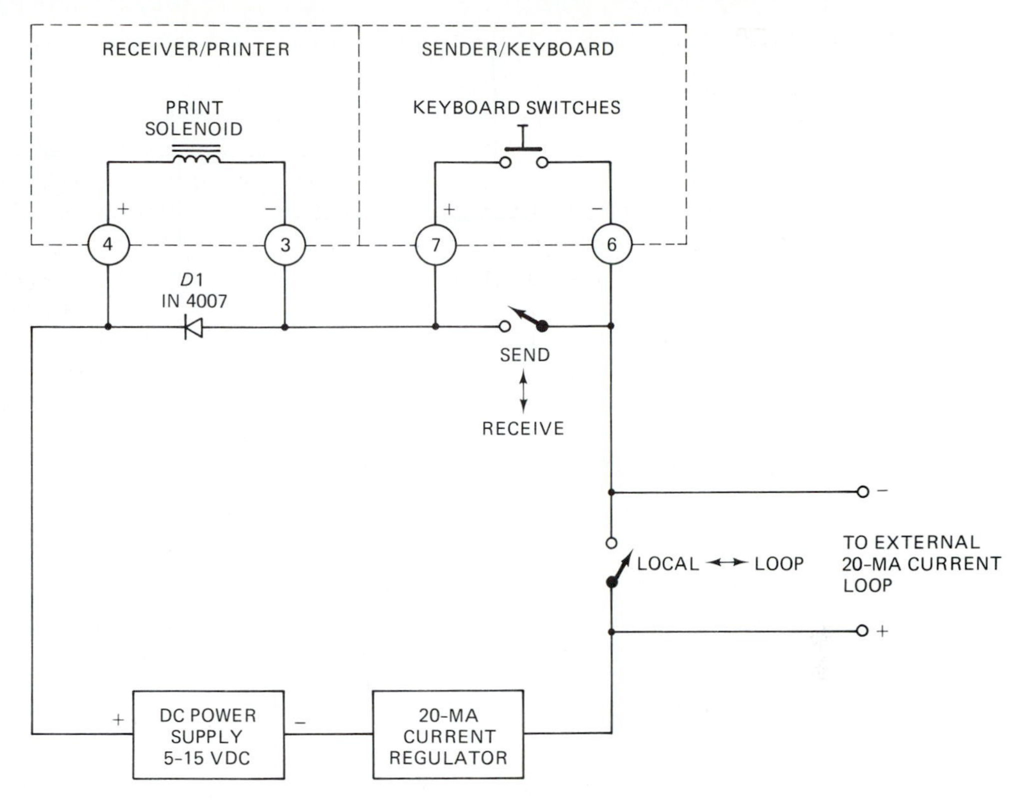


FIGURE 17-1

state equipment does not move the mechanism directly with the 20-mA loop, but rather with electronic drivers. The solid-state components can be damaged by the high voltage spike, so it is recommended that a 1N4007 be used even if it was not in the original design.

When the loop is closed, the circuit of Fig. 17-1 will produce a readable signal. Another similarly designed teletypewriter will be able to read the current variations produced by the machine. The binary 1 (HIGH) condition, also called a MARK in teletypewriter terminology, is defined as a current of 16 mA to 20 mA. The binary 0 (LOW), also called a SPACE, is defined as a current between 0 mA and 2 mA.

Figures 17-2 and 17-3 show how to interface 20 mA current loop equipment to TTL-compatible serial outputs on digital computers. The circuit in Figure 17-2 shows the transmitter arrangement. The assumption is that there is a single TTL-compatible bit from either a serialized or parallel output, or a UART IC. The TTL level is applied to an open collector TTL inverter, which has as its collector load an LED inside of an IC optoisolator. When the LED is turned on, the phototransistor is turned on hard. This transistor operates as an electronic switch in series with the 20 milliampere current loop. Thus, when the TTL bit is HIGH, the LED is on and the transistor is saturated. In that condition, the current loop transmits a MARK sign (equivalent to a logical 1 in binary).

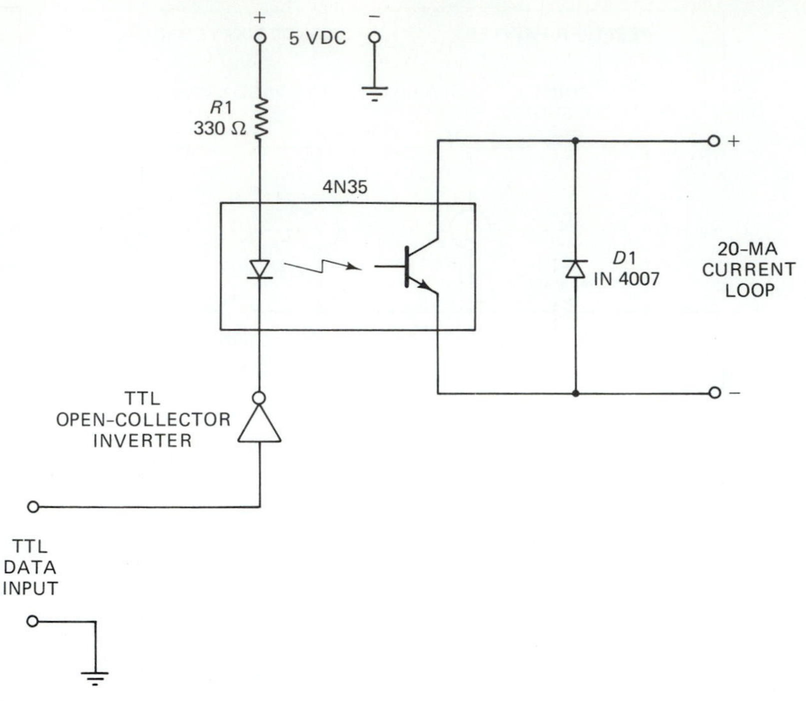


FIGURE 17-2

The receive end of the current loop-to-TTL interface is shown in Fig. 17-3. In this case, the optoisolator is still used, but in reverse. Here the LED is connected in series with the current loop. Thus, when a MARK is transmitted, the LED will be turned on. When a SPACE is transmitted the LED is turned off. During the MARK periods, the optoisolator phototransistor is saturated, and the input to the TTL inverter is LOW. This condition results in a HIGH on the output to the computer. The 0.01 μF capacitor is used for noise suppression.

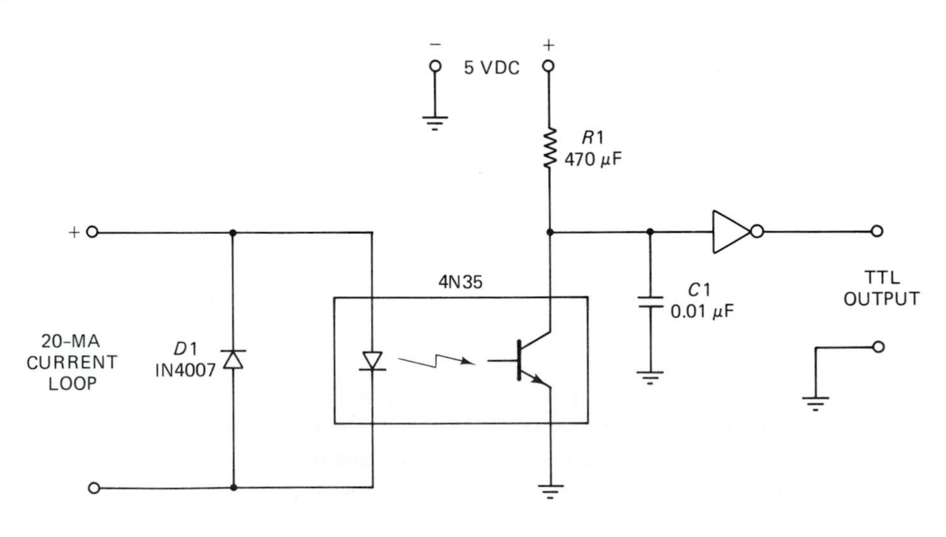


FIGURE 17-3

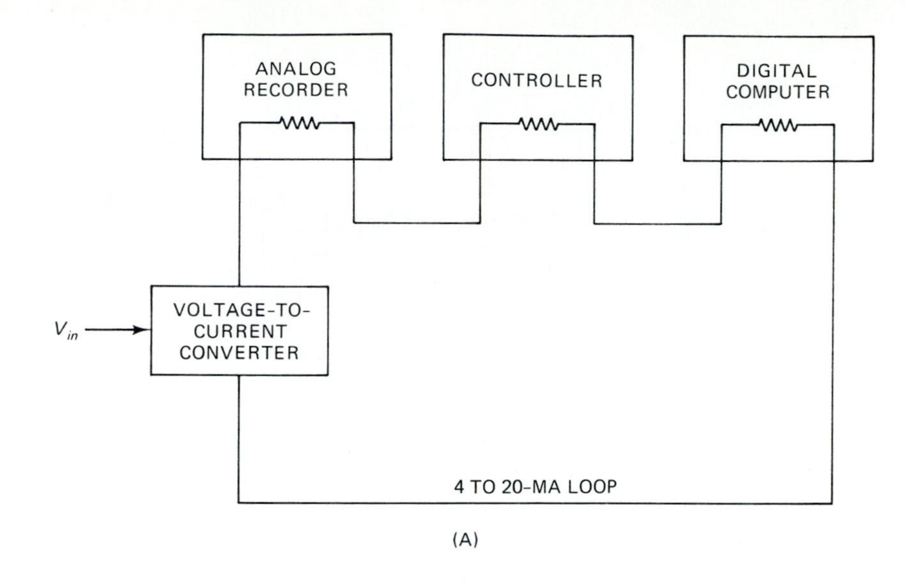


(A)

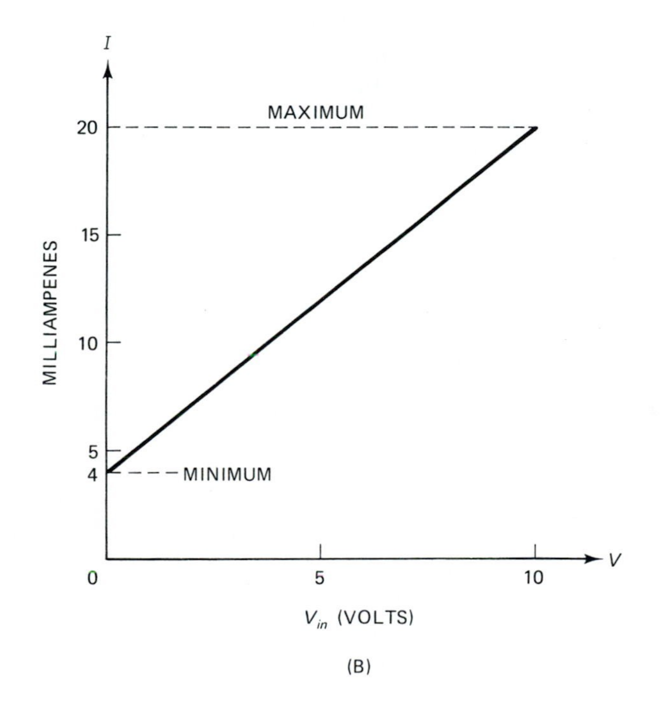


(B)

FIGURE 17-4

## 17-3.2 Four to Twenty Milliampere Current Loops

Industrial process control technology uses a current loop system in some instrumentation communications applications. Figure 17-4A shows a typical system in which three different devices are served by the same current loop. The current has a range

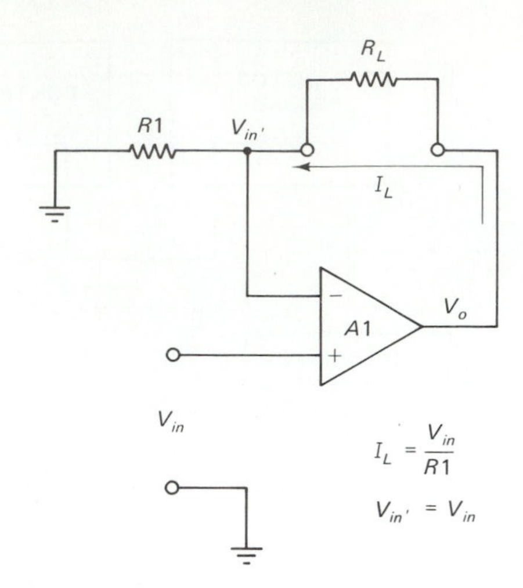


FIGURE 17-5

of permissible values of 4 mA to 20 mA (see Fig. 17-4B). The current loop transmits a range of values which represent parameters being measured. The overall dynamic range of the current is 16 mA, but the 4-mA offset raises the maximum current to 20 mA. The reason for the offset, and one of the advantages of the 4-to-20 mA instrumentation loop over the teletypewriter loop, is the zero input condition is represented by a known current (4 mA), not 0 mA. Thus, the 4-to-20 mA loop can distinguish between a zero parameter value (represented by 4 mA) and a zero current condition brought on by a failure in the circuitry or an open communications line, or if the equipment is turned off.

Sometimes the 4-to-20 mA current loop seems mystifying. Students may ask "What does the range represent?" It doesn't represent anything in general, it's just an available dynamic range that can be appropriated to represent anything. For example, in Fig. 17-4B the 4-to-20 mA current range is used to represent a 0 to 10 volt signal range. Thus, when the signal is 0 volts, a 4-mA current is transmitted, while a +10-volt signal produces a 20-mA current. The designer can redefine the meanings of the 4-mA minimum current and the 20-mA maximum current according to the needs of the system.

A *voltage-to-current converter* is needed to make the current loop work with ordinary input devices. Unlike the teletypewriter, where the current was either on or off depending on whether a 1 or 0 was transmitted, the analog 4-to-20 mA current loop requires a circuit which will produce an output load current proportional to the input voltage (Fig. 17-5). The current loop, represented by load resistor $R_L$, is the feedback current of a noninverting follower operational amplifier circuit. The current is proportional to the input voltage according to the following rule

$$I_L = \frac{V_o - V_{in}}{R_L} \tag{17-1}$$

Output voltage $V_o$ is found from

$$V_o = \frac{V_{in}R_L}{R_{in}} + V_{in} \tag{17-2}$$

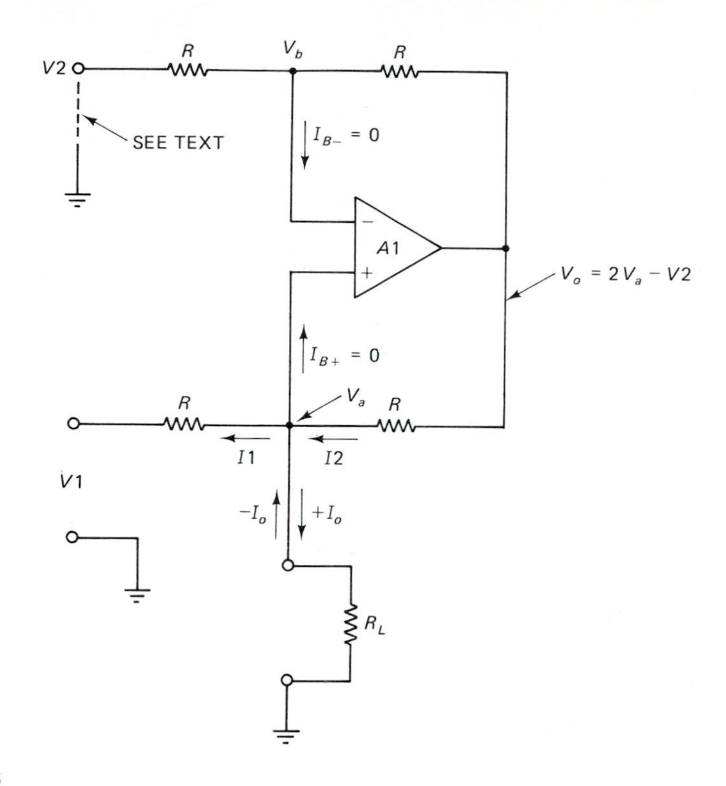


FIGURE 17-6

By algebra one can demonstrate substituting Eq. (17-2) into Eq. (17-1) produces

$$I_L = \frac{V_{in}}{R1} \tag{17-3}$$

The circuit of Fig. 17-5 operates over a wide range of load resistance values, but suffers because the load must float with respect to ground. A grounded load circuit is shown in Fig. 17-6. This circuit is called the Howland current pump after B. Howland of the MIT Lincoln Laboratories. The Howland current pump is unique because load current is independent of load resistance. The current is proportional to the difference between voltage $V1$ and $V2$. In the generalized case of the Howland current pump, $V2$ is not zero, although in many practical applications $V2$ is zero and the associated input resistor is grounded. Figure 17-7 is a graphical solution of the circuit evaluation for the case $V2 = 0$.

Another distinction of the Howland current pump is it can either sink or source current depending upon the relationship of $V1$ and $V2$. If $I_L > 0$, the current flows out of the circuit ("down" in Fig. 17-6), but if $I_L < 0$ the current flows into the circuit ("up" in Fig. 17-6). For simplicity the case where $V2 = 0$ is evaluated.

There are two output voltages in this circuit. Voltage $V_a$ is the load voltage and appears across the load resistance

$$V_a = I_L R_L \tag{17-4}$$

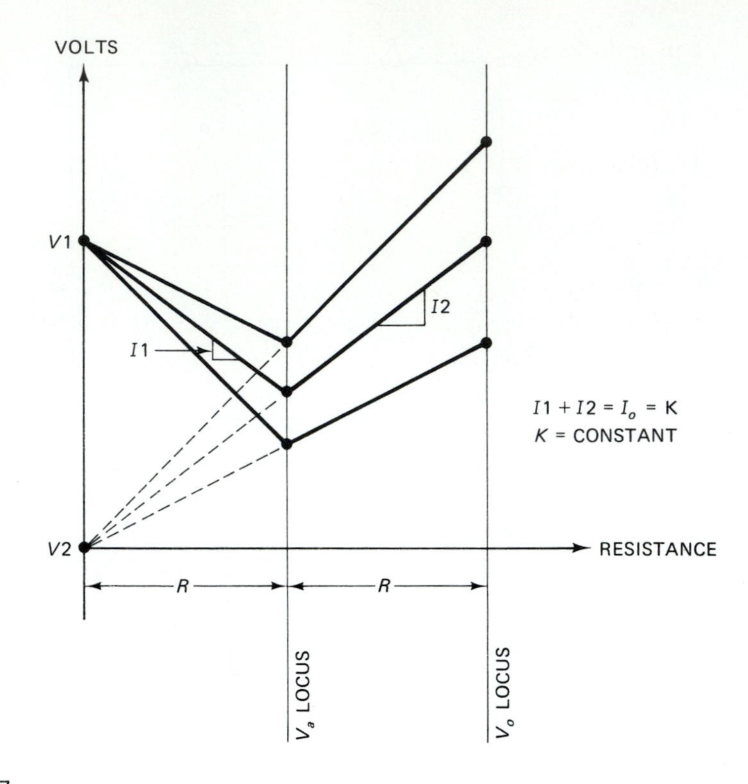


FIGURE 17-7

According to Kirchoff's Current Law (KCL)

$$I_L = I1 + I2 \tag{17-5}$$

By Ohm's law

$$I1 = \frac{V1 - V_a}{R} \tag{17-6}$$

and

$$I2 = \frac{V_o - V_a}{R} \tag{17-7}$$

Substituting Eq. (17-6) and Eq. (17-7) into Eq. (17-5)

$$I_L = \frac{V1 - V_a}{R} + \frac{V_o - V_a}{R} \tag{17-8}$$

This reduces to

$$I_L R = V1 + V_o - 2V_a \tag{17-9}$$

Therefore, we may write

$$V_a = \frac{V1 + V_o - I_L R}{2} \tag{17-10}$$

For voltage $V_a$ the operational amplifier is connected in the noninverting follower configuration. Because the two feedback network resistors are equal, the gain of the circuit is two. Therefore

$$V_o = 2V_a \tag{17-11}$$

or

$$V_o = V1 + V_o - I_L R \tag{17-12}$$

combining terms

$$0 = V1 - I_L R \tag{17-13}$$

$$I_L = \frac{V1}{\text{R}} \tag{17-14}$$

Equation (17-14) shows the output load current $I_L$ is proportional to input voltage $V_{in}$ and inversely proportional to the feedback resistance (assuming all four resistors are equal).

In the more general case where $V2$ is nonzero, Eq. (17-14) becomes

$$I_L = \frac{V1 - V2}{R} \tag{17-15}$$

To make the Howland current pump produce the 4 mA current which represents the 0 volts signal level, we must set $V2$ to a value which will force Eq. (17-15) to evaluate to 4 mA when $V1 = 0$.

### EXAMPLE 17-1

A Howland current pump is to be used to make a 4-to-20 mA current loop transmitter. What value of bias voltage $V2$ will force $I_L$ to be 4 mA when $V1 = 0$; assume $R = 1$ kohm.

### Solution

$$I_L = \frac{V1 - V2}{R}$$

$$4\text{ mA} = \frac{(0\text{ volts}) - V2}{1\text{ kohm}}$$

$$4\text{ volts} = 0 - V2$$

$$V2 = -4\text{ Vdc}$$

■

## 17-4 RS-232 SERIAL INTERFACING

Because serial data communications require only one channel, either a pair of wires, or a single radio or telephone channel, it is less costly than parallel data transmission. The benefit is especially noticeable on long line systems where the extra cost of wire and telecommunications channels becomes most apparent. The current loops discussed in the previous section were the earliest form of interface to peripherals and are still in use (although declining in popularity). In this section, we will discuss what is probably the most common form of voltage-operated serial communications channel, the RS-232C port.

The Electronic Industries Association (EIA) RS-232C is used for serial data transmission with voltage levels representing 1 and 0, rather than current levels. The RS-232C calls for a 25-pin D shell standard connector (the type DB-25) which is always wired in the same manner (see Table 17-1) using the same voltage levels. If all signals are defined according to the standard, it is possible to interface any two RS-232C devices without any problems.

A large collection of peripherals (modems, printers, video terminals) are fitted with DB-25 RS-232C connectors. Unfortunately, some manufacturers also use the DB-25 series of connectors in exactly the same gender as RS-232C, but not in the RS-232C manner. Thus, the existence of an RS-232C connector is not adequate proof an RS-232C is used.

**TABLE 17-1** RS-232 Pin Assignments for DB-25 Connector

| *Pin No.* | *RS-232 Name* | *Function* |
|---|---|---|
| 1 | AA | Chassis ground/common |
| 2 | BA | Data from terminal |
| 3 | BB | Data received from MODEM |
| 4 | CA | Request to send |
| 5 | CB | Clear to send |
| 6 | CC | Data set ready |
| 7 | AB | Signal ground |
| 8 | CF | Carrier detection |
| 9 | (x) | |
| 10 | (x) | |
| 11 | (x) | |
| 12 | (x) | |
| 13 | (x) | |
| 14 | (x) | |
| 15 | DB | Transmitted bit clock (internal) |
| 16 | (x) | |
| 17 | DD | Received clock bit |
| 18 | (x) | |
| 19 | (x) | |
| 20 | CD | Data terminal ready |
| 21 | (x) | |
| 22 | CE | Ring indicator |
| 23 | (x) | |
| 24 | DA | Transmitted bit clock, external |
| 25 | (x) | |

(x) = unassigned

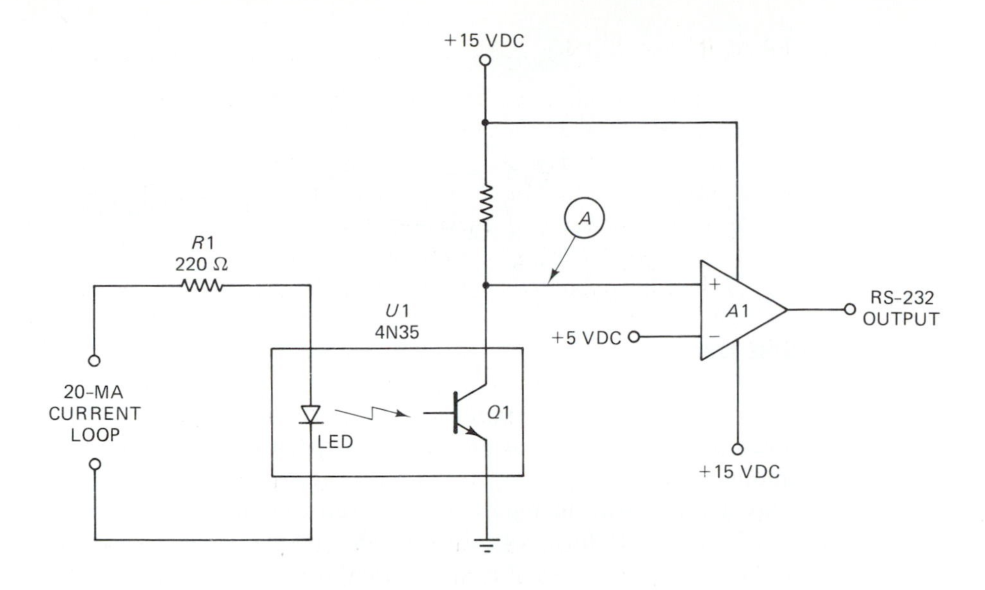


FIGURE 17-8

The RS-232C is older than most of our present-day digital devices, so it uses an obsolete set of voltages. In RS-232C format, the logical 1 (HIGH) is represented by a potential between −5 and −15 volts, while a logical 0 (LOW) is represented by a potential between +5 and +15 volts. Since most digital equipments today are based on TTL-compatible formats (0 = 0 volts, 1 >= +2.4 volts), some translation is needed. Perhaps the most common conversion method is to use the Motorola 1488 line driver and 1489 line receiver chips.

The circuit of Fig. 17-8 is used to convert a 20 mA current loop signal, as might be found with older design teletypewriters or printers, into an RS-232C signal. The circuit is designed around an optoisolator (*U*1). The optoisolator is an IC device in which light from an LED shines on the base region of a phototransistor. This is especially desirable if the current loop experiences high voltage transient spikes from the inductive kick of print solenoids. The output of the optoisolator is connected to the noninverting input of an operational amplifier (*A*1). The inverting input is biased to +5 Vdc. When the LED is turned off (during the 0 condition on the loop), the transistor is also turned off so the voltage at point A is high (essentially $V+$). This voltage applied to the noninverting input means the differential input voltage is effectively negative, which forces the output of *A*1 to saturate at $+V_o$ (about +12 Vdc).

The LED in the optoisolator is turned on when the 20 mA current flows. Resistor *R*1 limits the current in the LED to a safe value to prevent overburdening the loop with an excessive voltage drop. The 20 mA current turns on the LED, which in turn illuminates the phototransistor (*Q*1) and drives it to saturation. When this occurs the

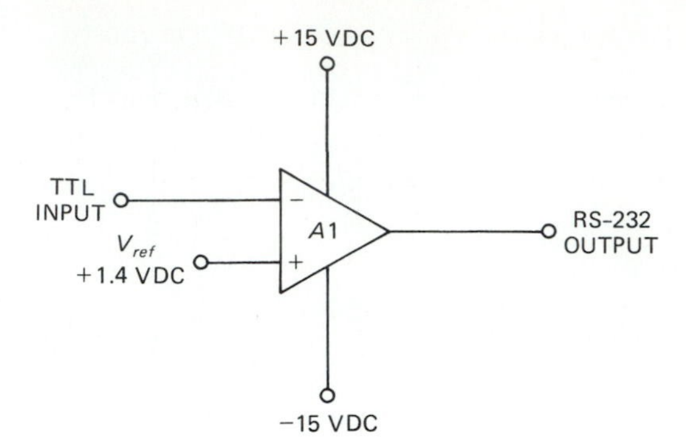


FIGURE 17-9

voltage on the collector (point A) drops to near zero. Under this condition the differential input voltage of the op-amp is positive, so the output saturates to $-V_o$, or about $-12$ Vdc.

The circuit shown in Figure 17-9 is used to convert TTL levels from a computer to RS-232C levels for transmission. This circuit is also based on the operational amplifier ($A1$). The $-15$ Vdc and $+15$ Vdc power supplies used for the operational amplifier are the source of the RS-232C levels. The TTL input is biased for noise band immunity by a reference voltage ($V_{ref}$) of 1.4 Vdc. The TTL levels are, HIGH (1) = +2.4 to +5 volts and LOW (0) = 0 to 0.8 volts. If a 1 is applied to the input of Fig. 17-9, the differential input voltage ($+2.4 - 1.4$) is positive, so the output of $A1$ is forced to $-V_o$. Similarly, when the input is 0, the differential input voltage of $A1$ is $0 - 1.4$, so the output is forced to $+V_o$.

## 17-5 MODULATION AND DEMODULATION

*Modulation* is the process of using one signal to alter a characteristic of another signal in order to impart information. In the context of communications, the altered signal is called the *carrier signal.* The other signal is called the *modulating signal.* The carrier signal usually has a much higher frequency than the modulating signal. For example, in an AM band radio signal the carrier will have a frequency between 550 KHz and 1620 KHz, and the audio modulating signal will have a frequency less than 5 KHz. Alternatively, in a wire carrier telephony system, the carrier might have any frequency between 5 KHz and 200 KHz, while the modulating signal is limited to the range 300 Hz to 3000 Hz.

Any of several carrier parameters can be altered by the modulating signal. In an *amplitude modulation* (AM) system the modulating signal varies the amplitude of the carrier signal. There are also two forms of angular modulation, *phase modulation* (PM) and *frequency modulation.* In these systems the carrier amplitude remains constant, but either its frequency or phase is varied. The two forms of angular modulation are similar enough so the terms are sometimes used interchangeably. For example, many FM transmitters used in landmobile radios are actually PM devices. Although interchanging PM and FM works in most practical situations, it is technically an error.

AM, PM, and FM usually involve a sinusoidal modulating signal altering a sinusoidal carrier. If the carrier is a square wave or pulse train, however, we can use *pulse amplitude modulation* (PAM), *pulse width modulation* (PWM), and *pulse position modulation* (PPM).

### 17-5.1 Amplitude Modulation (AM)

In AM systems a low frequency modulating frequency (with frequency $f_m$ and amplitude $v_m$) is used to alter the amplitude of a higher frequency carrier signal (frequency $f_c$ at amplitude $v_c$). These signals are defined for the sinusoidal case as

$$v_m = V_m \text{SIN}(2\pi f_m t) \tag{17-16}$$

and

$$v_c = V_c \text{SIN}(2\pi f_c t) \tag{17-17}$$

where

$v_m$ = the instantaneous amplitude of the modulating signal

$V_m$ = the peak amplitude of the modulating signal

$v_c$ = the instantaneous amplitude of the carrier signal

$V_c$ = the peak amplitude of the carrier signal

$f_c$ = the carrier frequency in hertz (Hz)

$f_m$ = the modulating frequency in hertz (Hz)

$t$ = time in seconds

Before examining amplitude modulation, let's first discuss what AM is not. Simply combining the two signals, $v_m$ and $v_c$, in a linear network does not produce AM. Figure 17-10A shows $v_m$ and $v_c$ in separate traces. In Fig. 17-10B the two signals are combined in a linear resistor network. The result is the *additive* waveform shown in Fig. 17-10B. The instantaneous amplitude of this signal is given by

$$v = v_m + v_c \tag{17-18}$$

$$v = V_m \text{Sin}(2\pi f_m t) + V_c \text{Sin}(2\pi f_c t) \tag{17-19}$$

Compare Fig. 17-10B with Fig. 17-10C. The waveform in Fig. 17-10C is a sinewave carrier amplitude modulated with lower frequency sinewave. The AM process is *multiplicative*, so the instantaneous amplitude is

$$v = (V_c + v_m)\text{Sin}(2\pi f_c t) \tag{17-20}$$

The instantaneous envelope is expressed in the form

$$v_e = V_c + v_m \tag{17-21}$$

or, to rewrite Eq. (17-20) in terms of the envelope expression

$$v = v_e \text{Sin}(2\pi f_c t) \tag{17-22}$$

and, accounting for $v_m$

$$v = (V_c + V\text{Sin}(2\pi f_m t))(\text{Sin}(2\pi f_c t)) \tag{17-23}$$

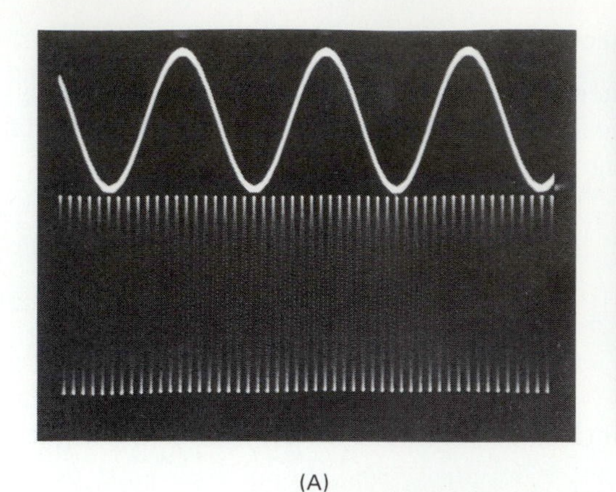

(A)

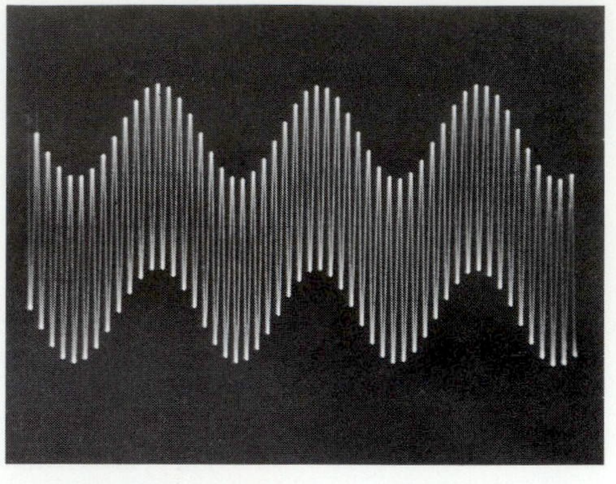

(B)

(C)

FIGURE 17-10

The ratio of modulating signal to carrier signal is called the *modulation index* ($m$)

$$m = \frac{V_m}{V_c} \tag{17-24}$$

$(0 < m <= 1)$

Equation (17-23) can be expressed in terms of $m$

$$v = V_c[1 + m\mathrm{Sin}(2\pi f_m t)(\mathrm{Sin}(2\pi f_c t))] \tag{17-25}$$

or when normalized

$$v = [1 + m\mathrm{Sin}(2\pi f_m t)(\mathrm{Sin}(2\pi f_c t)] \tag{17-26}$$

$$v = \mathrm{Sin}(2\pi f_c t) + [(m\mathrm{Sin}(2\pi f_m t)(2\pi f_c t)] \tag{17-27}$$

A well-known trigonometric identity gives some insight into the nature of amplitude modulation as derived from Eq. (17-27) and the preceding mathematical argument:

$$\mathrm{Cos}(A \pm B) = \mathrm{Cos}A\mathrm{Cos}B \mp \mathrm{Sin}A\mathrm{Sin}B \tag{17-28}$$

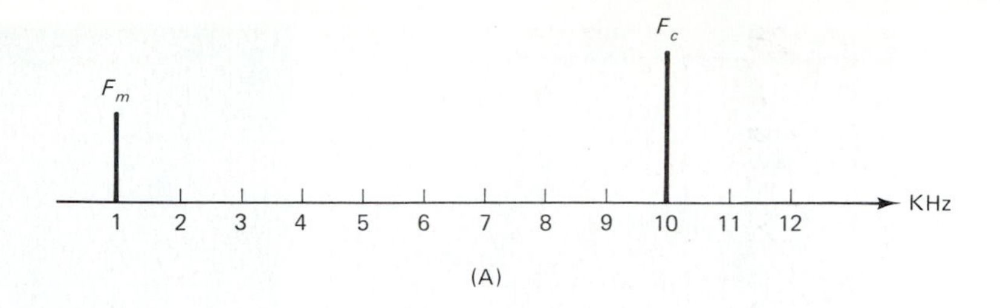


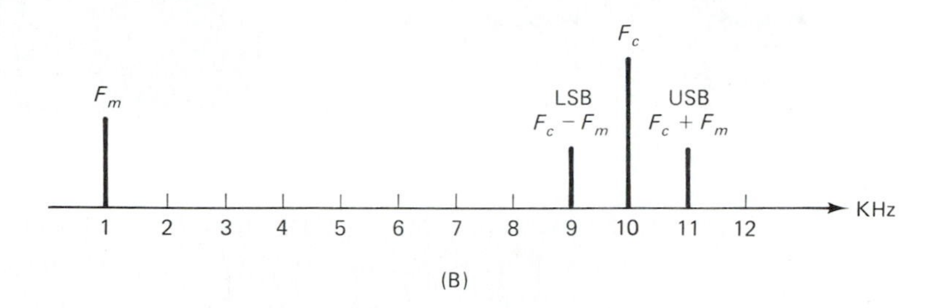


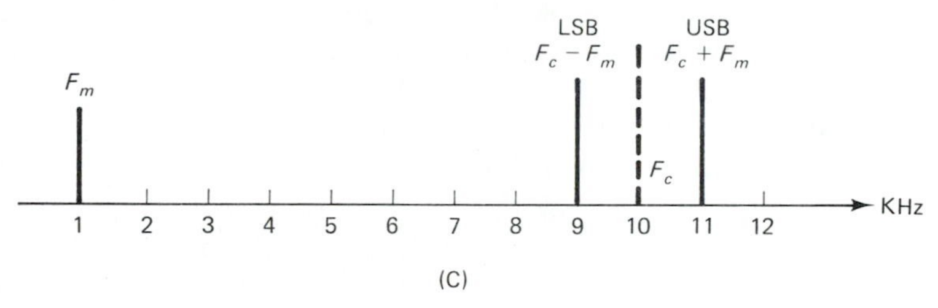


FIGURE 17-11

Applying the identity of Eq. (17-28) to Eq. (17-27) will result in

$$v = \text{Sin}(2\pi f_c t) + (m/2)[\text{Cos}(2\pi f_c - 2\pi f_m)t - \text{Cos}(2\pi f_c + 2\pi f_m)t] \tag{17-29}$$

Notice the terms $(2\pi f_c - 2\pi f_m)$ and $(2\pi f_c + 2\pi f_m)$ in Eq. (17-26). These terms tell us that sum and difference frequencies are generated. The sum frequency is called the *upper sideband* (USB) and the difference frequency is called the *lower sideband* (LSB). Figure 17-11 shows the result of amplitude modulation on the frequency domain as two separate spikes. In Fig. 17-11A the carrier $(f_c)$ and modulating $(f_m)$ signals are shown in frequency domain as two spiles. The result of amplitude modulation is shown in Fig. 17-11B. The carrier frequency is 10 KHz. The modulating frequency is 1 KHz. The USB is (10 + 1) KHz, or 11 KHz, and the LSB is (10 − 1) KHz, or 9 KHz. In other words, the sidebands vary from the carrier frequency by an amount equal to the modulation frequency. Figure 17-12A shows a simple modulator which is based on a single PN junction diode and a resistor combining network. The small-signal diode provides the nonlinearity over cyclic excursions of the input signal which is required to produce AM.

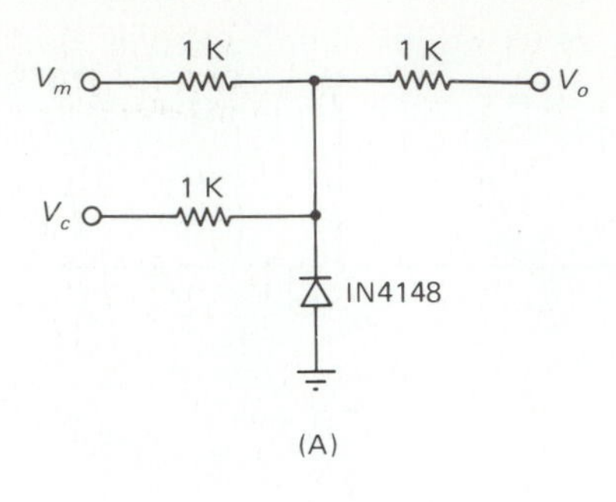


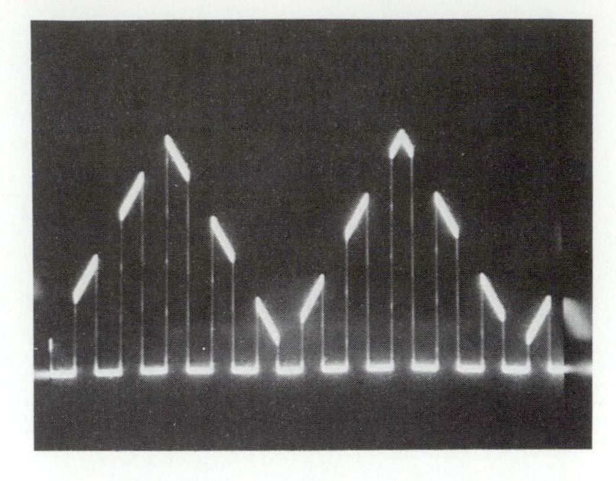

FIGURE 17-12 (B)

In Fig. 17-12B the result of using a 1 KHz triangle waveform to modulate a 5 KHz square wave is shown. The modulating signal is superimposed on the carrier in a manner quite different from the linear mixing previously seen in Fig. 17-10B.

Another form of simple AM modulator is shown in Fig. 17-13A, with the resultant waveform shown in Fig. 17-13B. Again a triangle wave was used to modulate a higher frequency square wave signal. When the carrier signal is modulated by an active modulator the result is the bipolar waveform seen in Fig. 17-10C. A point-for-point explanation of this type of waveform is shown in Fig. 17-13C, but with a triangle modulating signal and a square wave carrier for simplicity of illustration.

In Fig. 17-13A the modulation effect is obtained from a CMOS or MOS electronic switch shunted across the signal line. A high-pass filter passes only the carrier and sidebands to the output circuit. The modulating signal is passed along the line, but it is chopped at the carrier frequency because the control terminal on the switch is driven by the carrier signal ($v_c$). Demodulation of this signal is accomplished through a similar process, but with a low-pass filter instead (see Fig. 17-13C). This process is called *synchronous demodulation*. Another form of demodulation is *asynchronous demodulation* or *envelope detection*. This method uses a rectifier in series with the signal line followed by a low-pass filter circuit.

One final form of amplitude modulator is the analog multiplier shown in Fig. 17-14. Recall the AM signal is the product of the carrier and modulating signals. Thus, it is suitable to use an analog multiplier circuit AM (Chapter 18). The output of this circuit is

$$V_o = \frac{v_m v_c}{10} \tag{17-30}$$

A *balanced modulator* circuit is used to produce an output signal which reduces or suppresses the carrier signal, while retaining the sidebands. This type of signal is shown in Fig. 17-10C. It is frequently used in commercial, amateur, and military radio communications. The carrier signal ($f_c$) is suppressed a large amount (ideally to zero, but more likely $-40$ to $-60$ dB of suppression), leaving only the upper and lower sidebands. The advantage of this system is redistribution of the available power to only the sidebands. This form of amplitude modulation is usually called *double side-*

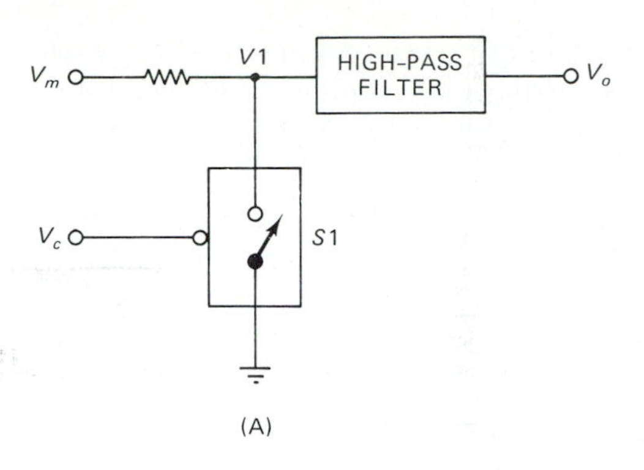


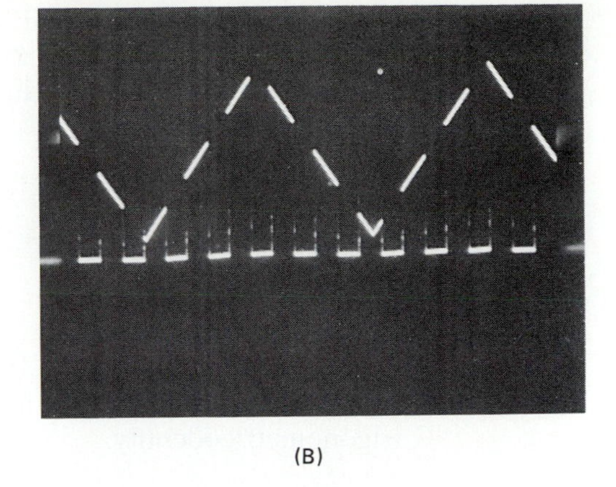

(B)

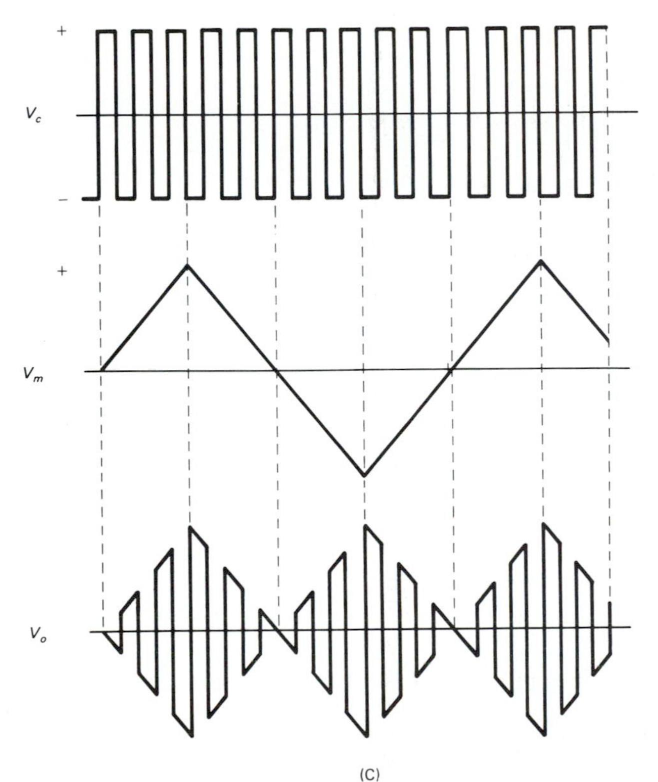


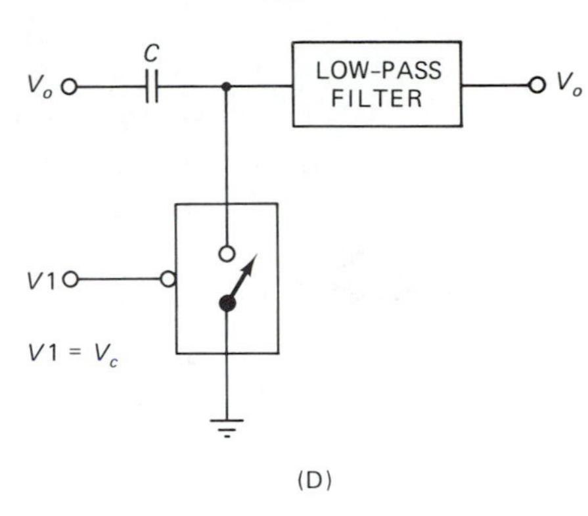


FIGURE 17-13

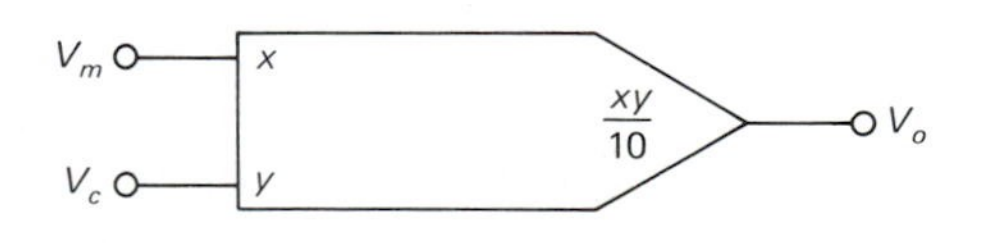


FIGURE 17-14

*band suppressed carrier* (DSBSC). A related and more common form of AM is *single sideband suppressed carrier* (SSBSC or simply SSB). The suppression of the carrier in DSBSC and in most forms of SSBSC is the result of the balanced modulator. Consider the following

$$v = v_m \times v_c \tag{17-31}$$

$$v = V_m \text{Sin}(2\pi f_m t) \times V_c \text{Sin}(2\pi f_c t) \tag{17-32}$$

$$v = V_m V_c [\text{Sin}(2\pi f_m t)][\text{Sin}(2\pi f_c t)] \tag{17-33}$$

Another form of Eq. (17-33) is sometimes used because it reflects the real situation in AM systems. This other form of the AM equation takes advantage of a standard trigonometric identity

$$\text{Sin}A\text{Sin}B = \frac{[\text{Cos}(A-B) - \text{Cos}(A+B)]}{2} \tag{17-34}$$

Let

$$A = V_c = V_{cp}\text{Sin}(2\pi f_c t) \tag{17-35}$$

and

$$B = V_m = V_{mp}\text{Sin}(2\pi f_c t) \tag{17-36}$$

Substituting Eq. (17-35) and Eq. (17-36) into Eq. (17-34)

$$V = (1/2)V_{mp}V_{cp}\text{Cos}[2\pi(f_c - f_m)] - (1/2)V_{mp}V_{cp}\text{Cos}[2\pi f_c + f_m)] \tag{17-37}$$

Compare Eq. (17-37) to Eq. (17-29). Note the carrier term $\text{Sin}(2\pi f_c t)$ is missing, indicating the carrier is suppressed. The signal of Eq. (17-37) is a DSBSC signal. If filtering or phasing is used to remove either upper or lower sideband, the signal becomes a single-sideband (SSB) signal. In communications, the SSB signal is probably the most widely used form of AM except in the VHF aviation and 27-MHz citizen's bands.

The SSB or DSBSC signal cannot be demodulated in the same manner as a standard AM signal. If an envelope detector is used, the recovered modulating signal will be distorted beyond recognition. The SSB and DSBSC signals are demodulated in a *product detector*. That is, a detector circuit that nonlinearly mixes a signal which represents a reconstructed version of the carrier with the DSB or SSB modulated signal to recover the modulation. For example, if a 455 KHz carrier is modulated with a 1 KHz sinewave in a balanced modulator, the two signals produced will be the difference (454 KHz) and sum (456 KHz). Only one of these needs to be transmitted. If the transmitted signal is then mixed with another 455 KHz carrier at the receiver end, it will produce the difference frequencies, 456 KHz − 455 KHz = 1 KHz, and 455 KHz − 454 KHz = 1 KHz. The modulation is thus recovered.

Figure 17-15A shows an IC balanced modulator, the LM-1496. An LM-1596 is also available with the military temperature range. The 14-pin DIP package is shown in

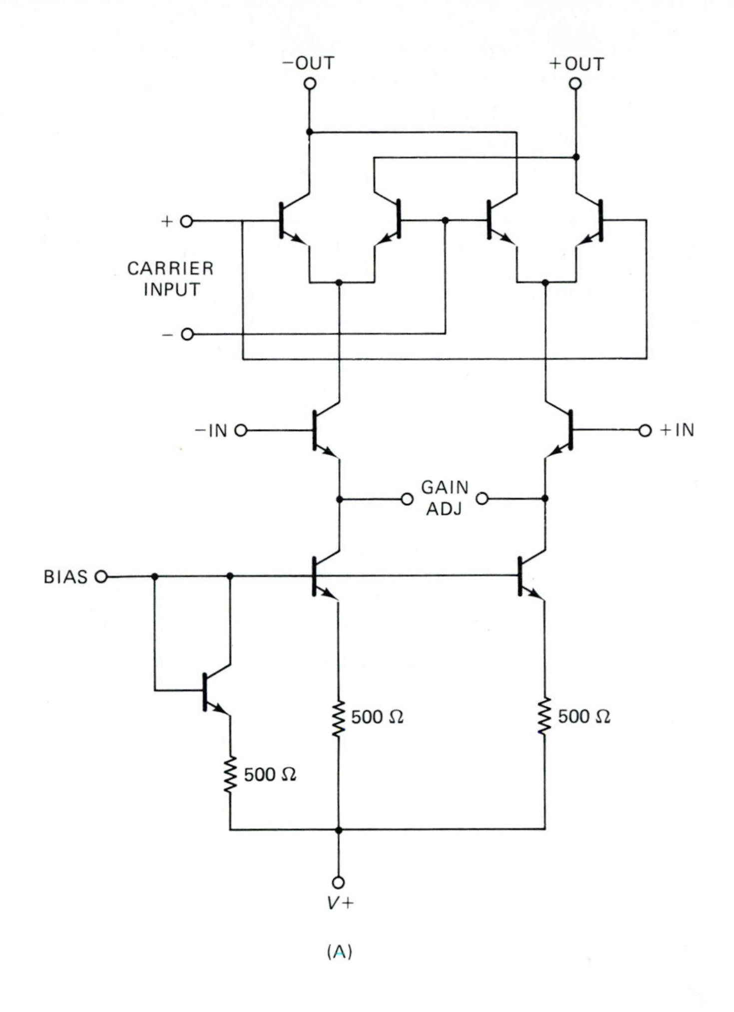


(A)

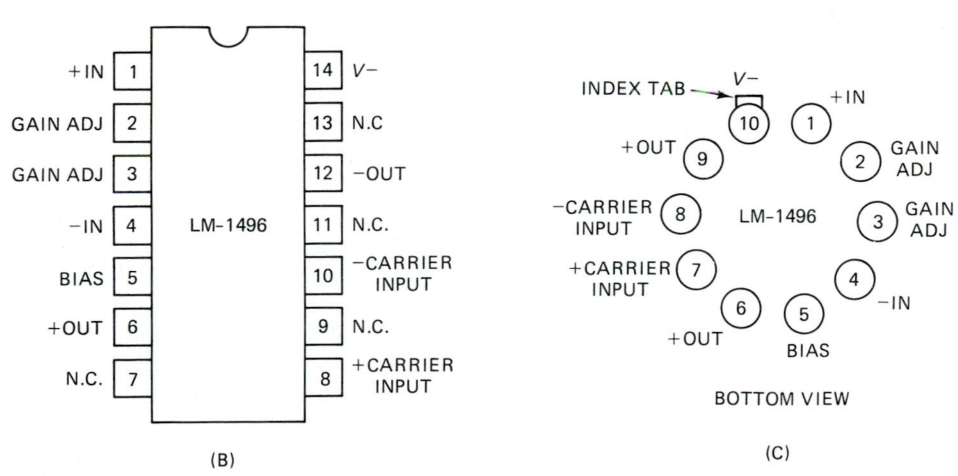


(B)

(C)

FIGURE 17-15

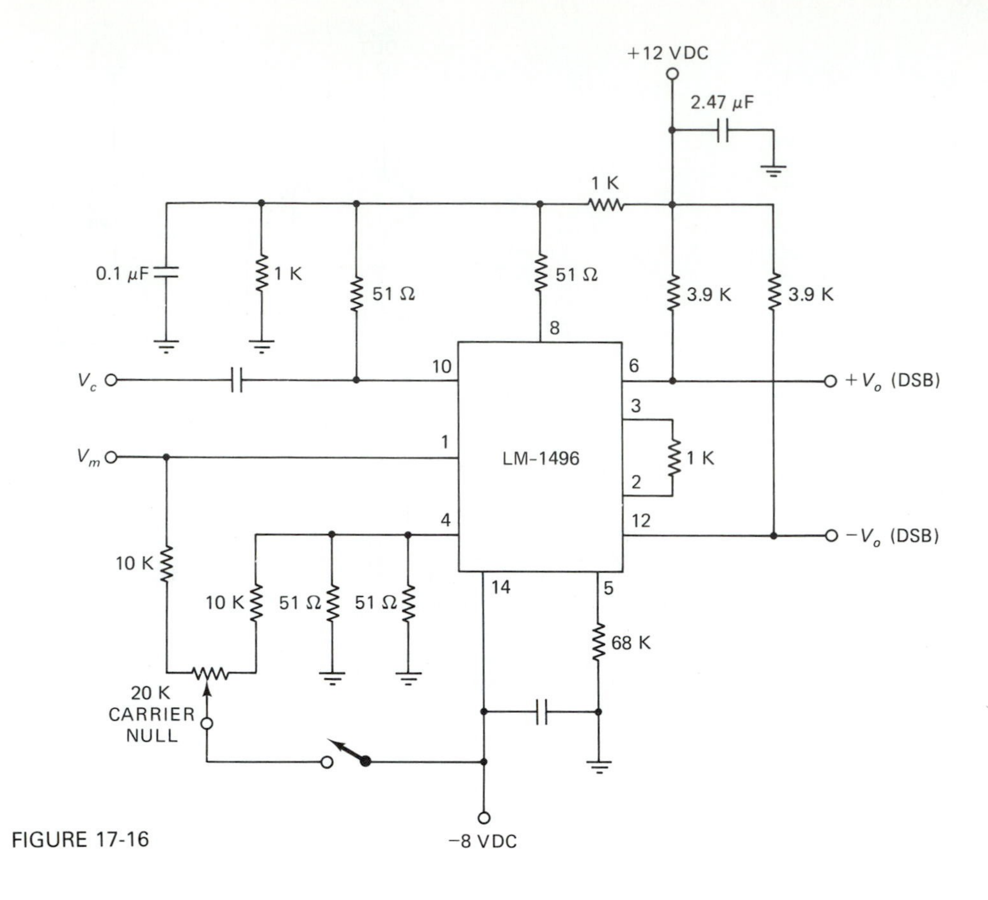


FIGURE 17-16

Fig. 17-15B. The 10-pin metal can package is shown in Fig. 17-15C. These IC balanced modulators will suppress the carrier 65 dB at 500 KHz and 50 dB at 10 MHz. The circuit for a balanced modulator based on the LM-1496 and LM-1596 is shown in Fig. 17-16A. The carrier signal is applied to the +CARRIER input (pin 8) and the signal is applied to the +IN input (pin 1). The −IN and −CARRIER inputs are terminated. The optimum signal level is 300 mV, although signals up to 500 mV can be accommodated. In SSB and DSBSC radio systems it is common to generate the sideband signal at a single frequency and then frequency translate it to the operating frequency in a heterodyne mixer.

A product detector circuit for the LM-1496 and LM-1596 is shown in Fig. 17-17. Here, the SSB or DSBSC signal is applied to the +IN and the carrier is applied to the −CARRIER input. The demodulated audio appearing on pin 12 will contain a residual of the carrier and RF signal. This must be stripped off in an RC low-pass filter. The DC level on the output is stripped off by the coupling capacitor.

## 17-5.2 Pulse Width Modulator Circuits

A *pulse width modulator* (PWM) produces a square wave output in which the duration of each cycle varies with the modulating signal. Another name for this type of circuit, which is equally descriptive, is *duty cycle modulator*. The circuit for a basic PWM is

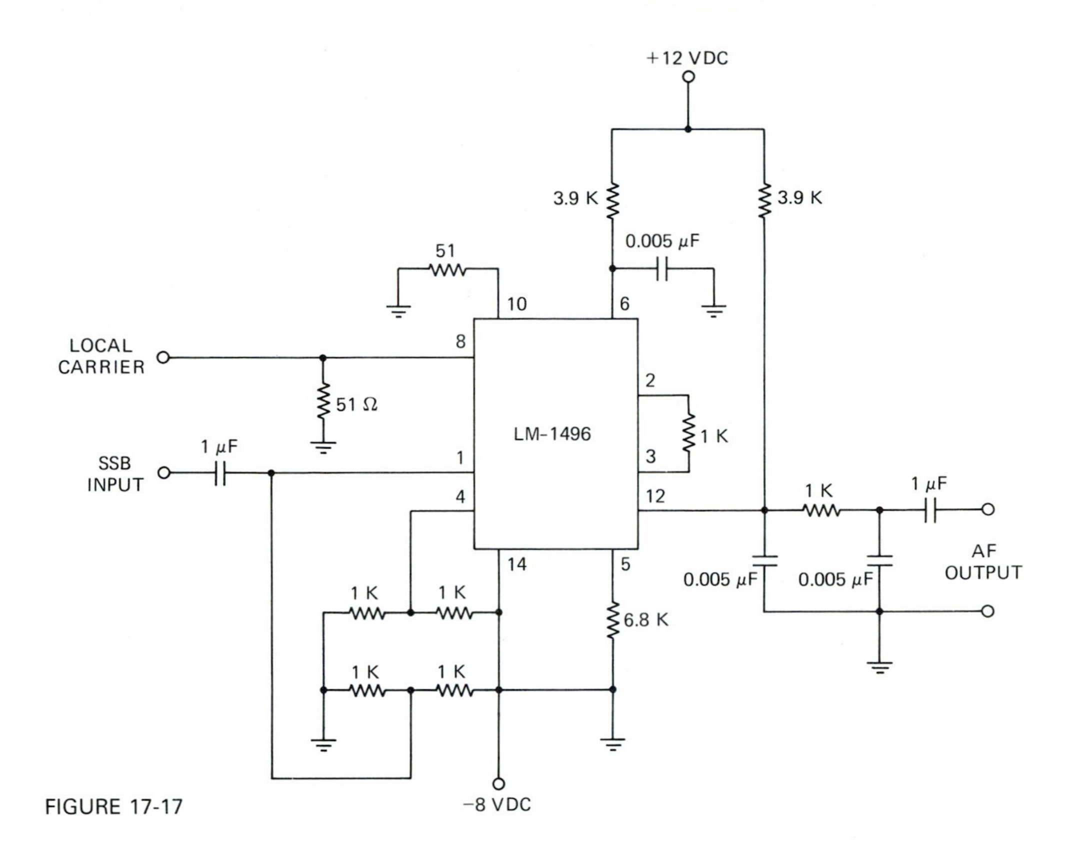


FIGURE 17-17

shown in Fig. 17-18A. The PWM is simply a voltage comparator used so the modulating signal ($V_m$) is applied to one input of the comparator, while the carrier signal ($V_c$) is applied to the alternate input of the comparator. In the case shown in Fig. 17-18A the modulating signal is applied to the noninverting input, while the carrier is applied to the inverting input of the comparator. This PWM circuit is therefore an inverting pulse width modulator. The carrier signal should be a triangle or sawtooth waveform.

The timing waveforms for the PWM are shown in Fig. 17-18B. The upper plot shows the triangle wave carrier ($V_c$) and the modulating signal ($V_m$). In electronic instrumentation applications the varying modulating signal might be an audio carrier, or more likely a time-varying DC potential which represents a physical parameter such as temperature or pressure.

According to the normal rules for a voltage comparator, output $V_o$ is LOW whenever $V_c < V_m$ and HIGH whenever $V_c > V_m$. Thus, the output pulse will snap HIGH on the peaks of the carrier signal, and will remain HIGH at time $t1$ which is inversely proportional to the value of the modulating signal.

The duty cycle of the output pulse produced by Fig. 17-18A (see lower plot in Fig. 17-18B) is given by

$$D = \frac{T1}{T2} \tag{17-38}$$

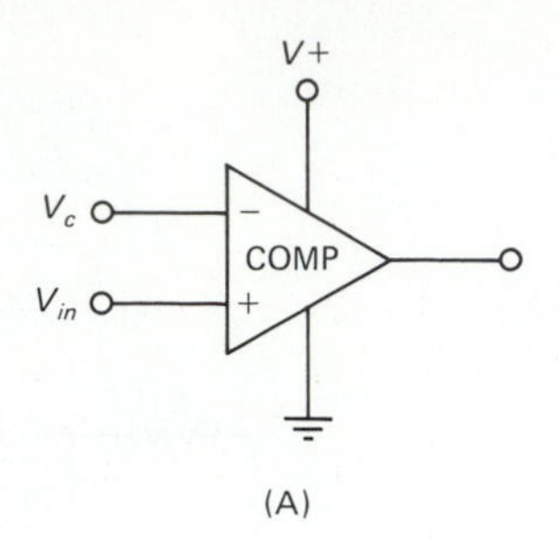


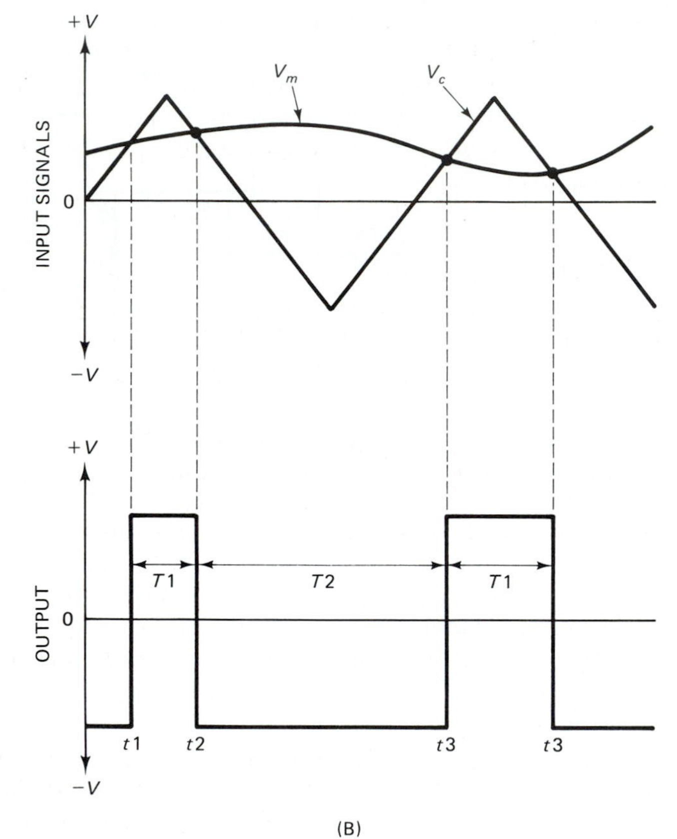


FIGURE 17-18

or related to the applied voltages

$$D = \frac{1 - (V_m/V_c)}{2} \tag{17-39}$$

## EXAMPLE 17-2

A pulse width modulator like the one in Fig. 17-18A uses a 0 to +6 volt triangle waveform carrier signal. Calculate the duty cycle ($D$) when a constant DC potential of +3.5 is applied to the modulation input.

**Solution**

$$D = \frac{1 - (V_m/V_c)}{2}$$

$$D = \frac{[1 - (3.5 \text{ volts}/6 \text{ volts})]}{2}$$

$$D = \frac{[1 - 0.583]}{2} = 0.21 = 21\%$$

■

In the example above a constant DC voltage was used as the modulating signal. In some instrumentation applications the modulating signal will be a DC level, or slowly varying DC level (as in the case of a temperature signal). It may also be a more dynamic signal. In the latter case Eq. (17-38) can be used to calculate the instantaneous duty factor if the instantaneous values of the carrier signal and modulating signal are used.

### 17-5.3 FM Demodulators

Over the years several methods have been used to demodulate FM and PM signals. Of these, however, only two are in widespread use in IC-based FM/PM receivers, *pulse counting detectors* and *IC quadrature detectors* (ICQD).

The FM/PM receivers typically receive a VHF/UHF RF signal and frequency translate it down to an intermediate frequency (IF) of 10.7 MHz (or something similar). A few two-way radio receivers further down-convert the 10.7 MHz IF to a second IF in the 455 KHz range. The latter are called *dual conversion receivers*. The former are called *single conversion receivers*. Both are examples of *superheterodyne receivers*.

The pulse counting FM detector is shown in Fig. 17-19A. An input stage ($A1$) is responsible for squaring the input FM IF sinewave signals. Either a voltage comparator, high-gain limiting amplifiers, or a Schmitt trigger may be used. The output of this stage is a squarewave containing the same frequency or phase variations as the input signal.

The pulses from $A1$ are used to trigger a dual output, one-shot multivibrator ($OS1$). The duration of $OS1$ output pulse must be much shorter than the minimum width of the input signals. The $Q$ and NOT-$Q$ outputs of $OS1$ are applied to a differential amplifier level translator. The output of the differential amplifier is a train of square wave pulses which have a constant amplitude and constant duration, but vary in repetition rate with changes of input frequency. If these pulses are integrated, they produce a varying output voltage which is the reconstructed modulating audio.

The IC quadrature detector is shown in block form in Fig. 17-19B and as a complete IC circuit in Fig. 17-19C.

In Fig. 17-19B the input signal is applied first to a high-gain cascade chain of wideband amplifiers. This has the effect of squaring the input sinewave signal. The clipping action is caused by the gain which is excessive for sinewaves. The clipping serves to remove amplitude peaks where the majority of noise rides. Most noise signals amplitude modulate the signal.

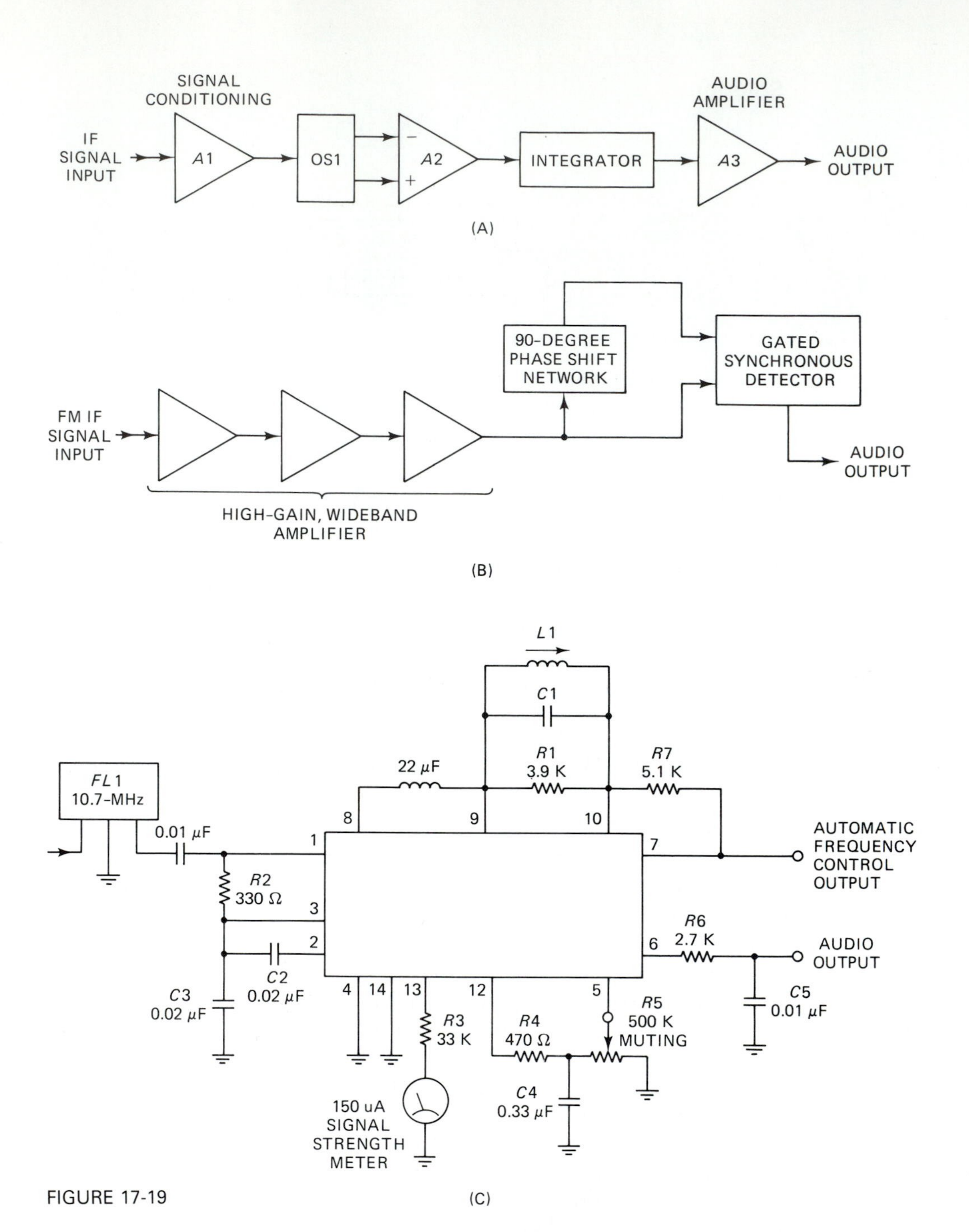

FIGURE 17-19

The square wave signal output from the wideband amplifier is split into two components. One component is fed directly to the gated synchronous quadrature detector stage and is called the 0° signal. The other signal is phase shifted −90° in an external RLC network. The −90° signal is also fed to the detector.

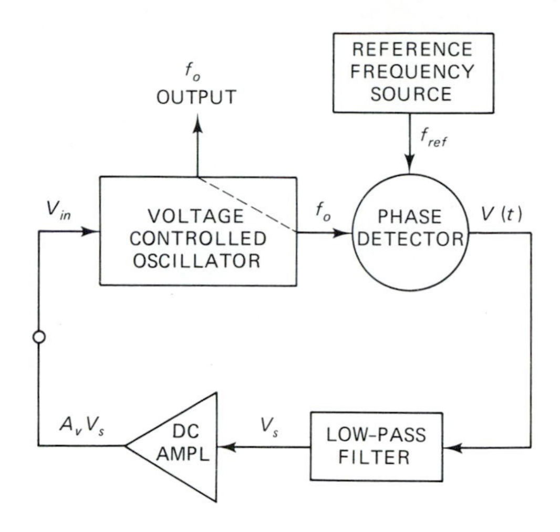


FIGURE 17-20

The gated synchronous detector serves as a quadrature phase detector in order to recover the modulating audio.

Figure 17-19C shows the circuit for an IC quadrature detector (ICQD) based on the CA-3089 (also called LM-3089). The input side of the ICQD is tuned by a ceramic or crystal bandpass filter (*FL*1) to the 10.7-MHz IF frequency. The filter bandpass is selected to accept the deviation of the FM carrier under modulation.

The phase shift network (*R*1/*C*1/*L*1) is tunable so it can be adjusted to account for tolerances in the IF center frequency. These tolerances are mostly due to variations in filter *FL*1.

## 17-6 PHASE LOCKED LOOP (PLL) CIRCUITS

The *phase locked loop* (PLL) circuit was invented in the 1930s as a synchronous AM demodulator. Oddly enough, its first intended use never caught on except in a few Voice of America shortwave relay receiver sites. Few, if any, other AM detectors are based on the PLL circuit. A host of other applications for the PLL are found, however: touchtone decoding, frequency shift keying (FSK) decoding, FM demodulation, FM/FM telemetry data recovery, FM multiplex stereo decoding (to reconstruct the pilot signal), motor speed control, and transmitter frequency control.

As a transmitter frequency controller the PLL has been very popular in cheap CB transmitters to the most expensive broadcast models. The PLL now dominates the frequency control sections of radio transmitters. The PLL are popular in frequency synthesizers because they allow multichannel fixed and discrete variable frequency control, yet the output frequency has the same stability of the crystal oscillator used as the PLL reference frequency.

Figure 17-20 shows the block diagram for a basic phase locked loop circuit. The main elements of the PLL circuit are a *voltage controlled oscillator* (VCO), *phase detector*, *reference frequency source*, and a *low-pass filter* (LPF). A *DC amplifier* may also be used for scaling or level translation of the DC control voltage from the output of the low-pass filter.

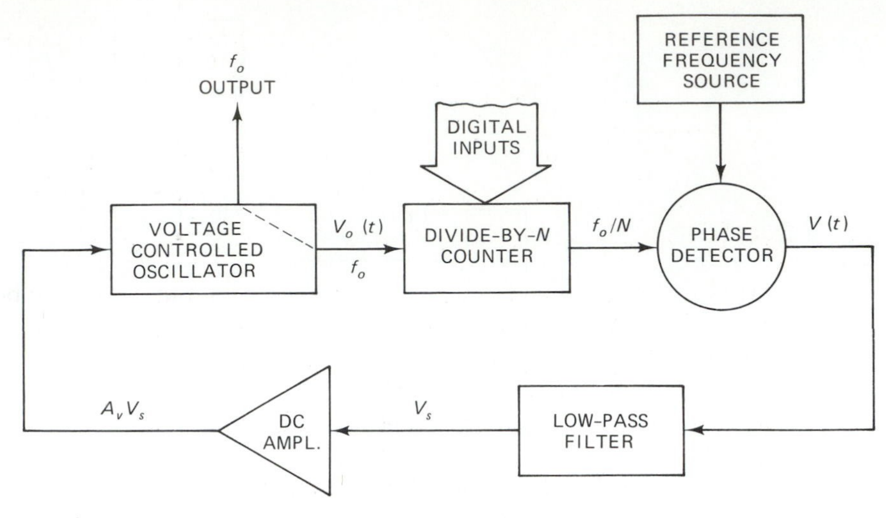


FIGURE 17-21

The VCO is a special form of variable frequency oscillator in which the output frequency $f_o$ is a function of the input control voltage ($V_{in}$). In Fig. 17-20 the VCO input voltage is also the output voltage from the DC amplifier.

The reference frequency source is a stable oscillator operating on a fixed frequency. The reference frequency ($f_{ref}$) in Fig. 17-20 is equal to or less than $f_o$.

The phase detector compares two signals and generates an output proportional to the phase difference between them. In a PLL circuit, the phase detector output ($v(t)$) will either be a pulse train (in digital phase detector) or a DC voltage (analog phase detectors). In both the digital and analog phase detector circuits the output must be processed in a low-pass filter to remove residuals of $f_o$ and $f_{ref}$. In the digital case, the low-pass filter also serves to create the DC control voltage by integrating the pulse train produced by the phase detector.

A modified form of phase detector is shown in Fig. 17-21. This circuit is most common in transmitter frequency control, signal generator, and similar applications where presettable, discrete frequencies are needed. The basic difference between these circuits is the *divide-by-N counter* between the VCO output and the phase detector input port. The reference frequency must be an integer subharmonic of the desired output frequency which will be produced by the VCO, $f_{ref} = f_o/N$.

The divide-by-N digital counter can be used to change the frequency of the VCO. If the division ratio of the counter is changed, the VCO will be forced to the new frequency which maintains the equality $f_{ref} = f_o/N$.

There are three modes of operation in a PLL, *free-running*, *capture* (also called *search*), and *locked*. In the free-running mode the VCO is not under control, and operates on an essentially random frequency within its range. The PLL is typically in the free-running mode for a brief period after turn/on, but then will be free-running only if a defect is present. In the capture mode, the PLL is attempting to lock onto the correct frequency. The VCO frequency tends to converge onto the desired frequency. When the VCO reaches the correct frequency and remains there, the PLL is said to be in the locked mode.

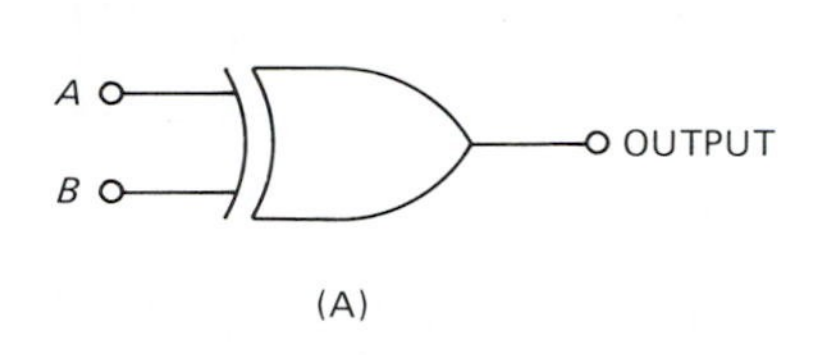


(A)

| A | B | OUTPUT |
|---|---|---|
| 0 | 0 | 0 |
| 0 | 1 | 1 |
| 1 | 0 | 1 |
| 1 | 1 | 0 |

1 = HIGH
0 = LOW

(B)

FIGURE 17-22

The reference frequency source controls the output frequency because it is compared with the VCO output in the phase detector. When there is a difference between the VCO output and the reference frequency, a DC control voltage is generated which pulls the VCO onto the correct frequency. Thus, the PLL is a form of feedback control system, or electronic servomechanism. The reference frequency sets the minimum step between discrete VCO frequencies.

The stability of the PLL is set by the stability of the reference frequency source. In the most stable systems, such as signal generators or transmitter channel controllers, the reference frequency will be a crystal oscillator which is either temperature compensated or operated inside of a stabilization oven. The output frequency of the crystal oscillator may be divided in a divide-by-N chain of digital counters to produce a low frequency such as 5 KHz, 1 KHz, or 100 Hz which are not easily obtained in crystal oscillators. At least one signal source locks the reference frequency to the extremely accurate and stable 60-KHz WWVB signal broadcast by the U.S. Government from a station at Fort Collins, CO.

## 17-6.1 Phase Sensitive Detector Circuits

The job of the phase sensitive detector (PSD) circuit is to generate an output proportional to the difference in phase between two input frequencies. Several different forms of PSD circuit are typically used in PLL circuits, but in this brief discussion only the digital form will be covered. Two basic phase sensitive detectors are used in most digitally controlled IC PLL circuits, the *exclusive-OR gate detector* (Fig. 17-22A) and the *R-S flip-flop edge detector* (Fig. 17-24A). Although both of these circuits are somewhat more sensitive to harmonics, noise, and reference source jitter, they are sufficiently easy to implement (and sufficiently useful) to cause them to see widespread application.

The exclusive-OR (XOR) gate, shown in Fig. 17-22A, is designed to output a HIGH only when the two inputs are different from each other (one is HIGH and the other is LOW). When the two inputs are at the same level, either both LOW or both HIGH, the output of the XOR gate is LOW. Figure 17-22B shows the truth table for the XOR gate. In the digital PLL circuit, the inputs of the XOR gate are excited with the square wave $f_o$ and $f_{ref}$ signals.

Figure 17-23 shows several cases for the relationship between the reference and output frequencies as applied to inputs A and B of the XOR gate. Despite the fact these signals are square waves, the convention of using radian notation is followed. The start of the cycle is designated with zero or $2\pi$, while the point half wave through the cycle is designated $\pi$. In Fig. 17-23A the two input signals are completely out of

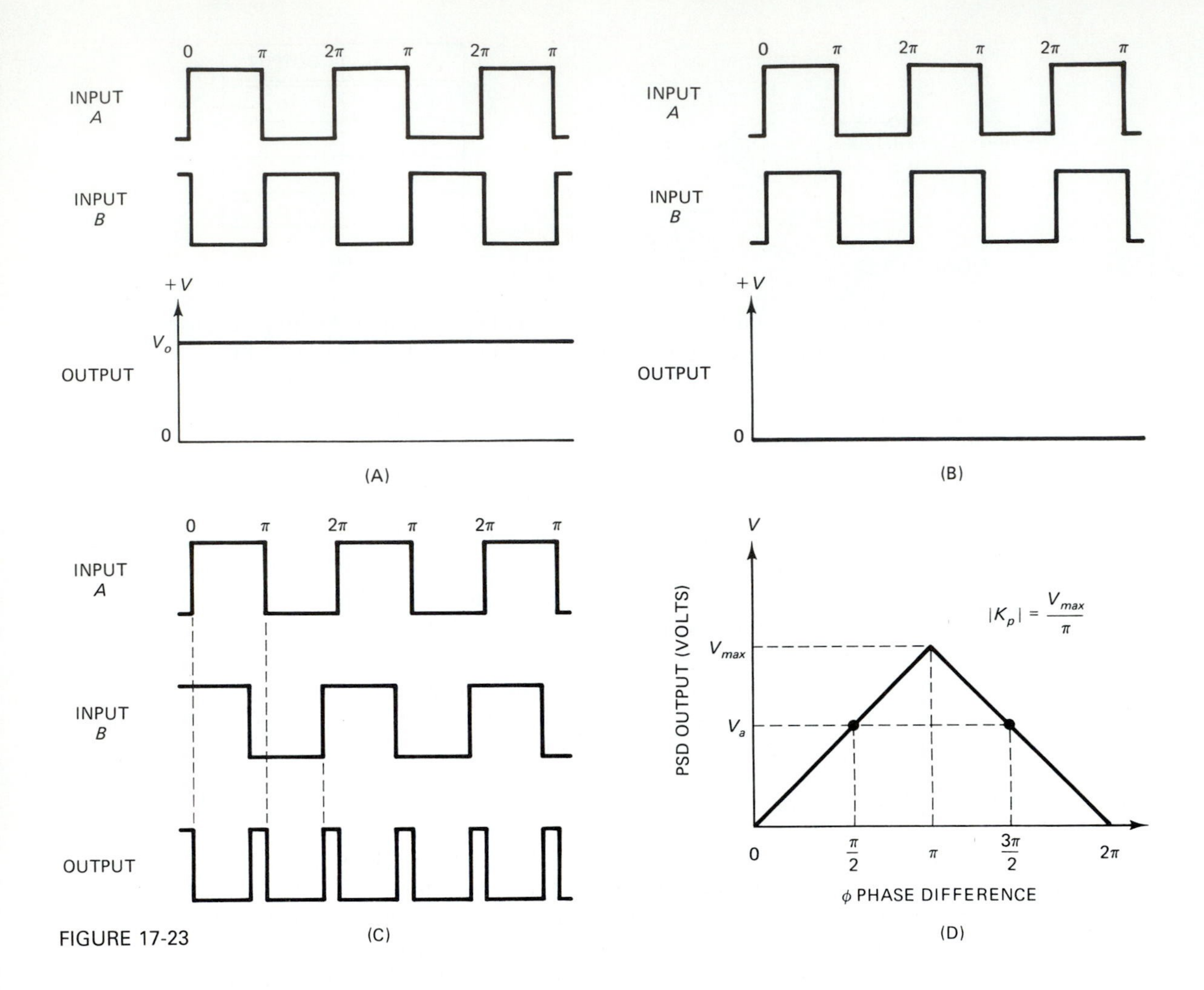


FIGURE 17-23

phase ($\phi = \pi$). This condition forces one of the inputs HIGH all of the time, but neither is HIGH at the same time. In other words, the two inputs always respond to a different level at the two inputs. If one is HIGH, the other is LOW and vice versa. Thus, the output of the XOR gate PSD is always HIGH for this condition. It is essentially a DC level.

The opposite condition is shown in Fig. 17-23B. Here the two signals are identical or in phase. In this condition, the two inputs of the XOR gate PSD are always the same, either HIGH or LOW. Thus, the output of the PSD is perpetually at zero. It is established that the output of the XOR gate PSD will be zero for in-phase signals, and maximum for 180° ($\pi$) out-of-phase signals. The condition for phase differences less than $\pi$ is shown in Fig. 17-23C. The output of the XOR gate PSD is a train of short duration pulses with a width proportional to the difference in phase between the two input signals.

A plot of the output of the XOR gate PSD is shown in Fig. 17-23D. Note the output voltage is maximum at $\phi = 0$, and zero at both $\phi = 0$ and $\phi = 2\pi$. The slope

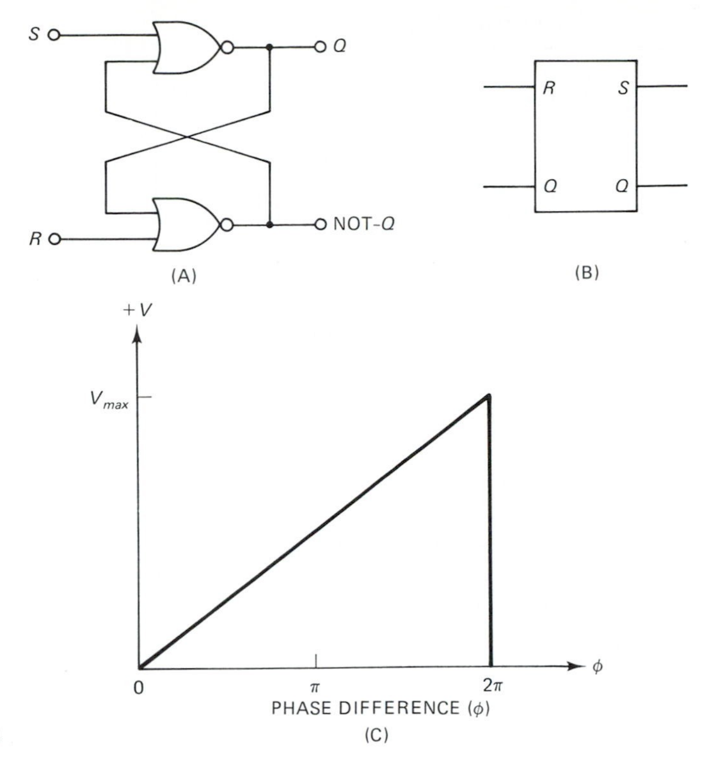


FIGURE 17-24

between the maxima and minima is the *conversion gain* ($K_p$). A problem with this detector can be easily seen in this graph. From a center phase difference of π (180°), where the voltage output is maximum, the voltage drops off linearly to zero. But the graph is symmetrical about 180°. The same voltage ($V_a$) therefore represents two different phase differences, in this example π/2 and 3π/2. The edge detector PSD circuit eliminates this problem at the cost of slightly greater complexity.

The *reset-set flip-flop* (R-S FF) circuit shown in Fig. 17-24 can be used for phase detection. The R-S FF can be constructed of either NAND or NOR gates, but in PSD circuits NOR-logic R-S FF is used. The two NOR gates in Fig. 17-24A are cross-coupled so the output of one becomes the input of the other. The schematic symbol shown in Fig. 17-24B is sometimes used to represent the R-S FF, especially in cases where the gate is found in a TTL or CMOS digital IC.

The operation of the NOR-logic R-S FF is simple. When a brief positive-going pulse is applied to the SET input, the Q-output is forced to HIGH and the NOT-Q output goes LOW (Q and NOT-Q are always opposite each other because they are complementary). The Q-output will remain HIGH until a second positive-going pulse is applied to the RESET input.

When used as a phase sensitive detector the two inputs are connected to the two different frequency sources. The SET input is connected to the reference frequency ($f_{ref}$) and the RESET input to the VCO frequency ($f_o$). The effect of using the NOR R-S FF is shown in Fig. 17-24C. Note the voltage rises linearly from zero to $V_{max}$ over the range 0 to 2π, which represents the entire potential 360° of the phase difference. Above the 2π point, however, the voltage drops to zero.

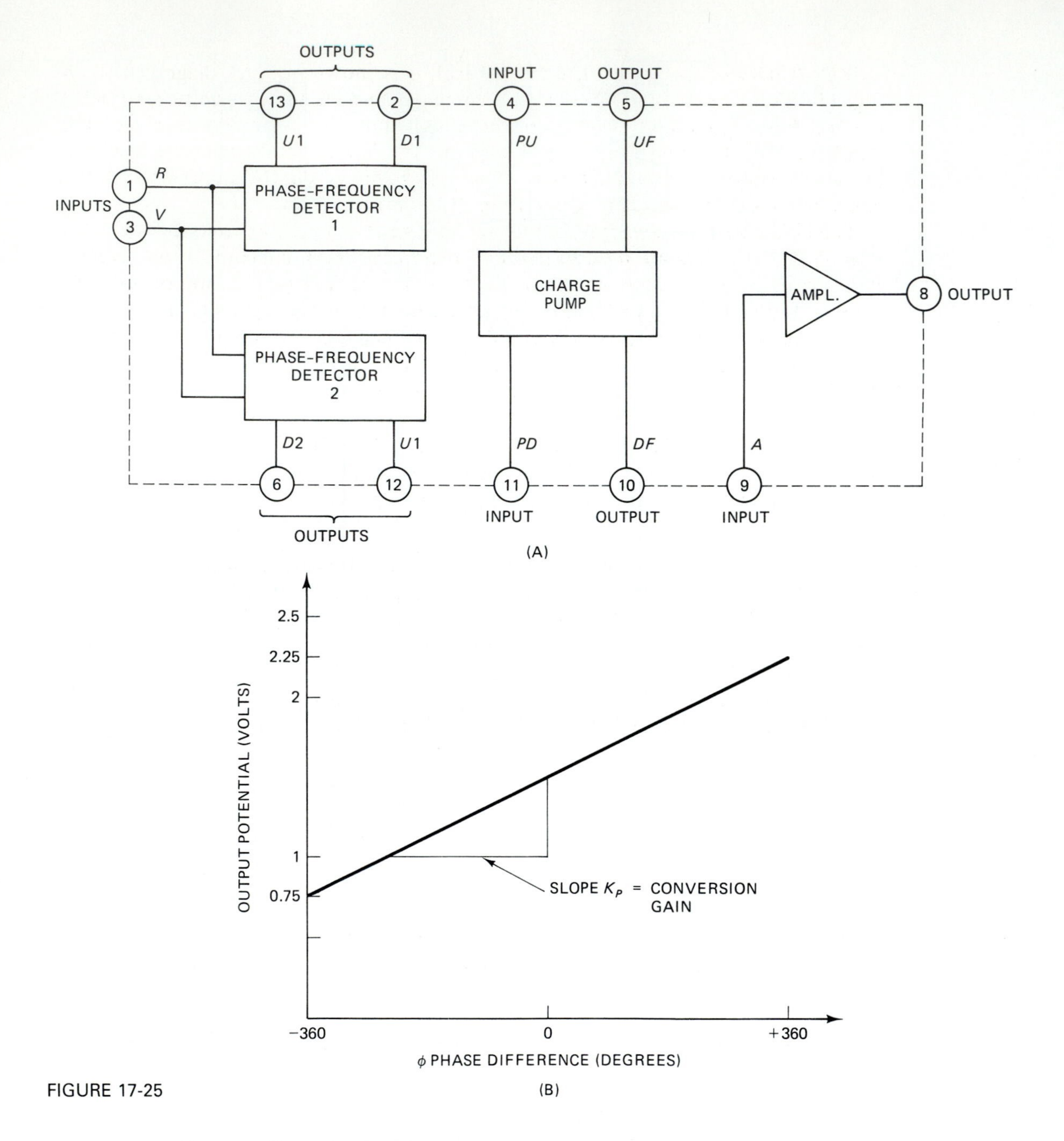


FIGURE 17-25

## 17-6.2 Phase Locked Loop IC Devices

Although the phase locked loop (PLL) circuit was invented in the 1930s, it required the modern integrated electronics technology to make the PLL practical enough for a large range of applications. In this section we will examine several of the more popular PLL IC devices.

**MC-4044/4344 PLL Chip.** The MC-4044 PLL is shown in block diagram form in Fig. 17-25A. It is used in a wide variety of applications including transmitter frequency control. In most cases, the 4044 is used in conjunction with other devices of the same family in order to extend its range of applications. The internal circuitry of the 4044 includes a pair of identical PSD circuits, a charge pump, and a DC amplifier. External circuitry links these elements to form the PLL circuit.

The phase sensitive detectors are digital and are fed in parallel from the same pair of inputs. Two PSDs are used to permit quadrature phase detection. If an ordinary PSD circuit is needed, only one phase detector is used. The quadrature form of PSD is less sensitive to noise and rejects certain harmonics of the input signals.

The charge pump is used to drive an external capacitor which is used as part of the low-pass filter network. The voltage across the charge pump capacitor is used as the DC loop control voltage. The DC amplifier is used to scale the DC control voltage as needed.

The 4044 has a transfer function as shown in Fig. 17-25B. The conversion gain ($K_p$) is on the order of 120 mV/radian when ordinary supply voltages are used. Note the phase detectors are linear over the range $-360°$ to $+360°$. This represents a 720° capture range.

**MC-145152 PLL IC.** Another PLL chip, the MC-145152, is diagrammed in Fig. 17-26. This IC contains three divide-by-integer counters. Counter $R$ is programmable to divide by 8, 64, 128, 256, 512, 1024, or 1160. Counter $N$ is programmable to any value between 3 and 1023. Counter $A$ is programmable to any value between 0 and 63. The $R$ counter is used to process the reference signal ($f_{ref}$). The others process the VCO output signal. The chip also contains a phase sensitive detector with differential outputs and a control logic section which governs the "housekeeping" of the device.

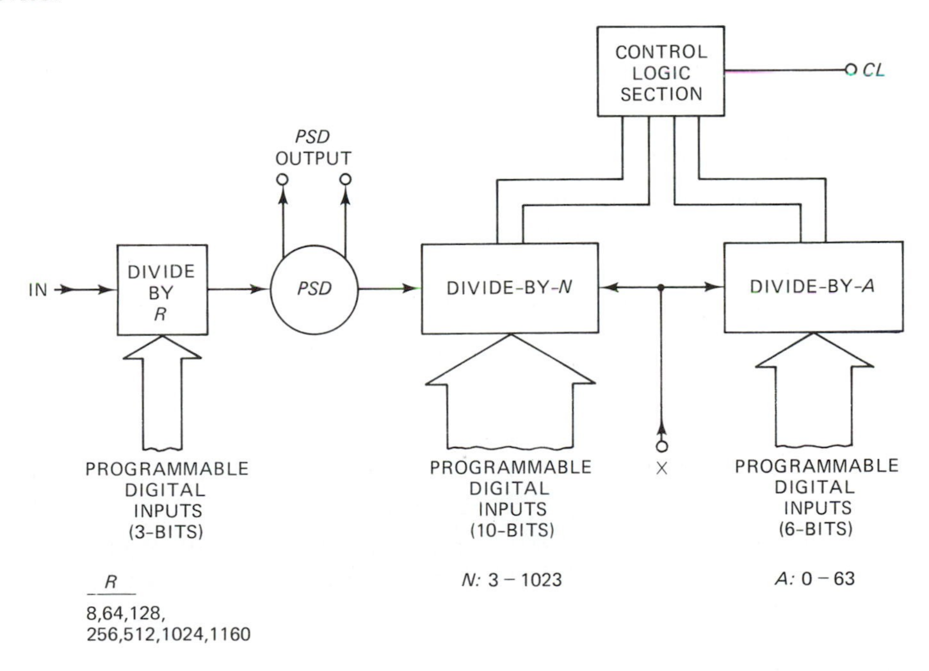


FIGURE 17-26

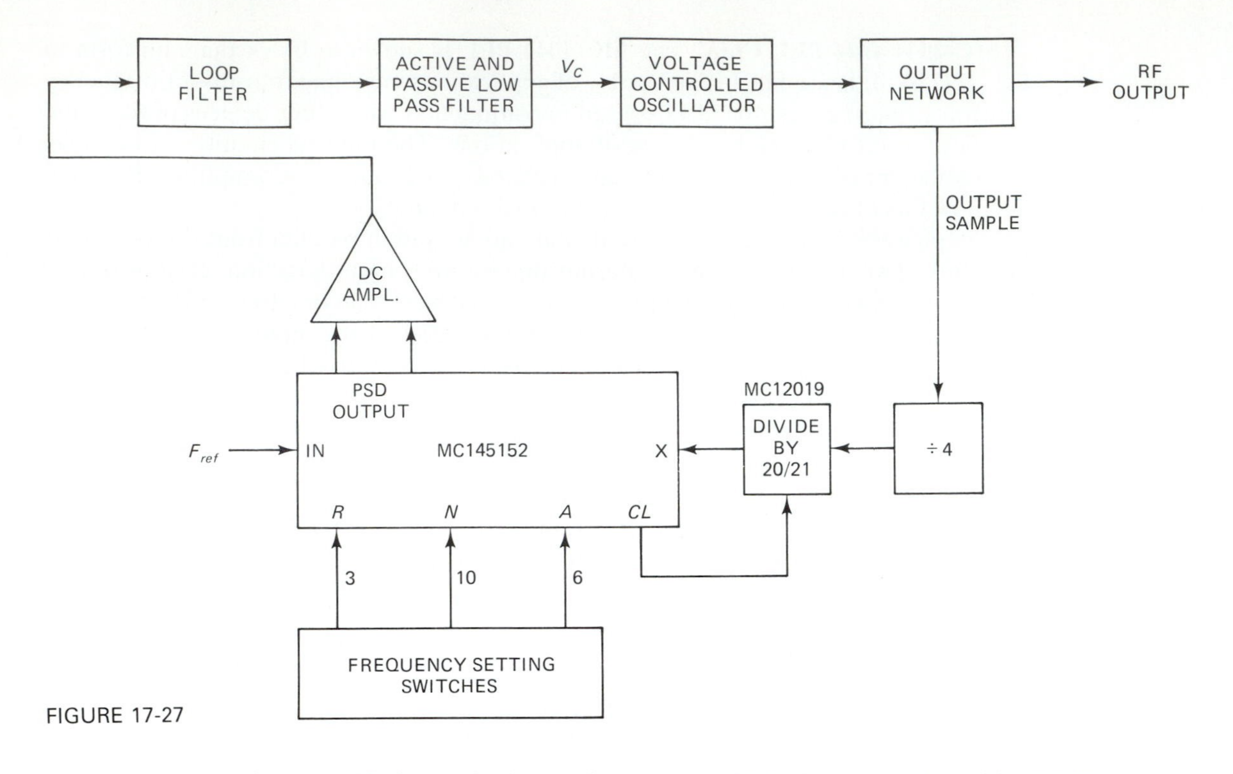


FIGURE 17-27

Figure 17-27 shows the block diagram of a PLL circuit developed by RF Prototype Systems. This PLL can be designed to operate at any frequency between 35 MHz and 920 MHz. The specific frequency range and spacing between channels depend on the design of the circuit and the particular voltage controlled oscillator (VCO) used.

The division ratios of the R, N, and A dividers inside the MC-145152 are set by external switches. The setting of these switches determines the operating frequency of the overall circuit. In practical circuits these switches may be any of several forms, DIP switches, front panel switches, a numerical ASCII touchpad, or a digital register loaded from a computer or other digital source.

The differential outputs of the PSD are used to drive a DC differential amplifier, which in turn drives the loop filter and a cascade series of active and passive low-pass filters. The output of these filters is the DC control voltage ($V_c$) which actually controls the VCO output frequency. The output network isolates and impedances matches the VCO and the RF output. The output network also samples the RF output signal to form the signal compared to the reference frequency. This output sample may be frequency divided (as shown) in two or more stages, or it may be applied directly to the input of the MC-145152 inputs.

**CMOS 4046 PLL IC.** The 4046 CMOS device (Fig. 17-28A) is a phase locked loop. Do not confuse the CMOS 4000 series with the MC-4044. In the CMOS 4000 series, the 4044 is a quad NAND-logic R-S flip-flop, not a PLL. The MC-4044, which is a PLL, uses a different technology and is not part of the CMOS series of chips.

The 4046 contains two phase-sensitive detectors. $\phi 1$ is an XOR-gate based PSD, and is somewhat sensitive to harmonics in the input signals. It is, however, also simple.

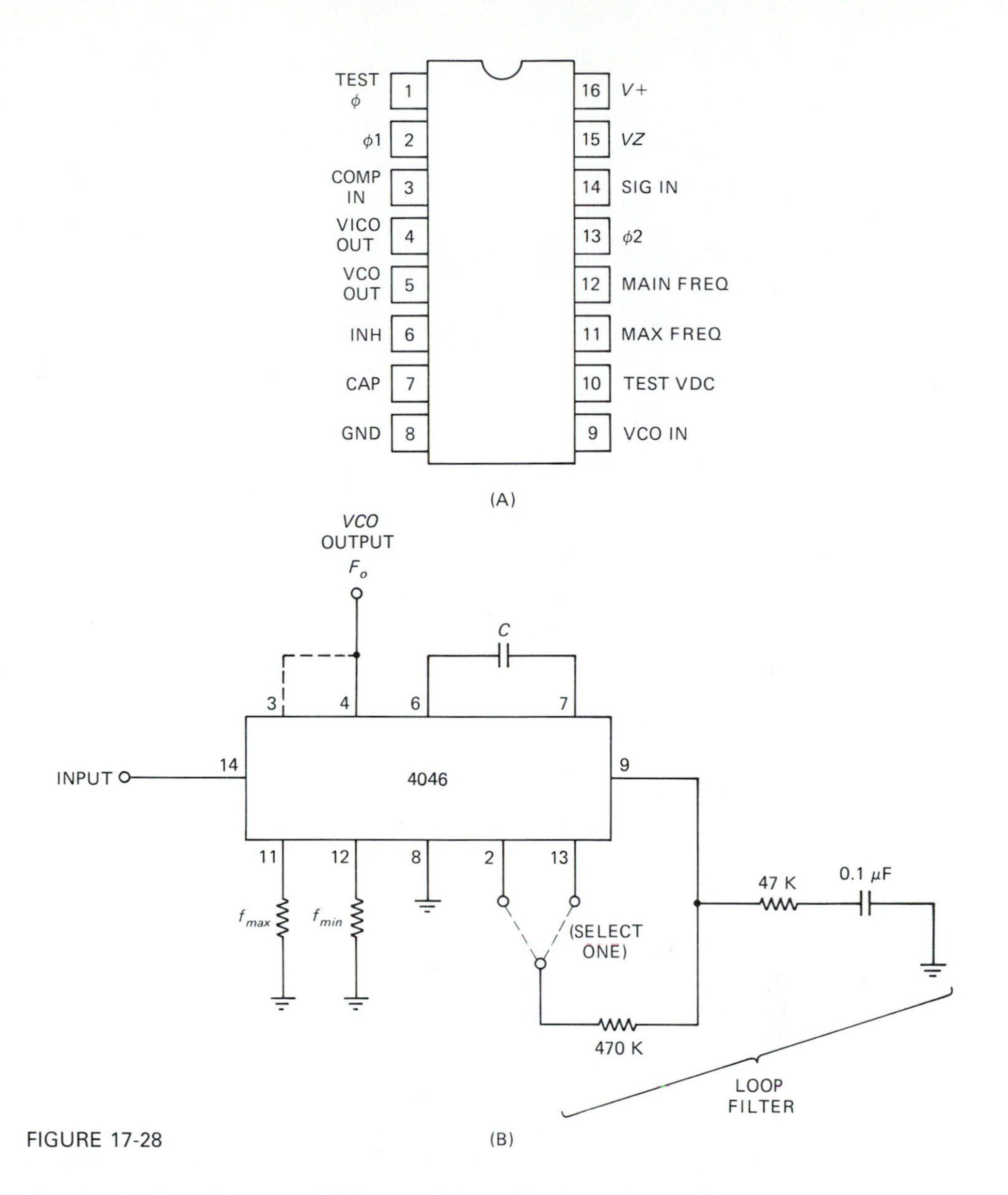


FIGURE 17-28

Phase detector ϕ2 uses a different form of logic. It is capable of a frequency range up to 2000:1. It will also accept any duty factor, while ϕ1 is limited.

The PSD is routed to the VCO input (pin 9) through an RC loop filter (see Fig. 17-28B). The operating frequency of the 4046 PLL device is set by a combination of the input voltage applied to pin 14, and the parameter setting components applied to pins 6, 7, 11, and 12. A capacitor ($>=$ 50 pF) is connected between pins 6 and 7. The maximum frequency resistor is connected to pin 11. It must be between 10 kohms and 1 megohms. The minimum frequency resistor is connected to pin 12 and must be larger than the maximum frequency resistor. The minimum resistor must be $>10$ kohms, but there is no upper limit.

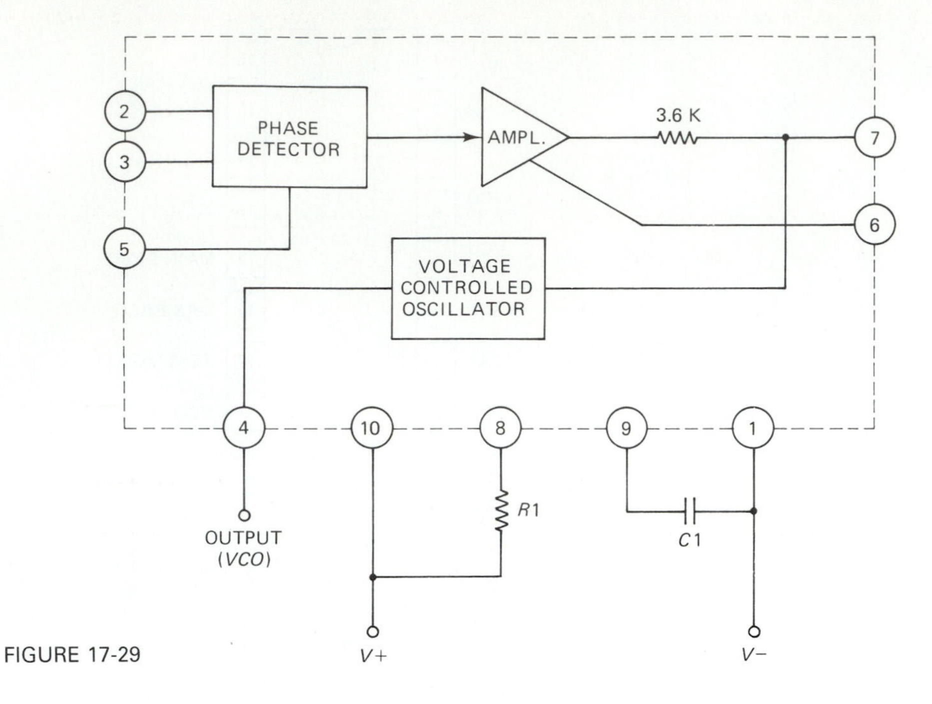


FIGURE 17-29

**NE/SE-565 PLL IC.** The NE/SE-565 was one of the earliest IC PLL devices. The 565 is part of a family of PLL devices which includes other PLL ICs and a VCO IC (NE/SE-566). The NE-565 operates over the commercial temperature range (0 to +70°C); the SE-565 operates over the wide commercial and military temperature range (−40°C to +125°C). The operating voltage can be ±6 Vdc to ±12 Vdc. The operating frequency range is 0.001 Hz to 550 KHz. The 565 contains a single phase detector, a DC amplifier, and a voltage controlled oscillator. The loop filter consists of a pair of components, resistor $R1$ and capacitor $C1$.

### 17-6.3 Loop Filter Circuits

The loop filter in a PLL circuit is a low-pass filter which has three basic functions. First, it produces the DC feedback control voltage that is used to set the VCO frequency. Second, it sets the responsiveness of the PLL. A very heavily damped loop filter will force the PLL to be sluggish and hard to change (which might be an advantage if there is a large amount of noise present). Third, the loop filter also removes any residual components of the VCO and reference frequency that are present in the PSD output. Several different types of circuit are used for the loop filter (Fig. 17-30).

Figures 17-30A through 17-30C show several passive RC network low-pass filters often used as loop filters in PLL circuits. In Fig. 17-30D and 17-30E are two active loop filters based on operational amplifiers. The −3 dB cut-off in the frequency response of these first-order low-pass filters is set by

$$f_{-3\text{dB}} = \frac{1}{2\,\pi\,R1\,C1} \tag{17-40}$$

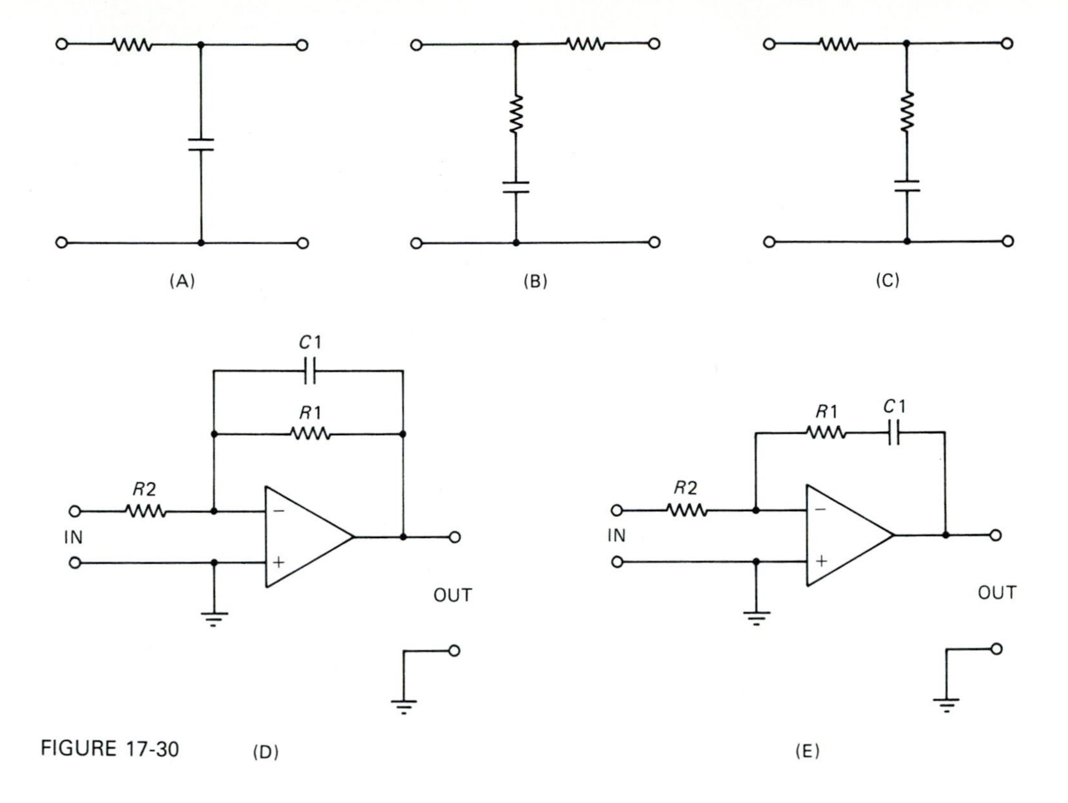


FIGURE 17-30 (D) (E)

These filter circuits have a relatively poor slope in the frequency response beyond −3 dB, on the order of −6 dB/octave. Second-order and higher active filters may also be used for the loop filter application (see Chapter 18).

## 17-6.4 A PLL Application: Sub-Audio Frequency Meter

Subaudio frequencies are below the range of human hearing, $f < 20$ Hz. In practical circuits the range of subaudio frequencies may be on the order 0.0001 Hz to 20 Hz. These frequencies are notoriously difficult to measure in regular digital frequency counters because the gate times are too long. In some models, short gate times (1 second) were used and the result extrapolated to the actual frequency based on the count during the gate time. This method produces relatively large errors. A better (but more complex) method used in some models is to measure the *period* of the input signal and take its reciprocal in an arithmetic circuit. This method was once extensively used in very low frequency counters. A final method, which has certain advantages, is to multiply the input frequency X100 or X1000 in a phase locked loop.

Figure 17-31 shows the block diagram of a PLL-based subaudio frequency meter. The VCO operates over a range which is either 100 or 1000 times the input frequency range. The VCO output is divided in a divide-by-N (where $N$ = 100 or 1000) counter before being applied to one input port of a phase sensitive detector (PSD). The subaudio input frequency is applied to the alternate input port of the PSD. If $N$ = 1000, a 1-Hz signal will read 1000 Hz on the output counter.

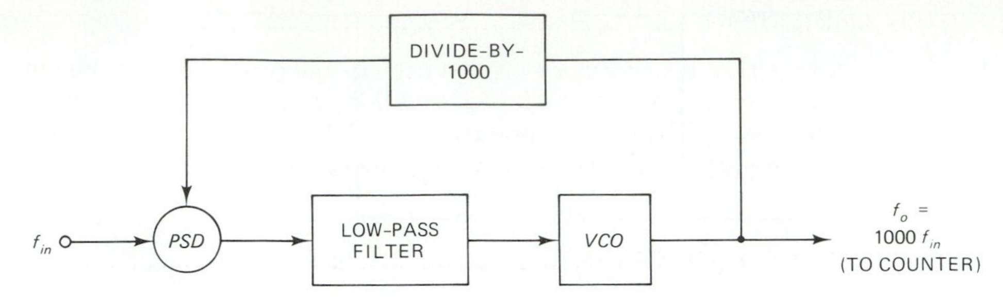


FIGURE 17-31

## 17-7 SUMMARY

1. Current loop circuits can be used in communications applications. The classical teletypewriter uses a 20 mA current loop (obsolete machines used 60 mA). Some instrumentation applications use a 4-to-20 mA current loop for communications.
2. The 4-to-20 mA current loop offers several advantages over other systems, easy daisy chaining of instruments, nonambiguous zero, and relative freedom from noise impulses.
3. Both 20 mA and 4-to-20 mA current loops can use voltage to current converter circuits (Howland current pumps, for example) on the transmit end, and optoisolators on the receive end. This flexibility makes the circuits relatively simple.
4. The EIA RS-232C standard for serial communications ports uses voltage levels to represent 1 and 0 binary numbers. The 1 level is represented in RS-232C by a voltage between $-5$ and $-15$ Vdc. The 0 level is represented by a voltage of $+5$ to $+15$ Vdc. In each case, potentials between 0 and |5 volts| are regarded as undefined.
5. Modulation is the process of altering a characteristic of one signal, called the *carrier*, in proportion to another signal, called the *modulating signal*, for the purpose of transmitting information or data. Modulation may vary either the frequency, phase or amplitude of the carrier signal.
6. Demodulation is the process of decoding the modulated carrier signal in order to recover the information represented by the original modulating signal.
7. Amplitude modulation can be performed by analog multiplier circuits because AM is a multiplication process. The output signal is the product of carrier and modulating frequencies.
8. These are two forms of *angular modulation*: *phase modulation* and *frequency modulation*. Forms of amplitude modulation include *amplitude modulation*, *double sideband suppressed carrier* (DSBSC), and *single sideband suppressed carrier* (SSBSC or SSB). The SSBSC form is sub-divided into *upper sideband* (USB) and *lower sideband* (LSB). Pulse modulation forms include *pulse position modulation* (PPM), *pulse width modulation* (PWM) and *pulse amplitude modulation* (PAM).
9. Phase locked loop (PLL) circuits create an output frequency from a voltage controlled oscillator (VCO) by comparing the VCO output to a stable reference source. The principal elements of a PLL circuit include the VCO, a phase sensitive detector, a low-pass loop filter, and (in some cases) a DC scaling amplifier.

## 17-8 RECAPITULATION

Now return to the objectives and Prequiz questions at the beginning of the chapter and see how well you can answer them. If you cannot answer certain questions, mark them and review the appropriate parts of the text. Next, try to answer the questions and work the problems below, using the same procedure.

## 17-9 STUDENT EXERCISES

1. Design, build, and test a 20 mA current loop communications system in which MARK is 16 mA to 20 mA and SPACE is 0 to 2 mA.
2. Design, build, and test a 4-to-20 mA current loop instrumentation communications circuit in which a 2 Vdc signal is represented by 4 mA and 10 Vdc is represented by 20 mA. Optional: Include a 1-bit test circuit which detects the off/fault state where $I = <<4$ mA when $V >= 2$ Vdc.
3. Design, build, and test a subaudio frequency meter based on the 4046 CMOS PLL IC.
4. Design, build, and test a pulse width modulator (PWM) based on (a) NOR-logic R-S FF and (b) XOR-gate.
5. Design, build and test an FM demodulator based on the pulse counting technique which will decode FM signals between 500 Hz and 5000 Hz.

## 17-10 QUESTIONS AND PROBLEMS

1. In a 20 mA current loop communications system, such as a teletypewriter, the MARK (logical 1) is represented by a current of ____________ to ____________ mA. A SPACE (logical 0) is represented by a current of ____________ to ____________ mA.
2. Some process instrumentation communications systems use a ____________ to ________ mA current loop.
3. In the question above, a current of 0.2 mA may indicate a ____________ .
4. A floating load current source (Fig. 17-5) has a 10 kohm load resistor ($R_L$). Assuming $R_L >> R1$, calculate the current when the output voltage is 10 Vdc and the input voltage is 2 Vdc.
5. A Howland current pump (Fig. 17-6) has two input voltages $V1 = +6$ Vdc and $V2 = +1$ Vdc. Assuming $R = 10$ kohms, calculate the output current in $R_L$.
6. A Howland current pump is used to make a 4-to-20 mA current loop transmitter. What value of bias voltage $V2$ will force $I_L$ to be 4 mA when input voltage $V1 = +2$ Vdc? Assume $R = 1000$ ohms.
7. List the voltage levels which represent logical 1 and logical 0 in the standard EIA RS-232C serial communications ports.
8. Define modulation.
9. List several characteristics of a carrier signal which can be varied in a modulation process.
10. List the forms of angular modulation.
11. List the forms of amplitude modulation.
12. List the forms of pulse modulation.

13. What type of sinusoidal modulation produces a sum and difference output frequency, but no carrier frequency?
14. Write the equation for *modulation index* in AM systems.
15. An ordinary amplitude modulation system has a 500 KHz sinusoidal carrier, and is modulated by a 10 KHz sinusoidal modulating signal. Assuming that no frequency selective output filtering is used, what frequencies are present in the output signal?
16. A voltage ______________ IC can be used as a pulse width modulator.
17. List two forms of FM demodulator commonly used in IC circuits.
18. List the elements of a phased locked loop (PLL), including any optional elements.
19. Draw the block diagram of a PLL in which the output frequency can be set by applying a digital control signal.
20. List two forms of digital phase sensitive detector circuit. Draw the transfer function $V_o$ compared to phase for both types.
21. What is the purpose of $D1$ in Fig. 17-1? What PIV rating value should it have?
22. Sketch the circuit of an op-amp based TTL-to-RS-232C converter.
23. Sketch the circuit of an RS-232-to-TTL converter.
24. Sketch the frequency spectrum of the following signals: (a) unmodulated, unkeyed CW, (b) AM, and (c) DSBSC.
25. Sketch the waveform of a modulated signal consisting of a squarewave carrier and a triangle modulating signal.

CHAPTER 18

# Analog Multipliers and Dividers

## OBJECTIVES

1. Learn the applications of the analog multiplier/divider.
2. Understand how multipliers and dividers work.
3. Learn other multiplier/divider applications, devices, and circuits.

## 18-1 PREQUIZ

These questions test your prior knowledge of the material in this chapter. Try answering them before you read the chapter. Look for the answers (especially those you answered incorrectly) as you read the text. After you have finished studying the chapter, try answering these questions again and those at the end of the chapter (see Section 18-23).

1. List three different forms of analog multiplier circuit.
2. An analog multiplier has a scaling factor of 1/20. Calculate the output voltage when $V_x = +1.5$ volts and $V_y = +2$ volts.
3. Draw the circuit for using an AD-533 multiplier square rooter.
4. Prove an analog multiplier connected as a squarer will operate as a frequency doubler when a sine wave signal is applied.

## 18-2 ANALOG MULTIPLIER AND DIVIDER IC DEVICES

Analog multiplier and divider circuits are available in both monolithic integrated circuit and hybrid circuit forms. Analog multipliers produce an output voltage $V_o$ which is the product of two input voltages, $V_x$ and $V_y$. The general form of the multiplier transfer function is

$$V_o = KV_xV_y \tag{18-1}$$

where

$V_o$ = the output potential in volts

$V_x$ = the potential (in volts) applied to the $X$ input

$V_y$ = the potential (in volts) applied to the $Y$ input

$K$ = a constant (usually 1/10)

If the proportionality constant $K$ is 1/10, Eq. (18.1) becomes

$$V_o = \frac{V_xV_y}{10} \tag{18-2}$$

There are several basic designs for analog multiplier circuits. The logarithmic amplifier was discussed in Chapter 11. Its use as a multiplier will be briefly reviewed here. When the outputs of two logarithmic amplifiers are first summed together and applied to an antilog amplifier, the output of the antilog amplifier is proportional (via $K$) to the product of the two input voltages. Transconductance amplifiers (Chapter 8) can also be used to make an analog multiplier. In Chapter 8 we discussed a multiplier based on the operational transconductance amplifier (OTA) IC device. There is also a *transconductance cell* analog multiplier. Other varieties of multiplier circuit will also be examined in this chapter.

## 18-3 ANALOG MULTIPLIER CIRCUIT SYMBOLS

Figure 18-1 shows typical symbols used to represent analog multiplier and divider circuits in schematic diagrams. Although there are standards for circuit symbols, the multiplier is one type of device in which corporate, IEEE, and military standards all are simultaneously used. The symbols shown in Fig. 18-1 are those commonly found. Other symbols (as well as variations of these) may also be used in actual practice.

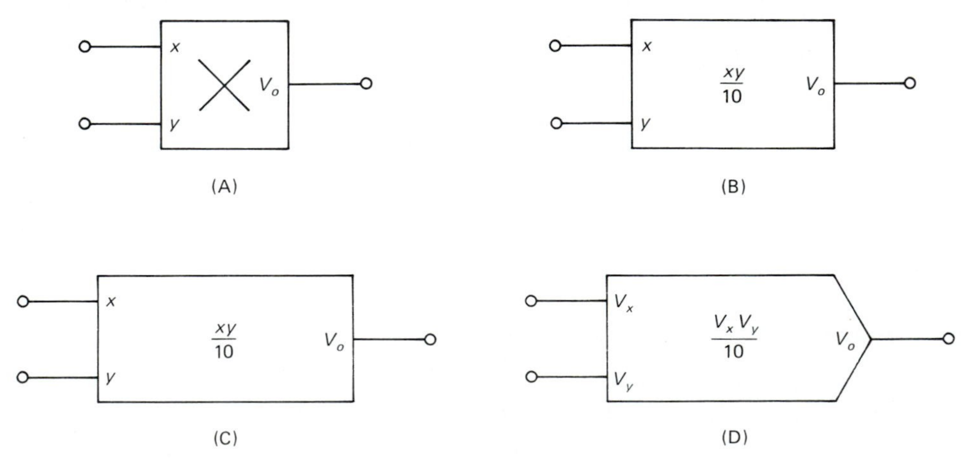


FIGURE 18-1

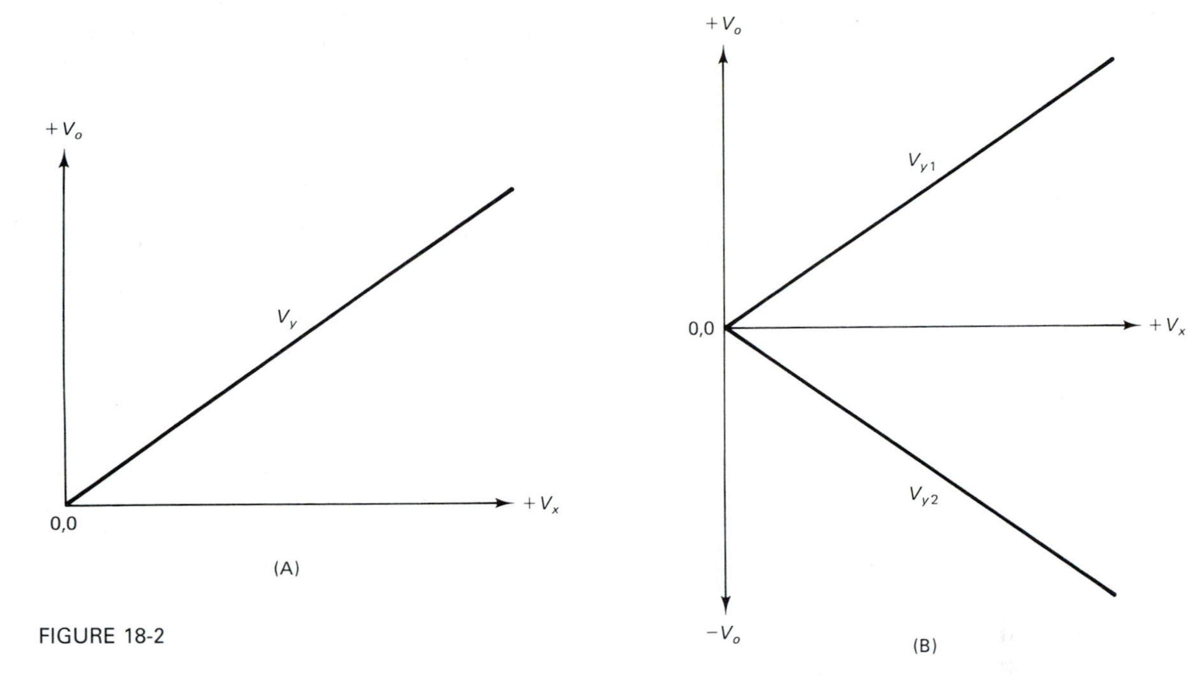


FIGURE 18-2

## 18-4 ONE-, TWO-, AND FOUR-QUADRANT OPERATION

Analog multipliers and dividers are classified according to the number of quadrants in which they operate. The quadrants are the four quadrants of the Cartesian coordinate system. Figure 18-2A illustrates one-quadrant operation. Here, both input voltages must be positive ($V_x \geq 0$), ($V_y \geq 0$). The only possible output voltage polarity is positive. At least one commercial hybrid multiplier operates in one quadrant, but with both input voltages negative. Again, the only permissible output voltage was positive. This type of operation is a rarity, however. The least complex multipliers based on logarithmic amplifiers are normally one-quadrant devices.

The two-quadrant multiplier (Fig. 18-2B) allows the output voltage to be either positive or negative, but there are constraints on the allowable input voltage polarities. One input voltage will be limited to positive values only and the other can be either positive or negative.

Four-quadrant operation (Fig. 18-3) is the most flexible because it allows operation with any combination of input signal polarity. The output signal can be either positive or negative, as can either (or both) input signal voltages. Figure 18-4 shows the relationship between input and output polarities. These limits are summarized as follows

$$QI\text{: } V_x \geq 0,\ V_y \geq 0,\ 0 < V_o < +V.$$

$$QII\text{: } V_x \leq 0,\ V_y \geq 0,\ -V < V_o < 0.$$

$$QIII\text{: } V_x \leq 0,\ V_y \leq 0,\ 0 < V_o < +V.$$

$$QIV\text{: } V_x \geq 0,\ V_y \leq 0,\ -V < V_o < 0.$$

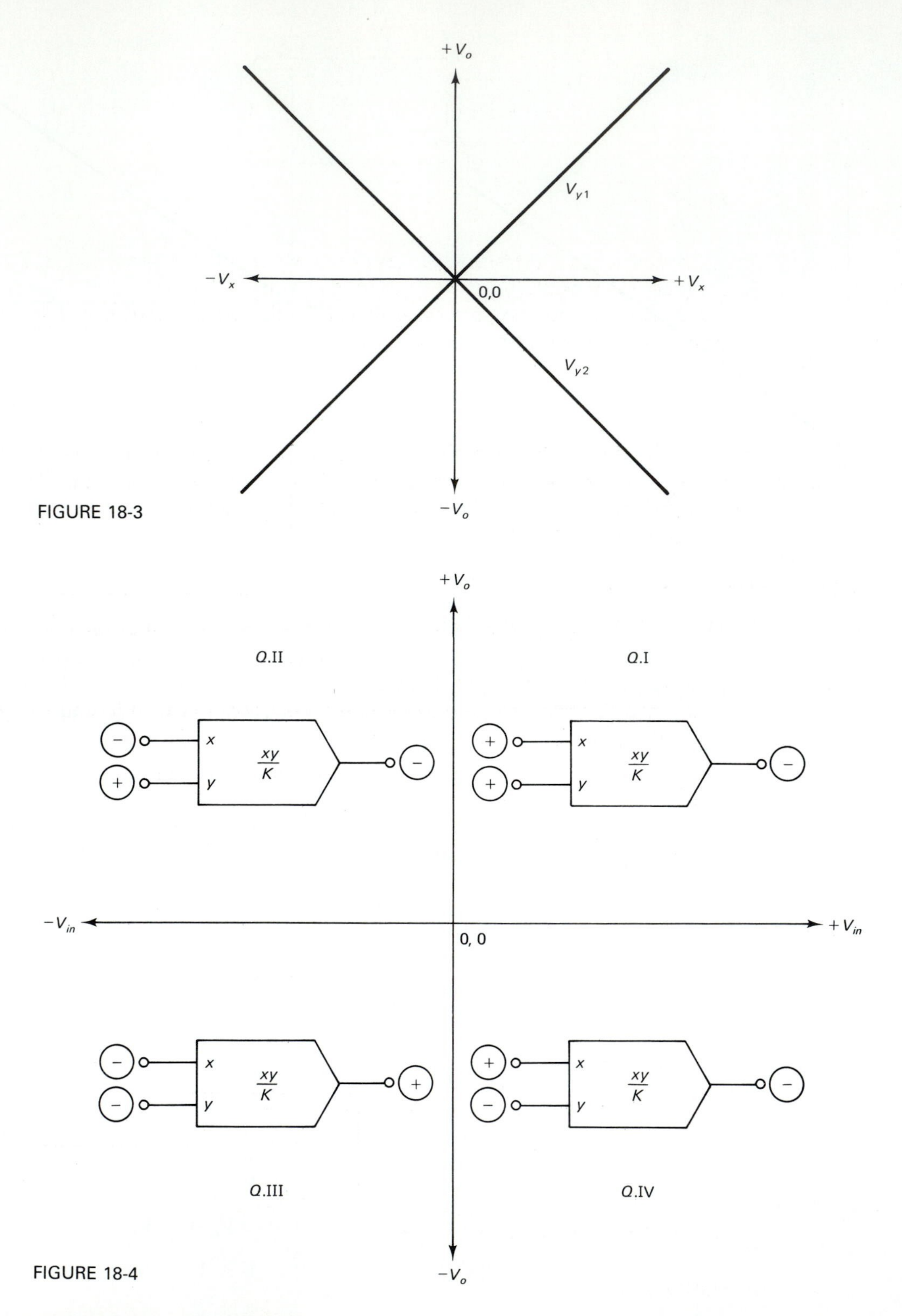


FIGURE 18-3



FIGURE 18-4

In most cases, $K = 1/10$ so these expressions become

$$V_o = \frac{V_{dx}V_{dy}}{10} \tag{18-8}$$

$$V_o = \frac{(X1-X2)(Y1-Y2)}{10} \tag{18-9}$$

The differential input multiplier is particularly useful in at least two different situations, where a Wheatstone bridge or other balanced output signal source supplies one or both multiplier input signals and where signals processing requires an input to the multiplier be the difference between source signals.

For example, if a fluid pressure transducer measures a pressure in which a variation rides on top of a certain large offset minimum pressure (the human blood pressure, in which the varying pressure wave rides atop the diastolic static pressure), the input signal will be a slowly varying DC wave on top of a large DC static offset potential. If the DC offset can be subtracted prior to being applied to the multiplier, the overall dynamic range of the system is improved.

Either input of Fig. 18-6 can be made single-ended by the simple expedient of grounding the unwanted input.

## 18-6 TYPES OF MULTIPLIER CIRCUIT

In this section we will examine several popular approaches to multiplier design. Some of them are currently used in state-of-the-art analog multipliers and others are obsolete. The reason for including the obsolete circuits is to demonstrate the process of multiplication and the range of possibilities. It is generally an error to disdain older methods because the same fundamentals are often resurrected when newer methods can make their implementation better. For example, the incandescent lamp multiplier circuit below is now obsolete, but the JFET manifestation of the same concept is currently used. It is nonetheless profitable to examine the lamp version because it is inherently easier to understand.

## 18-7 MATCHED RESISTOR/LAMP MULTIPLIER DIVIDERS

A simple and effective multiplier and divider circuit can be constructed from a pair of operational amplifiers and a special gain setting block (see Fig. 18-7A) consisting of a pair of matched photoresistors ($R1$ and $R2$), both of which are illuminated by the same incandescent lamp (IL1). Because $R1$ and $R2$ are matched for any given light level, $R1 = R2 = R$. Because $R1$ is in the feedback loop of amplifier $A2$, nonlinearities in the *voltage vs. brightness* ratio of IL1 are servoed out.

In past chapters we have used an analysis method based on Ohm's law and Kirchoff's Current Law (KCL). That practice will be continued here. In Fig. 18-7A, the input bias current of each amplifier ($I_{b1}$ and $I_{b2}$) are each zero. In addition, the noninverting inputs of both amplifiers are grounded, so the two summing junctions ($A$ and $B$) are at zero volts (ground) potential. By Ohm's law and considering $R1 = R2 = R$

$$I1 = \frac{V_z}{R} \tag{18-10}$$

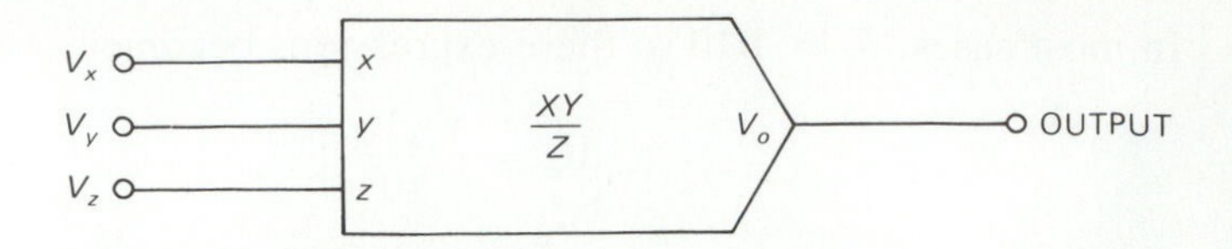


FIGURE 18-5

In most cases, the voltage ranges of the two inputs are symmetrical. For example, a typical range is $V_{x(\max)} = V_{y(\max)} = \pm 10$ volts.

Another category of device is the multiplier/divider shown in Fig. 18-5. These devices have a transfer function of the form

$$V_o = \frac{V_x V_y}{V_z} \tag{18-3}$$

In the multiplier mode, the $X$ and $Y$ inputs are used and the scaling factor $K$ is set by applying a voltage to the $Z$ input. In the division mode, either the $X$ and $Z$ or $Y$ and $Z$ inputs are used for signal voltages, while the scaling factor is set by applying a fixed voltage to the remaining input.

## 18-5 DIFFERENTIAL INPUT MULTIPLIERS AND DIVIDERS

Most of the multiplier and divider circuits presented in this chapter are single-ended devices. That is, the input signals are measured between the input terminal and ground, and there is only one input line each for $V_x$ and $V_y$. A differential input multiplier and divider circuit is shown in Fig. 18-6. In this circuit the differential input voltages $V_{dx}$ and $V_{dy}$ are defined as

$$V_{dx} = X1 - X2 \tag{18-4}$$

$$V_{dy} = Y1 - Y2 \tag{18-5}$$

The multiplier transfer function is

$$V_o = KV_{dx}V_{dy} \tag{18-6}$$

or

$$V_o = K(X1 - X2)(Y1 - Y2) \tag{18-7}$$

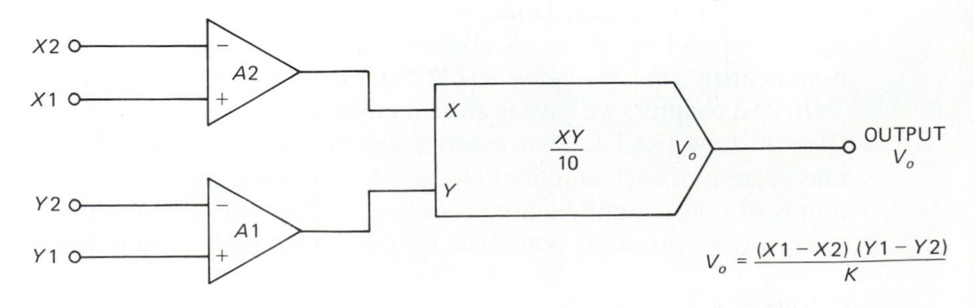


FIGURE 18-6

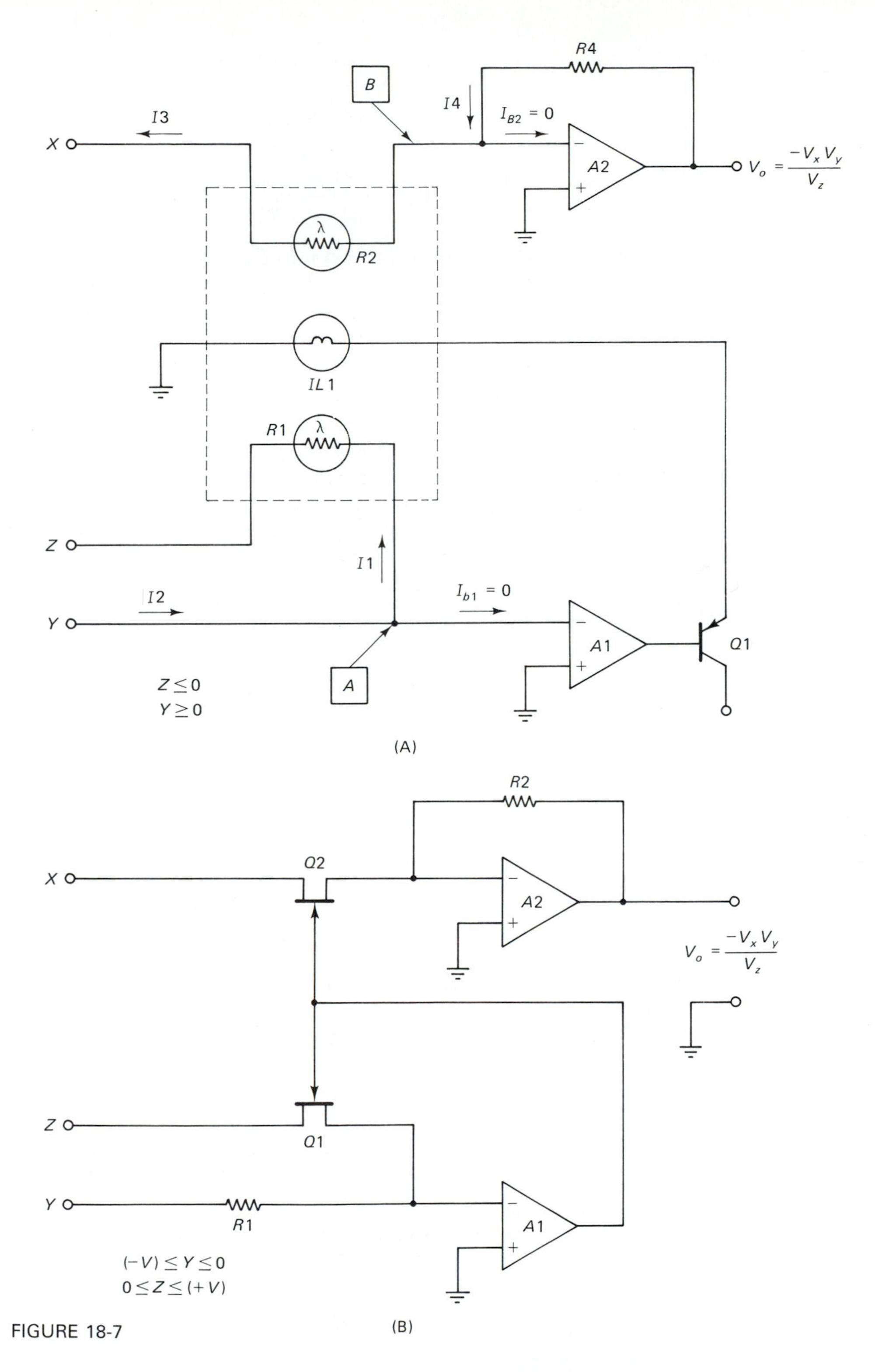
R4
B
I4
I3
$I_{B2} = 0$
X
A2
$V_o = \frac{-V_x V_y}{V_z}$
R2
IL1
R1
Z
I1
I2
$I_{b1} = 0$
Y
A1
Q1
A
$Z \leq 0$
$Y \geq 0$
(A)
R2
Q2
X
A2
$V_o = \frac{-V_x V_y}{V_z}$
Z
Q1
Y
R1
A1
$(-V) \leq Y \leq 0$
$0 \leq Z \leq (+V)$
(B)

FIGURE 18-7

and

$$I2 = \frac{V_y}{R3} \tag{18-11}$$

By KCL

$$I1 = -I2 \tag{18-12}$$

Substituting Eq. (18-10) and Eq. (18-11) into Eq. (18-12)

$$\frac{V_y}{R3} = \frac{-V_z}{R} \tag{18-13}$$

and rearranging terms

$$R = \frac{-V_z R3}{V_y} \tag{18-14}$$

Current $I3$ is by Ohm's law

$$I3 = \frac{V_x}{R} \tag{18-15}$$

So, by substituting Eq. (18-14) into Eq. (18-15)

$$I3 = \frac{V_x}{\left[\dfrac{-V_z R3}{V_y}\right]} \tag{18-16}$$

and then rearranging terms

$$I3 = -\frac{V_x V_y}{V_z R3} \tag{18-17}$$

Because $I_{b1} = 0$ and because point B is at ground potential, by KCL

$$I3 = -I4 \tag{18-18}$$

From Ohm's law

$$I4 = \frac{V_o}{R4} \tag{18-19}$$

By substituting Eq. (18-17) and Eq. (18-19) into Eq. (18-18)

$$-\frac{V_o}{R4} = -\frac{V_x V_y}{V_z R3} \tag{18-20}$$

or by rearranging terms

$$V_o = \frac{V_x V_y R4}{V_z R3} \tag{18-21}$$

Accounting for the fact that $V_z$ is restricted to negative values

$$V_o = -\frac{V_x V_y R4}{V_z R3} \tag{18-22}$$

Placing Eq. (18-22) into the standard form

$$V_o = -\frac{K V_x V_y}{V_z} \tag{18-23}$$

where

$$K = R4/R3.$$

The circuit of Fig. 18-7A is impractical today, but it was once used in discrete form in analog circuitry. Figure 18-7B shows a modern version in which the lamp and resistors have been replaced with *junction field effect transistors* (JFET). The circuit of Fig. 18-7B works on the basis of the channel resistance of the JFETs. At gate potentials below the pinch-off voltage, the JFET drain-source resistance is a function of the gate voltage. Thus, $Q1$ and $Q2$ operate as voltage controlled resistors.

## 18-8 LOGARITHMIC AMPLIFIER-BASED MULTIPLIERS

Many popular forms of analog multiplier-divider circuits are based on the properties of logarithmic and anti-log amplifiers (Chapter 11). A logarithmic amplifier has a transfer function of the form either

$$V_o = k\text{LN}\ (V_{\text{in}}) \tag{18-24}$$

or

$$V_o = k\text{LOG}\ (V_{\text{in}}) \tag{18-25}$$

Logarithms convert multiplication operations into addition and division operations into subtraction. Therefore

$$XY = \text{LOG}\ X + \text{LOG}\ Y \tag{18-26}$$

and

$$\frac{X}{Y} = \text{LOG}\ X - \text{LOG}\ Y \tag{18-27}$$

Equation (18-26) and Eq. (18-27) give us the basis for designing an analog multiplier-divider circuit. Figure 18-8 shows the block diagram of such a circuit. The X input ($V_x$) is applied to the input of LOGAMP-A to produce signal A = LOG($V_x$). Similarly, the Y input ($V_y$) is applied to the input of LOGAMP-B to produce a signal B = LOG($V_y$). These signals are each applied to a summer amplifier (in the multiplication case) or a difference amplifier (in the division case). The output of the summer circuit is $C = A + B$ or $C = \text{LOG}(V_x) + \text{LOG}(V_y)$. By passing this signal through an antilog amplifier it is possible to construct the product of $X$ and $Y$

$$V_o = \text{LOG}^{-1}(C) \tag{18-28}$$

$$V_o = \text{LOG}^{-1}[\text{LOG}(V_x) + \text{LOG}(V_y)] \tag{18-29}$$

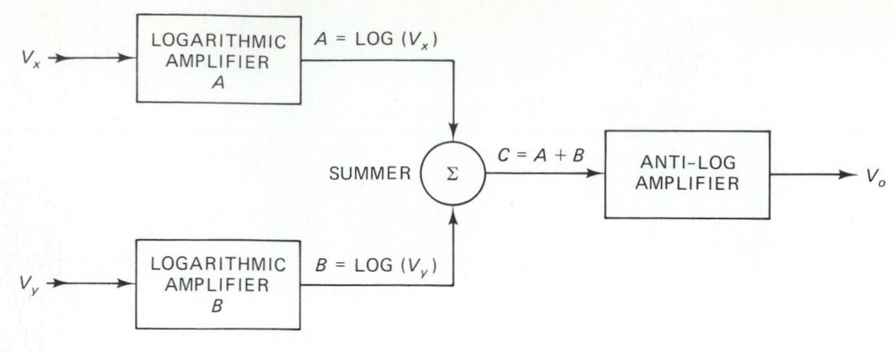


FIGURE 18-8

$$V_o = V_x V_y \tag{18-30}$$

An uncompensated LOG/ANTILOG amplifier exhibits a strong temperature dependence (Chapter 11). In order to prevent error, temperature compensation must therefore be incorporated into the multiplier circuit.

## 18-9 QUARTER SQUARE MULTIPLIERS

Figure 18-9 shows the block diagram of a quarter square multiplier. Consider the following expression

$$V_o = \frac{(X + Y)^2 - (X - Y)^2}{4} \tag{18-31}$$

Expanding this polynomial results in

$$V_o = \frac{(X^2 - X^2) + (Y^2 - Y^2) + 2XY + 2XY}{4} \tag{18-32}$$

$$V_o = \frac{2XY + 2XY}{4} \tag{18-33}$$

$$V_o = XY \tag{18-34}$$

Figure 18-9 shows the block diagram of a circuit which will implement Eq. (18-31). There are two chains of circuits in this multiplier. Each chain contains a two-port input amplifier. A1 is a summation amplifier and produces an output voltage $V_a = (V1 + V2)$. A2 is a difference amplifier and produces an output $V_b = (V1 - V2)$.

Each signal processing chain also contains an *absolute value amplifier* and a *squaring circuit*. The absolute value circuit is also called a *fullwave precise rectifier*. The only difference between these two chains is A is noninverting and B is inverting. Thus

$$V_C = \text{ABS}(V_a) \tag{18-35}$$

$$V_C = \text{ABS}(V1 + V2) \tag{18-36}$$

and

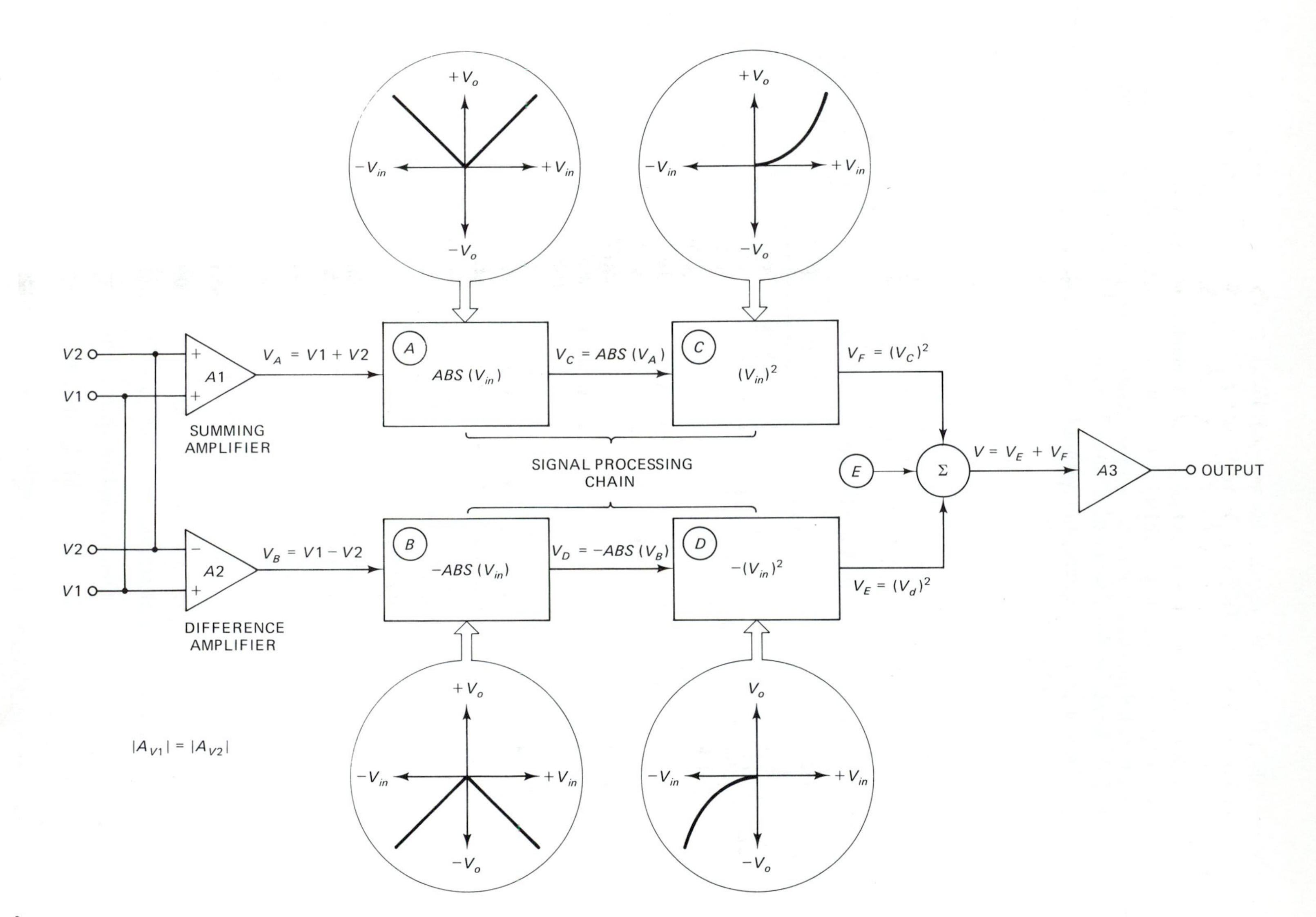

V2
V1
V2
V1
A1
A2
SUMMING AMPLIFIER
DIFFERENCE AMPLIFIER
$V_A = V1 + V2$
$V_B = V1 - V2$
A
$ABS (V_{in})$
B
$-ABS (V_{in})$
$V_C = ABS (V_A)$
$V_D = -ABS (V_B)$
C
$(V_{in})^2$
D
$-(V_{in})^2$
SIGNAL PROCESSING CHAIN
$V_F = (V_C)^2$
$V_E = (V_d)^2$
E
Σ
$V = V_E + V_F$
A3
OUTPUT
$+V_o$
$-V_o$
$V_o$
$-V_{in}$
$+V_{in}$
$|A_{V1}| = |A_{V2}|$


FIGURE 18-9

$$V_D = -\text{ABS}(V_B) \tag{18-37}$$

$$V_D = -\text{ABS}(V1 - V2) \tag{18-38}$$

The squarers (C and D) produce an output which is the square of the input voltage ($V_o = (V_{\text{in}})^2$). In analog multipliers the squarers are *diode breakpoint generators.* Ordinary PN junction diodes have a square law region in their operating characteristic. Using as few as ten PN junction diodes, each biased to slightly different points, produces a breakpoint generator with a linearity approaching $\pm 0.1\%$ of full-scale. The outputs of the squarers are

$$V_E = (V_D)^2 \tag{18-39}$$

$$V_F = (V_C)^2 \tag{18-40}$$

Combining these voltage in a summer produces

$$V_o = (V_C)^2 + (V_D)^2 \tag{18-41}$$

Relating Eq. (18-41) to Eq. (18-31) is a good exercise.

## 18-10 TRANSCONDUCTANCE MULTIPLIERS

The most commonly used form of IC analog multiplier is the *transconductance* type. The word "transconductance" indicates an output current ($I_o$) is controlled by an input voltage ($V_{\text{in}}$)

$$G_m = \frac{I_o}{V_{\text{in}}} \tag{18-42}$$

In older texts the unit of conductance was a *mho* (ohm spelled backwards), but today the *siemen* is used to denote conductance (it is only a name change, 1 siemen = 1 mho).

Figure 18-10 shows the circuit for a simple op-amp based transconductance multiplier. The basis for the circuit is a dual NPN transistor such as the MAT-01, MAT-02, or LM-114. It is important the two transistors be part of the same substrate in order to maintain thermal tracking between the two devices. For very low signal levels (10 mV or so), the following relationship is true

$$I_{ca} = kI_eV_x \tag{18-43}$$

where

$I_{ca}$ = the collector current of $Q1A$

$I_e$ = the emitter current

$V_x$ = the voltage applied to the base-emitter junction of Q1A

$k$ = q/2KT

$q$ = the electronic charge $1.6 \times 10^{-19}$ coulombs

$K$ = Boltzmann's constant ($1.38 \times 10^{-23}$ $J$/°K)

$T$ = the temperature in Kelvins (°K)

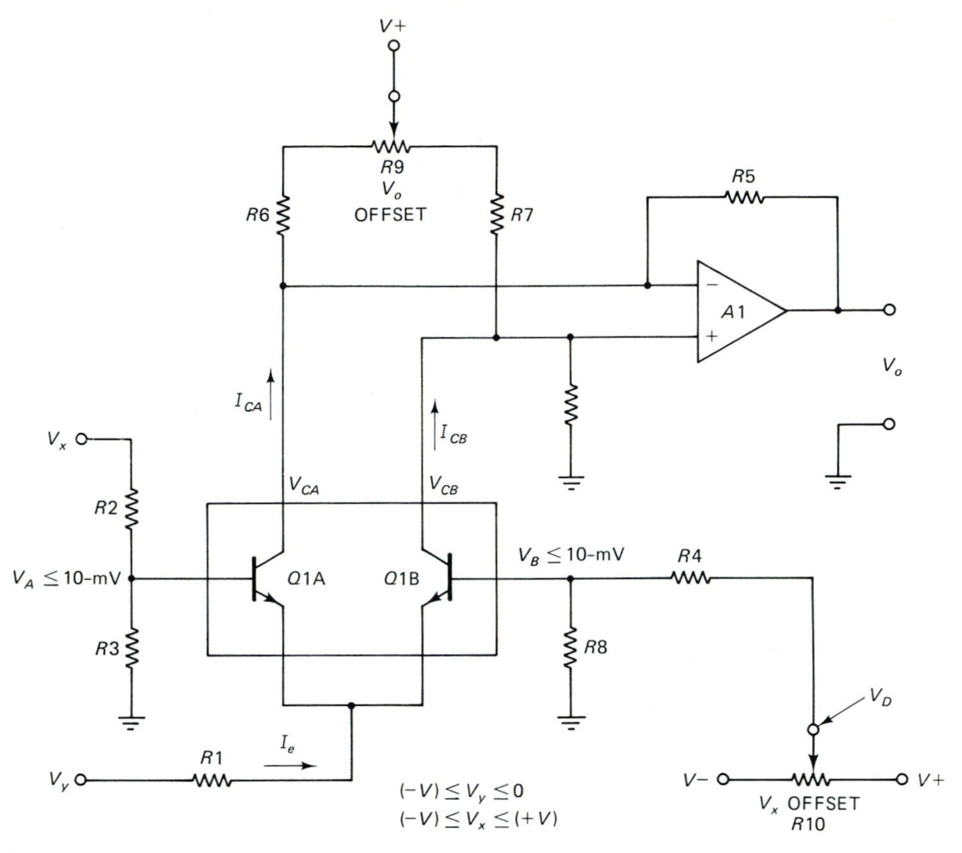


FIGURE 18-10

Output voltage $V_o$ is

$$V_o = I_{ca}R5 \tag{18-44}$$

The emitter current is supplied from input voltage $V_y$ so

$$I_e = \frac{V_y}{R1} \tag{18-45}$$

By substituting Eq. (18-43) and Eq. (18-45) into Eq. (18-44)

$$V_o = kV_xV_y(R5/R1) \tag{18-46}$$

or because $R5/R1$ is also a constant

$$V_o = k_{\text{tot}}V_xV_y \tag{18-47}$$

The transconductance cell forms the basis for most easily available IC multiplier-divider circuits. Figure 18-11 shows a typical IC multiplier circuit based on the transconductance cell. The output is a current, so it must be transformed into an output voltage in a *current difference amplifier*.

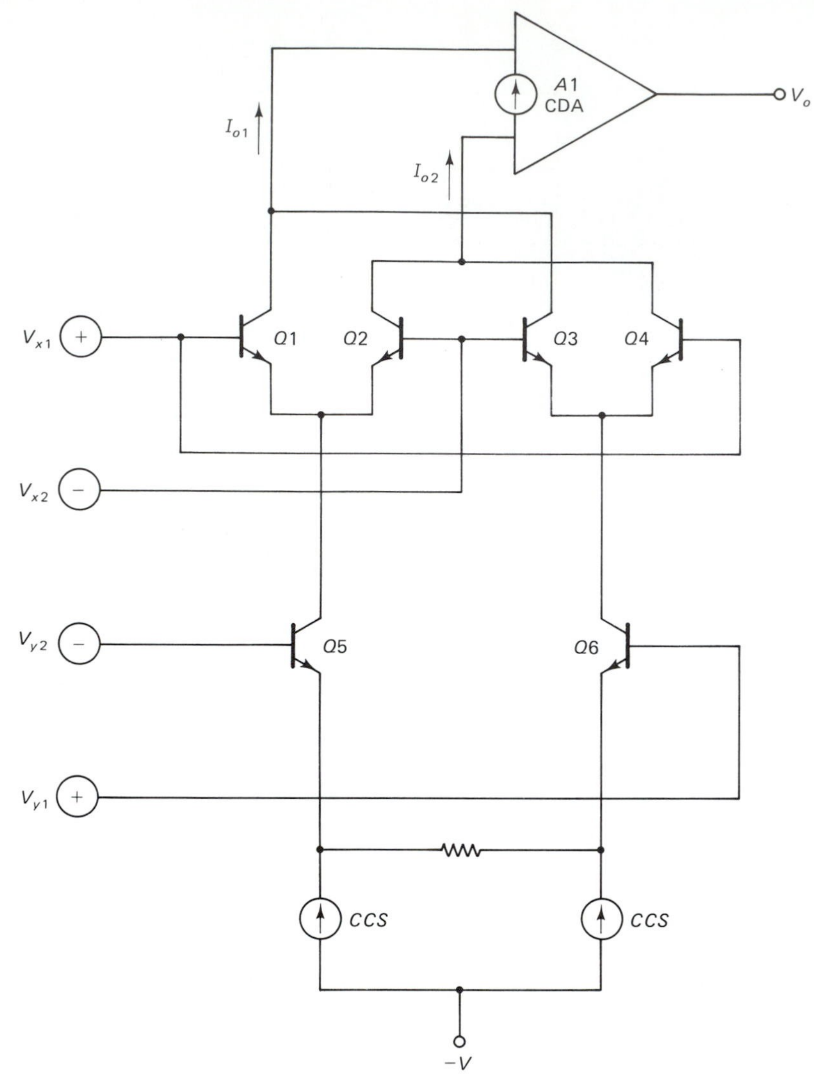


FIGURE 18-11

## 18-11 THE AD-533: A PRACTICAL ANALOG MULTIPLIER/DIVIDER

The Analog Devices AD-533 is a low-cost analog multiplier divider circuit in integrated circuit form. These chips tend to be more expensive than the discrete circuits discussed above, but they are often more cost effective because they tend to work better and require less tweaking to make them operate properly.

The pinouts for the 14-pin DIP version of the AD-533 are shown in Fig. 18-12. The AD-533 contains a transconductance multiplier for the $X$ and $Y$ inputs, and a summing junction at an operational amplifier for a $Z$ input. The AD-533L has a full-scale linearity error of only 0.5%. In addition, a low temperature coefficient of

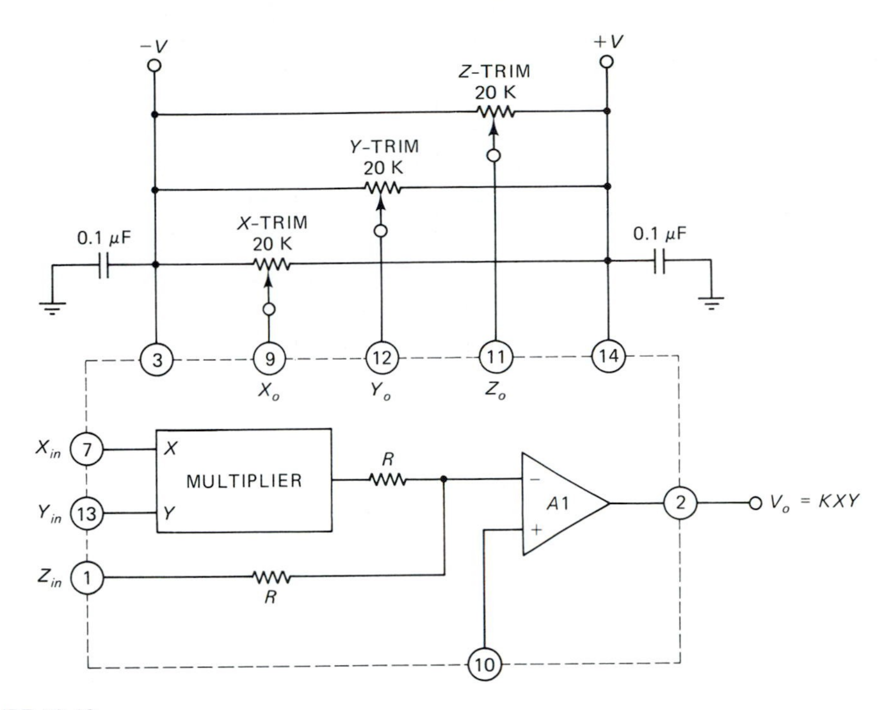


FIGURE 18-12

0.01%/°C is provided. The AD-533 is capable of operating to a small-signal bandwidth of 1 MHz, a full-power bandwidth of 750 KHz and a slew rate of 45V/μS. The AD-533 will multiply in four-quadrants with a transfer function of

$$V_o = \frac{V_x V_y}{10} \text{ volts} \tag{18-48}$$

For division, the AD-533 will operate in two-quadrants with a transfer function of

$$V_o = \frac{10V_z}{V_x} \text{ volts} \tag{18-49}$$

## 18-12 MULTIPLIER OPERATION OF THE AD-533

Figure 18-13 shows the connection of the AD-533 for straight multiplier operation. The $V_x$ and $V_y$ inputs are used for the operands. The $V_z$ input is fed back to the output. A scaling control circuit ($R1$ and $R2$) is used to set the output scale factor. The trimming circuits shown in Fig. 18-12 are also used, but not shown. The range of $V_x$, $V_y$, and $V_o$ is ±10 volts.

The trim procedure for the AD-533 multiplier is simple. It involves only three external multiturn potentiometers. The procedure is

1. Set $V_x = V_y = 0$ volts, and then adjust Z-TRIM for $V_o = 0$ volts.
2. Set $V_z = 0$ volts, and then apply a signal of 50 Hz at 20 volts peak-to-peak. Adjust X-TRIM for minimum signal output.

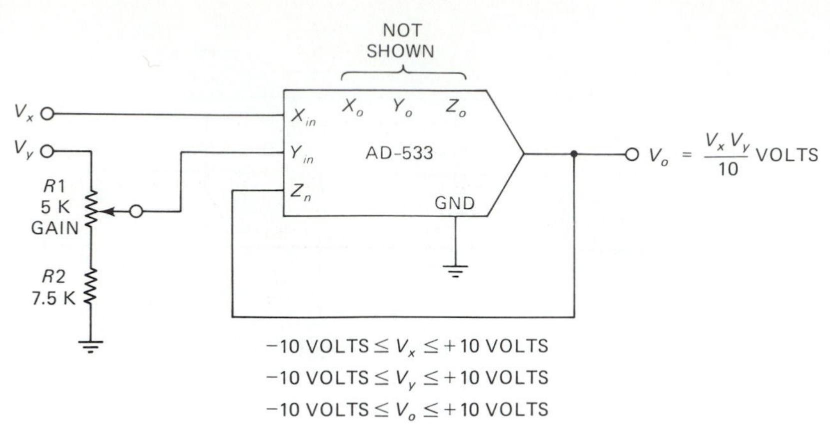


FIGURE 18-13

3. Set $V_y = 0$ and apply a signal of 50 Hz at 20 volts peak-to-peak; adjust Y-TRIM for minimum signal output.
4. Readjust Z-TRIM for $V_o = 0$ Vdc output.
5. Set $V_x = +10$ volts and $V_y$ to 50 Hz at 20 volts peak-to-peak; adjust $R1$ for $V_o = V_y$.

Rearranging the connection of the AD-533 will render it a divider, squarer, square-rooter, and other function circuits.

## 18-13 DIVIDER OPERATION OF THE AD-533

The AD-533 can also be used as an analog divider circuit with the transfer function

$$V_o = \frac{10V_z}{V_x} \text{ volts} \tag{18-50}$$

The circuit for the analog divider configuration is shown in Fig. 18-14. The X and Y inputs are used, with feedback going to the Y input. As was true in the previous case, the trimming circuits are not shown for simplicity, but are used nonetheless. The trim procedure for the analog divider is

1. Set all trimmer potentiometers at mid-scale.
2. Set $V_z = 0$, adjust Z TRIM so $V_o$ remains constant as $V_x$ is varied from $-10$ Vdc through $-1$ Vdc.
3. With $V_z = 0$, $V_x = -10$ Vdc, set Y TRIM for $V_o = 0$ Vdc.
4. Set $V_z = V_x$ and adjust X TRIM for the minimum variation of $V_o$ as $V_x$ is adjusted from $-10$ Vdc to $-1$ Vdc.
5. Repeat steps 2 and 3 for minimum interaction.
6. Set $V_z = V_x$ and trim the gain control ($R1$) for $V_o$ to have the closest approach to $\pm 10$-Vdc output as $V_x$ is varied over the range $-10$ Vdc to $-3$ Vdc.

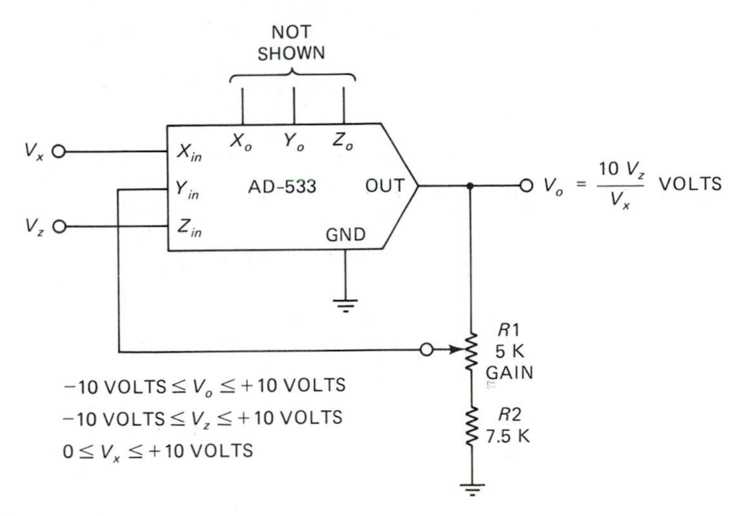


FIGURE 18-14

## 18-14 OPERATION OF THE AD-533 AS A SQUARER

A squarer outputs a voltage which is the square of the input voltage. Before the invention of good analog multiplier circuits, the standard means for squaring a voltage in analog circuits was in a diode breakpoint generator. In those circuits, a series of diodes operated in the square law region of their characteristic are used to fit an output voltage to the square of its input voltage. While the method worked, it suffered from a severe temperature sensitivity, and also from the fact that only a limited number of breakpoints were possible. In the analog multiplier version of a squarer circuit, however, it is possible to simply apply the same voltage to both $V_x$ and $V_y$ inputs. If $V_x = V_y = V$, then $V_xV_y = V^2$.

Figure 18-15 shows a circuit for a squarer based on the AD-533 analog multiplier-divider. In this circuit the $Y_o$ input, which is normally connected to the Y-TRIM potentiometer, is grounded. The X TRIM and Z TRIM potentiometers are the same, but are deleted from Fig. 18-15 for simplicity. When properly trimmed, the output

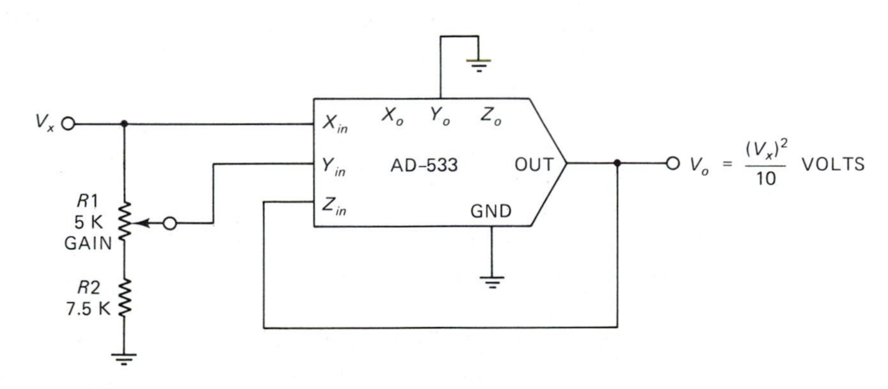


FIGURE 18-15

of the circuit in Fig. 18-15 is

$$V_o = \frac{(V_x)^2}{10} \tag{18-51}$$

### EXAMPLE 18-1

Find the output voltage of the squarer in Fig. 18-15 when the input voltage is $-5$ volts DC.

**Solution**

$$V_o = \frac{(V_x)^2}{10}$$

$$V_o = \frac{(-5 \text{ Vdc})^2}{10}$$

$$V_o = \frac{25 \text{ Vdc}}{10} = 2.5 \text{ Vdc}$$

■

The analog squarer circuit will operate in the two quadrants where the input voltage is positive and negative. The output voltage is only found in one quadrant ($Q$I).

## 18-15 OPERATION OF THE AD-533 AS A FREQUENCY DOUBLER

A special case in the operation of the squarer will allow frequency doubling. When a sine wave signal is applied to the input of a squarer, the output frequency will be doubled. The AD-533 makes a nearly ideal frequency doubler because it does not require a tuned input or output network. The tuning circuits used on other analog frequency multipliers make them single-frequency only. The frequency doubler based on the AD-533 is able to operate on any frequency within its range.

Consider the trigonometric identity

$$\text{Sin}(A)\text{Sin}(B) = \frac{1}{2}[\text{Cos}(A-B) - \text{Cos}(A+B)] \tag{18-52}$$

If A and B each represent the same signal, we can set $B = A$, and the identity takes the form

$$\text{Sin}(A)\text{Sin}(A) = \frac{1}{2}[\text{Cos}(A - A) - \text{Cos}(A + A)] \tag{18-53}$$

or collecting terms and redefining A as $2\pi ft$

$$\text{Sin}(2\pi ft)^2 = \frac{1}{2} - \frac{\text{Cos}(4\pi ft)}{2} \tag{18-54}$$

Rewriting Eq. (18-54) in a form that accounts for the standard IC multiplier which has a transfer function of

$$V_o = \frac{(V_x)^2}{10} \tag{18-55}$$

and because $V_x = V2\pi ft$

$$V_o = \frac{(V2\pi ft)^2}{20} \quad (18\text{-}56)$$

Using the identity of Eq. (18-54)

$$V_o = \frac{V^2}{20}[1 - \text{Cos}(4\pi ft)] \quad (18\text{-}57)$$

### EXAMPLE 18-2

A sinewave signal of 5 KHz at 10 volts peak is applied to the input of a squarer circuit. Calculate the output signal parameters.

#### Solution

The applied signal is defined at 10Sin($2\pi 5000t$).

1. The output of the squarer will be

$$V_o = \frac{(V_x)^2}{10}$$

$$V_o = \frac{[10\text{Sin}(2\pi 5000t)]^2}{10}$$

$$V_o = \frac{100}{10}\text{Sin}(2\pi 5000t)^2$$

$$V_o = 10\text{Sin}(2\pi 5000t)^2$$

2. The last expression is of the form suitable for using the trigonometric identity in Eq. (18-53)

$$V_o = 10\left[\frac{1}{2} - \frac{\text{Cos}(4\pi 5000t)}{2}\right]$$

$$V_o = 5.00 - 5\text{Cos}(2\pi 10000t)$$

The final expression shows a DC term of 5 volts and an AC term of 5Cos($2\pi 10000t$) or twice the input frequency. ■

## 18-16 OPERATION OF THE AD-533 AS A PHASE DETECTOR/METER

A multiplier driving a low-pass filter (Fig. 18-16) will operate as a phase meter. The output of this circuit is a DC level proportional to the difference in phase ($\phi$) between two different signals of the same frequency. Given a multiplier with a transfer function $V_o = V_xV_y/10$ and a low-pass filter which passed only signals with frequencies below the input frequency, only the DC level remains in the output and is proportional to phase difference. The input signals to the circuit of Fig. 18-16 are

$$V_x = V1\text{Sin}(2\pi ft) \quad (18\text{-}58)$$

and

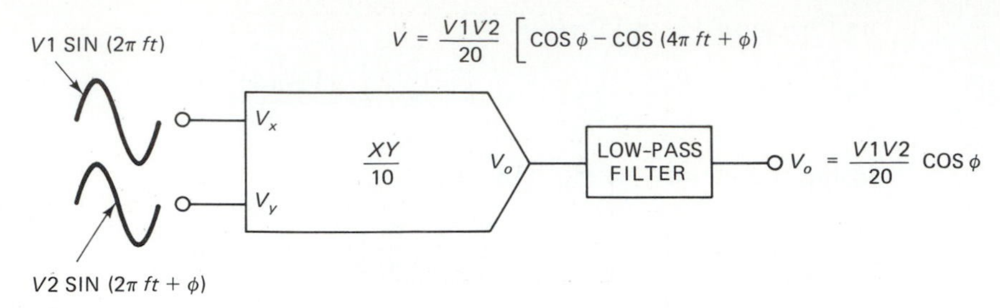


FIGURE 18-16

$$V_y = V2\text{Sin}(2\pi\text{ft} + \phi) \tag{18-59}$$

Appealing to the identity of Eq. (18-52)

$$\text{Sin}(V_x)\text{Sin}(V_y) = \frac{1}{2}[\text{Cos}(V_x - V_y) - \text{Cos}(V_x + V_y)] \tag{18-60}$$

and for a multiplier with the transfer function $V_xV_y/10$ and substituting Eq. (18-58) and Eq. (18-59)

$$V_o = \frac{V1V2}{20}[\text{Cos}(\phi) - \text{Cos}(4\pi ft + \phi)] \tag{18-61}$$

Equation (18-61) represents the output of the analog multiplier. When the low-pass filter removes the AC term, the equation becomes

$$V_o = \frac{V1V2}{20}\text{Cos}(\phi) \tag{18-62}$$

Solving Eq. (18-62) for the phase angle ($\phi$) results in

$$\phi = \text{Cos}^{-1}\left[\frac{20V_o}{V1V2}\right] \tag{18-63}$$

The phase detector operation described above assumes sinusoidal input signals. The circuit will also work with square waves or pulses, however, and in that case the output of the multiplier will be a series of pulses of width proportional to the phase difference. When the pulses are passed through the low-pass filter, the effect is to time average them and thereby produce a DC output that is zero when the signals are in quadrature, and maximum positive at 0° and maximum negative at 180° phase difference.

## 18-17 OPERATION OF THE AD-533 AS A SQUAREROOTER

A squarerooter will produce an output voltage which is the square root of the input voltage. Figure 18-17 shows the connection of the AD-533 as a squarerooter. In this circuit, the $Y_o$ input is grounded, as was true with the squarer, but the X TRIM and Z TRIM external potentiometers are used (but not shown). The input signal is applied to the Z input, while the Y and X inputs are fed back via a diode (*D*1). The circuit

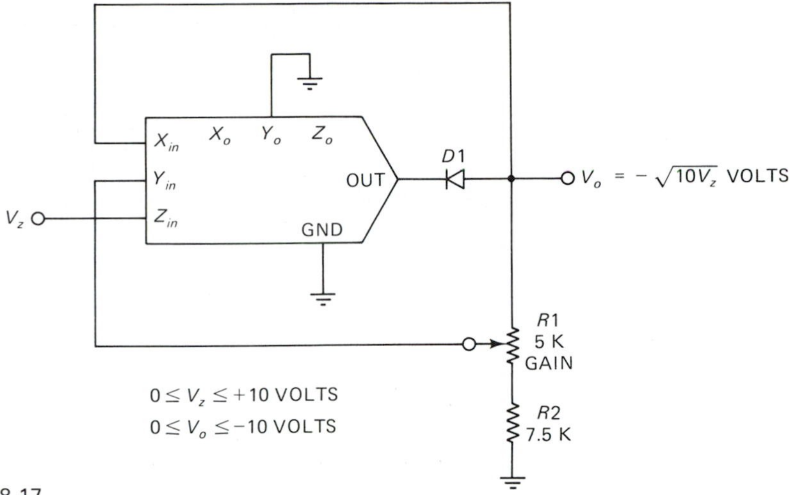


FIGURE 18-17

of Fig. 18-17 will accommodate input signals of 0 to +10 volts, but it outputs a negative voltage in the range of 0 to −10 volts. Therefore, the transfer function for this circuit is

$$V_o = -\sqrt{10V_z} \tag{18-64}$$

If a positive output voltage is needed, the output of Fig. 18-17 can be inverted in an operational amplifier circuit.

## 18-18 THE MULTIPLIER AS AN AUTOMATIC GAIN CONTROL (AGC)

An analog voltage multiplier can be used as an *automatic gain control* (AGC) if the right external circuitry is added. An AGC is designed to maintain a constant output signal level despite changes in the amplitude of the input signal. Examples of the use of AGC circuits include radio receivers and signal generators. The receiver uses the AGC (also sometimes called *automatic volume control* or AVC) to maintain a level output signal despite the fact that various radio stations being received vary considerably in strength. A signal generator might use the AGC circuit to maintain a level output even if variable tuned oscillators tend to output different signal levels at different frequencies.

An AGC circuit based on an analog multiplier is shown in Fig. 18-18. In this circuit the multiplier acts as a gain controlled amplifier. The output of the multiplier is $V_o = V_xV_y/10$. The input signal is applied to the X input. A DC feedback signal is applied to the Y input. The feedback signal is formed by sampling the output signal ($V_o$) rectifying it in a precise fullwave rectifier, and filtering the rectifier output to form a DC signal ($V_a$) proportional to the output signal amplitude. This DC control signal is applied to the input of a Miller integrator where it is compared to a reference level ($V_{\text{ref}}$) which determines the set-point of the circuit. The integrator output is applied to the Y input of the multiplier. The Y input of the multiplier acts as a gain control signal. If an external level-set is needed, the reference voltage $V_{\text{ref}}$ can be made variable.

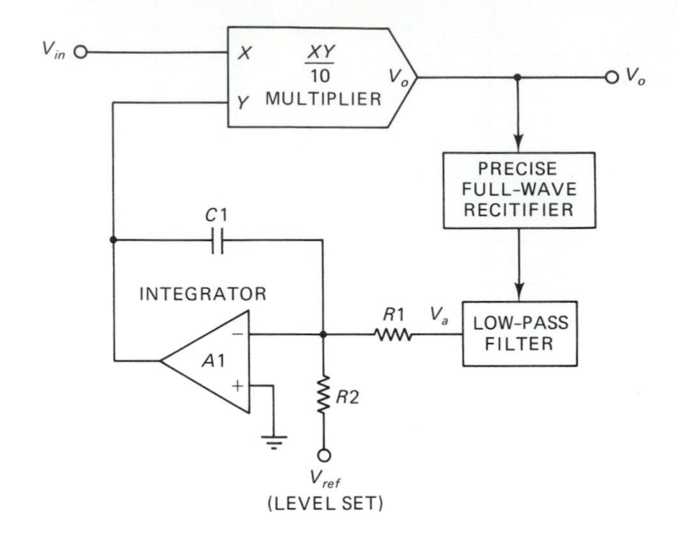


FIGURE 18-18

## 18-19 VOLTAGE CONTROLLED EXPONENTIATOR CIRCUIT

As part of the discussion of logarithmic amplifiers in Chapter 11 the Burr-Brown 4301/4302 multifunction modules were discussed. These hybrid circuits have a transfer function of the form

$$V_o = \left[\frac{V_y V_z}{V_x}\right]^m \tag{18-65}$$

In the transfer equation of the multifunction module (above) the exponent $m$ is normally set by an external resistor voltage divider connected to the $V_a$, $V_b$ and $V_c$ inputs (see Fig. 18-19A). If $m = 1$, the circuit operates as an ordinary multiplier or divider, depending on which inputs are used. For example, when $V_x$ is set to a constant reference of 10 volts, the multifunction module ($m = 1$) will operate as an XY/10 multiplier. Similarly, when $V_z$ is set to a constant 10 volts, the device functions as a divider of the form 10Y/X. If $m < 1$, the module operates as a rooter (a squarerooter when $m = 1/2$). If $m > 1$, the module operates as an exponentiator, including a squarer. If $m$ is set by a dynamic external voltage source, however, the multifunction module operates as a *voltage controlled exponentiator.*

In Fig. 18-19A the $V_b$ and $V_c$ inputs are driven by the outputs of a pair of XY/10 multipliers, which are in turn driven by control voltage $V1$ and $V2$. The $V_a$ input is connected to the alternate inputs of the multipliers. The transfer function of the circuit in Fig. 18-19A is

$$V_o = \left[\frac{V_y V_z}{V_x}\right]^{(V1/V2)} \tag{18-66}$$

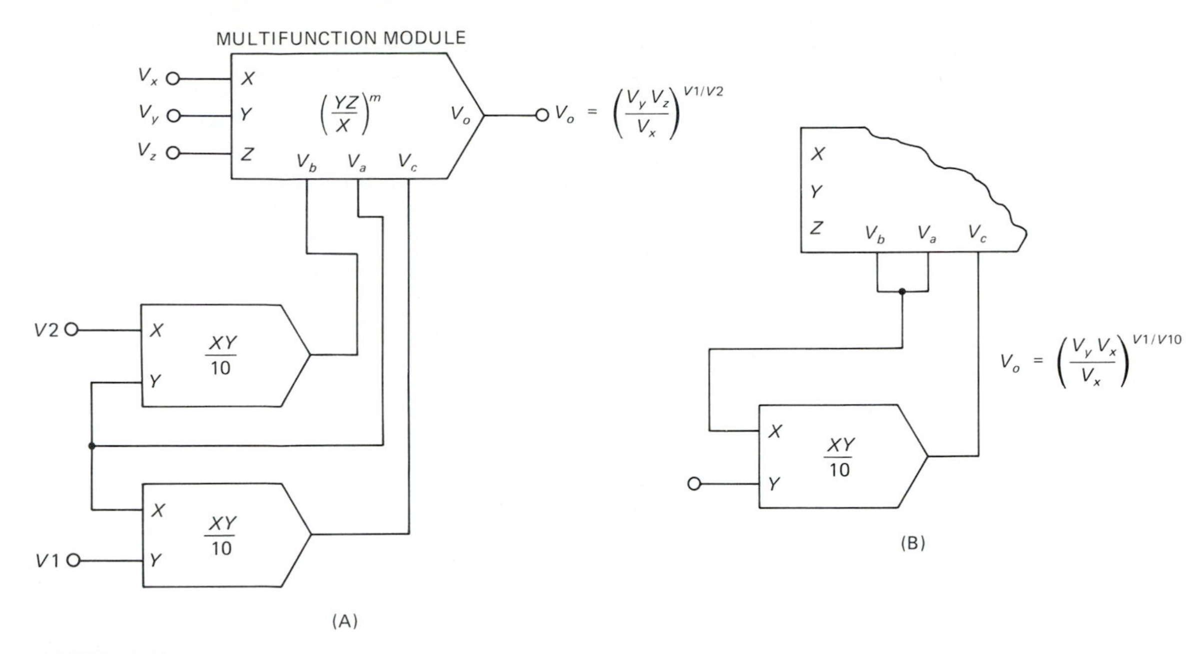


FIGURE 18-19

A single control voltage version of the exponentiator circuit is shown in Fig. 18-19B (it is a partial version of Fig. 18-19A). $V_a$ and $V_b$ are connected and applied to the X input of an external multiplier. The Y input of the multiplier becomes the control voltage, forming a transfer function of

$$V_o = \left[\frac{V_y V_z}{V_x}\right]^{(V1/10)} \tag{18-67}$$

Similarly, if the output of the multiplier is connected to the $V_b$ input and $V_a$ and $V_c$ are connected to the X input of the multiplier, the transfer function becomes

$$V_o = \left[\frac{V_y V_z}{V_x}\right]^{(10/V1)} \tag{18-68}$$

## 18-20 SUMMARY

1. Multipliers have a transfer function $KV_xV_y$, while dividers have a transfer function $KV_x/V_y$. In most cases, $K = 1/10$.
2. Different forms of multipliers may operate in one, two, or four quadrants.
3. Several different methods have been used for making multiplier circuits, matched resistor/lamp, matched JFET resistors, quarter square circuit, logarithmic amplifier, and transconductance cells. Of these, the log-amp and transconductance cells are the most commonly used today.
4. Typical applications of IC multiplier divider circuits include (in addition to multiplication and division), squarer, squarerooter, exponentiator, frequency doubler, phase detector/meter, and automatic gain control.

## 18-21 RECAPITULATION

Now return to the objectives and Prequiz questions at the beginning of the chapter and see how well you can answer them. If you cannot answer certain questions, place a check mark to each and review the appropriate parts of the text. Next, try to answer the questions and work the problems below, using the same procedure.

## 18-22 STUDENT EXERCISES

1. Design and build a transconductance analog multiplier based on an operational amplifier and a dual NPN transistor.
2. Design and build a quarter squares analog multiplier.
3. Connect and test the following AD-533 circuits (a) multiplier, (b) divider, (c) squarer.

## 18-23 QUESTIONS AND PROBLEMS

1. List four different forms of analog multiplier circuit.
2. An analog multiplier has a scaling factor of 1/10. Calculate the output voltage when $V_x = +2.5$ volts and $V_y = +3$ volts.
3. An analog divider with a scaling factor of 1/10 is exposed to the following input voltages, $V_x = +1.75$ volts and $V_y = +5.00$ volts. Calculate the output voltage.
4. Draw the circuit for using an AD-533 as a phase detector/meter.
5. Demonstrate mathematically that the phase detector works.
6. Demonstrate mathematically that the squarer works as a frequency doubler when a sine wave signal is applied.
7. Find the output voltage of the squarer in Fig. 18-15 when the input voltage is $-7.6$ volts DC.
8. A sinewave signal of 10 KHz at 5 volts peak is applied to the input of a squarer circuit. Calculate the output signal parameters.
9. Write the basic equation for a quarter square multiplier and evaluate it to prove it can form a multiplier.
10. Draw the circuit for a logarithmic amplifier based multiplier, and define the voltages at the outputs of each stage.
11. Draw the circuit for a logarithmic amplifier based divider and define the voltages at the output of each stage.
12. Sketch the circuit for an AD-533 connected as a multiplier circuit; as a divider.
13. Write the output expression for an analog multiplier when both inputs are connected together and a sine wave of $10\text{Sin}(2\pi ft)$ is applied.
14. An AD-533 is connected as a squarerooter. A $+4$ volt signal is applied to the input. What is the output voltage?

CHAPTER 19

# Active Filter Circuits

## OBJECTIVES

1. Recognize the response curves for low-pass, high-pass, bandpass, and notch filters.
2. Know the differences between Butterworth, Chebyshev, Cauer, and Bessel filters.
3. Be able to design first, second, and third order filters.

## 19-1 PREQUIZ

These questions test your prior knowledge of the material in this chapter. Try answering them before you read the chapter. Look for the answers (especially those you answered incorrectly) as you read the text. After you have finished studying the chapter, try answering these questions again and those at the end of the chapter (see Section 19-13).

1. A ____________ -pass filter passes only those frequencies above the cutoff frequency.
2. A cut-off slope of $-6$ dB/octave represents ____________ dB/decade.
3. A ____________ filter has a flat response within the passband and a heavily damped rolloff characteristic.
4. The ____________ filter has a very rapid rolloff near the cutoff frequency, but only at the expense of less attenuation at frequencies far removed from the cutoff frequency.

## 19-2 INTRODUCTION TO ACTIVE FILTERS

An electronic frequency selective filter favors some frequencies and discriminates against other frequencies (or bands of frequencies). In other words, *a filter circuit will pass some frequencies and reject or sharply attenuate others*. Frequencies that pass through the filter with little attenuation are said to be in the *passband*, while frequencies that are heavily attenuated are said to be in the *stopband*.

Filter circuits can be classified several ways: passive versus active, analog versus digital versus software, by frequency range (audio, RF, or microwave), or by passband characteristic. *Passive filters* are made of various combinations of passive components such as resistors ($R$), capacitors ($C$) and inductors ($L$). In general, passive filters are lossy and not very flexible. An *active filter*, on the other hand, is based on an active device such as a transistor or an operational amplifier along with passive components ($R$, $C$, and occasionally $L$) that determine frequency. In most cases, the passive components are resistors and capacitors (although a few inductor-based circuits are available).

Active filters use linear circuit techniques like those found throughout this textbook. Digital filters use digital IC devices, and are often based on capacitor switching techniques. Software filters implement solutions to frequency selective equations using computer programming techniques. The emphasis here will be on analog active filters.

Filters can also be classified by frequency range. "Audio" filters operate from the sub-audio to ultrasonic ranges (near DC to about 20 KHz). RF filters operate at frequencies above 20 KHz, up to about 900 MHz. Microwave filters operate at frequencies >900 MHz. These range designations are not absolute, but do indicate approximate points at which a change of design techniques generally takes place. For example, filters can be made frequency-selective using inductors.

But in the audio range the inductance values are large, so inductors are bulky, costly, and lossy. In addition, inductors produce stray magnetic fields which interfere with other nearby circuits. On the other hand, inductors are the elements of choice in the RF region. But once frequencies approach several hundred megahertz, the inductance values required become too low for practical use so other techniques are required. In the microwave and high UHF region transmission line and cavity techniques are used. The circuits discussed in this chapter are "audio" active filters, but can have a passband between subaudio and the low ultrasonic region.

Finally, filters may be classified by the nature of their frequency response characteristics. This method of categorizing filters takes note of the filter's passband and stopband. In this chapter we will examine *low-pass filters*, *high-pass filters*, *bandpass filters*, and *stopband filters*. We will also examine a related circuit called the *all-pass phase shifter*.

### 19-2.1 Filter Characteristics

Figure 19-1A shows in general terms the characteristics of theoretically ideal filters. These curves will be discussed in greater detail later. A *low-pass filter* has a passband from DC to a specified cutoff frequency ($F1$). All frequencies above the cutoff frequency are attenuated, so are in the stopband. A *bandpass filter* has a passband between a lower limit ($F2$) and an upper limit ($F3$). All frequencies lower than $F2$ or greater than $F3$, are in the stopband. A *high-pass filter* has a stopband from DC to a certain lower limit ($F4$). All frequencies $>F4$ are in the passband.

A *stopband filter* response is shown in Fig. 19-1B. This filter severely attenuates frequencies between lower and upper limits ($F5$ to $F6$), but passes all others. When the stopband is very narrow, the stopband filter is called a *notch filter*. Such filters are often used to remove a single, unwanted frequency. An example of such an application is removal of unwanted 60 Hz interference caused by proximity to local power lines.

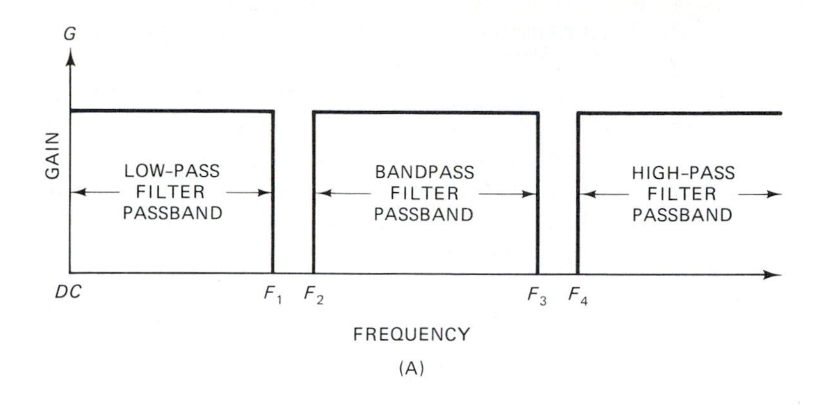


(A)

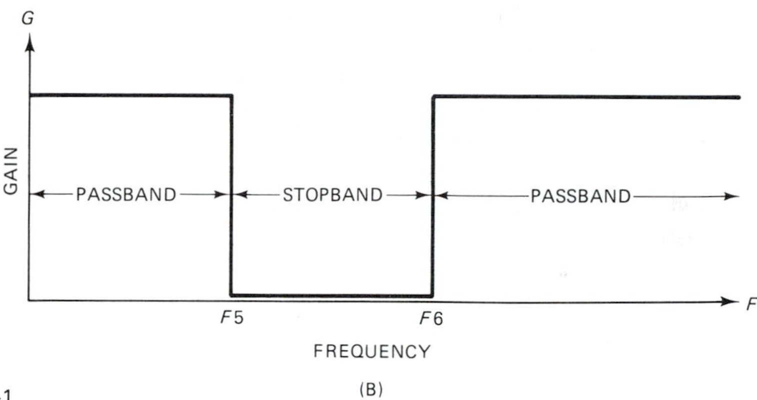


(B)

FIGURE 19-1

**"Ideal" Compared to Practice Filter Response Curves.** The response curves shown in Fig. 19-1 are unrealistic "ideal" generalizations. It might seem that a step-function cutoff is desirable, but in actual practice such a response is neither attainable nor desirable. The reason why it is undesirable is the "ideal" response actually causes a problem because the filter may ring when a fast risetime signal is applied. This is especially likely to occur on narrow bandpass filters. While sinusoidal signals do not usually pose a problem, noise impulses or transient step functions can easily cause unwanted ringing.

There are several common filter responses, *Butterworth*, *Chebyshev*, *Cauer* (also sometimes called *elliptic*), and *Bessel*. The basic Butterworth response is shown in Fig. 19-2A. The noteworthy properties of the Butterworth filter are both the passband and stopband are relatively flat, as is the transition region slope between them.

It is standard practice in filter design to specify the passband between points where the response falls off $-3$ dB from the mid-passband gain. Therefore, for the low-pass filter shown in Fig. 19-2A the cutoff frequency is where gain ($A1$) falls off to 0.707 times the low frequency gain ($A1$), or the $-3$ dB point.

At frequencies $f > f_c$, the gain falls off linearly at a rate which depends on the order of the filter. The slope ($S$) of the fall off is measured in either *decibels per octave* (a 2:1 frequency change) or *decibels per decade* (a 10:1 frequency change).

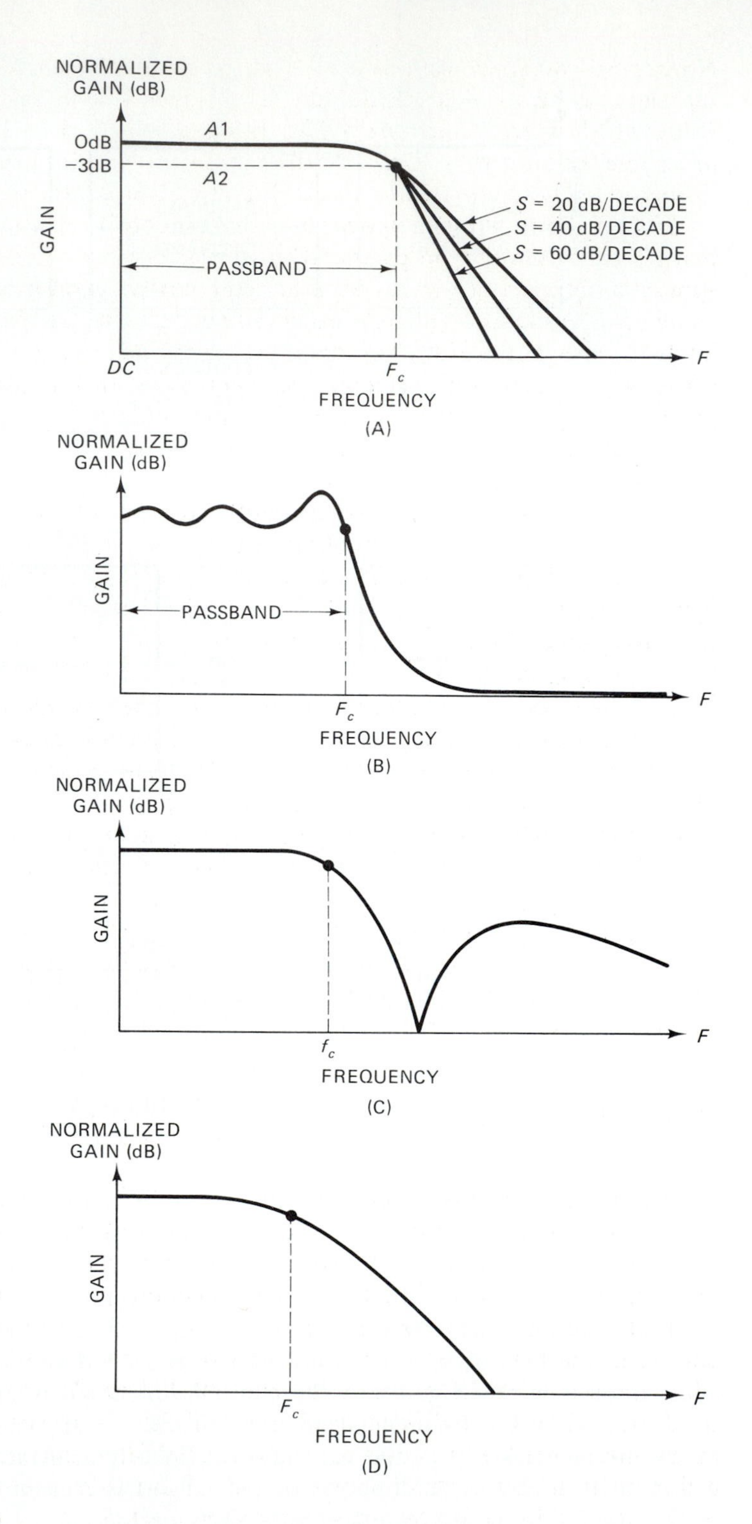
NORMALIZED
GAIN (dB)
OdB
−3dB
A1
A2
GAIN
S = 20 dB/DECADE
S = 40 dB/DECADE
S = 60 dB/DECADE
PASSBAND
DC
Fc
F
FREQUENCY
(A)
NORMALIZED
GAIN (dB)
GAIN
PASSBAND
Fc
F
FREQUENCY
(B)
NORMALIZED
GAIN (dB)
GAIN
fc
F
FREQUENCY
(C)
NORMALIZED
GAIN (dB)
GAIN
Fc
F
FREQUENCY
(D)

FIGURE 19-2

Note these two specifications can be scaled relative to each other, −6 dB/octave is the same slope as −20 dB/decade. The slopes shown in Fig. 19-2A cover three Butterworth cases. A *first-order filter* offers a roll-off of −20 dB/decade, a *second-order filter* offers a roll-off of −40 dB/decade, and a *third-order filter* rolls off at −60 dB/decade. These correspond to 6, 12, and 18 dB/octave, respectively.

On first glance, it might appear only third-order filters would be used because they make the transition from passband to stopband more rapidly. But higher order response is obtained at the cost of more complexity, greater sensitivity to component value error, and more difficult design. Some higher order filter designs are also more likely to oscillate than lower order equivalents. The selection of filter order is a trade-off between system requirements and complexity. If a large, undesired signal is expected at a frequency up to, say, the third harmonic of $f_c$, a high order response might be indicated. Conversely, weaker stopband signals in that region may permit the use of a lesser order filter.

The steepness and shape of the roll-off curve is a function of the filter *damping factor.* As a class, Butterworth filters tend to be heavily damped, which explains the gradual roll-off in the response curve. The Chebyshev filter response (Fig. 19-2B) is lightly damped, so has a variation (or ripple) within the passband. The Chebyshev filter offers a generally faster roll-off than the Butterworth filter, but at the cost of less flatness within the passband.

The *Cauer* or *elliptic* filter response curve (Fig. 19-2C) offers the fastest roll-off for frequencies close to the cutoff frequency, as well as relatively good flatness within the passband. Notches of −40 dB to −60 dB can be achieved close to $f_c$, but only at a cost of less attenuation further into the stopband. A typical Cauer filter response has a deep notch close to $f_c$, rises to a peak at some high frequency removed from $f_c$, and then gradually falls off for even higher frequencies at a rate of −20 dB to −40 dB/decade.

The response of the *Bessel* filter is shown in Fig. 19-2D. Although it appears similar to the Butterworth response, it is not maximally flat within the passband, and falls off somewhat less rapidly. The benefit of the Bessel filter is a flat phase response across the passband.

### 19-2.2 Filter Phase Response

Most frequency-selective circuits exhibit a phase change over the passband. The responses for two different filters, Butterworth and Bessel, are shown in Fig. 19-3. Note the maximally flat Butterworth exhibits a decidely nonlinear phase response in both passband and stopband. The frequency dependent phase shift of a low-pass filter is −45° at $f_c$ and increases by a factor of −45° for each additional increase of −20 dB/decade in the roll-off slope. Thus, a first order low-pass filter phase shift is −45°, while the second-order phase shift (for −40 dB/decade roll-off) is −90°, and for a third-order filter (−60 dB/decade) it is −135°. The high pass response is of the same magnitude as the low-pass case for each order of filter, but the sign is opposite. Thus, a first-order high-pass filter shows a +45° phase shift, a second-order filter shows +90°, and a third-order high-pass filter shows +135°.

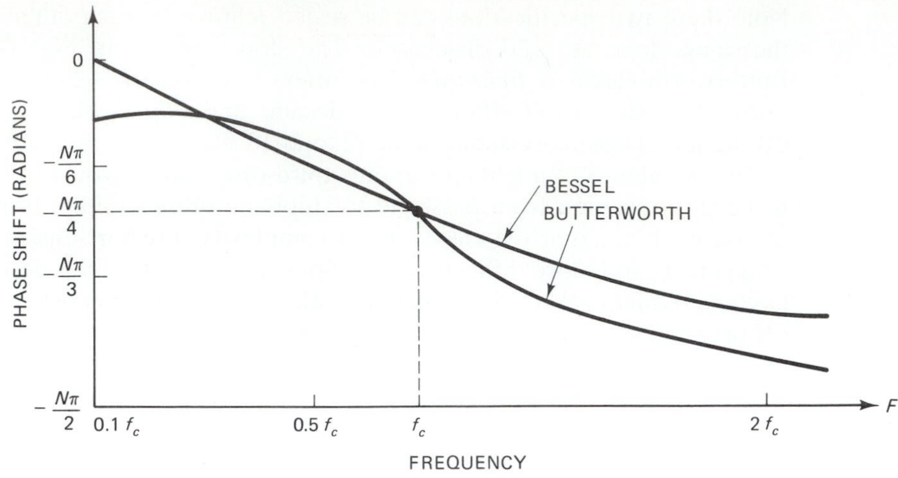


FIGURE 19-3

The Bessel filter also shows a phase shift over the passband, but it is nearly linear. A useful feature of this characteristic is it allows a uniform time delay all across the passband. As a result, the Bessel filter offers the ability to pass transient pulse waveforms with minimum distortion. For the Bessel filter, the phase shift maximum is

$$\Delta\phi_{max} = \frac{-n\ \pi}{2} \qquad (19\text{-}1)$$

where

$\Delta\phi$ = the maximum phase shift

$n$ = the order of the filter (i.e., number of poles)

In a properly designed Bessel filter, the cutoff frequency ($f_c$) occurs at a point where the phase shift is half the maximum phase shift, or

$$\Delta\phi_{fc} = \frac{-n\ \pi}{4} \qquad (19\text{-}2)$$

The Bessel filter is said to work best at the frequency where $f = f_c/2$.

## 19-3 LOW-PASS FILTERS

A model for the low-pass filter is shown in Fig. 19-4. This filter is called the *voltage controlled voltage source* (VCVS) filter or *Sallen-Key* filter. The basic configuration is a noninverting follower operational amplifier ($A1$). The op-amp selected should have a high gain bandwidth product, relative to the cutoff frequency, in order to permit the filter to operate properly. The gain of the circuit is given by

$$A_v = \frac{R_f}{R_{in}} + 1 \qquad (19\text{-}3)$$

So, if $R_{in} = R$, we may deduce

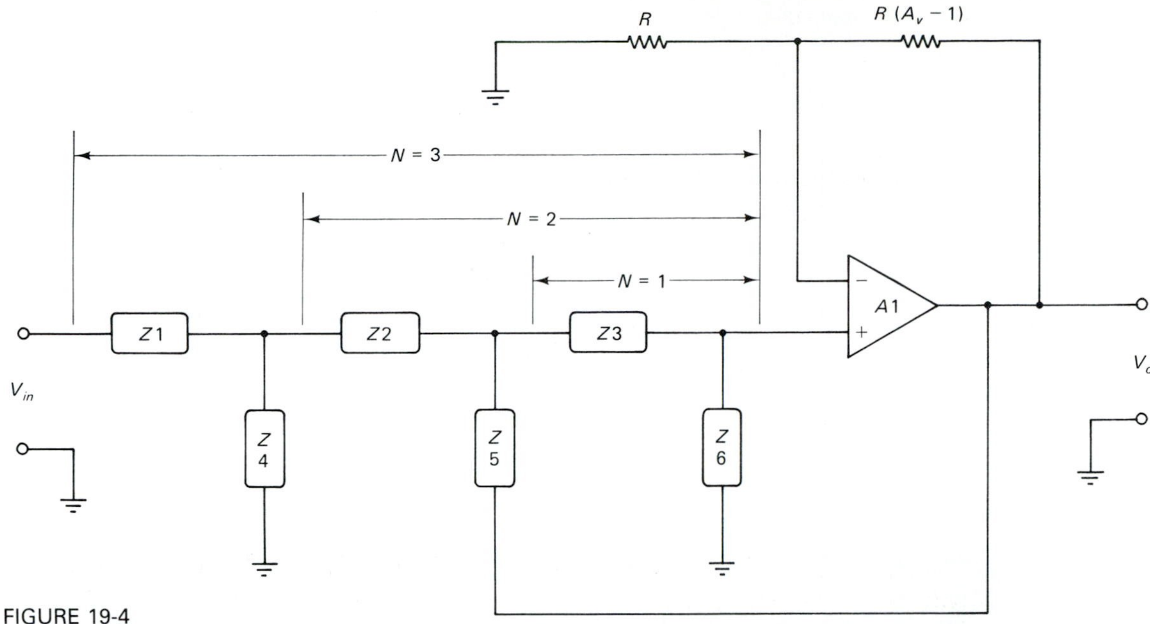


FIGURE 19-4

$$R_f = R(A_v - 1) \tag{19-4}$$

In some circuits, the gain may be unity. In those circuits the resistor voltage divider feedback network is replaced with a single connection between output and the inverting input.

The input circuitry of the generic VCVS filter consists of a network of impedances labelled $Z1$ through $Z6$. Each of these blocks will be either a resistance ($R$) or a complex capacitive reactance ($-jX_c$). Which element becomes which type of component is determined by whether the filter is a low-pass or high-pass type.

The *order* of the filter, denoted by $n$, refers to the number of poles in the design, or in practical terms, the number of RC sections. A first-order filter ($n = 1$) consists of $Z3$ and $Z6$, a second-order filter ($n = 2$) consists of $Z2$, $Z3$, $Z5$, and $Z6$, and a third-order filter ($n = 3$) consists of all six impedances ($Z1$–$Z6$). Higher order filters ($n > 3$) can also be built, but are not discussed here.

In a low-pass filter $Z1$ through $Z3$ are resistances, while $Z4$ through $Z6$ are capacitances. The component roles are reversed in high-pass filters. Now let's review the properties of the low-pass filter and then learn to design first-, second-, and third-order low-pass VCVS filters.

By way of review, Fig. 19-5 shows the low-pass Butterworth filter response curve. This type of filter is maximally flat within the passband, and passes all frequencies below a certain critical frequency ($F_c$). The breakpoint between the passband and the stopband is where the gain of the circuit has dropped off $-3$ dB from its lower frequency value. Above the critical frequency, the gain falls off at a certain rate indicated by the slope of the curve. The steepness of the slope is usually specified in terms of decibels (dB) of gain per octave of frequency and dB/decade is sometimes used.

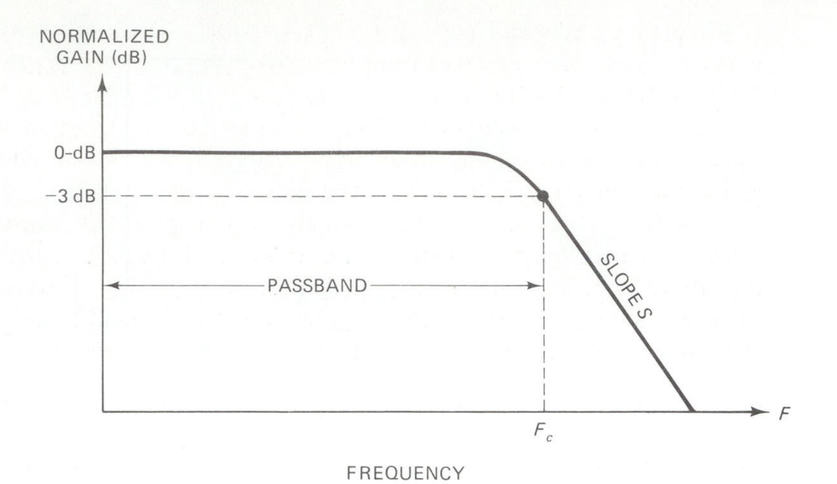


FIGURE 19-5

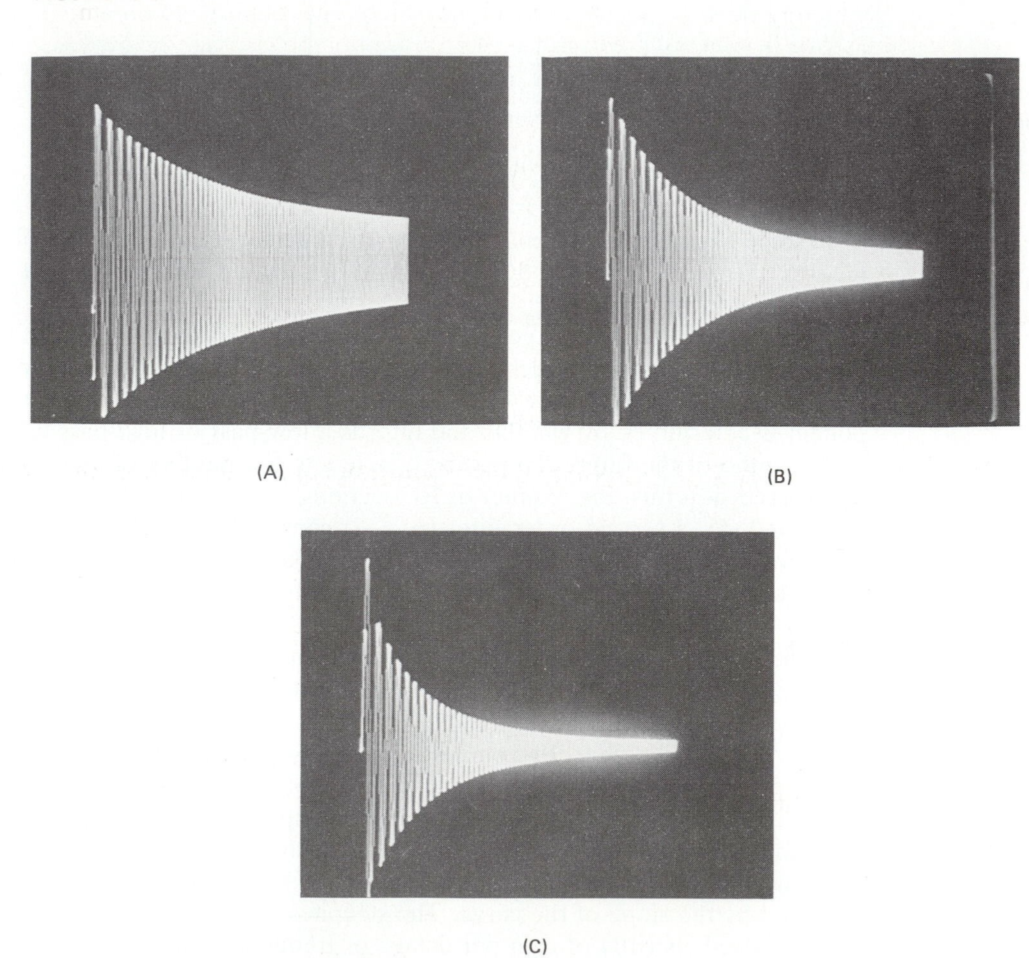


FIGURE 19-6

Frequency sweeps of a set of low-pass filters are shown in Fig. 19-6. These oscilloscope traces were created using a sweep generator that varies a sine wave signal from 1 KHz to 10 KHz. The −3 dB frequency of the filters was 3 KHz. The Y input of the oscilloscope displays the output amplitude of the filter. The X input of the oscilloscope was externally swept using the same sawtooth used to sweep the signal generator. Thus, the X-Y oscilloscope trace is a plot of the frequency response of the filter under test. Figure 19-6A shows the response of a first-order filter in the area from just below $f_c$ up to 10 KHz. The response of the second-order filter is shown in Fig. 19-6B, and the third-order filter in Fig. 19-6C. All three traces were taken under the same conditions, so one can see how the attenuation of frequencies more than $f_c$ is greater in the higher order filters.

### 19-3.1 First-Order Low-Pass (−20 dB/Decade) Filters

The first-order low-pass filter is shown in Fig. 19-7A. Its response curve is shown in Fig. 19-7B. The filter consists of a single-section RC low-pass filter driving the noninverting input of an operational amplifier. The gain of the op-amp is [(R2/R3) + 1]. The high input impedance of $A1$ prevents loading of the RC network. The general form of the transfer equation for the amplitude vs. frequency response for the first-order filter is

$$A_{dB} = 20\mathrm{LOG}(A_v) - 20\mathrm{LOG}(1 + (\omega_o)^2)^{1/2} \tag{19-5}$$

where

$A_{dB}$ = the gain of the circuit in decibels

$A_v$ = the voltage gain within the passband

LOG = the base-10 logarithms

$\omega_o$ = the ratio of the input frequency to the cutoff frequency ($f_o = f/f_c$)

The voltage at the output of the RC network ($V_a$) is found from the voltage divider equation

$$V_a = \frac{-jX_c V_{in}}{R - jX_c} \tag{19-6}$$

where

$-jX_c = 1/j2\pi fC$

$j$ = the imaginary operator ($\sqrt{-1}$)

Substituting the value for $-jX_c$

$$V_a = \frac{\dfrac{V_{in}}{j2\pi fC}}{R + \dfrac{1}{j2\pi fC}} \tag{19-7}$$

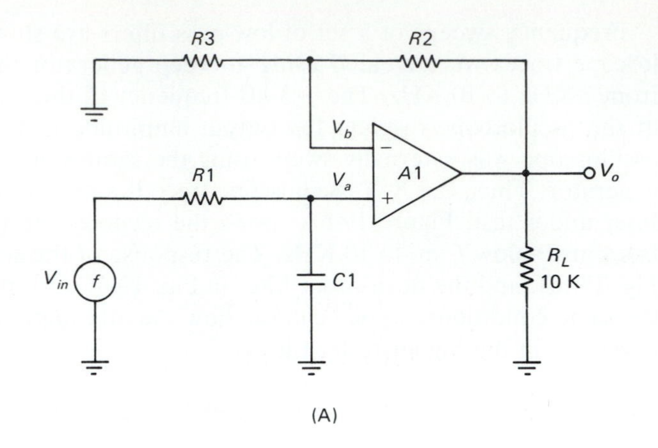


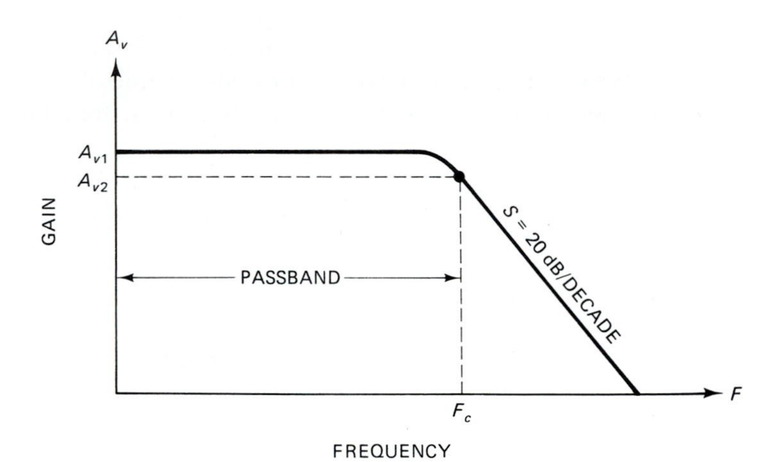


FIGURE 19-7

This simplifies to

$$V_a = \frac{V_{in}}{1 + 2\pi fCR} \tag{19-8}$$

If the transfer function of the noninverting follower is

$$V_o = V_{in}\left[\frac{R2}{R3} + 1\right] \tag{19-9}$$

and since $V_{in} = V_a$ [Eq. (19-7)]

$$V_o = \left[\frac{V_{in}}{1 + 2\pi fCR}\right]\left[\frac{R2}{R3} + 1\right] \tag{19-10}$$

Equation (19-9) can be put into a more general transfer equation

$$\frac{V_o}{V_{in}} = \frac{A_v}{1 + j(f/f_c)} \tag{19-11}$$

where

$V_o$ = the output signal voltage

$V_{in}$ = the input signal voltage

$A_v$ = the pass-band gain $[(R2/R3) + 1]$ [see Eq. (19-8)]

$f$ = the signal frequency

$f_c$ = the $-3$-dB frequency $(1/2\pi RC)$

The filter parameters are required to define the operation of any particular circuit. The *gain magnitude* and *phase shift* are found from the following equations

Gain magnitude

$$\left|\frac{V_o}{V_{in}}\right| = \frac{A_v}{[1 + (f/f_c)^2]^{1/2}} \tag{19-12}$$

and phase shift angle in radians

$$\phi = -\text{Tan}^{-1}(f/f_c) \tag{19-13}$$

### EXAMPLE 19.1

A first-order low-pass VCVS filter is built with the following component values, $R1 = 10$ kohms, $R2 = 2$ kohms and $R3 = 1$ kohms, $C1 = 0.01\ \mu F$. Characterize this filter if a 1200-Hz sine wave signal is applied.

### Solution

1. Calculate the $-3$-dB frequency:

$$f_c = \frac{-1}{2\pi R1C1}$$

$$f_c = \frac{-1}{(2)(\pi)(10{,}000\text{-ohms})(0.01 \times 10^{-6}\text{F})}$$

$$f_c = 1/0.000628 = 1{,}592\text{-Hz}$$

2. Passband gain $(A_v)$

$$A_v = \left[\frac{R2}{R3} + 1\right]$$

$$A_v = \left[\frac{(2 \text{ kohms})}{(1 \text{ kohms})} + 1\right] = 3$$

3. Calculate gain magnitude

$$\left|\frac{V_o}{V_{in}}\right| = \frac{A_v}{[1 + (f/f_c)^2]^{1/2}}$$

$$\left|\frac{V_o}{V_{in}}\right| = \frac{3}{[1 + (1200/1592)^2]^{1/2}}$$

$$\left|\frac{V_o}{V_{in}}\right| = \frac{3}{[1 + 0.057]^{1/2}}$$

$$\left|\frac{V_o}{V_{in}}\right| = \frac{3}{1.25} = 2.4$$

4. Phase shift angle (in radians)

$$\phi = -\mathrm{Tan}^{-1}(f/f_c)$$

$$\phi = -\mathrm{Tan}^{-1}(1200/1592)$$

$$\phi = -\mathrm{Tan}^{-1}(0.075) = 0.65 \text{ radians}$$

Because the filter characterization depends in part on the ratio $f/f_c$, the equations take different forms at different values of $f$ and $f_c$. These can be reduced to

At low frequencies well within the passband ($f < f_c$)

$$\left|\frac{V_o}{V_{in}}\right| = A_v = \frac{R_2}{R_3} + 1 \qquad (19\text{-}14)$$

At the $-3$ dB cutoff frequency ($f = f_c$)

$$\left|\frac{V_o}{V_{in}}\right| = 0.707A_v \qquad (19\text{-}15)$$

At a high frequency well above the $-3$ dB cutoff frequency ($f > f_c$)

$$\left|\frac{V_o}{V_{in}}\right| < A_v \qquad (19\text{-}16)$$

Table 19-1 shows the characteristics of first-order filters at several different ratios of $f/f_c$.

### 19-3.2 Design Procedure for a First-Order Low-Pass Filter

There are two basic ways to design a low-pass filter, *ground-up* and *frequency scaling*. In this section the ground-up method is discussed.

Procedure

1. Select the $-3$ dB cutoff frequency ($f_c$) from the circuit requirements and applications.
2. Select a standard value capacitance ($C \leq 1$ μF).
3. Calculate the required resistance from

$$R1 = \frac{1}{2\pi f_c C}$$

**TABLE 19-1** Filter Characteristics

| *Applied Frequency (Hz)* | *Gain Magnitude* | | *Phase Shift (degrees)* |
|---|---|---|---|
| | *Ave.* | *dB* | |
| 10 | 2.00 | 6.02 | −0.57 |
| 20 | 1.999 | 6.018 | −1.15 |
| 50 | 1.998 | 6.009 | −2.86 |
| 80 | 1.994 | 5.993 | −4.57 |
| 100 | 1.990 | 5.977 | −5.71 |
| 200 | 1.961 | 5.850 | −11.31 |
| 500 | 1.789 | 5.052 | −26.57 |
| 800 | 1.561 | 3.872 | −38.66 |
| 1000 | 1.414 | 3.010 | −45.00 |
| 2000 | 0.894 | −0.969 | −63.43 |
| 5000 | 0.392 | −8.129 | −78.69 |
| 8000 | 0.248 | −12.109 | −82.87 |
| 10000 | 0.199 | −14.023 | −84.29 |
| 20000 | 0.0998 | −20.011 | −87.14 |
| 50000 | 0.0399 | −27.960 | −88.85 |
| 80000 | 0.0249 | −32.042 | −89.28 |
| 100000 | 0.0199 | −33.980 | −89.43 |

4. Select the passband gain for $f < f_c$.
5. Select a value for resistor $R$, and
6. Calculate $R_f$ from

$$R_f = R(A_v - 1)$$

## EXAMPLE 19-2

A low-pass filter is needed for a biological transducer. The cutoff frequency should be 100 Hz, and the gain should be 5.

### Solution

1. $F_c = 100$ Hz
2. Select trial value for $C1$, 0.1 μF
3. Calculate $R1$

$$R1 = \frac{1}{2\pi f_c C1}$$

$$R1 = \frac{1}{(2)(3.14)(100 \text{ Hz})(0.1 \times 10^{-6} \text{ farads})}$$

$$R1 = 1/6.28 \times 10^{-5} \text{ ohms} = 15{,}923 \text{ ohms}$$

4. Select a trial value for $R1$, 10 kohms
5. Calculate $R_f$

$$R_f = R(A_v - 1)$$

$$R_f = (10 \text{ kohms})(5 - 1)$$

$$R_f = (10 \text{ kohms})(4) = 40 \text{ kohms}$$

■

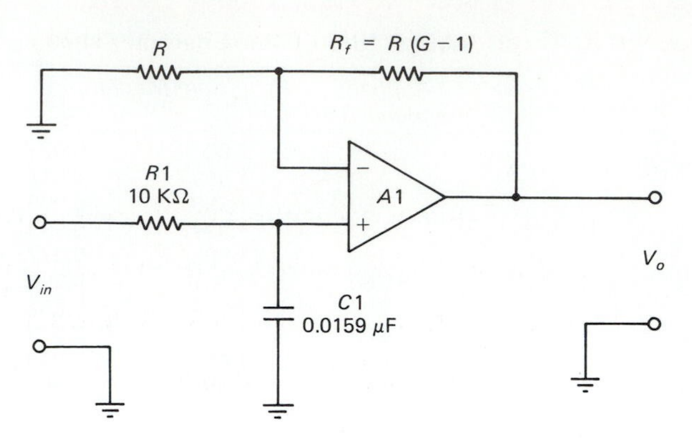


FIGURE 19-8

### 19-3.3 Design by Normalized Model

The filter design can be simplified by using a normalized model. First, design a filter for a standardized frequency (1 Hz, 10 Hz, 100 Hz, 1000 Hz, or 10,000 Hz) and list the component values. The values for any other frequency can then be computed for any other frequency by a simple ratio and proportion. An example of a first-order low-pass Butterworth filter is shown in Fig. 19-8. The component values shown in Fig. 19-8 are normalized for 1 KHz. The actual required component values ($R1'$ and $C1'$) are found by dividing the normalized values shown by the desired cutoff frequency in kilohertz.

$$C1' = \frac{(C1)(1\ \text{KHz})}{F} \tag{19-17}$$

or

$$R1' = \frac{(R1)(1\ \text{KHz})}{F} \tag{19-18}$$

Leave one of the values alone and calculate the other. In general, it is easier to obtain precision resistors in unusual values (or the value obtained by a potentiometer), so it is common practice to select a standard capacitance and calculate the new resistance.

**EXAMPLE 19-3**

Change the frequency of the normalized 1-KHz filter to 60 Hz (0.06 KHz).

**Solution**

$$C1' = \frac{(C1)(1\ \text{KHz})}{F}$$

$$C1' = \frac{(0.0159\ \mu\text{F})(1\ \text{KHz})}{0.06} = 0.265\ \mu\text{F}$$

■

A generalized form of the equations is given below

$$A' = A(f_c/f) \tag{19-19}$$

where

$A'$ = the new component value for either $C1$ or $R1$

$A$ = the original component value for either $C1$ or $R1$

$f_c$ = the filter $-3$ dB cutoff frequency

$f$ = the new design frequency

### 19-3.4 Second-Order Low-Pass (−40 dB/Decade) Filters

The circuit for a second-order low-pass filter is shown in Fig. 19-9A, while the response curve is shown in Fig. 19-9B. Note that this circuit is similar to the first-order filter, but with an additional *RC* network in the frequency selective portion of the circuit.

The particular version of this circuit shown in Fig. 19-9A is connected in the unity gain configuration. The purpose of *R*3 is to help counteract the DC offset at the output of the operational amplifier which is created by input bias currents charging the capacitors in the frequency selective network. The value of *R*3 in the unity gain case is 2*R*, where *R* is the value of the resistors in the frequency selective network. In cases where DC offset is not a problem, resistor *R*3 can be replaced with a short circuit between the op-amp output and the inverting input. If passband gain is required, resistors *R*3 and *R*4 are used.

The second-order VCVS filter is the most commonly used type. Its −40 dB/decade roll-off, coupled with a high degree of stability, results in a generally good trade-off between performance and complexity.

The general form of the second-order filter transfer equation is similar to the expression for the first-order filter

$$A_{dB} = 20\text{LOG}(A_v) - 20\text{LOG}[(\omega_o)^4 + (a^2 - 2)(\omega_o)^2 + 1]^{1/2} \tag{19-20}$$

where $a$ = the *damping factor* of the circuit and other terms are as defined earlier for the first-order case.

The damping factor term ($a$) is determined by the form of the filter circuit. For the Butterworth design, which is used in most of the examples of filters in this chapter, the value of $a$ is $[2]^{1/2}$, or 1.414.

The passband gain for this circuit is the normal gain for any noninverting follower/amplifier. If the output is strapped directly to the inverting input, or if *R*3 (but not *R*4) is used in the feedback network, the gain is unity ($A_v = +1$). For gains greater than unity ($A_v > 1$), the following is true

$$A_v = \frac{R3}{R4} + 1 \tag{19-21}$$

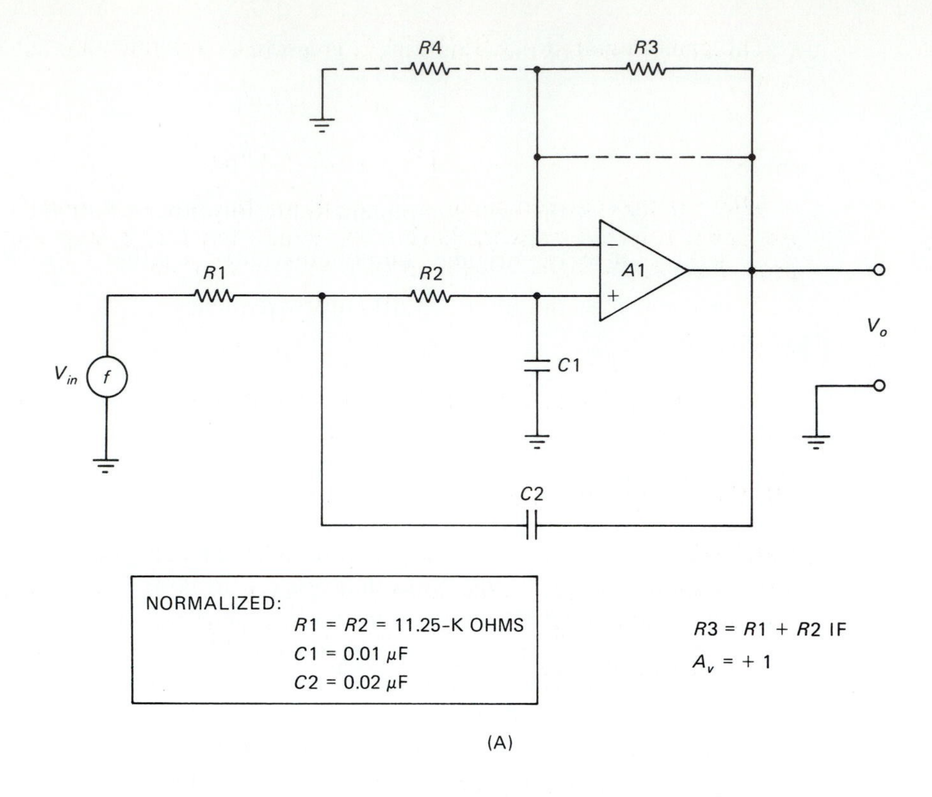


(A)

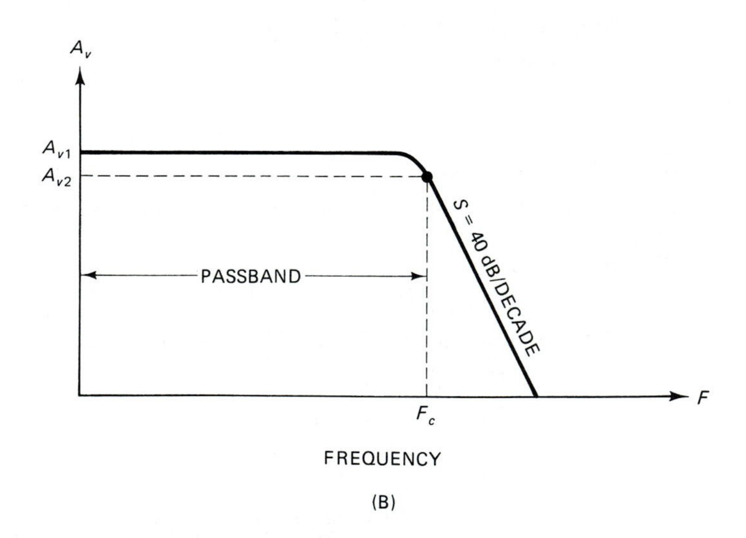


FIGURE 19-9 (B)

The cutoff frequency ($f_c$) is the frequency at which the voltage gain drops $-3$ dB from the passband gain [Eq. (19-21)]. This gain is found from

$$A_v = \frac{1}{2\pi\,[R1R2C1C2]^{1/2}} \tag{19-22}$$

The gain magnitude (ABS($V_o/V_{in}$)) is found in a manner similar to the first-order case

$$\left|\frac{V_o}{V_{in}}\right| = \frac{A_v}{[1 + (f/f_c)^4]^{1/2}} \tag{19-23}$$

There is no requirement in VCVS filters that like components ($R$ or $C$) in the frequency selective network have to be equal, but such a step simplifies the design procedure. If $R1 = R2 = R$, and $C1 = C2 = C$, then

$$f_c = \frac{1}{2\pi RC} \tag{19-24}$$

A constraint on this simplification is the Butterworth response is guaranteed only if $A_v \leq 1.586$.

### 19-3.5 Design Procedure

Procedure

1. Select the −3 dB cutoff frequency ($f_c$) from the circuit requirements and applications.
2. Select a standard value capacitance (30 pF $\leq C \leq$ 1 μF).
3. Calculate the required resistance from

$$R1 = \frac{1}{2\pi f_c C} \tag{19-25}$$

4. Select the passband gain for $f < f_c$.
5. Select a value for resistor $R4$, and
6. Calculate $R3$ from

$$R3 = R4(A_v - 1) \tag{19-26}$$

**EXAMPLE 19-4**

Design a 1 KHz second-order low-pass filter with a unity gain based on Fig. 19-9.

**Solution**

1. Select $f_c$, 1 KHz (given)
2. Select a trial value for $C1$ and $C2$, $C$ = 0.0056 μF
3. Calculate $R1 = R2 = \mathrm{R}$

$$R = \frac{1}{2\pi f_c C}$$

$$R = \frac{1}{(2)(3.14)(1000\ \text{Hz})(0.0056 \times 10^{-6}\ \text{farads})}$$

$$R = \frac{1}{3.52 \times 10^{-5}}\ \text{ohms} = 28{,}435\ \text{ohms}$$

4. Select $R3$:

$$R3 = 2R = (2)(28{,}435\ \text{ohms}) = 56{,}870\ \text{ohms}$$

■

The normalized 1 KHz trial values for doing scaling design of the second-order low-pass filter are shown in Fig. 19-9 as an inset. The design here is based on a more complex arrangement whereby $C2 = 2C1$. Some authorities maintain this is the superior design. The same scaling rule is applied to the second-order filter as was used in the first-order.

### EXAMPLE 19-5

Design a 3,000 Hz second-order low-pass filter with a gain of one using the scaling technique.

### Solution

$$R1' = \frac{R1 f_c}{f}$$

$$R1' = \frac{(11.25 \text{ kohms})(1000 \text{ Hz})}{3000 \text{ Hz}} = 3.75 \text{ kohms}$$

■

## 19-3.6 Third-Order Low-Pass (−60 dB/Decade) Filters

A third-order filter has a frequency roll-off slope of −60 dB/decade, or −18 dB/octave. There are two main forms of third-order filter. One type is similar to the first- and second-order filters, except an extra low-pass RC filter is in the frequency selective network. The other method is to cascade first- and second-order filters. Figure 19-10A shows the circuit for the former type, along with the response curve shown in Fig. 19-10B. This circuit is the normalized version. In past examples, filters were normalized using frequency $f_c$ as the determining factor, so in this example we will use the radian form in which frequency $\omega_c$ is the cutoff frequency in radians per second, and is equal to $2\pi f_c$. The filter circuit shown in Fig. 19-10A is normalized to one radian per second.

The values assigned to the capacitor values in Fig. 19-10A are parametric and are used in a two-step process for finding the actual final values $C1$, $C2$, and $C3$. The parameters are $C1' = 1.39$, $C2' = 3.55$, and $C3 = 0.20$ (for Butterworth filters).

### EXAMPLE 19-6

Design a −60 dB/octave low-pass filter for a frequency of 100 Hz. Use the scaling method, and the normalized values shown in Fig. 19-10A.

### Solution

1. Calculate $\omega_c$

$$\omega_c = 100 \text{ Hz} \times 6.28 = 628 \text{ radians/second}$$

2. State the parameter values

$$C1' = 1.39$$

$$C2' = 3.55$$

$$C3' = 0.20$$

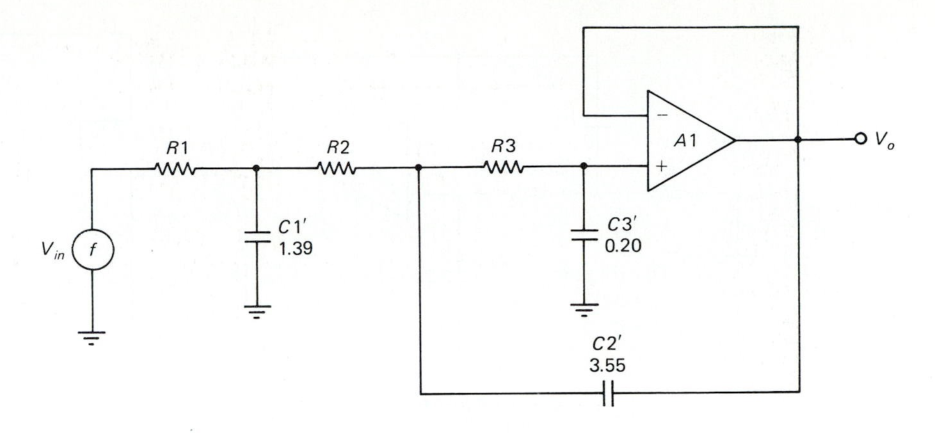


NORMALIZED TO 1 RADIAN/SECOND (1-Hz)
$R1 = R2 = R3 = R$

(A)

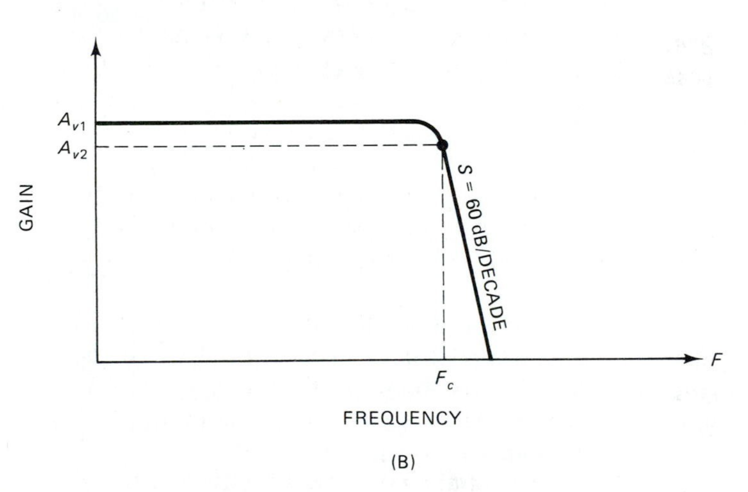


FREQUENCY

(B)

FIGURE 19-10

3. Perform frequency scaling:

$$C1a = \frac{C1'}{\omega_c} = \frac{1.39}{628} = 2.2 \times 10^{-3}$$

$$C2a = \frac{C2'}{\omega_c} = \frac{3.55}{628} = 5.65 \times 10^{-3}$$

$$C3a = \frac{C3'}{\omega_c} = \frac{0.20}{628} = 3.19 \times 10^{-4}$$

3. Select a trial value for $C2$, 0.1 μF ($1 \times 10^{-7}$ farads).

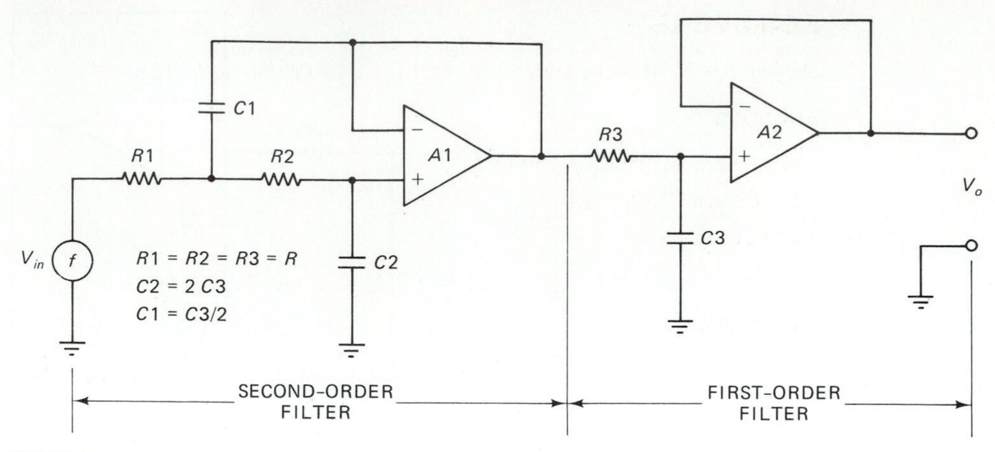


FIGURE 19-11

4. Solve for the value of $R1 = R2 = R3 = R$

$$R = \frac{C2a}{C2} = \frac{5.65 \times 10^{-3}}{1 \times 10^{-7}} = 56{,}500 \text{ ohms}$$

$$C1 = \frac{C1a}{R} = \frac{2.2 \times 10^{-3}}{56{,}500}$$

$$= 3.9 \times 10^{-8} = 0.0039\ \mu\text{F}$$

$$C3 = \frac{C3a}{R} = \frac{3.19 \times 10^{-4}}{56{,}500}$$

$$= 5.7 \times 10^{-9} \text{ farads} = 0.0057\ \mu\text{F}$$

■

A second method for designing a third-order low-pass filter is to cascade first-order and second-order filters of the same cutoff frequency (Fig. 19-11). The roll-off of the first-order filter is −20 dB/decade. For the second-order filter it is −40 dB/decade. The combined roll-off is −(20 + 40) = −60 dB/decade. The components in this circuit have the following relationships

$$R1 = R2 = R3 = R \tag{19-27}$$

$$R = \frac{1{,}000{,}000}{2\pi f_c C3_{\mu F}}$$

$C3$ is selected for convenient value

$$C1 = \frac{C3}{2} \tag{19-28}$$

$$C2 = 2 \times C3 \tag{19-29}$$

The design procedure calls for selecting a reasonable trial value for $C3$ and calculating $C1$ and $C2$. If these values are not standards, set another trial value of $C3$ and try again. Repeat the procedure until a good set of values is obtained. Finally, calculate $R$.

## EXAMPLE 19-7

Design a −60 dB/decade cascade filter for a frequency of 500 Hz.

### Solution

1. Select frequency $f_c$, 500 Hz
2. Select C3, 0.01 μF
3. Calculate C1

$$C1 = C3/2 = 0.01\ \mu F/2 = 0.005\ \mu F$$

4. Calculate C2

$$C2 = 2 \times C3 = (2)(0.01\ \mu F) = 0.02\ \mu F$$

5. Calculate $R1 = R2 = R3 = R$

$$R = \frac{1{,}000{,}000}{2\pi f_c C3_{\mu F}}$$

$$R = \frac{1{,}000{,}000}{(2)(3.14)(500\ Hz)(0.01\ \mu F)}$$

$$R = \frac{1{,}000{,}000}{31.4} = 31{,}850\ \text{ohms}$$

■

## 19-4 HIGH-PASS FILTERS

The *high-pass filter* is the inverse of the low-pass filter, so one can reasonably expect its frequency response characteristic to mirror that of the low-pass filter. Figure 19-12 shows the basic high-pass filter response with roll-off slopes of −20, −40, and −60

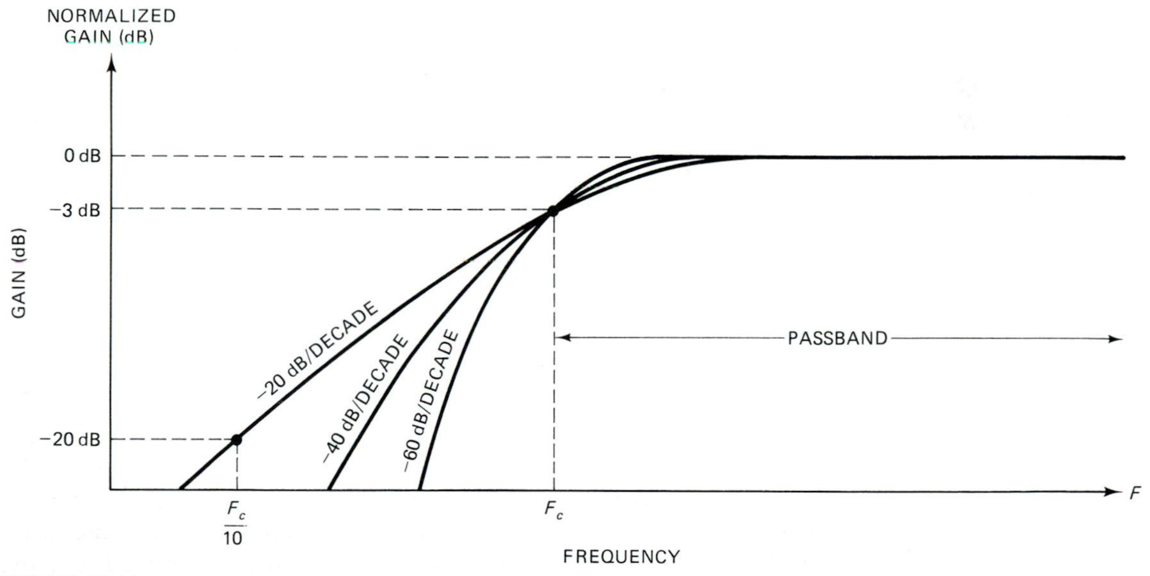


FIGURE 19-12

dB/decade. The passband of the high-pass filter are all frequencies above the cutoff frequency $f_c$. As in the low-pass case, $f_c$ is the frequency at which passband gain drops $-3$ dB, $A_{vc} = 0.707A_v$.

The cutoff frequency phase shift in a high-pass filter has the same magnitude as the low-pass case, but the sign is opposite. At $f_c$, the high-pass filter exhibits a phase shift of $+45°$ per 20 dB/decade of roll-off. Put another way, the phase shift is $(n \times 45°)$, where $n$ is the order of the filter.

The high-pass versions of the VCVS filters are of the same form as the low-pass filter (Fig. 19-4). In the case of the high-pass filter, however, impedances $Z1$ through $Z3$ are capacitances, while $Z4$ through $Z6$ are resistances. In the high-pass filter the roles of the resistors and capacitors are reversed.

### 19-4.1 First-Order High-Pass VCVS Filters (−20 dB/Decade)

The circuit for a first-order high-pass filter is shown in Fig. 19-13. This circuit is identical to the first-order low-pass filter in which the roles of $C1$ and $R1$ are interchanged. The filter shown here is normalized for 1 KHz. Passband gain of this circuit is

$$A_v = \frac{R_f}{R_{in}} + 1 \tag{19-30}$$

The voltage at the noninverting input of the operational amplifier ($V_a$) is developed across resistor $R1$, and is given by

$$V_a = \frac{j2\pi f R1C1V_{in}}{1 + j2\pi f R1C1} \tag{19-31}$$

The transfer equation for the circuit is

$$V_o = A_vV_a \tag{19-32}$$

$$V_o = \left[\frac{R_f}{R_{in}} + 1\right]\left[\frac{j2\pi f R1C1V_{\text{in}}}{1 + j2\pi f R1C1}\right] \tag{19-33}$$

In the traditional form, the equation becomes

$$\frac{V_o}{V_{in}} = \frac{A_v j(f/f_c)}{1 + j(f/f_c)} \tag{19-34}$$

As in the previous cases, the cutoff frequency $f_c$ is found from

$$f_c = \frac{1}{2\pi R1C1} \tag{19-35}$$

The gain magnitude of this circuit is the absolute value of the traditional form of the transfer equation:

$$\left|\frac{V_o}{V_{in}}\right| = \frac{A_v(f/f_c)}{[1 + (f/f_c)^2]^{1/2}} \tag{19-36}$$

The VCVS high-pass filter shown in Fig. 19-12 is normalized to 1 KHz. The same scaling technique is used for this circuit as was used for the low-pass filters.

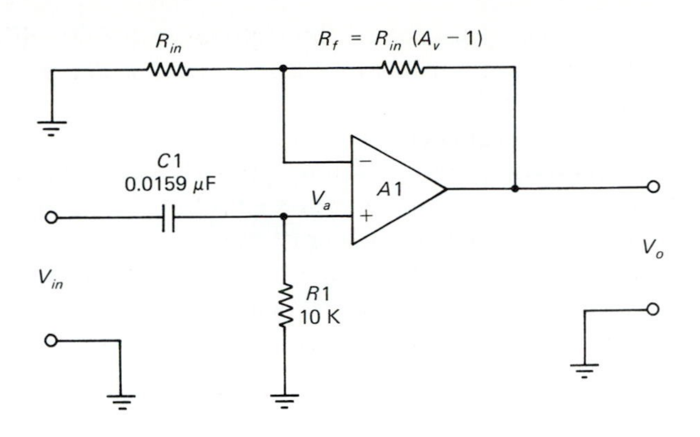


FIGURE 19-13

### 19-4.2 Second-Order High-Pass Filter (−40 dB/Decade)

The second-order high-pass filter has a roll-off slope of −40 dB/decade. This VCVS filter circuit is, like its low-pass counterpart, probably the most commonly used form of high-pass filter. The circuit is similar to the low-pass design except for a reversal of the roles of capacitors and resistors. The cutoff frequency is the frequency at which gain falls off −3 dB and is found from:

$$f_c = \frac{1}{2\pi[R1R2C1C2]^{1/2}} \tag{19-37}$$

or, in the case where $R1 = R2 = R$, and $C1 = C2 = C$

$$f_c = \frac{1}{2\pi RC} \tag{19-38}$$

The gain magnitude of the circuit is found from

$$\left|\frac{V_o}{V_{in}}\right| = \frac{A_v}{[1 + (f_c/f)^4]^{1/2}} \tag{19-39}$$

**EXAMPLE 19-8**

Calculate the cutoff frequency of a filter such as Fig. 19-14 in which $C1 = C2 = 0.0056$ μF and $R1 = R2 = 22$ kohms.

**Solution**

$$f_c = \frac{1}{2\pi RC}$$

$$f_c = \frac{1}{(2)(3.14)(22{,}000 \text{ ohms})(0.0056 \times 10^{-6} \text{ farads})}$$

$$f_c = \frac{1}{7.74 \times 10^{-4}} = 1{,}293 \text{ Hz}$$

■

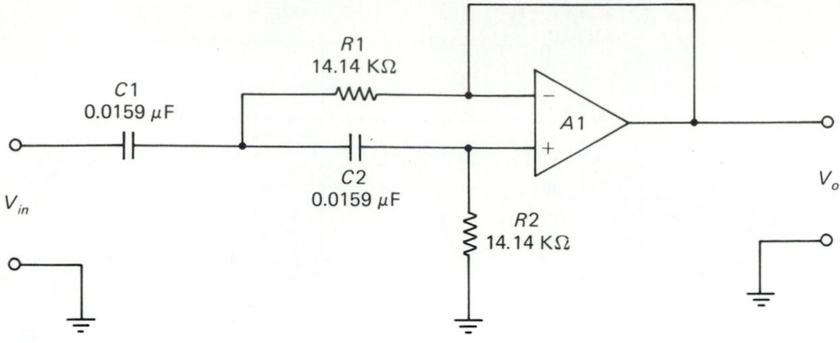


FIGURE 19-14

### 19-4.3 Multiple Feedback Path Filters

The Multiple Feedback Path (MFP) circuit shown in Fig. 19-15 is another form of active filter. The low-pass version is shown in Fig. 19-15A and the high-pass in Fig. 19-15B. The values are normalized for 1 KHz. The actual values are found in the manner described above. Change either the capacitor or the resistor values, but not both.

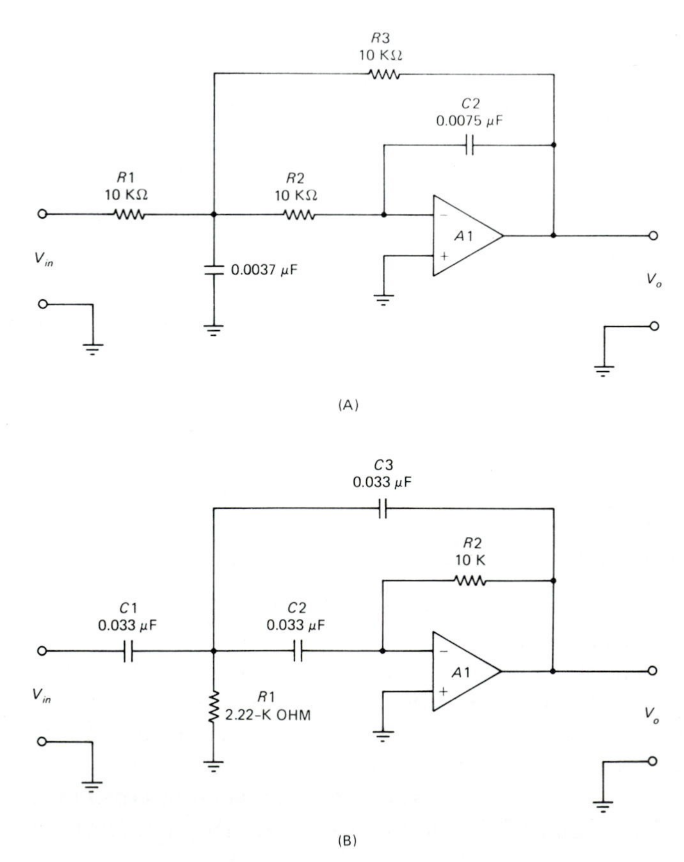


FIGURE 19-15

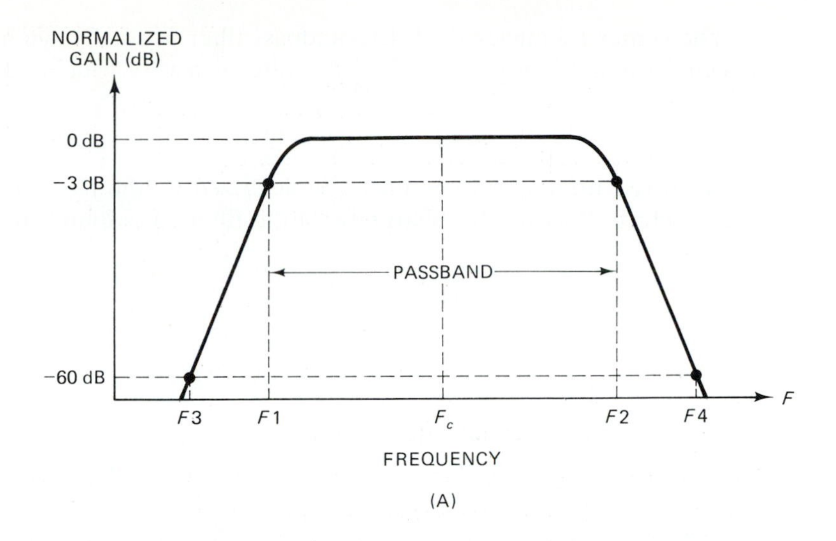


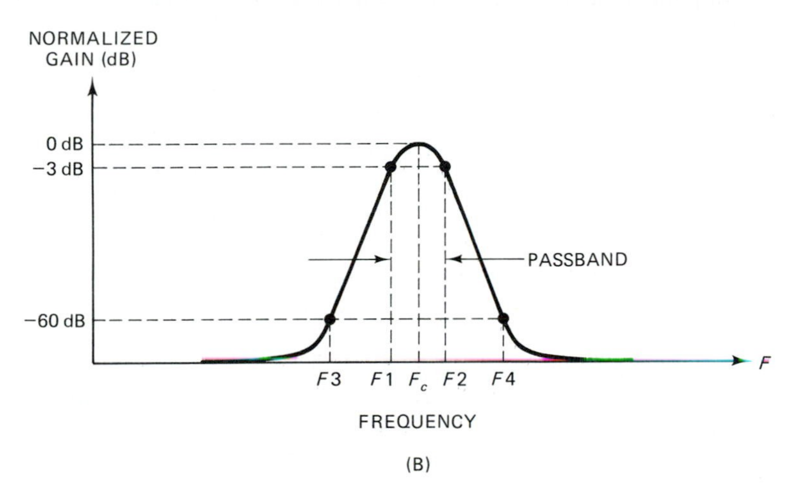


FIGURE 19-16

## 19-5 BANDPASS FILTERS

The bandpass filter has a passband between an upper limit and a lower limit. Frequencies above and below these limits are in the stopband. There are two basic forms of bandpass filter, *wide band pass* and *narrow band pass*. These two types are sufficiently different to offer different responses. The wide band pass filter may have a passband wide enough to be called a bandpass amplifier rather than a filter. The wide band pass filter response is shown in Fig. 19-16A. The narrow response is shown in Fig. 19-16B. The passband is defined as the frequency difference between the upper $-3$ dB point ($F2$), and the lower $-3$ dB point ($F1$). The bandwidth $BW$ is

$$BW = F2 - F1 \tag{19-40}$$

The center frequency $f_c$ of the bandpass filter is usually symmetrically placed between $F1$ and $F2$ or $(F2\text{-}F1)/2$. If the filter is a very wideband type, however, the center frequency is

$$f_c = [F1F2]^{1/2} \tag{19-41}$$

Bandpass filters are sometimes characterized by the *figure of merit* or $Q$. $Q$ is a factor which describes the sharpness of the filter. It is found from

$$Q = \frac{f_c}{BW} \tag{19-42}$$

$$Q = \frac{f_c}{F2 - F1} \tag{19-43}$$

The $Q$ of the filter tells us something of the pass band characteristic. Wideband filters generally have a $Q < 10$. Narrow band filters have a $Q > 10$.

### EXAMPLE 19-9

Determine the $Q$ of a filter which has a center frequency of 2,000 Hz and $-3$ dB frequencies of 1,800 Hz and 2,200 Hz.

#### Solution

$$Q = \frac{f_c}{F2 - F1}$$

$$Q = \frac{2{,}000 \text{ Hz}}{(2200 \text{ Hz} - 1800 \text{ Hz})}$$

$$Q = \frac{2{,}000 \text{ Hz}}{400 \text{ Hz}} = 5$$

■

The filter in the above example is clearly a wideband type. It has a bandpass of 400 Hz and a $Q$ of 5. Now compare the $Q$ of the narrowband filter in the next example.

### EXAMPLE 19-10

Determine the $Q$ of a filter which has a center frequency of 2,000 Hz, and $-3$ dB frequencies of 1,975 Hz and 2,125 Hz.

#### Solution

$$Q = \frac{f_c}{F2 - F1}$$

$$Q = \frac{2{,}000 \text{ Hz}}{(2125 \text{ Hz} - 1975 \text{ Hz})}$$

$$Q = \frac{2{,}000 \text{ Hz}}{50 \text{ Hz}} = 40$$

■

The *shape factor* of the filter characterizes the slope of the roll-off curve so it is related to the order of the filter. The shape factor is defined as the ratio of the $-60$ dB bandwidth to the $-3$ dB bandwidth:

$$S.F. = \frac{BW_{-60\text{dB}}}{BW_{-3\text{dB}}} \tag{19-44}$$

### 19-5.1 First-Order VCVS Bandpass Filters

A wide band first-order bandpass filter response is obtained by cascading first-order high-pass and low-pass filter circuits, as shown in Fig. 19-17A. This arrangement overlays or superimposes the frequency response characteristics of both filter stages. Figure 19-17B shows a situation when cascade high and low-pass filters are used. The low-pass filter response (solid line) is from DC to the $-3$ dB point at $F2$. The high-pass filter response is from the highest possible frequency within the range of the circuit down to the $-3$ dB point at $F1$. The passband is the intersection of the two sets, high and low-pass characteristics, which falls between $F1$ and $F2$.

The gain of the overall bandpass filter within the pass band is the product of the two individual gains, $A_{vt} = A_{VL} \times A_{VH}$. The gain magnitude term of this form of filter is found by

$$\left|\frac{V_o}{V_{in}}\right| = \frac{A_{vt}(f/F1)}{[1 + (f/F1)]^2[1 + (f/F2)^2]^{1/2}} \tag{19-45}$$

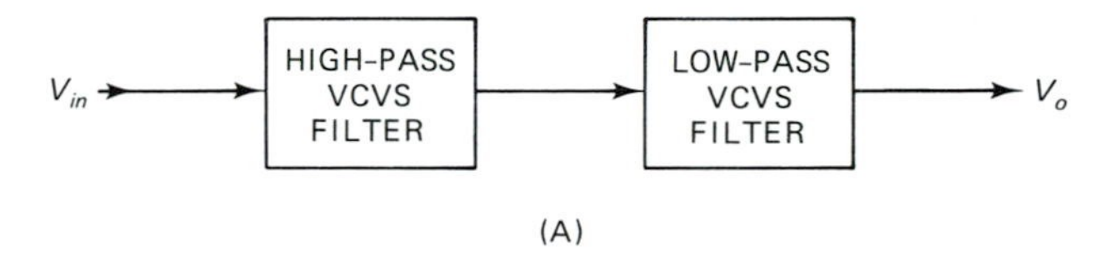


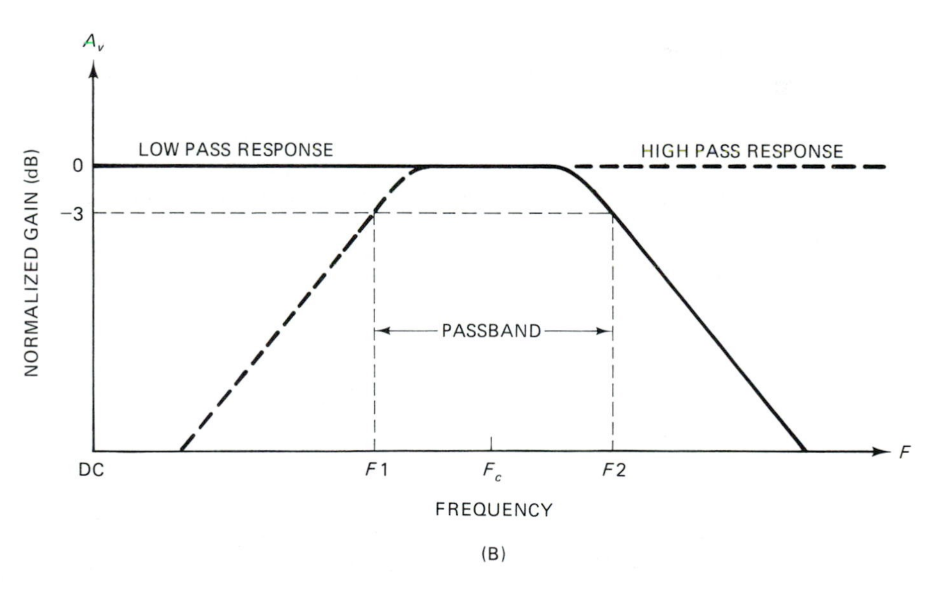


FIGURE 19-17

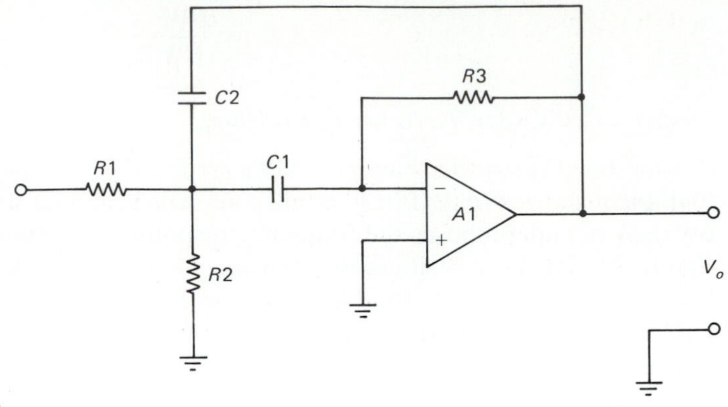


FIGURE 19-18

where

$$V_o = \text{the output signal voltage}$$

$$V_{in} = \text{the input signal voltage}$$

$$f = \text{the applied frequency}$$

$$F1 = \text{the lower } -3 \text{ dB point frequency}$$

$$F2 = \text{the upper } -3 \text{ dB point frequency}$$

$$A_{vt} = \text{the total cascade gain of the filter}$$

Cascading low- and high-pass filter sections can be used to make wideband filters, but because of component tolerance and problems it becomes less useful as $Q$ increases above about 10. For narrow band filters a Multiple Feedback Path (MFP) filter circuit like Fig. 19-18 can be used. This filter circuit offers first-order performance and relatively narrow bandpass. The circuit will work for values of $10 \leq Q \leq 20$, and gains up to about 15. The center frequency of the MFP bandpass filter is

$$f_c = \frac{1}{2\pi}\left[\frac{1}{R3C1C2}\left(\frac{1}{R1} + \frac{1}{R2}\right)\right]^{1/2} \tag{19-46}$$

In order to calculate the resistor values, it is necessary to first set the passband gain ($A_v$) and the $Q$. It is the general practice to select values for $C1$ and $C2$, and calculate the required resistances for the specified values of $f_c$, $A_v$, and $Q$. The resistor values are

$$R1 = \frac{1}{2\pi A_v C2 f_c} \tag{19-47}$$

$$R2 = \frac{1}{2\pi f_c(2Q^2 - A_v)C2} \tag{19-48}$$

$$R3 = \frac{2Q}{2\pi f_c C2} \tag{19-49}$$

and the gain

$$A_v = \frac{R3}{R1\left[1 + \frac{C2}{C1}\right]} \tag{19-50}$$

These equations can be simplified if the two capacitors are made equal ($C1 = C2 = C$), and assuming $Q > (A_v/2)^{1/2}$

$$R1 = \frac{Q}{2\pi f_c A_v C} \tag{19-51}$$

$$R2 = \frac{Q}{2\pi f_c C(2Q^2 - A_v)} \tag{19-52}$$

$$R3 = \frac{2Q}{2\pi f_c C} \tag{19-53}$$

$$A_v = \frac{R3}{2\,R1} \tag{19-54}$$

**EXAMPLE 19-11**

Design an MFP bandpass filter with a gain of 5 and a $Q$ of 15 when the center frequency is 2,200 Hz. Assume $C1 = C2 = 0.01\ \mu F$

**Solution**

$$R1 = \frac{Q}{2\pi f_c A_v C}$$

$$R1 = \frac{15}{(2)(3.14)(2200\text{ Hz})(5)(0.01 \times 10^{-6}\text{ farads})}$$

$$R1 = \frac{15}{0.00069} = 217{,}140\text{ ohms}$$

$$R2 = \frac{Q}{2\pi f_c C(2Q^2 - A_v)}$$

$$R2 = \frac{15}{(2)(3.14)(2200\text{ Hz})(0.01 \times 10^{-6}(2(15)^2 - (5))}$$

$$R2 = \frac{15}{0.062} = 240\text{ ohms}$$

$$R3 = \frac{2Q}{2\pi f_c C}$$

$$R3 = \frac{(2)(15)}{(2)(3.14)(2200\text{ Hz})(0.01 \times 10^{-6}\text{ farads})}$$

$$R3 = \frac{30}{0.000138} = 217{,}140\text{ ohms}$$

■

The MFP bandpass filter can be tuned using only one of the resistors. If $R2$ is varied, the center frequency will shift, but the bandwidth, $Q$, and gain will remain constant. To scale the circuit to a new center frequency using only $R2$ as the change element, select a new value of $R2$ according to

$$R2' = R2\left[\frac{f_c}{f_{c'}}\right]^2 \tag{19-55}$$

### EXAMPLE 19-12

Calculate the new value of $R2$ which will force the MFP filter in the previous example from 2,200 Hz to 1,275 Hz.

**Solution**

$$R2' = R2\left[\frac{f_c}{f_{c'}}\right]^2$$

$$R2' = (240 \text{ ohms})\left[\frac{2200 \text{ Hz}}{1275 \text{ Hz}}\right]^2$$

$$R2' = (240 \text{ ohms})(1.726) = 414 \text{ ohms}$$

■

## 19-6 BAND REJECT (NOTCH) FILTERS

A *band reject* or *notch* filter is used to pass all frequencies except a single frequency (or small band of frequencies). An application for this circuit is to remove 60 Hz interference from electronic instruments. The medical electrocardiograph (ECG) machine, for example, often has 60 Hz interference because the input leads are unshielded at the tips. These machines often include a switch selectable 60 Hz notch filter to remove the 60 Hz artifact which could result from interfering local electrical fields.

Figure 19-19A shows a typical active notch filter. Figure 19-19B shows the frequency response for the circuit. Note the gain is constant throughout the frequency spectrum except in the immediate vicinity of $f_c$. The depth of the notch is infinite in theory, but in practical circuits precision matched components will offer $-60$ dB of suppression. Ordinary bench-run components (not precision) can offer $-40$ to $-50$ dB of suppression. The resonant frequency of this notch filter is found from

$$f_c = \frac{1}{2\pi RC} \tag{19-56}$$

The gain of the circuit is unity, but the $Q$ can be set according to the following equations

$$Q = \frac{R_a}{2R} \tag{19-57}$$

or

$$Q = \frac{C}{C_a} \tag{19-58}$$

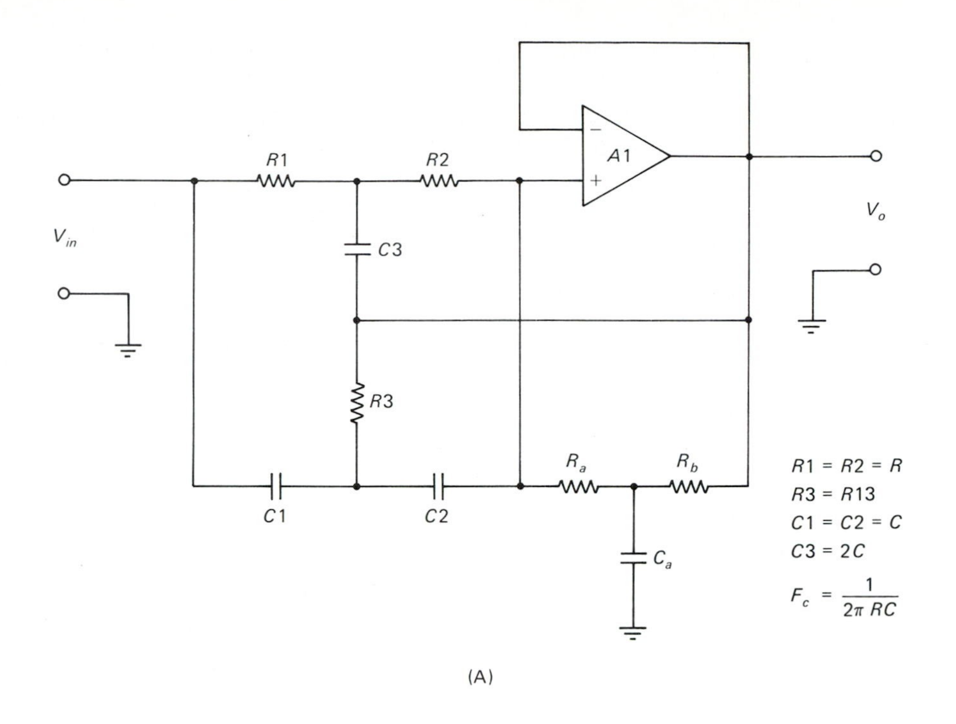


(A)

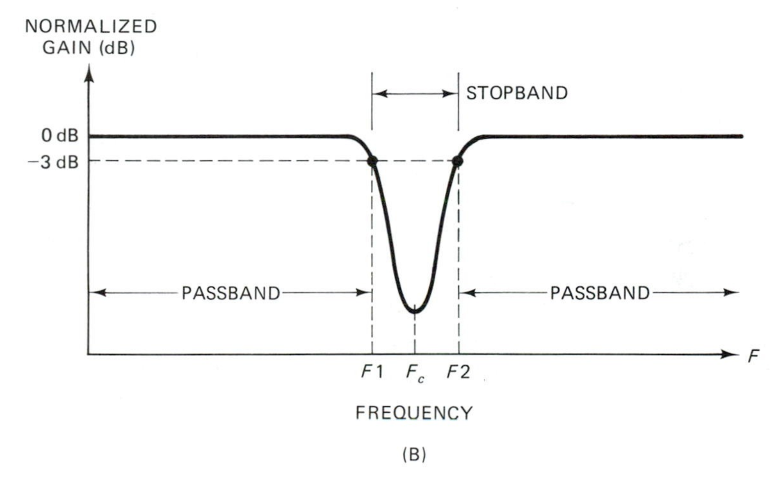


(B)

FIGURE 19-19

## EXAMPLE 19-13

Design a notch filter with a $Q$ of 8 for a frequency of 60 Hz.

### Solution

1. Select a trial value of $C1 = C2 = C$, 0.01 μF
2. Calculate the value of $R1 = R2 = R$

$$R = \frac{1}{2\pi f C}$$

$$R = \frac{1}{(2)(3.14)(60 \text{ Hz})(0.01 \times 10^{-6} \text{ farads})}$$

$$R = 1/3.768 \times 10^{-6} = 265{,}392 \text{ ohms}$$

3. Select $R3$

$$R3 = R/2$$

$$= \frac{265{,}392 \text{ ohms}}{2} = 132{,}696 \text{ ohms}$$

4. Select $C3$

$$C3 = 2\,C$$

$$C3 = (2)(0.01\ \mu\text{F}) = 0.02\ \mu\text{F}$$

5. Select $R_a$

$$R_a = 2QR$$

$$R_a = (2)(8)(265{,}392) = 4.24 \text{ megohms}$$

6. Select $C_a$

$$C_a = C/Q$$

$$C_a = 0.01\ \mu\text{F}/8 = 0.0013\ \mu\text{F}$$ ■

A bandstop filter is a notch filter with a wide stopband. Just as the wide bandpass filter can be made by cascading high- and low-pass filter, the wide band notch (or stopband) filter can be made by placing high- and low-pass filter sections in parallel. Figure 19-20 shows a band stop filter in which the outputs of a high-pass filter and a low-pass filter are summed together in a two-input unity gain inverting follower am-

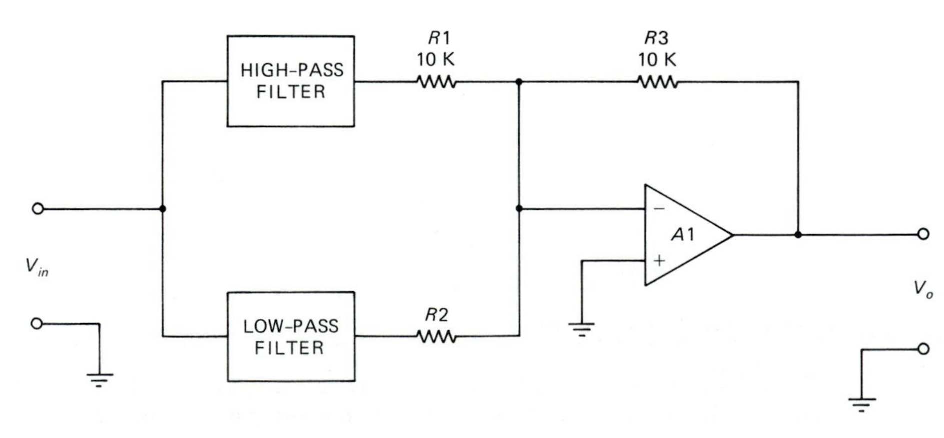


FIGURE 19-20

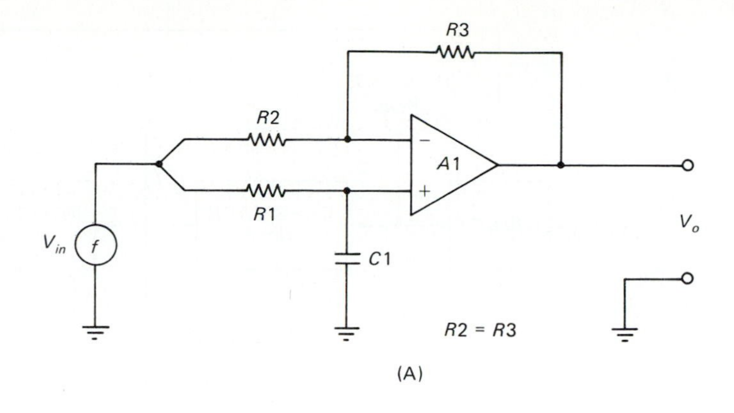


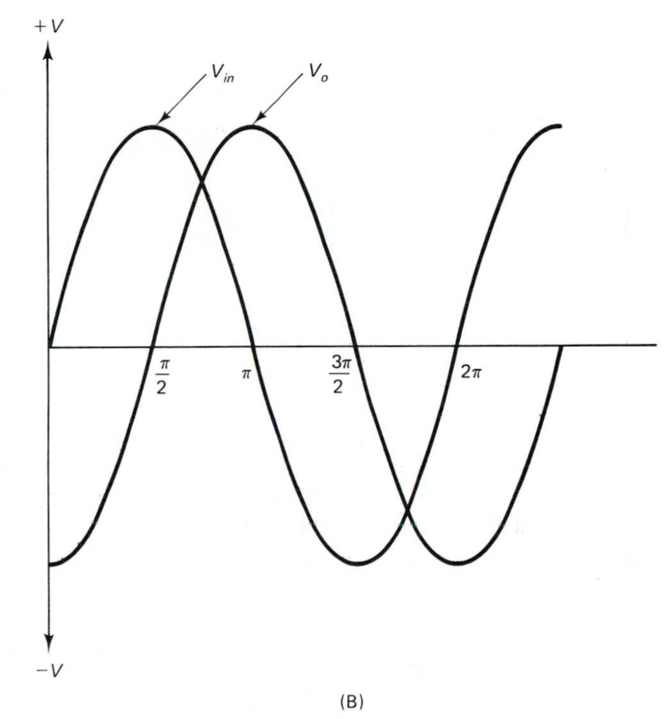


FIGURE 19-21

plifier circuit. The frequency response curves of the two filter sections are superimposed to eliminate the undesired band. Make the $-3$ dB point of the high-pass filter equal to the upper end of the stopband, and the $-3$ dB point of the low-pass filter equal to the lower end of the stopband.

## 19-7 ALL-PASS PHASE-SHIFT FILTERS

The *all-pass phase-shifter* (APPS) is a special category of filter in which all frequencies (within the ability of the op-amp) are passed, but are shifted in phase a specified amount. Figure 19-21A shows the circuit for an APPS that will exhibit phase shift

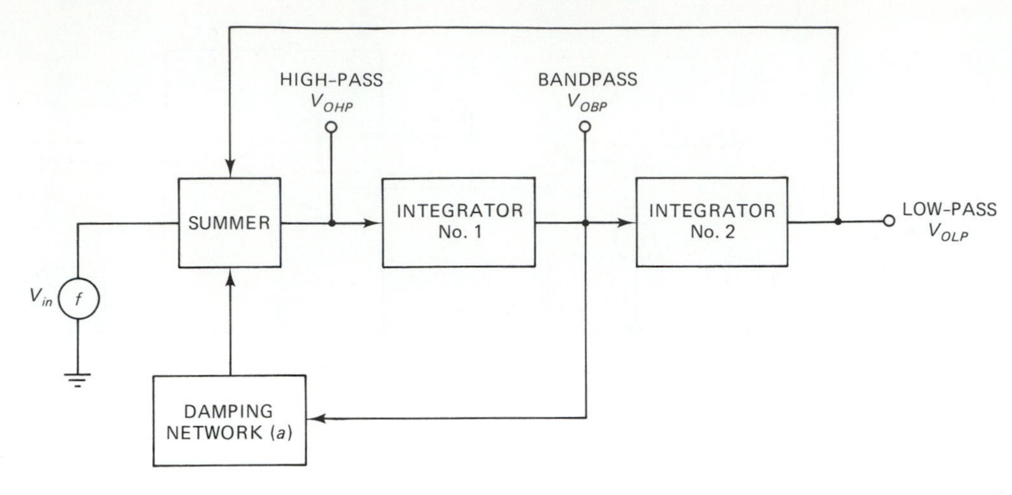


FIGURE 19-22

between input and output of −180° to 0°. If the roles of $R1$ and $C1$ are reversed, the phase shift will be −360° to −180°. The gain response of this circuit is

$$A_v = \frac{1 - 2\pi f R1C1}{1 + 2\pi f R1C1} \tag{19-59}$$

and the amount of phase shift

$$\Delta\phi = -2\ \mathrm{TAN}^{-1}(2\pi f R1C1) \tag{19-60}$$

## 19-8 STATE-VARIABLE ANALOG FILTERS

The *state-variable filter* is a variation on the multiple feedback path design using three operational amplifiers. The circuit is more complex than the other circuits presented in this chapter but it is also more versatile. The state variable filter is capable of *simultaneously* providing bandpass, low-pass and high-pass responses. Figure 19-22 shows the block diagram of the state variable filter. The constituent parts include a summing amplifier, two integrators, and a damping network. The integrators are inverting Miller integrators (Chapter 11). The summing amplifier is also a simple op-amp summer, and the damping network is a simple resistor voltage divider.

The damping factor ($a$) is the same for all three outputs. It is defined as the reciprocal of $Q$

$$a = \frac{1}{Q} \tag{19-61}$$

The state variable filter shown in Fig. 19-23 is normalized for 1 KHz, $C$ = 0.0159 μF and $R$ = 10 kohms. Designing for other frequencies is a matter of scaling these values to the new values required for the new frequency. As in the previous examples, change either $R$ or $C$, but not both. The value of components for the damping network are found from

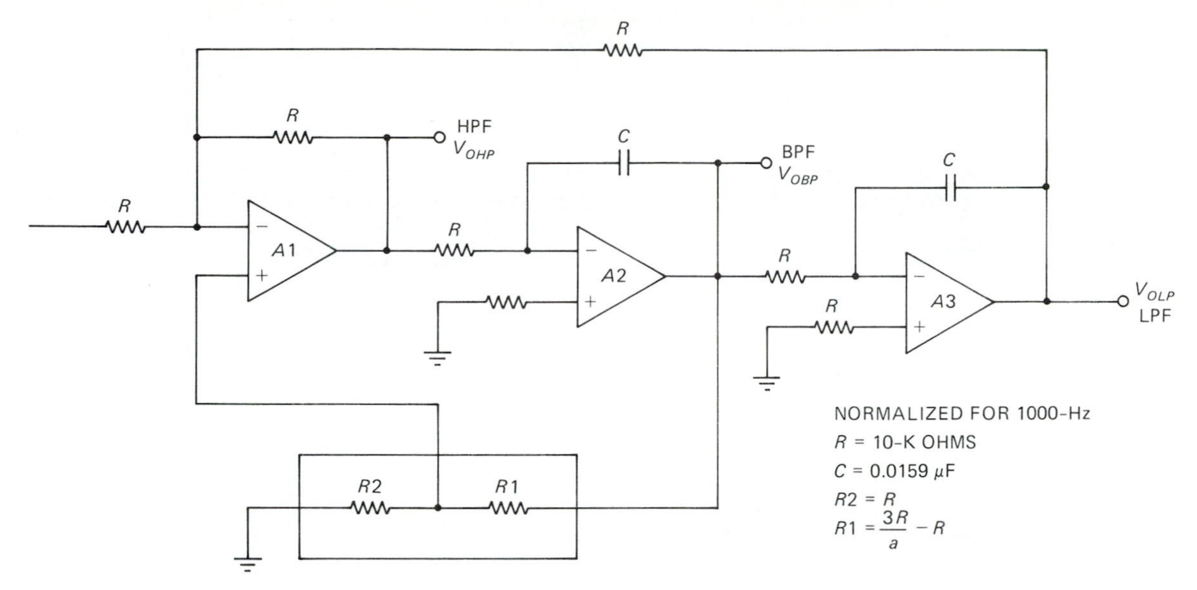


FIGURE 19-23

$$R1 = \frac{3R}{a} - R \tag{19-62}$$

and

$$R2 = R \tag{19-63}$$

## 19-9 VOLTAGE-TUNABLE FILTERS

A voltage tunable filter is one in which the −3 dB response point is a function of an input control voltage. The analog multiplier and divider can be used to make a simple voltage tunable filter in either low-pass (Fig. 19-24A), high-pass (Fig. 19-24B), or bandpass (Fig. 19-24C) versions.

The low-pass filter is shown in Fig. 19-24A. This circuit consists of a Miller integrator (Chapter 11) with an analog multiplier in a second negative feedback path through $R2$. The $X$ input of the multiplier is the output of the integrator, while the $Y$ input is the control voltage ($V1$) which tunes the filter. The output voltage of this circuit is

$$V_o = \frac{-V_{in}}{1 + \dfrac{2\pi f R1C1}{V1}} \tag{19-64}$$

The high-pass filter circuit is shown in Fig. 19-24B. This circuit consists of an analog divider driving an $RC$ differentiator ($R1C1$). The output of the differentiator is summed with the input signal. For $R2 = R3 = R$, $C1 = C$, and $R1 < R$, the following transfer equation obtains

$$V_o = -V_{in}\left[1 + \frac{2\pi f RC}{V1}\right] \tag{19-65}$$

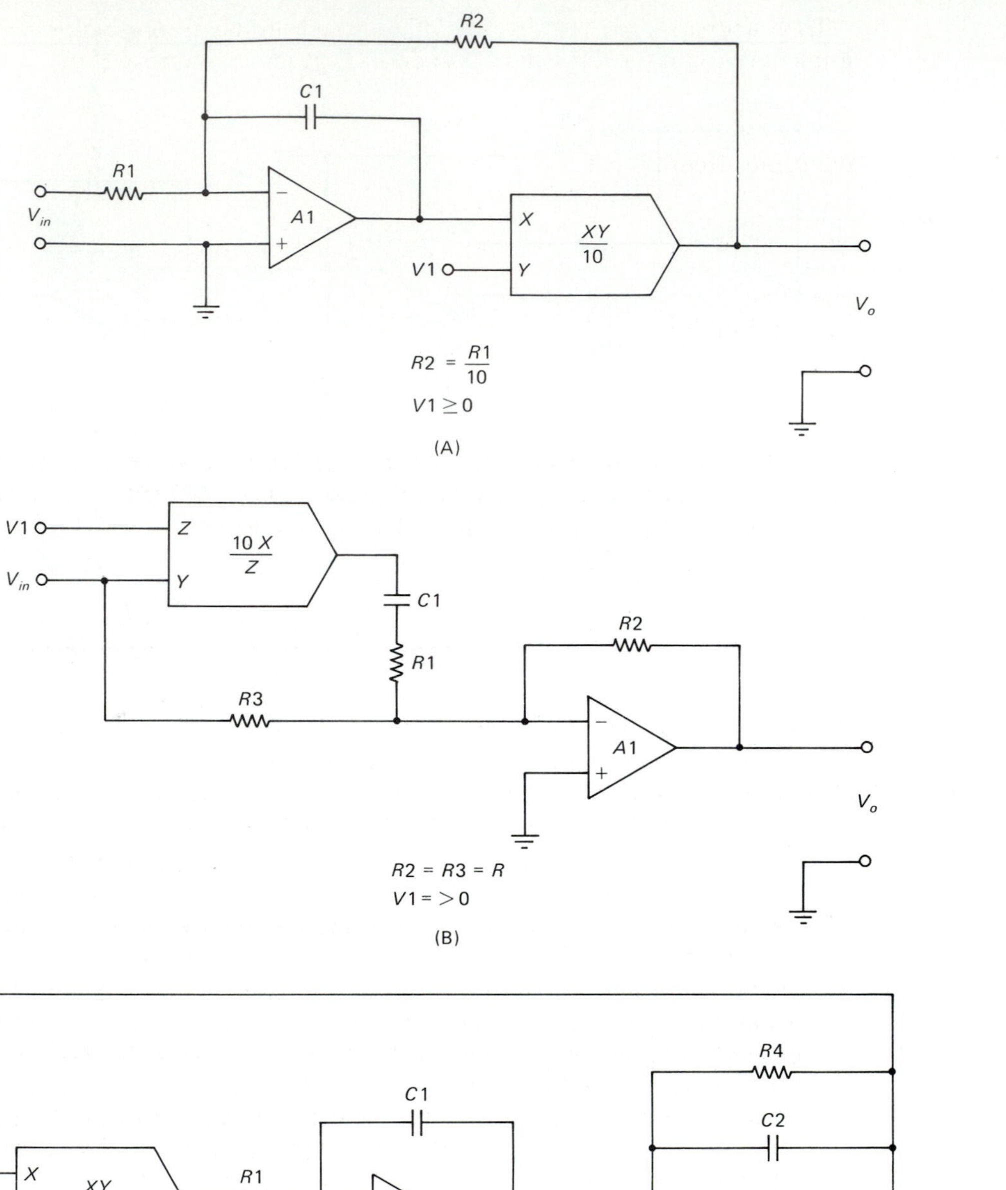
R2
C1
R1
V_in
A1
X
XY/10
V1
Y
V_o
R2 = R1/10
V1 ≥ 0
(A)
V1
Z
10 X/Z
V_in
Y
C1
R2
R1
R3
A1
V_o
R2 = R3 = R
V1 = > 0
(B)
R4
C1
C2
X
XY/10
R1
R2
V1
Y
A1
A2
R3
V_o
R1 = R2 = R
R3 = R4 = 10R
V1 ≤ 0
V_in
(C)

FIGURE 19-24

The bandpass circuit of Fig. 19-24C is based on the analog multiplier and a pair of integrator circuits. This circuit produces a gain of $-1$ and a $Q$ of

$$Q = [-10V1]^{1/2} \tag{19-66}$$

The center frequency is

$$f_c = \frac{1}{2\pi RC}\left[\frac{-V1}{10}\right]^{1/2} \tag{19-67}$$

(assuming a control voltage $V1 \leq 0$)
The bandwidth is

$$B.W. = \frac{1}{20\pi RC} \tag{19-68}$$

All three of these multiplier-of-divider based circuits are dependent on the properties of the device used. The multiplier gain, error, linearity, and response time determine the properties of the filter and the tuning rate.

## 19-10 SUMMARY

1. A filter circuit will pass some frequencies (called the passband), and either reject or severely attenuate other frequencies (called the stopband).
2. Passive filters are made from some combination of $R$, $L$, and $C$ components. Active filter may use these same components, but also add an active device such as an amplifier.
3. Filters can be categorized according to passband, *low-pass*, *high-pass*, *bandpass*, *stopband* (or *notch*), and *All-Pass Phase-Shifter.*
4. Another method of categorizing filters is according to passband shape. Common forms include: *Butterworth*, *Chebyshev*, *Cauer* (or *elliptic*), and *Bessel.*
5. The transition from passband to stopband is not abrupt, but falls off gradually. The slope of the roll-off is measured in either dB/octave (2:1 frequency change) or dB/decade (10:1 frequency change). Typical roll-off rates: first-order = $-20$ dB/decade, second-order = $-40$ dB/decade, and third-order = $-60$ dB/decade.
6. Filters can be designed by either the ground-up method or by the frequency scaling method. In the latter, a filter is first designed using the ground-up method for a standard frequency such as 1 Hz or 1000 Hz, and then simple ratios ($f_c/f$) are used to scale either the resistor or capacitor values.

## 19-11 RECAPITULATION

Now return to the objectives and Prequiz questions at the beginning of the chapter and see how well you can answer them. If you cannot answer certain questions, mark them each and review the appropriate parts of the text. Next, try to answer the questions and work the problems below, using the same procedure.

## 19-12 STUDENT EXERCISES

*In each of the following exercises you should measure the gain and the frequency response characteristics. Plot your work on graph paper, or tabulate the results if no graph paper is available.*

1. Design, build, and test a second-order low-pass filter circuit based on the Sallen-Key (VCVS) circuit (Fig. 19-9A). Use frequency scaling to make the −3 dB cutoff frequency of 500 Hz. Measure the response for at least 25 points from 10 Hz to 10 KHz.
2. Design, build, and test a high-pass filter of the same specifications as in the previous exercise.
3. Design, build, and test a notch filter with a $Q$ of 5 and a rejection frequency of 100 Hz.
4. Design, build, and test a narrow bandpass filter with a $Q$ of 15 for a center frequency of 1500 Hz.
5. Design, build, and test an all-pass phase-shifter for 1000 Hz which offers a phase change of approximately 80°. How may this circuit be made variable?

## 19-13 QUESTIONS AND PROBLEMS

1. A filter circuit has a response which has a gain of 6 dB +/1 0.25 dB from DC to 500 Hz, where the gain drops rapidly to 3 dB. The gain then rolls off to −23 dB at 5 KHz. This is a ____________-order, ____________ -pass filter.
2. A low-pass filter has a −40 dB/decade roll-off from a −3 dB frequency of 1000 Hz. If the output level is 6 volts peak-to-peak at 1000 Hz, what is the output voltage at 10 KHz?
3. A ____________ -pass filter passes only those frequencies below the cut-off frequency.
4. A cutoff slope of −12 dB/octave represents ____________ dB/decade.
5. A cutoff slope of −60 dB/decade represents ____________ dB/octave.
6. A third-order Bessel filter has a phase shift maxima of ____________ .
7. A frequency of 1500 Hz is applied to a first-order low-pass filter which has a −3 dB cutoff frequency ($f_c$) of 1000 Hz. Calculate response gain ($A_{dB}$) at this frequency if the mid-passband voltage gain is 15.
8. For the filter in the previous problem, calculate the gain magnitude and phase shift.
9. A first-order low-pass VCVS filter is built with the following component values, $R1$ = 15 kohms, $R2$ = 2.7 kohms, and $R3$ = 1.8 kohms, $C1$ = 0.015 μF. Characterize this filter if a 1200-Hz sine wave signal is applied.
10. A low-pass filter is needed for a biomedical transducer. The cutoff frequency should be 200 Hz and the gain should be 10.
11. Change the frequency of the normalized 1 KHz first-order low-pass filter in Fig. 19-8 to a new frequency of 200 Hz (a) by changing $R1$ and (b) by changing $C1$.
12. A second-order VCVS low-pass filter has the following component values, $R1$ = $R2$ = 10 kohms and $C1$ = $C2$ = 0.01 μF. Calculate the −3 dB frequency ($f_c$).
13. Design a 500 Hz second-order low-pass filter with unity gain based on the circuit of Fig. 19-9.
14. Design a 2,000 Hz second-order low-pass unity gain filter using the scaling method.
15. Design a −60 dB/decade low-pass filter for a frequency of 40 Hz. Use the scaling method and the normalized values shown in Fig. 19-10A.

16. Design a −60 dB/decade cascade filter for a frequency of 800 Hz.
17. Calculate the cut-off frequency for a filter such as Fig. 19-14 in which $C1 = C2 = 0.0047$ μF, and $R1 = R2 = 27$ kohms.
18. Find the $Q$ of a filter in which the center frequency is 1200 Hz and the −3 dB points are 1250 and 1175 Hz. Is this a narrow bandpass or wide bandpass filter?
19. Find the $Q$ of a filter in which the center frequency is 800 Hz, and the −3 dB points are 600 and 1000 Hz. Is this a narrow bandpass or wide bandpass filter?
20. Design an MFP bandpass filter with a gain of 6 and a $Q$ of 16 when the center frequency $f_c$ is 1000 Hz. Assume $C1 = C2 = 0.015$ μF.
21. Scale the filter in the previous problem for a new frequency of 1500 Hz. Change only *one* component value.
22. Design a notch filter with a $Q$ of 10 and a notch frequency of 100 Hz.
23. An all-pass phase shifter uses the following components, $R1 = 10$ kohms and $C1 = 0.1$ μF. What is the phase shift at 1000 Hz? Express your answer in degrees.
24. Design a 100 Hz state variable filter with a $Q$ of 3. Use the scaling method.

CHAPTER 20

# Design and Construction of DC Power Supplies

## OBJECTIVES

1. Learn the basic components required to make a low-voltage DC power supply.
2. Learn the required ratings, specifications, and margins for DC power supply circuits.
3. Be able to design common low-voltage DC power supplies.

## 20-1 PREQUIZ

These questions test your prior knowledge of the material in this chapter. Try answering them before you read the chapter. Look for the answers (especially those you answered incorrectly) as you read the text. After you have finished studying the chapter, try answering these questions again and those at the end of the chapter (see Section 20-15).

1. State the maximum operating current and output potential of an LM-340T-12 three-terminal IC voltage regulator.
2. A zener diode regulator must regulate a DC supply to +6.8 Vdc. The input DC source varies from 9 to 13 volts. Assume a constant load current of 90 mA. Find the value of series resistor, its power dissipation, and the power dissipation of the zener diode.
3. A "brute force" ripple filter in a 12 Vdc power supply is required to reduce the fullwave ripple to 0.8 when the load current is 1.2 amperes. Calculate the value of filter capacitance needed.
4. A 12 Vdc power supply must deliver 900 mA approximately 50% of the time. What package should be specified for a three-terminal IC regulator?

## 20-2 INTRODUCTION TO DC POWER SUPPLIES

The DC power supply is important to the success of any electronic circuit. The power supply converts the alternating current available from the power mains to the direct current needed to operate electronic circuits. The typical DC power supply consists of several different components, *transformer*, *rectifier*, *ripple filter*, and (in some designs) *voltage regulator*.

The transformer scales the AC voltage from the power lines up or down as needed for the particular application. The rectifier converts the bidirectional AC into unidirectional *pulsating DC*. The ripple filter smoothes the pulsating DC into nearly pure DC. The voltage regulator is used to stabilize the voltage in the face of changing load currents and AC input voltage.

Also part of some DC power supplies are functions such as *overvoltage protection*, and *current limiting*. These circuits, as well as the main components of the DC power supply are discussed in detail in this chapter. In this chapter you will also learn the fundamentals of DC power supply design, especially the regulated, low-voltage DC power supplies which are typically used with circuits containing linear integrated circuit elements.

## 20-3 RECTIFIERS

A rectifier in a DC power supply circuit removes impurities of the AC line current and make it right for electronic circuits which require pure, or nearly pure, direct current.

Before discussing the details of solid-state rectifiers, let's review the two basic forms of electrical current in the context of rectification. DC and AC (see Fig. 20-1). Direct current (DC) is graphed in Fig. 20-1A. The key feature of this form of electrical current is it is *unidirectional* (current flows through the circuit in only one direction). It will be zero until turned on (time *T*1) and will rise to a certain level and remain

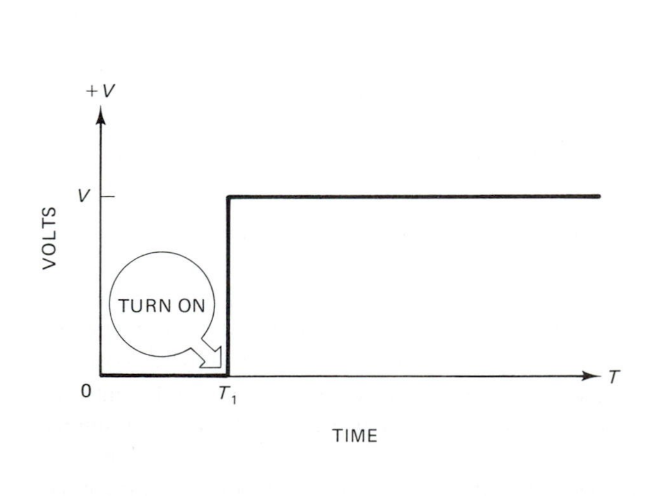


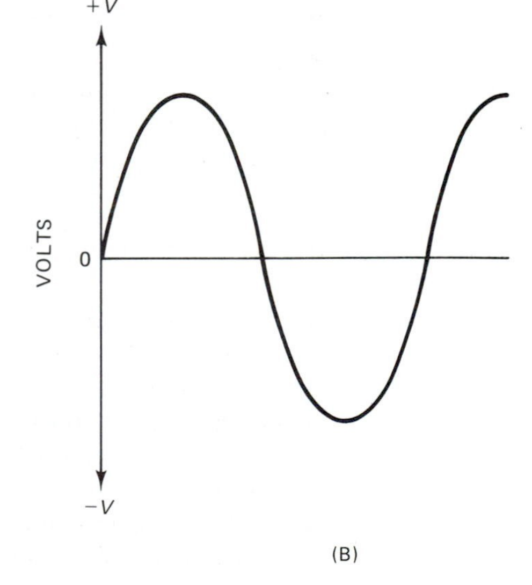


FIGURE 20-1 (A) (B)

there. Electrons flow from the negative terminal to the positive terminal of the power supply, and that polarity never reverses direction.

Alternating current, on the other hand, is *bidirectional* (see Fig. 20-1B). On one half-cycle the current flows in one direction. Then the power supply polarity reverses, so the current flows in the opposite direction. The electrons still flow from negative to positive, but since the positive and negative poles have switched places, the physical direction of current flow has reversed. In the normal AC power mains the voltage and current waveforms vary as a sinewave. By convention, flow in the positive direction is graphed above the zero volts (or zero amperes) line, and flow in the negative direction is graphed below the zero line.

AC is incompatible with nearly all electronic circuits, so it must be changed to DC by a *rectifier* and a *ripple filter.* The main requirement for a rectifier is that it *convert bidirectional AC into a unidirectional form of current.* Although rotary mechanical switches, synchronous vibrators, and vacuum tubes were used for rectification in industrial applications, all modern circuits rely on solid-state PN junction rectifiers.

### 20-3.1 PN Junction Diode Rectifiers

The modern solid-state rectifier really is not so new. Various versions of the rectifier date back to the dawn of the radio (indeed the electrical) age. All common rectifier diodes in use today, however, are *silicon PN junction diodes* (shown schematically in Fig. 20-2).

The PN junction diode rectifier (Fig. 20-2) consists of a silicon semiconductor material which is doped with impurities to form N-type material at one end and P-type material at the other end. The charge carriers (which form the electrical current) in the N-type material are negatively charged electrons. The charge carriers in the P-type material are positively charged "holes."

The reverse bias situation is shown in Fig. 20-2A. In this case the negative terminal of the voltage source ($V$) is connected to the $P$-type material, while the positive terminal is connected to the $N$-type material. Positive charge carriers are thus attracted away from the PN junction towards the negative voltage terminal, while negative charge carriers are drawn away towards the positive terminal. This leaves a charge free *depletion zone* in the region of the junction which contains no carriers. Under this condition, there is little or no current flow across the junction. Theoretically, the junction current is zero, although in real diodes there is always a tiny *leakage current* across the junction.

The forward-biased case is shown in Fig. 20-2B. Here the polarity of voltage source $V$ is reversed from Fig. 20-2A. The positive terminal is applied to the P-type material and the negative terminal is applied to the N-type material. Because like charges repel, the charge carriers in both P- and N-type material are driven away from the power supply terminals towards the junction. The depletion zone disappears, allowing positive and negative charges to get close to the boundary between regions. As these opposite charges attract each other across the junction, a current flows in the circuit.

From the above description it is apparent that a PN junction diode is able to convert bidirectional AC into unidirectional current because it allows current to flow in only one direction. Thus, it is a rectifier. However, the rectifier output current is not pure DC (as from batteries), but rather it is *pulsating DC.*

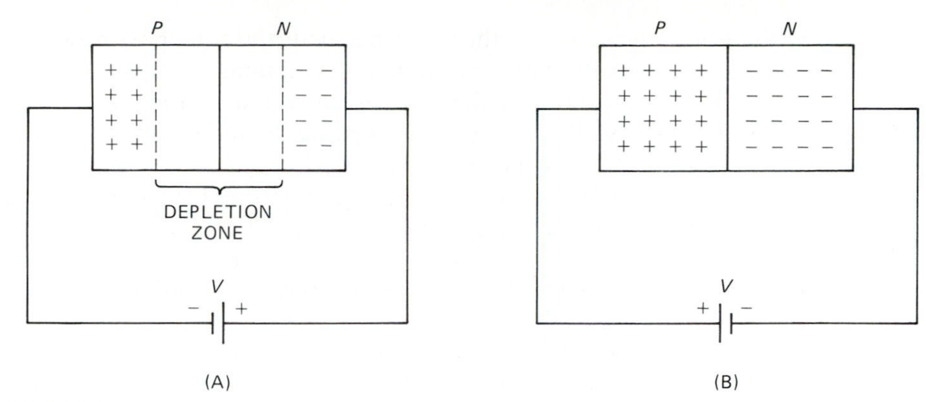

FIGURE 20-2

Figure 20-3 shows the standard circuit symbol for the solid-state rectifier diode (Fig. 20-3A) along with some common shapes of actual diodes. The input side where AC is applied is the anode. The DC output is the cathode. The diodes shown in Figs. 20-3B through 20-3G are positioned so the respective anodes and cathodes are aligned with those of the circuit symbol in Fig. 20-3A. Rectifiers 20-3B through 20-3E are epoxy package devices and are seen most often. The cathode end will be marked either with a rounded end (20-3B), a line (20-3C), a diode arrow (20-3D) or a plus sign (20-3E).

The diode shown in Fig. 20-3F is the older (now obsolete) tophat type. Unless otherwise specified, the tophat type can safely pass a current of 500 milliamperes, while those in Figs. 20-3B through 20-3E generally pass 1 ampere or more.

The stud-mounted type shown in Fig. 20-3G is a high current model. These diodes are rated at currents from 6 amperes and up (50- and 100-ampere models are easily obtained). These diodes are mounted using a threaded screw at one end which also forms one electrical connection. The other electrical connection is the solder terminal

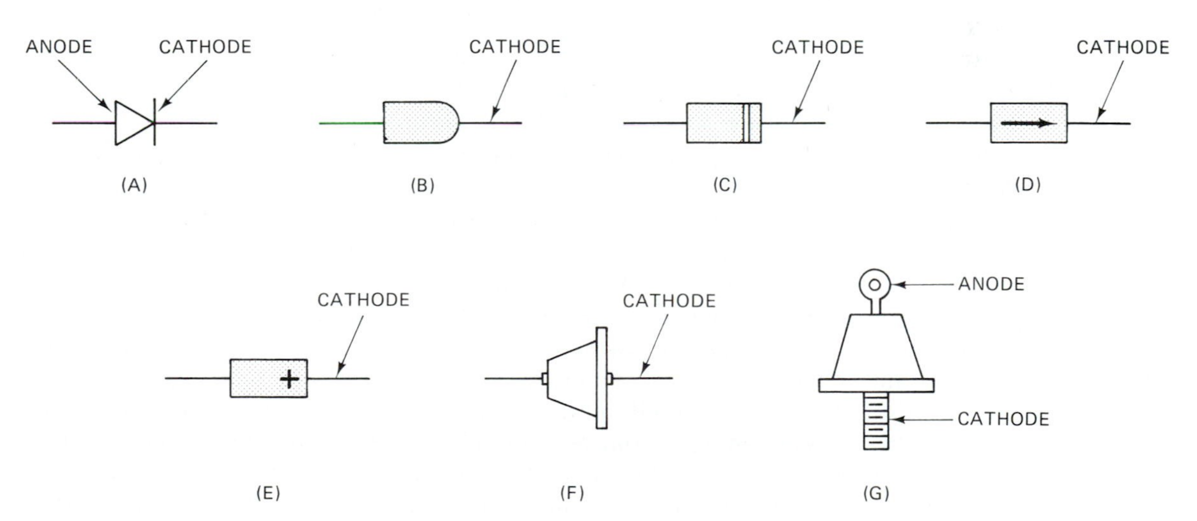

FIGURE 20-3

at the other end. Unless otherwise specified, the solder terminal is the anode, while the stud-mount is the cathode terminal. Exceptions to the polarity rule are sometimes seen. The reverse polarity diodes will have either an arrow symbol pointing in the opposite direction (the arrow always points to the cathode), or an R suffix on the type number (1NxxxxR instead of 1Nxxxx).

### 20-3.2 Rectifier Specifications

The proper use of solid-state rectifiers requires consideration of several key specifications, *forward current*, *leakage current*, *surge current*, *junction temperature*, and *peak inverse voltage* (PIV), also called *peak reverse voltage* (PRV).

The *forward current* is the maximum constant current the diode can pass without damage. For the 1N400x series of rectifiers the forward specified current is 1 ampere (1A). The *leakage current* is the maximum current that will flow through a reverse-biased junction. In an ideal diode, the leakage current is zero while in high quality practical diodes it is very low. The surge current is typically much larger than the forward current. It is sometimes erroneously taken to be the operating current of the diode. Surge current is defined as the maximum short duration current which will not damage the diode. "Short duration" typically means one AC cycle (1/60 second in a 60 Hz system). Don't use the surge current as if it were the forward current.

The specified *junction temperature* is the maximum allowable operating temperature of the PN junction. The actual junction temperature depends upon the forward current and how well the package (and environment) rids the diode of internal heat. Although typical maximum junction temperatures range up to +125°C, good design requires as low a temperature as possible. One reliability guide requires the junction temperature be held to a maximum of +110°C.

The *peak inverse voltage* (PIV) is the maximum allowable reverse bias voltage that will not damage the diode. This rating is usually the limiting rating in certain power supply designs.

### 20-3.3 Rectifier Circuits

Figure 20-4 shows a solid-state rectifier diode (D1) in a simple *halfwave rectifier* circuit. In Fig. 20-4A the diode is forward-biased, the positive terminal of the voltage source is connected to the anode of the rectifier. Current ($I$) flows through the load resistance ($R$). In Fig. 20-4B the opposite situation is found. The negative terminal of the voltage source is applied to the anode, so the diode is reverse-biased and no current flows.

The circuit in Fig. 20-4 is called a halfwave rectifier for reasons that become apparent when examining Fig. 20-4C. In this figure the output current through the load ($R$) is graphed as a function of time when an AC sinewave is applied. From time $T1$ to $T2$ the diode is forward biased, so current flows in the load (also from $T3$ to $T4$). But during the period $T2$ to $T3$ the diode is reverse-biased, so no current flows. Because the entire sinewave takes up the period $T1$ to $T3$, only half of the input sine wave is used. The output waveform shown in Fig. 20-4C is called a *halfwave rectified pulsating DC* wave.

The halfwave rectifier is low cost, but wastes energy due to its use of only one-half the input AC waveform. Efficiency is increased by using the entire waveform in

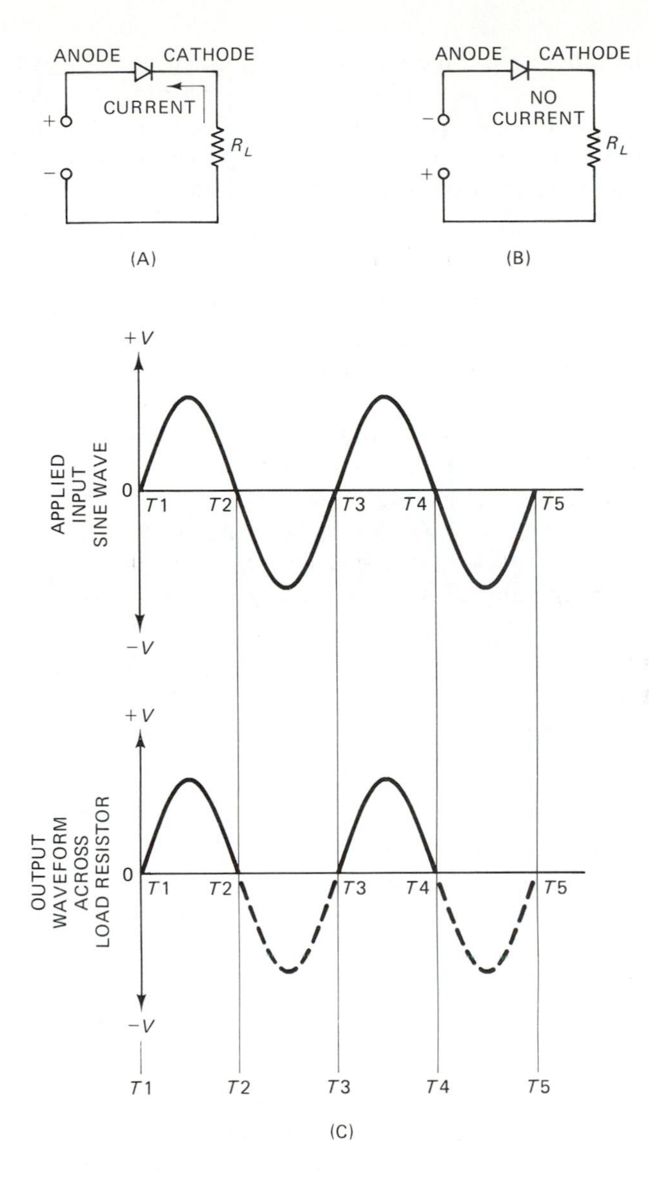


FIGURE 20-4

a *fullwave rectifier circuit.* Figure 20-5A shows the standard fullwave rectifier. This circuit uses a transformer which has a center-tapped secondary winding. Because the center tap (CT) is used as the zero volts reference (and in most circuits is grounded), the polarities at the ends of the secondary are always opposite each other (180°). On one halfcycle, point A is positive with respect to the CT and point B is negative. On the next halfcycle, point A is negative and point B is positive with respect to the CT. This situation makes $D1$ forward-biased on one halfcycle, while $D2$ is reverse-biased. Alternatively, on the next halfcycle, $D1$ is reverse-biased and $D2$ is forward-biased.

Follow the circuit of Fig. 20-5A through one complete AC cycle (times $T1$ through $T3$ in Fig. 20-5B). On the first half cycle ($T1-T2$), point A is positive, so $D1$ is

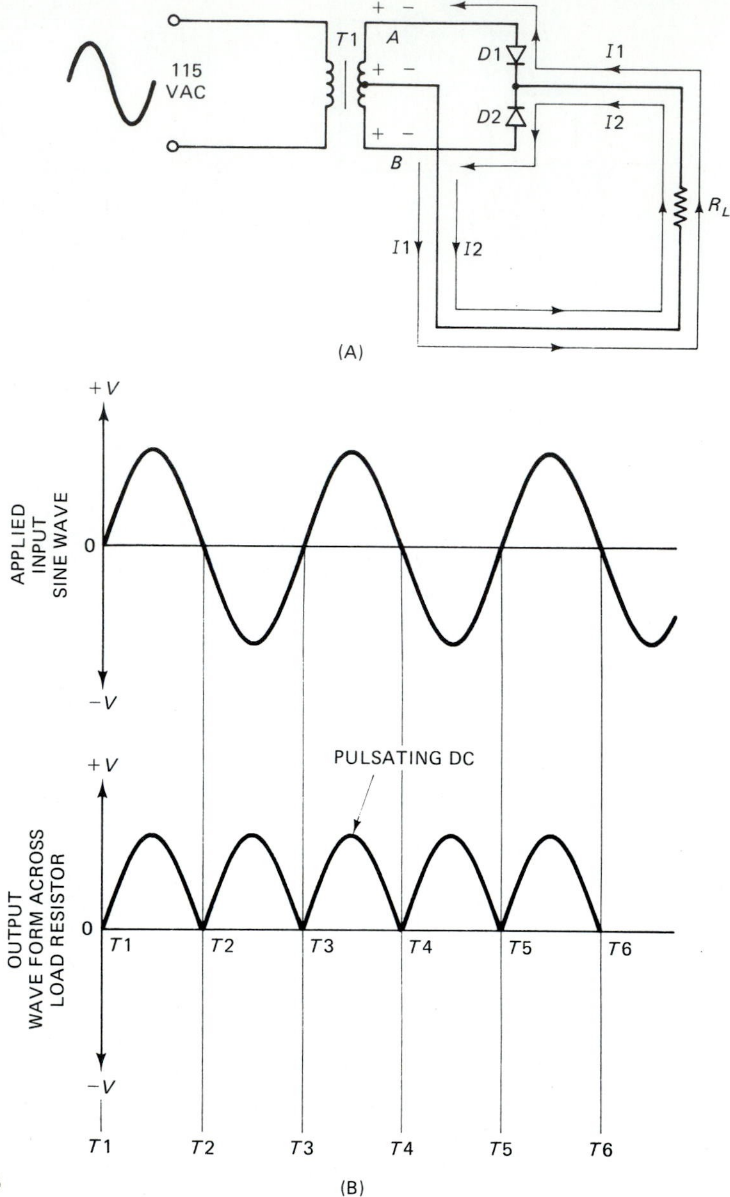


FIGURE 20-5

forward-biased and conducts current. $D2$ is reverse-biased. Current $I1$ flows from the CT, through load $R$, diode $D1$ and back to the transformer at point $A$. On the alternate half cycle, current $I2$ flows from the CT, through load $R$, diode $D2$ and back to the transformer at point B. Now notice what happened. $I1$ and $I2$ are equal currents, generated on alternate halfcycles, and they *flow in load R in the same direction.* Thus, a unidirectional current is flowing through load $R$ on both halves of the AC sinewave. The waveform resulting from this action is shown in Fig. 20-5B. It is called *fullwave rectified pulsating DC.*

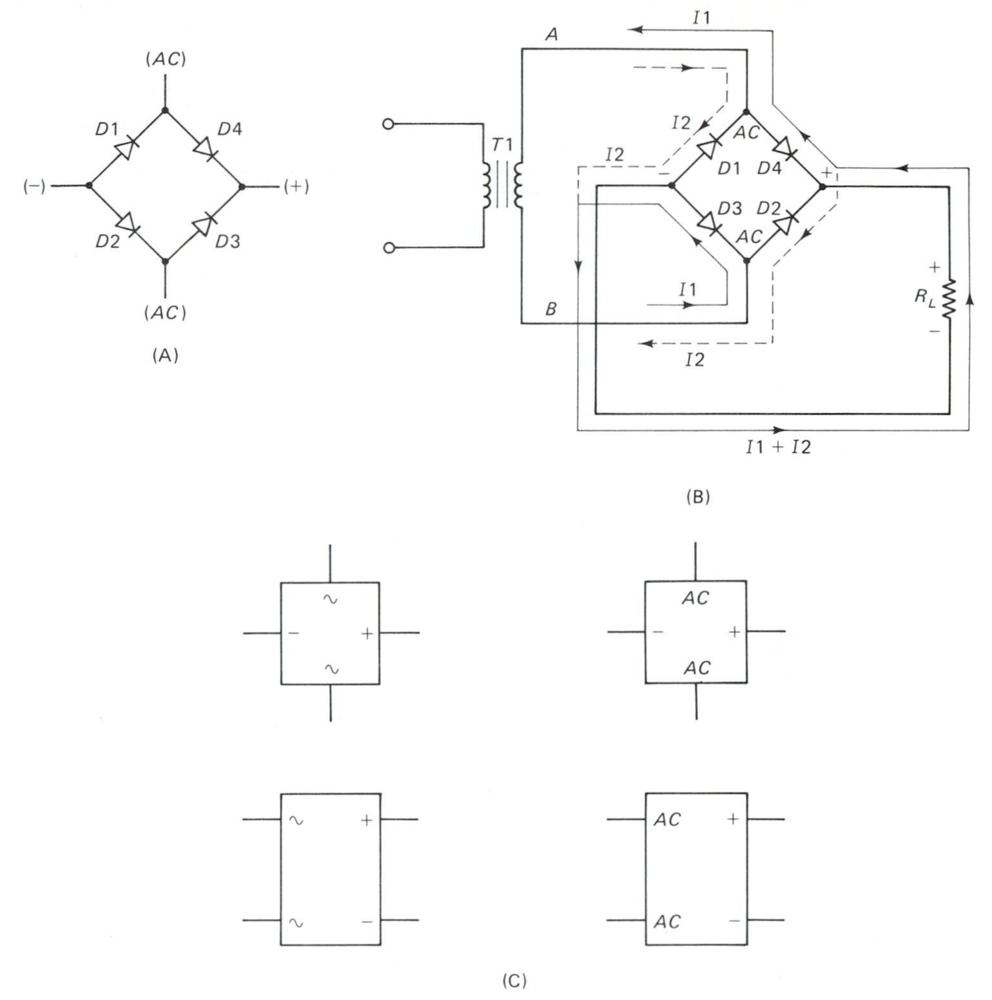


FIGURE 20-6

The center tap on the secondary can be eliminated by using the fullwave bridge rectifier circuit of Fig. 20-6A. This circuit requires twice as many rectifier diodes as the other form of fullwave rectifier, but allows the use of a simpler transformer (no center tap). The operation, however, is similar (see Fig. 20-6B). On one half cycle, point A is positive and point B is negative. Current *I*1 flows from the transformer at point B, through *D*4, load *R*, diode *D*1 and back to the transformer at point A. On the alternate half cycle, point A is negative and point B is positive. In this case, current *I*2 flows from point A, through diode *D*3, load *R* (in the same direction as *I*1), diode *D*2 and back to the transformer at point *B*.

A bridge rectifier can be built using four discrete diodes (*D*1–*D*4). In most modern equipment, however, a *bridge stack* is used. These parts are a bridge rectifier built into a single package with four leads. Figure 20-6C shows various alternate circuit symbols for bridge rectifier stacks.

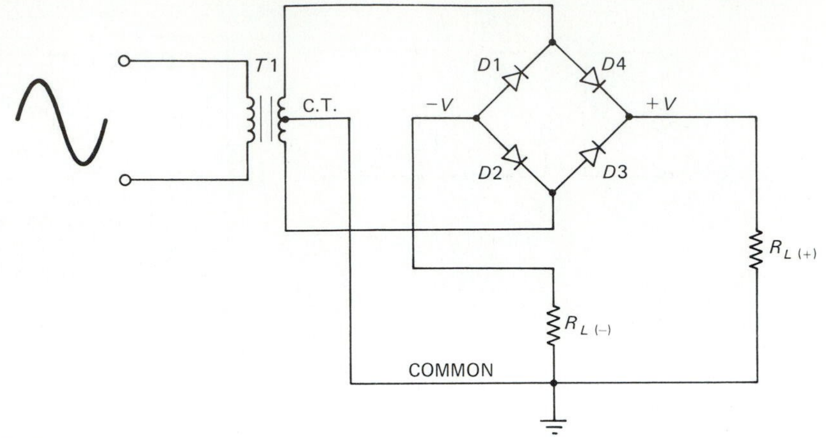


FIGURE 20-7

Figure 20-7 shows a halfbridge fullwave rectifier. This circuit is used more today because dual polarity power supplies are used in most equipment. Operational amplifiers and some CMOS devices typically require $\pm 12$ volt DC power supplies. A fullwave bridge rectifier stack coupled to a center-tapped transformer will make a pair of fullwave rectified DC power supplies. The CT is the common (or ground), while the bridge positive terminal supplies the positive voltage and the negative terminal supplies the negative voltage. Both outputs from the halfbridge rectifier are fullwave rectified pulsating DC.

## 20-3.4 Selecting Rectifier Diodes

The two parameters most often used to specify practical power supply diodes are the *forward current* and *peak inverse voltage.*

The forward current rating of the diode must be at least equal to the maximum current load the power supply must deliver. In practical circuits there is also a need for a safety margin to account for tolerances in the diodes and variations of the real load (as opposed to the calculated load). It is also true that making the rating of the diode larger than the load current will greatly improve reliability. A good rule is to select a diode with a forward current rating of 1.5 to 2 times the calculated (or design goal) load current (or more if available). Selecting a diode with a much larger forward current (100 amperes for a 1 ampere circuit) is both wasteful and likely to make the diode not work like a rectifier diode. The general rule is to make the rating as high as is feasible. The 1.5 to 2 times rule, however, should result in a reasonable margin of safety.

The peak inverse voltage (PIV) rating can be more complicated. In unfiltered, purely resistive, circuits the PIV rating need only be greater than the maximum peak applied AC voltage (1.414 times RMS). But if a 20% safety margin is desired, make it 1.7 times RMS voltage.

Most rectifiers are used in *ripple filtered* circuits (Fig. 20-8A). This makes the problem different. Figure 20-8B shows the simple halfwave rectifier capacitor filtered circuit redrawn to better illustrate the circuit action. Keep in mind that capacitor *C*1

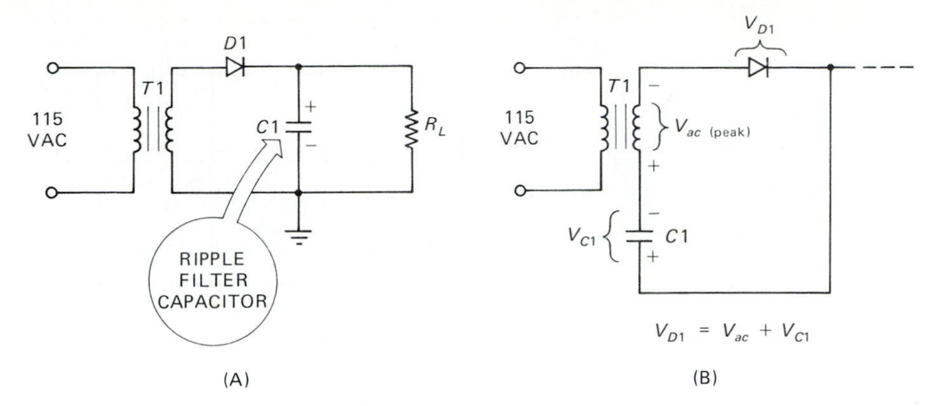


FIGURE 20-8

is charged to the peak voltage with the polarity shown. This peak voltage is 1.414 times the RMS voltage. The peak voltage across the transformer secondary ($V$) is in series with the capacitor voltage. When voltage $V$ is positive, the transformer voltage and capacitor voltage cancel out, making the diode reverse voltage nearly zero. But when the transformer voltage ($V$), is negative, the two negative voltages ($V$ and $V_c$) add up to twice the peak voltage

$$V_{D1} = V_{ac(\text{peak})} + V_{C1} \tag{20-1}$$

but

$$V_{C1} = V_{ac(\text{peak})} \tag{20-2}$$

so

$$V_{D1} = V_{ac(\text{peak})} + V_{ac(\text{peak})} \tag{20-3}$$

$$V_{D1} = 2V_{ac(\text{peak})} \tag{20-4}$$

Because $V_{ac(\text{peak})} = 1.414V_{ac(\text{rms})}$

$$V_{D1} = 2(1.414V_{ac(\text{rms})} \tag{20-5}$$

$$V_{D1} = 2.828V_{ac(\text{rms})} \tag{20-6}$$

The reverse voltage across the diode is approximately 2.83 times the rms voltage. Therefore, the absolute minimum value of PIV rating for the diode is 2.83 times the applied rms. If a 20% safety margin is preferred, the diode PIV rating should be 3.4 times the applied rms voltage (or more).

### 20-3.5 Using Rectifier Diodes

In most cases, especially low-voltage power supplies, diodes can be used as shown in the circuits above. In Fig. 20-9A, however, the proper way to use the solid-state diode rectifier is shown. The resistor in series with diode $D1$ ($R1$) is used to limit the forward current. Many circuits, especially those with capacitor-input filter circuits, have a large surge current at initial turn on. This current can sometimes destroy the diode, so $R1$

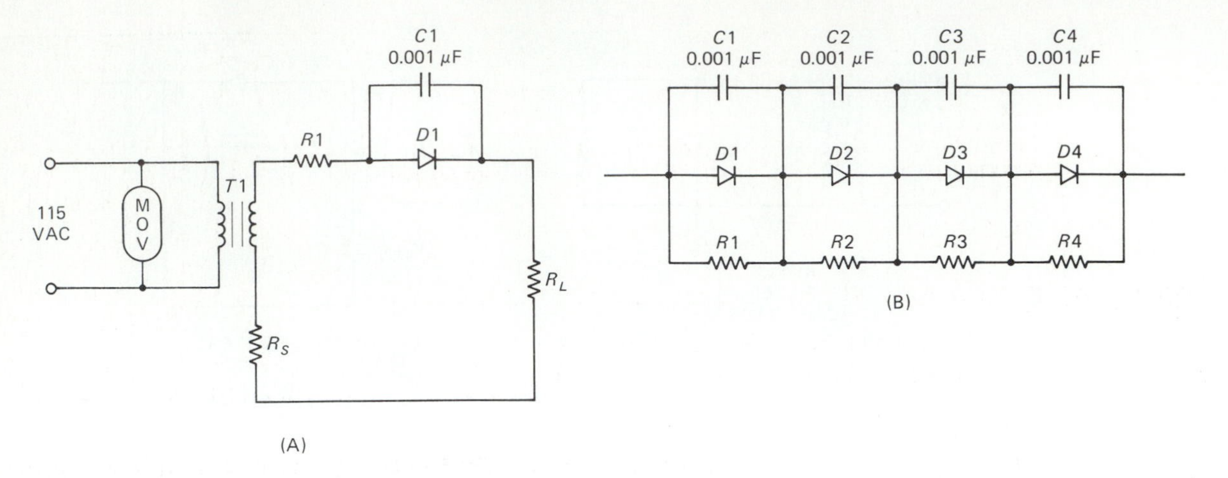


FIGURE 20-9

is used to limit the possible damage. The resistance value of $R1$ is typically 5 to 20 ohms. In most cases, however, $R1$ can be eliminated by using a diode with a current rating significantly larger than the load current (for example, two times). Also, the transformer secondary resistance ($R_s$) serves the current limiting function of $R1$ in many circuits.

Capacitor $C1$ in Fig. 20-9A is used to bypass high voltage transient spikes around the diode. These spikes could possibly blow the diode PN junction. In fact, high voltage line spikes are a frequent source of damage to rectifier diodes. Placing the capacitor in parallel with the diode will eliminate this problem. The DC working voltage (WVDC) of the capacitor should be equal to or greater than the PIV rating of the diode.

By use of 1000 volt PIV diodes (even in low-voltage circuits) much of the damage caused by transients is avoided. The capacitors can often be eliminated if a *metal oxide varistor* (MOV) or some other high voltage spike suppressor device is shunted across the AC supply voltage in the primary circuit of the power transformer (see Section 20-11).

Figure 20-9B shows the method for connecting several diodes in series to increase the PIV rating. Assuming the PIV ratings of the diodes are equal, the overall rating is four times the rating of one diode. If 1000 volt PIV diodes are used in this circuit, the total PIV rating of the assembly is 4000 volts.

The capacitors in Fig. 20-9B are used for the same purpose as in Fig. 20-9A. The resistors, however, are needed for a different purpose. They equalize the forward voltage drop across each diode. A 470 kohm, 2 watt resistor is typically used for 1000 volt PIV diodes. The 2 watt rating is required not because of the power dissipation of the resistors, but rather for their voltage rating. (Resistors *do* have voltage ratings.)

Figure 20-10 shows the proper method for mounting an axial lead rectifier on a prototyping perfboard or printed circuit board. This method is used anytime *except* where excessive vibration is expected, which includes most sedentary circuits or equipment. The space beneath the diode body allows air to circulate (keeping the diode cooler) and prevents diode heat from damaging the board.

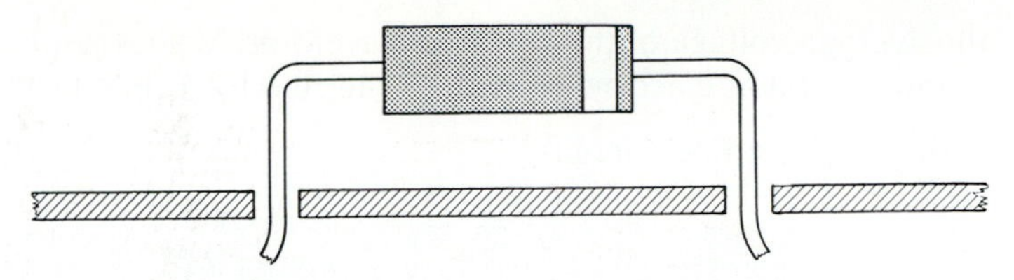

FIGURE 20-10

## 20-4 RIPPLE FILTER CIRCUITS

The pulsating DC output from either fullwave or halfwave rectifiers is almost as useless for some electronic circuits as the AC input waveform. A *ripple filter* circuit is used to smooth out the pulsating DC into a purer DC. Figure 20-11A shows the simplest form of filter circuit, a single capacitor ($C1$) connected in parallel with the load. Circuit action is shown in Fig. 20-11B. The job of the capacitor is to store electrical charge on voltage peaks and dump that charge into the load when the voltage drops between peaks. The shaded area of Fig. 20-11B shows the filling in caused by the capacitor charge. The output voltage is the sum of both rectifier and capacitor contributions. It is represented by the heavy line in Fig. 20-11B.

The value of the filter capacitor is determined by the amount of *ripple factor* which can be accepted. The ripple factor ($r$) is the ratio of the ripple voltage amplitude to

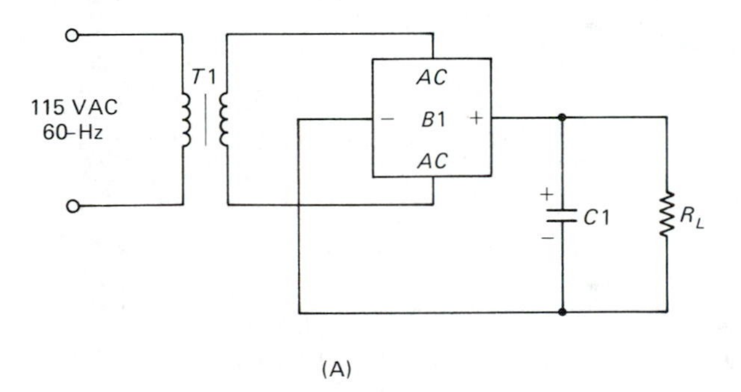


(A)

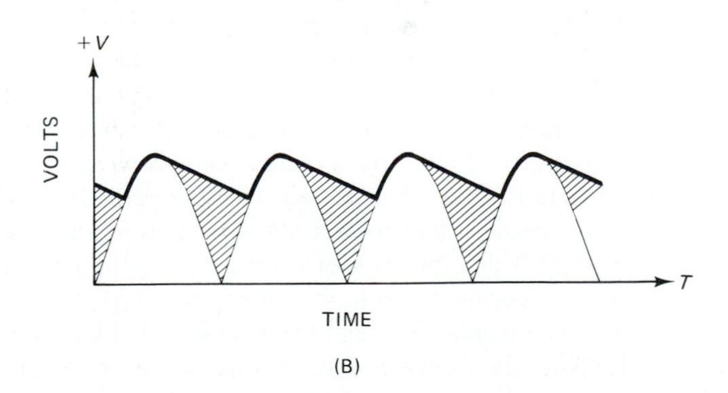


(B)

FIGURE 20-11

the average voltage of the rectified waveform. Values tend to be in the range 0.01 to 1.0 for common electronic circuits. The rule for ripple factor for 60 Hz circuits is

Halfwave Rectified Circuits

$$r = \frac{1{,}000{,}000}{208\ C1\ R_L} \tag{20-7}$$

Fullwave Rectified Circuits

$$r = \frac{1{,}000{,}000}{416\ C1\ R_L} \tag{20-8}$$

where

$C1$ = the capacitor value in microfarads

$R_L$ = the load resistance in ohms.

The load resistance is the ratio of the output voltage to the output current, $V_o/I_o$. Consider a fullwave example.

### EXAMPLE 20-1

A 12 Vdc fullwave rectified DC power supply delivers 0.5 amperes to the load. Calculate the ripple factor ($r$) if 1,000 μF is used for the filter capacitor.

#### Solution

$$r = \frac{1{,}000{,}000}{416\ C1\ R_L}$$

$$r = \frac{1{,}000{,}000}{(416)(1000\ \mu\text{F})(12\ \text{Vdc}/0.5\ \text{A})}$$

$$r = \frac{1{,}000{,}000}{(416)(1000\ \mu\text{F})(24)} = 0.1$$

■

So what do these figures mean in practical terms? A few oscilloscope waveform photos (Fig. 20-12) illustrate the effect of adding capacitance to the filter circuit. In all cases the AC applied to the rectifier was nominally 12 Vac, and the load resistance was 25 ohms. These photos were taken with the 'scope AC-coupled to permit expansion of the ripple in the presence of the large DC offset. The waveforms shown in Fig. 20-12 ride on top of the DC output applied to the load, but the DC offset is suppressed by the 'scope input circuit. The 'scope sensitivity was 5V/cm.

The unfiltered, fullwave rectified pulsating DC (Fig. 20-12A) was measured with a DC voltmeter and found to be 9.25 Vdc. This value is not the peak voltage (which was close to 12 Vdc), but rather an average value caused by the fact that the DC meter tends to average the reading. When a 150 μF capacitor is connected across the 25-ohm load the ripple reduces to 0.64 (Fig. 20-12B) and the output voltage rises to 10.86 Vdc. Note the filtering action is just beginning to take place. Connecting a 1000 μF capacitor across the load further drops the ripple factor to 0.096 (Fig. 20-12C).

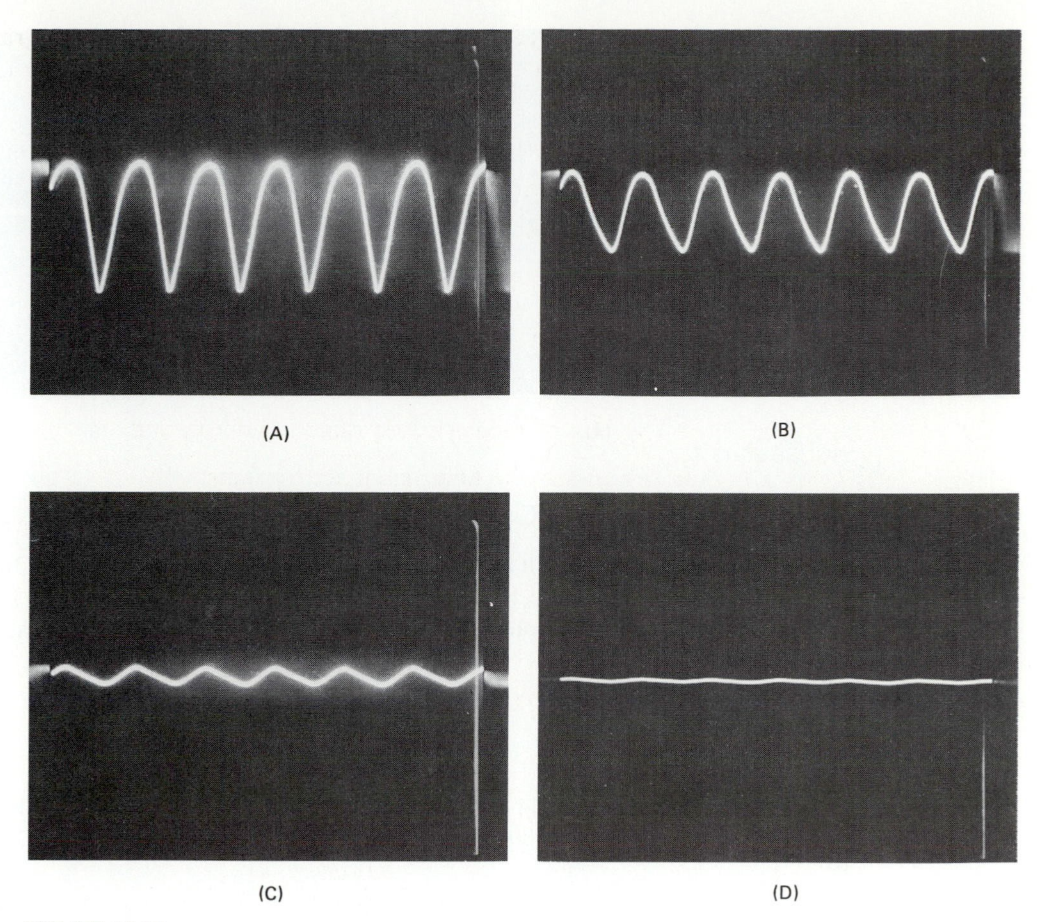

FIGURE 20-12

The voltage rises to 11.8 volts. Finally, a 6800 μF capacitor was connected across the load. The ripple factor dropped to 0.014 (Fig. 20-12D). This filtered DC is nearly pure, but still contains a small ripple factor.

Next, the sensitivity of the 'scope was increased from 5V/cm to 0.1V/cm (a 50X increase) and the same 6800-μF capacitor was used in parallel with the 25 ohm load. The ripple shows up a lot better under this condition (Fig. 20-13A). Recall from above this waveform represents a ripple factor $r = 0.014$.

A *voltage regulator* tends to smooth out the ripple considerably. Power supply makers sometimes claim their products have the "equivalent of one farad of output capacitance." What they mean is the ripple is reduced by a voltage regulator the same as a 1,000,000 μF capacitor across the load.

Examine Fig. 20-13B. Using the same 0.1V/cm 'scope sensitivity and the same power supply, this waveform shows a remarkable drop of ripple compared with Fig. 20-13A. The lesson here is to use a voltage regulator in circuits where it is necessary to reduce the ripple to a very low value.

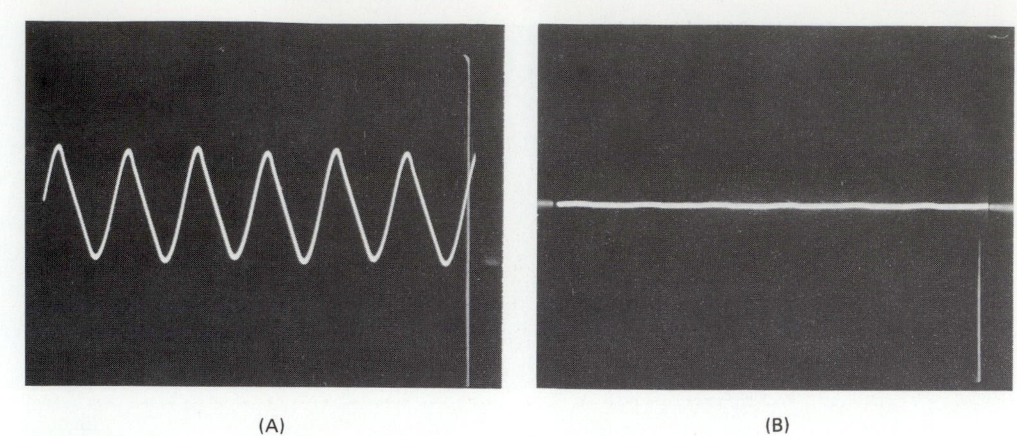

(A) (B)

FIGURE 20-13

### 20-4.1 Pi-Network Filter Circuits

Another form of filter circuit is the RC Pi-network shown in Fig. 20-14. Output voltage $V1$ represents a circuit like those above, but output voltage $V2$ has a lower voltage and substantially lower ripple factor. For fullwave circuits the ripple factor is

$$r = \frac{2.5 \times 10^{-6}}{C1\ C2\ R1\ R_L} \tag{20-9}$$

**EXAMPLE 20-2**

A DC power supply delivers 14 Vdc at 100 mA to a resistive load. Calculate the ripple factor of the output from the Pi-section RC filter if $R1 = 90$ ohms, $C1 = 1000\ \mu F$ and $C2 = 470\ \mu F$.

$$r = \frac{2.5 \times 10^{-6}}{C1\ C2\ R1\ R_L}$$

$$r = \frac{2.5 \times 10^{6}}{(1000\ \mu F)(470\ \mu F)(90\ \text{ohms})(14\text{Vdc}/0.100\ \text{A})}$$

$$r = \frac{2.5 \times 10^{6}}{5.92 \times 10^{9}} = 0.0004$$

■

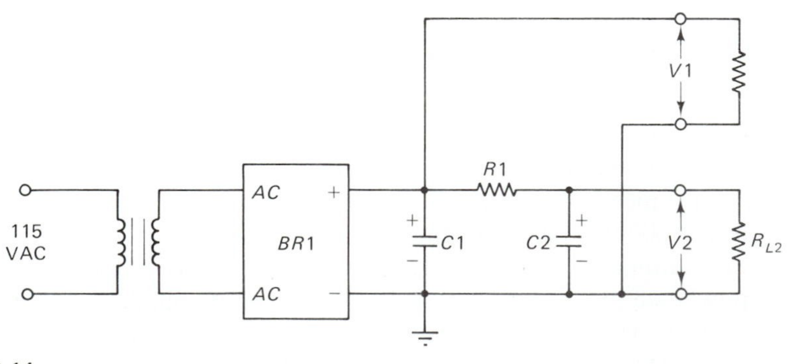


FIGURE 20-14

### 20-4.2 Working Voltage

The other significant rating for filter capacitors, besides capacitance, is the *DC working voltage* (WVDC). This rating specifies the maximum voltage the capacitor can safely sustain on a continuous basis (not a transient peak voltage). Because AC line voltages can vary $\pm 15\%$ and the WVDC rating tolerance may have a $\pm 20\%$ tolerance, it is prudent to use a capacitor with a WVDC rating which is at least half again higher than the normal output voltage expected. That is, the minimum WVDC rating of the capacitor should be 1.5 times the maximum output voltage of the rectifier. If a filter capacitor has a WVDC rating close to the power supply output voltage, a short life and a spectacular end may be the fate of the power supply.

## 20-5 VOLTAGE REGULATION

The output voltage of ordinary rectifier/filter DC power supplies is not stable. Rather, it may vary considerably over time. There are two main sources of variation in the output of this type of power supply. First, there is always a certain fluctuation of the AC input voltage. Ordinary commercial power lines vary from 105 to 120 volts AC (rms) normally, and may droop to less than 100 volts during power brownouts.

The second source of variation is created by load variation (see Fig. 20-15). The problem is caused by the fact that real DC power supplies are not ideal. The ideal textbook power supply has zero ohms internal resistance, while real power supplies have a certain amount of internal resistance (represented by $R_s$ in Fig. 20-15). When current is drawn from the power supply there is a voltage drop ($V1$) across the internal resistance, and this voltage is subtracted from the available voltage ($V$). In an ideal power supply, output voltage $V_o$ is the same as $V$, but in real supplies $V_o$ is equal to $(V - V1)$. Because $V1$ varies with changes in the load current $I_o$, the output voltage will also vary with changes in current demand.

The "goodness" or "badness" of a power supply can be defined in terms of its *percentage of regulation*. This specification is a measure of how badly the voltage

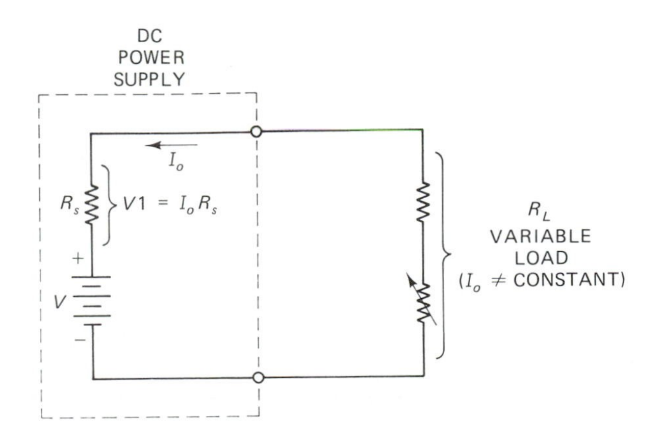


FIGURE 20-15

changes under changes of load current, and is found from

$$\%REG = \frac{(V - V_o)(100\%)}{V} \tag{20-10}$$

Where

$V$ = the open-terminal (no output current) output voltage

$V_o$ = the output voltage under full load current

$\%REG$ = the percentage of regulation

**EXAMPLE 20-3**

Find the percentage of regulation if a DC power supply output voltage drops from 15 to 13 volts as the output current is raised from zero to 2 amperes (which is the maximum allowable output current for this power supply).

$$\%REG = \frac{(V - V_o)(100\%)}{V}$$

$$\%REG = \frac{(15 - 13)(100\%)}{15}$$

$$\%REG = \frac{(2)(100\%)}{15} = 13\%$$

■

Many electronic circuits do not work properly under varying supply voltage conditions. Oscillators and some waveform generators, for example, tend to change frequency if the DC power supply voltage changes. Obviously, some means must be provided to stabilize the DC voltage. The zener diode is perhaps the simplest such voltage regulator device.

## 20-6 ZENER DIODES

The *zener* (pronounced zen-ner) *diode* is a special case of the PN junction diode. Figure 20-16 shows both the circuit symbol (Fig. 20-16A) and *I versus V* curve (Fig. 20-16B) for a zener diode.

In the forward bias region operation of the zener diode is the same as for other PN junction diodes. For this case, the anode is positive with respect to the cathode, so a forward bias current (+I) flows. For voltages greater than $V_g$, which is approximately 0.6 volts to 0.7 volts, the current flow increases approximately linearly with increasing voltage. At potentials less than $V_g$, the current increases from a small reverse leakage current ($I_L$) at $V = 0$, to a small forward current at $+V_g$. The diode is, like all other PN junction diodes, nonlinear in this low-voltage region.

The zener diode also acts like any other PN junction diode in the reverse bias region between $V = 0$ volts and the zener potential $-V_z$. In this region, only the small reverse leakage current flows.

At an applied potential of $-V_z$, or greater, the zener diode breaks down and allows

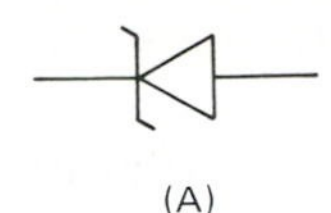
(A)

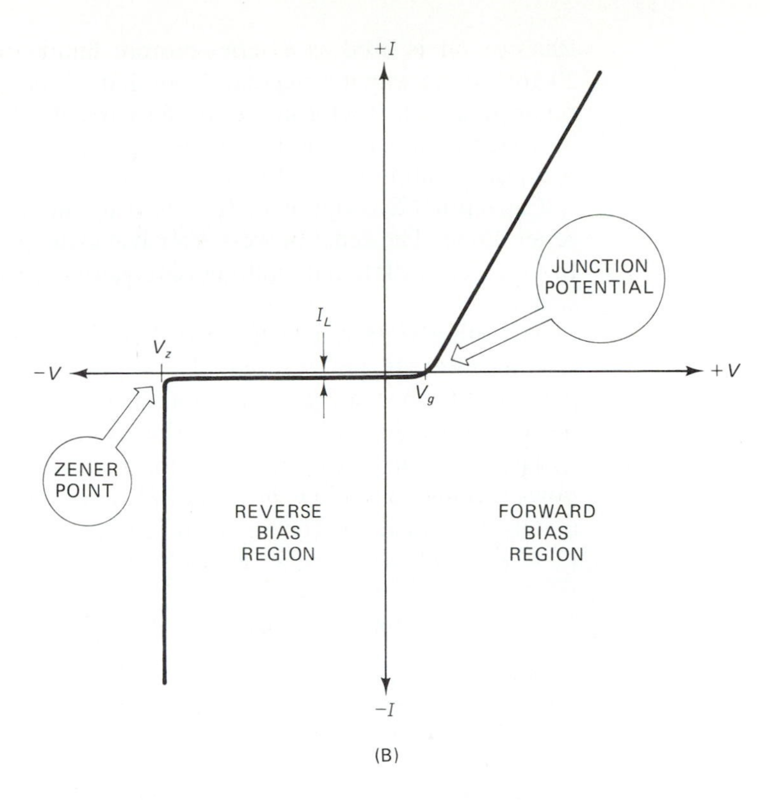

(B)

FIGURE 20-16

a large reverse current to flow. Note in Figure 20-16B that further increase in −V does not cause an increased voltage drop across the diode. Thus, the zener diode regulates the voltage to its zener potential by clamping action.

### 20-6.1 Zener Diode Regulator Circuits

A zener diode operates as a parallel or shunt regulator because it is connected in parallel with the load. It regulates by clamping the output voltage across the load to the zener potential. Figure 20-17 shows a typical zener diode regulator circuit which takes advantage of these attributes.

In Fig. 20-17, resistor $R_L$ represents the load placed across the power supply (the circuits that draw current from the supply). The value of the load resistance is $V_z/I_o$.

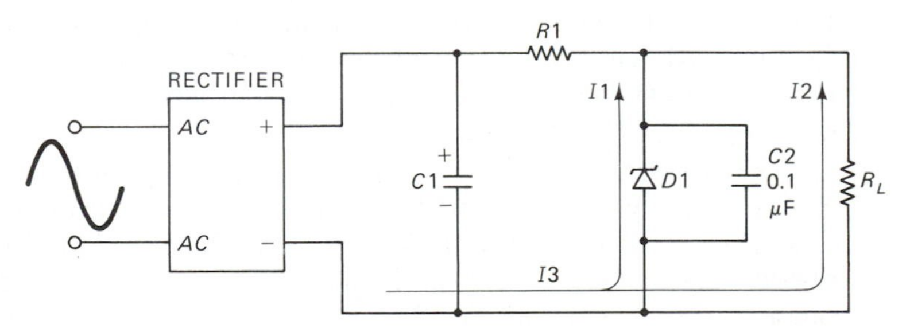


FIGURE 20-17

Resistor $R1$ is used as a series current limiter to protect the diode. Refer to Fig. 20-16B to see why it is needed. Note that $-I$ increases sharply when $-V$ reaches the zener potential. If $R1$ is not used, this current will destroy the diode. One of the tasks in designing a zener diode voltage regulator circuit is selecting the resistance value and power rating of resistor $R1$.

Capacitor $C2$ is optional. It is used to suppress the hash noise generated by the zener diode. The zener process is an avalanche phenomenon, so it is inherently noisy. In fact, certain RF and audio noise generators use a zener diode to create the noise signal.

The other capacitor in the circuit ($C1$) is the regular filter capacitor used in any rectifier/filter DC power supply. Its purpose is to smooth the pulsating DC into nearly pure DC. It does not really serve a function in the zener regulator circuit, except the power supply should be filtered prior to the regulator circuit.

The main current drawn from the rectifier ($I3$) is broken into two branches, $I1$ flows through the zener diode and $I2$ flows through the load. $I1$ usually is approximately 10% of $I2$. According to Kirchoff's law, the relationship between currents is $I3 = I1 + I2$.

### 20-6.2 Designing Zener Diode Voltage Regulator Circuits

When designing zener diode regulators it is necessary to know certain circuit conditions, and use them to specify (a) the resistance of $R1$, (b) the wattage rating of $R1$, and (c) the wattage rating of zener diode $D1$.

There are three circuit conditions, which are designated I, II and III. The properties of these three conditions are as follows:

**Condition I.** Variable supply voltage, with constant load current.

**Condition II.** Constant supply voltage, with variable load current.

**Condition III.** Variable supply voltage, with load current also variable.

Table 20-1 shows the design equations for all three conditions. Note the power dissipation expressions for $R1$ and $D1$ are the same for all three conditions. $V_{in}$ and $V_{in(\max)}$ are the same for the constant supply voltage case.

### EXAMPLE 20-4

A 6.8 Vdc zener diode must regulate in the presence of a supply voltage that varies from +9 to +12 volts. The current remains constant at 120 milliamperes (0.12 amperes). In this case, $V_{in(\min)}$ is 9 Vdc and $I2$ is 0.120 amperes. Calculate the value of $R1$, the power dissipation of $R1$ and the power dissipation of $D1$. Recommend minimum values for the power ratings of $R1$ and $D1$.

$$R1 = \frac{V_{in(\min)} - V_z}{1.1\ I2}$$

$$R1 = \frac{(9 - 6.8)}{1.1(0.12)}$$

$$R1 = 2/0.132 = 16.7 \text{ ohms}$$

**TABLE 20-1**

**CONDITION I.** Variable $V_{in}$, Constant $I_o$

$$R1 = \frac{V_{in(min)} - V_z}{1.1\ I2}$$

**CONDITION II.** Constant $V_{in}$, Variable $I_o$

$$R1 = \frac{V_{in(max)} - V_z}{1.1\ I2_{(max)}}$$

**CONDITION III.** $V_{in}$ and $I_{in}$ both variable

$$R1 = \frac{V_{in(max)} - V_z}{1.1\ I2_{(max)}}$$

**POWER DISSIPATIONS FOR ALL THREE CONDITIONS**

*Diode dissipation*

$$P_{D1} = \frac{[V_{in(max)} - V_z]^2}{R1} - [I2V_z]$$

*Resistor dissipation*

$$P_{R1} = P_{D1} + (I2V_z)$$

Since 16.7 ohms is not a standard value, it is necessary to either series-parallel two or more resistors to make 16.7 ohms, or use the nearest standard value (15 or 18 ohms).

The power dissipated by $D1$ is given by

$$P_{d1} = \frac{[V_{in(max)} - V_z]^2}{R1} - [I2 \times V_z]$$

$$P_{d1} = \frac{[12 - 6.8]^2}{16.7} - [(0.12)(6.8)]$$

$$P_{d1} = \frac{[5.2]^2}{16.7} - (0.816)$$

$$P_{d1} = \frac{27.04}{16.7} - (0.816)$$

$$P_{d1} = 1.62 - 0.816 = 0.803 \text{ watts}$$

A power dissipation of 0.803 watts means a 1 watt or greater rating is required for the zener diode ($D1$). For best reliability, maintain at least 2:1 ratio between the rated and dissipated wattage. Recommend a rating >1.6 watts (which means 2-watt standard value) diode. For most intermittent or low duty-cycle applications, however, a 1 watt zener diode should suffice.

The required power rating of $R1$ is found from calculating the actual power dissipation of $R1$ and picking a higher standard wattage rating. The power dissipation of $R1$ is

$$P_{R1} = P_{d1} + (I2 \times V_z)$$

$$P_{R1} = (0.803 \text{ watts}) + ((0.012)(6.8))$$

$$P_{R1} = 0.803 + 0.816 \text{ watts} = 1.62 \text{ watts}$$

Since $P_{R1}$ is 1.62 watts, use at least 2 watts for the rating of $R1$. A 3 watt or 5 watt resistor would yield even greater reliability.

The component values for this example are 16.7 ohms at 2 watts (or more) for $R1$, and 6.8 volts at 1 watt (or more) for $D1$. ■

Solving for the other conditions is similar, so further examples are not needed here. However, you may wish to calculate these case as a matter of interest.

## 20-7 INCREASING VOLTAGE REGULATOR OUTPUT CURRENT

The output current that can be supplied by a zener diode voltage regulator is somewhat limited. In cases where a larger output current is needed, it is possible to amplify the effect of the zener diode by using it to control the base of a series-pass transistor ($Q1$ in Figure 20-18). The output voltage produced by this circuit is approximately 0.6 to 0.7 volts less than the zener potential. This reduction is accounted for by the base-emitter potential of the transistor, $V_{b-e}$. The output current rating is limited by the collector current rating of the transistor, with due regard for the collector dissipation. The collector will dissipate a power of $(V_{in} - V_o) \times I_o$. If there is a large difference between $V_{in}$ and $V_o$, it is possible to exceed the maximum collector dissipation rating of $Q1$ even if less than the maximum collector current flows in the circuit. The output current must be limited to a value less than that required to exceed the collector dissipation under the voltage difference conditions established in the circuit.

Another series-pass voltage regulator circuit is the feedback regulator shown in Fig. 20-19. In this circuit, a sample of the output voltage and a reference potential are applied to the differential inputs of a feedback amplifier ($A1$). When the difference between $V_A$ and $V_{ref}$ is nonzero, the amplifier drives the base of transistor $Q1$ harder thereby increasing the output voltage. The actual voltage will be stable at a point determined by $V_{ref}$.

The circuit in Fig. 20-19 shows a highly useful feature in high current DC power supplies, especially where the power supply must be operated more than a few inches from the load. The voltage divider $R1/R2$ takes the sample of output voltage $V_o$ which drives $A1$. The lines from the positive output and negative output to the voltage divider are separate from the main current-carrying lines. This arrangement makes it possible to place these sense lines at the points in the actual circuit where the value of $V_o$ must be maintained at a precise value. For example, in a microcomputer that uses high current TTL devices, it matters little that +5 Vdc is maintained at the output of the DC power supply. It matters a lot, however, that the voltage at the microcomputer printed circuit board is +5 Vdc. If the (+)SENSE line is connected to the +5 VDC bus of the computer and the (−)SENSE line is connected to the ground bus, the feedback power supply will keep the voltage at the rated value at the PCB, not at the power supply. This method servos out $I \times R$ drop in the power supply lines.

## 20-8 THREE-TERMINAL IC VOLTAGE REGULATORS

Voltage regulators for low current levels (up to 5 amperes) are reasonably simple to build now that simple three-terminal IC regulators are available. The circuit used with positive three-terminal regulators is shown in Fig. 20-20. Typical package styles are

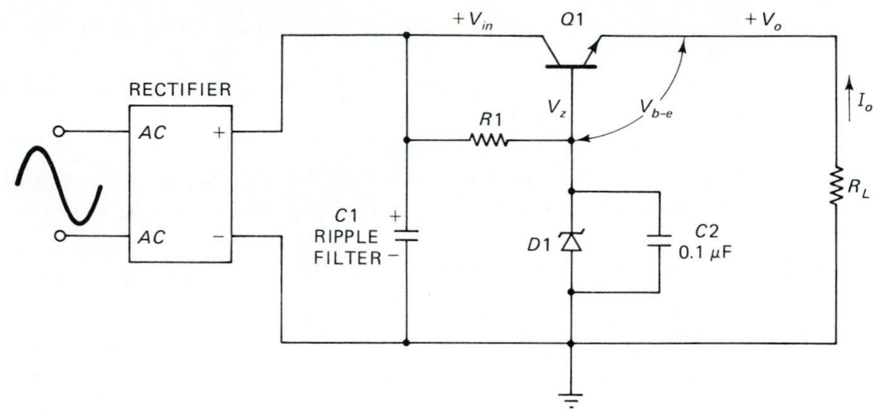


FIGURE 20-18

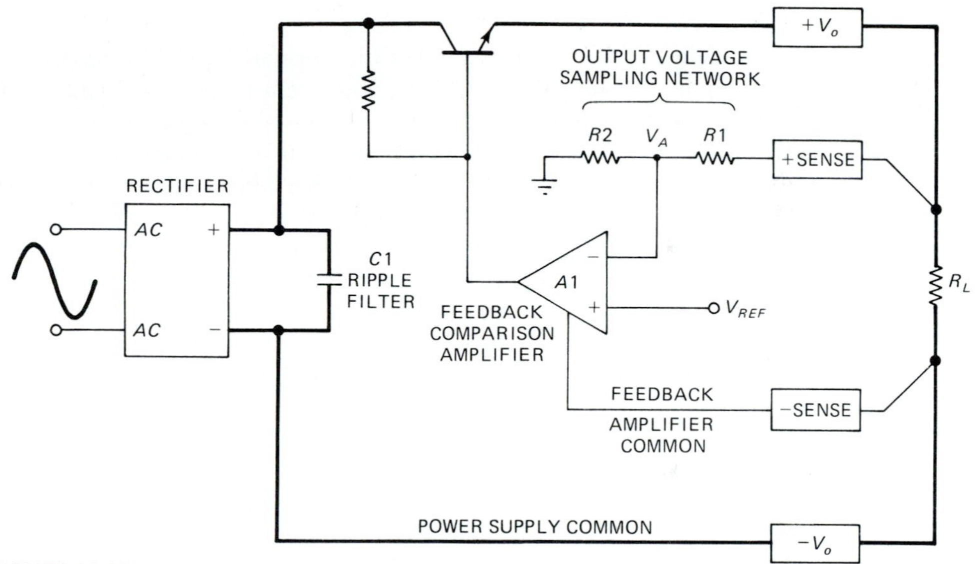


FIGURE 20-19

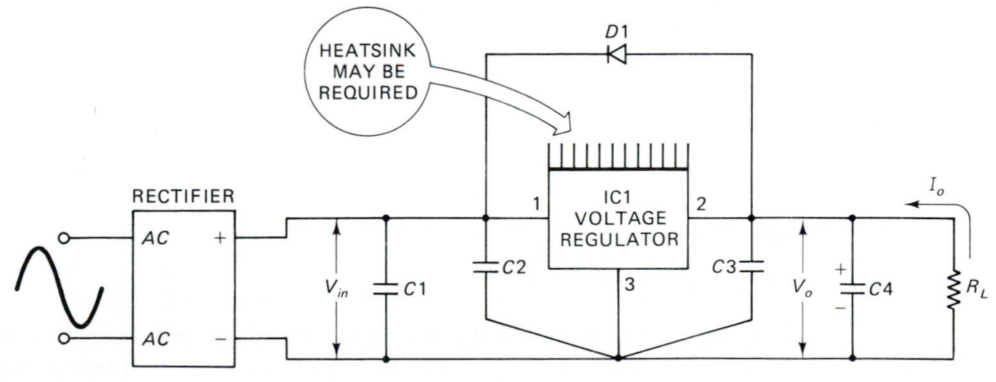


FIGURE 20-20

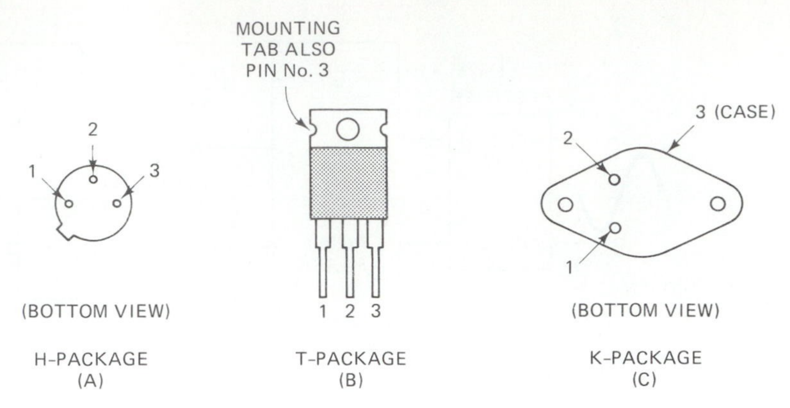


FIGURE 20-21

shown in Fig. 20-21. Capacitor *C*1 is the normal ripple filter capacitor and it should have a value of 1000 μF/ampere of load current (some authorities insist on 2000 μF/ampere). Capacitor *C*4 is used to improve the transient response to sudden increases in current demand (something that happens in digital circuits). Capacitor *C*4 should have a value of approximately 100 μF/ampere load current. Capacitors *C*2 and *C*3 are used to improve the immunity of the voltage regulator to transient noise impulses. These capacitors are usually 0.1 μF to 1 μF, and are to be mounted as close as possible to the body of the voltage regulator *IC*1.

Diode *D*1 is not shown in many circuits, but it is highly recommended for applications where *C*4 is used. If the diode is not present, charge stored in *C*4 would be dumped back into the regulator when the circuit is turned off. That current has been implicated in poor regulator reliability. The mechanism of failure is the normally reverse-biased PN junction formed by the IC regulator substrate and the circuitry become forward-biased under these conditions. This situation allows a destructive current to flow. The diode should be a 1 ampere type at power supply currents up to 2 amperes, and larger for larger current levels. For most low voltage (1 ampere or less) supplies a 1N400x is sufficient.

Several three-terminal IC voltage regulator packages are shown in Fig. 20-21. The H package (Fig. 20-21A) is used at currents up to 100 mA, the TO-220 T package (Fig. 20-21B) at currents up to 750 mA, and the TO-3 K package (Fig. 20-21C) at currents to 1 ampere.

There are two families of IC regulator. One is designated 78*xx*, in which the *xx* is replaced with the fixed output voltage rating. Thus, a "7805" is a 5 volt regulator, while a "7812" is a 12 volt regulator. The LM-340*y*-*xx* series is also used. The *y* is the package style (*H*, *K*, or *T*). The *xx* is the voltage. Thus, an "LM-340K-05" is a 1 ampere, 5 volt regulator in a similar-to-TO3 type-K package. An LM-340T-12 is a 12 volt, 750 mA regulator in a plastic TO-220 power transistor package.

Negative versions of these regulators are available under the 79*xx* and LM-320*y*-*xx* designations. Figure 20-22 shows the typical circuit symbol. Note that the pin-outs on the voltage regulator device are different from those of the positive regulator.

The minimum input voltage to the three-terminal IC voltage regulator is usually

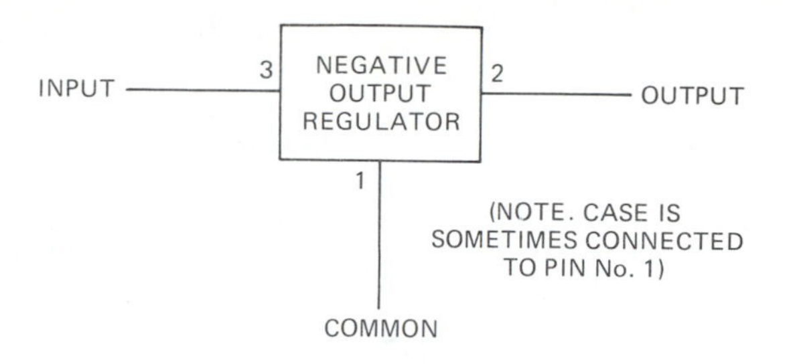


FIGURE 20-22

2.5 volts higher than the rated output voltage. Thus, for a +5 volt regulator, the minimum allowable input voltage is 7.5 Vdc. The power dissipation is proportional to the voltage difference between this input potential and the output potential. For a 1 ampere regulator the dissipation will be 2.5 watts if the minimum voltage is used, and considerably higher if a higher voltage is used. It is recommended that an input voltage which is close to the minimum allowable voltage be used. For +5 volt supplies used in digital projects, a standard 6.3 Vac transformer is sufficient. When fullwave rectified and filtered with 1000 μF/ampere or more, the output voltage will be approximately +8 Vdc. The student is encouraged to work out the arithmetic to prove this statement.

Figure 20-23 shows a dual-polarity DC power supply such as might be used in operational amplifier circuits, some microcomputers, and other applications. The voltage regulator portion of the circuit is a combination of positive and negative output versions of Fig. 20-20. The transformer/rectifier section bears some explanation, however. The rectifier is a 1 ampere bridge stack, but is not used as a regular bridge. The center-tap on the secondary of transformer *T*1 establishes a zero-reference, so the bridge actually consists of a pair of conventional full-wave bridges connected to the same AC source. Thus, the (−) terminal of the bridge stack drives the negative voltage regulator, and the (+) terminal drives the positive voltage regulator. This rectifier is sometimes called a half-bridge rectifier.

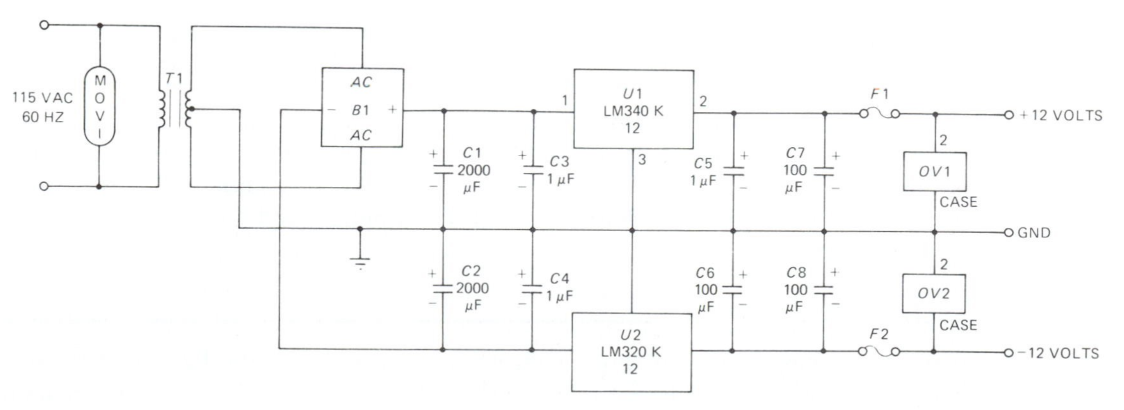


FIGURE 20-23

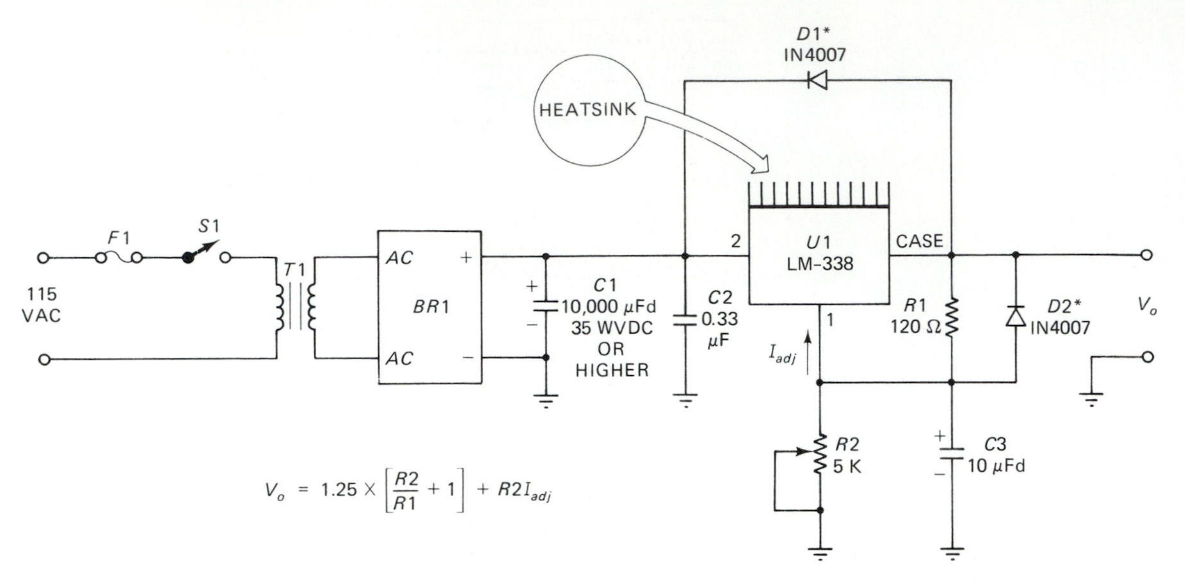


FIGURE 20-24

### 20-8.1 Adjustable IC Voltage Regulators

The IC voltage regulators discussed above are fixed types. They offer an output voltage that is predetermined and unchangeable without extraordinary effort. A variable output voltage regulator, on the other hand, can be programmed to any voltage desired within its range. These devices can be used for either variable DC power supplies, or to supply custom output voltages other than those allowed by the standard fixed voltages. Two similar models are considered here as examples.

The LM-317 and LM-338 are variable DC voltage regulators capable of delivering up to 1.5 amperes and 5 amperes, respectively, at voltages up to +32 volts DC. Figure 20-24 shows a typical circuit for these regulators. The input voltage must be 3 volts higher than the maximum output voltage. The output voltage is set by the ratio of two resistors, $R1$ and $R2$, according to the equation

$$V_o = [1.25 \text{ volts}] \left[\frac{R2}{R1} + 1\right] \tag{20-11}$$

An example from the National Semiconductor *Linear Databook* shows 120 ohms for $R1$, and a 5 kohm potentiometer for $R2$. This combination produces a variable output voltage of 1.2 Vdc to 25 Vdc, when $V_{in}$ is >28 Vdc. Diode $D1$ can be any of the series 1N4002 through 1N4007 for LM-317 supplies, and any 3 ampere type for LM-338 supplies.

## 20-9 OVERVOLTAGE PROTECTION

If the series-pass transistor shorts collector-to-emitter, or if the base control circuit fails, the input voltage will be applied to the output terminal of the voltage regulator. Because this voltage is often considerably higher than the regulated output voltage,

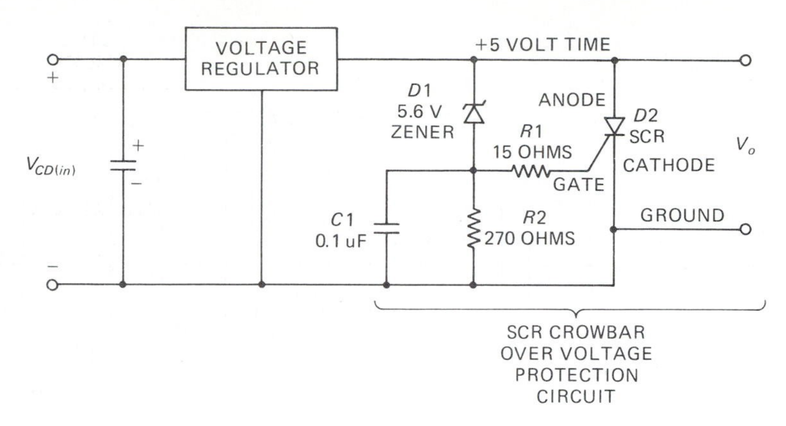


FIGURE 20-25

serious damage can result to the electronic circuitry powered by the regulator. A protective circuit called an *SCR crowbar* provides *overvoltage protection* to such circuits.

The SCR crowbar (Fig. 20-25) is shunted across the output of the power supply ($V$ is the power supply output voltage). A +5 Vdc supply is used as an example. Diode $D1$ is a zener diode which has a zener potential a little higher than the rated output voltage. Zener voltage from 5.6 to 6.8 Vdc can be used to protect a +5 Vdc power supply. Power supplies with other output voltages than +5 Vdc can be protected by scaling the zener diode voltage proportionally.

Diode $D2$ is a silicon controlled rectifier (SCR). This type of diode is open-circuited (a high resistance in both directions) until a current is caused to flow in the gate terminal. When this occurs, the SCR breaks over and operates as any ordinary PN junction diode.

The gate terminal of diode $D2$ is controlled by the network around diode $D1$. When the power supply voltage V exceeds the zener potential of $D1$, a current is conducted through $D1$ creating a voltage drop across $R2$. This voltage drop becomes the source for the gate current that flows in $R1$ and the gate of $D2$. At this instant, the SCR becomes conductive and shorts out the power supply line. Either a fuse or fuse resistor can be connected in series with the DC line, and will open-circuit when $D2$ conducts. The fuse should be in series with the positive input line of the voltage regulator. In a few cases, there is no fuse. In those circuits, the SCR ($D2$) must have a large current rating because it must carry the short-circuit current of the power supply. It will clamp the output to near ground level until the circuit is turned off.

*Lambda Electronics, Inc.* offers overvoltage protection integrated circuits. Current levels available are 2, 6, 12, 20, and 35 amperes, with available voltage levels being 5, 6, 12, 15, 18, 20, 24, 28, and 30 volts. Several package styles are used, which also indicate the ampere level. A TO-66 power transistor case is used for 2 amperes. The TO-3 power transistor case is for 6 amperes. Higher current levels are packaged in epoxy cases. The Lambda overvoltage protection modules are designated with a type number of the form L-*y*-OV-*xx*, in which *y* indicates ampere level and *xx* indicates voltage. Thus, an L-6-OV-5 device is a 5 volt, 6 ampere model.

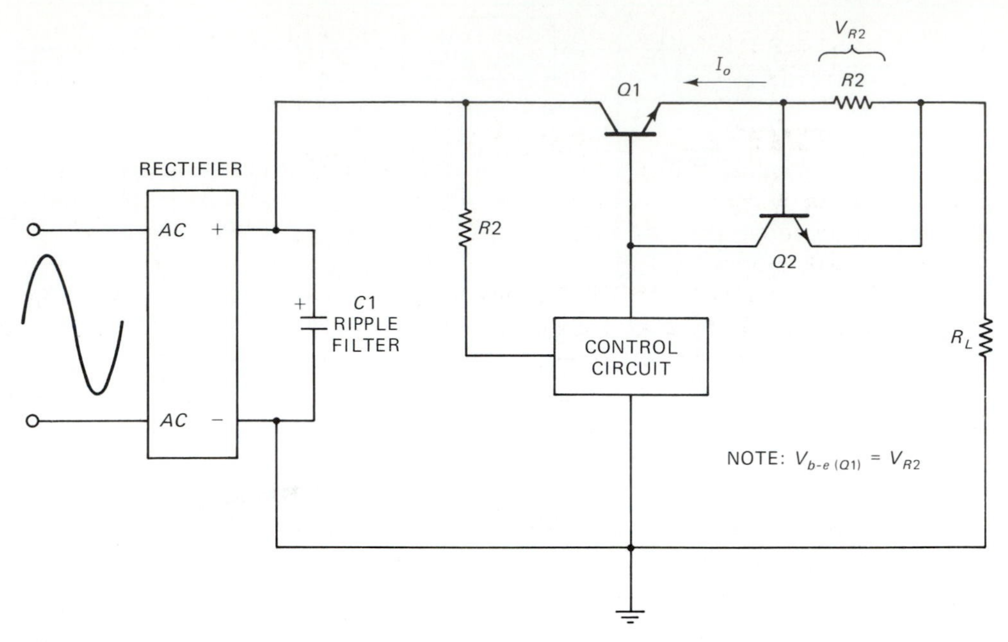


FIGURE 20-26

## 20-10 CURRENT LIMITING

Another catastrophe that can befall a DC power supply is an output short circuit. For unprotected power supplies, this will result in destruction of the circuit. It is possible to place a circuit in the power supply which will provide a "current knee" to shut down the supply above a certain level. Figure 20-26 shows a representative circuit.

Transistor *Q*1 in Fig. 20-26 is the series-pass transistor in a regulated power supply, while *Q*2 is the sense transistor that determines when the current flow is too high. Some IC voltage regulators can also be used with this circuit (and some three-terminal types have the circuit built-in) if they have a sense terminal or some other provision.

Resistor *R*2 is used to sense the level of current flow. The voltage drop across this resistor provides forward bias to *Q*2, and is proportional to the current flow:

$$V_{R2} = I_o \times R2 \qquad (20\text{-}12)$$

For silicon transistors, the forward bias voltage required to saturate the transistor is approximately 0.6 volts. When $V_{R2}$ exceeds this critical voltage, transistor *Q*2 is heavily forward-biased so its $V_{c-e}$ drops to a very low value, essentially shorting the base-emitter terminals of *Q*1. This action turns off the power transistor. The value of *R*2 is therefore

$$R2 = 0.6 \text{ Vdc}/I_{o(\max)} \qquad (20\text{-}13)$$

Consider a practical example. Suppose a computer power supply delivers 10 amperes maximum. The value of *R*2 would be

$$R2 = 0.6 \text{ Vdc}/I$$

$$R2 = 0.6 \text{ Vdc}/10 \text{ amperes}$$

$$R2 = 0.06 \text{ ohms}$$

A value of 0.06 ohms (60 milliohms) seems difficult to achieve, but such a resistor can be made from fine wire. Alternatively, several wirewound power resistors or fuse resistors can be connected in parallel to form the low value required. For example, a 0.33 ohm resistor is often used as a "fusistor" in auto radios or the emitter resistor in audio power amplifiers. Five of these resistors in parallel produce nearly 60 milliohms. Various values of fusistors are available from 0.09 to 1.5 ohms. These can be paralleled in assorted combinations to produce the required resistance.

## 20-11 HIGH VOLTAGE TRANSIENT PROTECTION

Experts warn that 20 μS to 500 μS transient pulses of 1500 volts or more strike residential and small business power lines several times per day. In industrial facilities that number may be considerably greater because of the heavy electrical machinery often in use. Until digital electronics devices, including computers, were widespread, however, this fact was interesting but trivial. But high voltage transient pulses can seriously disrupt digital circuits. The circuit may simply fail to operate correctly or can be damaged by the transient pulse. If a computer seems to occasionally "bomb out" while executing a program that ran properly only a short while ago, suspect these transient pulses as the root cause.

Figure 20-27 shows *metal oxide varistor* (MOV) devices shunted across the power

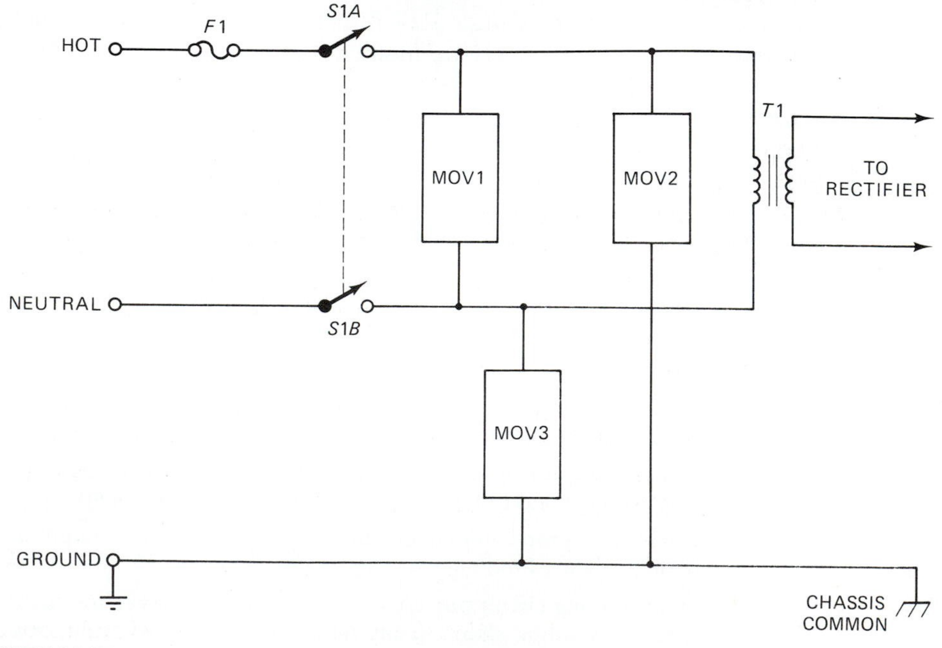


FIGURE 20-27

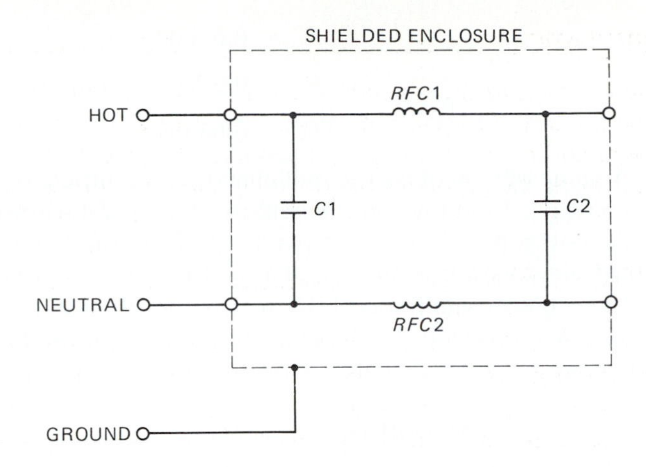


FIGURE 20-28

lines. Normally, only MOV1 will be needed, but MOV2 and MOV3 are recommended in serious cases. These devices can be modeled as a pair of back-to-back zener diodes with a $V_z$ rating of about 180 volts. The purpose of the MOV devices is to clip transient pulses over 180 volts.

Some applications require an LC low-pass filter on the AC power lines. In some very severe transient cases where the MOV is not sufficient, or, in cases in which a strong RF field is present (as in a radio transmitter), or, in cases where the digital device creates RFI, then the LC filter of Fig. 20-28 might be indicated. This filter should be mounted as close as possible to the point where the AC power cord enters the equipment.

Several manufacturers offer RFI filters which are shielded and especially suited for this service. Some models are molded inside of a chassis mounted AC receptacle.

## 20-12 SUMMARY

1. The DC power supply consists of a *transformer* to scale voltage levels, a *rectifier* to convert bidirectional AC to unidirectional pulsating DC, a *ripple filter* to smooth pulsating DC to nearly pure DC, and (in some circuits) a *voltage regulator* to stabilize the output voltage.
2. The rectifier is a PN junction diode. It should have a peak inverse voltage rating of $>2.83$ times the applied AC (RMS) voltage, and a forward current rating high enough to handle the full required load current.
3. The ripple filter may be a single capacitor shunted across the load, or an RC pi-network. The filter smooths the ripple to nearly pure DC.
4. The voltage regulator is used to stabilize the output voltage in spite of fluctuation of the input AC voltage and changes in the load current requirements.
5. An overvoltage protection circuit turns off the DC power supply in the event a fault occurs that produces too high an output voltage level.
6. A current limiting circuit provides protection against either too high an output current demand, or an output short circuit. When the output current increases above a certain point the power supply shuts down.

## 20-13 RECAPITULATION

Now return to the objectives and Prequiz questions at the beginning of the chapter and see how well you can answer them. If you cannot answer certain questions, mark them and review the appropriate parts of the text. Next, try to answer the questions and work the problems below, using the same procedure.

## 20-14 STUDENT EXERCISES

WARNING. Some of the exercises in this section involve working with 115 volt AC power from the mains. This type of power is extremely dangerous if not handled properly. Ideally, you should be working on a bench equipped with isolation transformers. Do not under any circumstances touch exposed AC power line wires, and don't even think about working on the exercises "hot" (with power applied). Ask your instructor for information on safely working with AC power circuits.

1. Use a low-voltage AC power transformer (12.6 Vac) rated at a current of at least 500 mA, and a silicon rectifier diode (1N4002 through 1N4007) to produce a half-wave rectifier circuit. Connect a 100 ohm, 5 watt (or more) resistor across the output of the rectifier. Measure the DC output voltage across the resistor, and compare with the calculated values for peak and RMS voltage.
2. Perform the exercise above, and then repeat the experiment using 100 μF/50 WVDC, 1000 F/50 WVDC, and 10,000 μF/50 WVDC filter capacitors. Compare the results.
3. Build the circuit of a DC power supply (see two previous exercises). Add a 12 volt three-terminal IC voltage regulator. Measure the output voltage, output current, and input voltage to the regulator.
4. Build a negative output voltage regulated DC power supply.
5. Design and build a ±12 Vdc/1 ampere bipolar DC power supply with a common shared between the two polarities.

## 20-15 QUESTIONS AND PROBLEMS

1. What is the principal job of a rectifier? A ripple filter?
2. Draw the circuits of (a) halfwave rectifier, (b) fullwave rectifier, and (c) fullwave bridge rectifier.
3. A rectifier is connected across a load resistor and a shunt capacitor that serves as a ripple filter. What is the minimum peak inverse voltage rating if the transformer delivers 25.6 Vac rms to the rectifier?
4. List the key specifications for a rectifier diode.
5. State the maximum operating current and output potential of an LM-340K-15 three-terminal IC voltage regulator.
6. A zener diode regulator must regulate a DC supply to +5.6 Vdc. The input DC source varies from 8 to 10 volts. Assume a constant load current of 60 mA. Find the value of series resistor, its power dissipation, and the power dissipation of the zener diode.
7. A single capacitor ripple filter in a 12 Vdc power supply is required to reduce the fullwave

ripple to 0.9 when the load current is 10 amperes. Calculate the value of filter capacitance needed.

8. Calculate the ripple factor in a 60 Hz halfwave rectifier circuit if a 12.6 Vac (rms) transformer is used to supply a current of 0.800 amperes; the filter capacitor is 2200 μF.
9. A 12 Vdc power supply must deliver 500 mA approximately 50% of the time. What package should be specified for a three-terminal IC regulator which will do this job?
10. A 270 Vdc power supply is filtered by a 60 μF/350 WVDC capacitor. The voltage can vary ±15%, and the filter capacitor has a ±20% tolerance in the WVDC rating. Calculate the "worst case" applied voltage and WVDC rating to determine if the capacitor selection was appropriate.
11. A DC power supply delivers 10 Vdc at 250 mA to a resistive load. Calculate the ripple factor of the output from the Pi-section RC filter if $R1 = 50$ ohms, $C1 = 2000$ μF, and $C2 = 220$ μF.
12. Find the percentage of regulation if a DC power supply output voltage drops from 14 to 10 volts as the output current is raised from zero to 1.5 amperes (which is the maximum allowable output current for this power supply).
13. A 9.1 Vdc zener diode must regulate in the presence of a supply voltage that varies from +16 to +22 volts. The current remains constant at 140 milliamperes (0.14 amperes). Calculate the value of current limiting resistor $R1$, the power dissipation of $R1$, and the power dissipation of $D1$. Recommend minimum values for the power ratings of $R1$ and $D1$.
14. An LM-338 variable three-terminal IC voltage regulator is connected with the following resistor values: $R1 = 120$ ohms and $R2 = 3300$ ohms. Find (a) the output voltage, (b) the minimum allowable input voltage.
15. A current limiting circuit is being designed for a 22 ampere DC power supply. Calculate the value of series resistor needed to sense the current level.

CHAPTER 21

# Troubleshooting Discrete and IC Solid-State Circuits

## OBJECTIVES

1. Learn the type of test equipment typically needed for troubleshooting solid-state circuits.
2. Understand the difference between AC-path and DC-path troubleshooting.
3. Learn the DC method for troubleshooting circuits.
4. Learn the *signal injection* and *signal detection* methods for troubleshooting cascade circuits.

## 21-1 PREQUIZ

These questions test your prior knowledge of the material in this chapter. Try answering them before you read the chapter. Look for the answers (especially those you answered incorrectly) as you read the text. After you have finished studying the chapter, try answering these questions again and those at the end of the chapter (see Section 21-10).

1. An NPN transistor common emitter amplifier is operated with a +12 Vdc potential applied through a load resistor to the collector. The 680 ohm emitter resistor is connected to ground. Calculate the emitter voltage drop. Is this transistor stage conducting?
2. The ____________ ____________ method of troubleshooting cascade circuits involves connecting a signal generator to the input, and then systematically looking for the signal at the output of each stage in succession with an oscilloscope.
3. A 0.01 Hz to 5 MHz function generator has an output impedance of 600 ohms. When troubleshooting a standard 4 MHz RF amplifier ($Z_{in} = Z_{out} = 50$ ohms) it is noted the signal voltage is considerably lower than one would expect. The fault could be traced to improper selection of test equipment. *True or false?*
4. List the types of test equipment typically used in solid-state troubleshooting.

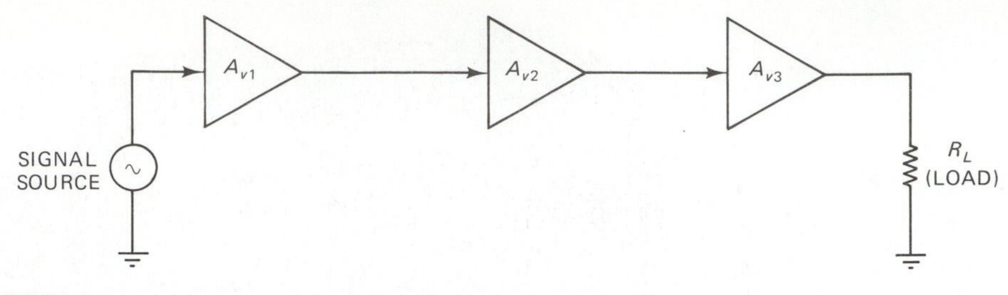


FIGURE 21-1

## 21-2 INTRODUCTION

It is one of those unfortunate facts of life that electronic circuits do not always work as expected. Whether in older equipment, that worked well for a long time before failing, or in newly constructed circuits that need to be debugged, certain troubleshooting skills are needed. The skills involved in troubleshooting electronic circuits are easily learned and with practice become almost second nature. These skills are based on easily learned logical processes. You can, in many cases, follow step-by-step procedures to find a fault in a circuit. For sake of illustration a multistage cascade amplifier (Fig. 21-1) is used as a general model of how to troubleshoot. Keep in mind that nearly all troubleshooting problems reduce to finding one of the following two conditions:

1. A *missing*, but required, path for current.
2. An added, undesired, path for current.

These current paths can be in either the AC path or the DC path of the circuit. Examples of the DC path include bias and load resistors. The AC path consists of transformers and coupling capacitors. It is also possible for a single component to be in both the AC and DC paths. Examples of the latter situation include active devices (such as ICs), diodes, and transformers.

There are several approaches to troubleshooting circuits, and we will examine them in detail below. But first, let's examine the matter of typical test equipment needed to troubleshoot solid-state circuits.

## 21-3 TEST EQUIPMENT

Unlike the old days of radio, when a wet finger and a good ear were "test equipment," electronic troubleshooting must be done with a variety of test instruments. The range of required instruments is wide because the ranges of possible circuits and applications are wide. As a result, it is impossible in a short textbook section to lay out all of the instruments that will ever be used. It is possible, however, to make some general statements about the classes and types of instruments needed. The instruments will fall into several different classes, signal sources, voltmeters, and oscilloscopes.

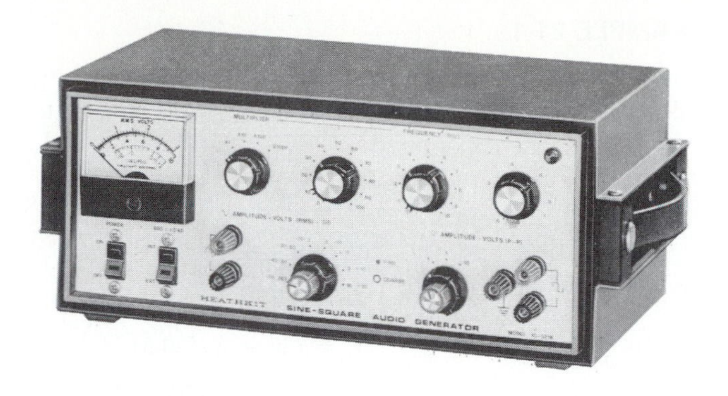


FIGURE 21-2

### 21-3.1 Signal Generators

A signal generator produces a controllable output waveform which can be used to excite circuits under test or troubleshooting. Although there is an immense variety of signal generators available, only a few very general categories are recognized, *audio generators*, *function generators*, and *RF generators*. While it can be argued successfully that these classes are too restrictive in the general sense, they are reasonable for our present discussion.

**Audio Signal Generators.** The audio generator produces a controllable output signal from approximately 20 Hz to 20,000 Hz. All audio generators produce a sinewave output signal. Most also produce a squarewave signal. If any particular model produces more than these two waveforms, it actually belongs in the *function generator* class.

Many commercial and broadcast audio circuits are based on a system impedance of 600 ohms. As a result, the standard audio generator is designed to have an output impedance of 600 ohms.

There are two basic forms of audio generators. They are different in the manner in which frequency is selected. The continuously tuned variety allows the operator to turn a single knob to select all frequencies within the set range. A separate range control is used to set the frequency range covered by the frequency setting knob. The precision to which this instrument can be set to a particular frequency is limited by the dial scale calibration, the width of the pointer or indicator, as well as by electronic factors such as the frequency control mechanism. Although exceptions exist, the continuously tunable generator is used on low resolution, low accuracy applications and noncritical troubleshooting. Its use in troubleshooting is limited to those applications where accuracy is either not needed or can be augmented with an external frequency counter.

The other category of signal generator is the stepped variety shown in Fig. 21-2. In this type of generator the range switch is calibrated in powers of ten (X0.1, X1, X10, X100, and X1000). There are two step selectable sub-range switches calibrated X10 through X100 and X1 through X10 respectively. A continuously variable vernier sets the exact frequency within the range and is calibrated 0 to 1. The set frequency is determined from the multiplication factor and adding all switch settings.

### EXAMPLE 21-1

A step selectable audio generator has four controls $A$, $B$, $C$, and $D$. These controls are set as follows, $A = 1000$, $B = 30$, $C = 6$, and $D = 0.5$. Calculate the output frequency.

$$A = 1000$$

$$B = 30$$

$$C = 6$$

$$D = 0.5$$

$$E = (30 + 6 + 0.5) \times 1000$$

$$F = 36.5 \times 1000 = 36{,}500 \text{ Hz}$$

■

The decibel scale normally used in audio work (called the *VU scale*) measures level in volume units in which the reference 0 dB (or 0-VU) is defined as 1 milliwatt at 1000 Hz dissipated in a 600 ohm load. The term *volume unit* is used instead of decibels to differentiate the modern scale from an archaic telephone scale which was based on a reference level definition of 0 dB as 6 mW at 1000 Hz dissipated in a 500 ohm load. The VU scale is nonetheless a decibel scale, and the power decibels equation (VU = 10LOG[P1/P2]) is used for VU scale work.

One of the principal differences between low-cost and high-cost audio signal generators is in the output metering and control circuitry. A low-cost instrument will have a poorly calibrated continuously adjustable output control and may or may not have an output meter. A higher grade instrument uses a precision attenuator with both *coarse* and *vernier* controls. The course control sets the range. The vernier offers fine control continuously over the range. The output meter may be calibrated in either volts (rms), VU, or decibels; some instruments combine two or all three of these scales into one meter. The advantage of the precision version is it allows exact output voltages to be produced for purposes of measuring gain or other circuit parameters.

**Function Generators.** The function generator is much like the audio generator, except that it outputs at least sinewave, squarewave, and triangle waveforms. Some function generators also output sawtooth and/or pulse waveforms. One difference between the function generator and the audio generator is the function generator usually has a wider frequency range. While audio generators cover from 20 Hz to 20,000 Hz, with some offering wider ranges of 1 Hz to 100 KHz (as in the case of Fig. 21-2), the function generator typically offers 0.1 Hz (or less) to 2 MHz. At least one function generator is available with an 11 MHz output frequency.

Most function generators do not have the precision attenuators or output metering that better audio generators offer. Most function generators have a 600 ohm output impedance, but some also offer in addition to the standard 600 ohm output impedance other common values, 50 ohms for RF circuits and TTL-compatible (LOW = 0 volts and HIGH > 2.4 volts) for use in digital circuits. The output metering, if used, is usually calibrated to the 600 ohm output.

The output impedance of any signal generator creates a problem in some circuits. For ideal power transfer, the source impedance of the signal generator must equal the load impedance it is driving. But when several impedances are available, it is

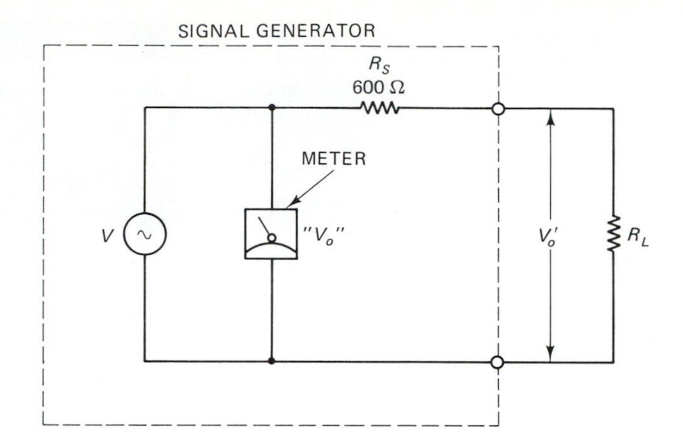


FIGURE 21-3

possible for an erroneous output to occur. Consider, the example, the case where a 600 ohm signal generator is used to drive a 50 ohm load. The output meter assumes that the load is 600 ohms, so the actual output voltage will depart from the metered value by an amount determined by the voltage divider equation

$$\text{Error} = \frac{R_L}{R_L + R_s} \tag{21-1}$$

Consider Fig. 21-3. The oscillator inside the signal generator produces a voltage $V$, which is passed to the output through source impedance $R_s$. The meter will read $V_o$ accurately only if $R_L = R_s$. But when these impedances are not equal, the output meter (or setting, if no meter is used) will be in error by the amount described in Eq. (21-1).

## EXAMPLE 21-2

A 600 ohm output signal generator is connected to the input of a 50 ohm circuit. Calculate the error in the output reading if this occurs.

### Solution

$$\text{Error} = \frac{R_L}{R_L + R_s}$$

$$\text{Error} = \frac{50 \text{ ohms}}{(50 \text{ ohms}) + (600 \text{ ohms})}$$

$$\text{Error} = 50/650 = 0.077$$

or, expressed as a percent, the output will be 7.7% of the meter reading. ■

**RF Generators.** The RF generator generally produces signals above the audio range, and maybe well into the gigahertz range. Although linear IC devices only

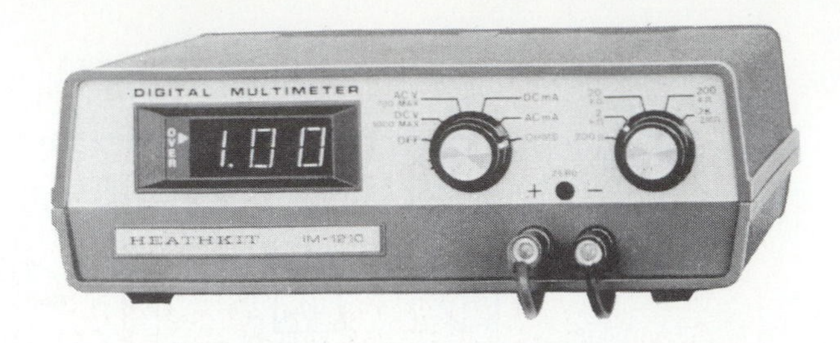


FIGURE 21-4

operated to about 100 MHz (and most only in the audio region) only a few years ago, there are now microwave ICs (MMICs) on the market. One might, therefore, see gigahertz level RF signal generators specified to test linear IC devices. The RF signal generator may also produce signals down to the upper end of the audio range. One model, for example, produces signals down to 10 KHz. Very Low Frequency (VLF) radio stations use these frequencies. There is no difference between a 10 KHz audio signal and a 10 KHz radio signal inside the electrical circuits. They differ in the environment, however, because audio signals generate air pressure variations (an acoustic wave), while RF signals produce an electromagnetic (or radio) wave. The difference between an RF and an audio generator operating on the same frequency is usually the output impedance. RF generators typically produce a 50 ohm output impedance, while audio generators produce 600 ohms.

### 21-3.2 Multimeters and Voltmeters

The multimeter (Fig. 21-4) is an instrument that will measure DC voltage, AC voltage (RMS), current, and resistance. Some models will also measure certain other parameters, such as capacitance, or will offer either decibel, VU or peak reading versions of the AC voltage function in addition to the normal RMS reading. For purposes of troubleshooting either analog or digital models are sufficient. It is important, however, to be able to measure down to tenths of a volt in order to correctly determine whether transistors are correctly biased. Only a few applications require either extreme precision or accuracy for troubleshooting purposes.

### 21-3.3 Oscilloscopes

If you are unable to have more than one instrument on a troubleshooting bench, then be sure to select an oscilloscope! The 'scope is the single most useful instrument for troubleshooting. Figure 21-5 shows a typical two-channel, high frequency oscilloscope designed for troubleshooting purposes. Although single-channel models are useful also, it is best to select a two-channel model because it is often necessary to compare two waveforms to inspect their time relationship.

## 21-4 METHODS OF TROUBLESHOOTING

There are two basic procedures for troubleshooting the circuit on a stage-by-stage basis, *signal tracing* and *signal injection*. Both of these methods involve methodically tracking a signal through successive stages in order to find where it is successfully getting through and where it is not.

FIGURE 21-5

*Signal injection* is shown in Fig. 21-6. In this case the output indicator is either the loudspeaker, or an oscilloscope, AC voltmeter or signal tracer across a "dummy load" used in place of the loudspeaker. The signal source is a signal generator capable of producing signals within the amplifier passband. In most receivers the generator must be capable of producing the kind of signals the circuit responds to (sine wave, square wave, pulse).

In signal injection, start at the output stages and inject a signal into the input of each stage in succession. If that stage and the stages beyond it are working, the signal will be detected on the output indicator. The signal generator output lead is then moved one stage closer to the receiver input and the test is tried again. When the bad stage is found, injecting a signal either fails to produce an output indication or the output is considerably weaker than before.

The input of the amplifier may require termination when using the signal injection method. Open amplifier inputs often lead to either self-oscillation or DC latchup of the output. In order to avoid these problems it is necessary to either short the input(s) to ground or terminate it in a specified resistance.

*Signal tracing* technique is shown in Fig. 21-7. This method differs from the signal injection technique in that it uses a fixed signal source at the input. The detector is then moved from stage to stage to find the stage where signal is either lost or becomes very weak. Using the reverse logic of signal injection, we can just as easily home in

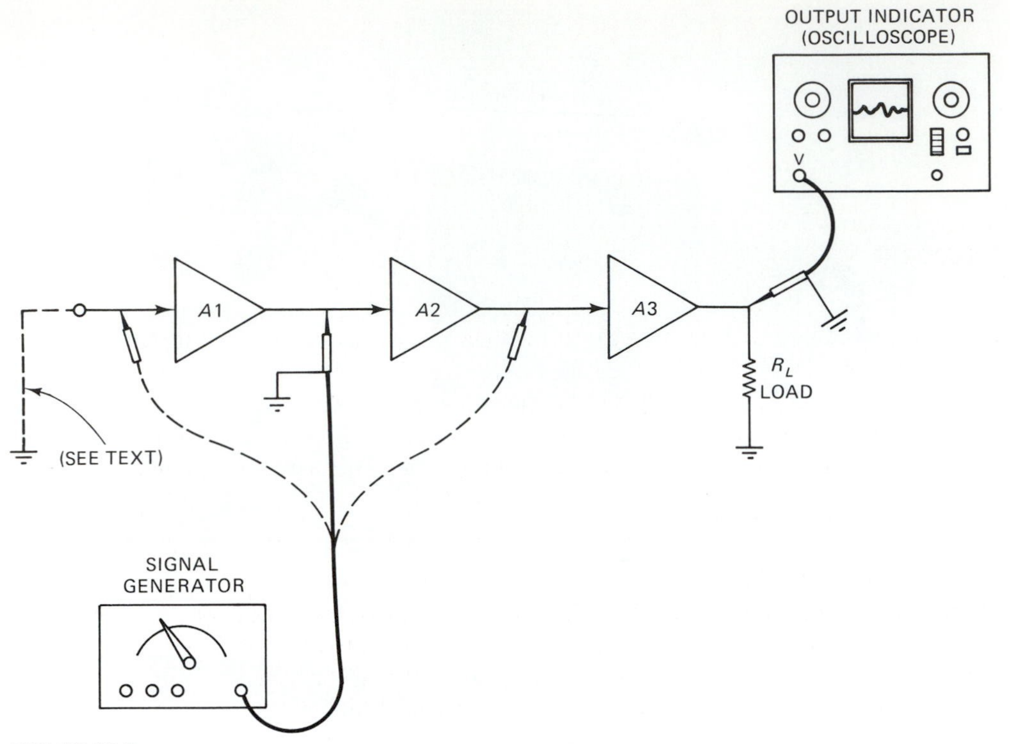


FIGURE 21-6

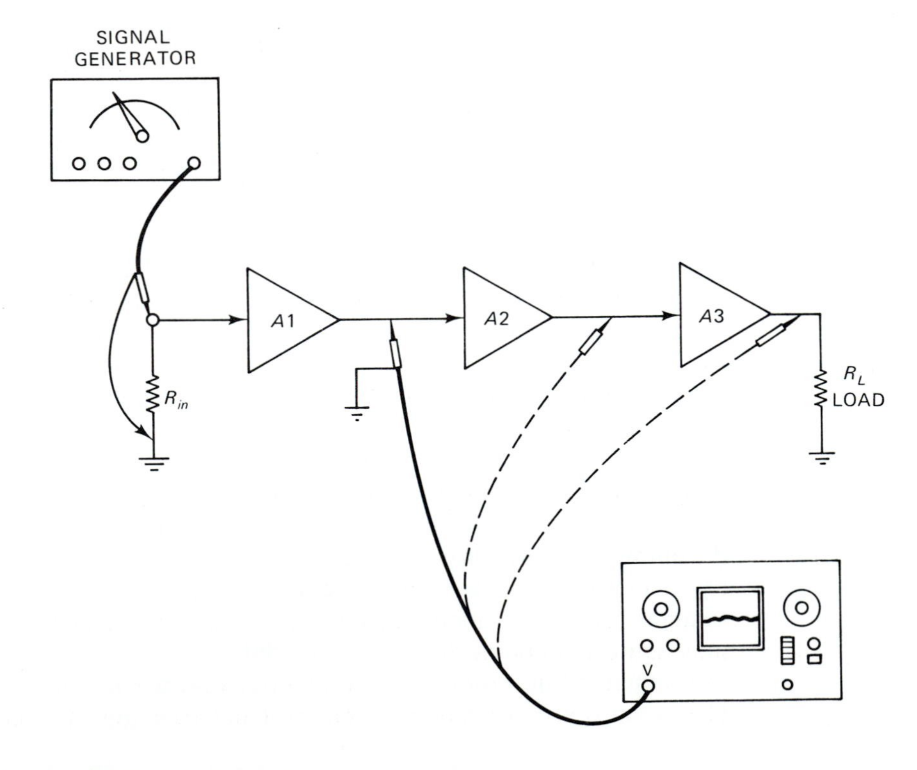


FIGURE 21-7

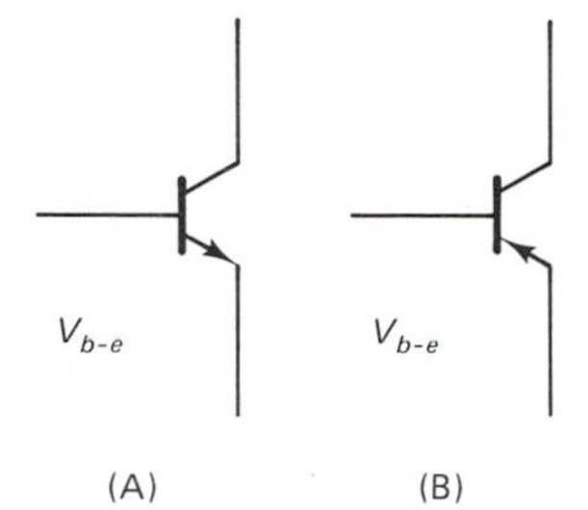


FIGURE 21-8

on the dead stage. One note of caution, however, signal tracers are high gain audio amplifiers (with a demodulator probe for detecting R.F. if necessary). Setting the *gain* control too high yields false indications when the signal is weak but present.

Once the dead stage is located other methods are used to pinpoint the component at fault. For example, DC methods can locate a bad transistor or linear IC.

## 21-5 TROUBLESHOOTING SOLID-STATE CIRCUITS

The sections above discussed methods for isolating the dead stage in a typical cascade amplifier. The methods of signal injection and signal tracing are capable of showing us which stage is bad. Then we can do further troubleshooting to isolate the bad component. In this section we will discuss a method for troubleshooting solid-state circuits. This method uses a DC voltmeter to isolate the bad stage. The method is not foolproof, but is well suited to many applications, especially when combined with other methods.

First examine Figure 21-8. NPN and PNP transistors are shown. In most amplifier circuits the base-emitter voltage will be 0.2 to 0.3 volts for normal germanium transistors (used in older equipment) and 0.6 to 0.7 volts for silicon transistors. The following relationships also exist

1. In a PNP transistor the base is more negative than the emitter.
2. In an NPN transistor the base is more positive than the emitter.

Further, the collector will be more positive than either base or emitter in NPN transistors, and more negative in PNP transistors. Keep in mind that the term "more negative" can be interpreted as "less positive" in some cases. Consider Fig. 21-9. Here we have a cascade chain of three stages. Each stage consists of a PNP transistor powered from a positive DC power supply. The collectors of the transistors are near ground potential, while the emitters and bases are closer to the +10.5 volt "B+" line. If you measure the collector voltage with respect to ground you will find it slightly positive, while the emitters are at a much higher potential. Thus, the collector being "less positive" acts exactly like it was "more negative."

Voltmeter A will measure the potential between the points being examined (in Fig. 21-9 an emitter) and ground. If the voltage is near normal at each stage, we can assume there are no massive short circuits, but we cannot get a hint of whether the stage is working properly.

Voltmeter B is connected with its positive electrode on the B+ distribution line. The other is used to probe the emitters of the stages. The B+ line sometimes can

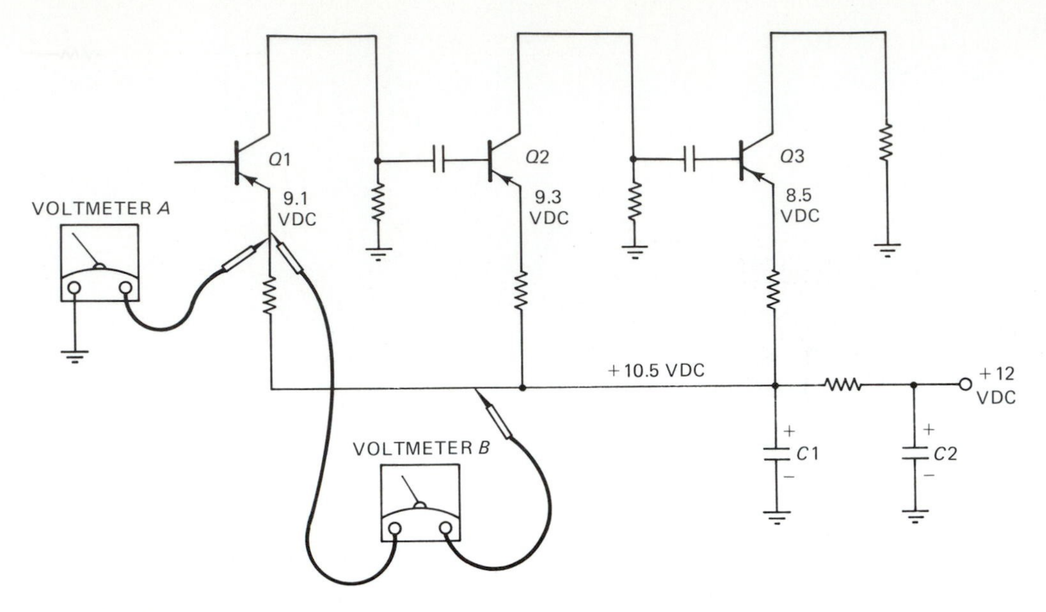


FIGURE 21-9

be located from the circuit diagram or service manual, but in some cases you will have to find it. Look for the electrolytic filter capacitor used to decouple the B+ line (C1 in Figure 21-9). This filter capacitor will denote the proper line (unless you accidentally selected C2) and usually has enough of a solder tab to allow connection of the voltmeter probe.

The voltage drop across the emitter resistor of each stage indicates the current conduction of the stage. If the service manual does not give the normal voltage drop, calculate it as the difference between the emitter potential printed on the schematic and the B+ voltage. In the case of the amplifier of Figure 21-9, for example, the emitter voltage is 9.1 volts, while the B+ voltage is 10.5 volts. The normal conduction of this stage will be 10.5 − 9.1, or 1.4 volts. Any radical departure from this value indicates a problem. For example, a shorted transistor would cause that conduction voltage to increase to nearly 10 volts, while a leaky transistor would place the voltage somewhat lower but still larger than 1.4 volts. Similarly, an open emitter (or other condition that cuts off the stage) will reduce the voltage across the emitter resistor to either zero or nearly zero.

It is possible to isolate the defective stage by looking at each emitter voltage in turn. Most often, the defective stage will show up from an anomaly in the emitter conduction voltages.

Radios with NPN transistors in the stages are similarly treated. Figure 21-10 shows a circuit with NPN transistors powered by a positive-to-ground DC power supply. The collectors of these transistors will be close to the B+ potential, while the emitter and base voltages will be much lower. The values shown in Figure 21-10 are typical, but they are not absolute. There are a lot of design choices that could alter the values, so buy and consult the service manual for the particular equipment.

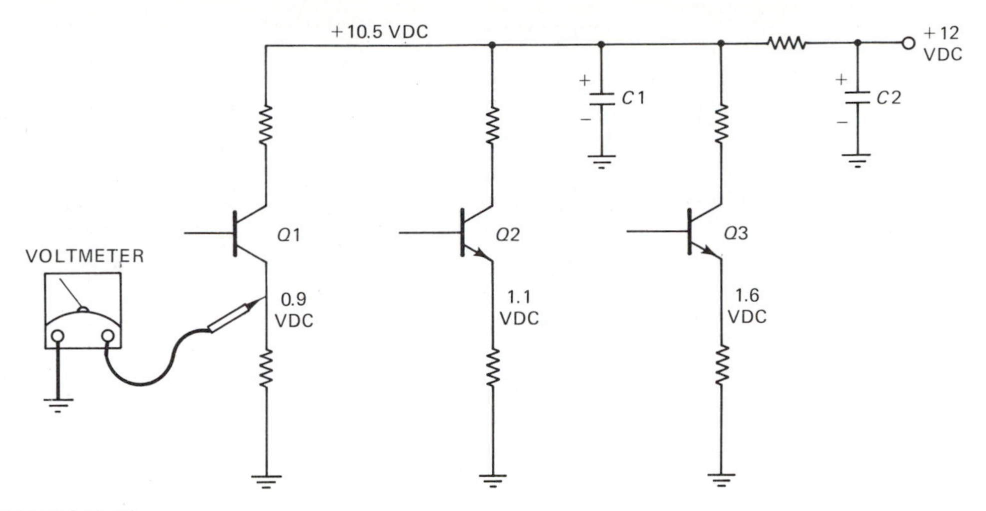


FIGURE 21-10

As in the case of the PNP transistor stages, the NPN emitter conduction voltage denotes the stage activity. In this type of circuit, however, the reference is ground instead of the B+ line. The principle is the same, however. Check each conduction voltage in turn and determine if any of them are incorrect. Fortunately, there are fewer calculations to make in this type of circuit. The emitter voltage on the schematic is the conduction voltage.

Variable frequency oscillators will behave a little differently than amplifiers. In most cases, the DC conduction of the oscillator transistor varies with frequency setting. Typically, the voltage will be higher at the low end of the range, and lower on the high end of the range (with a smooth transition as the frequency is varied). A sudden discontinuity in this transition might indicate a sudden cessation of oscillation (or a parasitic developing) at that point on the dial.

## 21-6 A SIMPLE "SEMICONDUCTOR TESTER"

With a little knowledge, the analog or digital multimeter becomes a transistor and diode tester. If you have an ohms scale, you can measure the diode's front-to-back resistance. From that determine whether or not the diode is leaking. You will even be able to tell which end of an unmarked diode is the cathode or anode.

## 21-7 OHMMETER ORIENTATION

To understand our test method let's first look at the test equipment. The question is "how do analog and digital ohmmeters work?" In analog ohmmeters a battery or electronic DC power source is connected in series with the meter movement and some calibration resistors (Fig. 21-11A). Battery current pushes its way through the meter movement into an external circuit and back to the battery. If we make it difficult for the current to flow by placing a resistance in series, the meter pointer will not swing

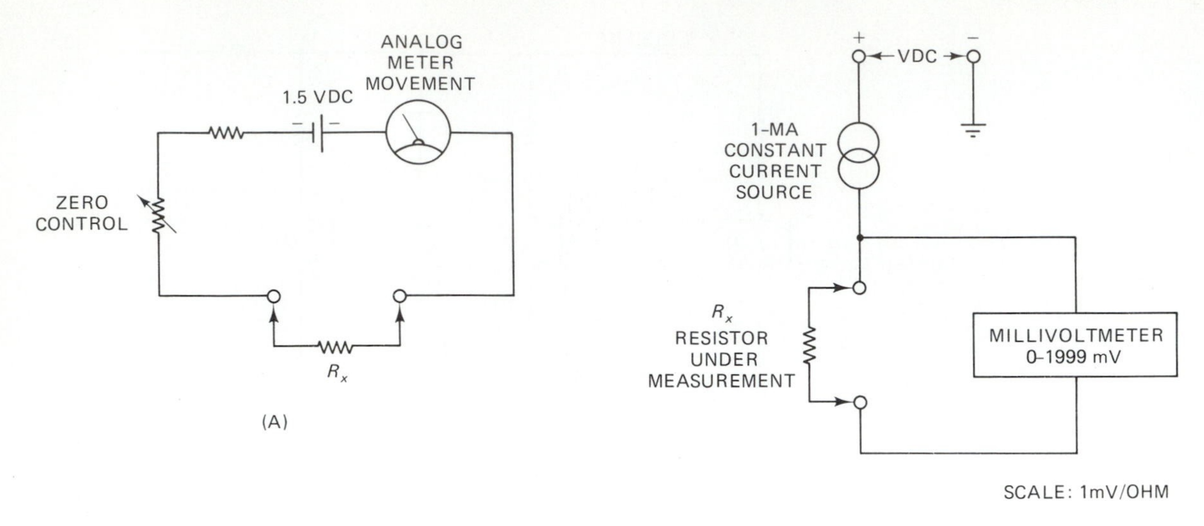


FIGURE 21-11

as far along the dial as when the probes are simply shorted together. That is why you will find high precision resistors in any decent multimeter: they limit the meter pointer travel (and the current) so the pointer always comes to rest at a specific point on the dial. Short an ohmmeter's leads together, and that place will be zero ohms.

When we connect our multimeter across an external resistance (with the multimeter in the *ohms* mode) the meter receives even less current than it did before; so there's less swing on the pointer's part.

Digital multimeters (DMM) use a different tactic for measuring resistance. In Fig. 21-11B we see that a constant current source is used to provide a precision reference current to the unknown resistance ($R_x$). The main circuit of the DMM is a millivoltmeter with a range of 0 to 1999 millivolts. By passing a constant current through the resistance, we can measure the resistance by measuring the voltage drop. For example, a 1 mA constant current produces a voltage drop of 1 mV/ohm. Typically, the constant current will be different for each range but the principle is the same.

### 21-7.1 Testing Diodes

Diodes are the easiest semiconductor devices to test. This method of transistor testing depends on the fact that the diode also represents the base-emitter and base-collector junctions of a transistor. We rely on diodes to pass electrical current *unidirectionally*. This unique ability forms the basis for our test. There are four states of being for any diode, they can be *open*, *shorted*, *leaky*, or *OK*. Before you start testing, it would be wise to have a sheet of paper and a pencil or pen to jot down the readings.

Let's first assume that we are testing a rectifier diode. With the ohmmeter set to the RX1 scale (Fig. 21-12A), place the probes across the diode leads. Record your readings and then swap probes (Fig. 21-12B) and take a second resistance reading. After you take both readings you're ready to interpret what you've discovered.

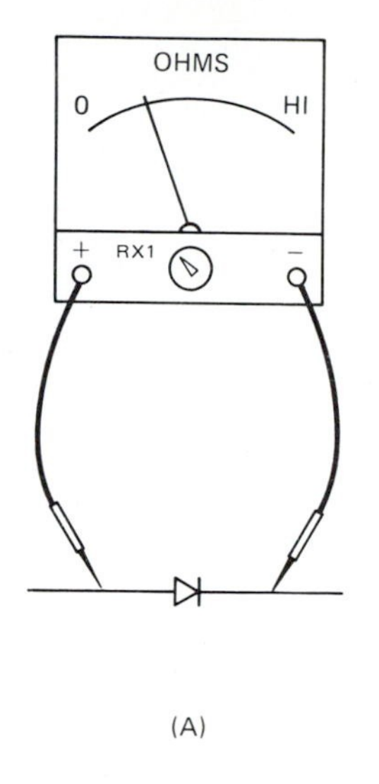


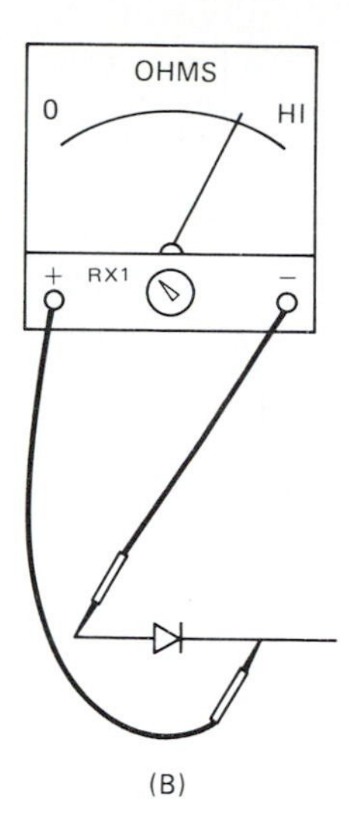


FIGURE 21-12

Let's assume for simplicity the first diode under test checked out OK. How did you know? If the diode is healthy, then one of your resistance readings will be much higher than the other. The actual resistance readings are not terribly important. It is the *ratio* between the readings that counts. Consider five-to-one (5:1) or higher normal for older-style rectifier diodes, and ten-to-one (10:1) for newer rectifier diodes and small signal diodes. For example, if the low reading is 500 ohms (a typical value), the second reading should be 5000 ohms or more on a good diode. Diodes of very recent manufacture may read a very high ratio, or even infinite.

Small signal diodes are tested in exactly the same manner as rectifier diodes, except the meter is set to the "RX100" scale rather than RX1. The lower RX1 scale produces a higher current flow in the external circuit, and can thus blow some small signal diodes.

Dead-shorted diodes often have a zero ohms resistance regardless of probe polarity (Fig. 21-13). Suppose, however, your first and second readings are almost identical (but not zero ohms). That diode is leaky, and is almost useless.

Shorted and open diodes test exactly as you might expect. Open diodes show a very high (or sometimes infinite) resistance, as in Fig. 21-14. On analog meters the pointer will not budge off the "infinity" symbol, while on digital meters the display will flash "1999." A shorted diode will show either zero ohms in both directions, or a very low resistance (less than 100 ohms). Note it is not always true that a shorted diode exhibits the same low resistance in both directions. In some cases, the resistances might be 50 and 80 ohms. This is very leaky and therefore to be considered "dead shorted" for all practical purposes.

The ohmmeter can also tell us which end of an unmarked diode is the anode or cathode. Here's where you must know the relative polarity of your ohmmeter probes. One way of determining polarity is to measure the voltage across the probes with a

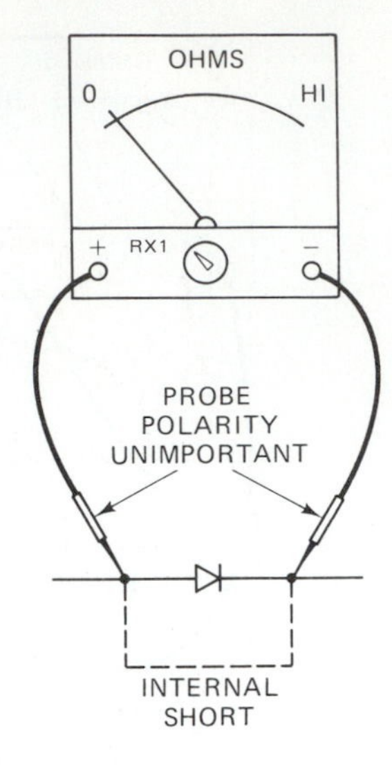


FIGURE 21-13

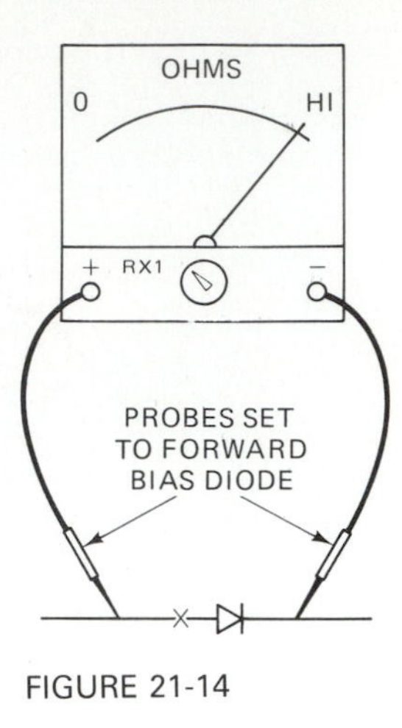


FIGURE 21-14

DC voltmeter and note the polarity of the reading. Another method is to take a known good diode which is marked as to cathode and anode (the cathode end is usually marked somehow: stripe of paint, bullet shape) and connect it to the meter in a direction that produces a low resistance. In this state the ohmmeter positive lead is applied to the diode anode and could be marked. The red ohmmeter lead is usually positive.

Once you know which probe of the ohmmeter is positive, you can use that information to tag any diode as to anode and cathode. Connect the diode across the leads to produce the low DC resistance reading. The positive lead is always connected to the anode end.

### 21-7.2 Testing Transistors

Transistor types can be (and often are) categorized in terms of family behavior. In any transistor catalogs many consecutively numbered transistor types share like characteristics. You will find NPN and PNP types identical in performance, but have equal and opposite polarities. These are called "complementary pairs."

Let's start by examining a small-signal PNP unit. The method for checking transistors with VOM, VTVM, TVM, or DMM is the same as checking a diode. Perform your tests with the ohmmeter on the RX100 scale for small transistors and the RX1 scale for power transistors. Connect your negative ohmmeter probe to the transistor's base lead. Separately touching the collector and emitter leads with the positive lead you will detect either a high-resistance or a low-resistance on each junction. Reversing the probe connections (placing the positive probe on the base) will show up the opposite reading (see Figs. 21-15A through 21-15D).

What do these resistance readings have to do with transistor testing? First, you found out if the transistor was leaky, shorted or open between elements. This would

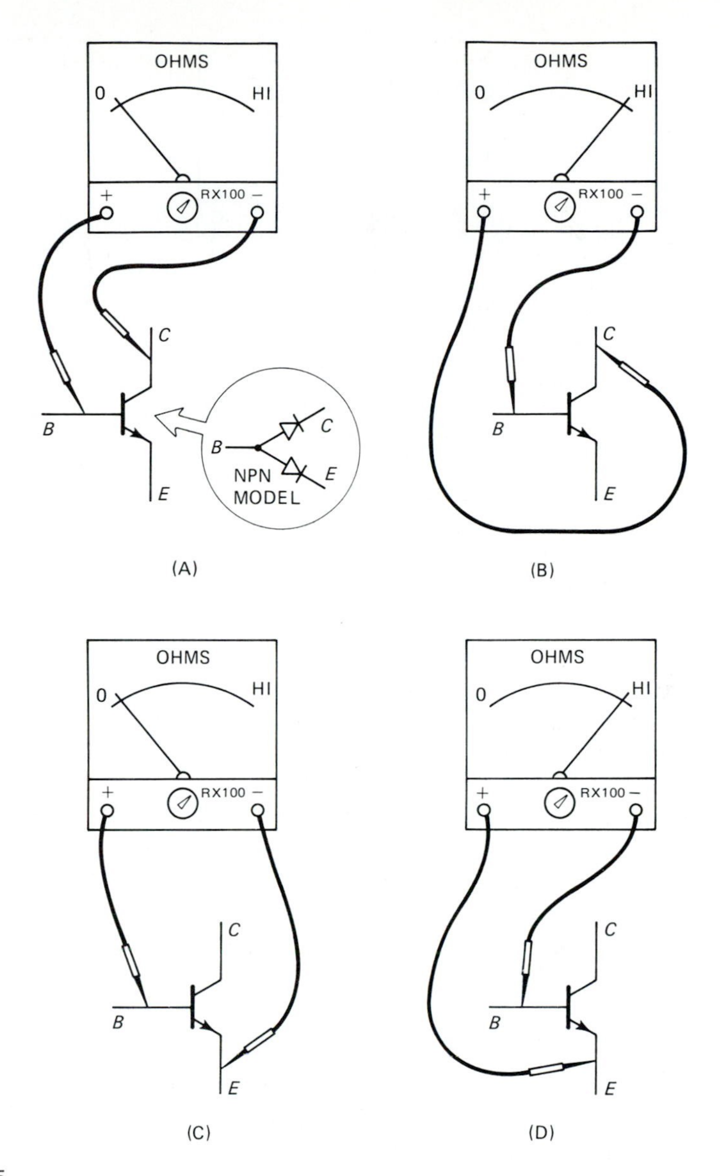


FIGURE 21-15

be apparent as you made your resistance measurements and found undiode-like behavior across the junctions. Next you either confirmed or discovered the transistor's polarity, whether it was PNP or NPN.

Suppose the base-emitter junction tests open. This situation is revealed as a very high resistance, whether or not the positive probe is connected to the base. If the base-emitter junction tests shorted, you will see this condition shows up as a very low resistance, no matter which way the ohmmeter probes are placed. Leaky transistors give you a real run for your measurement money. Indicating abnormal base-emitter

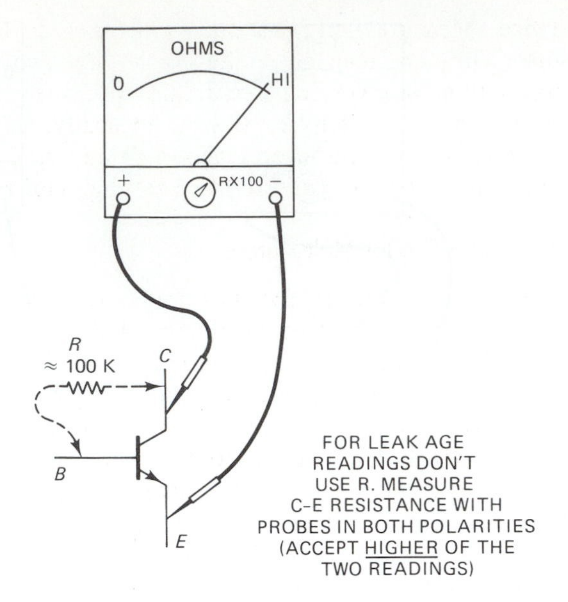


FIGURE 21-16

resistance values, they will not always test either open or shorted. Normal resistance characteristics for a particular transistor family simply will not show.

Finding out whether a transistor is PNP or NPN merely amounts to noting lowest resistance values as you check base-emitter junctions (refer again to Fig. 21-15). For PNP units, the lowest reading occurs when the positive probe is connected to the emitter and the negative lead is hooked to the base. For NPN transistors it is just the opposite. That is, you will get a low reading when the positive lead is connected to the base and the negative is connected to the emitter.

### 21-7.3 Testing for "Gain"

No transistor is worth much if its base terminal does not control collector-emitter current flow. All of the test methods listed above will give us some gross indications of failure, an open emitter or shorted C-E junction and so forth. But they don't tell us anything about whether or not the transistor will amplify.

Figure 21-16 shows a method used by some technicians to determine whether or not a transistor is good. Connect the ohmmeter across the collector-emitter leads in the polarity that would normally be found in circuit. For PNP units connect the negative probe to the collector and the positive probe to the emitter. For NPN units connect the positive probe to the collector and the negative probe to the emitter. If the transistor is good, the reading will be very high or infinity. Next, connect the base lead to the collector lead through a 100-kohm resistor. If the transistor works, the resistance reading across C-E terminals will drop.

### 21-7.4 Checking Leakage

Leakage in a transistor is the unwanted flow of current from collector to emitter (or vice versa). Checking a transistor for leakage is simply measuring resistance across

C-E twice. Measure first with one polarity, then switch the ohmmeter leads and measure again. The leakage reading is the *higher* of the two resistances (because of junction action, one reading is normally quite low). For germanium transistors the leakage will be lower than for silicon. Typically, silicon transistors will show nearly infinite resistance on the three scales normally used, RX1, RX10 and RX100. Germanium transistors may have 100K of leakage and still be good enough for use.

### 21-7.5 Using Digital Multimeters

The digital multimeter (DMM) has largely (but not entirely) replaced the old-fashioned VOM/VTVM. Earlier we discussed the method used in DMMs to measure resistance. Unfortunately, that method does not lend itself to our semiconductor test method because the voltage produced across the probe tips is not sufficient to forward bias PN semiconductor junctions. Although this feature makes it easy to make in-circuit resistance measurements without removing the semiconductors, it also prevents us from testing semiconductors. However, most recent DMMs are designed to overcome that problem. There will be a special ohms scale which will forward bias diode junctions. These scales are marked with either words such as "High Power," or more commonly with the diode symbol.

### 21-7.6 Safety Rule

Diodes and small transistors can be damaged by ohmmeter currents. Always start at the higher scales (RX100) and then drop down to lower scales (RX10 or RX1) only if necessary to get a readable deflection of the meter. In any event, stay on the same scale for both readings. The typical ohmmeter circuit changes currents on different ranges, so comparing readings taken on different readings means interpreting the results of two different bias levels. Also, when using an old VTVM (one made before about 1965) be certain the ohmmeter battery is 1.5 Vdc, not 22.5 Vdc (which will blow out most semiconductors).

## 21-8 TESTING LINEAR IC CIRCUITS

Linear IC devices are both easier and harder to troubleshoot than ordinary transistor circuits, depending on the situation. They are sometimes harder because they are seen as block functions in the schematic, so you may not have the insight into internal circuit workings that is normal with transistor circuits. This is one good reason to keep the data books and specifications sheets for linear devices nearby when troubleshooting. At the same time, however, the very fact the device is a block makes it easier to rapidly determine whether or not it is faulty.

There are two paths in any circuit, AC and DC. In the sample circuit of Fig. 21-17 the capacitors (*C*1 through *C*4) and the operational amplifier (*A*1) are in the AC path for current. Capacitors *C*1 and *C*2 are coupling capacitors. *C*3 and *C*4 are decoupling or bypass capacitors. The DC path consists of the DC power supply terminals (pins 4 and 7), the input pins (2 and 3) and the output pin (6). Let's examine the troubleshooting logic under several different scenarios.

**Condition 1.** The operational amplifier will not pass signal when a 1000-Hz AC

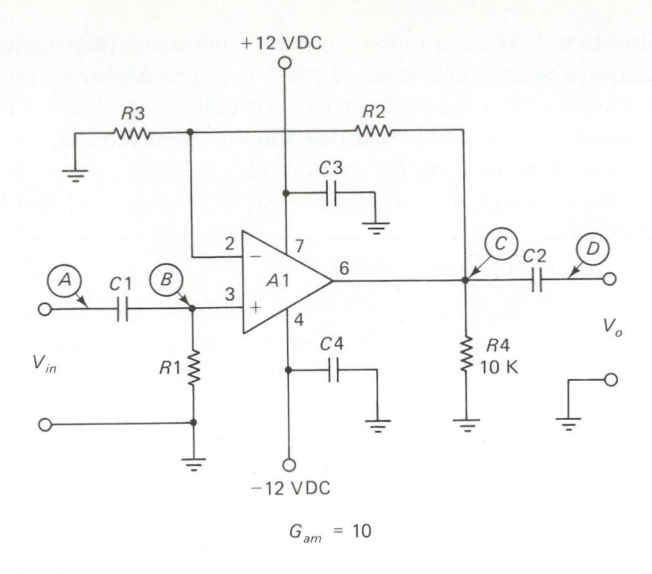


FIGURE 21-17

signal at 1 volt p-p is applied to the input. If an oscilloscope is available, examine the input (point A) to see if the signal generator is supplying signal (many a headache is caused by defective test equipment). Next, move the probe to point B and make sure the signal is passing through input coupling capacitor *C*1. If the signal is passing through *C*1, move the 'scope probe to the output (point C) and check for signal. If signal appears at point C, but not at the output (point D), the output coupling capacitor is probably open. Check by substitution.

**Condition 2.** The same scenario as Condition 1, but in this case there is signal at points A and B, but not at point C. This situation suggests that the operational amplifier is not passing the signal. Now it is time to examine the various operational amplifier terminals for DC levels. Check the V− and V+ power supplies (pins 4 and 7, respectively). The voltage should be within tolerance (usually ±15% if unregulated, or less if regulated).

Assuming the DC supplies are present, next check the output terminal (pin 6). Assume a high DC voltage close to either *V*− or *V*+ is found. It is possible to determine whether or not the operational amplifier is working by determining whether or not the inputs are capable of controlling the output. Short +IN and −IN together while monitoring the output pin. If the op-amp is good, then the voltage will drop to zero when pins 2 and 3 are shorted together. If the voltage does not change, then assume a fault and replace the IC.

**Condition 3.** No signal passes, and output pin 6 is found to be at a potential of +10 volts. Measurement of the power supply pins reveals that pin 7 is at +12 Vdc, while pin no. 4 is at ground potential. The problem here is loss of the *V*− DC power supply.

**Condition 4.** A high frequency oscillation is found riding on the output waveform. The DC voltages are found to be normal or very nearly so. To find the cause of this problem shunt the known good capacitors across $C3$ and $C4$ while monitoring the output on an oscilloscope. If the oscillation ceases when a good capacitor is shunted across the old capacitors, one or both capacitors are probably bad. It must be noted, however, other problems can cause oscillation, especially in circuits with feedback networks more complex than this one.

## 21-8 SUMMARY

1. Most circuit troubleshooting involves finding either (a) a lost path for current or (b) an undesired path for current.
2. Circuits contain both AC and DC paths for currents, and the troubleshooter must analyze both.
3. Cascade amplifiers can be approached using either signal injection or signal tracing.
4. An ohmmeter can be used to test diodes and transistors, although it is less useful for linear IC devices.

## 21-9 RECAPITULATION

Now return to the objectives and Prequiz questions at the beginning of the chapter and see how well you can answer them. If you cannot answer certain questions, mark them and review the appropriate parts of the text. Next, try to answer the questions and work the problems below, using the same procedure.

## 21-10 QUESTIONS AND PROBLEMS

1. List two different approaches to troubleshooting a cascade amplifier.
2. Can an ordinary digital ohmmeter be used to check a PN junction diode?
3. You can afford only one test instrument on your workbench. Select from the following: signal generator, multimeter, or oscilloscope.
4. A transistor amplifier uses a PNP transistor in a circuit with a negative ground DC power supply. The collector DC path returns to ground through a 1000 ohm resistor. The 680 ohm emitter resistor is connected to a +11 Vdc power supply line. The voltage from emitter to ground is 10.2 Vdc. Calculate the emitter resistor voltage drop, emitter current, and approximate DC voltage that appears from collector to ground.

# Index